Advances in Science, Technology & Innovation

IEREK Interdisciplinary Series for Sustainable Development

Advances in Science, Technology & Innovation (ASTI) is a series of peer-reviewed books based on important emerging research that redefines the current disciplinary boundaries in science, technology and innovation (STI) in order to develop integrated concepts for sustainable development. It not only discusses the progress made towards securing more resources, allocating smarter solutions, and rebalancing the relationship between nature and people, but also provides in-depth insights from comprehensive research that addresses the **17 sustainable development goals (SDGs)** as set out by the UN for 2030.

The series draws on the best research papers from various IEREK and other international conferences to promote the creation and development of viable solutions for a **sustainable future and a positive societal** transformation with the help of integrated and innovative science-based approaches. Including interdisciplinary contributions, it presents innovative approaches and highlights how they can best support both economic and sustainable development, through better use of data, more effective institutions, and global, local and individual action, for the welfare of all societies.

The series particularly features conceptual and empirical contributions from various interrelated fields of science, technology and innovation, with an emphasis on digital transformation, that focus on providing practical solutions to **ensure food, water and energy security to achieve the SDGs.** It also presents new case studies offering concrete examples of how to resolve sustainable urbanization and environmental issues in different regions of the world.

The series is intended for professionals in research and teaching, consultancies and industry, and government and international organizations. Published in collaboration with IEREK, the Springer ASTI series will acquaint readers with essential new studies in STI for sustainable development.

ASTI series has now been accepted for Scopus (September 2020). All content published in this series will start appearing on the Scopus site in early 2021.

Deepak Kumar · Nick P. Bassill · Khushboo Rani
Editors

Data Analytics and Remote Sensing-Based Analysis for Sustainable Development

Springer

Editors
Dr. Deepak Kumar
Atmospheric Sciences Group, Department of Geosciences, College of Arts and Sciences
Texas Tech University
Lubbock, Texas, USA

Center of Excellence in Weather and Climate Analytics, Atmospheric Sciences Research Center (ASRC)
University at Albany, State University of New York (SUNY)
Albany, New York, USA

Dr. Khushboo Rani
Bradley Department of Electrical and Computer Engineering
Virginia Tech
Blacksburg, Virginia, USA

Dr. Nick P. Bassill
Center of Excellence in Weather and Climate Analytics, Atmospheric Sciences Research Center (ASRC)
University at Albany, State University of New York (SUNY)
Albany, New York, USA

State Weather Risk Communication Center (SWRCC), Atmospheric Sciences Research Center (ASRC)
University at Albany, State University of New York (SUNY)
Albany, New York, USA

ISSN 2522-8714 ISSN 2522-8722 (electronic)
Advances in Science, Technology & Innovation
ISBN 978-3-031-98094-7 ISBN 978-3-031-98095-4 (eBook)
https://doi.org/10.1007/978-3-031-98095-4

This Springer imprint is published by the registered company Springer Nature Switzerland AG
The registered company address is: Gewerbestrasse 11, 6330 Cham, Switzerland

Dedicated to Beloved Parents

Preface

The effects of climate change are becoming more severe and prevalent with each year. This growing intensity highlights the need for innovative and flexible approaches to managing disasters. Data analytics and remote sensing are crucial tools in driving sustainable development. Remote sensing allows us to observe and monitor environmental changes on a global scale, while data analytics enables the interpretation of vast amounts of information to derive actionable insights. These technologies are essential for addressing climate change, managing natural resources, and mitigating disaster impacts. The integration of predictive analytics with high-resolution satellite imagery, decision-makers can craft strategies to promote environmental sustainability. Their integration supports real-time monitoring, enhances efficiency, and fosters informed policy-making. These empower communities to adapt to challenges and work towards a sustainable and resilient future.

Sustainable development is not just a goal but a pressing necessity, requiring a holistic understanding of environmental, social, and economic systems. With the advent of cutting-edge remote sensing technologies and advanced data analytics techniques, we now have an unprecedented opportunity to observe, analyse, and interpret our planet's dynamics with precision and foresight. From monitoring deforestation and urbanization to evaluating water resources and climate change impacts, these tools are reshaping how we approach sustainability challenges. Traditional approaches are no longer enough to address the increasing challenges caused by climate change. In an era of rapid technological advancement and escalating environmental challenges, the intersection of data analytics and remote sensing has emerged as a powerful tool for addressing the complexities of sustainable development. This book on *Data Analytics and Remote Sensing-Based Analysis for Sustainable Development* aims to present the role of the transformative potential of these disciplines in fostering a more sustainable future.

This book is intended for researchers, practitioners, policymakers, and students who seek to harness the power of data and technology for a positive global impact. It explores a range of topics, including the integration of big data with remote sensing, machine learning applications in environmental analysis, and case studies showcasing real-world implementations. Each chapter has been crafted to provide both theoretical foundations and practical insights, making this an essential resource for anyone passionate about the future of our planet.

As editors, we have had the privilege of collaborating with a diverse and passionate group of contributors, whose commitment to sustainability is evident in every chapter of this book. Our aim is that this work not only informs but also inspiresa valuable resource for academics, practitioners, and policymakers dedicated to advancing the collective mission of addressing climate change and promoting sustainable practices for future generations. We invite readers to join us on this journey, exploring the vast potential of geospatial technologies in fostering a more sustainable and equitable world. This publication seeks to go beyond highlighting the technical capabilities of these tools, emphasizing their critical role in saving lives, safeguarding resources, and driving sustainable development. It will equip communities with actionable geospatial intelligence to empower and adapt them to environmental challenges while building resilience and equity. We hope that this book provides insights that are both informative and transformative, sparking meaningful dialogue, innovative strategies, and collaborative partnerships in the pursuit of climate resilience. We hope this book serves not only as a guide to the technical aspects of data analytics and remote sensing but also as an inspiration to collaborate and innovate across disciplines. Together, we can leverage knowledge and technology to create solutions that ensure the well-being of both humanity and the environment. Let us embark on this journey of discovery and action, united by a shared vision of a sustainable world.

Chapter "Data Analytics and Remote Sensing for Sustainable Development"

This chapter explores the role of data analytics and remote sensing for sustainable development to offer critical insights into environmental, social, and economic systems. The integration of data analytics with remote sensing technologies enables the efficient monitoring and management of natural resources, climate change, and urban development. This reports the role of data analytics and remote sensing in achieving sustainable development goals and highlights key applications like agricultural optimization, disaster management, and environmental conservation. It emphasizes the importance of interdisciplinary collaboration and innovative approaches in harnessing these technologies to create a more sustainable and equitable future. It suggests that stakeholders make informed decisions to promote sustainability and resilience with the use of satellite imagery, sensor networks, and advanced data processing techniques.

Chapter "Mapping and Identification of Coconut Plantations Through a Pixel-Based Classification Algorithm with Sentinel 2 Datasets"

This chapter explores the mapping and identification of coconut plantations through a pixel-based classification algorithm with Sentinel 2 datasets. It highlights that agriculture is an important natural resource and coconut is one of the vital horticultural crops suitable for humid regions. Due to the impact of climate change and less awareness of farmers, major crop areas were converted into coconut plantations over the decades in the Western part of Tamil Nadu region. Even though numerous applications utilizing remote sensing tools have been attempted, regional mapping of coconut growing areas has not been studied at regular intervals. This study tries to recognize the dominance of coconut crop pattern changes in selected study areas. It uses multispectral Sentinel 2 data for the analysis with an ISODATA algorithm for unsupervised classification to separate the cluster of pixels. Based on their spectral signature, three clusters are isolated as vegetation for NDVI mapping. The training data are assigned to import the maximum likelihood algorithm of supervised classification to classify the heterogeneous coconut palm to measure the extent of the coconut plantation. The parallel ground truth observations are carried out for better accuracy of classifications. This finding shows changes in crop patterns to create awareness among farmers about their high-yield income crops over different climatic conditions. The study includes the importance and effectiveness of remotely quantifying the coconut plantation will help stakeholders, agricultural statistics, NGOs, and farmers to detect the crop rate per unit of land for implementation of appropriate agricultural practices to increase coconut palm production profits.

Chapter "Remote Sensing-Based Error Assessment of Mathematical Modelling for Predicted Urban Sprawl with Stochastic Approaches"

This chapter emphasizes the remote sensing-based error assessment of mathematical modelling for predicted urban sprawl with stochastic approaches. This highlights that the population explosion has become an extremely important theme for debates. Migration is also one of the most problematic issues which have become now an environmental issue. These two factors play a major role in case of expansion of urban sprawls. As we know to cure a disease, first we must diagnose the same. The same scenario may be applied here to make urban centres sustainable; first we must assess the sustainability. It is well known that economic, social, and environmental aspects play a major role in the assessment of overall sustainability. These urban sprawls are very important to evaluate the environmental well-being of urban life. These may affect the social and economic aspects of the centres negatively. The current study assesses the same with the help of mathematical modelling using stochastic methods and Geomatics.

Chapter "Data Analytics and Integration of CCME-Water Quality Index, Water Pollution Index (WPI), Contamination Index (CI) with GIS Approach for a Better Understanding of Drinking Water Quality"

This chapter examines the impact of Data Analytics and Integration of the CCME-Water Quality Index, Water Pollution Index (WPI), and Contamination Index (CI) with the GIS approach for a better understanding of drinking water quality. The current study determines the concentration of physiochemical parameters in surface water intended for drinking, together with any potential dangers to human health, and identifies any substantial anthropogenic pressures. Samples of water were gathered from nineteen locations, encompassing both urbanized and non-urbanized parts during 2022–2023 and its acquired specimens were analysed for twenty usual WQ variables. It is noticed that most of the WQ indicators were behind the guideline value of the World Health Organization (WHO) for surface water except TKN and Coliform. It is predominantly found in the overall region that all water samples show mild-alkaline behaviour. The TKN content is discovered to be inversely correlated with the presence of nitrogen fertilizer and shell debris.

Chapter "Comprehensive Study of Ground Experiences for Flood Resilience and Sustainable Development"

This chapter discusses a comprehensive study of ground experiences for flood resilience and sustainable development. Although it is questionable whether exposure to flooding always results in increased flood resilience, it is unclear what such a learning process entails. The flood experience leads to the development of flood-related knowledge and these are dependent on learning opportunity, learning motivation, and prior knowledge. Flood management and/or other actions are influenced by flood-related knowledge and constrained by restrictions such as lack of information and resources, attitudes, social capital, and policy restrictions. This flood-related information changes flood ability, recoverability, adaptability, and/or transformability, which in turn influences flood resilience. These analyse the various learning processes and their impacts on flood resistance in two environments. It implies that learning about flood mitigation is not possible in a setting that is well-protected by flood control infrastructure. The study advocates for learning-based flood mitigation to promote flood resilience.

Chapter "Habitat Suitability Modelling for Snow Leopards in Biosphere Reserve with Remote Sensing-Based Approaches"

This chapter deals with habitat suitability modelling for snow leopards in biosphere reserves with remote sensing-based approaches. The present work focuses on delineating the vulnerable zones with thematic mapping in the GIS platform. This study uses an extensive field survey to emphasize the role as a stronghold for elusive species like the snow leopard. The application of the maximum entropy model to predict habitat suitability is particularly noteworthy, offering a scientific basis for conservation efforts by identifying areas with the highest potential for snow leopard presence. This meticulous approach has revealed favourable for snow leopards, with a notable concentration to provide a foundation for targeted conservation strategies. The findings are instrumental in informing future preservation initiatives to ensure these regions remain a refuge for one of nature's most magnificent predators.

Chapter "A Comprehensive Review of Brick Kilns Mapping with Artificial Intelligence, Deep Learning and GIS Methods for Sustainable Development"

This chapter discusses the geospatial information services for climate-resilient natural resource management with a comprehensive review of Brick kiln mapping with artificial intelligence, deep learning, and GIS methods for sustainable development. Brick kilns play a significant role in meeting the global demand for construction materials, but their operations have raised environmental concerns, ranging from air pollution to deforestation and soil degradation. This review article provides a comprehensive analysis of various brick kiln mapping methods, from traditional approaches to state-of-the-art deep learning algorithms. The paper explores a wide array of mapping technologies, including aerial remote sensing, satellite imagery, machine learning, deep learning and geographic information systems (GIS), utilized to monitor brick kiln activities and their impact on the environment. The review highlights the strengths and limitations of different techniques and discusses recent advancements in the field. Early mapping methods which relied on manual surveys and basic tools transformed into visual inspection using aerial photography and satellite imagery which introduced more efficient mapping possibilities. The review further addresses the environmental consequences of brick kiln emissions, such as greenhouse gases, black carbon, and particulate matter, impacting air quality and climate change. The paper also highlights the socio-economic aspect, discussing the exploitation of workers and the health risks associated with brick kiln labour. This article is the first of its kind which aims to aid policymakers and researchers in selecting appropriate mapping techniques and promotes sustainable solutions for accurately mapping brick kilns. The critical analysis of the existing methodologies, results, and discussion explores the potential of emerging technologies with remote sensing and GIS. This review contributes to the advancement of brick kiln mapping practices, benefiting the environment, labour conditions, and the brick manufacturing industry as a whole.

Chapter "Assessment of Hydro-Meteorological Variables in Gandak River Basin with an Integrated Approach for Climate Resilient Sustainable Water Resource Management"

This chapter explains the assessment of hydro-metrological variables in the Gandak River basin with an integrated approach for climate-resilient sustainable water resource management. The area of the state is occupied with rich alluvial deposits with high-yielding aquifers sustaining the livelihood of millions of populations and river ecosystems. The high dependency on the growing population coupled with agricultural and industrial development imposes severe stress on the freshwater resources of the state. The present study uses an integrated approach of hydrochemical, isotopic, and remote sensing to assess the water resources and other hydrometeorological variables in the area. It was observed that both surface and groundwater are polluted at a few locations. Isotopic analysis suggests surface water is the admixture of glacier melt and precipitation. It was observed that land surface temperature is increasing in the basin showing an impact on vegetation and moisture in the basin.

Chapter "Remote Sensing of Active Tectonics and Geomorphic Signatures for Sustainable Development"

The current chapter describes climate change as one of the major problems of our day with remote sensing of active tectonics and geomorphic signatures for sustainable development. Everyone on the planet shares concerns about the ongoing research that aims to explore the tectonic influences shaping the development of streams. This investigation employs Geomorphic Indices of Active Tectonic (GIAT) parameters, such as the bifurcation ratio (Rb), basin elongation ratio (Re), hypsometric integral analysis (HI), basin asymmetry factor (AF), transverse topographic symmetry (T), stream gradient length ratio (SL), river longitudinal profile (RLP), lineaments, and abnormal stream behaviour. Hypsometric integral analysis indicates an early mature stage of basin evolution. The river's longitudinal profile value of 26.3% and stream length gradient ratio index of 6.73 suggest a second-order anomaly. Other parameters, including strike-slip faults in granite terrain, abnormal stream behaviour concerning the main river, trellis drainage pattern, lineament analysis, waterfall breaks, gorges, and quaternary sediment analysis, contribute to the GIAT analysis of watersheds. This extensive analysis sheds light on the tectonic influences shaping the river basin's evolution and helps comprehend the ongoing geological processes impacting stream behaviour and landscape formation.

Chapter "Exploring the Relationship Between Women's Health and WASH Conditions in Slum Areas Towards Sustainable Development"

This chapter discourses the relationship between women's health and WASH conditions in slum areas towards sustainable development. The relationship between women's health and Water, Sanitation, and Hygiene (WASH) conditions in slum areas is a critical yet underexplored facet of sustainable development. This study investigates how inadequate WASH infrastructure disproportionately affects the health and well-being of women in urban slums, exacerbating gender inequities and hindering progress towards global development goals. Slum areas often experience poor access to clean water, sanitation facilities, and hygiene resources, leading to heightened health risks such as waterborne diseases, reproductive health complications, and chronic illnesses. Women are particularly vulnerable due to biological and sociocultural factors, including their roles in caregiving and water collection. This study combines quantitative data analysis with qualitative insights gathered through fieldwork, focusing on slums in [specific regions/countries]. It evaluates the impact of WASH conditions on women's physical and mental health, as well as their ability to participate in economic and educational activities. Improved WASH conditions not only enhance health outcomes but also empower women by reducing time burdens, ensuring safety, and fostering greater equality. The findings reveal significant correlations between inadequate WASH infrastructure and adverse health outcomes, emphasizing the need for gender-sensitive interventions in slum development policies.

Chapter "Science of Data Analytics and Role of Microclimate in Crop Disease Prediction for Sustainable Development"

The purpose of this chapter is to explore the opportunities for the integration of the science of data analytics and the role of microclimate in crop disease prediction for sustainable development. The intersection of data analytics and microclimate science offers transformative potential in predicting crop diseases, a critical aspect of sustainable agricultural development. This research explores how advanced data analytics methodologies, combined with the understanding of microclimate dynamics, can enhance disease prediction accuracy and inform proactive measures in agriculture. Microclimates—localized weather conditions—are influenced by factors such as topography, vegetation, and human activities. Data analytics allows researchers to process extensive datasets with the microclimatic parameters, soil health indicators, and crop-specific vulnerabilities, to identify patterns that contribute to disease outbreaks. Machine learning algorithms, predictive modelling, and geospatial analysis serve as key tools in extracting actionable insights from complex, multi-dimensional datasets. This research highlights successful applications of data analytics in predicting diseases such as blight, rust, and powdery mildew across various crops. It emphasizes the need for integrating real-time sensor data, satellite imagery, and historical climatic information to develop robust predictive models. Furthermore, this study underscores the importance of translating predictions into localized, sustainable practices that mitigate disease impacts while promoting ecosystem health. The findings of this research demonstrate that combining data analytics and microclimate science not only empowers farmers and policymakers with precision tools but also advances global efforts towards sustainable agriculture. By reducing crop loss, improving resource efficiency, and minimizing chemical dependency, this approach supports the long-term goals of food security and environmental preservation.

Chapter "Mitigating Climate Change Effects on Food and Nutrition Security for a Sustainable Future"

The purpose of this chapter is to explore the opportunities for mitigating climate change effects on food and nutrition security for a sustainable future. Food must be produced, processed, distributed, prepared, eaten, and occasionally disposed of. Each of these processes produces greenhouse gases, which trap the heat of the sun and cause climate change. Food is linked to about one-third of all greenhouse gas emissions that are created by humans. People believe that the main effects of climate change are increasing temperatures. The storey does not begin with the temperature increase. Because the ecology is interrelated, changes in one area will therefore have a similar effect on others. The world's wealthiest nations will experience less change in their local climate and crop yields as a result of well-built information systems in place, while low-income or less developed nations will suffer more in terms of food security and food safety as a result of climate change and less resilient crop infrastructure. On the one hand, rising CO_2 levels are beneficial for crop growth, but on the other hand, CO_2 emissions are causing frequent climatic changes, such as extreme heat, severe weather, and droughts, which pose a serious threat to crops in high demand like wheat and maize. The worst effects of these consequences are likely to fall on nations already struggling with concerns like violence, pollution, deforestation, and other problems.

Chapter "Challenges and Future Trends of Data Analytics and Remote Sensing Based Analysis for Sustainable Development"

The purpose of this chapter is to explore the opportunities for the challenges and future trends of data analytics and remote sensing-based analysis for sustainable development. Issues like data quality and accuracy, high costs, data integration and interoperability, and a shortage of skilled professionals obstruct the effective utilization of remote sensing and data analytics. Besides, challenges related to data access, privacy concerns, and uncertainties in predictive modelling complicate long-term planning and decision-making hindering the utilization despite their enormous potential. Despite these obstacles, the integration of data analytics and remote sensing proposes significant opportunities to optimize resource management, enhance disaster response, and support policy development for achieving global sustainability goals. Future trends point towards advances in satellite technology, increased data availability, and improved machine learning algorithms will enhance the precision and accessibility of these tools. The continued evolution of these technologies, combined with interdisciplinary collaboration and capacity-building, holds the potential to overcome existing challenges and create a more sustainable, data-driven future. The growth of cloud computing and open-source platforms will facilitate broader access to remote sensing data, particularly in developing countries, empowering local communities and organizations to contribute to sustainable development initiatives.

This work represents the culmination of research, case studies, and technological advancements, crafted to inspire a new generation of sustainability-focused practitioners. It emphasizes not only the immense capabilities of geospatial data but also the critical need to integrate it into the broader framework of climate resilience and sustainable development.

We hope this compilation serves as a guiding resource for those engaged in the vital task of managing resources in ways that balance ecological preservation and development goals. As editors, we extend our deep gratitude to the contributors who have shared their knowledge and insights, fostering a collaborative effort to highlight the role of these technologies in addressing the environmental challenges of our time.

May this book catalyse innovation and action, encouraging the use of technology as a powerful ally in our shared endeavour to protect and nurture Earth's finite resources for future generations. Together, we can pave the way for a sustainable and resilient future.

Lubbock, USA
November 2025

Dr. Deepak Kumar
Dr. Nick P. Bassill
Dr. Khushboo Rani

Acknowledgements

The earth is always changing…readjusting to our existence. Each era is full of unique challenges
—Val Uchendu

The conception of this book, *Data Analytics and Remote Sensing-Based Analysis for Sustainable Development*, would not have been possible without the invaluable contributions and support of numerous individuals and organizations. As editors, we would like to acknowledge the help of all the people involved in this project and, more specifically, the authors and reviewers who took part in the review process. The completion of this edited volume would not have been possible without the invaluable contributions and support of numerous individuals and organizations.

We express our deepest gratitude to the contributing authors whose expertise, dedication, and scholarly rigour have shaped the contents of this book. Their commitment to advancing the field of geospatial technologies and climate resilience has been both inspiring and instrumental. We extend our appreciation to the reviewers and editors whose critical insights and constructive feedback have refined and enriched the quality of this work. Their meticulous efforts have ensured that this volume meets the highest standards of academic and professional excellence.

It is our sincere hope that this volume serves as a meaningful contribution to the field and inspires further innovation, collaboration, and progress in the pursuit of climate-resilient disaster management. We are grateful for all the facilities at the "Beautiful campus" for providing first-class infrastructure and offering research facilities, creative thinking, and scientific writing. This work could not have been completed without the great support and encouragement of several people. We are deeply grateful to our mentors and colleagues, whose guidance, feedback, and encouragement have been invaluable throughout this endeavour. Since at this moment it is difficult to memorize all of them, we would like to apologize for any person we forgot to mention. Regrettably, we cannot acknowledge by name all my fantastic friends, because the list would not fit in here, but we gratefully thank each one of them including wonderful friends at our workplace for the great and joyous moments we shared.

A special thanks is owed to the institutions, research organizations, and funding bodies that supported the development of this book, either through direct collaboration or by fostering the research that underpins the chapters herein. Their encouragement and resources have been invaluable. We are also deeply grateful to the global community of practitioners, policymakers, and researchers working tirelessly to mitigate the impacts of climate change and enhance disaster management practices. This book is a testament to their enduring efforts and a contribution to their ongoing mission. We are thankful to the institutions and organizations that have funded and supported studies in Geospatial Information and Services for Climate-Resilient Disaster Management. We would like to thank NOAA OAR for fellowship support under the Cooperative Agreement Grant # NA21OAR4590360. We would like to thank all colleagues working at the Center of Excellence in Weather and Climate Analytics, Atmospheric Science Research Center (ASRC), State University of New York at Albany and anonymous reviewers for their insightful comments and suggestions. Earnest appreciations are extended to the reviewers for their constructive comments improving the chapters.

Lastly, we sincerely express our special and heartfelt respect, gratitude, and gratefulness to my family members and parents for their endless support and blessings. We admire the almighty for being so kind to us during this period of trials and tribulations, whose grace has sustained us to reach this level of life. It has been a long journey when many obstacles stood in the way; yet the almighty sorted me through, turning every failure into successful and enjoyable moments.

Lubbock, USA Dr. Deepak Kumar
New York, USA Dr. Nick P. Bassill
Blacksburg, USA Dr. Khushboo Rani

About This Book

The book on *Data Analytics and Remote Sensing-Based Analysis for Sustainable Development* outlines the role data analytics and remote sensing for climate-resilient sustainable development. It provides a comprehensive and enlightening skill to utilize harnessing the power of data analytics with geospatial sciences for climate-resilient sustainable development. Remote sensing can be used to monitor the health of forests, and data analytics can be used to identify areas where deforestation is particularly high. Remote sensing involves the use of sensors to gather data about the environment from a distance, while data analytics refers to the use of statistical and computational techniques to extract insights and patterns from large datasets. Sustainable development with the use of remote sensing is used to monitor changes in the environment like deforestation, land use change, and water quality. This analysed data helps to identify trends and patterns to develop strategies for sustainable development. This information is used to develop targeted strategies for reducing deforestation to promote sustainable forestry practices and support reforestation efforts. These are also used to monitor water quality, and data analytics to identify sources of pollution and to develop strategies for improving water quality. This information is very much essential in developing countries, where access to clean water is a major challenge. Moreover, data analytics and remote sensing can be powerful tools for supporting sustainable development. We can better understand the challenges we face and develop effective strategies for addressing them by collecting and analysing data about the environment. This book is a timely exploration of how cutting-edge technology can shape our response to climate-resilient developments to avoid disasters. The book embarks on a journey through the intricate web of geospatial data, Data analytics and remote sensing to support sustainable development.

Contents

Data Analytics and Remote Sensing for Sustainable Development 1
Deepak Kumar, Nick P. Bassill, and Khushboo Rani

Mapping and Identification of Coconut Plantations Through a Pixel-Based Classification Algorithm with Sentinel 2 Datasets 15
B. Anand, R. Shanmathi Rekha, M. Vijaya Prabakaran, R. Amirtha Varshini, M. Kavya, B. Reshma, and K. Ramaswamy

Remote Sensing-Based Error Assessment of Mathematical Modelling for Predicted Urban Sprawl with Stochastic Approaches 25
Gaurav Kumar Mishra and Amit M. Deshmukh

Data Analytics and Integration of CCME-Water Quality Index, Water Pollution Index (WPI), and Contamination Index (CI) with GIS Approach for a Better Understanding of Drinking Water Quality 39
Abhijeet Das

Comprehensive Study of Ground Experiences for Flood Resilience and Sustainable Development 65
Puja Verma and Ambrina Sardar Khan

Habitat Suitability Modelling for Snow Leopards in Biosphere Reserve with Remote Sensing-Based Approaches 79
Vipul Maurya, Parag Madhukar Dhakate, Ashish Mani, Ashish Panda, and S. B. Lal

A Comprehensive Review of Brick Kilns Mapping with Artificial Intelligence, Deep Learning and GIS Methods for Sustainable Development 93
Yamini Agrawal, Hina Pande, Poonam Seth Tiwari, and Shefali Agrawal

Assessment of Hydro-Meteorological Variables in Gandak River Basin with an Integrated Approach for Climate Resilient Sustainable Water Resource Management 101
Rina Kumari, Ujjawal Kumar, Nang Swatie Manpoong, and Virendra Padhya

Remote Sensing of Active Tectonics and Geomorphic Signatures for Sustainable Development 117
Gurav Chandrakant and Md Babar

Exploring the Relationship Between Women's Health and WASH Conditions in Slum Areas Towards Sustainable Development 137
Renu Dhupper and Suhasini Jindal

Science of Data Analytics and Role of Microclimate in Crop Disease Prediction for Sustainable Development 157
Pritimoy Sanyal

Mitigating Climate Change Effects on Food and Nutrition Security for a Sustainable Future 169
Rosy Bansal, Gorakh Singh Teja, and Monika Hans

Challenges and Future Trends of Data Analytics and Remote Sensing Based Analysis for Sustainable Development 175
Deepak Kumar, Nick P. Bassill, and Khushboo Rani

About the editors

Dr. Deepak Kumar is an academic researcher with a multidisciplinary background in geospatial sciences, computational sciences, climate change, and sustainability. He is currently working as a research scientist in the Atmospheric Sciences Group at Texas Tech University USA. Before this, he served as a research scientist in the Atmospheric Sciences Research Center (ASRC) at the State University of New York at Albany, USA. He has been conducting interdisciplinary research on the urban-climate-energy nexus, focusing on policy making from the perspectives of humanities, social sciences, and technology. He has also served at Amity University, Delhi-NCR (India), as a full-time assistant professor with a wide experience in the research development-cum-implementation pipeline comprising idea conceptualisation, research design, data collection, processing, and analysis. He has implemented two government-sponsored research projects as the sole principal investigator and filed 7 patents with 1 copyright as inventor/co-inventor. He has mentored, advised, and supervised a cross-functional research group of 4 Ph.D. and 29 postgraduate students and conducted fifteen academic, research, and industrial field visits to students. As a young researcher, he enjoys developing skills, knowledge, and involvement through conference appearances, outreach activities, training services and contributions to the professional membership of scholarly associations. He has extensive experience in presenting research at international, national, regional, and local conferences and successful grant writing and published75+ research articles in high-impact Web of Science and Scopus-indexed international journals. He has participated in review activities for more than 230 research articles, project proposals, and grants as an invited/ad hoc/adjunct reviewer or as an editorial board member or associate/assistant editor. He has authored/edited/co-edited around 10 books with publishers like Elsevier, Springer, CRC Press, Taylor & Francis, and Wiley. He is dedicated to finding solutions for complex problems in the areas of climate change, urban planning, energy, the environment, and sustainability (SDGs 7, 10, 11, and 13).

Dr. Nick P. Bassill has several years of experience in climate computing research and its impact assessment. He is currently working as Director of the State Weather Risk Communications Center, along with Director of Research and Development at the Center of Excellence, State University of New York at Albany and Adjunct Professor at the Atmospheric Sciences Research Center, Department of Atmospheric and Environmental Sciences, the State University of New York at Albany, New York, USA. Dr. Bassill received his Ph.D. in Atmospheric and Oceanic Sciences from the University of Wisconsin-Madison, with a research focus on ensemble and parameterisation studies. Afterwards, he worked as a post-doctoral researcher at the University of Utah, where he studied aspects of Hurricane Sandy's unusual track (including what led some models to make an incorrect forecast). Afterwards, he was hired as a post-doctoral researcher with the New York State Mesonet to begin building an operational analysis system that would utilize Mesonet observations. Concurrently, he is actively involved in building meteorological products for and doing basic research with the New York State Mesonet. Currently, Nick spends his time working with the Center of Excellence on a variety of projects. He has published various research papers in international journals of high repute. He has also conducted sessions on Numerical weather prediction, with a specialisation in WRF modelling, Data assimilation, Weather forecasting, Tropical cyclones, Data analysis of both gridded model data and weather observations and the New York State Mesonet. He also has experience in providing scientific leadership to high-profile strategic computational sustainability and conservation initiatives with several years of research experience. Managed cross-sectoral and interdisciplinary teams of professionals to deliver on complex research projects. His core areas of interest cover financial markets, Energy and utilities, Emergency Management, Risk Management, Climate Computing, AI and Machine Learning, and Industry 4.0. He has presented several research papers at international and national conferences on these topics. He has experience in providing scientific leadership for high-profile strategic sustainability and conservation initiatives, backed by several years of research experience.

Dr. Khushboo Rani has over five years of experience in sustainable computing research and its impact assessment. She has published various research papers in international journals of high repute. She has also conducted FDPs on data analysis using contemporary tools and techniques. She received her Ph.D. in Computer Science and Engineering from the Indian Institute of Technology Guwahati (India). She has also conducted FDPs on data analysis using contemporary tools and techniques. She has academic experience teaching graduate and postgraduate students at the Indian Institute of Technology Guwahati (India). Her core area of research includes Network-on-Chip, Systems-on-Chip, Memory System Design, Embedded Systems, Secure Processor, Hardware Security and Trust, and Malware Detection. Along with that, her interest also lies in sustainability, sustainable computing, and machine learning. She has presented several research papers at international and national conferences on these topics. She also has experience in providing scientific leadership to high-profile strategic computational sustainability and conservation initiatives with several years of research experience. She managed cross-sectoral and interdisciplinary teams of professionals to deliver on complex research projects. Currently, she is working as a postdoctoral associate in the System Software Research Group (SSRG), Bradley Department of Electrical & Computer Engineering at Virginia Tech, prior to this, she was a post-doctoral researcher at the University of Florida. She has presented several research papers at international and national conferences on these topics. Also, she has experience in providing scientific leadership to high-profile strategic sustainability and conservation initiatives with several years of research experience.

Data Analytics and Remote Sensing for Sustainable Development

Deepak Kumar, Nick P. Bassill, and Khushboo Rani

1 Introduction

The advancement of technology has provided humanity with powerful tools to address the pressing challenges of sustainable development (Ahmad 2023; Verma and Bhalla 2024). Data analytics and remote sensing have emerged as pivotal in shaping our understanding and management of environmental, social, and economic systems among these tools (Ghisellini et al. 2016; Siraganyan et al. 2019). Data analytics can process and interpret huge amounts of information and complement remote sensing technologies that capture critical data from satellites, drones, and other sensor networks (Carretero-Peña et al. 2019; Corke et al. 2010). They offer unprecedented insights into the complexities of our planet, enabling more effective monitoring, assessment, and decision-making processes. Interdisciplinary collaboration and innovative approaches have the potential to foster a more resilient and equitable future (Mondol et al. 2025; Rijal et al. 2024). This offers the significance of integrating data analytics and remote sensing in the pursuit of sustainable development, exploring key applications such as environmental monitoring, disaster response, and natural resource management.

2 Significance of Data Analytics and Remote Sensing for Sustainable Development

Data analytics and remote sensing are pivotal in advancing sustainable development by providing actionable insights and facilitating informed decision-making. Remote sensing technologies including satellites and drones enable real-time monitoring of natural resources, land use, and environmental changes, while data analytics processes and interprets these datasets to identify trends, predict future outcomes, and optimize resource management (Chen 2019; Gutiérrez-Carrillo et al. 2021; Selvam and Al-Humairi 2025). These support sustainable development efforts for agriculture, climate change mitigation, urban planning, and biodiversity conservation (Basu 2014; Corburn 2009; Hendrickson et al. 2016). These improve the efficiency of resource utilization, enhance disaster management, and guide policy decisions to achieve environmental, social, and economic goals for a more sustainable future. Data analytics and remote sensing are critical in advancing sustainable development, as they provide critical insights and support decision-making across various sectors (Appel et al. 2018; Dasgupta et al. 2023; Le et al. 2024). These tools enable governments and organizations to monitor and manage ecosystems, optimize agricultural practices, and mitigate the impacts of climate change, ultimately supporting long-term sustainability and promoting effective policy formulation. The key significance includes:

D. Kumar (✉)
Atmospheric Sciences Group, Department of Geosciences, College of Arts & Sciences, Texas Tech University, Lubbock, Texas, USA
e-mail: deepakdeo2003@gmail.com

D. Kumar · N. P. Bassill
Center of Excellence in Weather & Climate Analytics, Atmospheric Sciences Research Center (ASRC), University at Albany, State University of New York (SUNY), Albany, New York, USA

N. P. Bassill
State Weather Risk Communication Center (SWRCC), Atmospheric Sciences Research Center (ASRC), University at Albany, State University of New York (SUNY), Albany, New York, USA

K. Rani
Bradley Department of Electrical & Computer Engineering, Virginia Tech, Blacksburg, Virginia, USA

D. Kumar et al. (eds.), *Data Analytics and Remote Sensing-Based Analysis for Sustainable Development*, Advances in Science, Technology & Innovation, https://doi.org/10.1007/978-3-031-98095-4_1

(a) **Environmental Monitoring and Management**

Remote sensing offers the collection of large-scale environmental data through satellites, drones, and sensors, providing real-time and accurate information about land use, deforestation, water bodies, air quality, and climate change patterns (Alam et al. 2024; Cooper et al. 2022; Jehanzaib and Kim 2020). Data analytics process this data to detect environmental changes, predict future trends, and assess the impact of human activities on the environment, allowing for timely intervention and sustainable resource management.

(b) **Agriculture and Food Security**

Remote sensing technologies can assess soil health, monitor crop growth, and detect early signs of disease or drought, enabling precision farming. Data analytics helps farmers optimize resource use (water, fertilizers, etc.), predict harvest yields, and reduce waste, improving food security and minimizing environmental impact (Lasco et al. 2014; Olawuyi and Mushunje 2019).

(c) **Urban Planning and Smart Cities**

Remote sensing aids in monitoring urban sprawl, land use, and infrastructure development, providing data essential for sustainable urban planning. Data analytics facilitates decision-making in transportation, waste management, energy consumption, and pollution levels, allowing for smarter, more sustainable cities (Bakhshipour et al. 2021; Kwak et al. 2020; Simões et al. 2022).

(d) **Disaster Risk Management**

Remote sensing allows for real-time monitoring of natural disasters like floods, wildfires, and hurricanes, and helps predict their occurrence and intensity. Data analytics models assess risks, optimize emergency response strategies, and minimize the economic and human impact of disasters, contributing to resilience and sustainability (Kumar and Bhaduri 2018; Liao et al. 2024; McNamara and Olson 2014).

(e) **Biodiversity Conservation**

Remote sensing tracks ecosystems, forests, and wildlife habitats, giving detailed insights into biodiversity and potential threats. Data analytics supports conservation strategies by assessing biodiversity trends, evaluating conservation efforts, and forecasting long-term ecological health (Turner et al. 2003; Wang et al. 2010).

(f) **Energy Management and Climate Change Mitigation**

Remote sensing monitors renewable energy sources like solar radiation and wind patterns, to determine the feasibility and efficiency of renewable energy projects. Data analytics optimize energy consumption, assess emissions reduction efforts, and guide policies to combat climate change, fostering a transition towards low-carbon economies (Capodaglio and Olsson 2020; Cecchini et al. 2020; Corburn 2009; Eckman et al. 2008; Hendrickson et al. 2016; Krajačić et al. 2018).

(g) **Water Resources and Management**

Remote sensing tracks changes in water bodies, groundwater levels, and rainfall patterns, which are crucial for managing water resources sustainably. Data analytics proposes efficient water allocation, predicting water scarcity, and identifying pollution sources, which is critical for ensuring equitable access to clean water (Botseva et al. 2022).

(h) **Policy Formulation and Decision Support**

Data analytics helps governments and organizations evaluate the effectiveness of policies and interventions by analysing large datasets from various sources. Remote sensing provides spatial context, while data analytics offers insights into socio-economic factors, guiding evidence-based policy decisions that support sustainable development goals (SDGs) (Aubrechtová et al., 2020; Böhner, 2006).

Sustainable development strategies lead to smarter, data-driven decisions to support environmental protection, social equity, and economic growth. These tools help manage resources efficiently, mitigate risks, and ensure that development initiatives are aligned with long-term sustainability goals.

3 Components of Data Analytics and Remote Sensing for Sustainable Development

The integration of data analytics and remote sensing for sustainable development comprises several essential components that collectively contribute to efficient data collection, processing, and decision-making. The first component, data collection, involves gathering data from various sources such as satellites, drones, and sensor networks. This raw data is then stored and managed using advanced database systems and cloud storage solutions to ensure accessibility and integrity (Alberto et al. 2016; Saah et al. 2019; Zhang et al. 2019). Preprocessing techniques include image correction and calibration to refine the data for comprehensive analysis. Data integration follows combining remote sensing data with other relevant datasets to provide a general view of the area (Bhattacharya et al. 2020; Duong and Nguyen-Thi 2021; Pramanik 2024). Advanced data analytics techniques like machine learning and statistical analysis are employed to extract meaningful insights and identify patterns. Visualization tools like geographic information systems (GIS) and interactive maps provide the analysed data in an easily understandable format to ease conversant decision-making. Decision support systems (DSS) enhance this process by integrating user-defined criteria and models to offer recommendations. Effective policy and governance frameworks and capacity building are also critical components to ensure the responsible use of technology and foster widespread adoption (Filimonau 2019; Jain et al. 2020). These components enable the effective application of data analytics and remote sensing to promote sustainability and resilience in various domains.

The key components of data analytics and remote sensing for sustainable development combine technology, processes, and methodologies to monitor, assess, and manage environmental, social, and economic resources. These components work together to support informed decision-making and drive sustainable practices across various sectors and the following key components:

(a) **Data Collection**

The primary component of remote sensing comprises the collection of spatial data from various platforms like satellites, drones, aircraft, or ground sensors. These tools capture detailed images and data on land cover, vegetation, water bodies, air quality, weather conditions, and other critical environmental parameters (Cai et al. 2017; Nichol and Wong 2009).These devices gather information across different wavelengths (e.g. visible, infrared, radar) to measure various factors such as temperature, humidity, vegetation health, soil moisture, and pollutant levels (Lay-Ekuakille et al. 2021; Toth and Jóźków 2016).

(b) **Data Processing and Preprocessing**

Raw remote sensing data typically requires preprocessing to correct for errors, such as atmospheric distortions, sensor inaccuracies, or cloud cover. This stage involves image correction, geometric rectification, and noise removal to enhance the quality and accuracy of data (Chen 2019; Forghani and Delavar 2014). Data analytics tools (like machine learning algorithms and statistical methods) organize, structure, and clean large datasets to ensure they are usable for analysis (Dasgupta et al. 2023; Renuka Devi and Kesavaraj 2019).

(c) **Data Integration and Fusion**

Remote sensing data is often integrated with data from other sources, such as geographic information systems (GIS), ground surveys, or environmental monitoring stations to create a comprehensive dataset and richer insights into the environment (Bhanumurthy et al. 2017; Lunetta et al. 1991; Pandey and Srivastava 2019). Data fusion techniques combine multiple data sources (e.g. satellite imagery, socioeconomic data, sensor data) to generate more accurate models and better understand the interactions between various factors impacting sustainable development.

(d) **Data Analysis and Modelling**

Data analytics encompasses large dataset analysis to identify patterns, trends, and correlations for informed decision-making. Advanced statistical methods, machine learning, and artificial intelligence (AI) algorithms predict future outcomes, optimize resource allocation, and identify potential risks or areas of concern (Ma et al. 2021; Sudmanns et al. 2019). These tools forecast future environmental changes, economic conditions, or resource demands for policymakers and organizations to implement proactive, sustainable strategies (Alhamwi et al. 2018; Tardioli et al. 2018).

(e) **Visualization and Interpretation**

Data visualization tools communicate complex results in an understandable and actionable format. Graphs, charts, heat maps, and geographic visualizations are commonly used to present the analysis of trends, environmental changes, and spatial patterns (Dao et al. 2012; Wang et al. 2012). Geographic information systems (GIS) integrate remote sensing data with spatial data to create detailed maps and 3D models for urban planning, disaster management, and land use decision-making.

(f) **Decision Support Systems (DSS)**

Decision support systems are interactive systems to combine data analytics and modelling to assist decision-makers in evaluating different scenarios, assessing risks, and optimizing outcomes (Böhner 2006; Kuznetsova et al. 2019). These systems help in planning for climate change, resource management, and infrastructure development to provide evidence-based recommendations. Remote sensing provides the spatial context in decision support systems to ensure actions (Lombardi et al. 2017; Rodriguez-Bachiller 2000).

(g) **Monitoring and Feedback**

Constant monitoring via remote sensing allows for real-time assessment of environmental conditions like monitoring deforestation, water quality, or urban growth. These observations inform timely interventions (Alotaibi and Nassif 2024; Coppin et al. 2004). Analytics tools track the effectiveness of interventions and policies over time, generating feedback to adjust strategies and improve future outcomes.

(h) **Sustainability Indicators and Metrics**

Remote sensing data track sustainability indicators, such as deforestation rates, greenhouse gas emissions, water scarcity, or biodiversity loss, for measuring progress towards sustainable development goals (SDGs). Data analytics quantifies these indicators and compares them against benchmarks or targets to provide insights into areas where improvements are needed (Abu-Rayash and Dincer 2021; Chauhan et al. 2023; Iannillo and Fasolino 2021; Michael et al. 2014).

(i) **Collaboration and Data Sharing**

Open-access satellite data repository platforms like Copernicus, and NASA Earth Observing System facilitate global collaboration, enabling countries, organizations, and researchers to access and share remote sensing data for sustainable development efforts. Cloud computing and data-sharing platforms provide collaboration across sectors for stakeholders to share findings, models, and recommendations for collective action towards sustainability (Doucet et al. 2024; Ekmen and Kocaman 2024).

The integration of data analytics and remote sensing makes a powerful ecosystem for sustainable development with the integration of accurate environmental data, advanced analytical tools, and decision support systems. These components help to manage natural resources, mitigate climate risks, optimize agricultural practices, protect biodiversity, and improve urban planning for a more sustainable future.

4 Challenges in Sustainable Development with Data Analytics and Remote Sensing

Several challenges hinder the effective application of data analytics and remote sensing in advancing sustainable development. One of the primary issues is the quality and accuracy of data, as remote sensing data can be affected by factors such as atmospheric interference, sensor calibration, or cloud cover, leading to unreliable results. The high costs of technology and infrastructure make it difficult for developing countries or smaller organizations to access advanced tools like high-resolution satellite imagery and powerful analytics platforms (Elkind et al. 2019; Wassie et al. 2017). Data integration and interoperability also pose challenges, as data from various sources may be incompatible or require extensive preprocessing to merge, hindering seamless analysis.

Moreover, there is often limited access to data, especially in resource-constrained regions, which can restrict the ability of governments and organizations to make data-driven decisions. The skills and expertise gap in both remote sensing and data analytics further complicates matters, as there is a shortage of trained professionals capable of interpreting complex data. Privacy and data security concerns also arise with the collection of sensitive geographic or personal data (Dev et al. 2023; Goumehei et al. 2019; Ma et al. 2021). Lastly, the lack of standardized policies and governance frameworks for data use, as well as the uncertainty in predictive models, can undermine the reliability and scalability of sustainable development efforts.

These challenges require investment in technology, capacity building, improved data-sharing frameworks, and better governance to fully leverage the potential of data analytics and remote sensing in promoting sustainability. The use of data analytics and remote sensing for sustainable development brings valuable insights and supports informed decision-making, but several challenges still obstruct their potential. These challenges can affect their applicability, scalability, and effectiveness across different regions and sectors. Some of the key challenges include:

(a) **Data Quality and Accuracy**

Inconsistent or incomplete data leads to flawed analysis, and it compromises the ability to make informed decisions. While satellite imagery and sensor data offer large-scale coverage, these data sources are often prone to interference from weather conditions, cloud cover, or atmospheric distortion. One of the key challenges is the quality and accuracy of data from remote sensing technologies. The inaccuracies arise from sensor calibration issues or low resolution of certain data, and make it difficult to extract precise and actionable information for sustainable development (Chen 2019; Forghani and Delavar 2014; Kumar 2003).

(b) **High Costs of Technology and Infrastructure**

The advanced technologies involved in remote sensing (such as high-resolution satellites, drones, and sensors) have significant costs for purchasing data and the infrastructure required to store, process, and analyse the data. The price of accessing high-quality, real-time data from private satellite companies is excessively expensive, especially for developing countries or small organizations with limited financial resources (Mohammed et al. 2020; Su et al. 2011; Zhang and Zhao 2024). Besides, data analytics often requires powerful computing systems, specialized software, and expert personnel, all of which add to the financial burden(Gogu et al. 2019; Shubin 2019; Simões et al. 2022).

(c) **Data Integration and Interoperability**

Remote sensing and data analytics rely on integrating data from various sources, including satellite imagery, sensor networks, geographic information systems (GIS), and socio-economic datasets. These data from different sources can vary in format, resolution, and scale, making it difficult to integrate and analyse seamlessly. Incompatible data formats, variable temporal scales, and different geospatial references often create challenges and require significant preprocessing and data cleaning for analysis (Lu et al. 2023; Oliveira 2021; Pinho et al. 2020). These barriers to data interoperability delay the efficiency and effectiveness of decision-making processes.

(d) **Limited Access to Data**

Remote sensing data from agencies like NASA or the European Space Agency (ESA) is publicly available but access to high-resolution satellite imagery or specialized data is restricted or expensive. This is mostly problematic for countries or regions that lack the financial means to purchase or access this data. The governments, NGOs, and other stakeholders may struggle to monitor and manage environmental issues effectively, hindering progress towards sustainable development goals (SDGs) due to the unavailability of comprehensive and timely data (DeLancey et al. 2019; Hird et al. 2017).

(e) **Skills and Expertise Gap**

Remote sensing and require specialized knowledge in fields like geospatial science, environmental management, machine learning, and statistics. Another significant challenge lies in the skills gap. There is a global shortage of skilled professionals capable of managing and analysing the large amounts of data generated by remote sensing technologies, which limits the impact of these technologies on sustainable development efforts. However, many regions, especially in low-income or developing areas, lack the trained personnel required to interpret complex datasets and apply them to real-world challenges (Kesavan et al. 2024; Maheswari et al. 2024).

(f) **Data Privacy and Security Concerns**

The collection of sensitive data through remote sensing technologies—such as high-resolution images of urban areas, infrastructure, or private land raises privacy and security concerns. Data protection regulations and laws are often underdeveloped or inconsistent across regions, raising questions about the ethical and legal implications of using such technologies. The potential for misuse of this data for surveillance or other unethical purposes creates resistance among stakeholders to the widespread adoption of these technologies (Chen et al. 2024; Kumar et al. 2018; Malhotra et al. 2021; Shaham et al. 2021).

(g) **Complexity of Interpreting Large Datasets**

Remote sensing generates vast amounts of data, and the complexity of analysing and interpreting these datasets is another significant challenge. The complexity of handling big data along with the potential for error amplifies if the models are not properly calibrated. The analysis of remote sensing data becomes challenging in fast-evolving environments where real-time decisions are crucial. Mining meaningful insights requires advanced data processing techniques, including machine learning, artificial intelligence, and statistical modelling (Ivey et al. 2004; Lancia et al. 2020).

(h) **Policy and Governance Issues**

Effective use of remote sensing and data analytics requires a strong governance framework and clear policies for data management, sharing, and privacy protection. Many countries lack consistent standards for data sharing, which makes it difficult to ensure that remote sensing data is used efficiently across different sectors. Moreover, unclear regulatory frameworks for the use of data—especially when dealing with private sector data or international data sources lead to legal and ethical issues, and block cross-border collaboration (Cunningham et al. 2019; Laurini et al. 2023; Weber et al. 2015).

(i) **Uncertainty in Predictive Models**

Predictive modelling based on remote sensing data provides valuable forecasts and insights into future environmental changes but these models often face uncertainty. Many factors like unpredictable weather patterns, human interventions, or unforeseen environmental changes affect the accuracy of predictions. This makes long-term planning for sustainable development more difficult and it is hard to account for all variables that could impact the outcomes of policies or interventions (Mishra and Nair 2024).

(j) **Scalability and Applicability**

Remote sensing data is excellent for large-scale monitoring, but application to local-level interventions can be challenging. Satellite imagery may not always capture local environmental conditions with the level of detail necessary for specific interventions. Scaling up small-scale solutions to larger areas or regions also be problematic due to differences in local conditions, resources, and needs. Thus, remote sensing provides valuable data at the global or regional level, but its application at smaller, localized scales can be more difficult (Makropoulos et al. 1999; Palmieri et al. 2018; Ulpiani et al. 2021).

Data analytics and remote sensing offer transformative potential for sustainable development but it still has several challenges—ranging from data accuracy and integration issues to high costs and limited expertise—that must be addressed for these technologies to reach their full potential. The development of robust governance frameworks and innovations in technology makes data more accessible and affordable. These challenges will allow remote sensing and data analytics to become indispensable tools for tackling global sustainability issues.

5 Advantages of Data Analytics and Remote Sensing

Data analytics and remote sensing offer substantial advantages in decision-making, and improvement in resource management to support sustainable development. One of the primary advantages of remote sensing is its ability to provide real-time, large-scale data from areas that are difficult or expensive to reach, such as forests, oceans, or remote regions. This allows for continuous monitoring of land use, vegetation, water bodies, and climate change, helping policymakers stay informed and responsive to environmental shifts. The value of remote sensing data is further increases when it is coupled with data analytics (Aleksandrowicz et al. 2019; Ali et al. 2021). Data analytics process vast amounts of satellite or sensor data, identify patterns and predict future trends, which helps in making data-driven decisions across sectors like agriculture, urban planning, disaster management, and environmental conservation. These analytics provide a prediction to crop yields, detect illegal deforestation, or forecast climate-related disasters, enabling early intervention. The analytics can optimize resource allocation with the identification of the most efficient land use, water, and energy for further sustainable practices (Ball et al. 2022; De Alban et al. 2020; Rahman and Sumantyo, 2010). The combination of remote sensing and data analytics enhances precision that allows interventions for more targeted and localized, rather than one-size-fits-all solutions. These technologies also enable the tracking of sustainability indicators, providing a way to measure and monitor progress towards global sustainability goals. Thus, the integration of remote sensing and data analytics offers a powerful toolkit for tackling persistent environmental challenges for better resource management, improved disaster resilience, and informed policy decisions to support long-term sustainability (Iannillo and Fasolino 2021; Nesticò and Maselli 2020).

6 Challenges for Sustainable Development with Data Analytics and Remote Sensing

There are several challenges which hold the widespread and effective use of data analytics and remote sensing for forward-moving sustainable development. One of the principal challenges is the quality and accuracy of data. Remote sensing data from satellites are affected by various factors such as cloud cover, atmospheric interference, and sensor limitations, leading to distorted or incomplete information (Kumar and Bassill 2024; Restrepo et al. 2021; Rosales Carreón and Worrell 2018; Zhao et al. 2008). The integration of diverse data sources from different platforms introduces issues of data compatibility and interoperability and these make it hard to combine and analyse data. Another noteworthy challenge is the high cost of both the technologies and infrastructure required to use remote sensing and data analytics. The purchase of high-resolution satellite imagery, maintenance of ground-based sensors, and powerful computing resources for data processing are financially unaffordable for developing countries or smaller organizations with limited budgets (De Alban et al. 2018; Holden et al. 2021; Konstantinov et al. 2015). There is also a skills gap in remote sensing and data analytics and it requires specialized knowledge in geospatial science, programming, machine learning, and environmental science. The shortage of skilled professionals limits the ability to fully leverage these technologies, particularly in regions that lack sufficient training resources. Likewise, the challenge of data access arises, as some critical datasets are restricted or expensive to obtain, preventing many stakeholders from using remote sensing for effective monitoring and planning. Privacy and ethical concerns related to data collection in sensitive regions lead to resistance to the use of remote sensing technologies (Gabriele et al. 2022; Lee et al. 2022). Data analytics provide valuable insights into future trends, and unforeseen variables including human behaviour or sudden natural events to affect the accuracy of predictions to make long-term planning for sustainable development. These predictive models based on remote sensing data often come with uncertainty. Thereby data analytics and remote sensing offer tremendous potential for sustainable development, challenges related to data quality, costs, accessibility, expertise, and uncertainty must be addressed to fully realize their benefits.

7 Impacts of Data Analytics and Remote Sensing on Sustainable Development

Data analytics with remote sensing have revolutionized the way we approach sustainable development. The integration of data analytics and remote sensing has substantial impacts on sustainable development for enhanced capabilities in monitoring, managing, and conserving resources. These technologies enable better decision-making to provide accurate, timely, and large-scale data for addressing global challenges such as climate change, resource depletion, and environmental degradation (Mohammed et al. 2020; Zhang and Zhao 2024). The most significant impact is the improvement in environmental monitoring for the continuous, real-time observation of ecosystems, land use, water bodies, and natural resources to provide the early detection of environmental changes. This helps policymakers and organizations to track critical indicators like deforestation, desertification, pollution, and biodiversity loss. These large datasets are processed and analysed to identify trends, predict future outcomes, and recommend targeted interventions. Analytics provide efficient resource management by optimizing agricultural practices, water use, and energy consumption (Horn 2020, 2021; Wooster et al. 2022). Remote sensing data monitor crop health and predict yields to make informed decisions on irrigation, fertilization, and pest control, leading to improved food security and reduced waste. It monitors water bodies for quality and availability with data analytics in predicting demand and distribution for sustainable use of water resources. This know-how plays a critical role in enhancing preparedness and response in the era of disaster management. It allows the monitoring of hazards like hurricanes, floods, wildfires, and earthquakes, providing real-time data to assess damage, plan evacuations, and allocate resources more effectively (Capodaglio and Olsson 2020; Croce and Vettorato 2021; Pioletti et al. 2018). Authorities can better anticipate and mitigate the impacts of natural disasters with predictive models to achieve sustainable development goals (SDGs) and track progress to identify gaps. They are used to evaluate improvement in SDG 13 (Climate Action), SDG 15 (Life on Land), and SDG 6 (Clean Water and Sanitation) with monitoring of climate patterns, land degradation, and water resources over time. These technologies support informed decision-making, promote efficiency in resource use, improve environmental conservation, and enhance disaster resilience—contributing to a more sustainable and resilient future.

8 Summary of Challenges and Opportunities

Data analytics and remote sensing have the potential to enhance efforts towards sustainable development, but their effective application has several challenges while offering a wide range of opportunities. These challenges and opportunities can summarized below:

8.1 Challenges for Data Analytics and Remote Sensing for Sustainable Development

(a) **Data Quality and Accuracy**

Remote sensing data are impacted by environmental factors such as cloud cover, atmospheric disturbances, and sensor calibration errors, leading to inaccurate or incomplete data. These issues limit the reliability of the information and make it difficult to generate precise insights for sustainable development.

(b) **High Costs**

The acquisition of high-resolution satellite imagery, maintaining sensor networks, and the infrastructure require expensive data storage and analysis. This is challenging for developing countries or smaller organizations with limited budgets to access these valuable technologies.

(c) **Data Integration and Interoperability**

Data from various sources (satellite imagery, drones, ground sensors) are collected in different formats, resolutions, and timescales. Integrating and harmonizing these diverse datasets are technically complex and require a significant effort to ensure the data can be combined effectively for analysis.

(d) **Lack of Skilled Workforce**

The use of remote sensing and data analytics requires a specific skill set in geospatial science, data processing, and environmental management. The skills gap is a major challenge for regions where access to education and training is limited.

(e) **Limited Access to Data**

Some high-quality remote sensing data is restricted or requires expensive licences. This limited access to real-time data or local-level datasets hinders the ability of governments and organizations to monitor and respond to environmental changes effectively.

(f) **Privacy and Ethical Concerns**

Remote sensing technologies capture high-resolution imagery and these raise concerns about privacy and data security in sensitive areas like residential zones or areas with proprietary land information. The ethical implications of data collection and usage limit the widespread adoption of these technologies.

(g) **Uncertainty in Predictions**

Predictive models based on remote sensing data are subject to uncertainty due to unpredictable environmental factors or human activities. These uncertainties destabilize the accuracy of forecasts and complicate long-term planning for sustainable development.

8.2 Opportunities for Data Analytics and Remote Sensing for Sustainable Development

(a) **Improved Environmental Monitoring**

Remote sensing offers an unparalleled ability to monitor large-scale environmental changes like deforestation, land degradation, and climate shifts. Data analytics process this information to detect trends, identify emerging risks, and provide actionable insights to protect ecosystems and biodiversity.

(b) **Optimized Resource Management**

Remote sensing helps farmers monitor soil moisture and health, optimize irrigation schedules, and predict crop yields, and it leads to increased food security with reduced environmental impact. These track water bodies to monitor water quality and predict demand for sustainable usage. Data analytics with remote sensing manages resources efficiently and optimizes agriculture practices, water and energy utilizations.

(c) **Enhanced Disaster Management and Response**

The analysis of remote sensing data helps to forecast disaster risks, plan evacuations, assess damage, and deploy resources more efficiently. The combination of real-time remote sensing data and predictive analytics provides better support to authorities to prepare and respond to natural disasters like floods, wildfires, hurricanes, and earthquakes.

(d) **Tracking Progress Towards SDGs**

Remote sensing and data analytics have a role in assessing sustainable development goals (SDGs). These enable governments and organizations to measure progress in areas like climate action, clean water and sanitation, and life on land to provide objective, data-driven assessments of environmental and social conditions.

(e) **Cost-Effective Monitoring**

Remote sensing offers a cost-effective means to monitor large areas and gather data without the need for physical presence and these are valuable in remote or inaccessible locations. The costs of satellite imagery and data processing have decreased over time and these technologies become more accessible and democratized access to critical information for sustainable development.

(f) **Support for Decision-Making and Policy Development**

Data analytics, when combined with remote sensing help in informed decision-making. It helps policymakers to understand the long-term implications of their decisions, optimize resource allocation, and identify sustainable alternatives. It guides urban planning, land use policies, and conservation efforts based on precise data on environmental changes and resource availability.

(g) **Empowering Local Communities**

Data analytics and remote sensing empower local communities with tools to monitor and manage their environment. These communities use low-cost remote sensing technologies to track environmental changes and advocate for more sustainable practices. This empowers local stakeholders to take an active role in shaping their development outcomes.

The intersection of data analytics and remote sensing offers substantial opportunities for sustainable development by providing powerful tools for monitoring, resource management, disaster response, and policy development. Nevertheless, challenges such as data quality, high costs, access limitations, and a lack of skilled expertise need to be addressed to answer their full potential. Overcoming these barriers through technological advancements, capacity building, and international collaboration will help ensure that these technologies are effectively applied to achieve global sustainability goals.

9 Summary

Data analytics and remote sensing have become essential tools for advancing sustainable development to monitor, manage, and conserve resources effectively across various sectors. These technologies provide real-time, large-scale data that supports informed decision-making, particularly in addressing global challenges such as climate change, environmental degradation, and resource depletion. Remote sensing allows for continuous monitoring of ecosystems, agriculture, and natural resources, offering insights that can help optimize resource management. Data analytics, on the other hand, processes and analyses vast amounts of data from remote sensing sources, uncovering patterns, trends, and actionable insights. They improve efficiency in areas like agriculture, water management, and disaster management, for more sustainable practices. They support the tracking and accomplishment of global sustainable development goals (SDGs) by offering real-time assessments of environmental and social conditions.

However, there are several challenges in applying these technologies effectively for sustainable development. Problems such as data quality, high costs, interoperability, and the skills gap in developing countries create barriers to widespread adoption. The concerns around data access, privacy, and the inherent uncertainty in predictive models complicate the use of remote sensing and data analytics for long-term planning. Despite these challenges, addressing them through improved infrastructure, training, and international collaboration could significantly enhance the impact of these technologies.

There is an extensive impact of data analytics and remote sensing on sustainable development. They provide valuable tools for environmental monitoring to track deforestation, land degradation, and climate change. These technologies support disaster preparedness and response for authorities to forecast and manage risks. They facilitate resource optimization and policy development for sustainable practices in agriculture, urban planning, and conservation. Even though data analytics and remote sensing face challenges related to cost, data quality, and expertise, their potential to enhance sustainable development is indisputable. These challenges can be addressed to utilize the advantages of these technologies by stakeholders to achieve significant progress in addressing the pressing environmental and societal issues of our time.

References

Abu-Rayash, A., & Dincer, I. (2021). Development of integrated sustainability performance indicators for better management of smart cities. *Sustainable Cities and Society*, *67*. https://doi.org/10.1016/j.scs.2020.102704

Ahmad, M. (2023). Spatially-aware artificial intelligence for sustainable development goals: Opportunities and challenges. In *Intelligent Engineering Applications and Applied Sciences for Sustainability*. IGI Global. https://doi.org/10.4018/979-8-3693-0044-2.ch024

Alam, A., Kumar, J., Sarkar, U. K., Jha, D. N., Sahu, S. K., Das, S. C. S., Srivastava, S. K., Kumar, V., & Das, B. K. (2024). Linking ecological characteristics with fish diversity, assemblage patterns and feeding guilds, and GIS applications along the temporal and spatial gradients in a large subtropical reservoir, India, for sustainable management. *Journal of Water and Climate Change*, *15*(2), 607–627. https://doi.org/10.2166/wcc.2023.513

Aleksandrowicz, L., Green, R., Joy, E. J. M., Harris, F., Hillier, J., Vetter, S. H., Smith, P., Kulkarni, B., Dangour, A. D., & Haines, A. (2019). Environmental impacts of dietary shifts in India: A modelling study using nationally-representative data. *Environment International*, *126*, 207–215. https://doi.org/10.1016/j.envint.2019.02.004

Alhamwi, A., Medjroubi, W., Vogt, T., & Agert, C. (2018). Modelling urban energy requirements using open source data and models. *Applied Energy*, *231*, 1100–1108. https://doi.org/10.1016/j.apenergy.2018.09.164

Ali, G., Abbas, S., Qamer, F. M., Wong, M. S., Rasul, G., Irteza, S. M., & Shahzad, N. (2021). Environmental impacts of shifts in energy, emissions, and urban heat island during the COVID-19 lockdown across Pakistan. *Journal of Cleaner Production*, *291*. https://doi.org/10.1016/j.jclepro.2021.125806

Alotaibi, E., & Nassif, N. (2024). Artificial intelligence in environmental monitoring: in-depth analysis. *Discover Artificial Intelligence*, *4*(1). https://doi.org/10.1007/s44163-024-00198-1

Appel, M., Lahn, F., Buytaert, W., & Pebesma, E. (2018). Open and scalable analytics of large Earth observation datasets: From scenes to multidimensional arrays using SciDB and GDAL. *ISPRS Journal of Photogrammetry and Remote Sensing*, *138*, 47–56. https://doi.org/10.1016/j.isprsjprs.2018.01.014

Aubrechtová, T., Semančíková, E., & Raška, P. (2020). Formulation matters! the failure of integrating landscape fragmentation policy. *Sustainability (Switzerland)*, *12*(10). https://doi.org/10.3390/SU12103962

Bakhshipour, A. E., Dittmer, U., Haghighi, A., & Nowak, W. (2021). Toward Sustainable Urban Drainage Infrastructure Planning: A Combined Multiobjective Optimization and Multicriteria Decision-Making Platform. *Journal of Water Resources Planning and Management*, *147*(8). https://doi.org/10.1061/(ASCE)WR.1943-5452.0001389

Ball, J. G. C., Petrova, K., Coomes, D. A., & Flaxman, S. (2022). Using deep convolutional neural networks to forecast spatial patterns of Amazonian deforestation. *Methods in Ecology and Evolution*, *13*(11), 2622–2634. https://doi.org/10.1111/2041-210X.13953

Basu, J. P. (2014). Agroforestry, climate change mitigation and livelihood security in India. *New Zealand Journal of Forestry Science*, *44*(Suppl 1), 1–10.

Bhanumurthy, V., Ram Mohan Rao, K., Jai Sankar, G., & Nagamani, P. V. (2017). Spatial data integration for disaster/emergency management: an Indian experience. *Spatial Information Research*, *25*(2), 303–314. https://doi.org/10.1007/s41324-017-0087-5

Bhattacharya, T. R., Bhattacharya, A., Mclellan, B., & Tezuka, T. (2020). Sustainable smart city development framework for developing countries. *Urban Research and Practice*, *13*(2), 180–212. https://doi.org/10.1080/17535069.2018.1537003

Böhner, C. (2006). Decision-support systems for sustainable urban planning. *International Journal of Environmental Technology and Management*, *6*(1–2), 193–205. https://doi.org/10.1504/ijetm.2006.008261

Botseva, D., Tanakov, N., & Nikolov, G. (2022). Intelligent Water Resources Management. *Environmental Science and Engineering*, 263–273. https://doi.org/10.1007/978-3-031-07500-1_23

Cai, M., Ren, C., Lau, K. K.-L., & Xu, Y. (2017). Spatial analysis on intra-urban temperature variation under extreme hot weather by incorporating urban planning and environmental parameters: A pilot study from Hong Kong. In B. L., R. S., & N. F. (Eds.), *Proceedings of 33rd PLEA International Conference: Design to Thrive, PLEA 2017* (Vol. 1, pp. 26–32). NCEUB 2017—Network for Comfort and Energy Use in Buildings.

Capodaglio, A. G., & Olsson, G. (2020). Energy issues in sustainable urban wastewater management: Use, demand reduction and recovery in the urban water cycle. *Sustainability (Switzerland)*, *12*(1). https://doi.org/10.3390/su12010266

Carretero-Peña, S., Blázquez, L. C., & Pinilla-Gil, E. (2019). Estimation of PM10 levels and sources in air quality networks by digital analysis of smartphone camera images taken from samples deposited on filters. *Sensors (Switzerland)*, *19*(21). https://doi.org/10.3390/s19214791

Cecchini, C., Magrini, A., & Morandotti, M. (2020). The energy-oriented management of public historic buildings: An integrated approach and methodology applications. *Sustainability (Switzerland)*, *12*(11). https://doi.org/10.3390/su12114576

Chauhan, A., Wasim, Md., Mohanty, S., Pandey, P. C., Pandey, M., Maurya, N. K., Rankavat, S., & Dubey, S. B. (2023). Earth observation applications for urban mapping and monitoring: research prospects, opportunities and challenges. In *Earth Observation in Urban Monitoring: Techniques and Challenges*. Elsevier. https://doi.org/10.1016/B978-0-323-99164-3.00007-0

Chen, P. (2019). Visualization of real-time monitoring datagraphic of urban environmental quality. *Eurasip Journal on Image and Video Processing*, *2019*(1). https://doi.org/10.1186/s13640-019-0443-6

Cooper, C. M., Troutman, J. P., Awal, R., Habibi, H., & Fares, A. (2022). Climate change-induced variations in blue and green water usage in U.S. urban agriculture. *Journal of Cleaner Production*, *348*. https://doi.org/10.1016/j.jclepro.2022.131326

Corburn, J. (2009). Cities, climate change and urban heat island mitigation: Localising global environmental science. *Urban Studies*, *46*(2), 413–427. https://doi.org/10.1177/0042098008099361

Corke, P., Wark, T., Jurdak, R., Hu, W., Valencia, P., & Moore, D. (2010). Environmental wireless sensor networks. *Proceedings of the IEEE*, *98*(11), 1903–1917. https://doi.org/10.1109/JPROC.2010.2068530

Croce, S., & Vettorato, D. (2021). Urban surface uses for climate resilient and sustainable cities: A catalogue of solutions. *Sustainable Cities and Society*, *75*. https://doi.org/10.1016/j.scs.2021.103313

Dao, Viet Nga; Ong, Hang See; Do, Nguyet Quang;Chee, Yung Xuen; Lai, L. C. (2012). Visualization Techniques in Smart Grid. *Smart Grid and Renewable Energy*, *03*(03), 175–185. https://doi.org/10.4236/sgre.2012.33025

Dasgupta, S., Li, B., & Chakraborty, S. (2023). Cost-effective soil sensors and data analytics: The future of soil characterization. In *Teaching in Land Management and Applied Geospatial Tools: The Use of Geographic Information Systems and Remote Sensing Data to Design Advanced Technical Interventions for Sustainable Land Management*. Nova Science Publishers, Inc.

De Alban, J. D. T., Connette, G. M., Oswald, P., & Webb, E. L. (2018). Combined Landsat and L-band SAR data improves land cover classification and change detection in dynamic tropical landscapes. *Remote Sensing*, *10*(2). https://doi.org/10.3390/rs10020306

De Alban, J. D. T., Jamaludin, J., Wong De Wen, D., Than, M. M., & Webb, E. L. (2020). Improved estimates of mangrove cover and change reveal catastrophic deforestation in Myanmar. *Environmental Research Letters*, *15*(3). https://doi.org/10.1088/1748-9326/ab666d

DeLancey, E. R., Kariyeva, J., Bried, J. T., & Hird, J. N. (2019). Large-scale probabilistic identification of boreal peatlands using Google Earth Engine, open-access satellite data, and machine learning. *PLoS ONE*, *14*(6). https://doi.org/10.1371/journal.pone.0218165

Dev, M., Krishna, V. V. S., Kumar, D., & Jha, M. K. (2023). Integrated geospatial database design for land use pattern analysis and its impact on local governance: A case study of Manesar Urban Complex, Gurugram, India. *Journal of Applied and Natural Science*, *15*(4), 1541–1556.

Doucet, T. C., Duinker, P. N., Charles, J. D., Steenberg, J. W. N., & Zurba, M. (2024). Characterizing Non-governmental Organizations And Local Government Collaborations In Urban Forest Management Across Canada. *Environmental Management*, *73*(1), 231–242. https://doi.org/10.1007/s00267-023-01889-8

Duong, H.-T., & Nguyen-Thi, T.-A. (2021). A review: preprocessing techniques and data augmentation for sentiment analysis. *Computational Social Networks*, *8*(1). https://doi.org/10.1186/s40649-020-00080-x

Eckman, R. S., Killough Jr., B. D., & Hilsenrath, E. (2008). Earth observations informing energy management: A CEOS and GEOSS perspective. *Proceedings of SPIE—The International Society for Optical Engineering*, *7151*. https://doi.org/10.1117/12.804931

Ekmen, O., & Kocaman, S. (2024). Remote sensing for UN SDGs: A global analysis of research and collaborations. *Egyptian Journal of Remote Sensing and Space Science*, *27*(2), 329–341. https://doi.org/10.1016/j.ejrs.2024.04.002

Elkind, K., Sankey, T. T., Munson, S. M., & Aslan, C. E. (2019). Invasive buffelgrass detection using high-resolution satellite and UAV imagery on Google Earth Engine. *Remote Sensing in Ecology and Conservation*, *5*(4), 318–331. https://doi.org/10.1002/rse2.116

Filimonau, V. (2019). Integrating the water-energy-food (Wef) nexus in the environmental governance of a metropolitan city in a developing economy: A case study of lima, peru. In *Policy and Governance in the Water-Energy-Food Nexus: A Relational Equity Approach*. Taylor and Francis. https://doi.org/10.4324/9780429427718-9

Forghani, M., & Delavar, M. (2014). A Quality Study of the OpenStreetMap Dataset for Tehran. *ISPRS International Journal of Geo-Information*, *3*(2), 750–763. https://doi.org/10.3390/ijgi3020750

Gabriele, M., Brumana, R., Previtali, M., & Cazzani, A. (2022). A combined GIS and remote sensing approach for monitoring climate change-related land degradation to support landscape preservation and planning tools: the Basilicata case study. In *Applied Geomatics*. Springer Berlin Heidelberg. https://doi.org/10.1007/s12518-022-00437-z

Ghisellini, P., Cialani, C., & Ulgiati, S. (2016). A review on circular economy: The expected transition to a balanced interplay of environmental and economic systems. In *Journal of Cleaner Production*. https://doi.org/10.1016/j.jclepro.2015.09.007

Gogu, C. R., Boukhemacha, M. A., Gaitanaru, D., & Moraru, I. (2019). Interaction Between City Subsurface Infrastructure and Groundwater. *Green Energy and Technology*, 219–223. https://doi.org/10.1007/978-3-319-99867-1_36

Goumehei, E., Tolpekin, V., Stein, A., & Yan, W. (2019). Surface Water Body Detection in Polarimetric SAR Data Using Contextual Complex Wishart Classification. *Water Resources Research*, *55*(8), 7047–7059. https://doi.org/10.1029/2019WR025192

Gutiérrez-Carrillo, M. L., Guerrero Delgado, M. C., Sánchez Ramos, J., Arco Díaz, J., Bestué Cardiel, I., & Álvarez Domínguez, S. (2021). Mitigating damage on heritage structures by continuous conservation using thermal real-time monitoring. Case study of Ziri Wall, city of Granada, Spain. *Journal of Cleaner Production, 296*. https://doi.org/10.1016/j.jclepro.2021.126522

Hendrickson, T. P., Nikolic, M., & Rakas, J. (2016). Selecting climate change mitigation strategies in urban areas through life cycle perspectives. *Journal of Cleaner Production, 135*, 1129–1137. https://doi.org/10.1016/j.jclepro.2016.06.075

Hird, J. N., DeLancey, E. R., McDermid, G. J., & Kariyeva, J. (2017). Google earth engine, open-access satellite data, and machine learning in support of large-area probabilisticwetland mapping. *Remote Sensing, 9*(12). https://doi.org/10.3390/rs9121315

Holden, P. B., Rebelo, A. J., & New, M. G. (2021). Mapping invasive alien trees in water towers: A combined approach using satellite data fusion, drone technology and expert engagement. *Remote Sensing Applications: Society and Environment, 21*. https://doi.org/10.1016/j.rsase.2020.100448

Horn, C. (2020). Biodiversity in cities. *Project Baikal, 67*, 64–71. https://doi.org/10.51461/projectbaikal.67.1756

Horn, C. (2021). Biodiversity in cities. *Plant Science Today, 8*(2), 64–71. https://doi.org/10.14719/PST.2021.8.2.1027

Iannillo, A., & Fasolino, I. (2021). Land-use mix and urban sustainability: Benefits and indicators analysis. *Sustainability (Switzerland), 13*(23). https://doi.org/10.3390/su132313460

Ivey, J. L., Smithers, J., De Loë, R. C., & Kreutzwiser, R. D. (2004). Community Capacity for Adaptation to Climate-Induced Water Shortages: Linking Institutional Complexity and Local Actors. *Environmental Management, 33*(1), 36–47. https://doi.org/10.1007/s00267-003-0014-5

Jain, M., Siva, V., Hoppe, T., & Bressers, H. (2020). Assessing governance of low energy green building innovation in the building sector: Insights from Singapore and Delhi. *Energy Policy, 145*(May), 111752. https://doi.org/10.1016/j.enpol.2020.111752

Jehanzaib, M., & Kim, T.-W. (2020). Exploring the influence of climate change-induced drought propagation on wetlands. *Ecological Engineering, 149*. https://doi.org/10.1016/j.ecoleng.2020.105799

Kesavan, H., Choudhary, S., Khan, S., Samhita, R., & Sharma, A. (2024). AI Based Urban Resilience Planning: Opportunities and Challenges. *Journal of Environmental and Earth Sciences, 6*(2), 200–214. https://doi.org/10.30564/jees.v6i2.6681

Konstantinov, P. I., Grishchenko, M. Y., & Varentsov, M. I. (2015). Mapping urban heat islands of arctic cities using combined data on field measurements and satellite images based on the example of the city of Apatity (Murmansk Oblast). *Izvestiya—Atmospheric and Ocean Physics, 51*(9), 992–998. https://doi.org/10.1134/S000143381509011X

Krajačić, G., Vujanović, M., Duić, N., Kılkış, Ş., Rosen, M. A., & Ahmad Al-Nimr, M. (2018). Integrated approach for sustainable development of energy, water and environment systems. *Energy Conversion and Management, 159*, 398–412. https://doi.org/10.1016/j.enconman.2017.12.016

Kumar, B., & Bhaduri, S. (2018). Disaster risk in the urban villages of Delhi. *International Journal of Disaster Risk Reduction, 31*(May), 1309–1325. https://doi.org/10.1016/j.ijdrr.2018.04.022

Kumar, N. (2003). Data Quality And Flow Of Errors In GIS. *Indian Cartographer*, 158–163.

Kumar, R., Son, L. H., Jha, S., Mittal, M., & Goyal, L. M. (2018). Spatial data analysis using association rule mining in distributed environments: a privacy prospect. *Spatial Information Research, 26*(6), 629–638. https://doi.org/10.1007/s41324-018-0207-x

Kuznetsova, E., Cardin, M.-A., Diao, M., & Zhang, S. (2019). Integrated decision-support methodology for combined centralized-decentralized waste-to-energy management systems design. *Renewable and Sustainable Energy Reviews, 103*, 477–500. https://doi.org/10.1016/j.rser.2018.12.020

Kwak, Y., Park, C., & Deal, B. (2020). Discerning the success of sustainable planning: A comparative analysis of urban heat island dynamics in Korean new towns. *Sustainable Cities and Society, 61*. https://doi.org/10.1016/j.scs.2020.102341

Lancia, M., Zheng, C., He, X., Lerner, D. N., & Andrews, C. (2020). Groundwater Complexity in Urban Catchments: Shenzhen, Southern China. *Groundwater, 58*(3), 470–481. https://doi.org/10.1111/gwat.12935

Lasco, R. D., Delfino, R. J. P., Catacutan, D. C., Simelton, E. S., & Wilson, D. M. (2014). Climate risk adaptation by smallholder farmers: The roles of trees and agroforestry. *Current Opinion in Environmental Sustainability, 6*(1), 83–88. https://doi.org/10.1016/j.cosust.2013.11.013

Laurini, R., Nijkamp, P., Kourtit, K., & Bouzouina, L. (2023). Knowledge management for regional policymaking. In *Knowledge Management for Regional Policymaking*. Springer International Publishing. https://doi.org/10.1007/978-3-031-15648-9

Lay-Ekuakille, A., Djungha Okitadiowo, J. P., Di Luccio, D., Palmisano, M., Budillon, G., Benassai, G., & Maggi, S. (2021). Image sensors for wave monitoring in shore protection: Characterization through a machine learning algorithm. *Sensors, 21*(12). https://doi.org/10.3390/s21124203

Le, T. T., Priya, J. C., Le, H. C., Le, N. V. L., Duong, M. T., & Cao, D. N. (2024). Harnessing artificial intelligence for data-driven energy predictive analytics: A systematic survey towards enhancing sustainability. *International Journal of Renewable Energy Development, 13*(2), 270 – 293. https://doi.org/10.61435/ijred.2024.60119

Lee, K., Kim, Y., Sung, H. C., Kim, S. H., & Jeon, S. W. (2022). Surface urban heat island in South Korea's new towns with different urban planning. *Environmental Monitoring and Assessment, 194*(5). https://doi.org/10.1007/s10661-022-09967-w

Liao, Y., Liao, W., Wu, D., Zhang, C., Xu, J., & Majumdar, S. (2024). Inspired Applications of Artificial Intelligence in Disaster-Resilient Regional Construction for Climate Risk. *Urban Sustainability, Part F3945*, 305–315. https://doi.org/10.1007/978-981-97-9658-8_16

Lombardi, P., Abastante, F., Torabi Moghadam, S., & Toniolo, J. (2017). Multicriteria spatial decision support systems for future urban energy retrofitting scenarios. *Sustainability (Switzerland), 9*(7). https://doi.org/10.3390/su9071252

Lu, X., Griebsch, C., Norris, J. M., & Ward, M. P. (2023). Landscape, Socioeconomic, and Meteorological Risk Factors for Canine Leptospirosis in Urban Sydney (2017–2023): A Spatial and Temporal Study. *Veterinary Sciences, 10*(12). https://doi.org/10.3390/vetsci10120697

Lunetta, R. S., Congalton, R. G., Fenstermaker, L. K., Jensen, J. R., McGwire, K. C., & Tinney, L. R. (1991). Remote sensing and geographic information system data integration: error sources and research issues. *Photogrammetric Engineering & Remote Sensing, 57*(6), 677–687.

Ma, Q., Li, H., & Thorstenson, A. (2021). A big data-driven root cause analysis system: Application of Machine Learning in quality problem solving. *Computers and Industrial Engineering, 160*. https://doi.org/10.1016/j.cie.2021.107580

Maheswari, E. A. P., Mernisi, F. A., Sidharta, S., & Puspitasari, C. (2024). Harnessing Artificial Intelligence for Effective Coastal Flood Disaster Management: A Systematic Literature Review. *Proceedings of 2024 International Conference on Information Management and Technology, ICIMTech 2024*, 530–535. https://doi.org/10.1109/ICIMTech63123.2024.10780929

Makropoulos, C., Butler, D., & Maksimovic, Ĉ. (1999). GIS supported evaluation of source control applicability in urban areas. *Water*

Science and Technology, 39(9), 243–252. https://doi.org/10.1016/S0273-1223(99)00238-3
Malhotra, P., Singh, Y., Anand, P., Bangotra, D. K., Singh, P. K., & Hong, W.-C. (2021). Internet of things: Evolution, concerns and security challenges. *Sensors, 21*(5), 1–35. https://doi.org/10.3390/s21051809
McNamara, K. E., & Olson, L. L. (2014). Moving towards risk? Intersections of disaster and mobility. *Proceedings of the 5th International Disaster and Risk Conference: Integrative Risk Management—The Role of Science, Technology and Practice, IDRC Davos 2014*, 465–468.
Michael, F. L., Noor, Z. Z., & Figueroa, M. J. (2014). Review of urban sustainability indicators assessment - Case study between Asian countries. *Habitat International, 44*, 491–500. https://doi.org/10.1016/j.habitatint.2014.09.006
Mishra, T., & Nair, P. S. (2024). Advancing Agriculture Predictive Models for Farming Suitability Using Machine Learning. *International Journal of Intelligent Systems and Applications in Engineering, 12*(5s), 494 – 502.
Mohammed, S. O., Razoumny, Y. N., Orlovskiy, A., Lobanov, V., & Vasyunin, Y. (2020). Space-based monitoring capabilities for the evaluation of inland water resources: Application to Nigeria. In G. A. D. C. J.-M. Razoumny Y.N. Graziani F. (Ed.), *Advances in the Astronautical Sciences* (Vol. 170, pp. 525–532). Univelt Inc.
Mondol, S. C., Adhikary, S. K., Nath, H., & Shuvo, S. P. (2025). A Deep Learning Approach Using Long Short-Term Memory Networks for Enhanced Prediction of Rainfall in the Northeastern Region of Bangladesh. In R. M., A. M.S., & M. M.H. (Eds.), *AIP Conference Proceedings* (Vol. 3262, Issue 1). American Institute of Physics. https://doi.org/10.1063/5.0247626
Nesticò, A., & Maselli, G. (2020). Sustainability indicators for the economic evaluation of tourism investments on islands. *Journal of Cleaner Production, 248*. https://doi.org/10.1016/j.jclepro.2019.119217
Nichol, J., & Wong, M. S. (2009). Mapping urban environmental quality using satellite data and multiple parameters. *Environment and Planning B: Planning and Design, 36*(1), 170–185. https://doi.org/10.1068/b34034
Olawuyi, S. O., & Mushunje, A. (2019). Determinants Of Adoption And Use-Intensity Of Soil And Water Conservation Practices Among Smallholder Farmers In Nigeria. *African Journal of Food, Agriculture, Nutrition and Development, 19*(3), 14571–14586. https://doi.org/10.18697/ajfand.86.17745
Oliveira, V. (2021). Urban form and the socioeconomic and environmental dimensions of cities. *Journal of Urbanism*. https://doi.org/10.1080/17549175.2021.2011378
Palmieri, M., Bozzella, S., Cascella, G. L., Bronzini, M., Torresi, M., & Cupertino, F. (2018). Wind micro-turbine networks for urban areas: Optimal design and power scalability of permanent magnet generators. *Energies, 11*(10). https://doi.org/10.3390/en11102759
Pandey, V., & Srivastava, P. K. (2019). Integration of microwave and optical/infrared derived datasets for a drought hazard inventory in a sub-tropical region of India. *Remote Sensing, 11*(4). https://doi.org/10.3390/rs11040439
Pioletti, M., Brigolin, D., & Pastres, R. (2018). The life cycle impact of energy final uses in small urban systems: Implications for emission accounting and EU sustainable local energy planning. *Sustainable Cities and Society, 42*, 252–258. https://doi.org/10.1016/j.scs.2018.07.003
Pramanik, S. (2024). AI's function in sustainable development's renewable energy planning. In *Next Generation Materials for Sustainable Engineering*. IGI Global. https://doi.org/10.4018/979-8-3693-1306-0.ch016
Rahman, M. M., & Sumantyo, J. T. S. (2010). Mapping tropical forest cover and deforestation using synthetic aperture radar (SAR) images. *Applied Geomatics, 2*(3), 113–121. https://doi.org/10.1007/s12518-010-0026-9
Renuka Devi, R., & Kesavaraj, G. (2019). Big data analytics in healthcare: Tools and application perspective. *International Journal of Advanced Science and Technology, 28*(17), 910–918.
Rijal, M., Luo, P., Mishra, B. K., Zhou, M., & Wang, X. (2024). Global systematical and comprehensive overview of mountainous flood risk under climate change and human activities. *Science of the Total Environment, 941*. https://doi.org/10.1016/j.scitotenv.2024.173672
Rodriguez-Bachiller, a. (2000). Geographical Information Systems and Expert Systems for Impact Assessment Part II : Expert Systems and Decision Support Systems. In *Journal of Environmental Assessment Policy and Management* (Vol. 2, Issue 3). https://doi.org/10.1016/S1464-3332(00)00042-4
Rosales Carreón, J., & Worrell, E. (2018). Urban energy systems within the transition to sustainable development. A research agenda for urban metabolism. *Resources, Conservation and Recycling, 132*, 258–266. https://doi.org/10.1016/j.resconrec.2017.08.004
Saah, D., Johnson, G., Ashmall, B., Tondapu, G., Tenneson, K., Patterson, M., Poortinga, A., Markert, K., Quyen, N. H., San Aung, K., Schlichting, L., Matin, M., Uddin, K., Aryal, R. R., Dilger, J., Lee Ellenburg, W., Flores-Anderson, A. I., Wiell, D., Lindquist, E., Goldstein J, Clinton N, Chishtie, F. (2019). Collect Earth: An online tool for systematic reference data collection in land cover and use applications. *Environmental Modelling and Software, 118*, 166–171. https://doi.org/10.1016/j.envsoft.2019.05.004
Selvam, A. P., & Al-Humairi, S. N. S. (2025). Environmental impact evaluation using smart real-time weather monitoring systems: a systematic review. *Innovative Infrastructure Solutions, 10*(1). https://doi.org/10.1007/s41062-024-01817-7
Shubin, R. (2019). How blue–green infrastructure can create liveable cities and address climate change. *SpringerBriefs in Architectural Design and Technology*, 55–63. https://doi.org/10.1007/978-981-13-0713-3_9
Simões, P., Matos, R. S., Costa, P. M., Castro, C., & Ferreira, P. T. (2022). Green and Blue Infrastructures as A Model of Sustainable Urban Planning—Landscape Design for Praça De Espanha in Lisbon—Portugal. Advances in Science, Technology and Innovation, 215–228. https://doi.org/10.1007/978-3-030-86499-6_19
Sudmanns, M., Tiede, D., Lang, S., Bergstedt, H., Trost, G., Augustin, H., Baraldi, A., & Blaschke, T. (2019). *Big Earth data: disruptive changes in Earth observation data management and analysis?* https://doi.org/10.1080/17538947.2019.1585976
Tardioli, G., Kerrigan, R., Oates, M., O'Donnell, J., & Finn, D. P. (2018). Identification of representative buildings and building groups in urban datasets using a novel pre-processing, classification, clustering and predictive modelling approach. *Building and Environment, 140*(May), 90–106. https://doi.org/10.1016/j.buildenv.2018.05.035
Toth, C., & Jóźków, G. (2016). Remote sensing platforms and sensors: A survey. In *ISPRS Journal of Photogrammetry and Remote Sensing*. https://doi.org/10.1016/j.isprsjprs.2015.10.004
Turner, W., Spector, S., Gardiner, N., Fladeland, M., Sterling, E., & Steininger, M. (2003). Remote sensing for biodiversity science and conservation. In *Trends in Ecology and Evolution* (Vol. 18, Issue 6, pp. 306–314).
Ulpiani, G., Ranzi, G., Feng, J., & Santamouris, M. (2021). Expanding the applicability of daytime radiative cooling: Technological developments and limitations. Energy and Buildings, 243. https://doi.org/10.1016/j.enbuild.2021.110990
Verma, S., & Bhalla, P. (2024). Sustainable development and AI: Navigating safety and ethical challenges. In *Demystifying the Dark Side of AI in Business*. IGI Global. https://doi.org/10.4018/979-8-3693-0724-3.ch011
Wang, D.-H., Hsieh, H.-C., Wu, C.-S., Honjo, T., Chiang, Y.-J., & Yang, P.-A. (2012). Visualization with Google Earth and gaming engine. *Proceedings—2012 IEEE 4th International Symposium on Plant Growth Modeling, Simulation, Visualization and Applications, PMA 2012*, 426–430. https://doi.org/10.1109/PMA.2012.6524868

Wang, K., Franklin, S. E., Guo, X., & Cattet, M. (2010). Remote Sensing of Ecology, Biodiversity and Conservation: A Review from the Perspective of Remote Sensing Specialists. In *Sensors* (Vol. 10, Issue 11, pp. 9647–9667).

Wassie, Y. A., Koeva, M. N., Bennett, R. M., & Lemmen, C. H. J. (2017). A procedure for semi-automated cadastral boundary feature extraction from high-resolution satellite imagery. *Journal of Spatial Science*, *8596*, 1–18. https://doi.org/10.1080/14498596.2017.1345667

Weber, S., Sadoff, N., Zell, E., & de Sherbinin, A. (2015). Policy-relevant indicators for mapping the vulnerability of urban populations to extreme heat events: A case study of Philadelphia. *Applied Geography*, *63*, 231–243. https://doi.org/10.1016/j.apgeog.2015.07.006

Wooster, E. I. F., Fleck, R., Torpy, F., Ramp, D., & Irga, P. J. (2022). Urban green roofs promote metropolitan biodiversity: A comparative case study. *Building and Environment*, *207*. https://doi.org/10.1016/j.buildenv.2021.108458

Zhang, L., Zhang, M., & Yao, Y. (2019). Multi-time scale analysis of regional aerosol optical depth changes in national-level urban agglomerations in China using MODIS collection 6.1 datasets from 2001 to 2017. *Remote Sensing*, *11*(2), 201. https://doi.org/10.3390/rs11020201

Zhang, L., & Zhao, H. (2024). Sustainable development mechanism: The role of natural resources, remittance and policy uncertainty. *Resources Policy*, *90*. https://doi.org/10.1016/j.resourpol.2023.104621

Zhao, J., Liu, H., & Dong, R. (2008). Sustainable urban development: Policy framework for sustainable consumption and production. *International Journal of Sustainable Development and World Ecology*, *15*(4), 318–325. https://doi.org/10.3843/SusDev.15.4:5a

Mapping and Identification of Coconut Plantations Through a Pixel-Based Classification Algorithm with Sentinel 2 Datasets

B. Anand, R. Shanmathi Rekha, M. Vijaya Prabakaran, R. Amirtha Varshini, M. Kavya, B. Reshma, and K. Ramaswamy

1 Introduction

Coconut is one of the main crops that safeguard the livelihood of countless people in the world (Cocos nucifera L.), and it is the worldwide most commercially significant palm species (Rouse et al. 1974; Frolking et al. 2002). The cultivation of coconut trees in tropical areas has a significant social and financial impact on coconut farmers. Worldwide over 94 nations cultivate coconut, and it produces 67.04 million coconuts. Therefore, mapping and precise estimation of coconut production has a vital role in boosting the economy (Sahay et al. 2017; Roxy 2010; Agarwal et al. 2013). India is the world's biggest producer of coconut, representing around 31.45% of worldwide produce (Sahay et al. 2017). Tamil Nadu is developed into a traditional region for coconut cultivation after Kerala and now produces more coconuts than any other state (Chen 2004; Ananthakrishnan 1988; Breiman 1996). Coconut has been demonstrated to thrive in a variety of agro-climatic environments despite its tropical origins.

The coconut has an incredible impact on the economy of rural areas, providing a means of subsistence for over 10 million people. The coconut trees generate a substantial amount of eco-friendly and cost-effective raw materials during their growth and processing (Dalponate et al. 2008). These comprise fluorescence for sugar, husks for coir mattings, vinegar production, coconut shells for crafting ornaments, activated charcoal, coconut leaves for broomsticks, and coconut stems for crafting furniture (Breiman et al. 1996, 2001; Wardlow et al. 2002; Abonales et al. 2003; Abou et al. 2003; Hatfield et al. 2010; Nayar 2017; Amara et al. 2017).

Six percent of the nation's edible oil supply comes from coconut oil. Additionally, the crop earns an annual contribution of 7000 crores to the GDP. As a result, coconut farming is becoming more prevalent in less conventional locales. Suitable soils for the coconut plantations are weak tropical soils that include alluvial, red sandy loam, laterite, coastal sandy, and reclaimed soils with pH levels between 5.2 and 8.0. However, this variety prefers a particular kind of soil for growth (Clausi 2002; Demmerers et al. 2012; Santillan et al. 2012; Bricher et al. 2013; Briggs et al. 2013; Carlson et al. 2013; Bocardo et al. 2015; Baruth et al. 2008; Kannan et al. 2017).

The cultivation of coconut is considered one of the most significant crops that influence the Indian economy. A comprehensive and large-scale assessment of the coconut plantation is necessary to understand its extent and existing cultivation scale. This helps in finding a market for the surplus coconuts, which can be challenging due to the rising trend in production (Zhang et al. 2011). Accurately measuring the area of a coconut plantation can be a difficult and time-consuming task. Traditional methods involve physically measuring the boundaries of the plantation, which can be challenging in areas with rugged terrain or dense vegetation (Walsh et al. 2001; Chen 2004; Boulesteix et al. 2012; Immitizer 2012; Hebbar 2014). Remote sensing technology has emerged as a valuable tool for mapping the area of coconut plantations in recent years. The study shows that the area of production and productivity of coconut in the country has increased. It was deemed necessary to investigate production and provide a fresh assessment of the evolving coconut area pattern due to the shifting market conditions in the industry.

B. Anand (✉) · R. Shanmathi Rekha
Department of Civil Engineering, Shri Vishnu Engineering College for Women, Vishnupur, Bhimavaram, Andhra Pradesh, India
e-mail: anandbalasubramanian06@gmail.com; dranandce@svecw.edu.in

R. Shanmathi Rekha
e-mail: drshanmathirekhace@svecw.edu.in

M. Vijaya Prabakaran · R. Amirtha Varshini · M. Kavya · B. Reshma · K. Ramaswamy
Department of Agriculture Engineering, KIT—Kalaignarkarunanidhi Institute of Technology, Coimbatore, Tamil Nadu, India

D. Kumar et al. (eds.), *Data Analytics and Remote Sensing-Based Analysis for Sustainable Development*, Advances in Science, Technology & Innovation, https://doi.org/10.1007/978-3-031-98095-4_2

Researchers have developed algorithms that can automatically detect and classify individual trees within a plantation using satellite imagery to overcome this challenge. These algorithms use a combination of spectral, spatial, and textural features to identify and classify trees based on their size, shape, and color. Once the trees have been classified, the total area of the plantation can be calculated using standard GIS software. The farmers could not get adequate profit for the coconut on many occasions due to the higher adoption rate of coconut farming. This work aims to compute the precision and efficiency of geospatial technologies in delineating coconut plantation areas. The objectives include (i) to utilize the sentinel 2 multispectral remote sensing data for meticulous measurement of coconut cultivation of the study area, (ii) to classify the landuse and land cover features and extracting the count of coconut trees in the selection region, and (iii) to estimate precise species-based statistics and comparative analysis of user and producer accuracy.

The coconut crop area is identified with remote sensing techniques and helps in estimating the cropping rate of the field (Parthasarathy et al. 2003; Son et al. 2012; Saheb et al. 2013; Ochoa 2019). The available fundamental information on the areas where coconut is farmed is not reliable. The investigation developed is to examine the issue and to fill the knowledge gap vis-à-vis the coconut-cultivated areas in the study area. Satellite imagery offers a high-resolution interpretation of the Earth's surface and allows researchers to identify and measure individual trees within a plantation. These data can be used to calculate the total area of the plantation. However, accurately identifying individual trees within a plantation can be challenging, especially in areas with dense vegetation. Reliable data are crucial for the coconut area to be planned and developed. The information and specifics of coconut-cultivated areas are required to increase the growth and production, as well as installing new agricultural measures to empower the profit and economy level of the block.

The spectral reflectance and other patterns of vegetation help in identifying the health and growth of the vegetation. They primarily use supervised and unsupervised classification takes the part, as well as multispectral and multitemporal data, for the identification and classification of crops. The spectral signatures differ for every vegetation or plantation in satellite images. It becomes difficult to identify with various classification methods due to the above reasons or spectral information. Identifying the coconut crop areas in the area of interest is the main aim of the study which includes data analysis, processing, and classification (Cross 2018; Mandal et al. 2011; Mellor et al. 2012; Xu et al. 2019). So selecting or exploring the improved classification methods is very important in crop identification (Kogan 1995; Bartolucci et al. 1977; Booth et al. 1989; Gad and Kusky 2007; Collen 2009; Singh et al. 2011; Kuenzer et al. 2013; Millioto et al. 2017; Kamalaris et al. 2018). The detection and estimation of crop rate per unit of land help in providing suitable agricultural measures to increase profit while growing coconut palm.

2 Study Area

The area selected for this research is the Coimbatore district in Tamil Nadu state of India. It is located near the coordinates 11.10168°N and 76.9558°E. It lies near the extreme northwestern area of Kerala in Tamil Nadu. The Nilgiris Biosphere Reserve surrounds the northern side of the mountains, while the reserve forest encloses the western side. The city and the eastern part of the district are primarily dry. The area has red soil covering around 60% of it. The average maximum temperature is between 35.9° and 29.2 °C, while the average lowest temperature is between 24.5° and 19.8 °C. Out of a total geographical area of 367,099 hectares, the district's net cultivated area is around 165,260 ha. The major plantation crop in the area is coconut with around 85,831 acres of area. The district of Coimbatore served as the intentional choice due to the reason that the Coimbatore district has been influenced by the significance of coconut. Udumalpet was selected (as shown in Fig. 1) among the 12 blocks in the district.

The state now grows coconut on 4.44 lakh hectares, producing 5128.1 million coconuts per hectare with a productivity of 11,526 coconuts per hectare. The state accounts for 21% of India's coconut farming area and 26% of coconut production, and it leads the country in coconut processing operations.

The most land being used is 8846.7 hectares for growing coconut in Coimbatore. India produced 19,247 million coconuts in 2021–2022, and it was 31% more than of the world's total production. The export value of coconut products, excluding coir items, surged by 41% to reach 3,236.83 crore in the year 2021–22. As a result, coconut production, productivity, and marketing have become integral components of the nation's economic growth.

3 Methodology

Crop classification and identification are essential for all countries, and generating distinct crop type maps is the responsibility of national/multinational agricultural agencies and other organizations with a focus on agriculture to an inventory of the crops. Figure 2 explains the comprehensive methodology of the study.

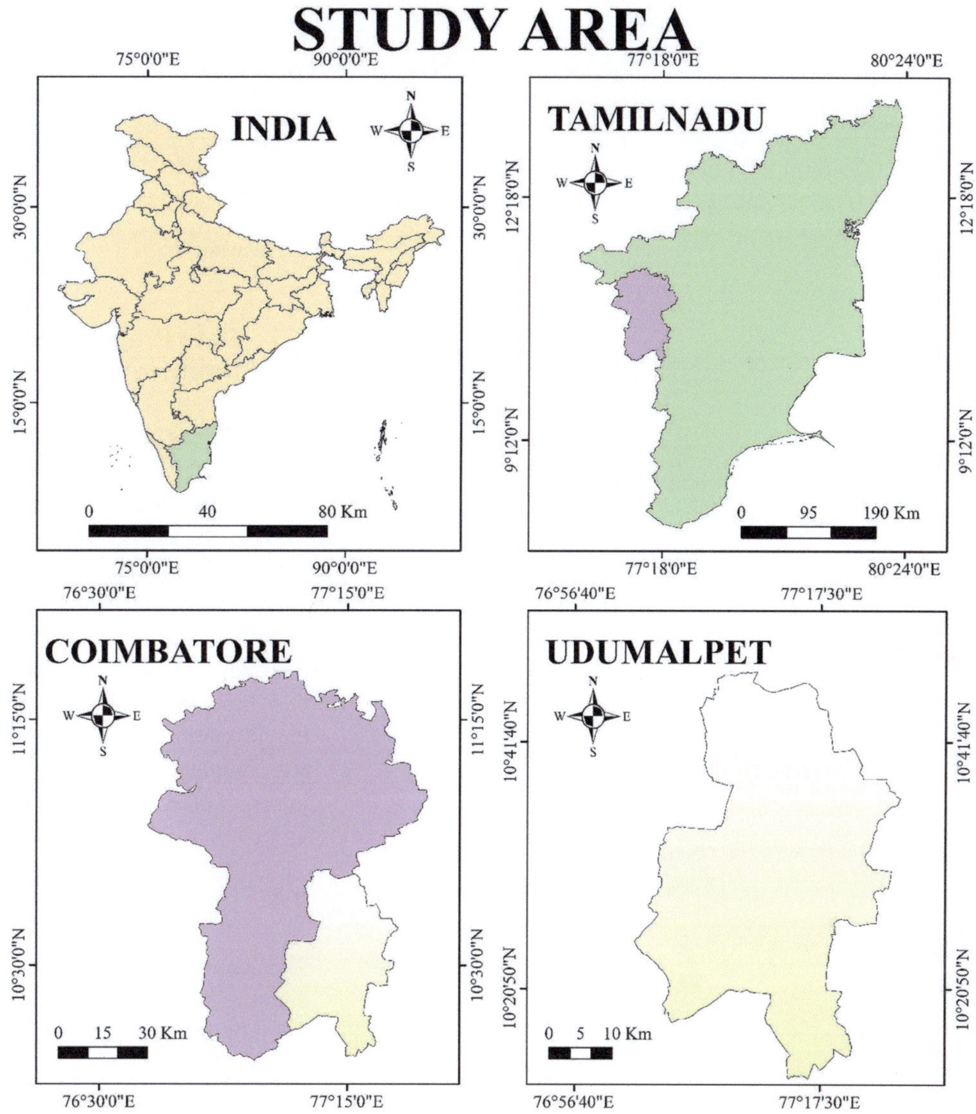

Fig. 1 Location map of the study area

3.1 Data Collection

High-resolution Sentinel 2 satellite data were employed to collect more accurate information and to differentiate coconut plantations from other species. The five-day worldwide S2 revisit generates multispectral imaging across the large regions.

The satellite carries a single multispectral instrument (MSI), which has 13 spectral bands as shown in Table 1 with a radiometric resolution of 12 bits in the visible/near-infrared (VNIR) and short-wave infrared (SWIR) spectral ranges with spatial resolutions of 4 bands at 10 m, 6 bands at 20 m, and 3 bands at 60 m. The orbital sweep measures 290 km in breadth. The Copernicus Scientific Data Portal provided Sentinel 2 photographs for the chosen blocks, which QGIS was used to access.

3.2 Data Selection

All relevant information was obtained precisely with the sensor details. This study uses a Sentinel 2 satellite imagery (three bands) for the year 2022 with mentioned band combinations as shown in Table 2.

The band combination plays a vital role for a better understanding of the detection of the features in images. We can easily identify the feature of interest from a precise image. There are band combinations to draw attention to geological, agricultural, or vegetative elements of an image. We can extract specific information from the image with the use of band combinations.

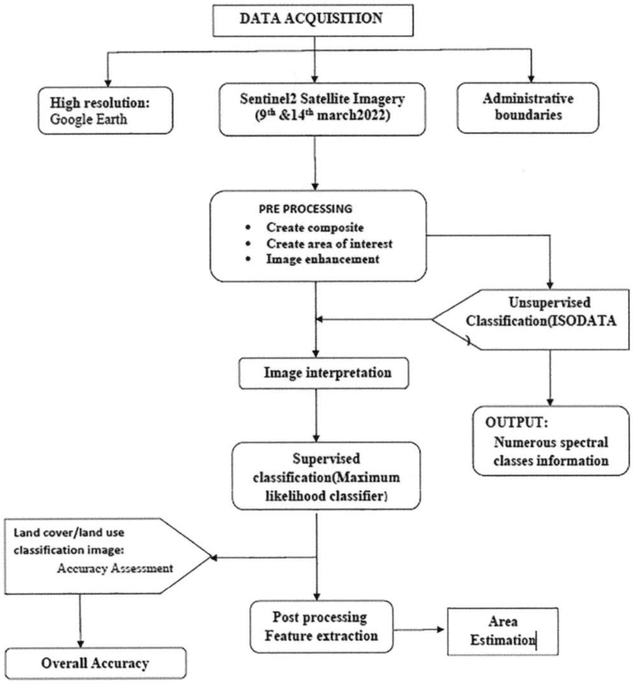

Fig. 2 Detailed methodology flow work of the study

Table 1 Combination of spectral bands of Sentinel 2 data

Sentinel2 bands	Central wavelength (nm)	Resolution (meters)	Band width (nm)
Band 1—coastal aerosol	443	60	20
Band 2—Blue	490	10	65
Band 3—Green	560	10	35
Band 4—Red	665	10	30
Band 5—Vegetation red edge	705	20	15
Band 6—Vegetation red edge	740	20	20
Band 7—Vegetation red edge	783	20	15
Band 8—NIR	842	10	15
Band8A—-Vegetation red edge	865	20	20
Band 9—Water vapor	945	60	20
Band10-SWIR—Cirrus	1375	60	30
Band 11—SWIR	1610	20	90
Band 12—SWIR	2190	20	180

Table 2 Different combinations and applications of spectral bands

Natural color	B4, B3, B2
Color infrared	B8, B4, B3
SWIR	B12, B8A, B4
Agriculture	B11, B8, B2
Geology	B12, B11, B2
Bathymetric	B4,B3,B1

3.3 Data Preprocessing and Image Interpretation

The collected satellite imagery will be processed before being inspected for cloud cover to reduce radiometric and geometric errors. The collected data are already geo-referenced. To increase the image's quality and make it easier to read, image enhancement is applied. It is accomplished with the contrast enhancement and histogram equalization approach. The input feature is used to clip off the selected region or area. The Sentinel 2 data are visually examined with the interpretation keys to determine the types of land cover present in the research region after image preprocessing. Most reliable details about features are easily inspected with image interpretation.

3.4 Improve Image Classification Methods

The identification and classification of various crops in an area are greatly aided by the image classification technique. Estimating crop acreage and crop productivity help in accurate crop identification. The unsupervised classification technique is applied to the sentinel 2 data that form the cluster of pixels as per their reflectance value. K-means clustering and ISODATA (Iterative Self-Organizing Data Analysis) are two familiar algorithm types of unsupervised machine learning methods. ISODATA algorithm is adopted over the K-means due to the adaptive number of clusters, dynamic cluster size, robustness to outliers, progressive refinement due to iteration, suitability for high dimensional data, tolerance to noise, and ability to handle unequal density clusters. ISODATA clustering algorithm gives the output of the classified image.

This image cluster is verified with the Google Earth image and the clusters as per their classification. The coconut plantation clusters are also identified and clipped as separate files. Later, this clipped coconut file is used in the next supervised maximum-likelihood algorithm. The supervised classifier is used to identify the pattern recognition with the spectral signature features. This algorithm used the training dataset to predict new pixels. This essential training data are obtained from the ISODATA algorithm. Using this training dataset, the maximum-likelihood algorithm classifies the satellite imagery. The coconut palm trees are extracted from this classification, and then the area and number of trees were computed using the GIS tool. The ground truth data were obtained through a field visit, and it was verified for its accuracy to validate the result.

4 Result and Discussion

It is feasible to classify identical pixels into distinct groups or classes based on their shared characteristics, allowing for the identification of various features within an image or dataset. This process relies heavily on comparing pixels to one another and assigning them to appropriate categories, which ultimately reveals the identities of these features. The success or failure of this classification process largely depends on the type of classifier used, as well as the quality and relevance of the training data. The classifier serves as the analytical tool that determines the grouping and accuracy of the classifications.

Numerous classification algorithms have been developed to address different challenges and applications. These techniques are generally divided into two primary categories: supervised and unsupervised classification methods. Supervised classification requires labeled training data to guide the classifier, whereas unsupervised classification identifies patterns without prior knowledge or labeled datasets.

This research uses a hybrid approach to merge both supervised and unsupervised classification algorithms. The strengths of each method are utilized to study and achieve a more robust and accurate classification process. The findings emphasize the effectiveness of combining these techniques to offer a valuable insights and meaningful outcomes for the analysis of complex pixel-based datasets.

4.1 Image Clustering Technique

ISODATA (Iterative Self-Organizing Data Analysis Technology) is a complete set of heuristic techniques combined into an iterative classification system. The algorithm includes several phases as a result of expertise obtained via experimentation. This ISODATA algorithm's ability to automatically erase, combine, and divide data using a mathematical function is one of its key benefits. The algorithm is very effective in cutting the clusters when the pixels in the cluster set values.

It separates the clusters which were formed early when they contained a high number of pixels and merges them while the spectral distance is very less between the clusters. The following algorithm was used to perform ISODATA classification:

$$\text{De} = \sqrt{\sum} xi = 1(\text{DNFi} + \text{DNGi})2$$

where

x	number of spectral bands used in a classification.
DNFi	digital number of pixel F in the ith band.
DNGi	digital number of pixel G in the ith band.

The result obtained with unsupervised classification is shown in Fig. 3, with 10 cluster classifications. The classified image compared with Google Earth's high-resolution satellite reveals a precise categorization was attained by this algorithm. From the result, the coconut cluster alone clipped as a region of interest. This region of interest was used as a training data set for the supervised classification.

4.2 Supervised Classification Technique

The image examiner regulates the pixel classification measure by devising the computation and mathematical creator relating to distinct types of land cover in a scene. The co-training approach uses samples from two free subsets to train two classifiers. It chooses unlabeled cases of high quality to get the other classifier. All cover types must be planned inside the described image for supervised classification.

The qualities of the interest-class indicators applied to the whole image are described with this data. Complex landscape settings typically enhance the likelihood of mixed pixels appearing in the images. Based on the Bayers rule, the maximum-likelihood classifier is given as

$$\Pr\{Ai|B\} = \Pr\{Ai\} * \Pr\{B|Ai\} \sum \Pr\{Ai\} * \Pr\{B|Ai\}$$

where

Ai—class, B—pixel response values.

$\Pr\{A|B\}$—the conditional probability of Class Ai occurring given the pixel response values, $\Pr\{A\}$ are prior probabilities of Class Ai occurring and is the probability that the pixel values are in Class Ai.

The extracted training dataset from the unsupervised classification is used in supervised classification for the accurate extraction of coconut from the sentinel data, and the obtained output is shown in Fig. 4 and Table 3.

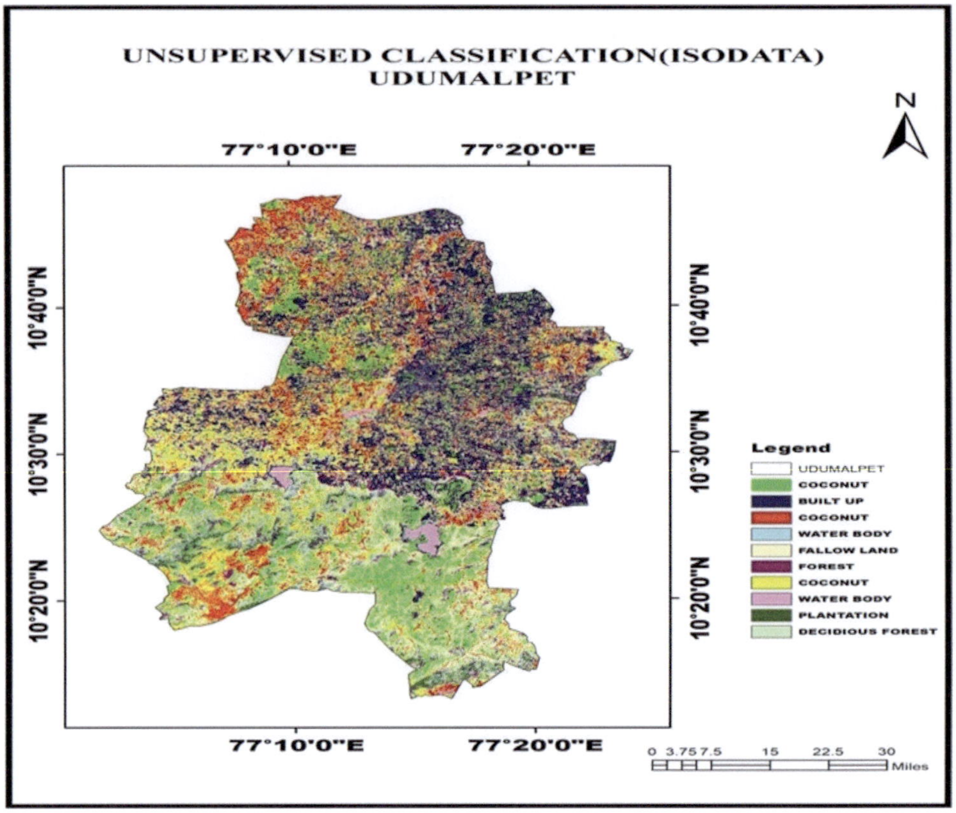

Fig. 3 Classification of cluster using unsupervised method

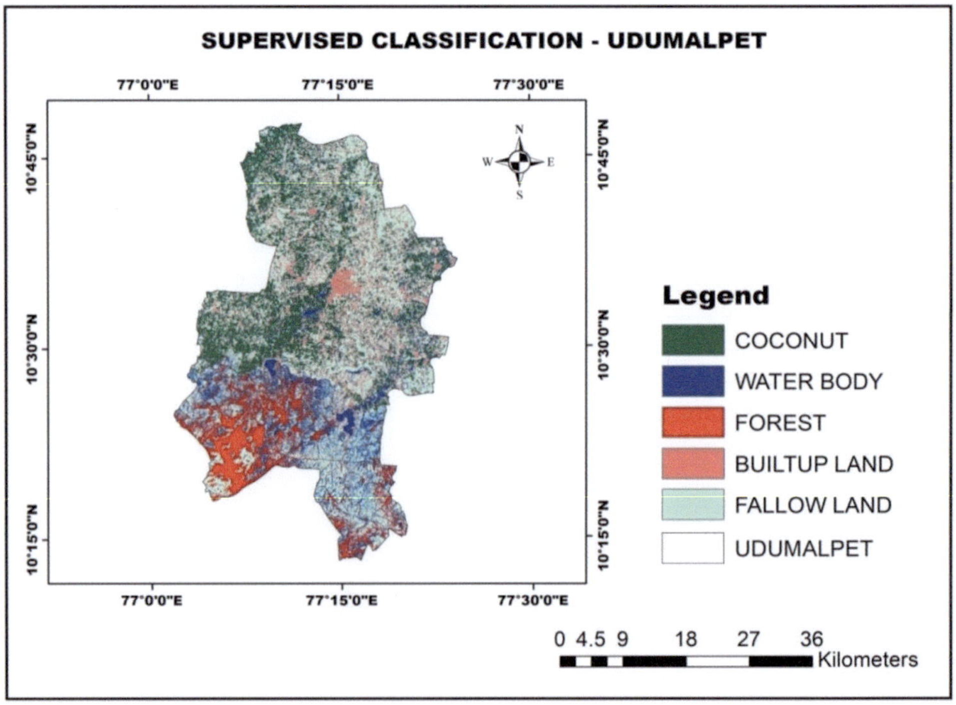

Fig. 4 Extraction of coconut using supervised classification method

Table 3 Classification of LULC classes with supervised classification

Classes	Pixel count	km^2
Coconut	3,109,906	310.9906
Water bodies	2,393,188	239.3188
Forest	14,410,117	144.1017
Build up land	2,724,967	272.4967
Fallow land	4,636,133	463.6133

4.3 Reliability of the Data

Several accuracy validation approaches are used for the categorized image. The accuracy of the individual unsupervised output was assessed by comparing high-definition satellite pictures. High-resolution satellite images are used from Google Earth to verify the coconut trees as seen in Fig. 5. One of the major difficulties with utilizing these images as a ground truth reference is the fact that the scenes comprising the research region had acquisition dates that were different from the date the analyzed picture was acquired and could only be used as an estimate of the study area. However, the output of the supervised classification is validated with the ground truth data.

The ground truth data were collected using the GPS, and a total of 70 points were obtained. These points were verified with the final result to check its precision level.

The confusion matrix was also used to compute the second accuracy assessment. The accuracy evaluation is completed by establishing arbitrary locations inside the research area. To validate the specific kind of land cover, several criteria are used. A confusion matrix that tabulates the relationship between real land cover classes and the map classes. The following equations are used to develop the assessment of detecting accuracy of the classification.

$$\text{user accuracy} = \frac{\text{Number of correctly classified in each category (Diagonal)}}{\text{Number of classified pixels in that category (The row total)}} * 100$$

$$\text{Producer accuracy} = \frac{\text{Number of correctly classified in each category (Diagonal)}}{\text{Number of classified pixels in that category (The row total)}} * 100$$

The main portion of accuracy measurement is overall accuracy (OA) which ensures the pixel is correctly classified to report the accuracy of a classification.

$$\text{Overall accuracy} = \frac{\text{Total number of correctly classified pixels in each category [Diagonal]}}{\text{Total number of reference pixel}} * 100$$

The kappa coefficient is a measure of how the classification results compare to sample data. It ranges from 0 to 1. If the kappa coefficient is equal to zero, there is no agreement between the classified and reference images. The kappa statistic, which controls the accuracy of an image classifier, is essentially a measurement of how closely the instances categorized by the machine learning classifier matched the data marked as actual data and random points.

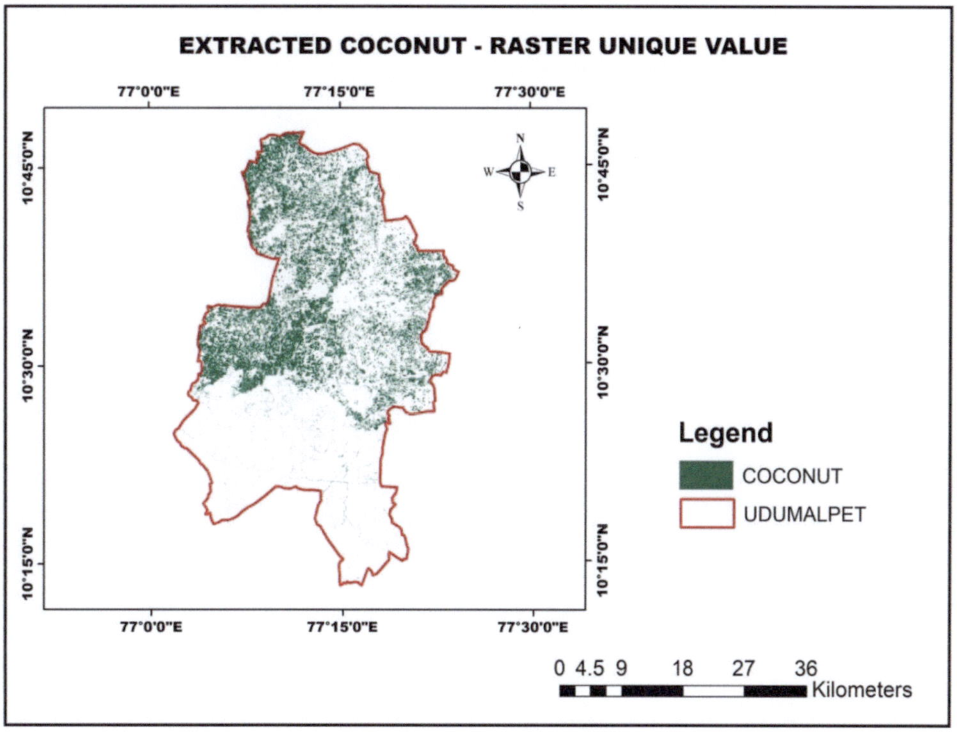

Fig. 5 Separation of coconut clusters from supervised classification

Table 4 Error matrix for the classified pixels versus field data

Class name	Coconut	Water bodies	Forest	Builtup	Fallow land	User total
Coconut	60	0	0	13	0	73
Water bodies	0	7	0	1	0	8
Forest	0	1	9	0	0	10
Built-up	0	0	0	12	1	13
Fallow land	1	1	0	4	16	22
Producer total	61	9	9	30	17	126

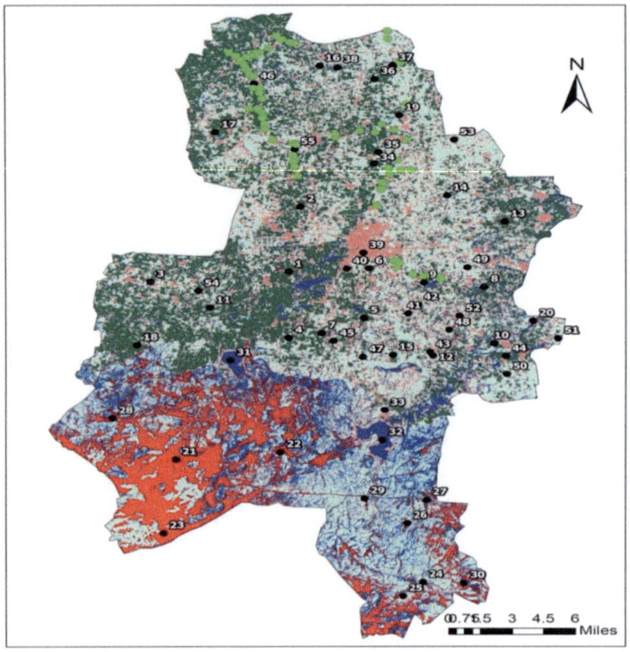

Fig. 6 Field-based ground truth points overlaid on the image

$$T = \frac{T_S - T_{CS} - \sum(\text{column total}*\text{row total})}{T_S^2 - \sum(\text{column total}*\text{row total})} * 100$$

An on-site evaluation of the research area was presented. From the field, further data and information were acquired as shown in Table 4. Throughout the site visit, a total of 70 GPS positions of the sample sites were recorded as shown in Fig. 6.

Accuracy evaluation results gave an 83% overall accuracy and 79% kappa value, which are considered very good are shown in Table 5. These values demonstrate the classification accuracy is reliable.

$$\text{Over All Accuracy} = 83\%$$

4.4 Computation of Area and Number of Coconut Trees

The coconut plantation area is measured using the GIS tool for area estimation of the coconut areas determined for the selected blocks using the supervised classification image. The field area in hectare is calculated to get the total area of pixels, and this gives the total area of the coconut distribution in study regions. According to the general square system, the identified coconut area was 310.99 km^2, and nearly 54,42,335 trees were found through the remotely sensed classification method. Overall, the study portrays that the selected area was classified as 45% coconut cultivation.

5 Conclusion

The total area under the coconut plantation in the Udumalpet block was detected and estimated using multispectral remote sensing data. This study demonstrates that remote sensing techniques are effective in classifying and detecting land cover features. The evaluation of coconut plantation areas was detected through the integration of remote sensing and GIS. With low-resolution satellite data, a dense coconut patch may blend with a forest and other plants. Sentinel 2 data

Table 5 Accuracy assessment of identified classes

Class name	Classified total	Reference total	Correctly classified	Users accuracy (%)	Producers accuracy (%)
Coconut	73	61	60	82	98
Waterbody	8	9	7	88	78
Forest	10	9	9	90	100
Built-up	13	30	12	92	40
Fallow land	22	17	16	73	94
Totals	126	126	126		

were used to prevent interclass mixing. These technologies are essential for processing massive volumes of geospatial data. The analysis of the high-resolution images showed how remote sensing may be used to generate spatial information on the arrangement of coconut and an evaluation of the amount of coconut found in the study area. In comparison with other techniques, remote sensing technology can contribute to the faster and more affordable provision of accurate and reliable landscape characteristics. Considering the geographical distribution of coconut cultivation, accurate data are available. The analysis of high-resolution images demonstrated how remote sensing and GIS may be used to offer a spatial representation of coconut movement and coconut assessment in the study area. Many such researchers utilize different kinds of remote sensing techniques to identify, classify, and analyze land use and land cover in any given region or geographic area. However, in supervised classification, the maximum-likelihood method is the most efficient. Because compared to other techniques, it provides more accuracy. The second crucial method for classification and identification is unsupervised methods. The generated land use and land map include the spatial variation of coconut trees. Remote sensing technology has revolutionized the way coconut plantation area is measured. By using satellite imagery and advanced algorithms, researchers can accurately map the boundaries and area of coconut plantations in a fraction of the time it would take using traditional methods. This technology has the potential to improve the management and productivity of coconut plantations, benefiting farmers and the wider economy.

Scope and Recommendation

Identification and suggestions using the estimation value can help in providing suitable agricultural policy measures to increase profit while growing coconut palm. Also, the entire zone can be assessed, and using the assessment, an app for farmers can be developed for efficient use.

References

Abou EL-Magd, I., Tanton, T. W., (2003). Improvements in land use mapping for irrigated agriculture from satellite sensor data using a multi-stage maximum likelihood classification. International Journal of Remote Sensing 24(21), 4197–4206.

Agarwal S., Vailshery L.S., Jaganmohan, M., Nagendra H. (2013). Mapping Urban Tree Species Using Very High Resolution Satellite Imagery: Comparing Pixel-Based and Object-Based Approaches. In: ISPRS International Journal of Geo- Information, Vol 2, pp. 220-236.

Amara, J., Bouaziz, B., and Algergawy, A. (2017). A deep learning-based approach for banana leaf diseases classification. In: BTW (Workshops). 79–88 Gesellschaft für Informatik, Bonn (2017).

Ananthakrishnan. R, Soman, M. K. (1988). The Onset Of The Southwest Monsoon Over Kerala: 190 1–1980, In: Journal Of Climatology, VOL. 8, 283–296.

Bartolucci, L. A., B. F. Robinson, and L. F. Silva. (1977). Field Measurements of the Spectral Response of Natural Waters. Photogrammetric Engineering and Remote Sensing 43 (5), 595–598.

Baruth, B., Royer, A., Klisch, A., Genovese, G. (2008). The use of remote sensing within the mars crop yield monitoring system of the European commission. The International Archives of the Photogrammetry, Remote Sensing and Spatial Information Sciences. Vol. XXXVII. Part B8. Pp: 936–939.

Boccardo, P.; Giulio Tonolo, F. (2015). Remote Sensing Role in Emergency Mapping for Disaster Response. In Engineering Geology for Society and Territory—Volume 5; Lollino, G., Manconi, A., Guzzetti, F., Culshaw, M., Bobrowsky, P., Luino, F., Eds.; Springer International Publishing: Cham, Switzerland, pp. 17–24.

Booth, D. J., and R. B. Oldfield. (1989). A Comparison of Classification Algorithms in Terms of Speed and Accuracy after the Application of A Post-Classification Modal Filter. International Journal of Remote Sensing 10 (7), 1271–1276. https://doi.org/10.1080/014311689089 03965.

Boulesteix, A. L., S. Janitza, J. Kruppa, and I. R. König. (2012). Overview of Random Forest Methodology and Practical Guidance with Emphasis on Computational Biology and Bioinformatics. Wiley Interdisciplinary Reviews: Data Mining and Knowledge Discovery 2 (6), 493–507. https://doi.org/10.1002/widm.1072.

Breiman, L. (1996). Out-of-Bag Estimation. Berkeley, CA: University of California Berkeley. https://www.stat.berkeley.edu/~breiman/OOB estimation.pdf

Breiman, L. (2001). Random Forests. Machine Learning 45 (1), 5–32. https://doi.org/10.1023/A:1010933404324.

Bricher, P. K., A. Lucieer, J. Shaw, A. Terauds, and D. M. Bergstrom. (2013). Mapping Sub-Antarctic Cushion Plants Using Random Forests to Combine Very High Resolution Satellite Imagery and Terrain Modelling. PLOS ONE 8 (8), 1–15. https://doi.org/10.1371/journal.pone.0072093.

Briggs, A. A., H. S. Young, D. J. McCauley, S. A. Hathaway, R. Dirzo, and R. N. Fisher. (2012). Effects of Spatial Subsidies and Habitat Structure on the Foraging Ecology and Size of Geckos. PLOS ONE 7 (8). https://doi.org/10.1371/journal.pone.0041364

Carlson, K. M., L. M. Curran, G. P. Asner, A. M. Pittman, S. N. Trigg, and J. M. Adeney. (2013). Carbon Emissions from Forest Conversion by Kalimantan Oil Palm Plantations. Nature Climate Change 3 (3), 283–287. https://doi.org/10.1038/nclimate1702.

Chai, Rhoda R; Use of GIS and remote sensing techniques to estimate coconut cultivation area: Case study of Kaloleni subcounty (2014).

Chen, Y., Xu, X., Zhang, D., et al., (2006). Correlation of vegetation distribution and terrain factors in northwestern of Sichuan Longmen mountain. Chinese Journal of Ecology 25(9), 1052-1055.

Chen, C., A. Liaw, and L. Breiman. (2004). Using Random Forest to Learn Imbalanced Data. Berkeley, CA: University of California Berkeley. https://statistics.berkeley.edu/sites/default/files/tec hreports/666.pdf.

Clausi, D. A. (2002). An Analysis of Co-Occurrence Texture Statistics as a Function of Grey Level Quantization. Canadian Journal of Remote Sensing 28 (1), 45–62. https://doi.org/10.5589/m02-004.

Collen, J. D., D. W. Garton, and J. P. A. Gardner. (2009). Shoreline Changes and Sediment Redistribution at Palmyra Atoll (Equatorial Pacific Ocean): 1874–Present. Journal of Coastal Research 25 (3), 711–722. https://doi.org/10.2112/08-1007.1.

Cross, M. D., T. A. Scambos, F. Pacifici, and W. E. Marshall. (2018). Validating the Use of Metre-Scale Multi-Spectral Satellite Image Data for Identifying Tropical Forest Tree Species. International Journal of Remote Sensing 39 (11), 3723–3752. https://doi.org/10.1080/014 31161.2018.1448482.

Dalponte, M., Bruzzone, L., Gianelle, D. (2008). Fusion of hyperspectral and lidar remote sensing data for classification of complex forest areas. IEEE Trans. Geosci. Rem. Sens. 46(5), 1416–1427.

Demmers, T., Yi Cao, Parsons, D., Wathes, C., Sophie Gauss. (2012). Simultaneous monitoring and control of pig growth and ammonia emissions. In: 2012 IX International Livestock environment Symposium (ILES IX), 3. American Society of Agricultural and Biological Engineers, Valencia (2012).

District Survey Report of Minor Minerals, Kozhikode District (2016). www.dmg.kerala.gov.in HAPIS Portal: https://aps.dac.gov.in

Elmer Abonales, R.T.S. Typhoon Yolanda Coconut DAMAGE Report; Philippine Coconut Authority (PCA): Leyte, Philippines, 2013.

Frolking, S., Qiu, J., Boles, S., Xiao, X., Liu, J., Zhuang, Y., Li, C., Qin, X. (2002).Combining remote sensing and ground census data to develop new maps of the distribution of rice agriculture in China. Global Biogeochem. Cycles. 16(4), 38–1.

Gad, S., & Kusky, T. (2007). ASTER spectral ratioing for lithological mapping in the Arabian–Nubian shield, the Neoproterozoic Wadi Kid area, Sinai, Egypt. Gondwana Research, 11(3), 326–335.

Govindasamy;"A Study on Production of Coconut in Coimbatore District, Tamilnadu." Shanlax International Journal of Economics, vol. 7, no. 1, 2018, pp. 27–35. https://doi.org/10.5281/zenodo.2528787

Hatfield, J.L., Prueger, J.H., (2010). Value of using different vegetative indices to quantify agricultural crop characteristics at different growth stages under varying management practices. Remote Sensing 2(2), 562-578.

Hebbar, R., Ravishankar, H. M., Trivedi, S., Subramoniam S.R., Raj, U., Dadhwal, V.K. (2014). Object Oriented Classification of High Resolution Data for Inventory of Horticultural Crops. In: The International Archives of Photogrammetry, Remote Sensing and Spatial Information Sciences, Vol XL-8, 2014.

Immitzer, M., Atzberger, C., Koukal, T. (2012). Tree Species Classification with Random Forest Using Very High Spatial Resolution 8-Band WorldView-2 Satellite Data. In: Remote Sensing, Vol-4, pp. 2661–2693.

Juepeng Zheng et.al.,; Coconut Trees Detection on the Tenarunga Using High- Resolution Satellite Images and Deep Learning; (July 2021). https://doi.org/10.1109/IGARSS47720.2021.9555008

Kamilaris, A., Prenafeta.Boldú, F.X. (2018). Deep learning in agriculture: a survey. Comput. Electron. Agric. 147, 70–90.

Kannan, B., Ragunath, K. P., Kumaraperumal, R., Jagadeeswaran, R., & Krishnan, R. (2017). Mapping of coconut growing areas in Tamil Nadu, India using remote sensing and GIS. Journal of Applied and Natural Science, 9(2), 771–773.

Kogan, F.N., (1995). Application of vegetation index and brightness temperature for drought detection. Advances in Space Research 15(11), 91- 100.

Kuenzer, C., Knauer, K., (2013). Remote sensing of rice crop areas. International Journal of Remote Sensing 34(6), 2101-2139.

Mandal, D.M., Mandal, S. (2011). Coconut (Cocos nucifera L.: Arecaceae): In health promotion and disease prevention. In: Asia Pacific Journal of Tropical Medicine, pp. 241–247.

Mellor, A., Haywood, A., Jones, S., Wilkes, P. (2012). Forest Classification Using Random Forest with Multisource Remote Sensing and Ancillary GIS Data. In: 16th Australian Remote Sensing and Photogrammetry Conference Proceedings.

Milioto, A., Lottes, P., Stachniss, C. (2017). Real-time blob-wise sugar beets vs weeds classification for monitoring fields using convolutional neural networks. ISPRS Ann. Photogramm. Remote Sens. Spatial Inf. Sci. 4, 41.

Motohka, T., Nasahara, K.N., Miyata, A., Mano, M., Tsuchida, S., (2009). Evaluation of optical satellite remote sensing for rice paddy phenology in monsoon Asia using a continuous in situ dataset. International Journal of Remote Sensing 30(17), 4343-4357.

Niti Singh et.al., comparison of different classification techniques to identify coconut growing areas in kozikhode district, kerala; The International Archives of the Photogrammetry, Remote Sensing and Spatial Information Sciences, Volume XLII-3/W6,2019.

Ochoa, K.S., Guo, Z. (2019). A framework for the management of agricultural resources with automated aerial imagery detection. Comput. Electron. Agric. 162, 53–69.

Parthasarthy, U., Das M.M., Kumaran P.M., Parthasarthy V.A. (2003). Identification of coconut plantation in Kasaragod district of Kerala with the help of satellite image. In: Indian Journal of Horticulture, Vol 60(4), 388–393, 200.

Parthipan; Digital Image Classification of Mango and Coconut for Natham Taluk, Dindigul District using Sentinel2a Optical Data; Indian Journal of Image Processing and Recognition (IJIPR) ISSN: 2582–8037(Online), Volume-1 Issue-1, December 2020.

Philippine Coconut Authority. Initial Report on Damage to Coconut by Tropical Storm Urduja & Typhoon Vinta. December 2017. Available online: https://pca.gov.ph/index.php/about-us/overview/10-news/110-initial-report-on-damage-to-coconut-bytropical-storm-urduja-typhoon-vinta (accessed on 30 June 2021).

Rouse, J.W., Haas, R.H., Scheel, J.A. and Deering, D.W. (1974). Monitoring vegetation systems in the Great Plains. In: Proceedings, 3rd Earth Resource Technology Satellite (ERTS) Symposium, Vol-1, pp 48–62.

Roxy, M.S., Sumithranand, V.B., Renuka, G. (2010). Variability of soil moisture and its relationship with surface albedo and soil thermal diffusivity at Astronomical Observatory, Thiruvananthapuram, south Kerala. In: Journel of Earth System Science, 119, No. 4, pp. 507–517.

Sahay, B., Chakraborty, A., Chaudhary K.K., Laxman, B., Murthy, C.S., Rao P.V.N. (2017). Mapping Multiple Horticulture Crops using Object Oriented Classification Techniques. In: 38th Asian Conference on Remote Sensing.

Santillan. J.R., M Santillan, M.M., Francisco. R. (2012). Using remote sensing to map the distribution of sago palms in northeastern Mindanao, Philippines: results based on Landsat ETM+ image analysis. In: The 33rd Asian Conference on remote Sensing.

Shaeb, B, Hareef. K, Joshi, A.K., Moharil, S.V. (2013). Surface reflectance retrieval from the high resolution multispectral satellite image using 6S radiative transfer model, In: International Journal of Remote Sensing and GIS, Vol-2(3), pp. 130–137.

Singh, N.J., Kudrat, M., Jain, K., Pandey, K., (2011). Cropping pattern of Uttar Pradesh using IRS-P6 (AWiFS) data. International Journal of Remote Sensing 32(16), 4511-4526.

Son, N.T., Chen, C.F., Cru, C.R. (2012). Mapping Major Cropping Patterns in Southeast Asia from Modis Data Using Wavelet Transform and Artificial Neural Networks. ISPRS-International Archives of the Photogrammetry, Remote Sensing and Spatial Information Sciences, 1, 421- 425.

Vetrivel, A., Kerle, N., Gerke, M., Nex, F., Vosselman, G. (2016). Towards automated satellite image segmentation and classification for assessing disaster damage using data-specific features with incremental learning. In Proceedings of the Geographic Object Based Image Analysis (GEOBIA), Enschede, The Netherlands, 14–16.

Walsh, S.J., Crawford, T.W., Welsh, W.F., Crews-Meyer, K.A. (2001). A multiscale analysis of LULC and NDVI variation in Nang Rong district, northeast Thailand. Agriculture, Ecosystems & Environment 85(1), 47-64.

Wardlow, B.D., Egbert, S.L. (2002). Discriminating cropping patterns in the US Central Great Plains region using time-series MODIS 250-meter NDVI data–Preliminary Results. In Proceedings, Pecora 15 and Land Satellite Information IV Conference, 10–15.

Xu, J.Z.; Lu, W.; Li, Z.; Khaitan, P.; Zaytseva, V. (2019). Building Damage Detection in Satellite Imagery Using Convolutional Neural Networks. arXiv:1910.06444.

Zhang, H., Lan, Y., Lacey, R., Hoffmann, W.C., Westbrook, J.K., (2011). Spatial analysis of NDVI readings with different sampling densities. Transactions of the ASABE 54(1), 349-354.

Remote Sensing-Based Error Assessment of Mathematical Modelling for Predicted Urban Sprawl with Stochastic Approaches

Gaurav Kumar Mishra and Amit M. Deshmukh

1 Introduction

Urban sprawl is the hysterical expansion of urban areas, which poses significant challenges to sustainable development. Accurate modeling offers critical insights into forecasting growth trends; however, inherent uncertainties necessitate rigorous error assessment. This research investigates remote sensing-based methods combined with stochastic approaches to improve the reliability of urban sprawl predictions. High-resolution satellite imagery supports real-time monitoring of spatial transformations, while stochastic techniques capture variability and randomness in urban evolution. The integrated framework aims to minimize prediction errors and facilitate data-driven decision-making for effective urban planning and management, thereby enhancing resilience in rapidly changing urban landscapes. This approach drives sustainable transformation.

Urban sprawl generally harms the overall ecosystem of a city (Bhatta 2010a, b). There are many regulatory urban planning theories which are given to minimize the effect of urban sprawl (Mitsova et al. 2011). Ebenezer Howard had given the concept of a Garden City where town, country, and town-country act together as three magnets and the benefits of each other can be shared (Howard 1965). For example, better employment opportunities will be in a town and a better environment will be in a country; so, if we adopt the town-country scheme, which talks about taking the benefits of towns and countries together, the problem would be solved (Howard 1965). Then, the neighborhood concept came in the early 1900s, which talked about participatory planning (Healey 2011). This concept was given by Clarence A Perry, an American urban planner (Patricios 2002). Later Radiant City was proposed by Le Corbusier (Wilson 1998). All these urban planning concepts were given to minimize the effect of environmental and social degradation due to the industrial revolution in those years (Berry and Rondinelli 1998). Generally, cities grow organically, and there are very less chances to set up new cities in a full-fledged manner (Montgomery 2007). So, the task of an urban planner is to look for managing these organically developed cities, so that they function properly (Montgomery 2007). While managing this task, he should be aware of all minimizing and mitigation measures which can control serious issues in the environment and society (Sharpley et al. 2015). It can be controlling the pollution or providing education and health facilities to the population (Perera et al. 2003). In the mitigation and management of these issues, one needs to follow some logical methods which are well-established (Bishop 2002).

Assessment of urban problems can address many issues which are related to the environment and society, and it is one the way of logical methods (Rashed and Weeks 2003). There are many examples which can be given to understand the urban sprawl assessment methods (Bhatta 2010a, b). In recent years, many studies have considered the sustainable development goals which had come into existence in 2015 (Blanc 2015). These studies are about simulating land use in different scenarios; these scenarios may be based on output values of social, economic, or environmental aspects of a city (Wang et al. 2022). Ecosystem service for optimization of land use was studied by some researchers for the Sichuan-Yunnan ecological barrier, in China (Li et al. 2021). Again, in China, for the assessment of the Wuhan Metropolitan area, the Markov-FLUS model was used (Ma and Wang 2022). A coupling of human and natural effects was considered for simulating multi-scenario simulation of land use in China (Liu et al. 2017). Remote sensing techniques have been extensively used for the assessment of urban sprawl. In India, Pune city was assessed by some researchers using remote sensing (Kantakumar et al. 2016). For monitoring the spatiotemporal assessment of Delhi

G. K. Mishra (✉) · A. M. Deshmukh
Visvesvaraya National Institute of Technology Nagpur, Nagpur, India
e-mail: ergaurav.iitr@gmail.com

A. M. Deshmukh
e-mail: amitdeshmukh@arc.vnit.ac.in

D. Kumar et al. (eds.), *Data Analytics and Remote Sensing-Based Analysis for Sustainable Development*, Advances in Science, Technology & Innovation, https://doi.org/10.1007/978-3-031-98095-4_3

city again, remote sensing satellite images were used (Tripathy and Kumar 2019). Expansion indicators were considered for assessing the Lucknow metropolitan area (Verma and Garg 2022). A patch-generating land-use simulation model was used to assess the drivers of sustainability (Liang et al. 2021) for Wuhan city in China. Fuzzy clustering and Cellular automata were coupled to modeling urban emergence and economic development zones in China (Liang et al. 2020). In India, urban sprawl around Bhubaneswar city was spatially modeled (Chatterjee et al. 2016). Assessment of eight Indian mid-sized cities was carried out using Geospatial tools (Chettry and Surawar 2021). Multi-scenario simulation was carried out in France to frame sustainable urban growth (Domingo et al. 2021). CA-based future land-use simulation model was proposed for a study in China (Liang et al. 2018). Multi-scenario landscape ecological risk simulation for Sustainable Development Goals was studied for the Central Mountainous Area of Hainan Island (Han et al. 2022). Multi-scenario simulation of land-use changes with ecosystem service value in the Yellow River Basin was carried out in this context (Lou et al. 2022). SDG-oriented multi-scenario land-use simulation was carried out under the background of urban expansion by some researchers (Zhou et al. 2022). Urbanization and its drivers for the expansion of Shanghai have been taken out by a few researchers (Yurui et al. 2014; Longyu et al. 2009). Some researchers have performed by modeling urban growth trajectories and spatiotemporal patterns for Lucknow city, India (Shukla and Jain 2019). At the regional level for the Yangtze River Basin, multi-scenario simulation of land use has been carried out (Hu et al. 2022). Urban slum detection was done in Pune city in India (Kohli et al. 2016). Landscape matrices were used to understand the urban sprawl dynamics in India (Sudhira et al. 2003). A rigors study was done by some Indian authors on framing smart city solutions for predicted urban sprawl in Delhi city, India (Mishra and Deshmukh 2023).

In all the above-mentioned studies, researchers usually talk about the qualitative aspect of urban sprawl. In this study, we see the quantitative aspects of urban sprawl, and we measure it with the help of probabilistic methods such as methods in stochastic processes. In this study, we observe that there are deviations in the predicted urban sprawl from the actual measured urban sprawl. The reason behind this deviation is not in the scope of this study. This study would focus only on the assessment part of the urban sprawl and not on the logic behind the deviation from the measured quantity of the urban sprawl. The study takes into consideration various satellite temporal datasets ranging from the year 1989 to 2019. The analysis of the year 2019 has been done based on predicted values, and the base year for prediction is taken as the year 2014. In the study, the analysis of measurement and prediction value is done for the interval of five years, starting from 1989 to 1994, 1994 to 1999,1999 to 2004, 2004 to 2009, 2009 to 2014, and 2014 to 2019.

For error assessment of modeled urban sprawl, we have considered three subsequent datasets, the first two datasets in ascending years were used for calculation of change in the area of different land uses and which type of transformation has taken place in succeeding years while comparing both the datasets by mean of geospatial interpretation. In this task, the location of different land uses attributed for the identification of its origin in the preceding dataset, and the probability of change was calculated for urban land-use class based on these two datasets. The probability of change is observed in the third dataset which would be subsequently five years later from the preceding year of our initial analysis task. The comparison is made between the projected land-use areas and the measured land-use areas of the targeted dataset which is the subsequent dataset in five years from the preceding dataset of our initial analysis task. For example, if we want to assess the error for the year 1999, so we will consider the three datasets in five years, i.e., 1989, 1994, and 1999. The three datasets for the years 1989, 1994, and 1999 were taken, and the change in the areas of land use were calculated between the datasets of the years 1989 and 1994. We also observed the attributed change in land uses from the dataset of the year 1989 to the year 1994. It means we observed the particular land-use change of the preceding year 1989 datasets to different land-uses in the succeeding year of 1994 and the areas were measured in square kilometers. Now based on these values, the probability of change in that land use was calculated, and the probability measure was applied to the dataset of the year 1994 for the projection of the land-use categories of year 1999. And, as we have already the land-use distribution of the dataset of the year 1999. So, the observed value through the probabilistic method was compared with the actual values of land-use distribution for the year 1999, and error estimation was carried out by differentiating both values.

Within the realm of urban sprawl studies, a noticeable dichotomy exists as researchers predominantly examine either quantitative or qualitative aspects. Recognizing the inherent interplay between these dimensions is crucial for a holistic comprehension of the subject. To bridge this gap, our research takes a comprehensive approach by simultaneously exploring both quantitative and qualitative facets of urban sprawl dynamics. The objective is to conduct class diversity analysis, which also provides values for class intensity analysis for all the classified classes in the land use and land cover (LULC) of the study area.

2 Literature Review

Effective natural resource management stands as a pivotal strategy in the pursuit of sustainable development. As urban landscapes continue to evolve, the proactive identification of potential changes becomes crucial for informed decision-making. In this context, the detection of land-use land-cover changes emerges as a valuable tool, enabling us to foresee and understand the dynamics of future urban sprawl. The complex relationship between natural resource management and land-use change detection underscores the interconnectedness of these domains in shaping the trajectory of sustainable urban development. Recognizing the significance of these factors lays the groundwork for our research, as the implications of land-use changes on urban sprawl dynamics and contributing to the broader discourse on sustainable urban planning were investigated. To comprehensively understand and analyze the evolution of urban landscapes, assessing changes over time is paramount, and spatiotemporal analysis of historical datasets emerges as a key method in this endeavor. Within this framework, remote sensing emerges as a powerful and indispensable tool for achieving diverse sustainability objectives within urban setups. By harnessing the capabilities of remote sensing technology, we can attain a holistic view of urban landscapes, enabling the assessment of environmental conditions, land-use changes, and the overall health of urban ecosystems. As we embark on our research journey, these connections emphasize the integral role of spatiotemporal analysis and remote sensing in unraveling the complexities of urban sustainability, providing a robust foundation for our investigation into the dynamics of urban sprawl.

In the Indian context, a surge of studies has been dedicated to reshaping urban policies and planning measures, particularly focusing on the quantification of land-use land-cover (LULC) changes and the analysis of influencing factors. For instance, Rahman conducted a study in the peri-urban areas of Delhi NCR, utilizing geomatics techniques to quantify LULC changes and dissect the factors steering these transformations. Ghosh took an integrative approach, employing GIS and statistical tools to map future urban sprawl in Midnapore town, India. This research contributes to the overarching narrative by emphasizing the importance of combining diverse methodologies for a comprehensive understanding of urban development. Similarly, Hasnine Rukhsana adopted the CA-Markov model to assess LULC changes in Kolkata, providing valuable insights into the dynamics of urban transformation over time. The utilization of advanced modeling techniques adds depth to the exploration of urban sprawl patterns. Angadi utilized landscape metrics and Shannon's entropy model for the assessment of urban sprawl in Barrackpore, India, offering a nuanced approach to understanding the spatial complexities of urban expansion. This study emphasizes the significance of incorporating landscape metrics for a holistic evaluation.

The city-specific focus is evident in the research conducted by Khan in Bhubaneswar, India, demonstrating a more detailed and localized perspective on urban sprawl dynamics. This refined approach allows for a more granular examination of the factors shaping urban development. Moving beyond city boundaries, Singh investigated peri-urban land-use changes in conjunction with agricultural growth in the Varanasi district, revealing the interconnectedness of urban and agricultural landscapes. This regional exploration contributes to our understanding of the multifaceted influences on urban sprawl. Mishra and Deshmukh (2023) undertook a rigorous study, framing an area-based development approach for predicting urban sprawl in Delhi city. This research contributes a predictive dimension, anticipating urban sprawl trajectories for effective urban planning.

Remote sensing methods and GIS applications have been recurrent themes in urban sprawl assessments, as demonstrated by studies like Omar and Inam, focusing on Belek, Turkey. This global perspective illustrates the versatility of these techniques in different geographical contexts. Expanding the scope, strategies explored coastal urban sprawls of the Mediterranean in the context of tourism promotion, showcasing the diverse applications of urban sprawl assessments across various domains. Some studies, such as Hoymann, emphasize the necessity of framing strict land-use policies to minimize residential land consumption. This policy-oriented perspective underscores the importance of regulatory measures in mitigating the impact of urban sprawl. Furthermore, researchers like Vargas, Sahoo, and Bakhti have investigated the impact of agricultural and urban land-use changes on plant and bird biodiversity. These ecological considerations highlight the far-reaching consequences of urban sprawl on the natural environment.

The evaluation of urban sprawl using Shannon's Entropy Index (SEI) and Fractal Dimensions (FD) has been a prevalent approach over the last two decades. In India, researchers applied Shannon's entropy model for assessing urban sprawl in Bangalore city, incorporating landscape metrics into the analysis. SEI was employed for the analysis of precipitation concentration characteristics in Iran. The expansion of Lilongwe city was determined using SEI, and SEI, along with GIS, was utilized for landslide susceptibility mapping in Western Rwanda, as well as for the assessment of Dejen District in Northwestern Ethiopia. In India, the determination of urban sprawl in Bardhaman city, West Bengal, was conducted using Shannon's entropy model in GIS. The assessment of urban growth in Jaipur city, India, was performed using SEI. SEI and FD were jointly employed to assess urban sprawl in Samsun, Turkey. For a comprehensive understanding, multidisciplinary papers have been published, such as one by Y. Chen et al. provide an overview of fractal dimensions and spatial measures of urban systems, exploring the connection from entropy to fractal dimension.

In India, researchers examined Tiruchirapalli city using SEI and FD. An equivalent relationship between normalized spatial entropy and fractal dimension was analyzed by another researcher. Modeling of urban growth and form with spatial entropy (SEI) was conducted by the same researcher. Shenzhen city was examined using remote sensing, SEI, and FD. Additionally, analyses of theoretical cities and Belgian cities were performed using FD.

In summary, this diverse array of studies collectively forms a rich collection of research endeavors, each contributing unique perspectives and methodologies to our understanding of urban sprawl. This chapter seeks to build upon these foundations, further exploring the intricate dynamics and implications of urban development in a specific context. Within the realm of urban sprawl studies, a noticeable dichotomy exists as researchers predominantly examine either quantitative or qualitative aspects. Recognizing the inherent interplay between these dimensions is crucial for a holistic comprehension of the subject. To bridge this gap, our research takes a comprehensive approach by simultaneously exploring both quantitative and qualitative facets of urban sprawl dynamics. This integrated approach contributes to a more robust understanding of the intricate dynamics at play, offering insights that can inform sustainable urban planning strategies.

3 Methods

For the assessment of urban sprawl, we have used 30 m of spatial resolution multispectral images from the United States Geological Society Earth Explorer's official website. The acquired images were for the years 1989, 1994, 1999, 2004, 2009, and 2014. ArcGIS software was used to classify the images into four classes, i.e., urban, forest, vegetation, and others. The same software was used to delineate the boundary of the study area. The area covered for each land-use class and change in each land-use was calculated using the ArcGIS tools. This comes at the end of the digital computing part of the study. Now mathematical modeling was used for prediction of error assessment of different land-use coverage for the years 1999, 2004, 2009, 2014, and 2019. For prediction purposes, a time-series analysis was carried out which is a stochastic process. For the prediction of area coverage of each land-use for the year 1999, the initial vector was determined by taking the probabilities of each land-use category for 1989, which was obtained by dividing the particular land-use area by the total study area. The transitional matrix was formed by taking the probabilities of land-use transformation from the year 1989 to 1994 for each land-use class. This way we can have the predicted land-use data for the year 1999 and actual satellite-measured land-use data for the same year. Now we can assess the error present in the predicted model of the land-use classes for that year. So, we can have an analysis of predicted urban sprawl and existing urban sprawl for each five-year gap from 1999 to 2019 (Figs. 1 and 2).

4 Results and Discussions

The LULC classes for datasets of years 1999–2014 were compared with the projected LULC class values through mathematical modeling and the error was assessed in these datasets. The data used in this calculation is given as follows for error assessment of the LULC class in the consecutive gap of five years. So, the projected and observed data have an interval of five years, which starts from 1999 to 2014 for assessment of the error. The supporting datasets range from 1989 to 2019, where for the years 1989 and 1994 we have only observed LULC values, for 2019 we have only projected LULC values, and for the years 1999, 2004, 2009, and 2014, we have both the LULC values.

Figure 3 denotes the LULC class area for the year 1989, whereas Table 1 denotes the probability vector for the same year which is obtained by dividing the LULC class area by the total study area.

Error Assessment for the LULC Class of Year 1999

For calculating the projected LULC class value, the following equation will be used, where $a_{1994}\ b_{1994}\ c_{1994}\ d_{1994}$ are the initial vectors, which denote the probability of LULC classes, viz. Urban, Vegetation, Water, and Others, respectively, for the observed dataset of the year 1994, and $a'_{1999}\ b'_{1999}\ c'_{1999}\ d'_{1999}$ are the projected probabilities of LULC, classes viz. Urban, Vegetation, Water, and Others respectively, for the projected dataset of the year 1999. The initial vectors of the year 1994 are multiplied with the Transitional Potential Matrix of the years 1989–1994, to get the projected LULC class probabilities of the year 1999.

$$\begin{bmatrix} a_{1994} & b_{1994} & c_{1994} & d_{1994} \end{bmatrix} \times \begin{bmatrix} 0.871601841 & 0.000438325 & 0.001818977 & 0.126140857 \\ 0.027326675 & 0.950665303 & 0.000309365 & 0.021698657 \\ 0.113857198 & 0.004185681 & 0.530704106 & 0.351253015 \\ 0.10943246 & 0.004604127 & 0.004007119 & 0.881956295 \end{bmatrix} = \begin{bmatrix} a'_{1999} & b'_{1999} & c'_{1999} & d'_{1999} \end{bmatrix}$$

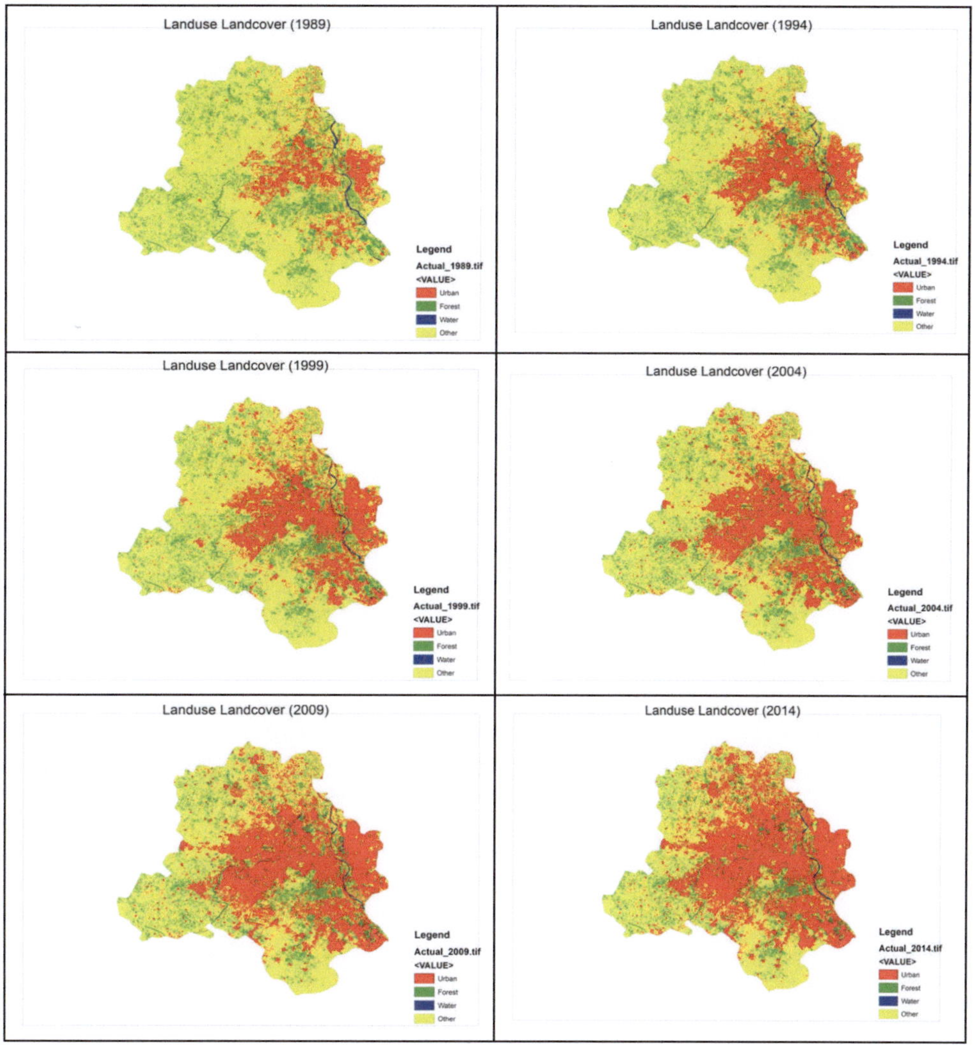

Fig. 1 Land-use land-cover classes for Delhi city

$$\begin{bmatrix} 0.180508597 & 0.128909447 & 0.008024418 & 0.682557538 \end{bmatrix} \times \begin{bmatrix} 0.871601841 & 0.000438325 & 0.001818977 & 0.126140857 \\ 0.027326675 & 0.950665303 & 0.000309365 & 0.021698657 \\ 0.113857198 & 0.004185681 & 0.530704106 & 0.351253015 \\ 0.10943246 & 0.004604127 & 0.004007119 & 0.881956295 \end{bmatrix} = \begin{bmatrix} 0.236460686 & 0.125804408 & 0.00736137122 & 0.630370042 \end{bmatrix}$$

Now the difference is calculated between projected probability values and observed probability values of the same year, i.e., 1999. The differences are multiplied by the total area of LULC classes, i.e., 1502 km^2, and error in LULC projection is observed in square kilometers for each LULC class.

$$\text{Error(Probability)} = \begin{bmatrix} a'_{1999} & b'_{1999} & c'_{1999} & d'_{1999} \end{bmatrix} \sim \begin{bmatrix} a_{1999} & b_{1999} & c_{1999} & d_{1999} \end{bmatrix}$$

$$\text{Error(Probability)} = \begin{bmatrix} 0.236460686 & 0.125804408 & 0.00736137122 & 0.630370042 \end{bmatrix} \sim \begin{bmatrix} 0.234146 & 0.126801 & 0.00549 & 0.633562 \end{bmatrix}$$

$$\text{Error(Probability)} = \begin{bmatrix} 0.002314686 & 0.000996592 & 0.00187137122 & 0.003191958 \end{bmatrix}$$

$$\text{Error(LULC Area)} = \begin{bmatrix} 3.476658372 & 1.496881184 & 2.810799572 & 4.794320916 \end{bmatrix}$$

In this section, Fig. 4 denotes the LULC class area for the year 1994, and Table 2 denotes the initial vector of the same year which is obtained by dividing the LULC area by the total study area. Figure 5 shows the LULC conversion from one LULC class to another LULC class or the same class like water to water or water to vegetation, etc., and Table 3 presents the probability of change which is obtained by dividing the converted area change class by the original class of the LULC area.

Error Assessment for the LULC Class of Year 2004

For calculating the projected LULC class value, the following equation will be used, where $a_{1999}\ b_{1999}\ c_{1999}\ d_{1999}$ are the initial vectors, which denote the probability of LULC

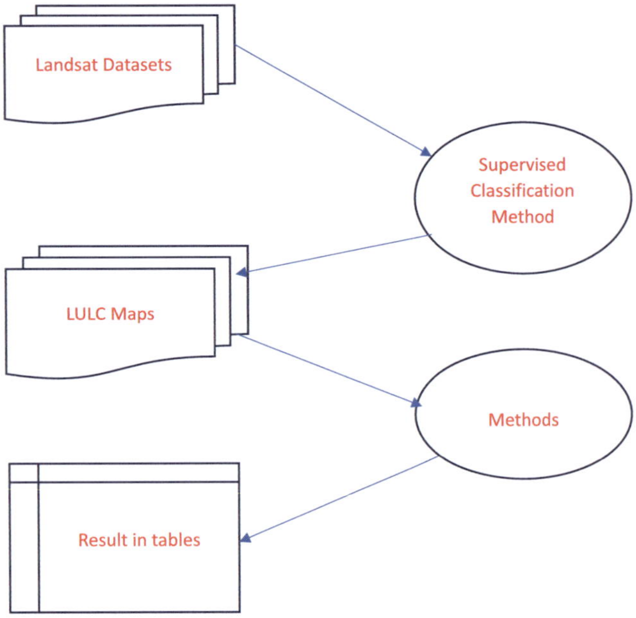

Fig. 2 Flowchart of the methodology adopted

classes, viz. Urban, Vegetation, Water, and Others, respectively, for the observed dataset of the year 1999, and $a'_{2004}\ b'_{2004}\ c'_{2004}\ d'_{2004}$ are the projected probabilities of LULC classes, viz. Urban, Vegetation, Water, and Others, respectively, for the projected dataset of the year 2004. The

Table 1 Class-wise area and probability for the year 1989

Class_1989	Area_1989	Probability (initial vector)
Urban	161.379965	0.107409771
Vegetation	198.130949	0.131870148
Water	13.521053	0.008999216
Others	1129.4381	0.751720865

initial vectors of the year 1999 are multiplied with the Transitional Potential Matrix of the years 1994–1999, to get the projected LULC class probabilities of the year 2004.

$$\begin{bmatrix} a_{1999} & b_{1999} & c_{1999} & d_{1999} \end{bmatrix} \times \begin{bmatrix} 0.986488682 & 0.000198415 & 0.000413385 & 0.012899518 \\ 0.018132799 & 0.972493578 & 0.000199622 & 0.009174002 \\ 0.094458203 & 0.005537074 & 0.412051761 & 0.487952962 \\ 0.077583862 & 0.001990001 & 0.003050243 & 0.917375894 \end{bmatrix} = \begin{bmatrix} a'_{2004} & b'_{2004} & c'_{2004} & d'_{2004} \end{bmatrix}$$

$$\begin{bmatrix} 0.234146 & 0.126801 & 0.00549 & 0.633562 \end{bmatrix} \times \begin{bmatrix} 0.986488682 & 0.000198415 & 0.000413385 & 0.012899518 \\ 0.018132799 & 0.972493578 & 0.000199622 & 0.009174002 \\ 0.094458203 & 0.005537074 & 0.412051761 & 0.487952962 \\ 0.077583862 & 0.001990001 & 0.003050243 & 0.917375894 \end{bmatrix} = \begin{bmatrix} 0.28295359 & 0.124650632 & 0.004316459787 & 0.588076317 \end{bmatrix}$$

Fig. 3 Land-use-distribution in the year 1989

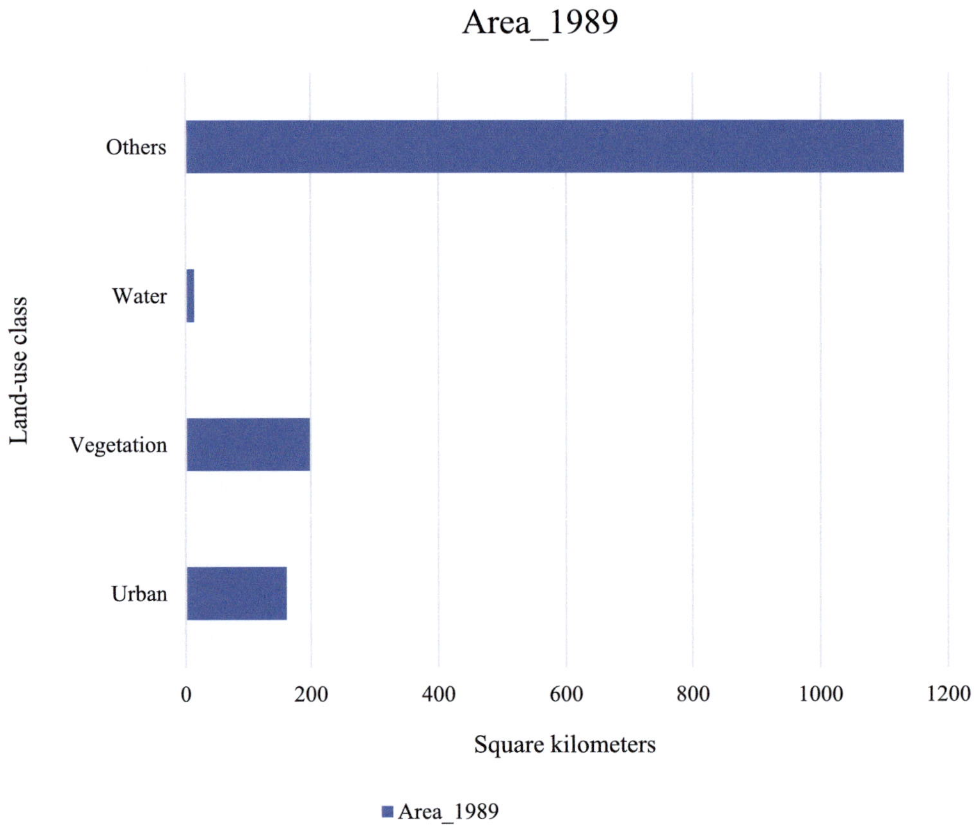

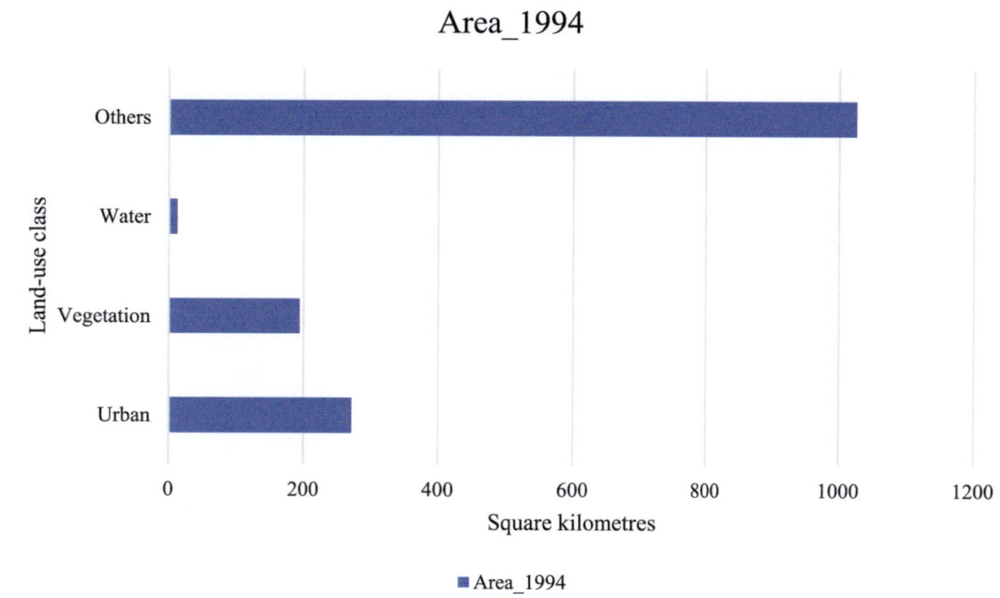

Fig. 4 Land-use-distribution in the year 1994

Table 2 Class-wise area and probability for the year 1994

Class_1994	Area_1994	Probability (initial vector)
Urban	271.21003707000	0.180508597
Vegetation	193.68349456600	0.128909447
Water	12.05650486170	0.008024418
Others	1025.52708540000	0.682557538

Now the difference is calculated between projected probability values and observed probability values of the same year, i.e., 2004. The differences are multiplied by the total area of LULC classes, i.e., 1502 km^2, and error in LULC projection is observed in square kilometers for each LULC class.

$$\text{Error(Probability)} = \left[a'_{2004}\ b'_{2004}\ c'_{2004}\ d'_{2004}\right] \sim \left[a_{2004}\ b_{2004}\ c_{2004}\ d_{2004}\right]$$

$$\text{Error(Probability)} = \left[0.28295359\ 0.124650632\ 0.004316459787\ 0.588076317\right] \sim \left[0.28211511\ 0.123853396\ 0.005286903\ 0.588744591\right]$$

$$\text{Error(Probability)} = \left[0.000142079\ 0.000797236\ 0.000970443213\ 0.000668274\right]$$

$$\text{Error(LULC Area)} = \left[0.213402658\ 1.197448472\ 1.457605706\ 1.003747548\right]$$

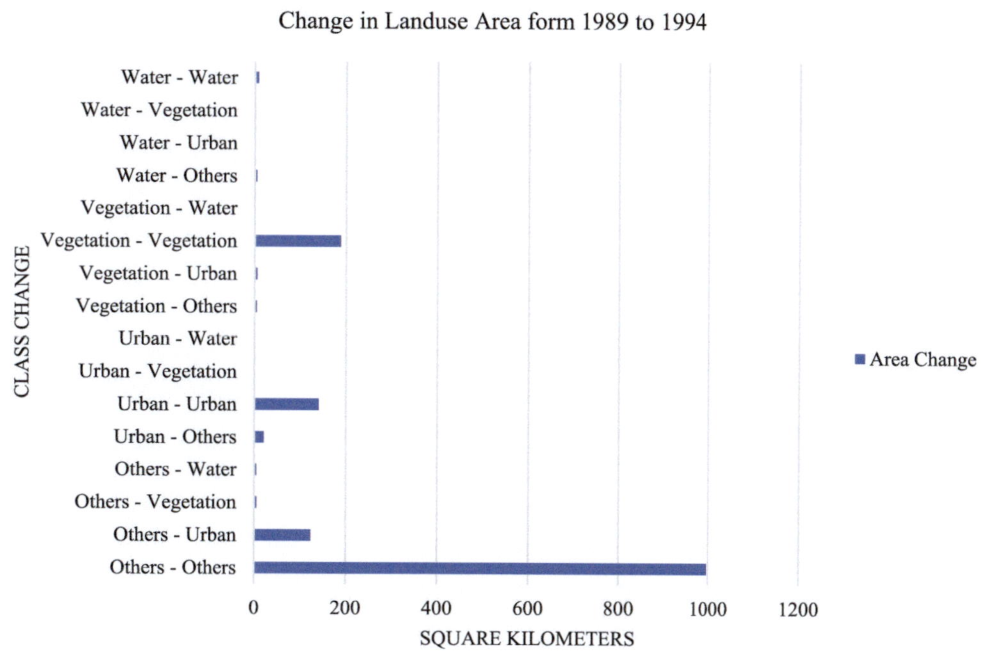

Fig. 5 Bar chart showing the change in land-use area from the year 1989 to 1994

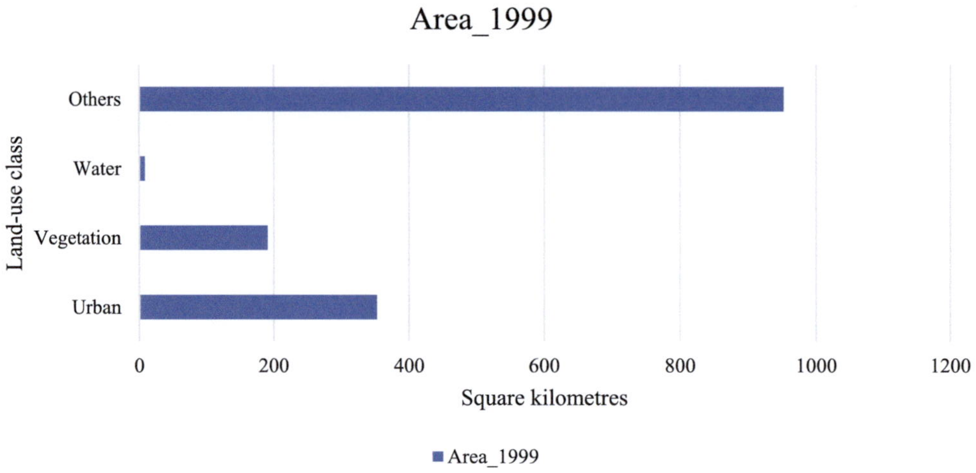

Fig. 6 Land-use-distribution in the year 1999

Table 3 Class-wise change in area and probability between the years 1989 and 1994

Change (1989–1994)	Area change	Probability of change
Others-others	996.019263	0.881956295
Others-urban	123.585305	0.10943246
Others-vegetation	5.199576	0.004604127
Others-water	4.525358	0.004007119
Urban-others	20.35519	0.126140857
Urban-urban	140.649283	0.871601841
Urban-vegetation	0.070732	0.000438325
Urban-water	0.293526	0.001818977
Vegetation-others	4.299116	0.021698657
Vegetation-urban	5.414185	0.027326675
Vegetation-vegetation	188.353611	0.950665303
Vegetatio initial vectors of the year 1999 are multiplied n-water	0.061294	0.000309365
Water-others	4.748403	0.351253015
Water-urban	1.539175	0.113857198
Water-vegetation	0.056584	0.004185681
Water-water	7.174307	0.530704106

Table 4 Class-wise area and probability for the year 1999

Class_1999	Area_1999	Probability (initial vector)
Urban	351.79718361300	0.234146
Vegetation	190.51512445600	0.126801
Water	8.24885330526	0.00549
Others	951.90660239200	0.633562

In this section, Fig. 6 denotes the LULC class area for the year 1999, and Table 4 denotes the initial vector of the same year which is obtained by dividing the LULC area by the total study area. Figure 7 shows the LULC conversion from one LULC class to another LULC class or the same class like water to water or water to vegetation, etc., and Table 5 presents the probability of change which is obtained by dividing the converted area change class by the original class of the LULC area.

Error Assessment for the LULC Class of Year 2009

For calculating the projected LULC class value, the following equation will be used, where $a_{2004}\ b_{2004}\ c_{2004}\ d_{2004}$ are the initial vectors, which denote the probability of LULC classes, viz. Urban, Vegetation, Water, and Others, respectively, for the observed dataset of the year 2004, and $a'_{2009}\ b'_{2009}\ c'_{2009}\ d'_{2009}$ are the projected probabilities of LULC classes, viz. Urban, Vegetation, Water, and Others, respectively, for the projected dataset of the year 2009. The initial vectors of the year 2004 are multiplied with the Transitional Potential Matrix of the years 1999–2004, to get the projected LULC class probabilities of the year 2009.

$$\begin{bmatrix} a_{2004} & b_{2004} & c_{2004} & d_{2004} \end{bmatrix} \times \begin{bmatrix} 0.992065774 & 0.00026844 & 6.21439E-05 & 0.007603642 \\ 0.025774579 & 0.967792671 & 1.54215E-05 & 0.006417329 \\ 0.041241393 & 0.000188027 & 0.956099311 & 0.002471269 \\ 0.073100309 & 0.001692533 & 3.35296E-05 & 0.925173628 \end{bmatrix} = \begin{bmatrix} a'_{2009} & b'_{2009} & c'_{2009} & d'_{2009} \end{bmatrix}$$

$$\begin{bmatrix} 0.28211511 & 0.123853396 & 0.005286903 & 0.588744591 \end{bmatrix} \times \begin{bmatrix} 0.992065774 & 0.00026844 & 6.21439E-05 & 0.007603642 \\ 0.025774579 & 0.967792671 & 1.54215E-05 & 0.006417329 \\ 0.041241393 & 0.000188027 & 0.956099311 & 0.002471269 \\ 0.073100309 & 0.001692533 & 3.35296E-05 & 0.925173628 \end{bmatrix} = \begin{bmatrix} 0.326323791 & 0.120936698 & 0.005093114452 & 0.547642799 \end{bmatrix}$$

Now the difference is calculated between projected probability values and observed probability values of the same year, i.e., 2009. The differences are multiplied by the total area of

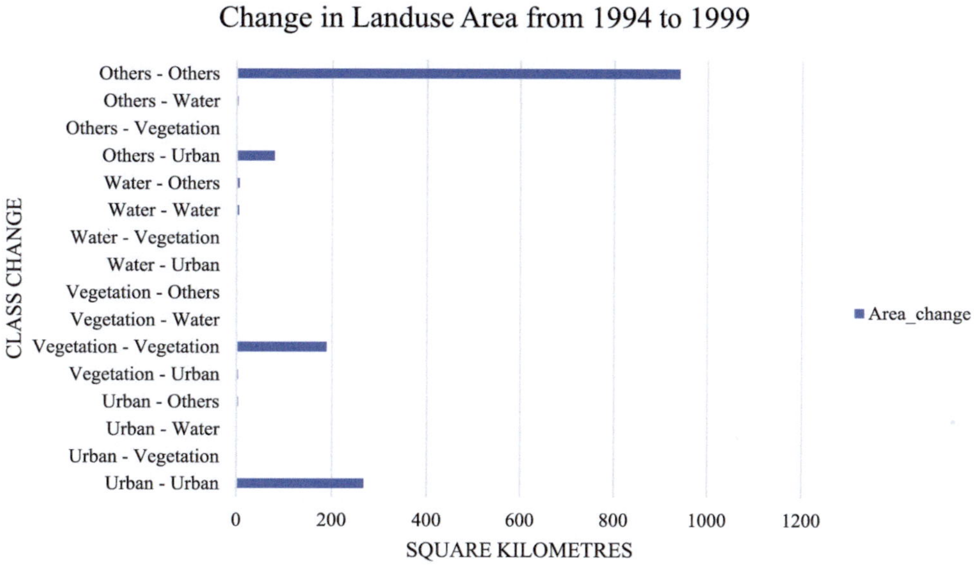

Fig. 7 Bar chart showing the change in land-use area from the year 1994 to 1999

Table 5 Class-wise change in area and probability between the years 1994 and 1999

Change (1994–1999)	Area change	Probability of change
Urban-urban	267.53456888600	0.986488682
Urban-vegetation	0.05381004174	0.000198415
Urban-water	0.11210942437	0.000413385
Urban-others	3.49833414575	0.012899518
Vegetation-urban	3.51193687146	0.018132799
Vegetation-vegetation	188.35128933400	0.972493578
Vegetation-water	0.03866246142	0.000199622
Vegetation-others	1.77680870122	0.009174002
Water-urban	1.13856731644	0.094458203
Water-vegetation	0.06674202702	0.005537074
Water-water	4.96673295426	0.412051761
Water-others	5.88162044138	0.487952962
Others-urban	79.55476241360	0.077583862
Others-vegetation	2.04055359491	0.001990001
Others-water	3.12773011883	0.003050243
Others-others	940.68044272500	0.917375894

LULC classes, i.e., 1502 km^2, and error in LULC projection is observed in square kilometers for each LULC class.

$$\text{Error(Probability)} = \left[a'_{2009}\ b'_{2009}\ c'_{2009}\ d'_{2009} \right] \sim \left[a_{2009}\ b_{2009}\ c_{2009}\ d_{2009} \right]$$

$$\text{Error(Probability)} = \left[0.326323791\ 0.120936698\ 0.005093114452\ 0.547642799 \right] \sim \left[0.345117984\ 0.117145197\ 0.007115399\ 0.53062142 \right]$$

$$\text{Error(Probability)} = \left[0.018794193\ 0.003791501\ 0.002022284548\ 0.017021379 \right]$$

$$\text{Error(LULC Area)} = \left[28.22887789\ 5.694834502\ 3.037471391\ 25.56611126 \right]$$

In this section, Fig. 8 denotes the LULC class area for the year 2004, and Table 6 denotes the initial vector of the same year which is obtained by dividing the LULC area by the total study area. Figure 9 shows the LULC conversion from one LULC class to another LULC class or the same class like water to water or water to vegetation, etc., and Table 7 presents the probability of change which is obtained by dividing the converted area change class by the original class of the LULC area.

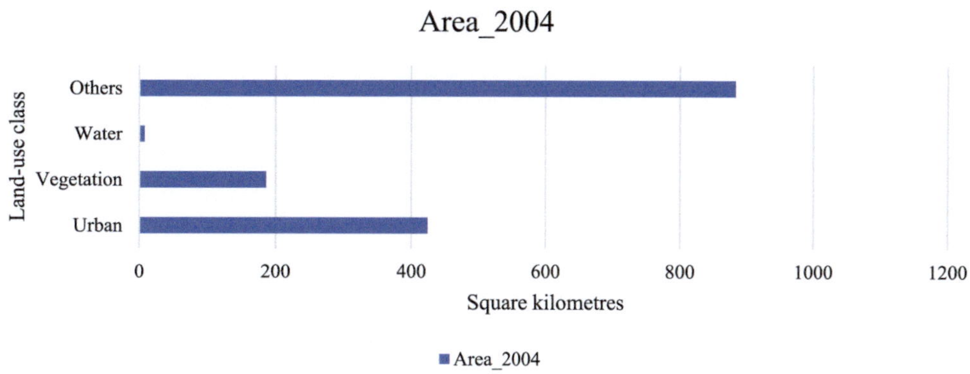

Fig. 8 Land-use-distribution in the year 2004

Table 6 Class-wise area and probability for the year 2004

Class_2004	Area_2004	Probability (initial vector)
Urban	423.86799709800	0.28211511
Vegetation	186.08535666000	0.123853396
Water	7.94338568536	0.005286903
Others	884.56797291600	0.588744591

Table 7 Class-wise change in area and probability between the years 1999 and 2004

Change (1999–2004)	Area change	Probability of change
Urban-urban	348.989285	0.992065774
Urban-vegetation	0.094432	0.00026844
Urban-water	0.021861	6.21439E-05
Urban-others	2.674812	0.007603642
Vegetation-urban	4.910398	0.025774579
Vegetation-vegetation	184.377297	0.967792671
Vegetation-water	0.002938	1.54215E-05
Vegetation-others	1.222586	0.006417329
Water-urban	0.340192	0.041241393
Water-vegetation	0.001551	0.000188027
Water-water	7.886672	0.956099311
Water-others	0.020385	0.002471269
Others-urban	69.580148	0.073100309
Others-vegetation	1.611029	0.001692533
Others-water	0.031915	3.35296E-05
Others-others	880.621696	0.925173628

Error Assessment for the LULC Class of Year 2014

For calculating the projected LULC class value, the following equation will be used, where $a_{2009}\ b_{2009}\ c_{2009}\ d_{2009}$ are the initial vectors, which denote the probability of LULC classes, viz. Urban, Vegetation, Water, and Others, respectively, for the observed dataset of the year 2009, and $a'_{2014}\ b'_{2014}\ c'_{2014}\ d'_{2014}$ are the projected probabilities of LULC classes, viz. Urban, Vegetation, Water, and Others, respectively, for the projected dataset of the year 2014. The initial vectors of the year 2004 are multiplied with the Transitional Potential Matrix of the years 2004–2009, to get the projected LULC class probabilities of the year 2014.

$$\begin{bmatrix} a_{2009} & b_{2009} & c_{2009} & d_{2009} \end{bmatrix} \times \begin{bmatrix} 0.990598366 & 0.000538334 & 0.000419607 & 0.008443692 \\ 0.059101444 & 0.928808769 & 0.00053539 & 0.011554397 \\ 0.043621909 & 0.004502952 & 0.69966442 & 0.252210719 \\ 0.098660462 & 0.003286066 & 0.005490521 & 0.892562951 \end{bmatrix} = \begin{bmatrix} a'_{2014} & b'_{2014} & c'_{2014} & d'_{2014} \end{bmatrix}$$

$$\begin{bmatrix} 0.345117984 & 0.117145197 & 0.007115399 & 0.53062142 \end{bmatrix} \times \begin{bmatrix} 0.990598366 & 0.000538334 & 0.000419607 & 0.008443692 \\ 0.059101444 & 0.928808769 & 0.00053539 & 0.011554397 \\ 0.043621909 & 0.004502952 & 0.69966442 & 0.252210719 \\ 0.098660462 & 0.003286066 & 0.005490521 & 0.892562951 \end{bmatrix} = \begin{bmatrix} 0.401457027 & 0.110766538 & 0.008098495248 & 0.479673931 \end{bmatrix}$$

Now the difference is calculated between projected probability values and observed probability values of the same year, i.e., 2014. The differences are multiplied by the total area of LULC classes, i.e., 1502 km^2, and error in LULC projection is observed in square kilometers for each LULC class.

$$\text{Error(Probability)} = \begin{bmatrix} a'_{2014} & b'_{2014} & c'_{2014} & d'_{2014} \end{bmatrix} \sim \begin{bmatrix} a_{2014} & b_{2014} & c_{2014} & d_{2014} \end{bmatrix}$$

$$\text{Error(Probability)} = \begin{bmatrix} 0.401457027 & 0.110766538 & 0.008098495248 & 0.479673931 \end{bmatrix} \sim \begin{bmatrix} 0.414121681 & 0.107615339 & 0.004804078 & 0.473458901 \end{bmatrix}$$

$$\text{Error(Probability)} = \begin{bmatrix} 0.012664654 & 0.003151199 & 0.003294417248 & 0.00621503 \end{bmatrix}$$

$$\text{Error(LULC Area)} = \begin{bmatrix} 19.02231031 & 4.733100898 & 4.948214707 & 9.33497506 \end{bmatrix}$$

Fig. 9 Bar chart showing the change in land-use area from the year 1999 to 2004

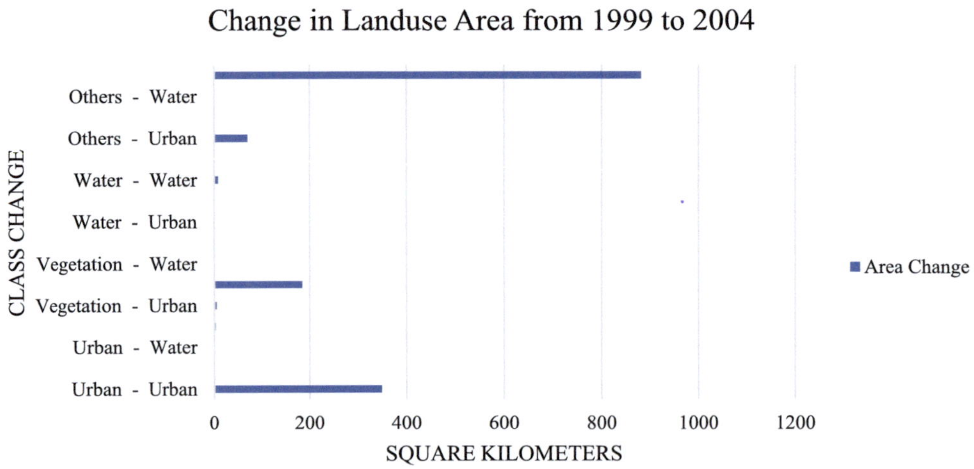

Fig. 10 Land-use-distribution in the year 2009

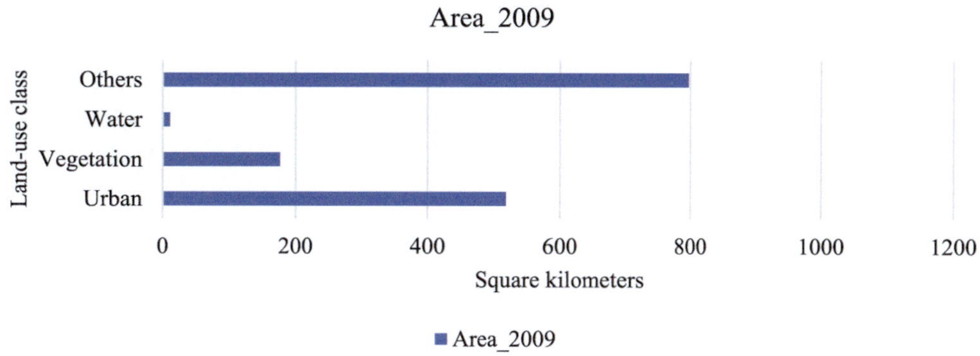

Table 8 Class-wise area and probability for the year 2009

Class_2009	Area_2009	Probability (initial vector)
Urban	518.528391	0.345117984
Vegetation	176.006796	0.117145197
Water	10.690652	0.007115399
Others	797.241187	0.53062142

Table 9 Class-wise change in area and probability between the years 2004 and 2009

Change (2004–2009)	Area change	Probability of change
Urban-urban	419.85802653100	0.990598366
Urban-vegetation	0.22816919669	0.000538334
Urban-water	0.17784754273	0.000419607
Urban-others	3.57879845863	0.008443692
Vegetation-urban	10.99749771140	0.059101444
Vegetation-vegetation	172.83118119100	0.928808769
Vegetation-water	0.09962456675	0.00053539
Vegetation-others	2.15002287330	0.011554397
Water-urban	0.34635708814	0.043621909
Water-vegetation	0.03575335001	0.004502952
Water-water	5.55532156060	0.69966442
Water-others	2.00254808749	0.252210719
Others-urban	87.26520694440	0.098660462
Others-vegetation	2.90652641573	0.003286066
Others-water	4.85636712437	0.005490521
Others-others	789.47218748200	0.892562951

In this section, Fig. 10 denotes the LULC class area for the year 2009, and Table 8 denotes the initial vector of the same year which is obtained by dividing the LULC area by the total study area. Figure 11 shows the LULC conversion from one LULC class to another LULC class or the same class like water to water or water to vegetation, etc., and Table 9 presents the probability of change which is obtained by dividing the converted area change class by the original class of the LULC area.

The Projection for the LULC Class of the Year 2019

For calculating the projected LULC class value, the following equation will be used, where a_{2014} b_{2014} c_{2014} d_{2014} are the initial vectors, which denote the probability of LULC

Fig. 11 Bar chart showing the change in land-use area from the year 2004 to 2009

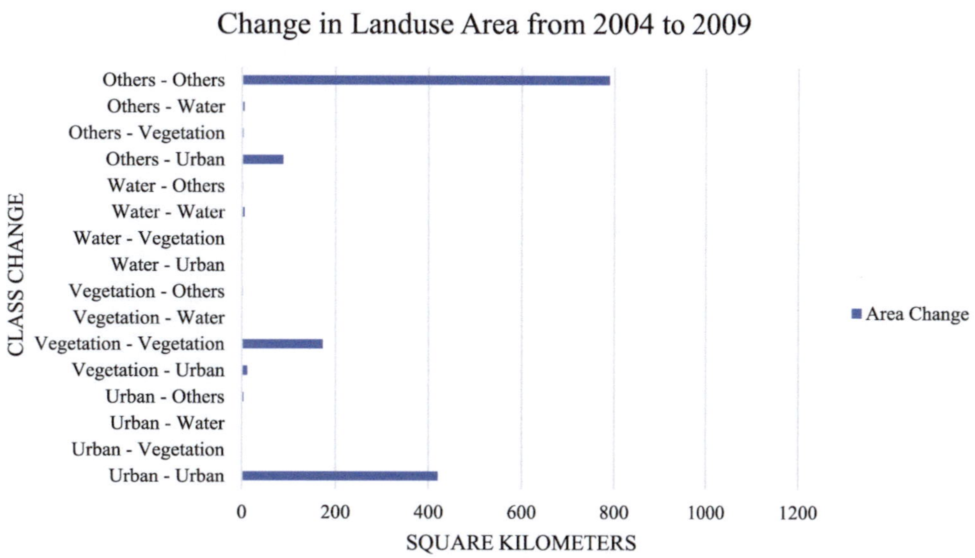

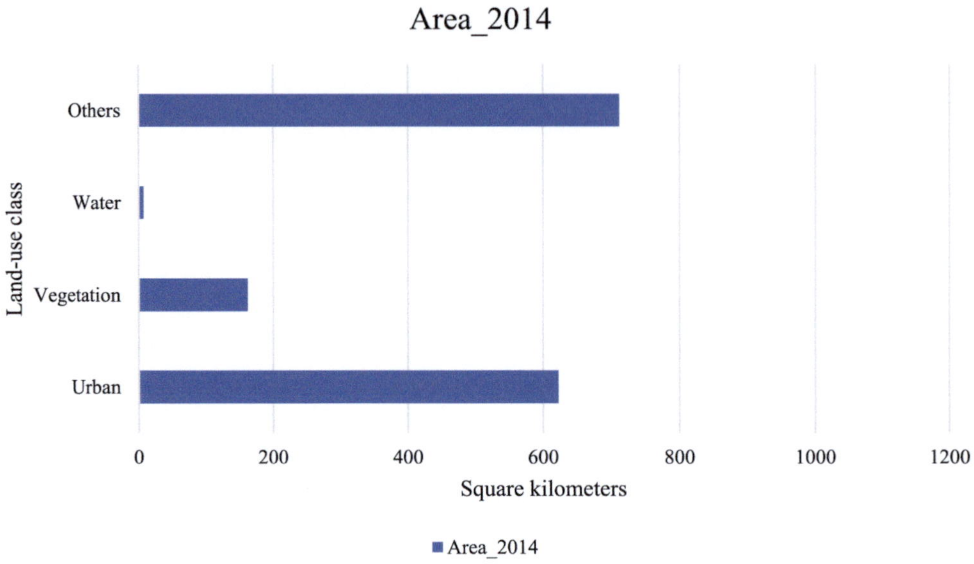

Fig. 12 Land-use-distribution in the year 2014

Table 10 Class-wise area and probability for the year 2014

Class_2014	Area_2014	Probability (initial Vector)
Urban	622.192232	0.414121681
Vegetation	161.685396	0.107615339
Water	7.217831	0.004804078
Others	711.342738	0.473458901

$$\begin{bmatrix} a_{2014} & b_{2014} & c_{2014} & d_{2014} \end{bmatrix} \times \begin{bmatrix} 0.989482712 & 0.000792355 & 0.000221204 & 0.009503729 \\ 0.087138059 & 0.896482511 & 0.000207966 & 0.016171464 \\ 0.114410016 & 0.008050497 & 0.508714525 & 0.368824962 \\ 0.116076132 & 0.004273229 & 0.002043061 & 0.877607579 \end{bmatrix} = \begin{bmatrix} a'_{2019} & b'_{2019} & c'_{2019} & d'_{2019} \end{bmatrix}$$

$$\begin{bmatrix} 0.414121681 & 0.107615339 & 0.004804078 & 0.473458901 \end{bmatrix} \times \begin{bmatrix} 0.989482712 & 0.000792355 & 0.000221204 & 0.009503729 \\ 0.087138059 & 0.896482511 & 0.000207966 & 0.016171464 \\ 0.114410016 & 0.008050497 & 0.508714525 & 0.368824962 \\ 0.116076132 & 0.004273229 & 0.002043061 & 0.877607579 \end{bmatrix} = \begin{bmatrix} .474649367 & .098864652 & .003524933796 & .422957519 \end{bmatrix}$$

classes, viz. Urban, Vegetation, Water, and Others, respectively, for the observed dataset of the year 2014, and $a'_{2019}\ b'_{2019}\ c'_{2019}\ d'_{2019}$ are the projected probabilities of LULC classes, viz. Urban, Vegetation, Water, and Others, respectively, for the projected dataset of the year 2019. The initial vectors of the year 2004 are multiplied with the Transitional Potential Matrix of the years 2009–2014, to get the projected LULC class probabilities of the year 2019.

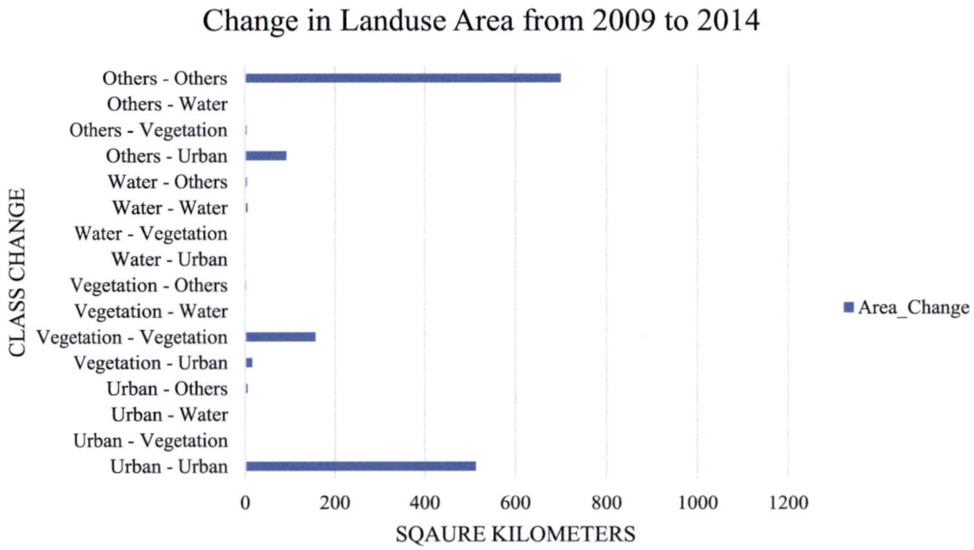

Fig. 13 Bar chart showing the change in land-use area from year 2009 to 2014

Table 11 Class-wise change in area and probability between the years 2009 and 2014

Change (2009–2014)	Area change	Probability of change
Urban-urban	513.032768	0.989482712
Urban-vegetation	0.410825	0.000792355
Urban-water	0.114691	0.000221204
Urban-others	4.927549	0.009503729
Vegetation-urban	15.336289	0.087138059
Vegetation-vegetation	157.780826	0.896482511
Vegetation-water	0.036602	0.000207966
Vegetation-others	2.846176	0.016171464
Water-urban	1.22279	0.114410016
Water-vegetation	0.086042	0.008050497
Water-water	5.437033	0.508714525
Water-others	3.941923	0.368824962
Others-urban	92.530049	0.116076132
Others-vegetation	3.406403	0.004273229
Others-water	1.628625	0.002043061
Others-others	699.584584	0.877607579

Now the probabilities are multiplied by the total area of LULC classes, i.e., 1502 km^2, and LULC projection is observed in square kilometers for each LULC class.

$$\text{LULC Area} = \begin{bmatrix} 712.9233492 & 148.4947073 & 5.294450562 & 635.2821935 \end{bmatrix}$$

In this section, Fig. 12 denotes the LULC class area for the year 2014, and Table 10 denotes the initial vector of the same year which is obtained by dividing the LULC area by the total study area. Figure 13 shows the LULC conversion from one LULC class to another LULC class or the same class like water to water or water to vegetation, etc., and Table 11 presents the probability of change which is obtained by dividing the converted area change class by the original class of the LULC area.

5 Conclusion

This study highlights the effectiveness of integrating remote sensing and stochastic approaches for error assessment in urban sprawl modeling. The improvement of predictive accuracy and the addressing uncertainties and this framework will support data-driven urban planning. The findings highlight the potential of this scheme to guide sustainable development and adaptive management in dynamic urban environments. The results illustrate that there is very little error which is found between the projected LULC class probabilities and observed LULC class probabilities, which are obtained from the satellite datasets, using this stochastic method.

These LULC class values are converted into class-wise areas and further analysis could take place. These LULC class area values help frame different sustainable development scenarios, e.g., economic development scenario, social progression scenario, and environmental protection scenario. These class-wise values are used as inputs in system dynamics modeling, where different variables can act with each other and generate final land-use demand. There are many process models which help in generating spatial maps through the land-use class values like cellular automata are used in these cases, where the spatial map is generated with the help of the evolution of land-use modeling. Evolution of land-use modeling is performed after getting the land-use demand or any other mode. The validation of land-use demand is generated with the model where these class-wise values are calculated through a stochastic process for the validation of land-use demand. This type of assessment could be useful in any type of creation of sustainable development scenarios, proposal of guidelines for land-use planning, and urban and regional planning.

References

Berry, M. A., & Rondinelli, D. A. (1998). Proactive corporate environmental management: A new industrial revolution. *Academy of Management Perspectives*.

Bhatta, B. (2010a). *Analysis of urban growth and sprawl from remote sensing data*.

Bhatta, B. (2010b). Causes and Consequences of Urban Growth and Sprawl. In *Analysis of Urban Growth and Sprawl from Remote Sensing Data* (pp. 17–36).

Bishop, S. R. (2002). What Do We Really Know About Mindfulness-Based Stress Reduction? *Psychosomatic Medicine*.

Blanc, D. L. (2015). Towards Integration at Last? The Sustainable Development Goals as a Network of Targets. *Sustainable Development*.

Chatterjee, N. D., Chatterjee, S., & Khan, A. (2016). Spatial modeling of urban sprawl around Greater Bhubaneswar city, India. *Model. Earth Syst. Environ.*

Chettry, V., & Surawar, M. (2021). Urban Sprawl Assessment in Eight Mid-sized Indian Cities Using RS and GIS. *Journal of the Indian Society of Remote Sensing*.

Domingo, D., Palka, G., & Hersperger, A. M. (2021). Effect of zoning plans on urban land-use change: A multi-scenario simulation for supporting sustainable urban growth. *Sustainable Cities and Society*.

Han, N., Yu, M., & Jia, P. (2022). Multi-Scenario Landscape Ecological Risk Simulation for Sustainable Development Goals: A Case Study on the Central Mountainous Area of Hainan Island. *Int. J. Environ. Res. Public Health*.

Healey, P. (2011). The universal and the contingent: Some reflections on the transnational flow of planning ideas and practices. *Planning Theory*.

Howard, E. (1965). *Garden Cities of To-Morrow.* MIT Press.

Hu, Y., Liu, Y., & Li, C. (2022). Multi-Scenario Simulation of Land Use Change and Ecosystem Service Value in the Middle Reaches of Yangtze River Urban Agglomeration. *Sustainability*.

Kantakumar, L. N., Kumar, S., & Schneider, K. (2016). Spatiotemporal urban expansion in Pune metropolis, India using. *Habitat International*.

Kohli, D., Sliuzas, R., & Stein, A. (2016). Urban slum detection using texture and spatial metrics derived from satellite imagery. *Journal of Spatial Science.*

Li, C., Wu, Y., Gao, B., Zheng, K., Wu, Y., & Li, C. (2021). Multi-scenario simulation of ecosystem service value for optimization of land use in the Sichuan-Yunnan ecological barrier, China. *Ecological Indicators.*

Liang, X., Guan, Q., Clarke, K. C., Liu, S., Wang, B., & Yao, Y. (2021). Understanding the drivers of sustainable land expansion using a patch generating land use simulation (PLUS) model: A case study in Wuhan, China. *Computers, Environment and Urban Systems.*

Liang, X., Liu, X., Chen, G., Leng, J., Wen, Y., & Chen, G. (2020). Coupling fuzzy clustering and cellular automata based on local maxima of development potential to model urban emergence and expansion in economic development zones. *International Journal of Geographical Information.*

Liang, X., Liu, X., Li, D., Zhao, H., & Chen, G. (2018). Urban growth simulation by incorporating planning policies into a CA-based future land-use simulation model. *International Journal of Geographical Information Science.*

Liu, X., Liang, X., Li, X., Xu, X., Ou, J., Chen, Y., ... Pei, F. (2017). Multi-scenario simulation of ecosystem service value for optimization of land use in the Sichuan-Yunnan ecological barrier, China. *Landscape and Urban Planning.*

Longyu, S., Guofan, S., Shenghui, C., Xuanqi, L., Tao, L., Kai, Y., & Jingzhu, Z. (2009). Urban Three-dimensional Expansion and Its Driving Forces: A Case Study of Shanghai, China. *Chin. Geogra. Sci.*

Lou, Y., Yang, D., Zhang, P., Zhang, Y., Song, M., & Huang, Y. (2022). Multi-Scenario Simulation of Land Use Changes with Ecosystem Service Value in the Yellow River Basin. *Land.*

Ma, B., & Wang, X. (2022). What is the future of ecological space in Wuhan Metropolitan Area? A multi-scenario simulation based on Markov-FLUS. *Ecological Indicators.*

Mishra, G. K., & Deshmukh, A. M. (2023). Framework for an Area-Based Development Approach for predicted Urban Sprawl in Delhi city. In *Geoinformatics for Sustainable Urban Development.* CRC Press.

Mitsova, D., Shuster, W., & Wang, X. (2011). A cellular automata model of land cover change to integrate urban growth with open space conservation. *Landscape and Urban Planning*, 141–153.

Montgomery, J. (2007). Making a city: Urbanity, vitality and urban design. *Journal of Urban Design.*

Patricios, N. (2002). Urban design principles of the original neighbourhood concepts. *Urban Morphology.*

Perera, F. P., Rauh, V., Tsai, W.-Y., Kinney, P., Camann, D., Barr, D., ... Whyatt, R. M. (2003). Effects of transplacental exposure to environmental pollutants on birth outcomes in a multiethnic population. *Environmental Health Perspectives.*

Rashed, T., & Weeks, J. (2003). Assessing vulnerability to earthquake hazards through spatial multicriteria analysis of urban areas. *International Journal of Geographical Information Science.*

Sharpley, A. N., Bergström, L., Aronsson, H., Bechmann, M., Bolster, C. H., Börling, K., ... Withers, P. J. (2015). Future agriculture with minimized phosphorus losses to waters: Research needs and direction. *Ambio.*

Shukla, A., & Jain, K. (2019). Modeling Urban Growth Trajectories and Spatiotemporal Pattern: A Case Study of Lucknow City, India. *Journal of the Indian Society of Remote Sensing.*

Sudhira, H., Ramachandra, T., & Jagadish, K. (2003). Urban sprawl: metrics, dynamics and modelling using GIS. *International Journal of Applied Earth Observation and Geoinformation.*

Tripathy, P., & Kumar, A. (2019). Monitoring and Modeling of Spatio-Temporal Urban Growth of Delhi, India using Cellular. *Cities.*

Verma, R., & Garg, P. K. (2022). Multi-temporal urban growth analysis with expansion indicators in. *Journal of Urban Management.*

Wang, Z., Li, X., Mao, Y., Li, L., Wang, X., & Lin, Q. (2022). Dynamic simulation of land use change and assessment of carbon storage based on climate change scenarios at the city level: A case study of Bortala, China. *Ecological Indicators.*

Wilson, M. O. (1998). DANCING IN THE DARK: The inscription of blackness in Le Corbusier's Radiant City. In *Places Through the Body.* Routledge.

Yurui, L., Jing, W., Yansui, L., & Hualou, L. (2014). Problem regions and regional problems of socio-economic development in China: A perspective from the coordinated development of industrialization, informatization, urbanization and agricultural modernization. *J. Geogr. Sci.*

Zhou, M., Ma, Y., Tu, J., & Wang, M. (2022). SDG-oriented multi-scenario sustainable land-use simulation under the background of urban expansion. *Environmental Science and Pollution Research.*

Data Analytics and Integration of CCME-Water Quality Index, Water Pollution Index (WPI), and Contamination Index (CI) with GIS Approach for a Better Understanding of Drinking Water Quality

Abhijeet Das

1 Introduction

Rivers act as the primary freshwater resource that humans need to preserve of around 2000 Km^3 water globally. In terms of both activities connected to human progress and survival, surface water is one of the most significant natural resources. Globally, the contamination of resources for surface water is turning into a serious ecological issue that necessitates ongoing assessment and supervising at all stages to ensure biosphere robustness (Das 2022a, b, c). Observations show that anthropogenic activities provide a significant and widespread threat to the biological population of the estuarine ecosystem (Liu et al. 2021). The maintenance of good water quality is crucial for both increased profitability and the protection of health of people. Chemical pollution is important in the context of global water pollution because home and agricultural discharges are linked to chemistry changes in the water that disrupt ecosystems. Additionally, it plays a significant role in the evolution of species; some of which are vulnerable to the effects of pollution. Chemical indicators may therefore be used to detect contamination in aquatic ecosystems. While doing so, water resource managers must manage their resources effectively and sustainably to satisfy the needs and expectations of both present and future generations. Nowadays, more people are aware of the effects of river and sediment contamination on the ecosystem. Examples of processes that are the result of human activity include sewage discharge, industrial effluent, agricultural runoff, and changes in riverine habitats (Qu et al. 2022). Long-term buildup of solids produces a soluble component in water that tends to stick to the inorganic materials and deposit as sediment. Additionally, metals can accumulate in species through the food chain, increasing health risks and harming estuarine environments. Large quantities of free ions can be supplied to rivers, which will work more effectively as a means of large parameter storage as well as a gauge for the level of contamination in aquatic ecosystems. Hence, surface water is being more fragile to contaminants from both organic and inorganic sources, namely industries, sewage, discarded household garbage, and irrigation drainage water, and a persistent decline in the quality of surface water has been seen in recent decades (Das et al. 2023a, b). The concept of sustainable water management has led to the publication of several hydrological studies that concentrate on the biological function of water in aquatic ecosystems. A multidimensional indicator approach called "water footprint" shows how much water is used by users and how much of it is polluted, along with what kind of pollution it causes. Although developed nations have adopted laws and strategies to address this issue, developing nations still lack the constitutional structure necessary to address the immediate effects of this threat. Monitoring water quality is therefore very important. In this context, the WQI model, which is crucial to the control of water resources, was specifically created for evaluating the temperature of river water in marine ecosystems. To simplify and make it replicable, this approach condenses voluminous data on water quality into a single composite number. To convey the general status of water quality and its suitability for various uses, it condenses a complicated set of facts into a numerical expression with fewer units (Kadam et al. 2021). Meanwhile, its use for monitoring and management of water resources has been growing quickly. Recent research has shown that when huge volumes of water quality variables data are converted into a numerical form, the existing WQI models yield a significant degree of uncertainty. It is crucial to pinpoint pollution-prone locations and those where providing protection will pay off in terms of social, economic, and technological factors. Realistic and efficient surface water management techniques

A. Das (✉)
Department of Civil Engineering, C.V. Raman Global University (CGU), Bhubaneswar, Odisha, India
e-mail: das.abhijeetlaltu1999@gmail.com;
abhijeetlaltu1994@gmail.com

D. Kumar et al. (eds.), *Data Analytics and Remote Sensing-Based Analysis for Sustainable Development*, Advances in Science, Technology & Innovation, https://doi.org/10.1007/978-3-031-98095-4_4

are adopted, such as Canadian Council of Ministers of the Environment (CCME)-WQI, contamination index (CI/C_d), and water pollution index (WPI). It combines the natural environment with engineering methods and provides a wide range of advantages, including water treatment, economic growth, enhanced public safety and health, habitat for flora and fauna, improved esthetics, and satisfaction of recreational needs (Sharma et al. 2022). The Canadian Council of Ministers of the Environment (CCME)-WQI is now among the most well-liked and one of the primary indicators that is advised for determining the water's quality for consumption by the United Nations Environment Programme Global Environment Monitoring System (GEMS). The benefits of this process are its adaptability, versatility to intake water performance measures and thresholds, and inclusion of random errors. However, these techniques focus on facts and are more precise (Haji et al. 2021). Geographic information system (GIS)-based geospatial framework is yet another extensively useful component that enables the visualization of geographic variability and the dispersion of water characteristics in the research region (Das et al. 2023a, b). It is a useful tool for mapping the vulnerability of cities to flooding, which allows for the spatial analysis of a lot of data about natural hazards. Additionally, the spatial results produced using GIS tools aid in a more accurate examination of the distribution of exposure, sensitivity, and overall vulnerability within the study area. Local and regional authorities will also be able to quickly obtain important data and choose the best measures to lower the danger of pollution owing to the spatial analysis. Contour mapping has been one of the best methods for visualizing spatial differences in water quality (Murugesan et al. 2020). Using kriging interpolation and inverted distance weighted (IDW) procedures, which are straightforward and understandable geostatistical techniques for mapping the spatial distribution patterns in water, physiochemical characteristics for water samples were mapped spatially. The IDW method, which uses a sequential pairing of well-known grid patterns to get an unknown grid value, is said to be the most widely used deterministic methods for interpolating several variables. Typically, this method makes use of the weighted average, which is intended for forecasting the value at each location using the nearest known points. While the inverse distance between the known and anticipated sample points is used to estimate the weight values, it illustrates an autocorrelation-based statistical model. To explain variation in the surface, this technique believes that a spatial correlation is indicated by the direction or distance between the sample points (Zhang et al. 2021). However, using municipal/industrial wastewater for drinking and irrigation over an extended period of time has negative effects on both the water's quality and human and environmental health. Thus, other widely utilized used tools are water pollution index (WPI) and contamination index (CI) which are quite skilled in classifying and analyzing the measured data. Additionally, these tools help to identify harmful pollution and its elemental concentrations in the selected parameters and provide an assessment of the overall extent of contamination of all samples that have been collected. However, this monitored data in combination with such testing assets that can provide information to aquatic governing bodies without any information being lost, while it promotes a clear and simple principle for quick decision-making in an emergency to cure the inappropriateness of water (Xiao et al. 2021). The Mahanadi River in Odisha receives a huge volume of effluent water, each year from a variety of sources, including sources at home, industrial waste, urban sprawl, and improper waste governance, which may have a negative impact on the water quality. Although being an interdistrict river, interstate, and worldwide significance, nothing is known about the situation with relation to the quality of its surface water. To the author's knowledge, no research has been done on the effects of urban growth on the geographic variance of various developmental activities take place in urban and nearby locations. Thus, the purpose of this investigation was to choose the best mix of quality indicators of water that lowers model certainty. The major goals of this study are to evaluate the effectiveness of various indexing strategies for forecasting WQI values and to find the key indicators of water quality while avoiding the loss or concealment of actual data using a variety of feature selection algorithms. Therefore, the research's findings will give readers a thorough grasp of different indicator optimization strategies and how well they function in terms of lowering WQI model certainty. The outcomes might also be helpful in figuring out the optimum algorithm to select the primary input of the water model in water research investigations that might be successful in improving the development of the WQI model.

2 Study Area

The Mahanadi River is the greatest waterway in the province of Odisha and the third-largest river on the Indian Peninsula. Its source is in Pharsiya village in the Dhamtari district of Chhattisgarh, at 442 m above mean sea level (MSL). The river is around 900 km long, while the coastline is about 200 km (Samanta et al. 2020). The area of study (Fig. 1) lies beneath the geographic coordinates of 80–86 °E and 19–23 °N, it is bordered by the plains of Chhattisgarh and Odisha, and it narrows even further to form a sizable deltaic zone near Puri and Cuttack, both in Odisha, before it meets the Bay of Bengal. The Mahanadi, the Brahmani, and the Baitarani are a network of three significant rivers that are transferring the sediments into the Bay of Bengal. So, while all these companies use a significant amount of the river's water, they also discharge pollutants and other effluents into the river's basin. Even though these industries typically have environmental

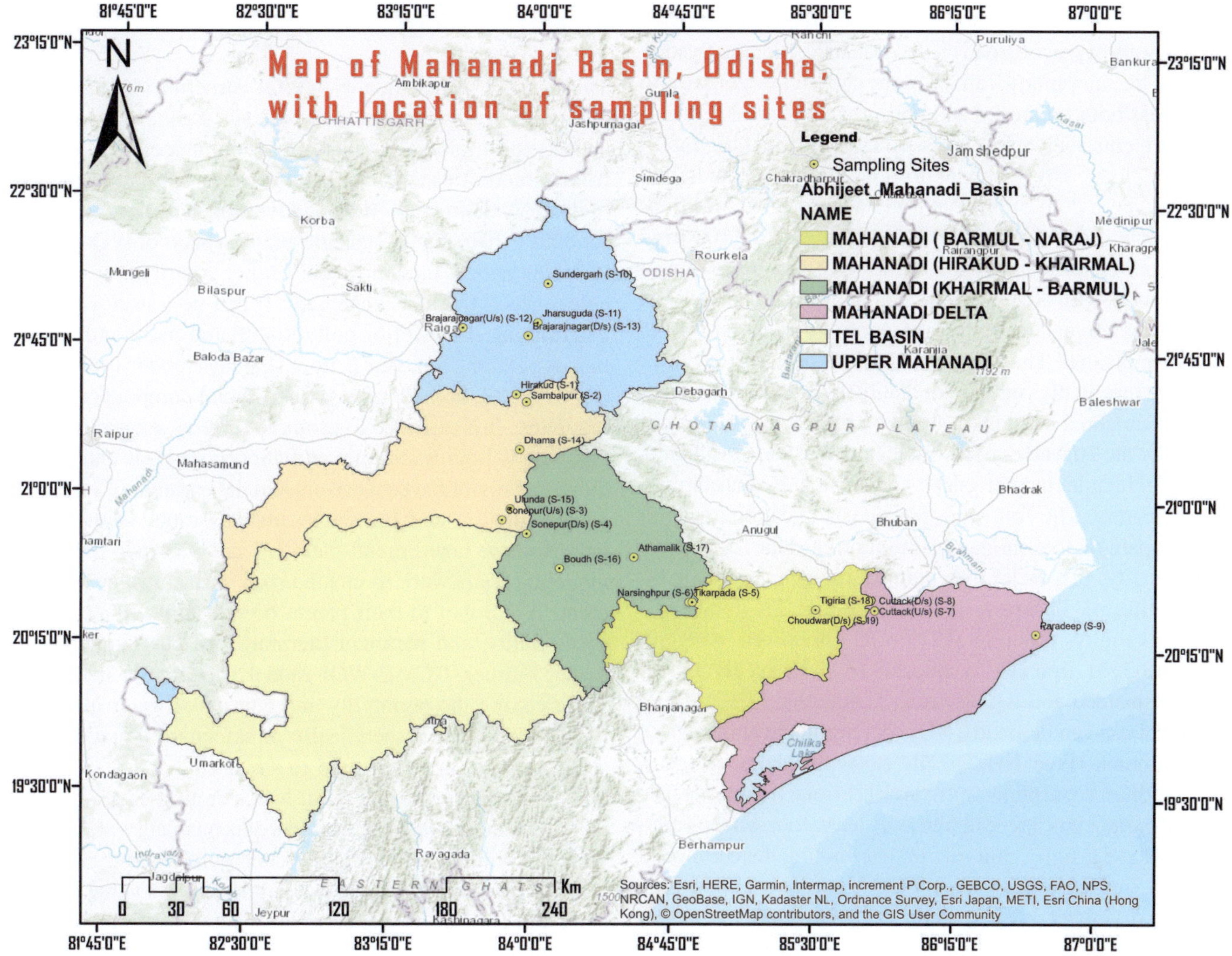

Fig. 1 Location of the study region (Mahanadi River basin, Odisha, India)

audit clearance, their effluents contain a number of harmful substances (Das 2022a, b, c, 2023a, b, c, d, e, f, g). Generally, in this basin, summers are hot and winters are frigid in the chosen locale, and it has a tropical climate, with an average annual rainfall of between 1200 and 1400 mm. The study area was dominated by a subtropical, moist temperate zone with thunderstorms and snowfall. This river always has stagnant water in it from January to May each year, which contributes to the buildup of pollution in the area. More than 90% of the total precipitation falls during the monsoon season, which starts in June and lasts until October, even if it occurs in spells of varying lengths and intensities. It is noticed that January is expressed as the coldest month, with average temperatures of 1–2 °C. However, the hottest month is June, with an average temperature, recorded as of 33 °C, and occasionally it rises to a maximum of 46 °C. Most of the locals work in the agricultural and farming industries, cultivating important crops including rice, wheat, maize, barley, grains, and vegetables. Additionally, the region is well known for its ability to produce rice through irrigation using surface water and additional surface water. Therefore, along the river, evaluation of surface waters and risk assessment for human health are crucial. In contrast, both marine and fluvial processes have an impact on the redistribution of deltaic sediments. The deltaic coast is in command during the southwest monsoon by waves, while beyond the non-monsoonal season, it is governed by the interaction of tides and waves.

3 Sampling Collection, Experimental Design, and Its Assay

Twenty water samples ($n = 20$) were retrieved in replicates from 19 different locations of the studied area during the chosen period of 2022–2023, following the standard guidelines (APHA 2005), in such a way that it ensures that each sample accurately represents the study region. The water samples were obtained during the monsoon season

(June–September). The sample location's registered latitude and longitude were recorded using Global Positioning System (GPS). The sampling locations were chosen horizontally to encompass both urbanized and non-urbanized areas, considering elements such as the site's accessibility, bathing facilities, ease of use, safety, stream flow, proximity to homes, and the area where domestic waste pollutants mingle. Additionally, samples were taken after an equal interval of 100 m in 250 ml pre-cleaned polythene bottles that had been soaked and cleaned with 10% HNO_3. Samples were dehydrated in an oven for a temperature of approximately 60 °C, to remove moisture. The samples that had been stirred after that were repeatedly rinsed with distilled water and decanted. The research area includes a summary of the analytical techniques for the 20 water quality indicators that were evaluated, some of which include physical, chemical, and microbiological indicators. The following physicochemical characteristics are part of the obligatory monitoring data for surface water quality: pH, BOD (biochemical oxygen demand), TC (total coliform), TDS (total dissolved solids), TH (total hardness), alkalinity, chloride (Cl^-), DO (dissolved oxygen), sulfate (SO_4^{2-}), iron (Fe^{2+}), fluoride (F^-), boron (B^+), TSS (total suspended solids), electrical conductivity (EC), COD (chemical oxygen demand), NH_3–N (ammoniacal nitrogen), free ammonia (Free NH_3), TKN (total Kjeldahl nitrogen), SAR (sodium adsorption ratio), and TH (total hardness). The chosen parameters are examined at least four times annually and have a significant role in determining the ecological health of surface water bodies. EC (electrical conductivity), pH, DO (dissolved oxygen), and TDS (total dissolved solids) were measured in situ with a portable multimeter, while the Cl^- (chloride), sulfate (SO_4^{2-}), fluoride (F^-), and nitrate ($NO_3{}^-$) were measured using spectrophotometric technique. Alkalinity and total hardness (TH) were analyzed by titration methods. A total of two metals, namely iron (Fe^{2+}) and boron (B^+), were analyzed by ion chromatograph (IC). Remaining tests of parameters were completed in compliance with World Health Standard (WHS) regulations. To prevent errors and preserve the quality, precision, and accuracy of the equipment, the dependability of the samples was often tested in between the samples by looking at the copies and established concentration guidelines. The instrument's calculated precision and accuracy values were found to be 5% and 10%, respectively. Additionally, MATLAB software was employed in the dataset analysis. Maps of the locations and the spatial distribution of the various metrics measured for the surface water of the research region were also created using ArcGIS 10.5 software.

4 Methodology

The surface water quality (WQ) structure is frequently employed in most studies to classify surface water's appropriateness for drinking. It functions as a unit score that derived from numerous water quality metrics and a substantial amount of data. Modeling tools are heavily utilized by specialists and scholars to better understand the many aspects of WQ. The use of mathematical modeling will further contribute to sustainable surface water management. It can be a useful tool for determining the level of pollution-related harm and recommends potential remedial actions (Panneerselvam et al. 2020). CCME, WPI, and CI analysis are crucial components of the modeling. It is intended to promote surface water infrastructure that is both traditional and conventional. It encourages the discovery of the cost drivers and the creation of improved cost prediction and life cycle analysis tools. These simulations show how crucial elements are incorporated into a modeling framework to create urban surface water management. According to their parent organization, functionality, accessibility, and pertinent literature, the classification was made. Further, CCME-WQI model is a crucial instrument for assessing the portability and control of water quality. It is an index created specifically to safeguard aquatic life in an ecosystem. It was created to streamline and present vast amounts of water quality data. It is a cutting-edge scientific communication tool that compares data on many water quality variables with predetermined water quality criteria (Wagh et al. 2017). The final WQI is a single unitless value ranging from 0 to 100, with 0 denoting the "worst" water quality and 100 denoting the "best" water quality and is computed from analyzed data as input into the dataset for a specific aquatic habitat. The six categories of water quality that can be separated based on the WQI model are as follows: WQI < 44, poor water quality; $45 <$ WQI < 64, marginal water quality; $65 <$ WQI < 79, fair water quality; $80 <$ WQI < 94, good water quality; and when WQI lies between 95 and 100, water was excellent for drinking purposes. This is based on three characteristics that describe how human activity affects the quality of the water. F_1 (Scope) indicates the proportion of water quality indicators ("failed variables") that do not reach the regulatory guideline values across all the variables used in the water quality evaluation. It is expressed as $F_1 =$ (Number of failed variables/Total number of variables) * 100. Further, F_2 (frequency) is the percentage of individual tests ("failed tests") which does not meet the objectives. It is expressed as $F_2 =$ (Number of failed tests/Total number of tests) * 100. Again, F_3 (amplitude) is the amount by which failed test values does not meet their objectives and is computed as in equation as excursion$_i =$ (Failed Test Value$_i$/Objective$_j$) $-$ 1, where the test value falls short of the goal, and excursion$_i$ $=$ (Objective$_j$/Failed Test Value$_i$) $-$ 1, where the test value exceeds the objective. Now, the normalized sum of excursions

(NSE), also known as the sum of individual test excursions from their objectives, is now calculated by summing (Σ) the individual test excursions from their objectives and dividing by the total number of tests in this research study. This is presented in equation as NSE = [(Σ excursion$_i$)/Number of tests]. Hence, F_3 is finally computed using an asymptotic function that scales the normalized sum of the excursions (NSE) from objectives to produce a range between 0 and 100. The factors F_1, $F_{2,}$ and F_3 are generated, by which three components are added together as vectors to determine the value. Further, three acquired factors are combined to calculate WQI in the manner described below: CCME-WQI $= 100 - \left(\left[F_1^2 + F_2^2 + F_3^2\right]/1.732\right)$. The above equation witnessed that the square of the index equals the sum of the squares of each element. Ultimately, this method views the index as a three-dimensional space with a single axis that is determined by each factor. With this model, changes in any one of the three elements will directly affect the index. In a later stage, WPI and CI are components of numerical methods that comprehensively transform a large amount of data of water quality into one output. The most recent WQI has been used in numerous research to evaluate the quality of Indian river water recently. A latest study has shown that this revised method is beneficial for evaluating surface water quality to lower model uncertainty. Therefore, three indexing approaches were employed in this research to evaluate river water quality (Masood et al. 2021). In addition, these methods are frequently employed for evaluating various dynamics in surface water bodies. The examination of these systems' chemical health condition may result from WPI. It was foremost implemented by the nutrient contamination index method that gets revised after Atique and An (2019). The identification of important water quality indicators has been done using a variety of methods, such as expert judgment, data accessibility, and environmental significance. The Delphi procedure was one of the most popular approaches in this regard for eliciting expert opinion via surveys or interviews. To quantify pollution in ecosystems, this model comprises of five metrics that are significant water quality contamination indices. The boundaries specified after the first three observed distribution of obtained values served as the basis for establishing the scoring criteria assigned to each metric (Hossain and Patra 2020). This approach depends on how each metric contributes to the appraisal of the overall environmental well-being of the model. The main purpose of this method is to identify the environment's characteristics as well as the main harmful elements and their impurities that are present in river water. This method was utilized in the current investigation to evaluate the water's quality and identify the physicochemical components responsible for water contamination. It is created by the following equation: WPI $= C_i/L_i$, where C_i is the revealed concentration of the ith parameter, and L_i is the allowable limit of the ith parameter. Consequently, when the WPI value is greater than 1.1, the presence of pollutants in the water requires treatment before use. WPI greater than 1.1 indicates a high concentration, and the water bodies located in certain areas may be contaminated by certain factors. It also shows that if the value is higher than 1.1, the surrounding area and any excess or concentrated amounts have the potential to contaminate nearby water bodies. Its index classes are arranged in the following hierarchy, such as < 1.1 indicates very pure, 1.1–2.1 represents moderately polluted, 2.1–4.1 indicates polluted, 4.1–6.1 represents impure, and > 6.1, highly impure water. Moreover, contamination index (C_d), which occurs most often used for evaluation of water contamination, was used to determine the combined impact of pollutants on drinking water (Backman et al. 1998). It also explains the degree of a parameter's contamination. Additionally, it describes how the combined impact of heavy or trace metals harms drinking water (Ren et al. 2021). The degree of contamination is measured as: $C_d = \Sigma Cf_i$ and $Cf_i = (C_{ai}/C_{ni}) - 1$, where Cf_i represents the contamination component, and C_{ai} and C_{ni} are the observed proportion and the highest permitted ceiling (HPLC) of the ith ion, suggested as per WHO (2004), respectively. The resulting index is compared to a common database that includes numbers in the range of 1–3 as: (< 1), (1–3), and (> 3) with a rating of "low," "medium," and "high," respectively. Finally, these models are employed for the important evaluation of water quality in several nations throughout the globe.

5 Results and Discussion

The essence of physiochemical parameters in this study cannot be underplayed because it is the primary determinant for drinking, industrial, domestic, and industrial uses. Understanding the impact of each parameter on a surface water prospect map requires water quality analysis. It helps to interpret the surface water potential index.

5.1 Water Quality for Drinking Purposes

For surface water used in irrigation and drinking, this study aims to analyze the effect of specific chemical parameters on agriculture commodities and public health (Krishnan et al. 2021). Presented in this study is a detailed overview of physiochemical results depicting the chemical composition of surface water in the study area. Further, in order to identify the vulnerable areas for the protection and management of surface water resources, the spatial distribution of values for the drinking surface water quality index was plotted using ArcGIS software using the IDW interpolation technique. The pH level is known to alter the quality of water for a variety of uses as well as the way minerals dissolve in surface water.

Because they have an impact on other physicochemical and biological factors, they are also seen to be important parameters for evaluating water quality. In addition, the pH of the water has an impact on biotic activities, and photosynthesis also raises the pH of the water. During the monitoring period, in this study, its value varied in the range from 7.739 to 7.899, not exceeding the permissible limit of 6.5 (H^+ has a strong acidic level dominance) to 8.5 (OH^- takes control at the alkaline phase) as per WHO standards. The data suggest that it exhibits a somewhat alkaline characteristic because temperature fluctuations might affect biological activities, particularly ecological biodiversity and chemical processes in water (Panneerselvam et al. 2020). However, the minimum values of pH were noted in site 11 and 12, and it was probably because of CO_2 and the oxidation of organic molecules, and pH is known to decrease when heat input increases. The study also showed that drinking water's pH is rarely a solution. When readings are above or below the advised range, however, it acts as a warning sign of potential problems, such as too much bicarbonate dissolved in irrigation and drinking water. This shows that, in terms of pH, all samples in the research area are safe for consumption. Microbes that reduce or ammonify nitrogen require DO concentrations of less than 5 mg/l. Microbes also use it as a part of the biodegradation process, which can change the oxygen content of water. Pressure and temperature, chemical composition, the breakdown of organic matter, primary production, biological activity, and respiration are a few of the elements that control it. While its consumption occurs during both the oxidation of nitrogen molecules and the metabolic decomposition of organic matter, the degree of oxygen saturation is directly correlated with both the temperature regime and the rate of photosynthesis in water (Nizam et al. 2022). The measured values in all samples, which varied from 7.27 to 8.2, greatly outperformed the WHO-recommended cutoff of 6 mg/l, indicating strong activity of oxidizing bacteria and optimum circumstances for ammonia production. The levels have increased in most places due to the addition of contaminants from metropolitan areas, as well as evidence of organic matter and other pollutants. The quantity of oxygen required in a surface water body to completely decompose organic matter is known as BOD. In other words, it displays the amount of waste that can be biologically broken down in the water. Due to competition for DO, excessive BOD levels in water can kill aquatic life and are a sign of high amounts of organic pollutants. Low organic matter is indicated by low BOD readings. It also results from annual cleanings of silt, soil, and other solid trash that take place between November and January. In the study region, its readings span between 1.04 and 2.38, which is within the allowable criteria of 5 mg/l. This can imply a low level of organic trash, and the ecosystem of rivers supports aquatic life. The low findings at every sampling site would suggest that the higher temperatures and increased rates of biological activity in the river environment are likely to be aggravating organic pollutants. According to the findings of the current study, an increase in DO levels at all sample locations is caused by the fact that there are very few people living there; thus, there are no large effluent discharges into the river. The rise in levels can be mostly attributed to reaeration along the river stretch due to its ability to purify itself (Balamurugan et al. 2020). Total coliform (TC) concentration is mostly derived from human feces, cattle, wild creatures, releases of wastewater into surface waterways, agricultural, urban and rainwater, soil, and other ecosystems for the purposes of further evaluating the fecal contamination situation. Its reported readings ranged from 1211 to 42,528 MPN/100 ml, and likewise, thereafter, level was inside the WHO-recommended range, except at sites 8, 9, and 19. It might have been caused by untreated sewage being dumped into the river. It may have been caused by the heavy use of fertilizers and pesticides on agricultural grounds, as well as waste detergents and sewage effluents (Ravi et al. 2020). Also, the raised level at the three places in the river system may have been influenced by drainage water discharged from the surrounding areas. Additionally, water can be severely contaminated by man-made activities such as farming and enterprises adjacent to streams, rivers, and canals. This contamination is especially harmful in locations with high pollution levels, such as unplanned big cities. Additionally, slum residents with poor sanitation systems who lived near rivers may have contributed to a larger bacterial load in surface water. The flow of untreated residential wastewater into the river and the unhygienic living circumstances of those living nearby may have raised the coliform burden in the river water. TSS levels show the amount of organic or mineral-based suspended particles in water. In a river system, high particulate matter (TSS) levels have an impact on habitat quality, productivity, and light penetration. Furthermore, the dataset's significant standard deviation in TSS, which consistently represents the regional heterogeneity in suspended solids, is another factor. The levels in the present study ranged from 28.61 to 74.88. Water with values less than 100 mg/l had no notable pollutant discharges, while such high pattern of distribution at sampling locations was also impacted by urban slum activities that take place near the riverbanks, including open defecation (Gao et al. 2021). Typically, variations in sewage intake, rainstorms, storm water, and point emissions of industrial and municipal effluents may have an impact on the shift in spatial dependency on the indicators of water quality in river water. The concentration of dissolved or suspended organic and inorganic compounds in water, which are caused by industrial effluents and agricultural discharges, is connected to COD. It is sometimes represented as the mass of oxygen consumed per liter of solution. It ranges from 6.746 to 22.90 mg/L in this river basin, which is under the threshold belt of 30 mg/l. Since more organic molecules can be oxidized chemically than biologically, it is not surprising that COD was also found to be higher than BOD. Higher COD often indicates that effluents from significant municipal and industrial applications have been released. NH_4–N concentrations ranged from 0.598 to 1.928 in the river that dropped below the 2 mg/l WHO-recommended threshold. This shows that there may not be as much anthropogenic activity in and around the river as there is NH_4–N input into the river. Thus, the river's low productivity may be caused by excessive NH_4–N levels inhibiting the growth of algae and other plants (Zhai et al. 2019). Additionally, precipitation or storm events may collect nutrients from farms, cities, and sewage treatment facilities and then transfer them into aquatic environments, leading to higher value concentrations. Additional factors for free NH_3 contamination, environmental degradation, and significant stress on the river and aquatic ecosystem include human and urban activities growing in the research area's vicinity. According to a current study, inadequate sanitation systems and poor waste management are two key contributors to river contamination (Karunanidhi et al. 2021). Due to the large amount of garbage and excrement that is discharged into the water because of the high concentration of people, the high population density, and the sewer system, the water quality and concentration level are ultimately impacted. Results from the ongoing research showed values ranging from 0.02 to 0.06, which is below the threshold proposed by WHO (2 mg/l). TKN would originate from the decomposition of the organic waste, accounting for components from domestic, industrial, and agricultural sources as well as fecal contamination. Initiating the surface water flow via garbage also allows dangerous chemicals taken away from the landfill to degrade and leak into the water. Trash water, excrement, household waste, human activities, and nitrogenous fertilizers from farms near the examined stations may all be contributing factors to the degradation of the related rivers. It also suggests that organic matter degradation processes are not fully complete, and it is a reliable sign that urban effluents are contaminating rivers. During the sampling period, the reported values ranged from 3.279 to 11.789. This also accords with the findings at most places, encompassing higher values, which is because of heavily populated and industrialized tropical watershed, claiming that untreated sewage has mostly been responsible for changes in water quality, which raises the possibility of ongoing sewer leaks or illegal releases. EC is a crucial factor in figuring out whether water is suitable for drinking. Since high salt concentrations make soil saline and interfere with plants' ability to absorb water via their roots, it is crucial to assess the danger of salinity in drinking and irrigation water. Additionally, salts in the water have a negative impact on crop productivity since they make the soil more saline, and it is unfit for human consumption. Additionally, it serves as a gauge for how well water conducts electricity and shows how much total dissolved solids are present (Aravinthasamy et al. 2021). Therefore, EC evaluated the salinity risk of the current investigation, and their concentrations ranged from 137.9 to 7779.29 μS/cm, which is considered suitable for drinking except at site 9. According to the EC level, 94.73% of the water samples collected were below the WHO (2250) acceptable limits for drinkable water. At site 9, the soils in the research area may get more salinized because of using this mildly salty water over an extended period. A high concentration of Na^+ reduces soil permeability and causes an infiltration issue because a portion of Na^+ can be exchanged for Ca^{2+} and Mg^{2+}. The selection of crops is severely constrained by saline soils, which also negatively impact germination and crop production (Das 2022a, b, c). However, extremely saline and low alkaline water can be used for crops that can withstand a lot of salt. Very high saline and somewhat alkaline water should not be utilized for drinking or irrigation, especially in soils with inadequate internal drainage. Concentration of SAR lies in between 0.408 and 16.58, which is below the 10 mg/l threshold suggested by the WHO. Higher values felt at site 9 suggest that noteworthy salt concentration in water is mostly caused by wastewater mixing, industrial wastes mixing, mineral deposits, detergents, other soaps, and machinery used in food processing. Large- and small-scale enterprises, as well as the high population density in these regions that results in high municipal waste production, could all be contributing factors to the greater sodium concentration in these areas. Major parameters like B^+ are the by-product of primary silicate weathering and are also human-sourced; they enter the subsurface aqueous system because of contaminated surface water. However, it enters the water system as small amounts of anthropogenic inputs (nitrogen fertilizers) from agricultural activity (Yousefi et al. 2018). Therefore, assessing the health status of this indicator will provide further insight on the space-based advancement of water quality. This variable is important in enhancing the quality of irrigation water for greater crop yields. The fertility of soils will decline during long-term agriculture without the use of B^+. As a result, this is a crucial component for improving irrigation. The amount of B^+ present in the river water ranged from below the detection limit to 2 mg/l. The suggested range for its concentration in drinking water, 0.029–0.55, along with the total number of surface water samples collected from the study area, indicates that the water is safe for consumption. It is preferable to build urban areas that cause an increase in the flow of pollutants into aquatic habitats, which are released from urban drainage networks and systems that control storm water runoffs, to limit pollution levels to lower levels. TDS is determined by weighing the deposits that remain after a sample has been evaporated to dryness, and it shows that the resistance of surface water samples rises as more dissolved inorganic ions are present. Due to extremely high or low TDS levels in the water, which also signal the presence of synthetic chemicals that might harm fish and other aquatic plants by producing an intolerable odor and color, the growth of aquatic species can be hampered or even result in their death. However, TDS in the study area ranged from 82.28 to 13,229.50 mg/l, with an average of 811.90 mg/l, it was found that approximately 94.7% of the samples were within the acceptable limits (WHO), thus reflecting the low level of dissolved solids, and it suggests that the water is classified to be fresh. A higher number implies that there is a greater influence of partially or completely untreated waste waters from agglomerations, industry, and agriculture on the water recipients and that site 9 has higher organic matter levels in its water. Total hardness (TH) is a measurement of the amount of calcium and magnesium ions present, represented in terms of the amount of calcium carbonate. They frequently regulate the water's hardness, and an excess of them can cause scaling and stomach problems in people. It is important to note that in water with high carbonate and bicarbonate levels, calcium and magnesium can precipitate as carbonates, which then can increase the levels. In most water bodies, calcium and magnesium are in a state of equilibrium, whereas crop production is decreased by higher magnesium water concentrations. Its findings in every sample that was analyzed ranged from 51.19 to 2194.90 mg/l. The WHO maximum limit of 300 mg/l was found to be violated by none of the concentrations in the 20 samples. At site 9, runoff from agricultural farmlands tainted with fertilizers could be one of the potential causes. From the investigated dataset, the ion that is most prevalent in the studied area seems to be chlorine. The value in all the water samples ranges from 9.648 to 4904, and the ideal Cl^- value should be < 250 mg/l as per WHO guidelines. This demonstrates that Cl values were within the acceptable range. According to recent observations, there is just one station in the research region where there is a risk from chloride. Therefore, recycled water with a Cl^- value of 0 to 250 may be used without limits; however, site 9 requires special attention. Higher presence of Cl^- at site 9 (4904) in water sample is due to farming operations and excessive aquifer exploitation. Consequently, rising concentrations could have a negative impact on human health (indigestion, kidney, and palatability). This further shows increasing seawater intrusion, increased leaching from sewage effluents, changes in evaporation rates, and human impacts (production and commercial processes, farming methods, and landfills). Sulfate (SO_4^{2-}) concentrations ranged from 4.968 to 376.06, and its values is within the 200 mg/l allowable limits. Gypsum dissolution and human-made activities like backflow from irrigation systems and clogged drain wastewater are blamed for the excessive sulfate concentration in some areas. A wide range of sulfate values at point 9 suggests that improper home wastewater disposal was the primary source in the area, which was then blamed for fertilizer applications in the agriculture sector (Shyam et al. 2022). Additionally, the weathering process and interaction between the rock and rainwater may have increased the temporary overall hardness. However, a low concentration of sulfate compared to chloride throughout the time studied shows that organic matter is breaking down, there is little nutrient use, and there is less agricultural runoff in the region. The constituents of F^- in the surface water are phosphate salts and the fluoride-bearing mineral like apatite. Dental fluorosis results if the concentration value exceeds the standard limit. When people eat leaves and vegetables, the amount of F^- in the irrigation water affects their health and causes fluorosis to spread. Fluoride levels that are greater than normal confirm the effects of human interferences including poor home wastewater disposal, industrial waste leaching, and runoff from agricultural areas. Its value in the present research span from 0.259 to 1, satisfying well within the prescribed threshold of 1 mg/l. The use of fertilizer application (chemical and organic) for agricultural activities is to blame for the rise in nitrate concentrations in surface water. Additionally, the level of nitrate content was impacted by the impact of wastewater effluents on the local surface water (Wu et al. 2018). It may cause hypertension, goiter, and gastric cancer. The oxidation of nitrogen-based fertilizers, human activity, animal excrement through microbial nitrification, and phytoplankton activity are the likely sources. When oxygen diffusion becomes a limiting factor, this parameter is likely formed, particularly in the deep levels of the biofilm where nitrates are converted to nitrites in an anoxic environment. As a result, this value rises as the biofilm gets older and more compact. Given that the reported amounts ranged between 1.287 and 2.609, the nitrate concentration was below the 45 mg/l WHO limit. This result at site 9 showed a potential risk, which is linked to urban discharges and anthropogenic activities. However, prolonged use of industrial and municipal sewage may cause heavy metals to build up in river water. It was anticipated that either natural sources (acid sulfate soils) or industrial wastewater would be the source of iron in the aquatic environment in the study area. Fe contamination of surface water is a widespread issue, and industrial and agricultural wastewater is a major source for other heavy metals as well. In brief, the experiment's findings varied between 0.59and 2.61. A higher value of 2.61 at site 9 ind icates a low-to-high cancer risk associated with long-term use of this surface water. So, before using it for drinking or irrigation, home and industrial sewage should be treated centrally. The removal of toxins from sewage can be accomplished using natural treatment techniques like reed beds and artificial/natural wetlands. In this case of Fe^{2+}, as per WHO recommendations, the readings in the river are within the permissible range, i.e., 3 mg/l. The most prevalent cation and anion, referred to be Fe^{2+} and Cl^-, respectively, are the main dissolved ion constituents in the river, which are typically low, except at site 9. Using the projected means, the river has a general pattern of ionic dominance of: $Fe^{2+} > B^+$ and $Cl^- > SO_4{}^{2-} > NO_3{}^- > F$. The increasing trend of Fe^{2+} concentration shows the rise in urban pollution and the industrial/agricultural practices that control surface water contamination. To avoid dumping untreated wastewater into streams and river courses, decentralized and functioning sewage treatment plants can also be constructed for the entire river basin. The data plotted on the spatial diagram for all parameters are shown in Fig. 2a–t.

5.2 CCME-WQI Method

Further, the surface water of Odisha is mostly polluted at some locations, and people of this region still rely on a few substitutes for household and recreational purposes; thereby, CCME method was used to measure the rate of decline of surface water quality (Das 2023a, b, c, d, e, f, g). The WQI's value over the sampling period, varying between 34.709 and 85.519, and the individual water sample grading according to WQI score are represented in Fig. 3. In the meantime, the categorization of samples created, based on WQI values, signified that 73.68%, 15.79%, and 10.53% of the samples classify good, fair, and poor categories, respectively. In addition, fair and poor type of classes represented around 26.32% and verified that the surface water is typically unfit for consumption and domestic usage. As a result, after receiving treatment, people should consume it. Organic and nutrient pollution are more significant in sites 8 and 9, possibly because these areas acquired more contaminants from wastewater and water than the other areas. The IDW interpolation approach was used in ArcGIS 10.5 software to zone for changes in water quality (Fig. 4). However, these findings demonstrate that water quality varies significantly over time and that it is typically different throughout the dry and rainy seasons. However, it is reflected from the spatial map that water quality was good at the upstream side as compared to downstream portion (Soujanya Kamble et al. 2020; Das 2023a, b, c, d, e, f, g). This outcome can be attributed to increased rainfall and water infiltration into the ground, which diluted the water system (Das 2023a, b, c, d, e, f, g). In contrast, the water quality downstream was caused by poor management of waste disposal sites and commercial mining effluents, as well as inorganic fertilizers and pesticide leakage in agricultural areas.

5.3 Estimation of Water Pollution Index (WPI)

As opposed to that, the results of WPI and CI agree with the computations' findings of the CCME. In the sampling period (Fig. 5), the WPI of river span between 0.29 and 15 encompasses a mean value of 2.44. As per taxonomy of WPI, this river stands as "very pure to highly impure water" even though each water point's water quality tended to be better with an approximate value of 63.16%, containing 12 out of 19 locations, falling under very pure water category. Regarding spatial location in Fig. 6, the WPI value was substantially higher at three locations, which falls under highly impure level of water quality. Taking this into account, it attains for three locations, out from 19 locations, which can reduce around 15.78% of monitoring costs, and these sites can be vulnerable zones (Das, A 2023; Abba et al. 2023). These findings demonstrated that there seems to be substantial variations in water quality between sampling locations, mostly because of industrial production, agricultural, transportation, and maritime economic activities (Egbueri et al. 2023). This method suggests that monitoring sites from 8, 9, and 19 can serve as fewer sampling sites due to similar aspects of water contamination. If not, it can be thought to switch to other potentially polluted places to quickly identify pollution sources. Additionally, these locations show that eutrophication was a major concern for the river.

5.4 Estimation of Contamination Index (CI)

In case of CI, in Fig. 7, the computed values were in between 0.2 and 4.0. About 63.16% of the samples ranked low in the hierarchy, demonstrating that this water is fit for human consumption. The remaining 21.05 and 15.79% of the locations grouped under medium- and high-polluted categories and, thus, are not advised to be consumed while drinking. This will make it easier to identify the river's most polluted section. In the entire part of the region, surface water with CI depicts low- to high-polluted zone is found. Around 12 samples represent low, four shows medium, and finally, three samples represent high-polluted water quality categories. The spatial variations across the region are shown in Fig. 8. Samples 8, 9, and 19 connect to the urban area, where high CI values indicate the significant role that human activities have played in contaminating surface water (Das, A 2023; Ramos-Pacheco et al. 2023). The recorded highest value of (4) of CI, recorded for the sampling location 8, is situated close to the handicraft and leather industry. This could have several origins, and it serves as an example of how human activity negatively affects surface water quality (Laltu 2023; Begum et al. 2023).

Furthermore, rapid urbanization, which arises from thermal runoff pollution, is one of the factors that may be cited from the results of WPI and CI, which highlight the main causes of pollution. This indicates that rainfall has reduced the amount of water that naturally seeps into the soil in urbanized areas, forcing the water to run onto the surface of the heat-absorbing material before eventually draining into the river.

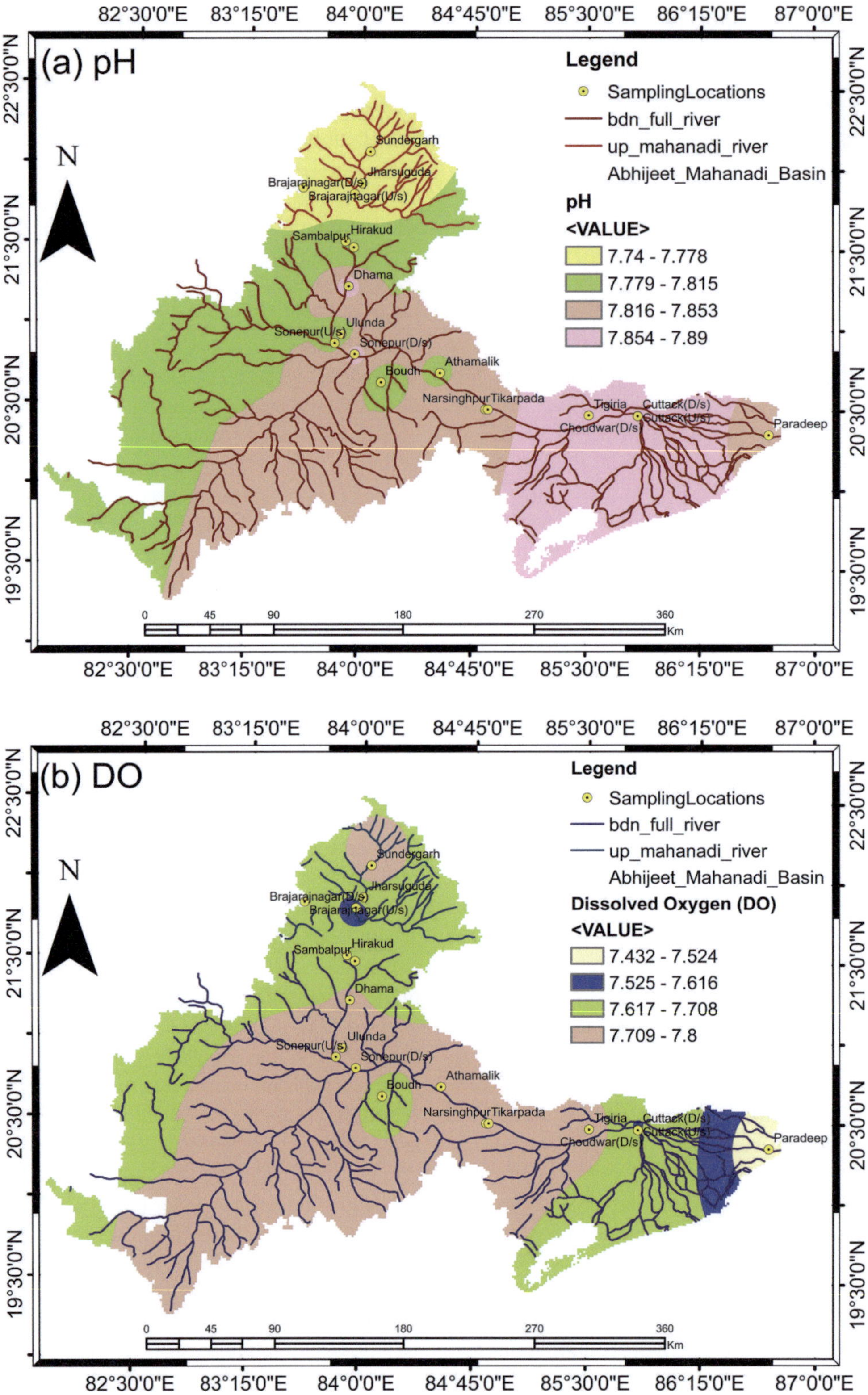

Fig. 2 Spatial variation of major parameters in river stretch, namely **a** pH, **b** DO, **c** BOD, **d** TC, **e** TSS, **f** Alkalinity, **g** COD, **h** NH_4–N, **i** free NH_3, **j** TKN, **k** EC, **l** SAR, **m** B^+, **n** TDS, **o** TH, **p** Cl^-, **q** SO_4^{2-}, **r** F^-, **s** NO_3^-, and **t** Fe^{2+}

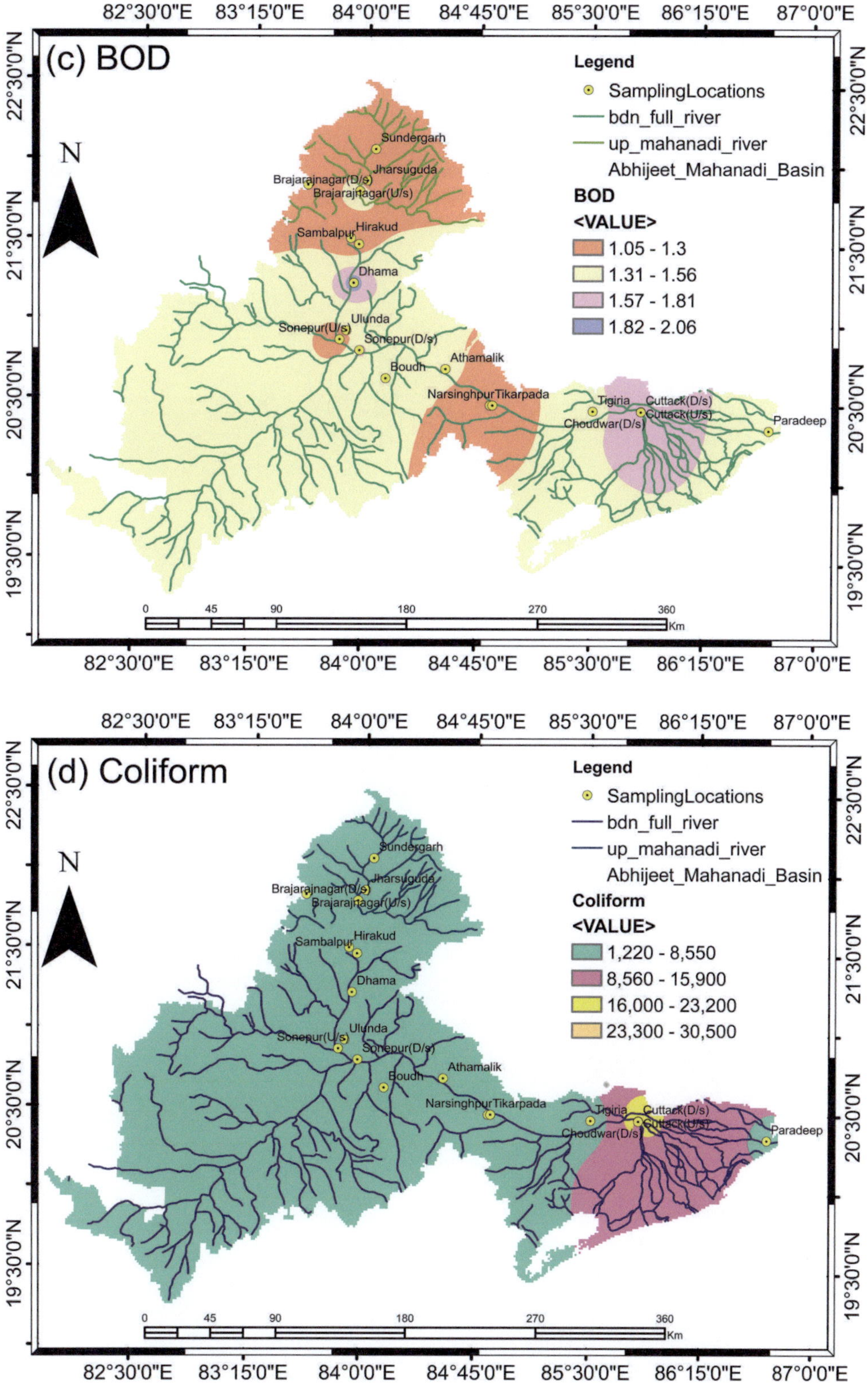

Fig. 2 (continued)

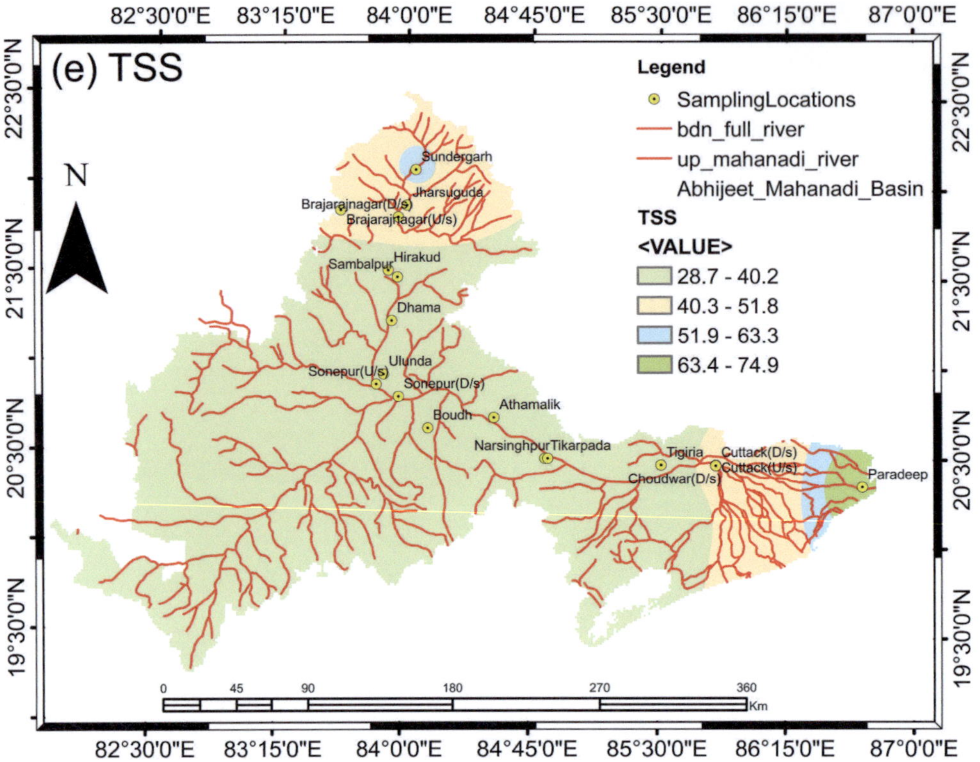

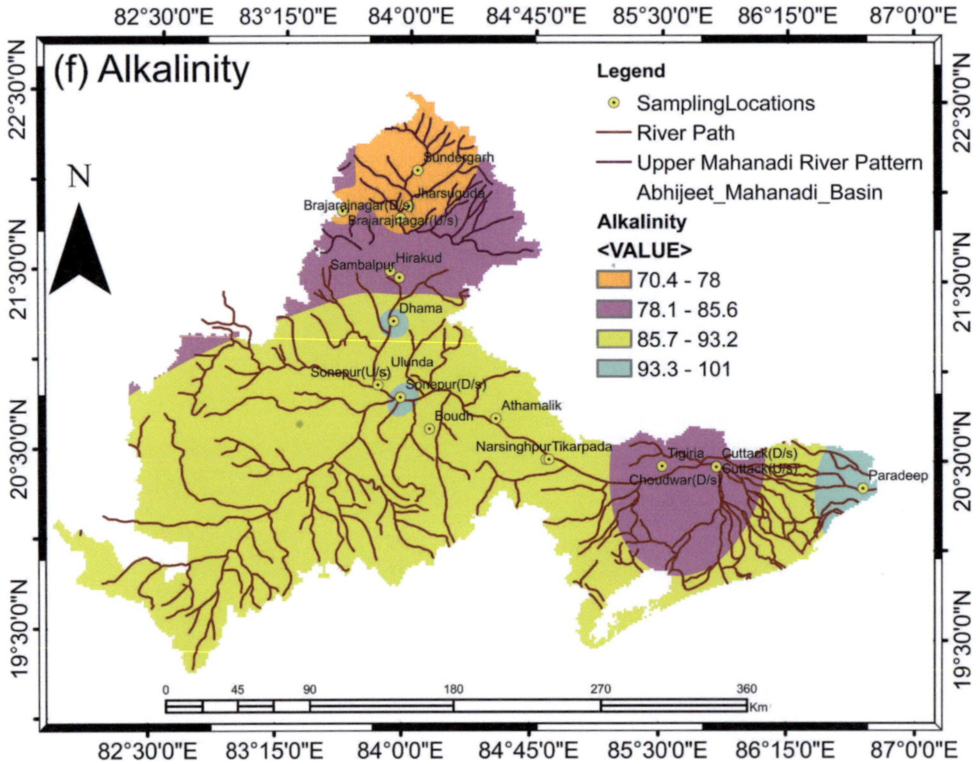

Fig. 2 (continued)

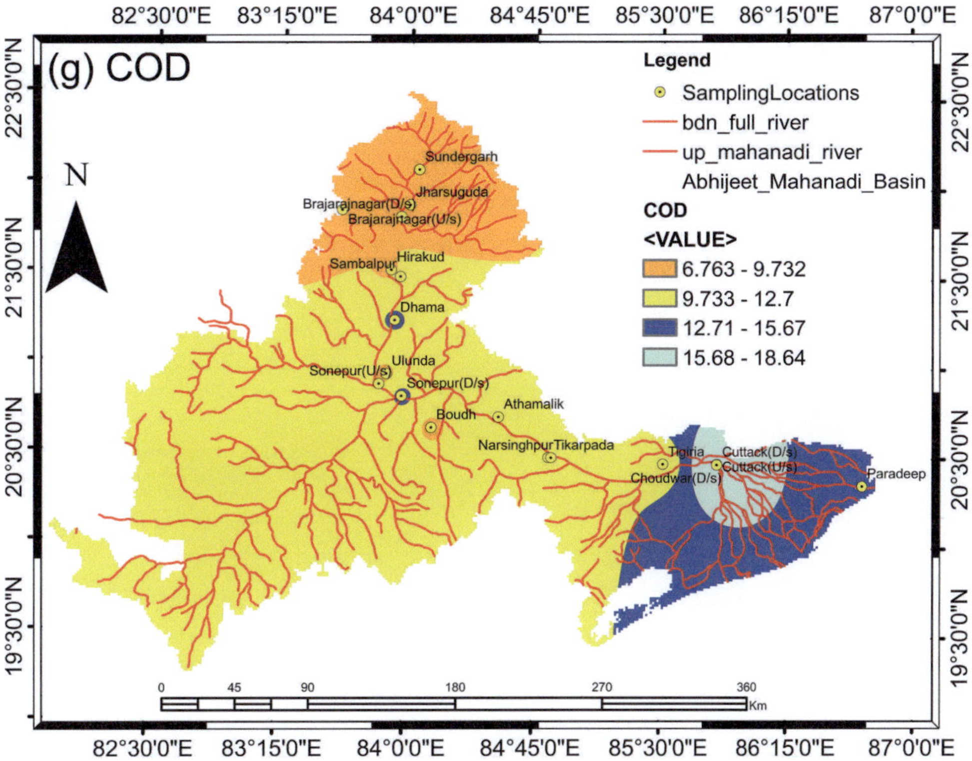

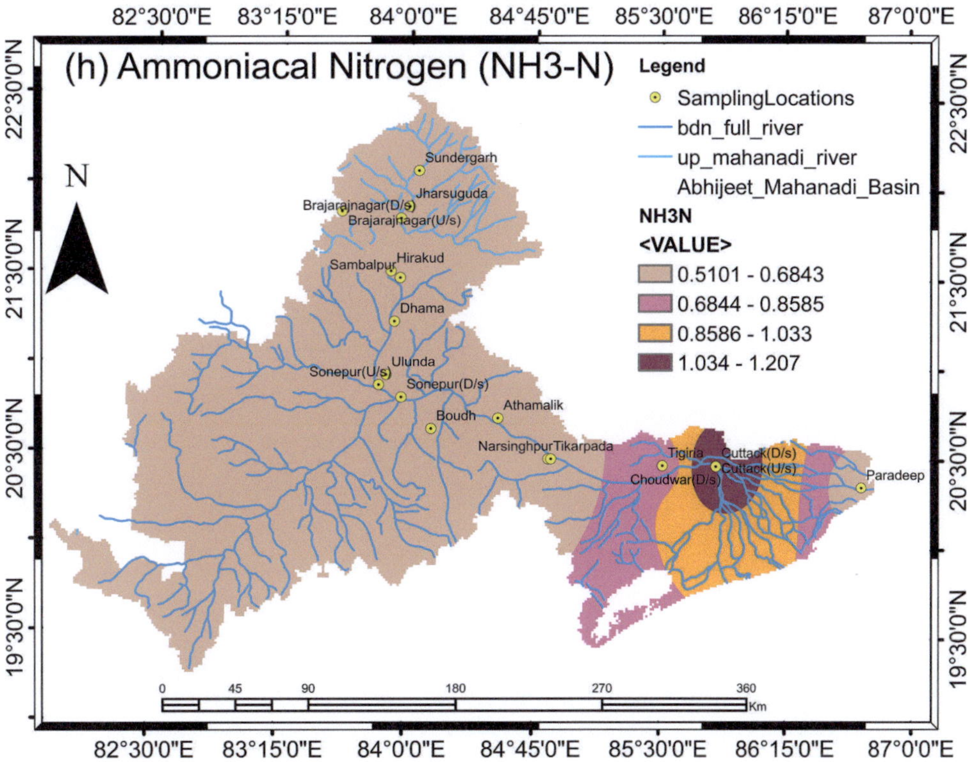

Fig. 2 (continued)

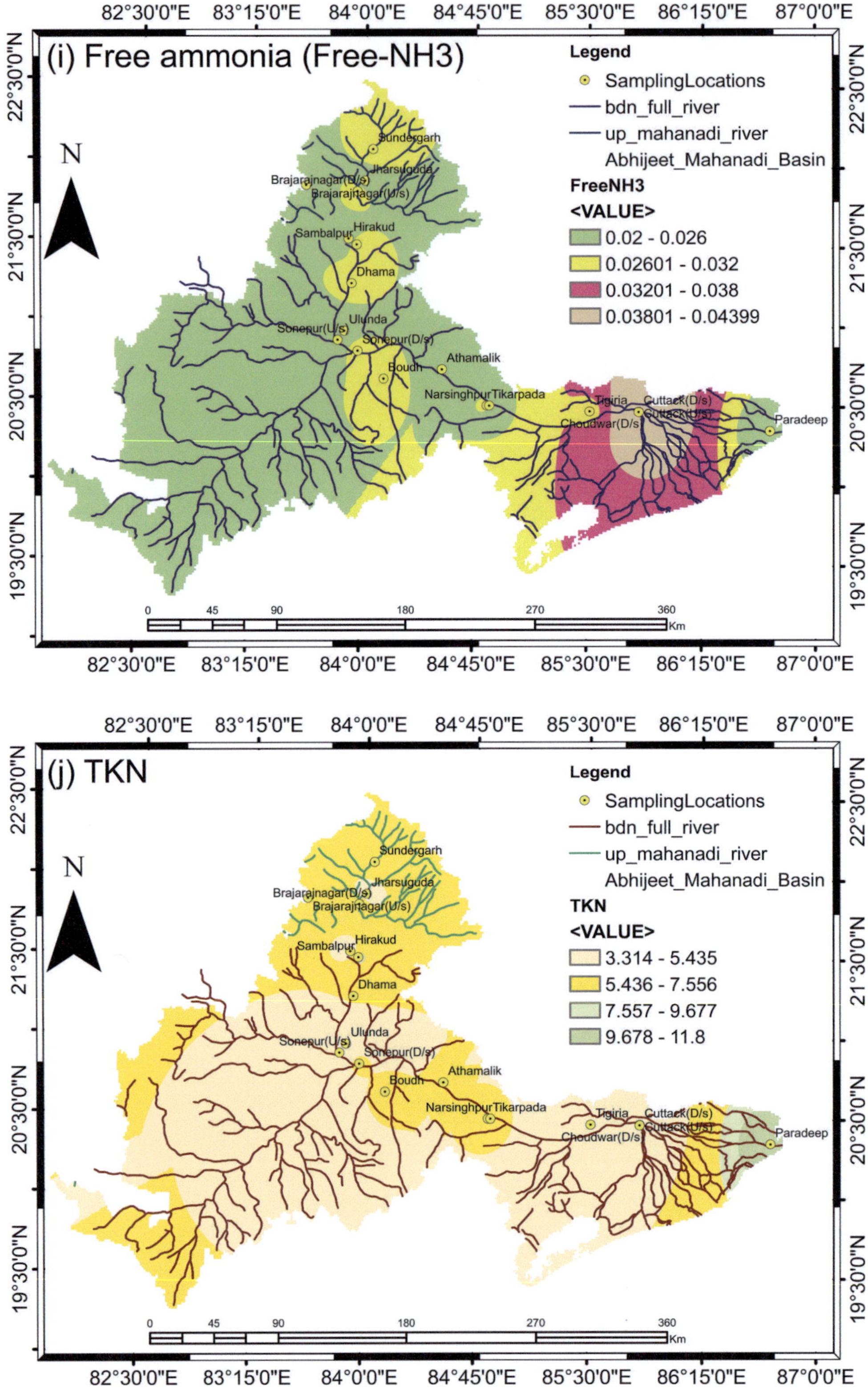

Fig. 2 (continued)

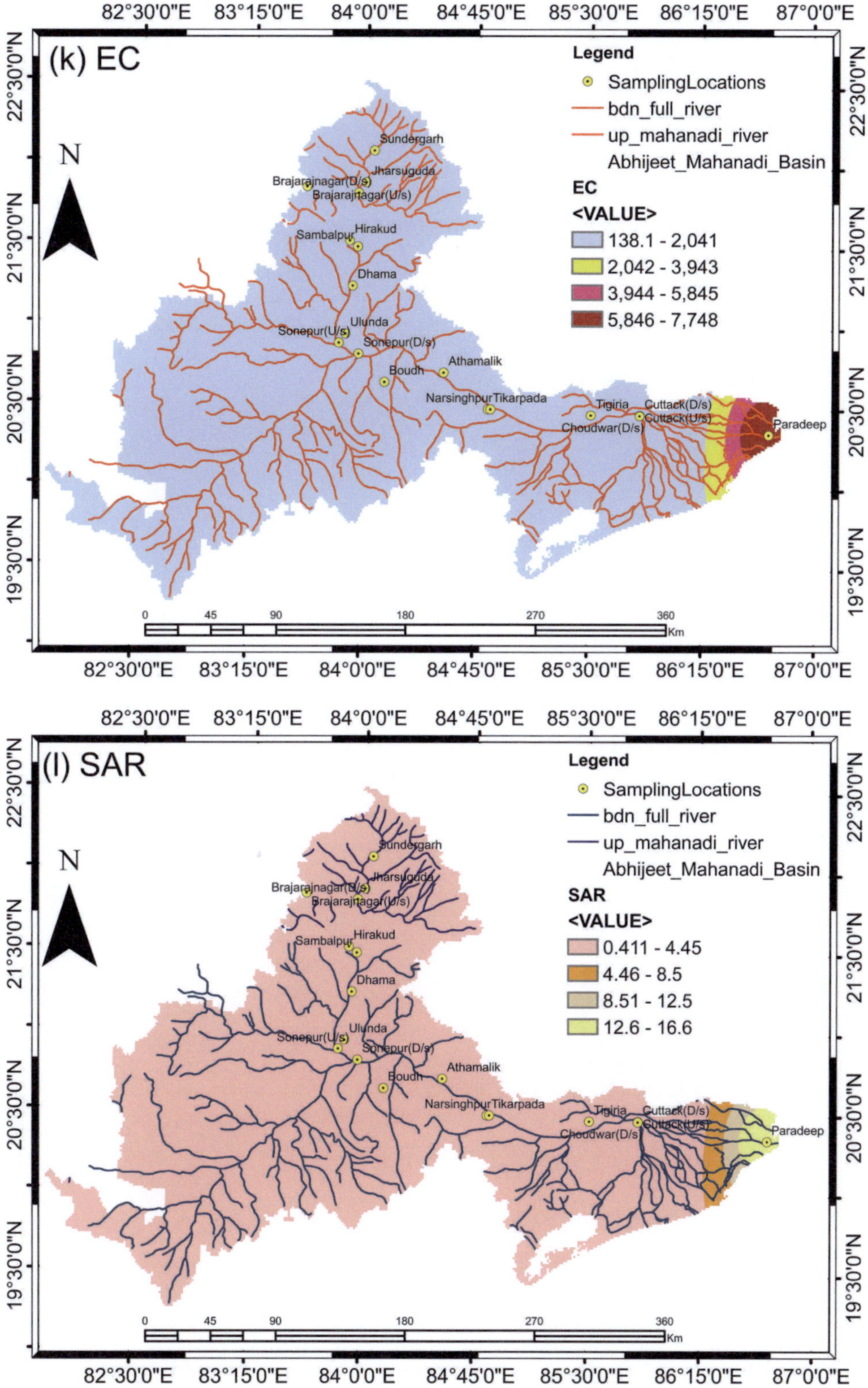

Fig. 2 (continued)

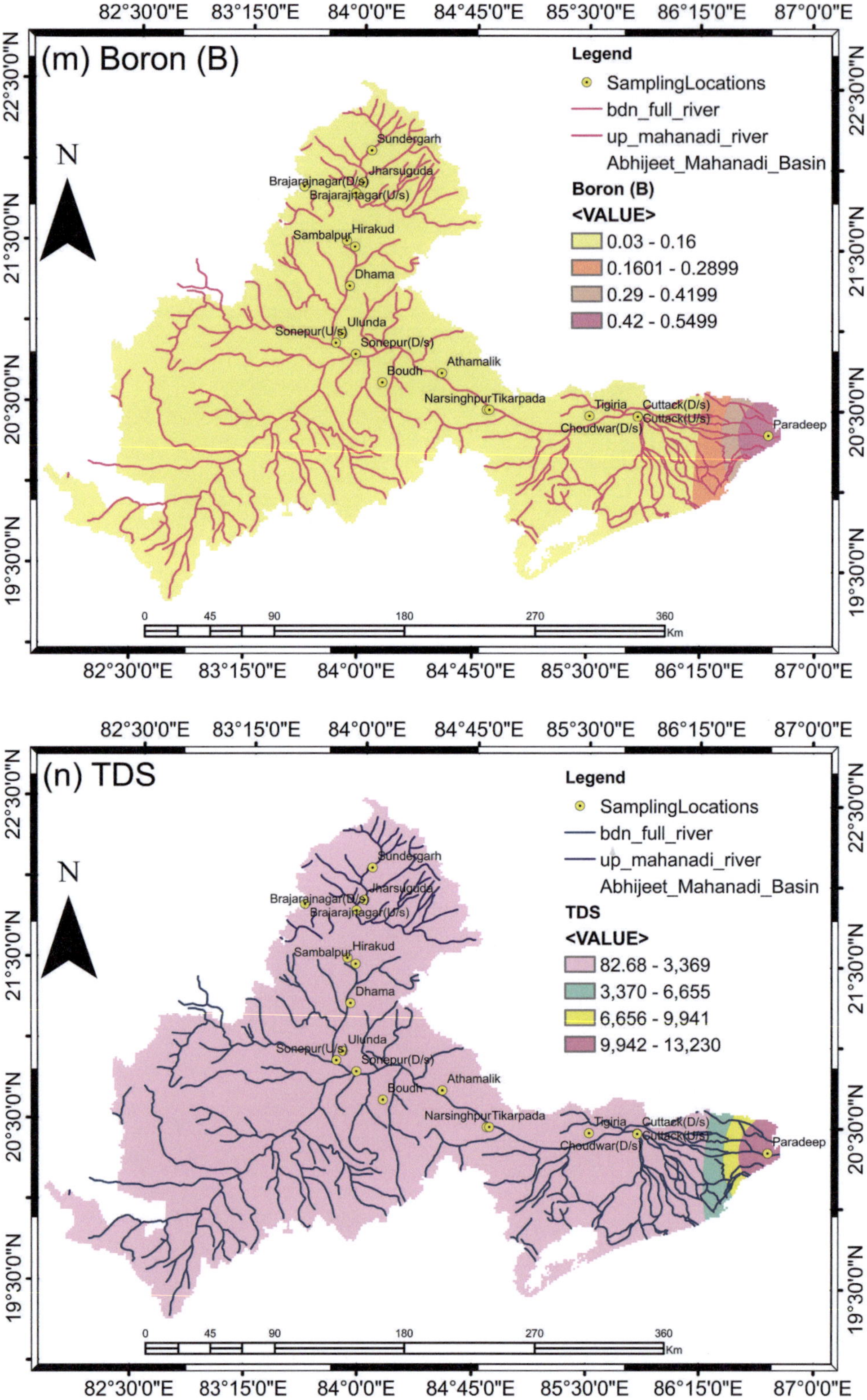

Fig. 2 (continued)

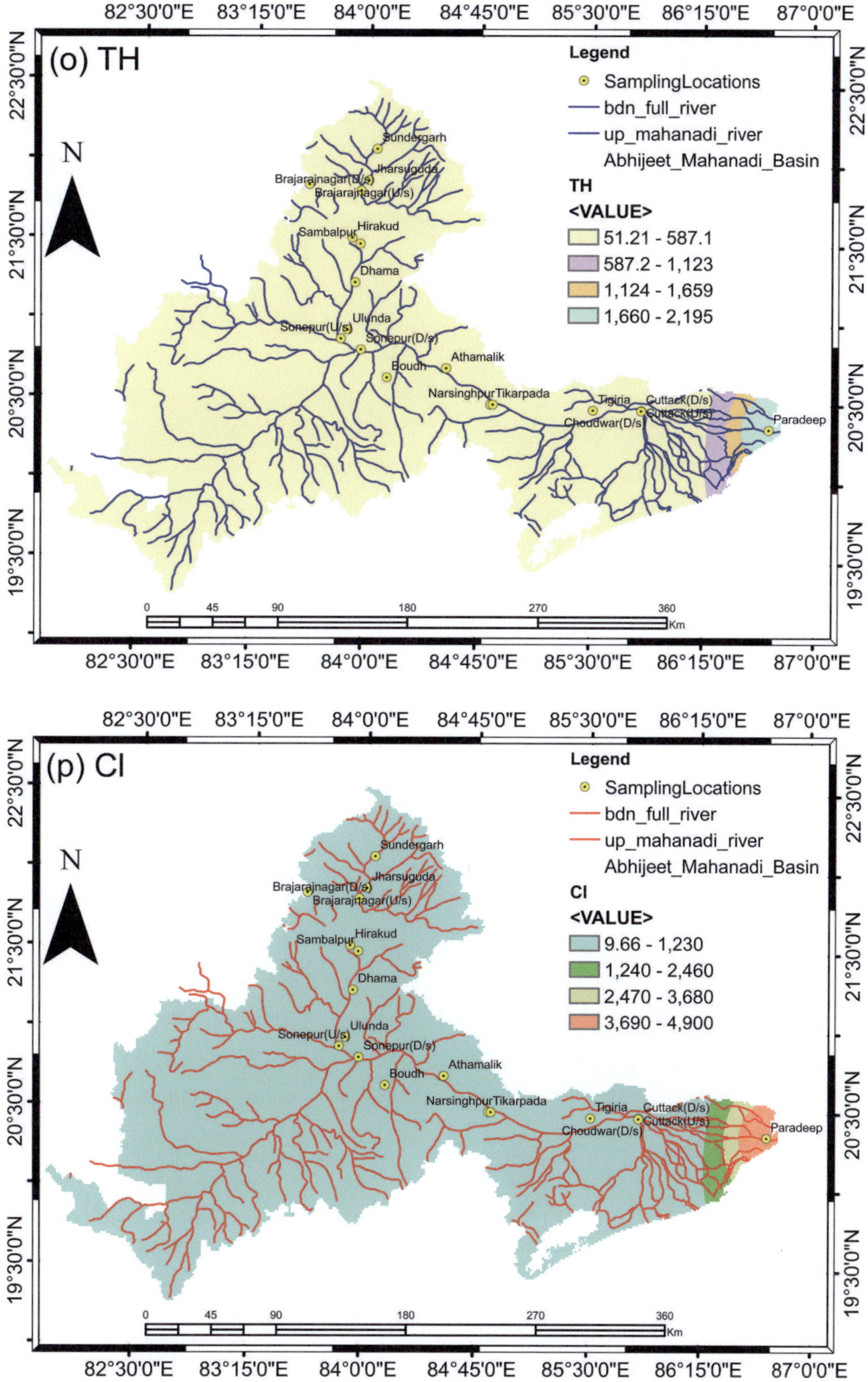

Fig. 2 (continued)

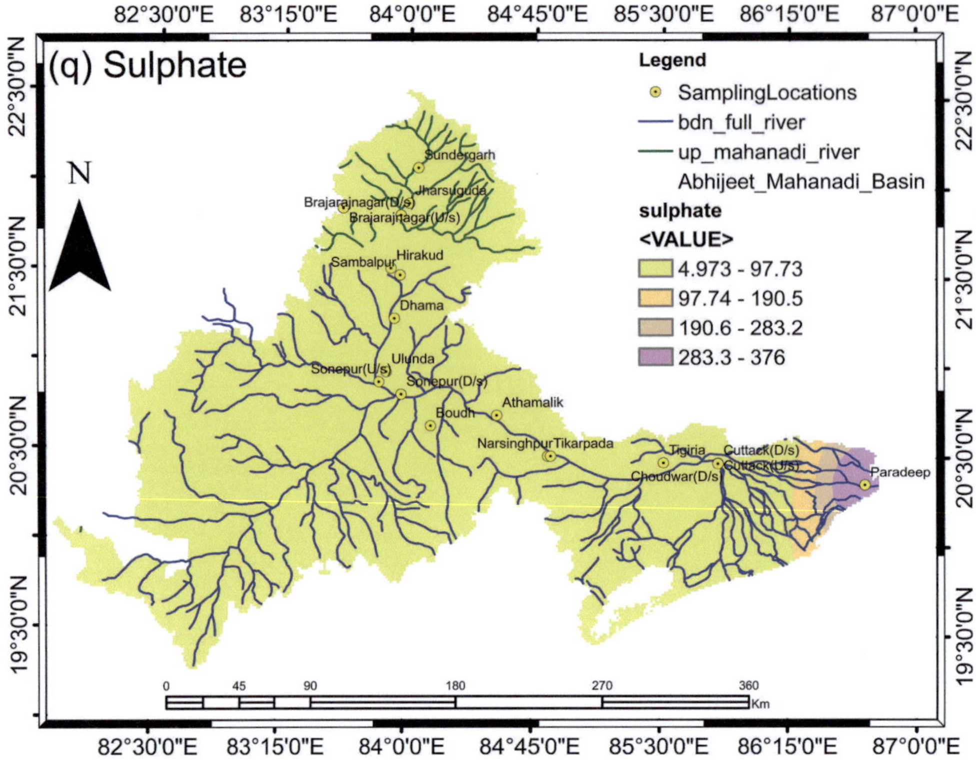

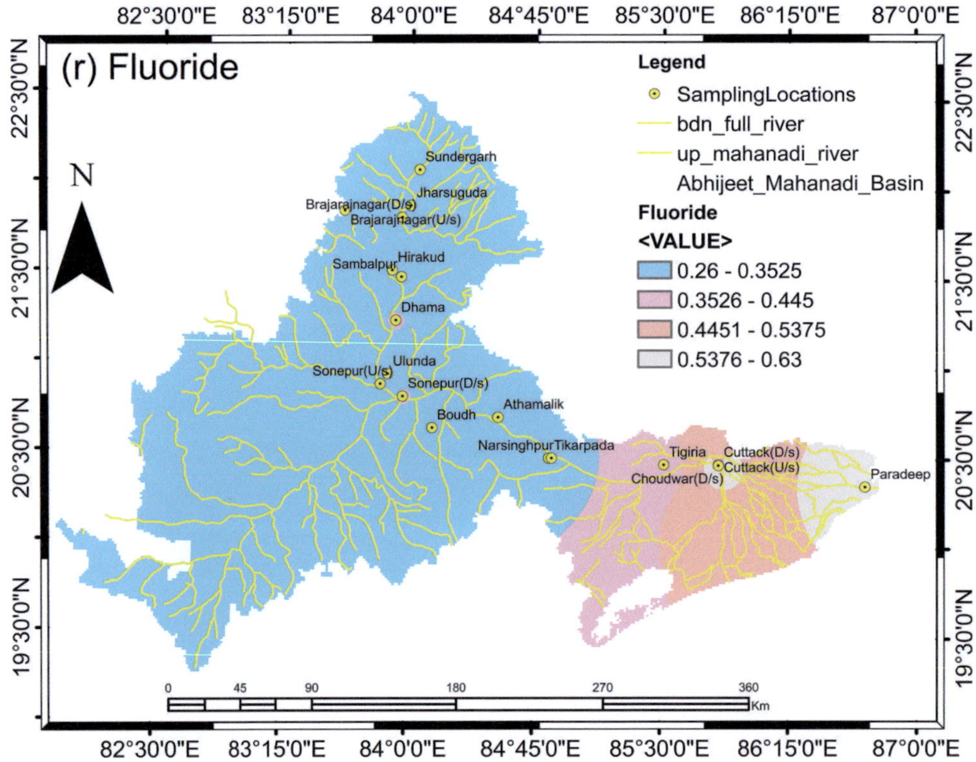

Fig. 2 (continued)

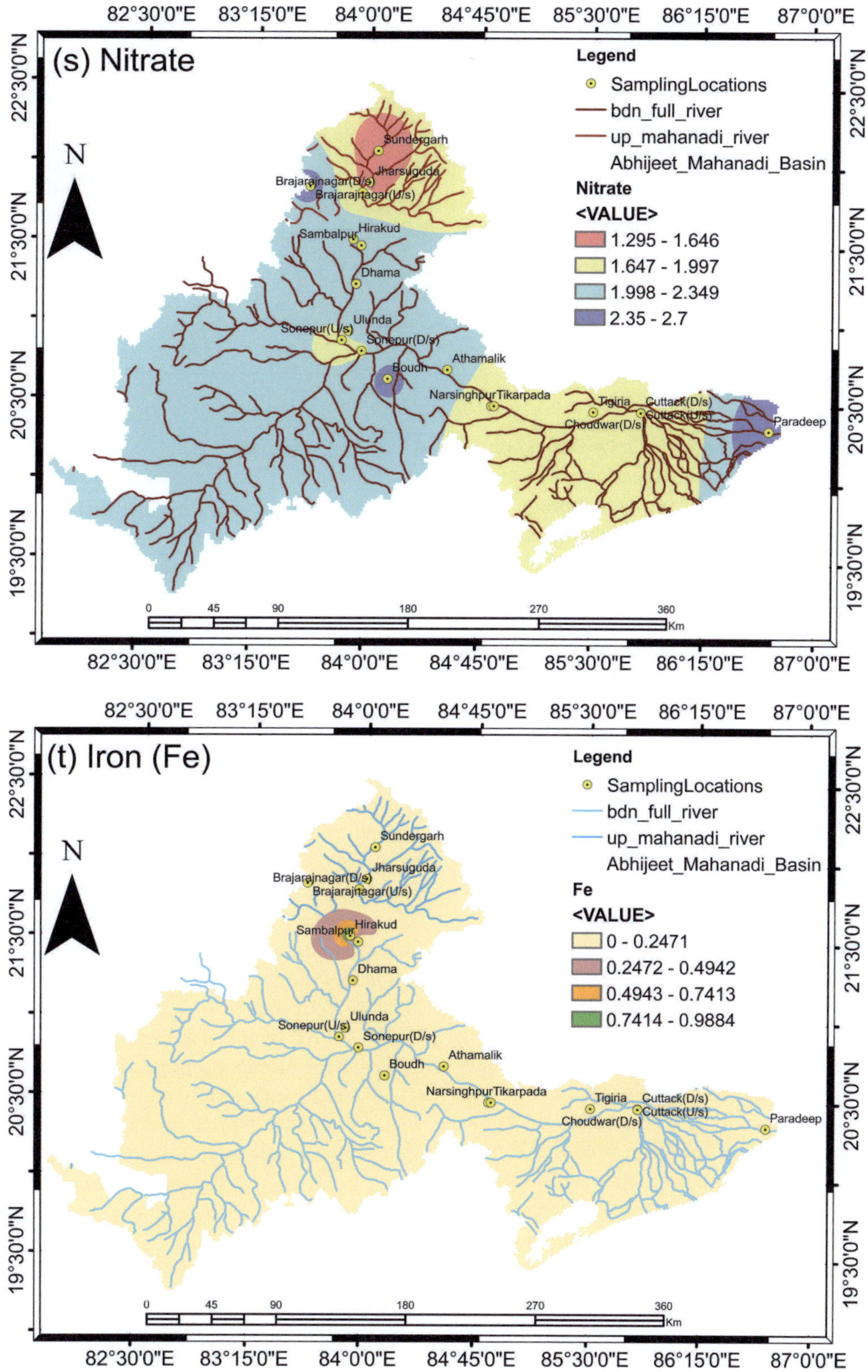

Fig. 2 (continued)

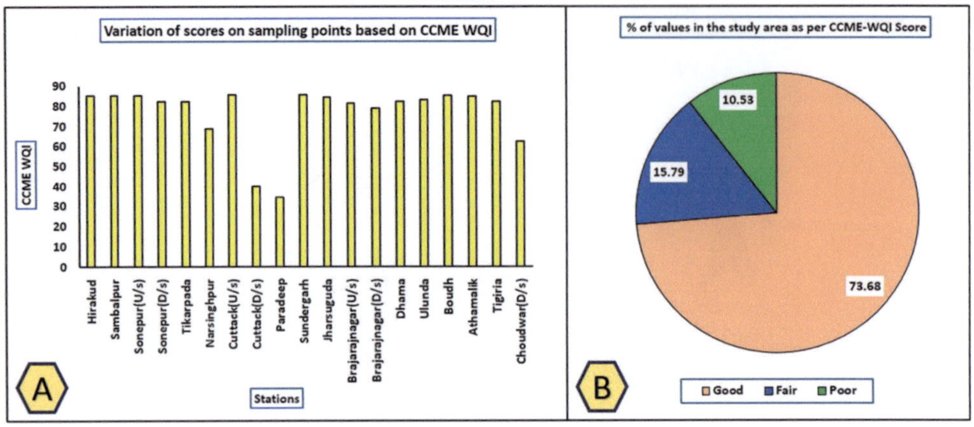

Fig. 3 Distribution of scores for the sampling sites according to CCME classification

According to the explanation, the region's two most significant economic drivers—agricultural cultivation and industrial production—also attracted people from neighboring states, making them two potential sources of pollution. The low density of pollution sources in the upstream could also help to explain the current state of the environment (Ahirvar et al. 2023). Agricultural, industrial, and populated areas put extra pressure on the midstream and downstream. In addition, the buildup of contaminants because of downstream water flow also increased the number of contaminants in the water, namely TC, TKN, EC, SAR, Cl^-, SO_4^{2-}, and Fe^{2+}, in this part of the river. All these factors lead to the augmentation of sodium content in the post-monsoon period due to dissolving activity, depending on the impact of water–rock processes and human-induced activity (Das 2023a, b, c, d, e, f, g). They increase the salts in the water that people drink, posing a risk to their health. As a result, the higher dilution capacity brought about by the increase in water volume and consequently better water quality felt at the upstream portion. From the different indexing approaches, CCME (10.53%), WPI (15.79%), and CI (15.79%) of the vulnerable zones (poor, highly impure, and high pollution surface water quality sites) were observed during the monitoring period, respectively. This discrepancy results from the topographically guided dilution of surface water by the intrusion of distinct recharge waters (Asomaku 2023). While the low and extremely poor surface water quality are only appropriate for household purposes, the excellent and good surface water grades are fit for human consumption. All water pollution index patterns were considered, and the outcomes of the WQI calculation are generally identical. From these created maps, hot spots can be identified that are the principal sources of the pollution load in the river, allowing for the adaptation of decentralized sewage treatment for that section of the river to enhance the level of water quality along that stretch.

6 Conclusions

Surface water plays a key importance role in the domain of Mahanadi Watershed of Odisha, India, used for drinking and irrigation purposes, due to shortage of surface water resources. Therefore, the current study region concentrated on the annual and spatial change of appropriateness for surface water, using drinking surface water quality index (CCME-WQI), CI, WPI, and GIS. Reliable methods are used in the current study to zone sample sites according to their overall pollution levels, which prioritizes decision-making. The following conclusions were summarized below: In line with the CCME model, the research field was categorized as 10.53% poor prospective, 15.79% as fair possibility, and 73.68% as good potential water. Further, the degree of contamination due to physiochemical parameters to human health was calculated by the CI and WPI. In a comparative analysis, the WPI approach grouped the region as follows: 63.16% (12 samples) as very pure water, 21.05% (four samples) as moderately polluted water, and 15.79% (three samples) as highly impure water. Calculations using CI for evaluating predictive abilities showed that the 63.16% (12 locations) demonstrates about low water class, whereas 36.84% (seven samples) belonged to medium- to high-polluted category. Hence, three sampling sites were found to be highly contaminated. In addition, a GIS-based method that combines many indexing strategies finds the most common regions that are acceptable or unsuitable for human consumption in accordance with all indexing techniques. The classification of sampling locations for restoration operations based on their pollution in various time periods and evaluation of the overall pollution levels of each sampling location were successful using all methods. Prior to entering the city limits, the river is still fresh and can be used for drinking and domestic purposes without any treatment, according to reports on drinking WQI and pollution analysis. The examined areas demonstrate that several variables, including industrial, household, sewage disposal, and agricultural runoffs,

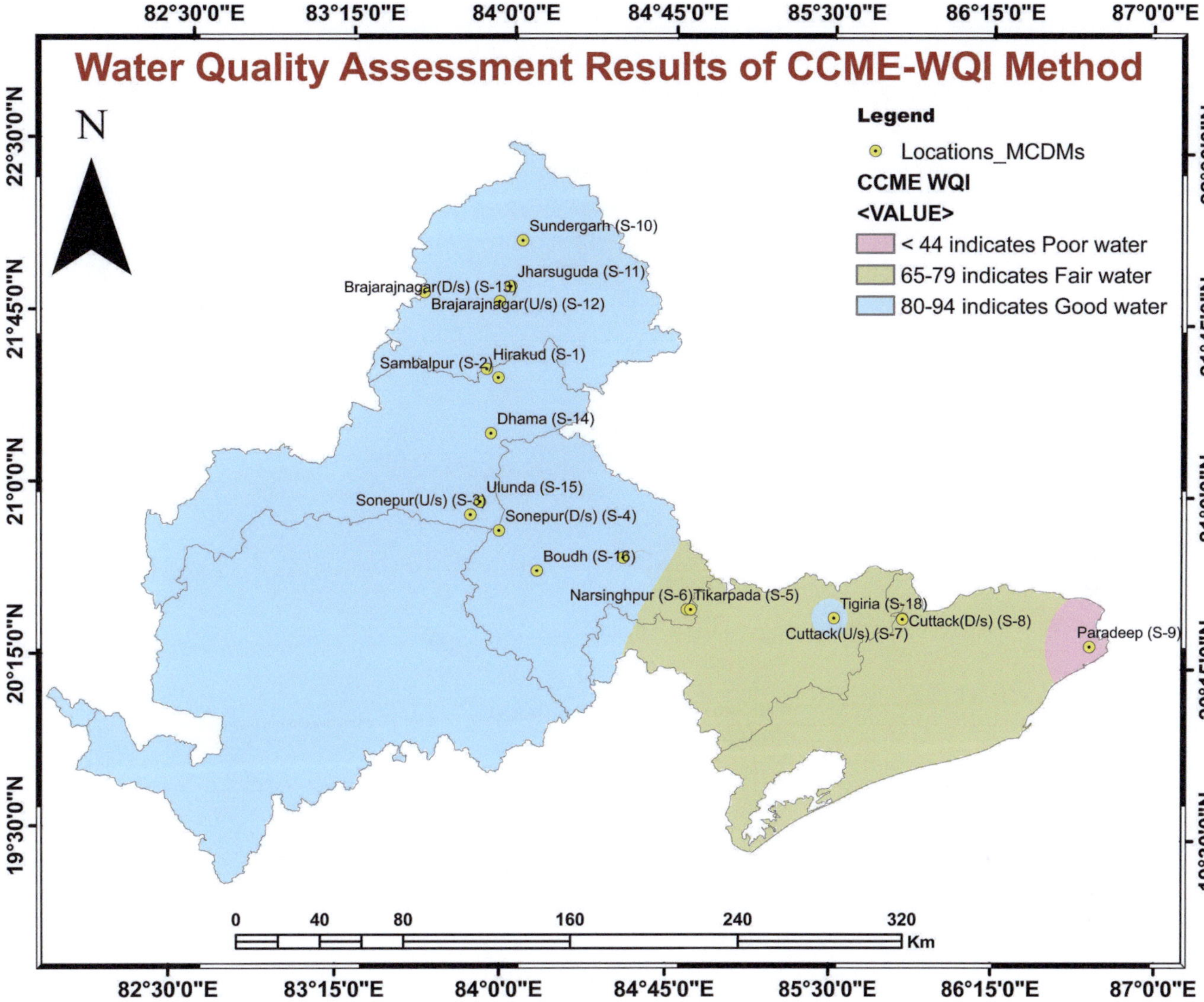

Fig. 4 Classification of water quality for drinking using CCME method

Fig. 5 Distribution of scores for the sampling sites according to WPI classification

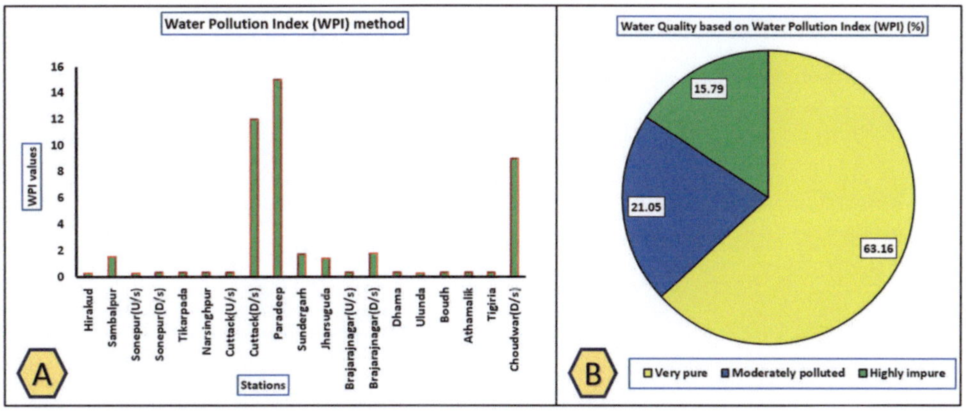

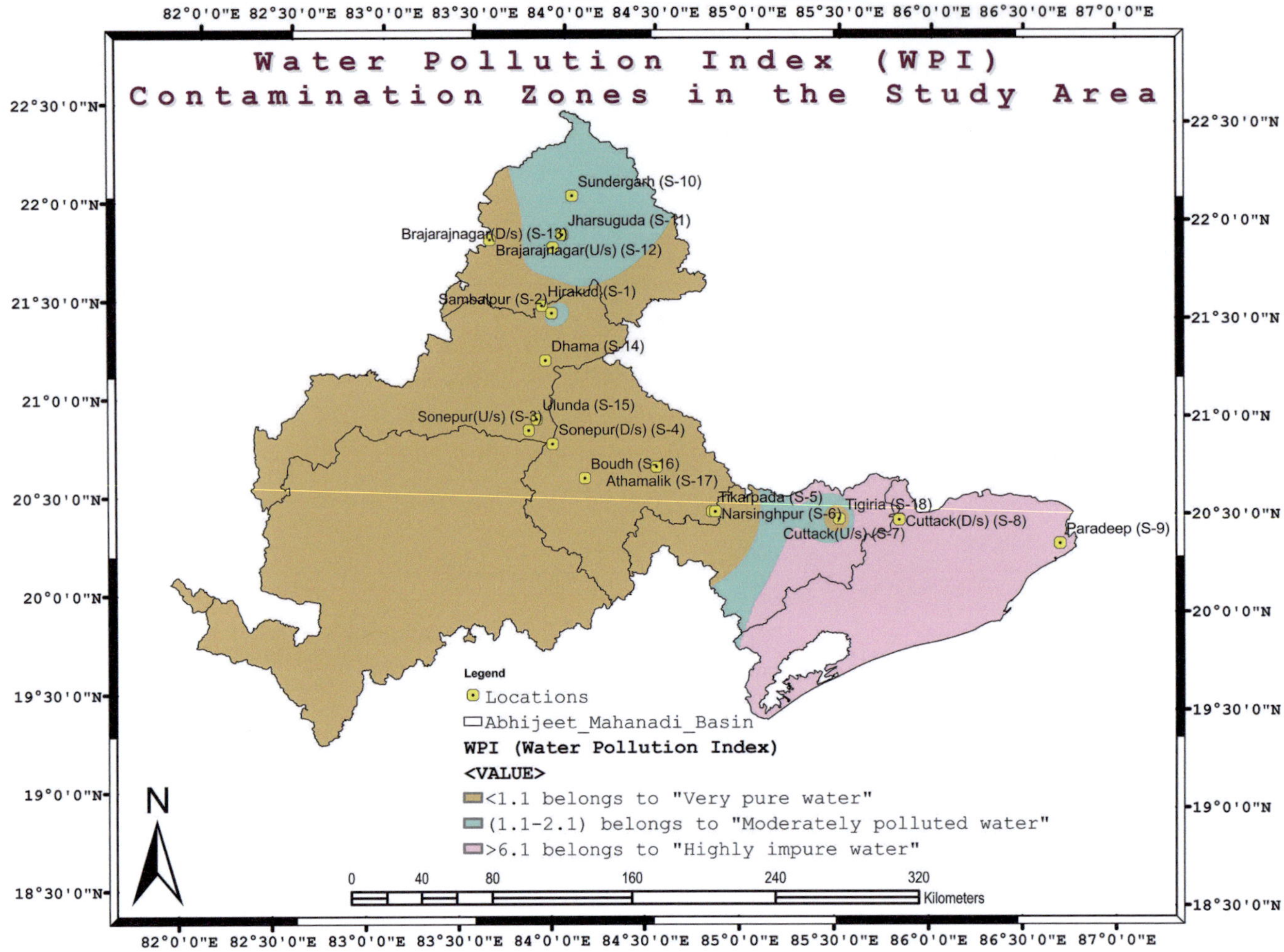

Fig. 6 WPI classification of surface water quality for drinking

Fig. 7 Distribution of scores for the sampling sites according to CI classification

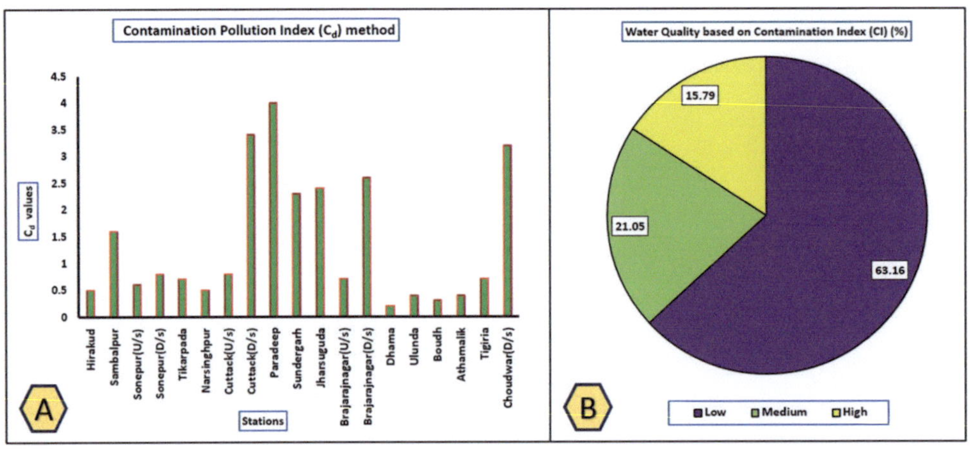

contribute to river contamination. These insights have a great deal of potential to increase the productivity of home water consumption, agricultural facilities and their output, and drinking water given the widespread agricultural activities in the research area. It is recommended that monitoring and management was given top priority to prevent pollution of water resources and to develop the necessary technology to make surface water fit for human use. Future research will assist in identifying the vulnerable areas for yearly and spatial variations in surface water quality, so that civic authorities can implement corrective actions to improve the water quality.

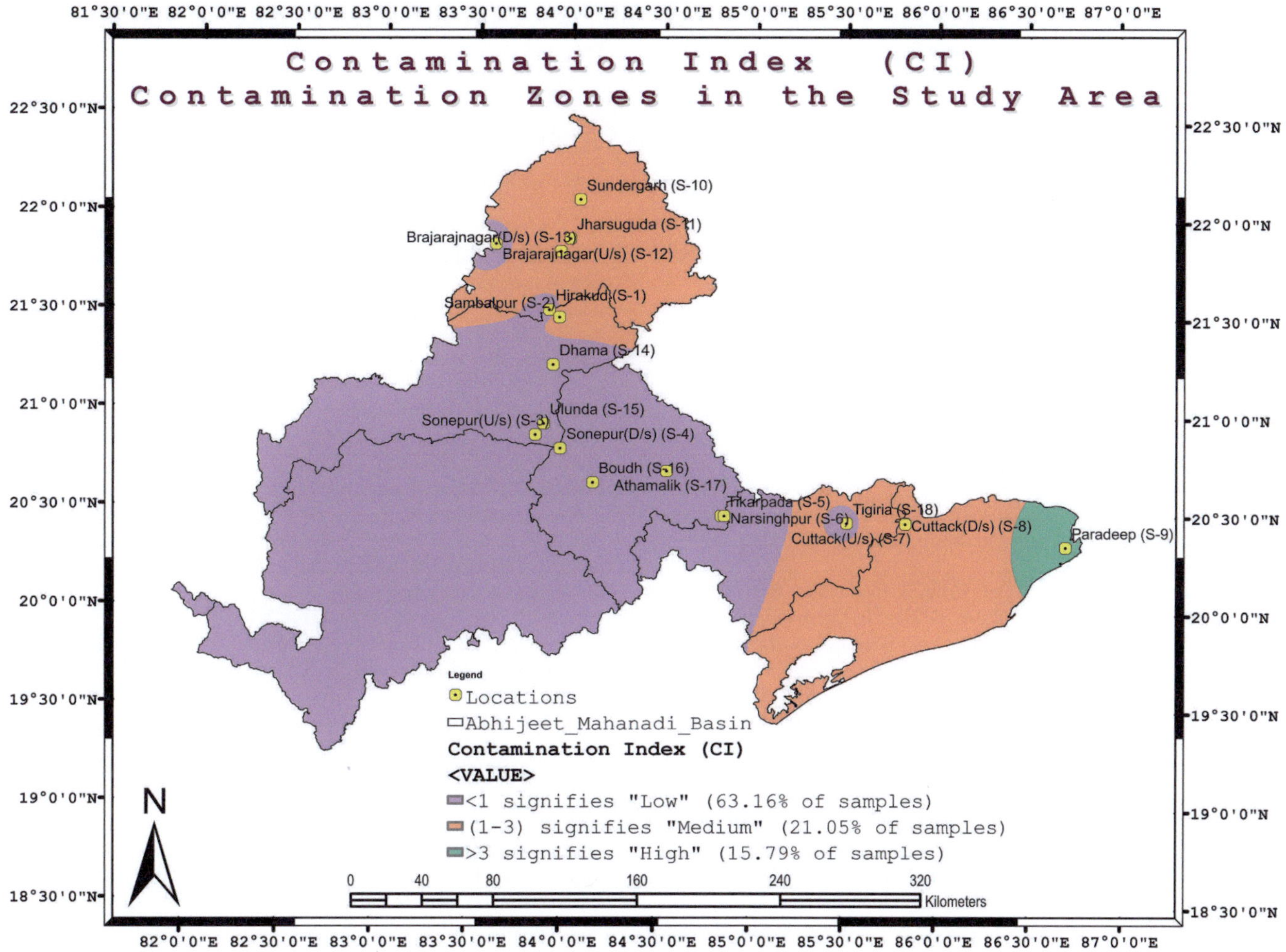

Fig. 8 Spatial distribution of suitability of surface water quality for drinking using CI

Additionally, it will help to concentrate on examining interactions between surface waters, enhancing data integration, adding socioeconomic elements, accounting for the consequences of climate change, and establishing long-term monitoring and modeling. This groundbreaking research will also aid in comparing water quality standards and identifying water quality changes.

Acknowledgements The authors acknowledge the financial support received from the Department of Civil Engineering, C.V. Raman Global University (CGU), Bhubaneswar, Odisha, Government of India, to carry out the present work. The authors are also thankful to Odisha State Pollution Control Board and Central Water Commission (CWC) for supplying the information required to carry out the current study. The author acknowledges the family members - Ashok Kumar Das (Father), Urmila Mohakud (Mother), and Sarita Das (Sister), that plays a key role in successful completion of this article. Special thanks go to the Ranu Prativa Research Group, Ruchika Square, Bhubaneswar, for providing necessary software and hardware support.

References

Abba, S. I., Yassin, M. A., Mubarak, A. S., Shah, S. M. H., Usman, J., Oudah, A. Y.,... & Aljundi, I. H. (2023). Drinking Water Resources Suitability Assessment Based on Pollution Index of Groundwater Using Improved Explainable Artificial Intelligence. Sustainability, 15(21), 15655.

Ahirvar, B. P., Das, P., Srivastava, V., & Kumar, M. (2023). Perspectives of heavy metal pollution indices for soil, sediment, and water pollution evaluation: An insight. Total Environment Research Themes, 6, 100039.

APHA, 2005. Standard Methods for the Examination of Water and Waste Water, Twenty-First Ed. American Public Health Association, Washington DC.

Aravinthasamy, P., Karunanidhi, D., Shankar, K., Subramani, T., Setia, R., Bhattacharya, P., Das, S., 2021. COVID-19 lockdown impacts on heavy metals and microbes in shallow groundwater and expected health risks in an industrial city of South India. Environ. Nanotechnol. Monit. Manag. 16, 100472.

Asomaku, S. O. (2023). Quality assessment of groundwater sourced from nearby abandoned landfills from Industrial City in Nigeria: Water pollution indices approach. HydroResearch, 6, 130-137.

Atique U, An K-G (2019) Reservoir water quality assessment based on chemical parameters and the Chlorophyll dynamics in relation to nutrient regime. Pol J Environ Stud 28(3):1043–1061

Backman, B., Bodiš, D., Lahermo, P., Rapant, S., Tarvainen, T., 1998. Application of a groundwater contamination index in Finland and Slovakia. Environmental Geology 36, 55–64.
Balamurugan, P., Kumar, P.S., Shankar, K., Nagavinothini, R., Kumar, P.S., 2020b. Impact of climate and anthropogenic activities on groundwater quality for domestic and irrigation purposes in Attur region, Tamilnadu, India. Desalination Water Treat. 208, 172–195.
Begum, W., Goswami, L., Sharma, B. B., & Kushwaha, A. (2023). Assessment of urban river pollution using the water quality index and macro-invertebrate community index. Environment, Development and Sustainability, 25(8), 8877-8902.
Das, A., Goya, A., & Soni, A. (2023a, July). Use of water quality indices and its evaluation to verify the impact of Mahanadi River basin, Odisha. In AIP Conference Proceedings (Vol. 2721, No. 1). AIP Publishing.
Das, A., Goyal, A., & Soni, A. (2023b, July). Deciphering surface water quality for irrigation and domestic purposes: A case study in Baitarani Basin, Odisha. In AIP Conference Proceedings (Vol. 2721, No. 1). AIP Publishing.
Das, A. (2022a). Multivariate statistical approach for the assessment of water quality of Mahanadi basin, Odisha. Materials Today: Proceedings, 65, A1–A11.
Das, A. (2022b, August). Water Criteria Evaluation for Drinking Purposes in Mahanadi River Basin, Odisha. In International Conference on Trends and Recent Advances in Civil Engineering (pp. 237–260). Singapore: Springer Nature Singapore.
Das, A. (2022c, December). Surface Water Quality Modelling Using Water Quality Index (WQI) and Geographic Information System (GIS) on the Mahanadi Basin, Odisha. In International Conference on Water Technologies (pp. 21–46). Singapore: Springer Nature Singapore.
Das, A. (2023a). 5 Identification of Surface Water Contamination Zones and its Sources on Mahanadi River, Odisha Using Entropy-Based WQI and MCDM Techniques. Optimization and Computational Case Studies in Civil Engineering, 67.
Das, A. (2023b). Anthropogenic Effects on Surface Water Quality Assessment in Baitarani River Basin, Odisha Using GIS and MCDM Techniques. Engineering Research Transcripts, 5, 37-64.
Das, A. (2023c). Assessing Surface Water Quality for Drinking Water Supply using Hybrid GIS-Based Water Quality Index (WQI) in Mahanadi River Basin (MRB), Odisha, India. AIJR Proceedings, 1–21.
Das, A. (2023d). Assessment of potability of surface water and its health implication in Mahanadi Basin, Odisha. Materials Today: Proceedings.
Das, A. (2023e). Modelling of Surface Water Quality and Spatial Mapping in Mahanadi Basin, Odisha. Journal of Mines, Metals & Fuels, 71(9).
Das, Abhijeet. (2023f). Abundance, Characterization, Understanding Water Quality Degradation and Strategies for Pollutant Reduction in Mahanadi River Basin, Odisha. https://doi.org/10.3233/ATDE23 0857
Das, Abhijeet. (2023g). Characterization of Surface Water Quality Using Water Evaluation Indices, EDAS and Geo-Statistics in Brahmani River Basin (BRB), Odisha, India. https://doi.org/10.3233/ATDE23 0802
Egbueri, J. C., Ayejoto, D. A., & Agbasi, J. C. (2023). Pollution assessment and estimation of the percentages of toxic elements to be removed to make polluted drinking water safe: a case from Nigeria. Toxin Reviews, 42(1), 146–160.
Gao, Z., Han, C., Xu, Y., Zhao, Z., Luo, Z., Liu, J., 2021. Assessment of the water quality of groundwater in Bohai rim and the controlling factors—a case study of northern Shandong peninsula, North China. Environ. Pollut. 285, 117482.
Haji, M., Karuppannan, S., Qin, D., Shube, H., Kawo, N.S., 2021. Potential human health risks due to groundwater fluoride contamination: a case study using multi- techniques approaches (GWQI, FPI, GIS, HHRA) in Bilate River basin of southern Main Ethiopian rift, Ethiopia. Arch. Environ. Contam. Toxicol. 80 (1), 277–293.
Hossain M, Patra PK (2020) Water pollution index—A new integrated approach to rank water quality. Ecol Ind 117:106668
Kadam, A., Wagh, V., Jacobs, J., Patil, S., Pawar, N., Umrikar, B., Sankhua, R., Kumar, S., 2021. Integrated approach for the evaluation of groundwater quality through hydro geochemistry and human health risk from Shivganga river basin, Pune, Maharashtra, India. Environ. Sci. Pollut. Res. 1–23.
Karunanidhi, D., Aravinthasamy, P., Deepali, M., Subramani, T., Shankar, K., 2021. Groundwater pollution and human health risks in an industrialized region of southern India: impacts of the COVID-19 lockdown and the monsoon seasonal cycles. Arch. Environ. Contam. Toxicol. 80 (1), 259–276
Krishan, G., Kumar, B., Sudarsan, N., Rao, M.S., Ghosh, N.C., Taloor, A.K., Bhattacharya, P., Singh, S., Kumar, C.P., Sharma, A., Jain, S.K., Sidhu, B.S., Kumar, S., Vasisth, R., 2021b. Isotopes (δ18O, δD and 3H) variations in groundwater with emphasis on salinization in the state of Punjab, India. Sci. Total Environ. 789, 148051 https://doi.org/10.1016/j.scitotenv.2021.148051.
Laltu, A. D. (2023). Drinking Suitability and Source Apportionment of Physical Parameters in Surface Water in Mahanadi Basin, Odisha. Odisha (April 14, 2023).
Liu, J., Peng, Y., Li, C., Gao, Z., Chen, S., 2021. An investigation into the hydrochemistry, quality and risk to human health of groundwater in the central region of Shandong Province, North China. J. Clean. Prod. 282, 125416.
Masood, A., Aslam, M., Pham, Q.B., Khan, W., Masood, S., 2021. Integrating water quality index, GIS and multivariate statistical techniques towards a better understanding of drinking water quality. Environ. Sci. Pollut. Res. 1–17.
Murugesan, B., Karuppannan, S., Mengistie, A.T., Ranganathan, M., Gopalakrishnan, G., 2020. Distribution and trend analysis of COVID-19 in India: geospatial approach. J. Geogr. Stud. 4(1), 1–9.
Nizam, S., Virk, H.S., Sen, I.S., 2022. High levels of fluoride in groundwater from northern parts of indo-Gangetic Plains reveals detrimental fluorosis health risks. Environ.
Panneerselvam, B., Karuppannan, S., Muniraj, K., 2020. Evaluation of drinking and irrigation suitability of groundwater with special emphasizing the health risk posed by nitrate contamination using nitrate pollution index (NPI) and human health risk assessment (HHRA). Hum. Ecol. Risk Assess.: Int. J. 27 (5), 1324–1348.
Qu, S., Duan, L., Shi, Z., Liang, X., Lv, S., Wang, G., Yu, R., 2022. Hydrochemical assessments and driving forces of groundwater quality and potential health risks of sulfate in a coalfield, northern Ordos Basin, China. Sci. Total Environ. 835, 155519.
Ramos-Pacheco, B. S., Choque-Quispe, D., Ligarda-Samanez, C. A., Solano-Reynoso, A. M., Choque-Quispe, Y., Aguirre Landa, J. P.,... & Choque-Quispe, K. (2023). Water Pollution Indexes Proposal for a High Andean River Using Multivariate Statistics: Case of Chumbao River, Andahuaylas, Apurímac. Water, 15(14), 2662.
Ravi, R., Aravindan, S., Shankar, K., Balamurugan, P., 2020. Suitability of groundwater quality for irrigation in and around the main Gadilam river basin on the east coast of southern India. Arch. Agric. Environ. Sci. 5(4), 554–562.
Ren, X., Li, P., He, X., Su, F., Elumalai, V., 2021. Hydrogeochemical Processes Affecting Groundwater Chemistry in the Central Part of the Guanzhong Basin, China. Archives of Environmental Contamination and Toxicology 80, 74–91.
Samanta S, Kumar V, Nag SK, Raman RK, Saha K, Bandyopadhyay S, Mohanty BP, Das BK (2020) Metal contaminations in sediment

and associated ecological risk assessment of river Mahanadi India. Environ Monit Assess 192(1):1–17
Sharma, M.K., Kumar, M., Malik, D.S., Singh, S., Patre, A.K., Prasad, B., Das, P.C., 2022. Assessment of groundwater quality and its controlling processes in Bemetara District of Chhattisgarh state, India. Appl Water Sci 12(5), 1–20.
Shyam, Radhe, Krishan, Gopal, Kheraj, Kumar, Amit, 2022. Evaluation of Groundwater quality for life-supporting activities: A case study of Haryana, India. Int. J. River Basin Manag. https://doi.org/10.1080/15715124.2022.2135518
Soujanya Kamble, B., Saxena, P.R., Kurakalva, R.M., Shankar, K., 2020. Evaluation of seasonal and temporal variations of groundwater quality around Jawahar Nagar municipal solid waste dumpsite of Hyderabad city, India. SN Appl. Sci. 2 (3), 1–22.
Wagh VM, Panaskar DB, Muley AA, Mukate SV (2017) Ground- water suitability evaluation by CCME WQI model for Kadava River Basin, Nashik, Maharashtra, India. Model Earth Syst Environ 3:557–565. https://doi.org/10.1007/s40808-017-0316-x
WHO (2004) Guidelines for drinking water quality. In: Recommendations, vol 1, 3rd edn. WHO, Geneva, p 515.
Wu, C., Wu, X., Qian, C., Zhu, G., 2018. Hydrogeochemistry and groundwater quality assessment of high fluoride levels in the Yanchi endorheic region, Northwest China. Appl. Geochem. 98, 404–417.
Xiao, J., Wang, L., Chai, N., Liu, T., Jin, Z., Rinklebe, J., 2021. Groundwater hydrochemistry, source identification and pollution assessment in intensive industrial areas, eastern Chinese loess plateau. Environ. Pollut. 278, 116930.
Yousefi, M., Ghoochani, M., Mahvi, A.H., 2018. Health risk assessment to fluoride in drinking water of rural residents living in the Poldasht city, northwest of Iran. Ecotoxicol. Environ. Saf. 148, 426–430.
Zhai, Y., Zheng, F., Zhao, X., Xia, X., Teng, Y., 2019. Identification of hydrochemical genesis and screening of typical groundwater pollutants impacting human health: a case study in Northeast China. Environ. Pollut. 252, 1202–1215.
Zhang, Y., Dai, Y., Wang, Y., Huang, X., Xiao, Y., Pei, Q., 2021. Hydrochemistry, quality and potential health risk appraisal of nitrate enriched groundwater in the Nanchong area, southwestern China. Sci. Total Environ. 784, 147186.

Comprehensive Study of Ground Experiences for Flood Resilience and Sustainable Development

Puja Verma and Ambrina Sardar Khan

1 Introduction

Flood resilience encompasses a community's ability to prepare for, withstand, and recover from flood events while minimizing disruption and damage. This concept involves not only physical infrastructure but also social, economic, and environmental aspects of a community's response to flooding. By examining past flood occurrences, communities can gain valuable insights into vulnerabilities, effective mitigation strategies, and areas for improvement in their flood management approaches. This chapter's introduction gives an overview of the significance of flood resilience, emphasises the value of drawing lessons from the past, and describes the goals and framework of the investigation.

According to Willner et al. (2018), flooding is the most common natural hazard in the world and annually results in significant losses in terms of both lives and property. The idea of flood resilience is gaining more attention as the frequency and severity of extreme storms rises and urbanisation continues to develop. The expanding literature on flood resilience has explored its definition and application (e.g. De Bruijn et al., 2017; Hegger et al., 2016; Liao, 2012), assessment measures (e.g. Keating et al., 2017; Kotzé and Reyers, 2016; Lhomme et al., 2013; Restemeyer et al., 2015), and practical strategies (e.g. Lennon et al., 2014; Schelfaut et al., 2011; Vis et al., 2003; Zevenbergen et al., 2008). However, the method through which flood resistance is increased or decreased is still hardly ever explicitly discussed.

In the literature on social-ecological resilience, it is suggested that natural disturbances, or innate environmental dynamics, play a significant role in social-ecological resilience (Berkes et al., 2000; Carpenter et al., 2001; Folke, 2006; Olsson et al., 2004). This idea is supported by the frequent use of the term "learning" in the literature. By improving the social-ecological system's ability to adapt, learning from one disturbance can boost resilience to that very disturbance (Carpenter et al., 2001; Gunderson and Holling, 2002). Since human societies that are susceptible to natural hazards can also be thought of as social-ecological systems, such an argument has been used to discuss hazard resilience (see Resilience Alliance, 2007). Godschalk (2003) contends that the ability to learn from hazard experience is necessary for risk resilience. According to Aldunce et al. (2015), community learning through experience and knowledge sharing is the foundations for enhancing resilience to hazards. Zevenbergen et al. (2008) contend that gaining knowledge from past flood experiences aids in the development of effective flood resilience strategies; Ten Brinke et al. (2008) contend that this is because flood events give the system the opportunity to repeatedly go through the learning cycles to lessen its vulnerability to subsequent events of the same type.

Although it is unknown what mechanisms such learning entails and if flood experience always results in flood resistance, existing literature seems to indicate that learning from flood experience adds to flood resilience.

In order to theoretically illustrate how distinct flood experience patterns resulting from various habitats can affect flood resilience, we apply the model to examine the learning processes in two fundamentally different situations.

P. Verma (✉)
Department of Architecture (FOAPD), Integral University, Lucknow, India
e-mail: pujaverma@iul.ac.in

A. S. Khan
Department of Environment Science, Integral University, Lucknow, India
e-mail: ambrina@iul.ac.in

D. Kumar et al. (eds.), *Data Analytics and Remote Sensing-Based Analysis for Sustainable Development*, Advances in Science, Technology & Innovation, https://doi.org/10.1007/978-3-031-98095-4_5

2 Natural Disruption and Learning

Natural hazards have been suggested to be crucial to hazard resilience as learning opportunities, much as natural disturbances are essential to social-ecological resilience. Aldunce et al. (2015), for instance, make the case that "accepting that disasters and changes happen opens up the possibility of framing them as opportunities for improvement and innovation". This is consistent with the idea of "bounce forward", in which natural disasters are seen as teaching opportunities for improving livelihoods (Manyena et al., 2011). According to the majority of the hazard resilience literature (e.g. Birkmann et al., 2010; Manyena et al., 2011), adaptations undertaken in the wake of a disaster may have an impact on resilience. To lessen the susceptibility of the system, such modifications could be made to hazard management actions, policies, or institutions (Aldunce et al., 2015). The majority of the research now in existence solely addresses how to learn from disasters; for instance, the disaster resilience of place model put forth by Cutter et al. (2008) suggests that learning only occurs when the impact of the hazard surpasses the community's capacity for absorption. Natural disasters do not necessarily result from risks, though. Flooding is a phenomenon that does not always result in a calamity. In this post, we make the supposition that every flood event, no matter how harmful, innocuous, or advantageous, offers possibilities for learning. What is taught makes a difference.

2.1 Face off Floods and Learning Experiences

As was already indicated, it has been argued that developing flood resilience requires learning from past flood experiences (Ten Brinke et al., 2008; Zevenbergen et al., 2008). The majority of the material now in print focuses on learning as it relates to modifications in flood management strategies and property-level flood mitigation.

In reaction to flood disasters, flood management policies are frequently revised or strengthened, a process known as "policy learning" (Brody, 2003; Nye et al., 2011). Large, regional floods could speed up the process and involve people who were previously indifferent to flood management policies (Johnson et al., 2005). A small, local flood might trigger a negotiation process among the stakeholders (local residents, local and state governments, etc.) to gradually change flood management policies. It is possible that policy learning will result in policies that support flood resilience, but it is uncertain if learning itself will always boost flood resilience because policy learning is significantly influenced by "policy legacies" (Brody, 2003; Van Buuren et al., 2016). In recent years, property-level flood mitigation has drawn more attention as flood management has moved its emphasis from flood prevention to "flood risk management" in acknowledgment of the fact that flooding can never be totally avoided (Bubeck et al., 2012; Osberghaus, 2017). Shared responsibility between the government and citizens is essential to the idea of managing flood risk (Bichard and Kazmierczak, 2012; Henstra et al., 2019). Citizens must also be responsible for minimising flood damages at the property level. Property-level flood mitigation measures include floodgates, air-vent covers, flood proofing, etc. (Bichard and Kazmierczak, 2012; Owusu et al., 2015). By enhancing individual families' capacity to avoid or minimise flood damage, they contribute significantly to the community's flood resilience (Begg et al., 2017). Empirical studies demonstrate that flood experience increases risk awareness (Burningham et al., 2008), attention to weather forecast and flood warning (Knocke and Kolivras, 2007), and willingness to take mitigation and response measures (Bubeck et al., 2013; Kreibich et al., 2005; Lawrence et al., 2014; Osberghaus, 2017). Learning is rarely explicitly discussed in the literature on property-level flood mitigation.

We see these behavioural adjustments as a clear example of learning. There is evidence that having experienced a flood can help one grasp its effects. For instance, Siegrist and Gutscher (2008) demonstrate that people who have never experienced a flood have trouble imagining how they would feel during one: while they can visualise the potential consequences, such as property damage, thanks to media reports, they are unable to picture the unfavourable emotions, such as fear, insecurity, and helplessness, which are frequently described by those who have actually experienced a flood.

In conclusion, research to far points to learning from environmental disturbance as a key strategy for food resilience, presuming that flood resilience is affected by learning from past flood experiences. There are still a lot of unanswered questions. What exactly can be learned from the experience of a flood, for instance? Do various groups' experiences with the same flood disaster result in different learning processes that have varied consequences on flood resilience? Fundamentally, does enhanced flood resistance result from learning from past flood experiences? Answering these queries is outside the purview of this chapter, because it is required to establish the theoretical connections between flood experience and flood resistance before they can be addressed. The LFF model proposed in this chapter serves this objective. But before we introduce the model, it is important to first define flood resilience.

3 Flood Resilience

Divergent ideas exist regarding flood resilience. There is widespread agreement that flood resilience refers to the capacity to recover swiftly from a flood disaster (De Bruijn,

2005; Gilbert, 2010; Lhomme et al., 2013), although there is disagreement over the contribution of flood resistance to flood resilience. Here, the term "flood resistance" refers to the ability to use flood control infrastructure, such as levees, floodwalls, dams, piped drainage networks, and pumping stations, as well as temporary emergency flood-fighting measures, to prevent flooding from happening in a region that is naturally prone to flooding. Some scholars explicitly argue that flood resistance is important to flood resilience (e.g. Dieperink et al., 2018; Hegger et al., 2016; Morrison et al., 2018; Restemeyer et al., 2015). For instance, Hegger et al. (2016, p. 3) view "resistance as an enhancing factor for resilience"; Restemeyer et al. (2015, p. 47) contend that the "resistant strategy" (technical measures such as dikes, dams, and sluices) "may also make a city more resilient" and that technical measures "must be an inherent part of a resilience strategy". Other academics define flood resilience independently of flood resistance. For instance, Liao (2012) defines flood resilience as the capacity to withstand flooding or to quickly reorganise after a flood disaster. Zevenbergen et al. (2008) define flood resilience as the capacity to adapt to and recover from the hazard. According to Sörensen et al. (2016), flood resilience refers to the capacity to prepare for, respond to, reorganise, and learn from floods. According to Liao et al. (2019), the idea that flood resilience depends on having strong flood resistance is related to the idea that flooding is frequently dangerous and should be avoided. But this impression is not full. Flooding can also be a vital development resource for rural populations in some regions of the world, including Bangladesh and the Mekong Delta (Thompson and Sultana, 1996; Ehlert, 2012). Without flood resistance, it is also possible to lessen the effects of flooding. Communities whose livelihoods depend on recurring floods have created adaptable methods, such as building temporary footbridges and living on stilts, to enable them to be flooded without suffering loss or harm (Liao et al., 2016).

It is crucial to understand the differences between these three closely related but distinct concepts. The hydrological phenomena known as a "flood" occurs when water temporarily covers the ground. Flooding that takes place in a human settlement and could be dangerous is referred to as a "flood hazard". Flooding that significantly damages property and disrupts social and economic operations is referred to as a "flood disaster". Flood disasters are not always caused by flooding. Only when human populations are unable to lessen the effects of flood threats can flood disasters occur.

Because flooding is not always dangerous, we define flood resilience as the capacity to prevent or reduce flood damage when flooding occurs; alternatively, flood resilience is the capacity to quickly recover from the flood disaster to a functional state in the event that flooding causes significant damage and severe socioeconomic disruption. Our definition, which is based on Holling's (1973) original concept of resilience, acknowledges that flooding is intrinsic to any human community near a river and that natural disturbance is inherent to system dynamics.

This definition is consistent with the frequently cited definition of social-ecological resilience, which is "the capacity of a system to absorb disturbance and reorganise while undergoing change so as to still retain essentially the same function, structure, identity, and feedbacks" (Walker et al., 2004, p. 2). It focuses on the ability to absorb the effects of flooding when they occur. Here, "floodability"—the capacity to withstand flooding without suffering damage or interruption (Liao, 2012); and "recoverability"—the capacity to quickly recover from a flood disaster—are key components of flood resilience to a specific flood event. However, adaptability—here defined as the capacity to continuously modify flood management to accommodate a changing flood regime—is also necessary to sustain flood resilience through several flood events throughout time. It should be noted that adaptability is a notion that is process-oriented and places a focus on incremental adjustment (Heazle et al., 2013; Keck and Sakdapolrak, 2013). Despite being currently flood-adapted, a community lacks adaptability if it does not keep changing its flood-adaptive measures in response to rising flood danger. To enhance flood resilience in a community that currently lacks it, transformability—here defined as the ability to substantially alter flood management and other systems—is necessary.

4 Flood Experience

Flooding Experience: The two modes of learning—direct knowledge acquisition and learning through experience—must first be distinguished (Geresdi & Horváth, 2000). The former involves acquiring knowledge that has been codified and pre-processed for the learner through the use of books, teachers, media, etc. This type of knowledge is frequently explicit. The LFF paradigm, on the other hand, is focussed on the latter—learned from experience. Learning from experience also creates tacit knowledge in addition to explicit information. While tacit knowledge gained via experience is typically contextualised, explicit knowledge from books and other sources simply provides second-hand information and offers generalised notions (Eraut, 2004). For instance, it is common knowledge that elevating a building above the flood level can prevent flood damage; however, experiencing a flood first-hand could add locally relevant information to that knowledge, such as the appropriate height of elevation or structural design.

Additionally, actually experiencing flooding aids in identifying the issue with flood management efforts since each flood tests their viability (see Kolb, 1984). In this context, the term "flood experience" refers to actual interaction with and observation of a flood. Flood impact and flood regime are factors that affect flood experience. We want to emphasise once more that experiencing a flood is not always a calamity. It might just be "nuisance flooding", which only causes small injury and inconvenience but "poses significant threats to public safety or causes major property damage" (Moftakhari et al., 2018, p. 4218), or it could be completely harmless and even beneficial for human livelihood. As a result, the impact of a flood on a person or a community could be either useful, inconvenient, or disastrous. Additionally, distinct flood regimes could result in various flood experiences. The term "flood regime" describes various aspects of flooding, including its frequency, diversity, magnitude, timing, length, rate of change, etc. For instance, experiencing lengthy flooding frequently differs greatly from experiencing flash flooding infrequently.

5 Learning in a Protected Setting

People are safeguarded in well-protected environments by flood control infrastructure that meets high design standards. An environment like this is typical; the majority of big cities throughout the world are well-protected, making flooding considerably less common than it would have been otherwise. It is well known that a sense of community connected to flood control infrastructure can result in decreased awareness of flood danger (Ludy and Kondolf, 2012; Pielke, 1999; Tobin, 1995). This suggests that the general public has a low motivation to learn about flood mitigation. In the Alpine region of Italy, for instance, it has been discovered that the construction of flood control equipment has caused locals to progressively lose their knowledge of self-defence and their willingness to make preparations for. Since most floods are avoided, there is also minimal opportunity for initial learning. According to research, people who have never experienced a flood are frequently overconfidently sure that they would be safe in the future (Lawrence et al., 2014). Although a safe atmosphere is widely desired, learning cannot take place there because there is little incentive and chance for learning to begin with. Additionally, even while flood control equipment can considerably lower the frequency of floods, flooding still cannot be completely eliminated. When flooding does occur, for which people with low risk awareness are often unprepared, it is often catastrophic because it means the failure of flood control infrastructure (Tobin, 1995). Due to its abrupt onset, tremendous velocity, and potential to flood places that would not normally be flooded, flooding linked with levee or dam breaches is far more dangerous than natural floods (Cenderelli, 2000).

Because of their existing understanding of flood mitigation, individuals are likely to seek increased flood control after a devastating flood experience (Birkland et al., 2003). For further information, Morrison et al. (2019). As previously noted, it is still widely accepted that flooding must be avoided in the first place for flood safety, despite significant conceptual changes in flood management. According to research (Terpstra, 2011), viewing flood control structures in person helps foster trust. Construction of new flood control infrastructure or reinforcement of existing flood control infrastructure is frequently the result of prior knowledge that flood resistance is crucial and that flood control infrastructure can be trusted, combined with the policy legacy of flood resistance (Barnett et al., 2015; Wiering et al., 2017).

This may be seen in New Orleans, where Hurricane Katrina's devastation prompted a significant update to the city's flood management system (Gotham et al., 2018). While improving flood resistance, such flood control measures do not improve floodability, or the capacity to lessen flood damage when flooding occurs. However, experiencing a disaster offers the chance to learn about recovery and may prompt the public to call for new policies or tools to improve recoverability. In terms of adaptation, a flood disaster might aid adaptability if it prompts actions other than boosting flood resistance. However, it would not promote transformability if the action simply strengthens the current flood management strategy, for example, if structural measures continue to be the main solution to flood safety.

6 Education in an Unprotected Setting

There are still rural populations that depend on and adapt to periodic flooding, notwithstanding the relative rarity of an entirely unprotected but frequently-flooded environment (Cuny, 1991). Since floods are needed for agriculture and fishery, households in these communities have developed measures, such as building stilt houses to stay above the flood level, using boats and footbridges to remain mobile, and selecting flood-tolerant crops for agriculture, to avoid damage and disruption during the flood season (Cuny, 1991; Ehlert, 2012; Paul, 1984). These flood adaptation strategies undoubtedly represent the outcomes of a lengthy process of learning by doing over many generations.

Unprotected, frequently-flooded environments are beneficial for learning, despite sounding uncomfortable. Because there is no infrastructure for centralised mitigation, such as flood control, people must take decentralised, property-level precautions to survive. It suggests a greater desire to learn about flood mitigation. The majority of villagers in the unprotected communities of the Vietnamese Mekong Delta believe that they alone—and not the government—are completely responsible for flood protection (Liao et al., 2016). Furthermore, in an unprotected area, floods happen naturally far more frequently than in a protected setting. This implies more flood experiences and thus more learning opportunities. However, because of the flood-adaptive measures, most floods experienced are harmless.

People are always learning by doing as they are obliged to make preparations for every flood. Each flood they encounter adds to their knowledge of flood mitigation, which is now dominated by property-level strategies. The knowledge of natural flood hydrology and flood ecology (more fish during the flood season, for example) that is impossible to learn in a well-protected environment is also generated by flood experience. However, not all floods prompt action; only a catastrophic flood would. For example, in the aforementioned villages in the Vietnamese Mekong Delta, a flood that is too high or too stormy can become a disaster, and most households would further raise their stilt houses soon after a disaster (Liao et al., 2016). They become even more flood-resistant as a result, increasing flood resistance. If the flood tragedy prompts action that could hasten future recovery, recoverability may also improve. However, depending on how well the households are able to adapt their flood protection measures, their experience with flooding may or may not be a factor in adaptability. As was already indicated, while the majority of households in the aforementioned villages are able to continue raising their homes after every flood that was higher than the height of the floor, some households are unable to do so due to financial difficulties (Liao et al., 2016).

7 Study Part Lucknow Area

India's Uttar Pradesh state has Lucknow as its capital. The city's water supplies are under stress due to urban sprawl and rising population, similar to most urban environments around the world (Figs. 1 and 2). Wetlands and lakes have begun to expand into the city as a result of the populace below a particular pay being unable to live in residential areas. In the low-lying residential parts of the city, the encroachment of these drains and lakes has led to regular flooding and the loss of key hydrological features like wetlands and lakes that were formerly the primary recharge source for the subterranean aquifer. Streets in Lucknow flooded after heavy rain, trees snapped. According to Singh et al. (2005), the absence of sewage treatment facilities in the city has also resulted in significant pollution of the river and the phreatic aquifer. A master plan is required for the management of sewage water, storm water, and existing water resources, which calls for the involvement of environmental agencies, the public works department, the city administration, the financial department, and, last but not least, the individual efforts of city (Table 1 and Figs. 3, 4 and 5).

7.1 Flood Problems and Issue Raised in Lucknow

The Indian state of Uttar Pradesh's capital, Lucknow, has frequently experienced flooding, which has seriously disrupted life and caused damage. There are multiple interconnected causes that contribute to Lucknow's flood problem:

Geographical Elements:

Topography: Because Lucknow is located along the banks of the Gomti River, it frequently floods when it rains a lot. The city's inherent slope may cause water to collect in lower regions.

River Overflow: During the monsoon season, the Gomti River frequently overflows its banks, causing flooding in nearby areas.

Encroachment on Water Bodies: As a result of rapid urbanisation, natural water bodies and drainage systems are being encroached upon and filled up. As a result, the city's ability to handle surplus water is diminished.

Fig. 1 Life in Lucknow is disrupted by heavy rain and water logging—Hindustan Times

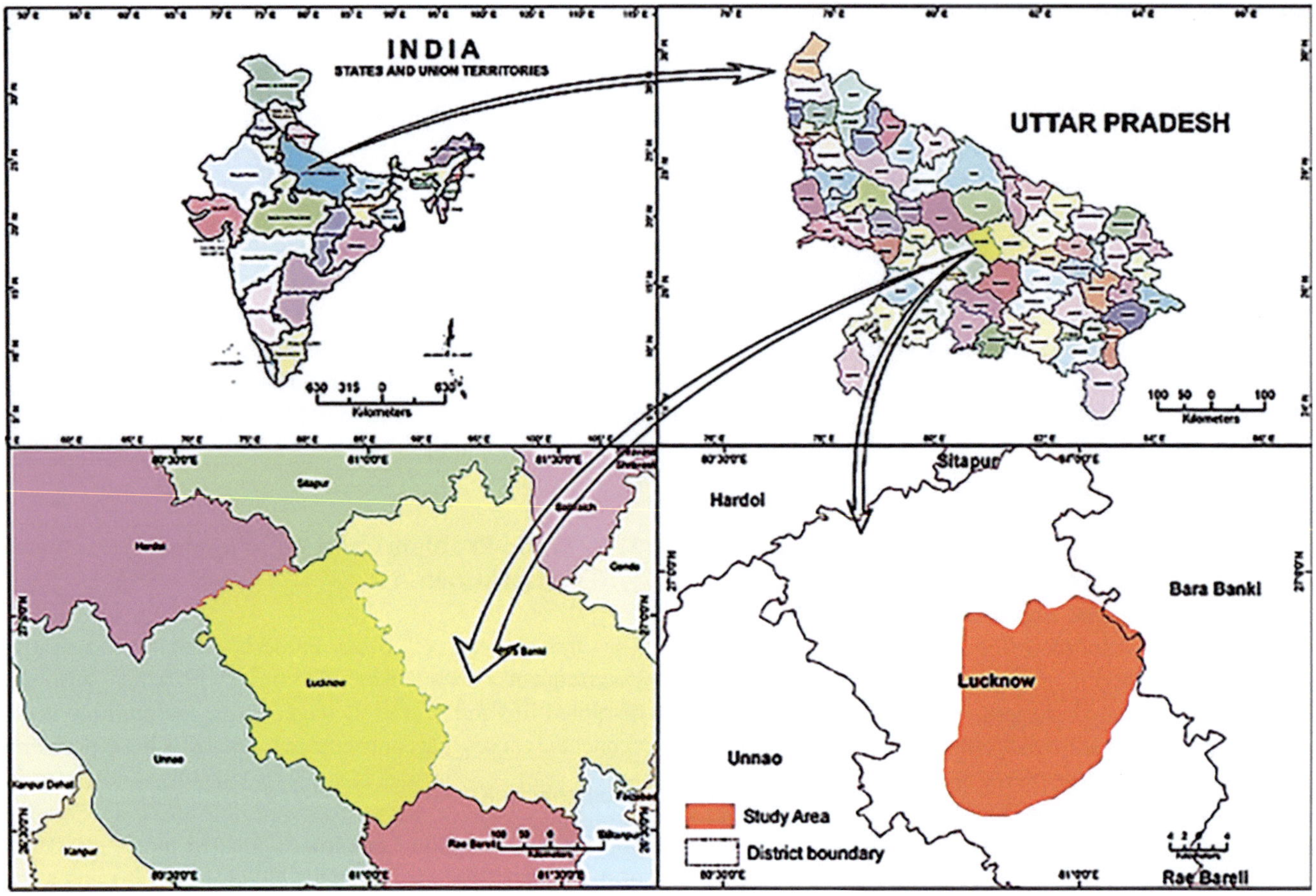

Fig. 2 Location map of Lucknow area

Fig. 3 Comparative study for average temperatures and precipitation in Lucknow

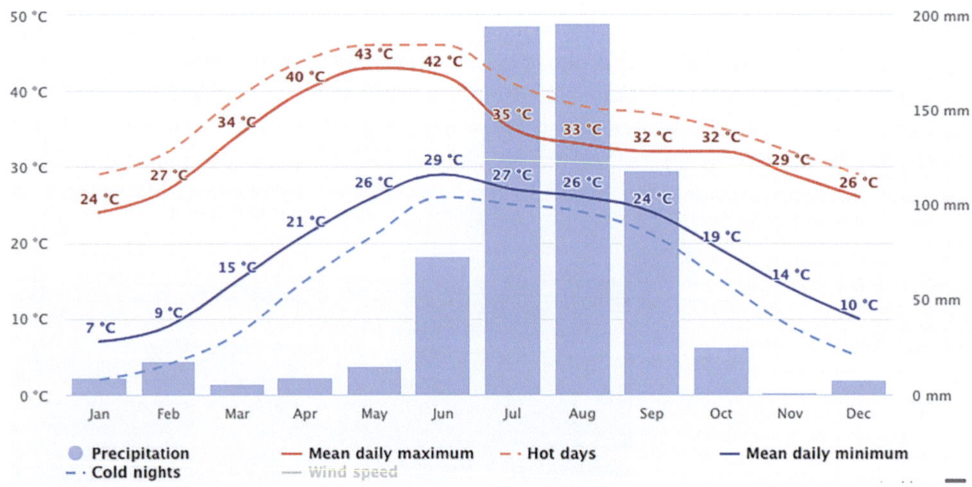

Construction and Infrastructure: When new structures, roads, and other infrastructure are developed, natural drainage patterns are frequently disregarded, increasing surface runoff and decreasing precipitation absorption. Inadequate Lucknow's current drainage infrastructure is frequently unable to withstand prolonged downpours. The efficacy of many drains is decreased because they are old, improperly maintained, or blocked with trash.

Insufficient Planning: To address the expanding population and shifting climate patterns, there has not been enough thorough planning done for drainage systems.

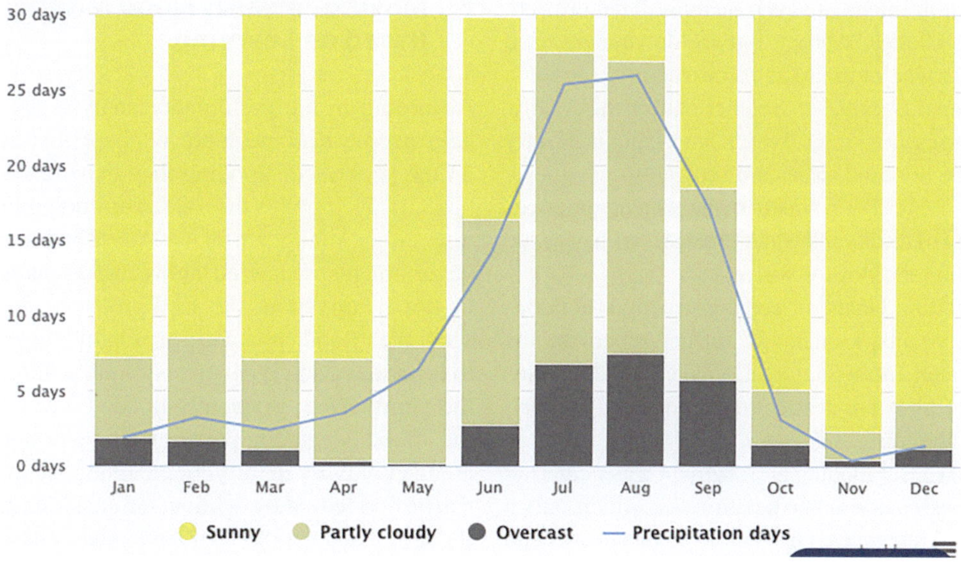

Fig. 4 Cloudy, sunny, and precipitation days in Lucknow

	January	February	March	April	May	June	July	August	September	October	November	December
Avg. Temperature °C (°F)	14.9 °C (58.9) °F	18.6 °C (65.4) °F	24.3 °C (75.7) °F	30.3 °C (86.6) °F	32.8 °C (91.1) °F	32.3 °C (90.2) °F	29 °C (84.1) °F	28.4 °C (83.1) °F	27.7 °C (81.8) °F	25.5 °C (77.9) °F	21.2 °C (70.1) °F	16.4 °C (61.6) °F
Min. Temperature °C (°F)	9.1 °C (48.3) °F	12.2 °C (54) °F	16.9 °C (62.4) °F	22.6 °C (72.6) °F	26.1 °C (78.9) °F	27.5 °C (81.5) °F	26.2 °C (79.2) °F	25.8 °C (78.4) °F	24.5 °C (76.2) °F	20 °C (68) °F	14.9 °C (58.8) °F	10.3 °C (50.6) °F
Max. Temperature °C (°F)	21.3 °C (70.3) °F	25.2 °C (77.4) °F	31.5 °C (88.8) °F	37.7 °C (99.9) °F	39.2 °C (102.6) °F	37.1 °C (98.9) °F	32.3 °C (90.2) °F	31.7 °C (89) °F	31.4 °C (88.5) °F	31.1 °C (88) °F	27.7 °C (81.9) °F	23.1 °C (73.5) °F
Precipitation / Rainfall mm (in)	19 (0)	25 (0)	14 (0)	9 (0)	17 (0)	129 (5)	310 (12)	265 (10)	165 (6)	34 (1)	2 (0)	10 (0)
Humidity(%)	67%	60%	44%	30%	38%	54%	79%	82%	80%	66%	58%	64%
Rainy days (d)	2	2	2	2	3	8	18	18	13	2	0	1
avg. Sun hours (hours)	8.4	9.6	10.6	11.5	11.9	10.7	8.4	8.0	8.4	9.6	9.6	9.0

Fig. 5 Data: 2015–2023 min. temperature °C (°F), max. temperature °C (°F), precipitation/rainfall mm (in), humidity, rainy days. Data: 2015–2022: avg. Sun hours in Lucknow city

Table 1 Settlements on both side of the Gomti River in lucknow

S. no.	Cis-Gomti area	Trans-Gomti area
1	Hazratganj	Nirala Nagar
2	Aminabad	Aliganj
3	Hussainganj	Daliganj
4	Lalbagh	Old and New Hyderabad
5	Gologanj	Mahanagar
6	Wazirganj	Nishatganj
7	Rajendra Nagar	Indira Nagar
8	Malviya Nagar	Gomti Nagar
9	Sarojni Nagar	Vikas Nagar
10	Aishbagh	Jankipuram
11	Rajajipuram	
12	Chowk	
13	Saadatganj	

Increased Rainfall: Intense and erratic rainfall patterns are a result of climate change. Abrupt, intense downpours have the potential to overwhelm drainage infrastructure.

Unseasonal Rains: Because of weather pattern variability, there have been unseasonal rains, which have caught the city off guard and made flooding issues worse.

Waste Management: Poor waste management practices lead to the clogging of drains and water channels with garbage and debris, impeding the flow of water.

Illegal Construction: Unauthorised construction on floodplains and along drainage channels further obstructs the natural flow of water. Infrastructure Damage: Roads, buildings, and public utilities are destroyed by flooding, necessitating expensive rebuilding and repair.

Public Health Issues: Flooding and stagnant water can lead to waterborne illnesses and other health risks, which puts a pressure on medical resources. Disruptions to businesses and livelihoods result in large financial losses. Recuperation and rehabilitation can come at a high cost. People may be uprooted and need temporary housing and assistance, especially in low-lying and vulnerable locations.

Enhancing Drainage Systems: Maintaining and modernising drainage systems to withstand intense downpours and avoid blockages.

Urban planning: Putting stronger restrictions on development and construction in place to protect naturally occurring water bodies and drainage routes. Creating all-encompassing plans for flood response and management that include effective evacuation procedures and early warning systems.

Green Infrastructure: Using green infrastructure to improve water absorption and lessen runoff, such as rain gardens, permeable pavements, and more green space.

Raising Community Participation and Awareness: Educating the public about waste management and enticing them to get involved in flood response and preparedness efforts.

Lucknow's flood issue calls for a multifaceted solution that includes proactive community involvement, efficient waste management, improved urban design, and infrastructural upgrades as during heavy floods the condition of the city doesnt allow the life to prosper, as depicted in the local news paper clippings in Figs. 6, 7, 8 and 9.

7.2 Lucknow Newspaper Report from (2018–2023)

Figs. 6, 7, 8 and 9.

8 Moving Towards Flood Mitigation Based on Learning

According to the preliminary analysis above, learning can help people become more resilient to flooding depending on the patterns of flooding they experience in various locations. It also implies that different flood mitigation strategies may either strengthen or weaken flood resilience. The two environments discussed above represent the two extremes of flood mitigation: the well-protected environment represents the "flood resistance paradigm", where flood safety is to be attained solely by preventing flooding in the first place; the unprotected, frequently-flooded environment represents the "flood adaptation paradigm", where flood safety is to be attained only by modifying the living environment, so that it can be flooded safely, without the at-risk of damage. Although the term "adaptation" can have many distinct meanings, for the sake of this study, it will be used to refer to the organism's adjustment to its environment. While disaster experience may improve recoverability, a component of flood resilience, prior knowledge, and policy legacies may lock in flood mitigation in the same direction as flood resistance to limit floodability (Van Buuren et al., 2016; Wiering et al., 2017). This makes learning about flood mitigation difficult in areas with flood control systems. Our LFF model reveals that in contrast to the claim that flood control infrastructure is a critical component of flood resilience, it may actually decrease flood resilience over time.

Our analysis, however, is focussed on the kind of environment that solely depends on large-scale flood control infrastructure, or environments where flood resistance is localised, meaning that only specific areas are protected by smaller-scale flood control infrastructure. However, some ecosystems are distinguished by specific combinations of flood resistance and flood adaption techniques. In order to further understand how different levels of flood resistance affect flood resilience

Fig. 6 Streets in Lucknow flooded after heavy rain

Fig. 7 Streets in Lucknow flooded after heavy rain

Fig. 8 Faizullahganj locality in Lucknow is flooded as a result of severe monsoon rains

by meditating the learning process from flood experience, empirical research is required. According to some (Aldunce et al., 2015; Berkes, 2007), resilience is a desirable trait for mankind's survival in the era of unpredictability brought on by climate change. A flood mitigation paradigm that encourages effective learning from flood experience—rather than suppressing it—is necessary for fostering flood resilience. In order to foster flood resilience, we urge "learning-based flood mitigation", a flood mitigation strategy that strives to encourage effective learning from innocuous floods by the general people. Distributed methods, such as localised flood retention areas and strategies for property-level flood proofing, are given priority in learning-based flood mitigation. It works to create a setting where flooding occurs in safe ways to offer educational opportunities. It is predicated on

Fig. 9 Faizullahganj locality in Lucknow is flooded as a result of severe monsoon rains

the idea that more frequent flooding will encourage people to think more deeply about the phenomenon of flooding and raise their awareness of the risk of flooding.

Flood mitigation that is based on learning encourages civic engagement to boost learning motivation. In the industrialised world, it has long been assumed that the government bears primary responsibility for flood control (Owusu et al., 2015). To be sure, the importance of shared responsibility between the government and citizens has been recognised in the concept of flood risk management (Henstra et al., 2019). But learning-based flood mitigation focuses more on the responsibility of citizens, as research has shown that the perception of self-responsibility can increase people's motivation to take flood mitigation and preparation measures (Motoyoshi, 2006). It is particularly important to modern cities, which can easily be devastated by an extreme storm event, despite being well-protected by flood control infrastructure.

However, there would be at least three obstacles in the way of the transition to learning-based flood mitigation. First, how people perceive flooding affects whether they will accept distributed flood prevention methods (Harries, 2012). Such perception is greatly influenced by mass media, which frequently exaggerates the negative effects of flooding (Kasperson and Kasperson, 1996) and tends to concentrate on the dramatic aspect of flood impacts (Birkland et al., 2003). This may be what led people to believe that floods should be resisted in the first place. Second, the absence of standardised methods to assess the costs and advantages of dispersed flood mitigation measures may make it more difficult for the government to promote them (Harries and Penning-Rowsell, 2011).

Thirdly, governments—particularly those in low-income countries—might not have additional financial capacity to subsidise low-income households after spending on expensive flood control infrastructure. Socioeconomically marginalised populations might not be able to afford to take property-level flood mitigation measures. Transformability is therefore necessary to move towards learning-based flood mitigation. It requires changing the current flood mitigation system into one where citizens understand their responsibility in flood management based on a more comprehensive understanding of flooding, which is not always harmful. It also requires a shift of public investment from centralised to distributed flood mitigation measures.

Undoubtedly, the focus of the conversation surrounding flood management has changed from flood resistance to flood risk management, where non-structural solutions are given more consideration (Butler and Pidgeon, 2011; Sayers et al., 2013). The idea of flood risk management, while acknowledging the limitations of flood control infrastructure and accepting a certain amount of risk (Challies et al., 2016), has not yet addressed the effect of flood resistance on the experience of flooding and the ensuing learning process. It is still believed that having some level of initial flood resistance is crucial (Driessen et al., 2016). A more critical analysis of conventional wisdom is necessary as the globe adjusts to a new normal of unpredictable extremes.

9 Concluding Remarks

The theoretical connections between flood experience, learning, and flood resistance are established by the LFF model. It can be used as an analytical framework to investigate flood resilience and the process of learning from past flood experiences. It would be fruitful to apply the LFF model to conduct empirical studies to compare the learning processes under different flood mitigation schemes with different degrees of flood resistance to understand their potential effects on flood resilience. It would also be important to study different cities or towns with the same degree of flood resistance to assess whether the learning process is consistent. Researchers could first characterise the flood experience of each case in terms of flood regime and flood impact, and then examine the influencing factors, i.e. learning motivation, learning opportunity, and prior knowledge. Subsequently, researchers could identify the newly generated flood-related knowledge, as well as the associated action taken. Attention should be paid to whether there is any barrier to prevent knowledge from turning into action. Finally, researchers could analyse how the action affects flood resilience in terms of floodability, recoverability, adaptability, and/or transformability. Such a study should provide insights into how to move towards learning-based flood mitigation.

Several flood disasters brought on by severe storm occurrences occurred solely in 2019; to mention a few, they occurred in Queensland (Australia), Iowa (USA), Jiangxi (China), and Nagano (Japan). We anticipate that the LFF model will spur additional study into learning, a critical component of building resilience, so that mankind can be better prepared for the era of climate change.

References

Aldunce, P., Beilin, R., Howden, M., Handmer, J., 2015. Resilience for disaster risk management in a changing climate: practitioners' frames and practices. Global Environ. Change 30, 1–11. https://doi.org/10.1016/j.gloenvcha.2014.10.010.

Baan, P.J., Klijn, F., 2004. Flood risk perception and implications for flood risk management in The Netherlands. Int. J. River Basin Manag. 2 (2), 113–122. https:// doi.org/https://doi.org/10.1080/15715124.2004.9635226.

Barnett, J., Evans, L.S., Gross, C., Kiem, A.S., Kingsford, R.T., Palutikof, J.P., Pickering, C. M., Smithers, S.G., 2015. From barriers to limits to climate change adaptation: path dependency and the speed of change. Ecol. Soc. 20. https://doi.org/10.5751/ES-07698-200305

Begg, C., Ueberham, M., Masson, T., Kuhlicke, C., 2017. Interactions between citizen responsibilization, flood experience and household resilience: insights from the 2013 flood in Germany. Int. J. Water Resour. Dev. 33, 591–608. https://doi.org/10.1080/07900627.2016.1200961.

Berkes, F., 2007. Understanding uncertainty and reducing vulnerability: lessons from resilience thinking. Nat. Hazards 41, 283–295. https://doi.org/10.1007/s11069-006-9036-7.

Berkes, F., Colding, J., Folke, C., 2000. Rediscovery of traditional ecological knowledge as adaptive management. Ecol. Appl. 10, 1251–1262. https://doi.org/10.2307/2641280.

Bichard, E., Kazmierczak, A., 2012. Are homeowners willing to adapt to and mitigate the effects of climate change? Climatic Change 112, 633–654. https://doi.org/10.1007/s10584-011-0257-8

Birkland, T.A., Burby, R.J., Conrad, D., Cortner, H., Michener, W.K., 2003. River ecology and flood hazard mitigation. Nat. Hazards Rev. 4, 46–54. https://doi.org/10.1061/(ASCE)1527-6988(2003)4:1(46)

Birkmann, J., Buckle, P., Jaeger, J., Pelling, M., Setiadi, N., Garschagen, M., Fernando, N., Kropp, J., 2010. Extreme events and disasters: a window of opportunity for change? Analysis of organizational, institutional and political changes, formal and informal responses after mega-disasters. Nat. Hazards 55, 637–655. https://doi.org/10.1007/s11069-008-9319-2.

Brody, S.D., 2003. Are we learning to make better plans? A longitudinal analysis of plan quality associated with natural hazards. J. Plan. Educ. Res. 23, 191–201. https://doi.org/10.1177/0739456X03258635.

Bubeck, P., Botzen, W.J., Kreibich, H., Aerts, J.C., 2013. Detailed insights into the influence of flood-coping appraisals on mitigation behaviour. Global Environ. Change 23, 1327–1338. https://doi.org/10.1016/j.gloenvcha.2013.05.009.

Bubeck, P., Botzen, W.J., Aerts, J.C., 2012. A review of risk perceptions and other factors that influence flood mitigation behavior. Risk Anal 32, 1481–1495. https://doi.org/ https://doi.org/10.1111/j.1539-6924.2011.01783.x.

Burningham, K., Fielding, J., Thrush, D., 2008. 'It'll never happen to me': understanding public awareness of local flood risk. Disasters 32, 216–238. https://doi.org/10.1111/j.1467-7717.2007.01036.x.

Butler, C., Pidgeon, N., 2011. From 'flood defence'to 'flood risk management': exploring governance, responsibility, and blame. Environ. Plann. C 29, 533–547. https://doi.org/10.1068/c09181j.

Carpenter, S., Walker, B., Anderies, J.M., Abel, N., 2001. From metaphor to measurement: resilience of what to what? Ecosystems 4, 765–781. https://doi.org/10.1007/s10021-001-0045-9.

Cenderelli, D.A., 2000. Floods from natural and artificial dam failures. In: Wohl, E.E. (Ed.), Inland Flood Hazards: Human, Riparian and Aquatic Communities. Cambridge University Press, Cambridge, UK, pp. 73–103.

Challies, E., Newig, J., Thaler, T., Kochsk€amper, E., Levin-Keitel, M., 2016. Participatory and collaborative governance for sustainable flood risk management: An emerging research agenda. Environ. Sci. Pol. 55, 275–280. https://doi.org/10.1016/j.envsci.2015.09.012

Cohen, W.M., Levinthal, D.A., 1990. Absorptive capacity: a new perspective on learning and innovation. Adm. Sci. Q. 35, 128–152. https://doi.org/10.2307/2393553. Colding, J., Elmqvist, T., Olsson, P., 2003. Living with disturbance: building resilience in

Colding,J.,Elmqvist,T.,Olsson,P.,2003.Livingwithdisturbance:building resiliencein social-ecological systems. In: Berkes, F., Colding, J., Folke, C. (Eds.), Navigating Social-Ecological Systems: Building Resilience for Complexity and Change. Cambridge University Press, Cambridge, UK, pp. 163–185.

Cuny, F.C., 1991. Living with floods: alternatives for riverine flood mitigation. Land Use Policy 8, 331–342. https://doi.org/10.1016/0264-8377(91)90023-C.

Cutter, S.L., Barnes, L., Berry, M., Burton, C., Evans, E., Tate, E., Webb, J., 2008. A place- based model for understanding community resilience to natural disasters. Global Environ. Change 18, 598–606. https://doi.org/10.1016/j.gloenvcha.2008.07.013.

De Marchi, B., Scolobig, A., 2012. The views of experts and residents on social vulnerability to flash floods in an Alpine region of Italy. Disasters 36, 316–337. https://doi.org/10.1111/j.1467-7717.2011.01252.x.

De Bruijn, K., Buurman, J., Mens, M., Dahm, R., Klijn, F., 2017. Resilience in practice: five principles to enable societies to cope with extreme weather events. Environ. Sci. Pol. 70, 21–30. https://doi.org/10.1016/j.envsci.2017.02.001.

De Bruijn, K.M., 2005. Resilience and Flood Risk Management: a Systems Approach Applied to Lowland Rivers. D thesis, Delft University of Technology).

Dieperink, C., Mees, H., Priest, S., Ek, K., Bruzzone, S., Larrue, C., Matczak, P., 2018. Managing urban flood resilience as a multilevel governance challenge: an analysis of required multilevel coordination mechanisms. Ecol. Soc. 23 https://doi.org/ https://doi.org/10.5751/ES-09962-230131.

Driessen, P.P.J., Hegger, D.L.T., Bakker, M.H.N., van Rijswick, H., Kundzewicz, Z.W., 2016. Toward more resilient flood risk governance. Ecol. Soc. 21 https://doi.org/ https://doi.org/10.5751/ES-08921-210453.

Ehlert, J., 2012. Beautiful Floods: Environmental Knowledge and Agrarian Change in the Mekong Delta, Vietnam, vol. 19. LIT Verlag, Berlin, Germany.

Ellis, S., Mendel, R., Nir, M., 2006. Learning from successful and failed experience: the moderating role of kind of after-event review. J. Appl. Psychol. 91, 669. https://doi.org/10.1037/0021-9010.91.3.669

Eraut, M., 2004. Informal learning in the workplace. Stud. Contin. Educ. 26, 247–273.https://doi.org/10.1080/158037042000225245

Eraut, M., 2000. Non-formal learning and tacit knowledge in professional work. Br. J. Educ. Psychol. 70, 113–136. https://doi.org/10.1348/000709900158001.

Fernandez-Gimenez, M.E., Ballard, H.L., Sturtevant, V.E., 2008. Adaptive management and social learning in collaborative and community-based monitoring: a study of five community-based forestry organizations in the western USA. Ecol. Soc. 13. https://www.ecologyandsociety.org/vol13/iss2/art4/

Folke, C., 2006. Resilience: the emergence of a perspective for social–ecological systems analyses. Global Environ. Change 16, 253–267. https://doi.org/10.1016/j.gloenvcha.2006.04.002.

Folke, C., Carpenter, S.R., Walker, B., Scheffer, M., Chapin, T., Rockstrom, J., 2010. Resilience thinking: integrating resilience. adaptability and transformability. Ecol. Soc. http://www.ecologyandsociety.org/vol15/iss4/art20/.

Fox-Rogers, L., Devitt, C., O'Neill, E., Brereton, F., Clinch, J.P., 2016. Is there really "nothing you can do"? Pathways to enhanced flood-risk preparedness. J. Hydrol. 543, 330–343. https://doi.org/10.1016/j.jhydrol.2016.10.009.

Gotham,K.F.,Campanella,R.,Lauve-Moon,K.,Powers,B., 2018. Hazard experience geophysical vulnerability, and flood risk perceptions in a postdisaster city, the case of New Orleans. Risk Anal. 38, 345–356. https://doi.org/10.1111/risa.12830

Gilbert, S.W., 2010. Disaster Resilience: A Guide to the Literature, 1117. U.S. Department of Commerce, National Institute of Standards and Technology (NIST), Special Publication. https://ws680.nist.gov/publication/get_pdf.cfm?pub_id1/4906887.

Godschalk, D.R., 2003. Urban hazard mitigation: creating resilient cities. Nat. Hazards Rev. 4, 136–143. https://doi.org/10.1061/(ASCE)1527-6988(2003)4:3(136).

Geresdi, I., & Horváth, Á. (2000). Nowcasting of precipitation type. Part I: Winter precipitation. Időjárás, 104(4), 241–252.

Gunderson, L.H., Holling, C.S. (Eds.), 2002. Panarchy: Understanding Transformations in Human and Natural Systems. Island Press, Washington, D.C., USA.

Gunderson, L.H., 2010. Ecological and human community resilience in response to natural disasters. Ecol. Soc. 15. http://www.ecologyandsociety.org/vol15/iss2/ar.

Haer, T., Botzen, W.W., de Moel, H., Aerts, J.C., 2017. Integrating household risk mitigation behavior in flood risk analysis: an agent-based model approach. Risk Anal. 37, 1977–1992. https://doi.org/10.1111/risa.12740.

Harries, T., Penning-Rowsell, E., 2011. Victim pressure, institutional inertia and climate change adaptation: the case of flood risk. Global Environ. Change 21 (1), 188–197. https://doi.org/10.1016/j.gloenvcha.2010.09.002.

Harries, T., 2012. The anticipated emotional consequences of adaptive behaviour—impacts on the take-up of household flood-protection measures. Environ. Plann. 44, 649–668. https://doi.org/10.1068/a43612.

Harries, T., 2013. Responding to flood risk in the UK. In: Cities at Risk. Springer, Dordrecht, pp. 45–72.

Heazle, M., Tangney, P., Burton, P., Howes, M., Grant-Smith, D., Reis, K., Bosomworth, K., 2013. Mainstreaming climate change adaptation: an incremental approach to disaster risk management in Australia. Environ. Sci. Pol. 33, 162–170. https://doi.org/10.1016/j.envsci.2013.05.009.

Hegger, D.L.T., Driessen, P.P.J., Wiering, M., van Rijswick, H., Kundzewicz, Z.W., Matczak, P., Crabbe, A., Raadgever, G.T., Bakker, M.H.N., Priest, S.J., Larrue, C., Ek, K., 2016. Toward more flood resilience: is a diversification of flood risk management strategies the way forward? Ecol. Soc. 21, 52. https://doi.org/10.5751/ES-08854-210452.

Henstra, D., Thistlethwaite, J., Brown, C., & Scott, D. (2019). Flood risk management and shared responsibility: Exploring Canadian public attitudes and expectations. Journal of Flood Risk Management, 12(1), e12346

Holling, C. S. (1973). Resilience and stability of ecological systems. Annual Review of Ecology and Systematics, 4, 1–23. https://doi.org/10.1146/annurev.es.04.110173.000245

Horváth, F., & Cloetingh, S. A. P. L. (1996). Stress-induced late-stage subsidence anomalies in the Pannonian basin. Tectonophysics, 266(1–4), 287–300.

Johnson, C.L., Tunstall, S.M. and Penning-Rowsell, E.C. (2005) 'Floods as Catalysts for Policy Change: Historical Lessons from England and Wales'. International Journal of Water Resources Development, 21(4): 561–575.

Kasperson, R. E., & Kasperson, J. X. (1996). The social amplification and attenuation of risk. The annals of the American academy of political and social science, 545(1), 95105.

Keating, A., Mechler, R., Campbell, K., Magnuszewski, P., Mochizuki, J., & Szönyi, M. (2017). Development and testing of a community flood resilience measurement tool. Natural Hazards and Earth System Sciences, 17, 77–101. https://doi.org/10.5194/nhess-17-77-2017

Keck, M., & Sakdapolrak, P. (2013). What is social resilience? Lessons learned and ways forward. Erdkunde, 5–19.

Knocke, E. T., & Kolivras, K. N. (2007). Flash flood awareness in southwest Virginia. Risk Analysis: An International Journal, 27(1), 155–169.

Kolb, D. A. (1984). Experiential learning: Experience as the source of learning and development. Prentice Hall.

Kotzé, I., & Reyers, B. (2016). Piloting a social-ecological index for measuring flood resilience: A composite index approach. Ecological Indicators, 60, 45–53. https://doi.org/10.1016/j.ecolind.2015.05.032

Kreibich, H., Thieken, A. H., Petrow, T., Müller, M., & Merz, B. (2005). Flood loss reduction of private households due to building precautionary measures–lessons learned from the Elbe flood in August 2002. Natural hazards and earth system sciences, 5(1), 117–126.

Lawrence, J., Quade, D., & Becker, J. (2014). Integrating the effects of flood experience on risk perception with responses to changing climate risk. Natural Hazards, 74(3), 1773–1794.

Lennon, M., Scott, M., & O'Neill, E. (2014). Urban design and adapting to flood risk: the role of green infrastructure. Journal of Urban Design, 19(5), 745–758.

Lhomme, S., Serre, D., Diab, Y., & Laganier, R. (2013). Analyzing resilience of urban networks: A preliminary step towards more flood resilient cities. Natural Hazards and Earth System Sciences, 13(1), 221–230. https://doi.org/10.5194/nhess-13-221-2013

Liao, K. (2012). A theory on urban resilience to floods: A basis for alternative planning practices. Ecology and Society, 17(4), Article 48. https://doi.org/10.5751/ES-05231-170448

Liao, K. (2016). Learning from the ecological wisdom of living with floods. Global Environmental Change, 36, 220–230. https://doi.org/10.1016/j.gloenvcha.2015.12.011

Liao, Y. J., Zhao, H. T., Jiang, Y., Ma, Y. K., Luo, X., & Li, X. Y. (2019). An innovative method based on cloud model learning to identify high-risk pollution intervals of storm-flow on an urban catchment scale. Water research, 165, 115007.

Ludy, J., & Kondolf, G. M. (2012). Flood risk perception in lands "protected" by 100-year levees. Natural hazards, 61(2), 829–842.

Manyena, S. B., O'Brien, G., O'Keefe, P., & Rose, J. (2011). Disaster resilience: a bounce back or bounce forward ability?. Local environment, 16(5).

Moftakhari, H. R., AghaKouchak, A., Sanders, B. F., Allaire, M., & Matthew, R. A. (2018). What is nuisance flooding? Defining and monitoring an emerging challenge. Water Resources Research, 54(7), 4218–4227.

Morrison, A., Noble, B. F., & Westbrook, C. J. (2019). Flood risk management in canada's prairie provinces: an analysis of decision-maker priorities and policy preferences. Environmental Management, 64(5), 608–625.

Morrison, A., Westbrook, C. J., & Noble, B. F. (2018). A review of the flood risk management governance and resilience literature. Journal of Flood Risk Management, 11(3), 291–304.

Motoyoshi, T. (2006). Public perception of flood risk and community-based disaster preparedness. A better integrated management of disaster risks: Toward resilient society to emerging disaster risks in mega-cities, 121–134.

Nye, M., Tapsell, S., & Twigger-Ross, C. (2011). New social directions in UK flood risk management: moving towards flood risk citizenship?. Journal of flood risk management, 4(4), 288–297.

Ogston, A. S., Cacchione, D. A., Sternberg, R. W., & Kineke, G. C. (2000). Observations of storm and river flooddriven sediment transport on the northern California continental shelf. Continental Shelf Research, 20(16), 2141–2162.

Olsson, P., Gunderson, L. H., Carpenter, S. R., Ryan, P., Lebel, L., Folke, C., & Holling, C. S. (2004). Shooting the rapids: Navigating transitions to adaptive governance of social-ecological systems. Ecology and Society, 9(2), 18. https://doi.org/10.5751/ES-00679-090218

Osberghaus, D. (2017). The effect of flood experience on household mitigation—Evidence from longitudinal and insurance data. Global Environmental Change, 43, 126–136.

Owusu-Manu, D. G., Badu, E., Holt, G. D., & Edwards, D. (2015). Determinants of management innovation in the Ghanaian construction consulting sector. Construction Economics and Building, 15(1), 75–88.

Paul, B. K. (1984). Perception of and agricultural adjustment to floods in Jamuna floodplain, Bangladesh. Human Ecology, 12(1), 3–19.

Pielke, R. A., Jr. (1999). The honest broker: Making sense of science in policy and politics. Cambridge University Press.

Resilience, R. A. U. (2007). Research Prospectus.

Restemeyer, B., Woltjer, J., & van den Brink, M. (2015). A strategy-based framework for assessing the flood resilience of cities–A Hamburg case study. Planning theory & practice, 16(1), 45–62.

Sayers, P., Yuanyuan, L., Galloway, G., Penning-Rowsell, E., Fuxin, S., Kang, W.,... & Le Quesne, T. (2013). Flood risk management: A strategic approach.

Schelfaut, K., Pannemans, B., Van der Craats, I., Krywkow, J., Mysiak, J., & Cools, J. (2011). Bringing flood resilience into practice: the FREEMAN project. Environmental Science & Policy, 14(7), 825–833.

Siegrist, M., & Gutscher, H. (2008). Natural hazards and motivation for mitigation behavior: People cannot predict the affect evoked by a severe flood. Risk Analysis: An International Journal, 28(3), 771–778.

Singh, V. P., Wang, S. X., & Zhang, L. (2005). Frequency analysis of nonidentically distributed hydrologic flood data. Journal of Hydrology, 307(1–4), 175–195.

Sörensen, J., Persson, A., Sternudd, C., Aspegren, H., Nilsson, J., Nordström, J.,... & Mobini, S. (2016). Rethinking urban flood management—time for a regime shift. Water, 8(8), 332.

Ten Brinke, W. B. M., Bannink, B. A., & Ligtvoet, W. (2008, August). The evaluation of flood risk policy in the Netherlands. In Proceedings of the Institution of Civil Engineers-Water Management (Vol. 161, No. 4, pp. 181-188). Thomas Telford Ltd.

Terpstra, T. (2011). Emotions, trust, and perceived risk: Affective and cognitive routes to flood preparedness behavior. Risk Analysis: An International Journal, 31(10), 1658–1675.

Thompson, P. M., & Sultana, P. (1996). Distributional and social impacts of flood control in Bangladesh. Geographical Journal, 1–13.

Tobin, G. A. (1995). The levee love affair: A stormy relationship? Water Resources Bulletin, 31(3), 359–367. https://doi.org/10.1111/j.1752-1688.1995.tb04064.x

Van Buuren, A., Ellen, G. J., & Warner, J. F. (2016). Path-dependency and policy learning in the Dutch delta: toward more resilient flood risk management in the Netherlands?. Ecology and Society, 21(4).

Vis, M., Klijn, F., De Bruijn, K. M., & Van Buuren, M. (2003). Resilience strategies for flood risk management in the Netherlands. International journal of river basin management, 1(1), 33–40.

Walker, B., Holling, C. S., Carpenter, S. R., & Kinzig, A. (2004). Resilience, adaptability and transformability in social–ecological systems. Ecology and Society, 9(2), 5. https://doi.org/10.5751/ES-00650-090205

Wiering, M., Kaufmann, M., Mees, H., Schellenberger, T., Ganzevoort, W., Hegger, D. L.,... & Matczak, P. (2017). Varieties of flood risk governance in Europe: How do countries respond to driving forces and what explains institutional change?. Global environmental change, 44, 15-26.

Willner, S. N., Otto, C., & Levermann, A. (2018). Global economic response to river floods. Nature Climate Change, 8(7), 594–598. https://doi.org/10.1038/s41558-018-0173-2

Zevenbergen, C., van Herk, S., van den Born, G. J., & Gouweleeuw, B. T. (2008). Towards flood resilient communities: The case for flood resilience indicators. Journal of Flood Risk Management, 1(1), 13–23. https://doi.org/10.1111/j.1753-318X.2008.00001.x

Habitat Suitability Modelling for Snow Leopards in Biosphere Reserve with Remote Sensing-Based Approaches

Vipul Maurya, Parag Madhukar Dhakate, Ashish Mani, Ashish Panda, and S. B. Lal

1 Introduction

The world's tallest and most recent mountain range is the Himalayas (Devan 1988). Around 70 million years ago, the Tethys Sea dried up due to the collision of the tectonic plates of the Indian subcontinent and the Eurasian continent, which also caused the Himalayan orogenesis to commence (Wadia 1966).

Programs for population monitoring are crucial to conservation efforts. They offer data on the condition of wildlife populations and can assist in assessing the success of conservation efforts, enabling adaptive management. Population monitoring programs can contribute to a better understanding of the biology of the species, including the basic processes of survival, reproduction, temporary emigration, and immigration, and the crucial rates associated with these processes, in addition to population trends (Sharma et al. 2014).

Previous research on the habitat utilisation of snow leopards has identified two major factors influencing this selection. Several abiotic factors, including terrain slope, aspect, elevation, snow cover, distance from rivers, and annual mean temperature are the first factors that predict the occurrence of snow leopards (Wolf and Ale 2009; Li et al. 2013; Aryal et al. 2014; Sharma et al. 2015; Alexander et al. 2016). The availability of prey and human activity are examples of biotic elements that make up the second component (Wolf and Ale 2009; Xu and Luo 2010; Aryal et al. 2014; Alexander et al. 2016).

One of the largest land mammals, snow leopards, is an indicator species for the Indian Himalayan ecology (Fox et al. 1988; Maheshwari 2016); protecting snow leopards also entails protecting high-altitude ecosystems (Bhatnagar et al. 2001; Maheshwari 2016). An estimated two million km^2 could serve as the prospective home for the 7463–7980 surviving snow leopards worldwide (McCarthy et al. 2016). Anonymous (2008) estimates that the potential habitat of snow leopards in India is 128,757 km^2, of which 92% is found in the western Himalayan Mountains, home to 400–700 snow leopards. Sadly, there is little data on the distribution and appropriate habitat of snow leopards in the western Himalayas (Maurya et al. 2016 and 2017).

Many areas of Uttarakhand are situated in the area that separates the Greater Himalayan and Trans-Himalayan biogeographic zones, even though just a small portion of the state is located within the Trans-Himalayan Biogeographic Zone, which offers the snow leopard the ideal habitat. Similar research on the snow leopard and wildlife values of the Trans-Himalayan regions of Himachal Pradesh, Lahaul-Spiti, and Pangi Valley was also done by Bhatnagar (1997), Bhatnagar et al. (1999, 2008), Saberwal (1996), Vinod and SathyaKumar (1999); however, some areas of Himachal Pradesh have scant information on snow leopards.

Considering that signs and camera traps in some areas of the NDBR in Uttarakhand, we look at the habitat that is appropriate for the species. It is crucial to comprehend the habitat distribution of snow leopards in the NDBR to enhance research findings and conservation strategies. Thus, the purpose of our study was to use the MaxEnt model to

V. Maurya (✉) · S. B. Lal
Department of Forestry, Sam Higginbottom University of Agriculture Technology and Sciences, Prayagraj, Uttar Pradesh, India
e-mail: vipulforestry@gmail.com

S. B. Lal
e-mail: sblal_123@shiats.edu.in

P. M. Dhakate
Chief Conservator of Forests, Uttarakhand Forest Department, Dehradun, India
e-mail: paragenetics@gmail.com

A. Mani
Amity School of Natural Resources and Sustainable Development (ASNRSD), Amity University, Noida, Uttar Pradesh, India
e-mail: ashish.mani@s.amity.edu

A. Panda
Academy of Scientific and Innovative Research (AcSIR), Ghaziabad, Uttar Pradesh, India
e-mail: pelochelyspanda@gmail.com

D. Kumar et al. (eds.), *Data Analytics and Remote Sensing-Based Analysis for Sustainable Development*, Advances in Science, Technology & Innovation, https://doi.org/10.1007/978-3-031-98095-4_6

evaluate the possible habitat distribution for snow leopards in the study area as well as biotic and abiotic elements of influence.

2 Material and Methods

2.1 Study Area

One significant protected area in the Himalayas at a high altitude is the Nanda Devi Biosphere Reserve (NDBR). According to Anonymous (1987) and Gupta (2002), it is one of the world's least disturbed protected areas. The NDBR covers an area of 624.62 km^2 in the core zone and 1612.12 km^2 in the buffer zone. It is located between 30°17 N and 30°41 N latitude and 79°40 E and 80°05 E longitude. For management and conservation purposes, the NDBR has been divided into three zones: core zones (national parks), buffer zones, and transition zones (Kumar 2002; Negi 2002). The NDBR now covers 5800 km^2, up from 2336.74 km^2, and includes portions of the districts of Bageshwar and Pithoragarh in Kumaun, Uttarakhand State, and Chamoli District in Garhwal. For management and conservation purposes, the NDBR has been divided into three zones: core zones (national parks), buffer zones, and transition zones (Kumar 2002; Negi 2002). There are no human habitations in the core zones, and they are under strict protection, while the buffer zone contains 47 villages. The region usually experiences three distinct seasons—long winter, short summer, and rainy—and a temperate, monsoonal climate. A 929 mm of rain falls on average. Approximately 47% of the yearly rainfall falls in two short months, i.e. July and August (Purohit et al. 2001). The temperature ranges from 3 to 7.5 °C at the minimum and from 11 to 24 °C at the maximum. The Trans-Himalayan region is between 4400 and 5500 m above sea level. It exhibits minimal rainfall and has all the features of a normally arid, cold climate (Rawat 2005) (Figs. 1 and 2).

2.2 Methods

The rigorous field surveys conducted in the Trans-Himalayan regions of Uttarakhand are essential for understanding the population structure, status, and distribution of prey species, which are critical for the conservation of apex predators like the snow leopard. In the survey, the Garhwal and the Kumaon regions of Uttarakhand covered a total of 37 transects to assess the presence of snow leopards. The deployment of camera traps was based on the evidence of signs indicating the elusive feline's presence. This method is a cornerstone in wildlife conservation efforts, providing invaluable data without disturbing the animals (Fig. 3a, b).

The use of trail survey methods, stratification into survey blocks, and the systematic use of camera traps in 2 × 2 km^2 grids based on the presence of snow leopard indicators, all contribute to a comprehensive dataset. Such data is invaluable for wildlife management and conservation efforts, ensuring that these elusive creatures are preserved for future generations. Based on information science, MaxEnt models have been proposed by Jaynes (1957a, b) and more recent studies

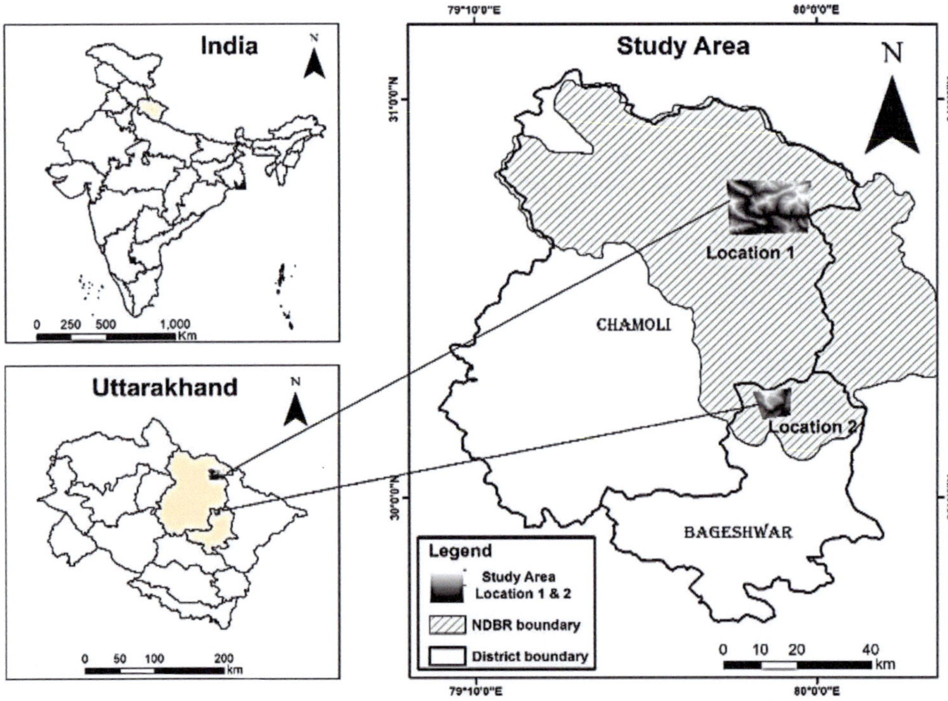

Fig. 1 Map of the study area

Fig. 2 Glimpses of the study area

(Jaynes and Bretthorst 2003; Xing and Hao 2011) and are widely employed in various academic disciplines. Based on ecological niche theory, the MaxEnt model investigates a target species' potential distribution within a study area using data from species "presence" surveys. The methods used for predicting habitat suitability for snow leopard in the Uttarakhand Himalaya followed the standard methods adopted by previous studies. MaxEnt is now a popular SDM technique used by conservationists to forecast a species' distribution based on a collection of data and environmental indicators. To assess the response function and contribution of environmental factors and to anticipate the possible geographic range of a species, the premise behind SDM is to match the known locations of a species with the environmental parameters of these places. These models calculate the basic ecological niche in the environmental domain (species reaction to abiotic environmental stimuli) and extrapolate it to the spatial domain to calculate the probability of presence for any given area. The Shutter Radar Topographical Mission (SRTM) Digital Elevation Model (DEM) data has been used to generate a topographic layer of slope, aspect, and elevation. Additionally, the ArcGIS Desktop Software have been used to prepare topography and study area maps.

3 Result and Discussion

3.1 Aspect

Aspect helps us in the identification of the downslope course of the most extreme pace of progress in an incentive from every cell to its neighbours. Aspect or viewpoint is a part of slope direction that helps in understanding the slope (Mani and Kumar 2020). The estimations of the yield raster will be towards the viewpoint. The aspect maps are divided into 10 classes, namely "flat (− 1°), north (0–22.5°), northwest (22.5°–67.5°), and so on till 360°, which indicates the slope direction. The aspect maps of Bageshwar and Chamoli south direction in the maximum areas is shown in Fig. 4a, b.

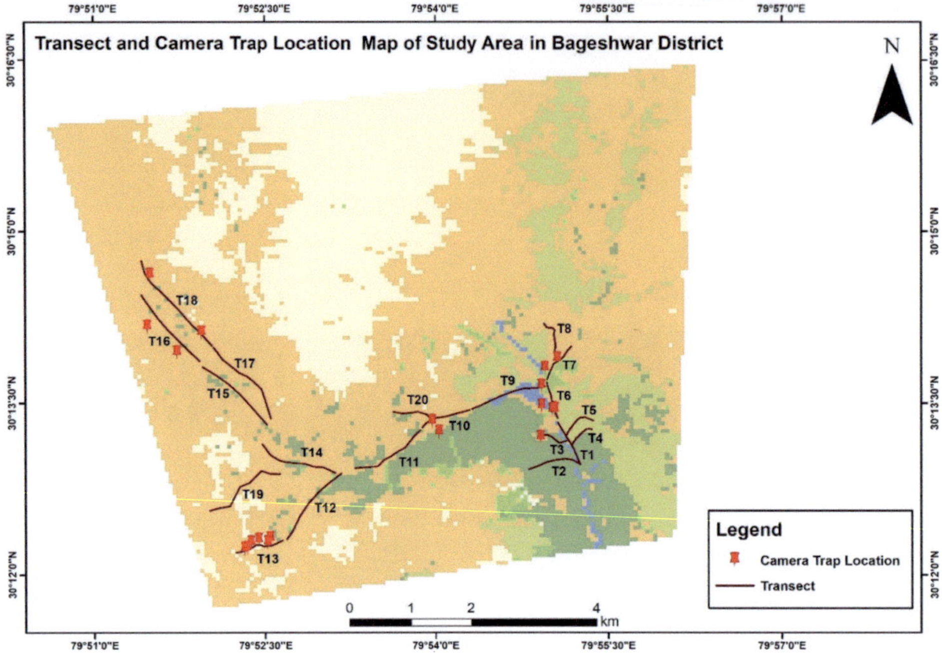

(a)

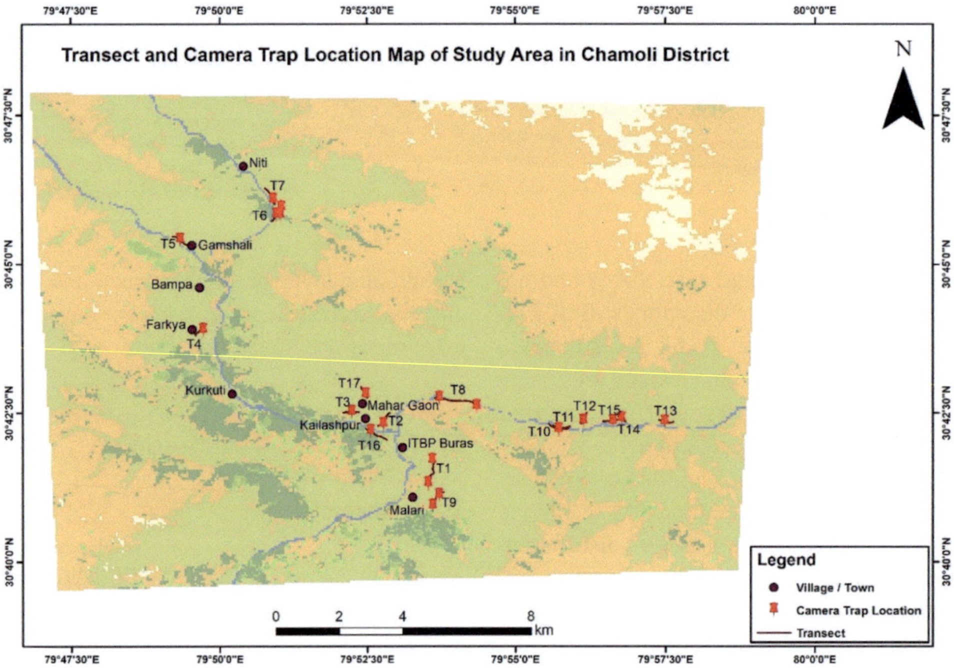

(b)

Fig. 3 **a** Transact map of Bageshwar district. **b** Transact map of Chamoli district

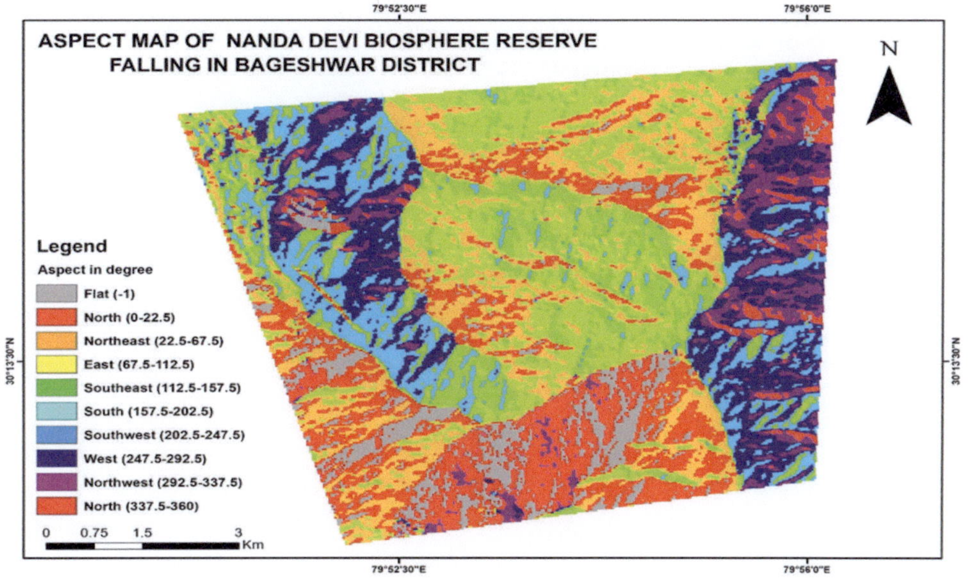

(a)

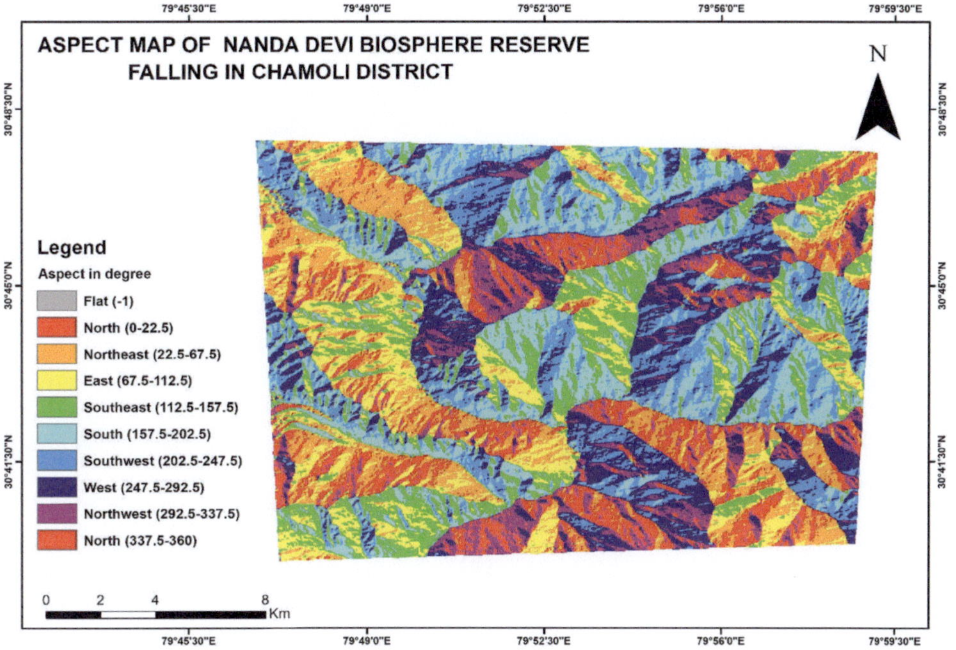

(b)

Fig. 4 **a** Aspect map of Bageshwar district. **b** Aspect map of Chamoli district

3.2 Slope

Slope or slant is the proportion of steepness or the level of the tendency of the land surface, which was a significant factor that imparted in this work (Mani and Kumar 2020). The slope which is measured in degrees was assembled into five classes 1°–15° (very gentle), 15°–30° (gentle), 30°–45° (moderate), 45°–60° (steep), and more than 60° (very steep). The slope maps display the maximum area having a moderate to steep slope in the Bageshwar region and a steep to very steep slope in the Chamoli region (Fig. 5a, b).

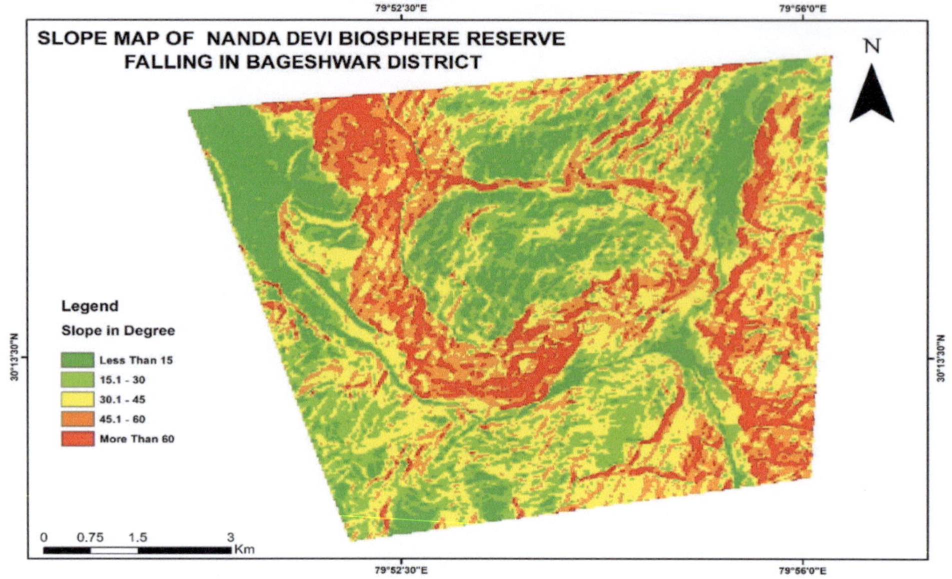

(a)

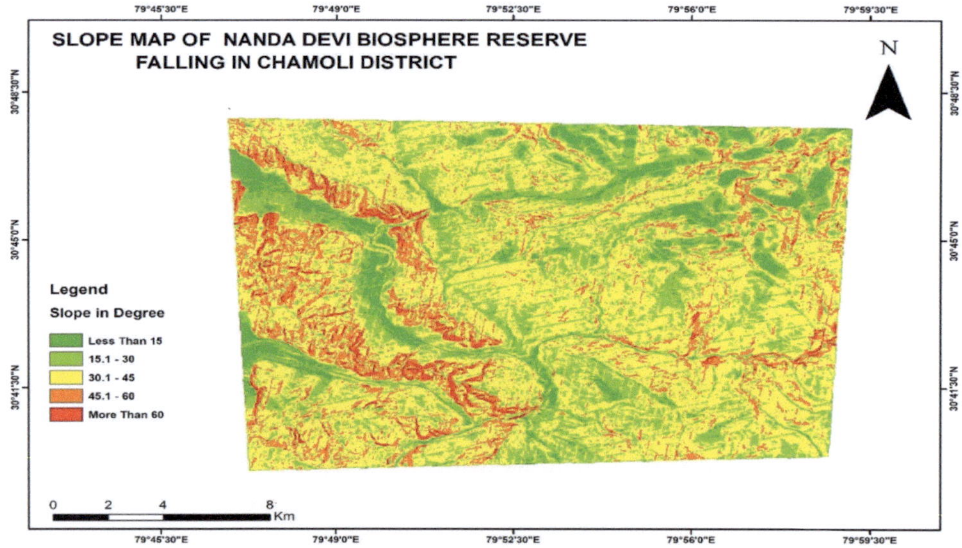

(b)

Fig. 5 **a** Slope map of Bageshwar district. **b** Slope map of Chamoli district

3.3 Elevation

The elevation is a significant land factor in the study because the habitat of snow leopards is in high-altitude regions. The elevation maps have been prepared in ArcGIS using the SRTM DEM data. The state of Uttarakhand is situated between 172 and 7764 m above sea level (Mani et al. 2023). The very high elevation is found in the maximum area. The study area in Bageshwar District has an elevation range between 2852 and 6578 m and the study area in Chamoli district has an elevation range between 2666 and 6014 m (Fig. 6a, b). The elevation maps have been prepared in ArcGIS using the SRTM DEM data. There is an obvious correlation between the area's slope and elevation.

3.4 Predicting Habitat Suitability for Snow Leopards in the NDBR

The upper Himalayan region district of Uttarakhand, which includes Uttarkashi, Tehri Garhwal, Rudraprayag, Chamoli, Bageshwar, and Pithoragarh, has been taken into

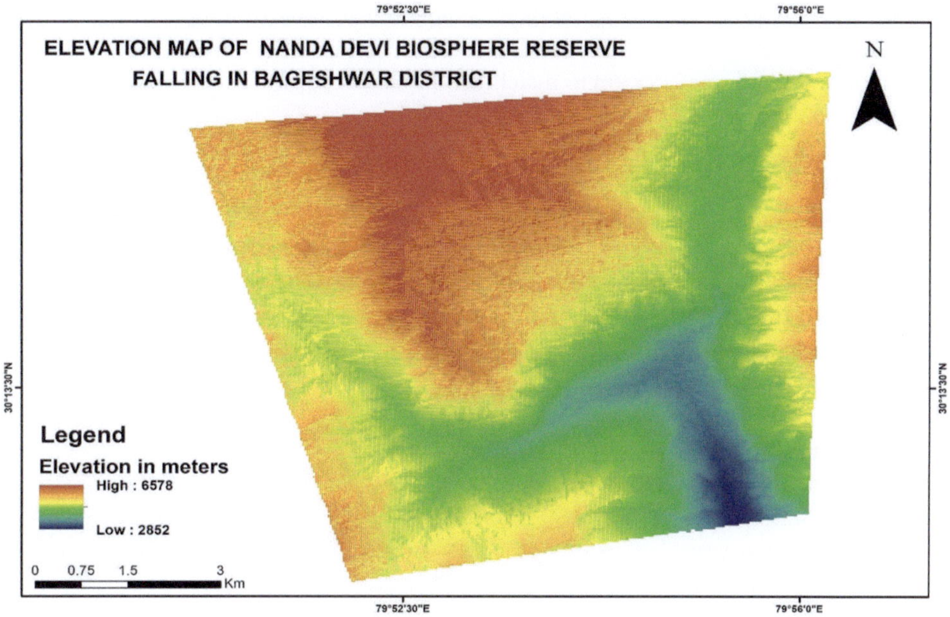

(b)

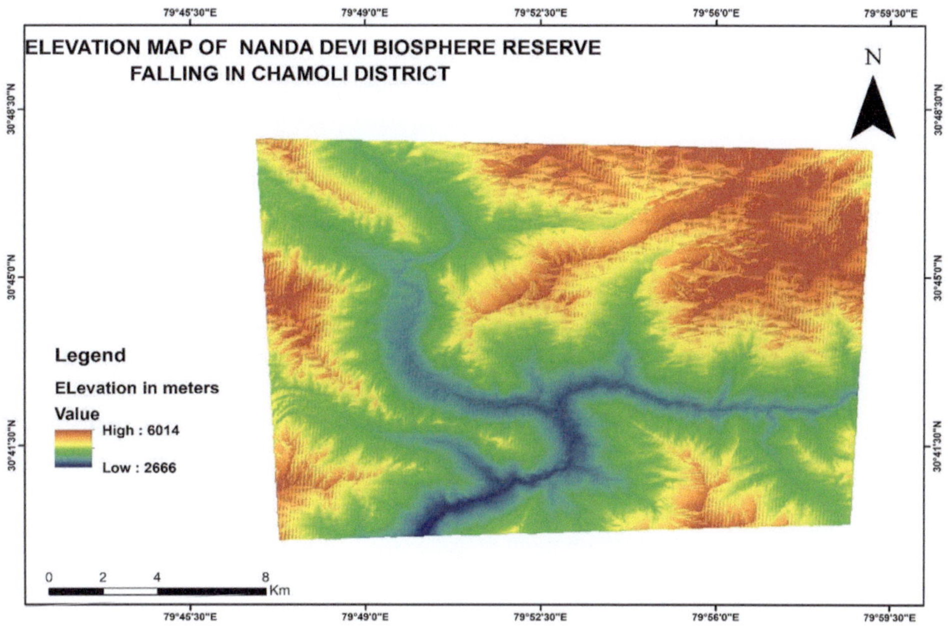

(a)

Fig. 6 **a** Elevation map of Chamoli district. **b** Elevation map of Bageshwar district

consideration for MaxEnt modelling due to the area's potential as a snow leopard habitat (Peterson and Robins 2003). A total of 484 snow leopard observations (direct and indirect evidence) were used to estimate the appropriate habitat in the NDBR (Maheshwari and Sharma 2010), Himachal Pradesh (Sharma 2009), and Uttarakhand (Maheshwari and Sharma 2010). The coordinates of each data point were located in ArcMap 10.5, either by using data from the literature or a geographical projection. The records' assigned geographic coordinates have a positional accuracy of 30 m. The chosen locations were given a spatial resolution of 1x1 km to match the environmental data's resolution (WorldClim). For each 1 km grid cell, only one location was selected if more than one observation was gathered in the grid (i.e. to avoid autocorrelation with low sample size; Phillips et al. 2006; Pearson et al. 2007).

3.4.1 Environmental Variables

One of the most important challenges facing academics, scientists, planners, and politicians worldwide is climate change. Out of all the climate variables, temperature and precipitation are the most important one for identifying changes in the climate that are mostly brought about by industry and urbanisation (Mani et al. 2021). A variety of bioclimatic environmental variables were chosen and retrieved from the WorldClim database (Version 1.4, http://www.worldclim.org/bioclim.htm; Hijmans et al. 2005). These metrics, which describe physiologically important elements for characterising a species' range, are derived from climatologies of monthly temperature and precipitation. Overall, 24 environmental parameters were used. All spatial data layers were resampled to a spatial resolution of 1 km using the closest neighbour resampling method in ERDAS 10.5 and ArcGIS 10.5 software. Using autocorrelation, the multi-collinearity of 19 bioclimatic predictor variables was investigated. $r > 0.7$-positive variables.

3.4.2 Model Performance

According to Phillips and Dudík (2008), models with AUV scores > 0.7 are regarded as potentially helpful and suggest outstanding prediction performance. The MaxEnt model's AUC score (SD) for the training data was high (0.991–0.001). The little AUC standard deviation (0.001) indicates that the present data was not overfitted. Nine variables were found to account for 99.5% of the model's variance according to the jackknife evaluation process (Table 1 and Fig. 1).

Average daily temperature range, Mean Diurnal Range (bio 2), which accounts for 42.7% of the snow leopard habitat suitability forecast, had the single greatest impact on model predictions. According to model estimates, the likelihood of seeing snow leopards increased with an elevation of over 3000 m and drastically dropped above 6700 m (Fig. 2a).

The contribution of bio 2 was just 50%, and bio 14 was 5.5 (Table 1). The likelihood of seeing a snow leopard increased when the amount of precipitation during the driest and coldest quarter climbed > 80 mm (Fig. 2b, c), while it reduced when seasonal precipitation variance increased (CV = 30%) (Fig. 2d). The likelihood of seeing a snow leopard also dropped as the temperature rose (Fig. 2e, f).

Table 1 Environmental variables used, i.e. predicting the potential distribution of snow leopards in NDBR

S. No.	Variable	Percent contribution	Permutation importance
1	bio_2	42.7	50
2	bio_14	34.8	5.5
3	forest_ classification	11.4	0
4	bio_15	5.8	25
5	bio_3	3.5	8.6
6	bio_19	0.9	10.2
7	Ruggedness	0.8	0.7
8	bio_5	0	0
9	prec_average	0	0
10	temp_avarage_	0	0
11	bio_1	0	0
12	Altitude	0	0

3.4.3 Environmental Data

For the modelling study, which is deemed to be most relevant to animal existence, 19 bioclimatic variables representing annual trends, seasonality, and extreme or limiting environmental factors were used. These variables were derived from globally interpolated datasets (source: http://www.worldclim.org); (Pearson et al. 2007). Since bioclimatic variables with finer than 30 m resolution are not available, all analyses were carried out at the 30 m pixels spatial resolution of the environmental datasets (Stockwell and Peterson 2002). To complete the modelling experiment, all environmental data layers were finally chopped for the study area.

3.4.4 Software Developer

19 bioclimatic variables obtained from globally interpolated datasets for environmental information (source: http:/ /ww Model of maximum entropy). By determining a probability distribution that is based on a distribution of maximum entropy (with regard to a collection of environmental factors), the maximum entropy (MaxEnt) technique assesses a species' environmental niche (Phillips et al. 2006). In our investigation, 10% of the data points were used as the random test percentage along with default values for various parameters, maximum iterations of 500, convergence threshold of 0.001, and other parameters. We describe annual trends using the maximum entropy (MaxEnt) model, version 3.4.4 from WorldClim.org.

3.4.5 Model Validation

Despite some recent arguments, receiver operating characteristic (ROC) analysis was used to measure the prediction

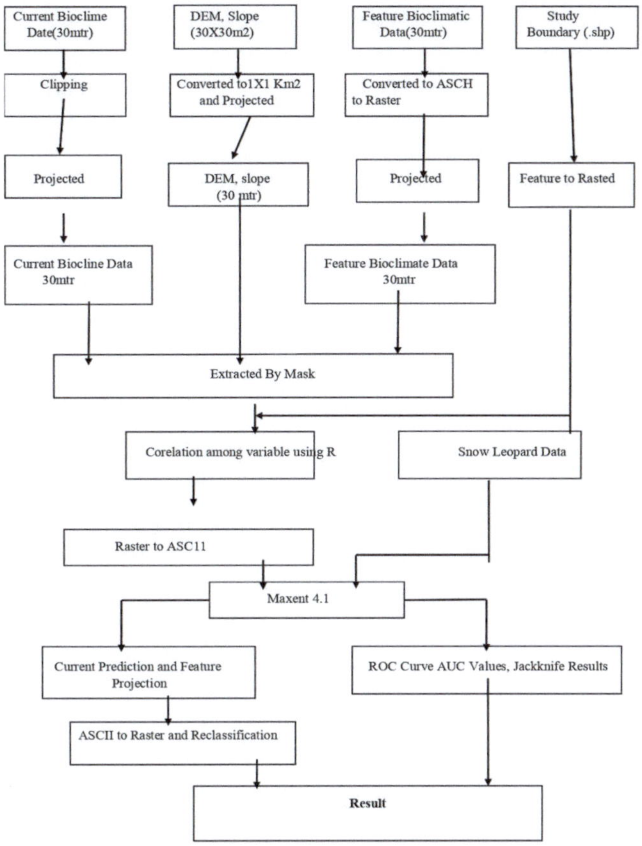

Fig. 7 Flowchart of methodology by MaxEnt

accuracy of maximum entropy model outputs due to its wider application in modelling studies (Boubli and de Lima 2009; Lobo et al. 2008; Vanderwal et al. 2009; Yates et al. 2010). The sensitivity values, the correct positive percentage, and the false-positive fraction for all accessible probability thresholds can be plotted on a ROC chart. A good model is one that maximises sensitivity against small false-positive fraction values, and this is measured by computing the area under the curve (AUC) that is (0.991). A good model is one with an AUC statistic that is closer to 1.0, which indicates complete agreement between the model and test data. AUC values closer to 0.5 are seen as being on par with randomness. The flowchart of MaxEnt methodology is shown in Fig. 7.

Using maximum entropy modelling to delineate the distribution patterns of snow leopard in the area threshold, the emission rate should be closer to the predicted omission. Here, the area under the curve is 0.991. This graph shows the receiver operating characteristics (ROC) for the data modelling. Specificity for habitat conformation can be defined by using predicted area rather than study area boundary (.shp) feature to raster average sensitivity versus 1-specificity for snow leopard.

The analysis of this graph reveals the omission rate and predicted area as a standard function of the cumulative threshold. The emission rate is calculated both on the training presence records and (if test data are used) on the test records. According to the definition of the cumulative threshold, the emission rate should be closer to the predicted omission.

The five environmental characteristics' impact on the distribution of the study area's target species is demonstrated using a jackknife graph analysis (Fig. 8). Because the snow leopard favours a habitat with good annual food availability reliant on the ecosystem 14, the forest type has a significant impact on the distribution pattern in this case. They decide to reside in a bio-class 2 and forest-class environment. Along with it, altitude and the human impact index have an impact on snow leopard distribution.

Unregularised training gain is 3.0873; regularised training gain is 3.4717; training AUC is 0.988. The testing gain that is irregular is 2.1266. The standard deviation is 0.003, and tests'

Fig. 8 Jackknife test of environmental variables for snow leopard in the NDBR

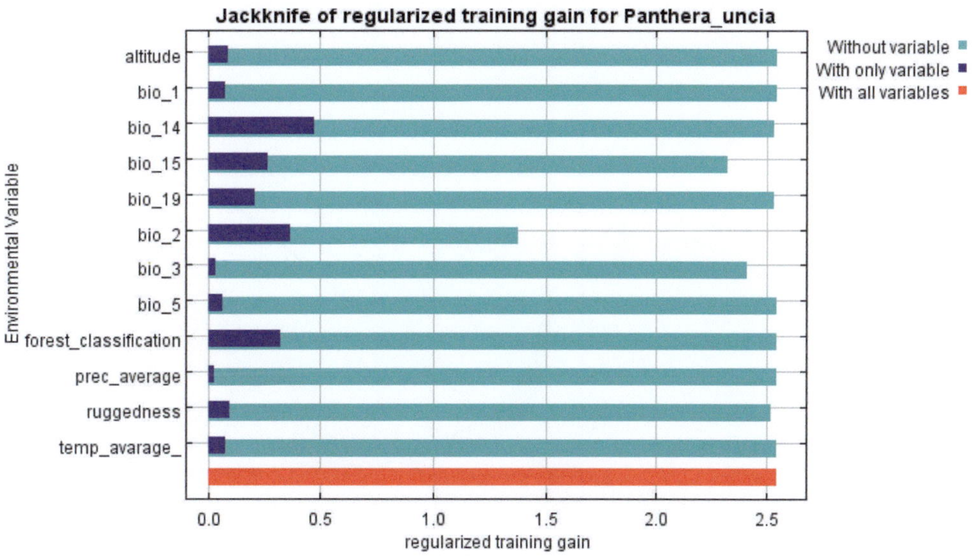

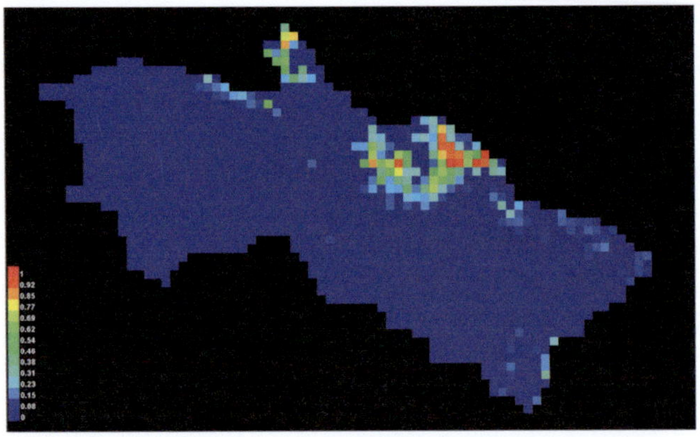

Fig. 9 Habitat suitability map of snow leopard in the study area

AUC is 0.991 (Fig. 9). After 500 iterations, the algorithm ended (Fig. 10).

The maximum entropy model has been successfully used to model the potential distributions of snow leopards in the study area. About 2.29% of the area is suitable for the snow leopard habitat. It has been found that the snow leopard is very intensively distributed in Malari, Niti area. The northern and eastern part has a non-habitable climate for the snow leopard—somewhat, the core region provides some suitable habitat for the snow leopard. Various environmental variables may be contributing to this spatial distribution of snow leopards. We also get that the environmental variable, i.e. bio 14, dramatically affects the distribution pattern of snow leopards following bio 2 and forest class (Fig. 11a–l). Any particular change in major contributing environmental parameters can lead to significant fluctuation in habitat and distribution patterns.

3.4.6 Potential Distribution Area

The majority (89.2%) of snow leopard presence records correlated with regions where the chance of occurrence was greater than 0.50 (Fig. 12). In the higher elevation of the NDBR, the model suggested that snow leopards with a high adaptability threshold would be found in greater numbers. Elith et al. (2011) argued that the logistic probability assigned 0.50% would indicate that the 42,432 km^2 of land in the western Himalayas would be appropriate for snow leopards. While some landscapes in the western Himalayas lack protected status, the prediction includes 11,247 km^2 (or 26.5%) within protected areas. 22,162 km^2 were deemed to be of moderate appropriateness (0.30–0.50%), while 42,037 km^2 were deemed to be of low suitability (0.30%).

Based on existing presence records, this study projected the possible distribution of snow leopards in India's western Himalayas (collected during the present study and published sources). The output of this model can assist conservation plans and serve as a benchmark for the collection of presence and absence data on snow leopards. MaxEnt models only identify regions with similar environmental conditions to occurrence localities across the species distribution range (Pearson et al. 2007). Mean Diurnal Range (Mean of monthly [max temp–min temp]) plays an important role in predicting the distribution of snow leopards, the distribution of snow leopards in higher elevations ranged from 3000 to 6700 m, characterized by a narrow range of precipitation (100 mm) and maximum temperatures during summer (15 °C), characterised by low vegetation cover (15% vegetation cover, grassland, and shrub cover), and this model output and previous field surveys revealed this (Chundawat 1992; McCarthy and Chapron 2003; Mishra et al. 2003; Forrest et al. 2012).

Fig. 10 Average omission and predicted area for snow leopard

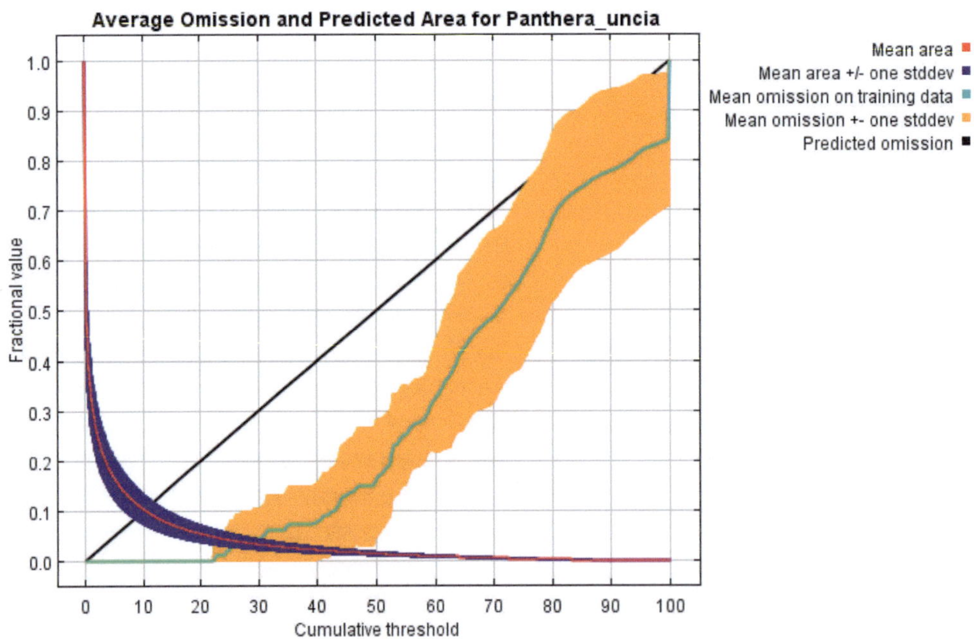

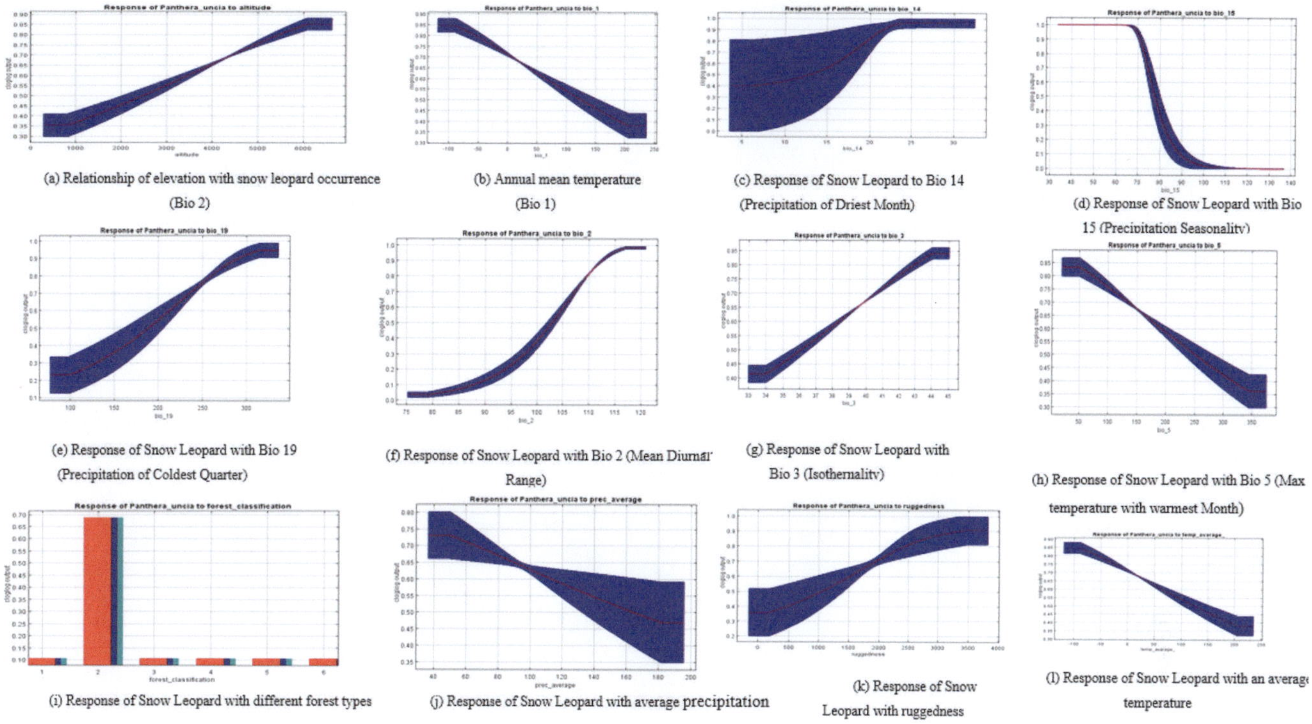

Fig. 11 a–l Graphical representation of MaxEnt model

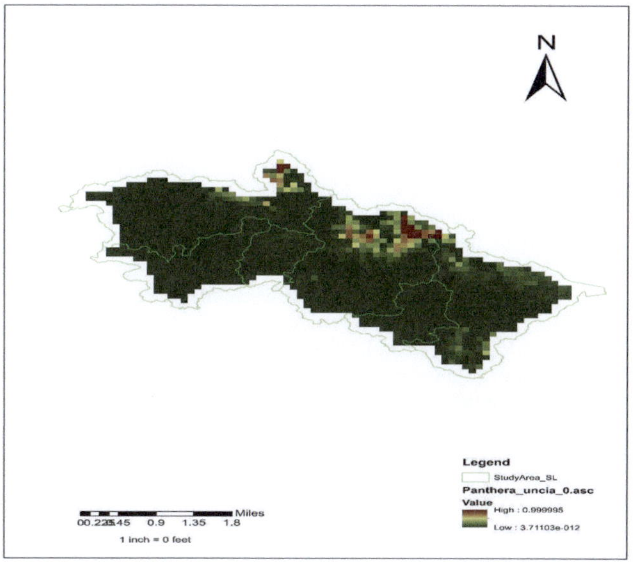

Fig. 12 Predicting potential distribution habitat for snow leopard in NDBR

The core zone (pariba) was assessed to have 717.33 km^2 of possible snow leopard protection; this model anticipated that the western Himalaya would have 106,631 km^2 of potential snow leopard habitat under high, moderate, and low appropriate habitat classes. Out of the entire potential snow leopard habitat anticipated in the western Himalayas, the model indicated that 39.8% of the land area (probability > 0.50%), which is located in Niti Valley, Malari, Gamsali, Pindari, Sunderdunga, and Kopkot, is extremely appropriate for snow leopards.

According to this model's predictions, the majority of the snow leopard's potential habitats (2.29%) are not protected areas, and roughly 80% of the species' population lives outside of PAs, which is consistent with earlier studies (Bhatnagar et al. 2001). Zonation of landscapes is the first step in creating regional-level conservation plans (i.e. identification of landscape at larger spatial scale units, which is > 1000 km^2) in order to advance the conservation initiative and make planning for snow leopards to occur beyond the PAs.

In the research region, only 74.2% is favourable for the presence of snow leopards, whereas 22.9% is somewhat suitable. Minor conservation units or core areas (National Park and Wildlife Sanctuary) with small buffer areas should be constructed inside these landscape units (Bhatnagar et al. 2001; Mishra et al. 2010) (Fig. 13).

4 Conclusion

The snow leopard remains one of the most mysterious and revered species in the high-altitude ecosystems of the Greater and Trans-Himalayas. Despite their elusive nature, recent research efforts have begun to shed light on the snow leopard's presence in the Indian Himalayas, particularly in

(a) (b)

Fig. 13 a Indirect evidence of snow leopard. b Camera trap images of snow leopard found in different area

Uttarakhand. The pioneering study by the WWF in 2010, although it did not find evidence of snow leopards in Bageshwar, has paved the way for subsequent research that has successfully captured photographic evidence of these majestic animals. The confirmation of snow leopards in the NDBR's buffer zone, along with sightings of other predators like Tibetan wolves and common leopards, underscores the rich biodiversity of the region. The presence of key prey species such as the Himalayan Tahr, Musk Deer, and Blue Sheep further supports the suitability of the habitat for snow leopards. This growing body of research is crucial for conservation efforts, providing valuable insights into the snow leopard's habits, movements, and the challenges they face in a rapidly changing environment.

References

Alexander JS, Gopalaswamy AM, Shi K, Hughes J, Riordan P. 2016. Patterns of snow leopard site use in an increasingly human-dominated landscape. PLoS One, 11(5): e0155309.

Anonymous, 2008. The Project Snow Leopard, Ministry of Environment and Forests, Government of India, New Delhi.

Anonymous. 1987. The Nanda Devi Biosphere Reserve. Indian National Man and Biosphere Reserve Committee. United Printing Press, New Delhi.

Aryal A, Brunton D, Ji WH, Karmacharya D, McCarthy T, Bencini R, Raubenheimer D. 2014. Multipronged strategy including genetic analysis for assessing conservation options for the snow leopard in the Central Himalaya. Journal of Mammalogy, 95(4): 871–881.

Bhatnagar Y. V. 1997. Ranging and Habitat Use by Himalayan Ibex (Capra ibex sibirica) in Pin Valley National Park.Ph.D. dissertation, Saurashtra University, Rajkot.

Bhatnagar, Y. V., Mathur, V. B., McCarthy, T. 2001 A regional perspective for snow leopard conservation in the Indian Trans-Himalaya. Presented at the National Workshop on Regional Planning for Wildlife Protected Areas, 6–8 Aug 2001, India Habitat Centre, New Delhi as part of GEF –India Eco-development Project Initiative, Wildlife Institute of India.

Bhatnagar, Y. V., Rana, B. S., Bhalla, K. K., Sharma, M. P., Rana, H. L., Singh, P. and Raghunath, R. 2008. Exploring the Pangi Himalaya, A preliminary Wildlife Survey in the Pangi region of Himachal Pradesh. Technical Report.

Bhatnagar, Y. V., Wangchuk, R. and Jackson, R. 1999 A survey of depredation and related wildlife-human conflicts in Hemis National Park, Ladakh, Jammu and Kashmir, India. Unpub. Rept., International Snow Leopard Trust. 20 pages.

Boubli, J.P. & De Lima, M.G., 2009: Modelling the Geographical Distribution and Fundamental Niches of Cacajao spp. and Chiropotes israelita in Northwestern Amazonia via a Maximum Entropy Algorithm. Int. J. Primatol. 30, 217–228.

Chundawat, R. S. 1992. The ecological studies of snow leopard and its associated prey species in Hemis National Park, Ladakh, India. PhD thesis, University of Rajasthan.

Devan, M.L. 1988. State of Himalayas - Ecology, Environment, Geography, Resource and Publications: A call for Action. Pp 183, In: Chadha, S.K.180 (ed) 1988. Himalayas: ecology and environment, Mittal Publications, Delhi, India.

Elith, J., Phillips, S. J., Hastie, T., Dudík, M., Chee, Y. E., & Yates, C. J. 2011. A statistical explanation of MaxEnt for ecologists. Diversity and distributions, 17(1), 43–57.

Forrest, J. L., Wikramanayake, E., Shrestha, R., Areendran, G., Gyeltshen, K., Maheshwari, A., & Thapa, K. 2012. Conservation and climate change: Assessing the vulnerability of snow leopard habitat to treeline shift in the Himalaya. Biological Conservation, 150(1), 129–135.

Fox, J. L., Sinha, S. P., Chundawat, R. S. and Das, P. K. 1988. A field survey of snow leopard presence and habitat use in northwestern India, in: Freeman, H. (Ed.) Proceedings of the fifth International Snow Leopard Symposium. International Snow Leopard Trust and Wildlife Inst. of India, pp. 99–112.

Gupta, S.K. 2002. Tourism and Heritage Resources in Garhwal Himalayas: An Approach to Planning and Management. Kaveri Books, New Delhi.

Hijmans, R. J., Cameron, S. E., Parra, J. L., Jones, P. G., and Jarvis, A. 2005. Very high resolution interpolated climate surfaces for global land areas. International Journal of Climatology, 25, 1965–1978.

Jaynes ET, Bretthorst GL. 2003. Probability Theory: the Logic Science. Cambridge, Unites Kingdom: Cambridge University Press.

Jaynes ET. 1957a. Information theory and statistical mechanics. Physical Review, 106(4): 620–630.

Jaynes ET. 1957b. Information theory and statistical mechanics. II. Physical Review, 108(2): 171–190.

Kumar, A. 2002. Nanda Devi Biosphere Reserve. In: Ramakrishna, P.S. (ed.), Traditional Leopard Panthera uncia in Tost Mountains, South Gobi, Mongolia. PLoS ONE 9(7): e101319. doi:https://doi.org/10.1371/journal.pone.0101319.

Li J, Schaller GB, McCarthy TM,Wang DJ, Jiagong ZL, Cai P, Basang LM, Lu Z. 2013. A communal sign post of snow leopards (Panthera uncia) and other species on the Tibetan Plateau, China. International Journal of Biodiversity, 2013: 370905.

Lobo, J.M., Jime'nez-Valverde, A. & Real, R., 2008: AUC: a misleading measure of the performance of predictive distribution models. Glob Ecol Biogeogr., 17,145–151.

Maheshwari, A., 2016. Conservation and management of Snow leopard and co-predators with special reference of large carnivore– human conflicts in the select areas of western Himalayas, PhD Thesis, Saurashtra University.

Maheshwari, A. and Sharma, D. 2010 Snow leopard conservation in Uttarakhand and Himachal Pradesh. WWF India, New Delhi.

Mani, A., Bansal, D., Kumari, M., & Kumar, D. (2023). Land Use Land Cover Changes and Climate Change Impact on the Water Resources: A Study of Uttarakhand State. In River Conservation and Water Resource Management (pp. 1–16). Singapore: Springer Nature Singapore.

Mani, A., & Kumar, D. (2020). Morphometric Analysis of Manali Watershed of Beas River Basin for Watershed Management. VayuMandal, 46, 21–29.

Mani, A., Kumar, D., Kumar, D. (2021). Evaluating Decadal Change in Green Cover of Dehradun City. In: Sharma, P. (eds) Geospatial Technology and Smart Cities. The Urban Book Series. Springer, Cham. https://doi.org/10.1007/978-3-030-71945-6_26

Maurya, V., Lal, S.B., Dhakate, P.M. 2017. "First picture of snow leopard in Kumaon Himalayas India, CAT news 66" Autumn 2017, IUCN, Cat Specialist Group.

Maurya, V., Meena, D. S., Singh, P., Dhakate, P. M., & Lal, S. B. 2016. First Photographic (Camera Trap) Evidence of Tibetan Wolf Canis lupus chanco in the Uttarakhand Himalaya, India. Journal of the Bombay Natural History Society (JBNHS), 113, 18–20.

McCarthy, T., & Chapron, G. (Eds.). 2003. Snow leopard survival strategy (Vol. 105). Seattle, Washington: International Snow Leopard Trust.

McCarthy, T., Mallon, D., Sanderson, E.W., Zahler, P., and Fisher, K., 2016. What is a snow leopard? Biogeography and status overview, in Snow Leopards Biodiversity of the World: Conservation from Genes to Landscapes, McCarthy, T.M. and Mallon, D.P., Eds., New York: Elsevier, p. 23. e42.

Mishra C, Bagchi S, Namgail T, Bhatnagar YV 2010. Multiple Use of Trans- Himalayan Rangelands: Reconciling Human Livelihoods with Wildlife Conservation.Pages 291–311 in J. T. Du Toit, R. Kock, and J. C. Deutsch, editors. Wild Rangelands: conserving wildlife while maintaining livestock in semi-arid ecosystems.Ist Edition. Blackwell Publishing.

Mishra, C., Allen, P., McCarthy, T. O. M., Madhusudan, M. D., Bayarjargal, A., & Prins, H. H. (2003). The role of incentive programs in conserving the snow leopard. Conservation Biology, 17(6), 1512–1520.

Negi, A.S. 2002. Nanda Devi Biosphere Reserve: Annual Plan Operation- 2001-02. Wildlife Preservation Organization, Uttaranchal.

Pearson, R., Raxworthy, C., Nakamura, M. and Peterson, A. 2007. Predicting species distributions from small numbers of occurrence records: a test case using cryptic geckos in Madagascar. Journal of Biogeography, 34:102–117.

Peterson, A.T. & Robins, C.R., 2003: Using ecological niche Modeling to predict Barred Owl invasions with implications for Spotted Owl conservation. Cons. Bio., 17, 1161–1165.

Phillips, S. J. and Dudík, M. 2008. Modeling of species distributions with Maxent: new extensions and a comprehensive evaluation. Ecography, 31: 161-175

Phillips, S. J., Anderson, R. P. and Schapire, R. E. 2006. Maximum entropy modeling of species geographic distributions. Ecological Modelling, 190: 231–259.

Purohit, A., Maikhuri, R. K., Rao, K. S., & Nautiyal, S. 2001. Impact of bark removal on survival of Taxus baccata L.(Himalayan yew) in Nanda Devi biosphere reserve, Garhwal Himalaya, India. Current Science, 586–590.

Rawat, G.S. 2005. Alpine meadows of Uttarakhand: ecology, land use and status of medicinal and aromatic plants. Bishen Singh Mahendra Pal Singh, Dehradun, 219 pp.

Saberwal, V. K. 1996. Pastoral politics: Gaddi grazing, degradation and biodiversity conservation in Himachal Pradesh, India. Conservation Biology, 10: 741–749.

Sharma, R. K. 2009. Can livestock benefit an endangered carnivore? Snow leopard population estimation along a density gradient of wild and domestic ungulates. Final report submitted to Snow Leopard Conservation Grants Program.

Sharma RK, Bhatnagar YV, Mishra C. 2015. Does livestock benefit or harm snow leopards? Biological Conservation, 190: 8–13.

Sharma, K., Bayrakcismith, R., Tumursukh, L., Johansson, O., Sevger, P., McCarthy, T., & Mishra, C. 2014. Vigorous dynamics underlie a stable population of the endangered snow leopard Panthera uncia in Tost Mountains, South Gobi, Mongolia. PloS one, 9(7), e101319.

Stockwell, D. R., & Peterson, A. T. 2002. Effects of sample size on accuracy of species distribution models. Ecological modeling, 148(1), 1–13.

Vanderwal, J., Shoo, L.P., Graham, C. & Williams, S.E., 2009: Selecting pseudo-absence data for presence-only distribution modeling: How far should you stray from what you know? Ecol. Model., 220, 589–594.

Vinod, T. R. and Sathyakumar, S. 1999. Ecology and conservation of mountain ungulates in Great Himalayan National Park, Western Himalaya. In: An ecological study of the conservation of biodiversity and biotic pressures in the Great Himalayan National Park Conservation Area-An ecodevelopment approach. Forestry Research Education and Extension Project –Great Himalayan National (FREE-GHNP), Final Project Report, Wildlife Institute of India, Dehradun.

Wadia, D. N. 1966.Geology of India. 4th Edition. Tata Megraw-Hill Publishing Company, New Delhi.

Wolf M, Ale S. 2009. Signs at the top: habitat features influencing snow leopard Uncia uncia activity in Sagarmatha National Park, Nepal. Journal of Mammalogy, 90(3): 604–611.

Xing DL, Hao ZQ. 2011. The principle of maximum entropy and its applications in ecology. Biodiversity Science, 19(3): 295–302. (in Chinese).

Xu WH, Luo C. 2010. Application of MAXENT model in Rhinopithecus roxllanae habitat assessment in Qinling Mountain. Forest Engineering, 26(2): 1–3, 26. (in Chinese).

Yates, C.J., Mcneill, A., Elith, J. & Midgley, G.F., 2010: Assessing the impacts of climate change and land transformation on Banksia in the South West Australian Floristic Region. Diversity Distrib., 16, 187–201.

A Comprehensive Review of Brick Kilns Mapping with Artificial Intelligence, Deep Learning and GIS Methods for Sustainable Development

Yamini Agrawal, Hina Pande, Poonam Seth Tiwari, and Shefali Agrawal

1 Introduction

Brick kilns have long been an integral part of urban and industrial sector, serving as crucial centers for brick production to meet the world's soaring construction demands. The process of brick making has evolved over time, with traditional techniques blending with modern technology to meet the growing demands of urbanization and infrastructure development.

In the developing world, the brick industry is a large, unorganized, unregulated, and polluting industry. As per the current scenario, globally more than 1500 billion bricks are manufactured each year, with 1300 billion bricks (87%) of these are produced in Asia (About the Global Brick Industry | Habla Zig Zag Kilns, n.d.). Annually, brick kilns worldwide consume 375 million tons of coal. Brick kilns utilize a range of fuels for firing bricks which includes coal, wood, local biomass, waste oil, used tires, sawdust, and even cow dung. The industry emits significant amounts of greenhouse gases (GHGs), including CO_2 emissions and short-lived climate pollutants (SLCPs) like black carbon (soot) due to fuel inefficiency, incomplete fuel combustion, outdated kiln technologies, and scavenged fuel usage. Brick kilns are the primary stationary emitter of atmospheric black carbon in South Asia. They account for 20% of the average black carbon (BC) production worldwide (Climate and Clean Air Coalition 2015; Environmental Impact of Brick Kilns | Habla Zig Zag Kilns n.d.; Maheshwari and Jain 2017). Brick kilns are also a substantial source of greenhouse gas emissions and carbon emissions, which influence climate change and global warming (Climate and Clean Air Coalition 2015; Guttikunda and Calori 2013).

Brick kiln operations leads to other environmental concerns as well. Excavating soil for clay leads to soil degradation, reduced water-holding capacity, and erosion, affecting agricultural productivity and contributing to desertification (Saha et al. 2021b). Additionally, substantial water usage in clay preparation, cooling, and dust suppression depletes local water resources. Deforestation results from the high demand for wood and biomass as fuel for firing bricks. Moreover, emissions, including black carbon, from numerous kilns have transboundary effects, such as the melting of glaciers in the Himalayas and alterations in rainfall patterns in the Indian subcontinent (Glaciers of the Himalayas World Bank Report, 2021). Such environmental consequences emphasize the urgency of adopting sustainable practices in the brick kiln industry.

Brick kilns have been characterized as a source for modern-day slavery by United Nations Sustainable Development Goals (UN SDG), in which SDG 8.7 (Goal 8 | Department of Economic and Social Affairs, n.d.) specifically targets the eradication of forced labor, modern slavery, and child labor. This is particularly prevalent in some regions of South Asia, where brick kilns play a significant role in the construction and building industry. The workforce predominantly comprises impoverished individuals, many of whom are migratory laborers from marginalized communities. The health and well-being of kiln workers, their families, and surrounding communities are adversely impacted by respiratory and cardiovascular diseases caused by air pollution from

Y. Agrawal (✉) · H. Pande · S. Agrawal
Photogrammetry and Remote Sensing Department, Indian Institute of Remote Sensing, Indian Space Research Organisation, Dehradun, Uttarakhand, India
e-mail: yamini_a@ce.iitr.ac.in

H. Pande
e-mail: hina@iirs.gov.in

S. Agrawal
e-mail: shefali_a@iirs.gov.in

Y. Agrawal · P. S. Tiwari
Geoweb Services, Information Technology and Distance Learning Department, Indian Institute of Remote Sensing, Indian Space Research Organisation, Dehradun, Uttarakhand, India
e-mail: poonam@iirs.gov.in

D. Kumar et al. (eds.), *Data Analytics and Remote Sensing-Based Analysis for Sustainable Development*, Advances in Science, Technology & Innovation, https://doi.org/10.1007/978-3-031-98095-4_7

brick kilns, resulting in thousands of premature deaths annually. According to (Climate and Clean Air Coalition 2015), approximately 15 million people are employed in Indian brick kiln industry.

Brick kilns operate informally without proper regulation. Mapping brick kilns helps in identifying their locations and monitoring their activities. It also enables agencies to implement targeted measures in order to mitigate the adverse impact of brick kiln pollution effectively. Brick kilns are spatially distributed across regions, making their mapping a challenging task. Brick kiln mapping is crucial for various purposes, including environmental monitoring, urban planning, and labor welfare. A comprehensive review of brick kiln mapping methods is essential to assess the various techniques and technologies employed in accurately identifying and monitor these crucial industrial facilities. Over time, researchers have explored diverse approaches, ranging from traditional image processing methods to state-of-the-art deep learning algorithms. This review aims to highlight the strengths and limitations of different mapping methods, the advancements made in recent years, and the potential for integrating emerging technologies, such as remote sensing and geographic information systems (GIS), to achieve more precise and efficient brick kiln mapping. By critically analyzing existing methodologies, this review seeks to contribute to the development of optimal and sustainable solutions for accurately mapping brick kilns, aiding policymakers, and promoting best practices in the brick manufacturing industry. As far as it is known, this review is first of its kind which analyses various brick kiln mapping techniques in a comprehensive manner.

2 Brick Kiln Mapping Techniques

2.1 Visual Interpretation Techniques for Brick Kiln Mapping

The history of brick kiln mapping can be traced back to traditional methods that relied on manual surveys and mapping techniques. The field surveys involved recording the locations and characteristics of each kiln, often using basic tools like compasses and measuring tapes. The resulting maps were typically hand-drawn and showed the spatial distribution of brick kilns in a particular region (Hammond 1977).

The introduction of aerial photography and satellite imagery provided new opportunities for more efficient and comprehensive brick kiln mapping. In early 2000s, it was recognized that remote sensing data can help in assessing the extent of damage caused by brick kilns in the South East Asian Countries, which are abundant in brick kilns. The utility of earth observation (EO) data for mapping brick kilns was first explored by (Yadav 2003). In this study, SPOT satellite data coupled with field observations gave insights into the number of operational kilns and their environment impacts on soil degradation inside the JNU Campus, New Delhi. The shifting pattern of brick kiln and rice cultivation in the same area was studied by (Haack and Khatiwada 2007) in Kathmandu Valley, Nepal. However, visual inspection of kilns in satellite imagery is time-consuming and subject to interpretation biases.

Efforts for mapping brick kilns using remote sensing data picked pace after 2018 when brick kilns were characterized as source of modern-day slavery in South-East Asian countries (Boyd et al. 2018, 2021; Foody et al. 2019; Lee et al. 2021; Li et al. 2019).

2.2 Satellite Image Processing Techniques for Brick Kiln Mapping

Image processing algorithms have been employed to differentiate brick kilns from their surroundings by analyzing their spectral characteristics and texture information or using statistical approaches. (Boyd et al. 2018) estimated a total number of 55,387 kilns in the 1,551,997 km^2 of "Brick-Kiln Belt" of South Asia using sampling approaches. (Hassan et al. 2019) utilized Landsat and Google Earth Imagery to examine the spatial patterns of brick fields development in Dhaka, Bangladesh between 1990 and 2015 using quadrat analysis. Statistical and classical image processing approaches, however, are subject to expert knowledge and manual interpretation, making them time-consuming and less adaptable to variations in data. Thus, these methods have replaced by more sophisticated and automated methods, particularly machine learning and deep learning algorithms, which have demonstrated superior performance in brick kiln mapping tasks.

2.3 Deep Learning Techniques for Brick Kiln Mapping

Machine learning algorithms are designed to handle complex data patterns and adaptability to handle diverse conditions can achieve higher accuracy in mapping brick kilns due to their ability to learn intricate patterns and relationships in the data. The algorithms can distinguish between subtle differences, leading to improved identification and delineation of brick kilns. With time-series data and continuous monitoring using machine learning and geographic information system, changes in brick kiln locations over time can be assessed. Temporal analysis helps understand the dynamics of the brick kiln industry and its expansion or contraction (Boyd et al. 2021; Li et al. 2019) developed a change detection method based on time-series Landsat imagery to monitor brick kilns' changing dynamics over time. Random forest regression analysis was utilized to identify the break points in the Landsat short-infrared bands over a time period of 35 years, where the first break corresponded to date of construction and second break corresponded to date of abandonment of the kiln. The difference between the second and first break gave the average time of the operation of the active kiln. Many studies have employed deep learning models to delineate brick kilns from surroundings. Deep learning methods use distinctive features of brick kilns, such as kiln structures, clay extraction pits, and stockpiles of raw materials to identify them in satellite imagery. (Nazir et al. 2019) introduced Tiny-Inception-ResNet-v2, a novel deep neural network designed to identify brick kilns in the brick kiln belt of South Asia. In a subsequent study, (Nazir et al. 2020) presented Kiln-Net, a two-staged gated neural network architecture that achieved course-to-fine resolution identification of brick kilns. (Foody et al. 2019) employed the region-based convolutional neural network (CNN) named Faster R-CNN (Ren et al. 2017) to classify brick kilns in Rajasthan, India. (Misra et al. 2020) proposed an integrated approach of CNN-based object detection and machine learning-based pixel classification to identify brick kiln locations in medium resolution Sentinel-2 Imagery. In this study, VGG-16 (Simonyan and Zisserman 2015) model for object detection and random forest-based Normalized Difference Vegetation Index (NDVI) classification was used to correctly identify brick kilns in the vicinity of Delhi-NCR and showed the utility of mapping brick kilns in courser resolution dataset using deep learning techniques. (Boyd et al. 2021) identified 66,455 kilns in the "Brick Kiln Belt" using YOLOv3 object detection model (Redmon and Farhadi n.d.). (Lee et al. 2021) created a low-cost scalable framework for extraction of brick kilns in Bangladesh using deep learning techniques and examining the proximity of kilns to schools and health facilities. (Tahir et al. 2021) employed deep learning techniques for the automatic detection and localization of brick kilns in satellite images. In a comparative analysis conducted by (Paul et al. 2022), three neural networks, namely Inception ResNet V2, ResNet 101, and ResNet 152, were assessed for the detection of brick kilns in the Indo-Gangetic plains of West Bengal. (Nazir et al. 2023) proposed a two-stage detection technique of brick kilns which first utilizes spectral indies-based classification of brick kilns in Sentinel-2 Imagery and an object detector YOLOv3 (Redmon and Farhadi n.d.) on fused high-resolution images to filter out false predictions and improved localization. (Imaduddin et al. 2023) compared different machine learning classification algorithms by utilizing the spectral response of land use and land cover features to classify brick kilns using Sentinel-2 imagery in Delhi-NCR.

Although deep learning models are scalable and can be trained to perform brick kiln mapping automatically, thus reducing the need for manual intervention and saving time and labor, they require large volumes of labeled training data to achieve optimal accuracy. Also, deep learning models might struggle to generalize well to regions with dynamic environment, architectural styles, or kiln technologies, leading to potential misclassifications.

2.4 GIS Based Techniques for Brick Kiln Mapping

Deep learning integrated with GIS can optimize the process of mapping brick kilns, enhance accuracy, and facilitate comprehensive spatial analysis, enabling the integration of brick kiln locations with other geographic information. (Hassan et al. 2019) performed proximity and cluster analysis to analyze spatial distribution of brick kilns with various land use classes. Average nearest neighbor (Clark and Evans 1954) and Ripley's K function (Ripley 1977) techniques were utilized to analyze the clustering pattern of brick kilns. (DeWitt et al. 2021) carried out a visual inspection and geospatial analysis of high-resolution multispectral and hyperspectral imagery to quantify total number of kilns, type of kiln, number of people employed and production capacity of brick kilns in Kabul, Afghanistan. (Purakayastha 2021) conducted a geo-statistical analysis of brick kilns in Kulpi, a small part of West Bengal. (Boyd et al. 2021) also carried out a spatiotemporal analysis to observe the increase and shifting patterns of brick kilns along the "Brick-Kiln Belt." Nighttime light data, coupled with geospatial techniques like proximity analysis were used to establish relationship between the supply and demand of bricks. A spatial correlation between channel mobility of river Ganga in lower Gangetic Plains of West Bengal and brick kilns was carried out by (Ghosh and Biswas 2022) using GIS techniques which indicated a decrease in channel width and mobility with expansion of brick kilns.

3 Brick Kiln Environmental Assessment

The environmental impacts of brick kilns have emerged as a pressing concern in recent times. From air pollution to deforestation, soil degradation to water contamination, the operations of brick kilns have left an indelible mark on the delicate balance of our ecosystem (Parvez et al. 2023; Skinder et al. 2014a, b). The following section delves into the multifaceted effects of brick kilns on the environment. There are several accounts of literatures conducted for studying the effects of brick kilns on environment, but only relevant literature conducted after 2000 have been cited in the present context.

The impact of brick kilns on air quality is a significant concern due to the emission of harmful air pollutants (Akinshipe and Kornelius 2018; Beard et al. 2022; Bhanarkar et al. 2002; Khan et al. 2019; Nasir et al. 2021; Öztaş and Yüksek 2022; Pangtey et al. 2004; Pervaiz et al. 2021; Skinder et al. 2014a, b; Suresh et al. 2016; Ullah et al. 2022; Weyant et al. 2014). These pollutants contribute to air pollution and have detrimental effects on both human health and the environment (Beard et al. 2022; Chen et al. 2017; Joshi and Dudani 2008; Khan et al. 2019; Motalib et al. 2015; Raza et al. 2021a, b; Thomas et al. 2014; Ullah et al. 2022). Firstly, the combustion process in brick kilns releases greenhouse gases like carbon dioxide (CO_2) (Motalib et al. 2015). These gases contribute to global warming and climate change. Brick kilns also release sulfur dioxide (SO_2) and nitrogen oxides (NO_2) (Bhanarkar et al. 2002; Skinder et al. 2014a, b) which cause acid rain when they react with atmospheric moisture. Acid rain has severe impact on soil quality, water bodies, and vegetation. Brick kilns also emit significant amounts of PM2.5 during the brick-making process, especially when using traditional, less efficient technologies (Arif et al. 2018; Guttikunda et al. 2013; Hamid et al. 2022). Inhalation of PM2.5 can lead to a range of health issues (Joshi and Dudani 2008; Sanjel et al. 2016, 2017). The incomplete combustion in brick kilns also leads to the release of CO into the atmosphere (Weyant et al. 2014). Brick kilns also release black carbon, a type of particulate matter that contributes to the melting of glaciers and snow surfaces (Arif et al. 2018). Apart from gaseous emissions, brick kilns produce significant amounts of dust and ash. These particles are carried by the wind over long distances, causing pollution in surrounding areas and affecting air quality. These harmful emissions lead to rising sea levels and other environmental disruptions. Traditional brick kilns, like fixed chimney kilns, have lower combustion efficiency and lack proper emission controls (Aniyikaiye et al. 2021). Transitioning from traditional kilns to more modern and energy-efficient technologies can significantly reduce air pollution (Bashir et al. 2023; Bruce et al. 2007; Mahboob et al. 2021; Pervaiz et al. 2022; Schmidt 2013; Seay et al. 2021).

Apart from harmful emissions and degradation of air quality, brick kilns have significant impacts on land, soil, vegetation, water, and forests, leading to environmental degradation and ecological imbalances.

Brick kilns require vast areas of land for their operations, leading to land use changes and potential deforestation or habitat destruction. The extraction of clay for brick production results in the removal of topsoil, which is crucial for supporting vegetation and agriculture (Hossain et al. 2019; Lone and Afroza 2008; Nath et al. 2018; Pangtey et al. 2004; Singh and Asgher 2005; Skinder et al. 2015). The various gaseous emissions from brick kilns affect the crop growth and lead to lower productivity of plants (Achakzai et al. 2017; Adrees et al. 2016; Ahmad et al. 2012; Nath et al. 2018).

The extraction of clay for brick-making destroys the soil structure, leading to soil erosion and reduced soil fertility (Chowdhury and Rasid 2020, 2021; Hamid et al. 2022; Saha et al. 2021a). The land surrounding brick kilns become compacted, making it difficult for vegetation to grow. Also, the harmful brick kiln emissions, such as particulate matter, contaminate the soil, and hinder its ability to support plant growth (Hamid et al. 2022; Haque et al. 2022; Kumar et al. 2021; Saha et al. 2021a).

Brick kilns consume huge amounts of water in clay preparation and cooling processes. This leads to water scarcity in areas where water resources are already limited. Moreover, the runoff from brick kilns, containing pollutants such as heavy metals and suspended solids, contaminate the nearby water bodies, affecting aquatic life and water quality (Ghosh and Biswas 2022).

The demand for fuelwood in brick kilns, especially in areas where other fuels are scarce or expensive, leads to deforestation. The extraction of wood for firing bricks leads to the loss of forest cover and disrupt wildlife habitats, affecting biodiversity and ecological balance (Komal et al. 2022).

Another aspect of concern is the treatment of working animals like donkeys, bullocks, and horses, which are used to transport heavy loads of raw materials, bricks, or other construction materials within the brick kiln premises. These animals are subjected to strenuous and often dangerous tasks without proper rest, nutrition, or medical care (Bukhari et al. 2021; Farhat et al. 2020; Rodrigues et al. 2020; Watson et al. 2020).

4 Impact of Brick Kilns on Livelihood of Labor

The impact of brick kilns extends beyond their environmental consequences but also affects the lives of the workers involved. In many regions across the world, brick kilns remain a source for exploitation, where vulnerable individuals are trapped in cycles of bonded labor, children are forced to work as labors (Larmar et al. 2017; Nasir et al. 2021; Pellenq et al. 2022; Roshania et al. 2022; Sinaga et al. 2022), facing harsh working conditions and meager wages (Gahlot et al. 2020; Larmar et al. 2017; Nazir 2018; Roshania et al. 2023). Brick kilns emit various harmful pollutants and continuous exposure to these airborne pollutants lead to respiratory problems, including coughing, wheezing, and exacerbation of preexisting respiratory conditions like asthma and bronchitis as well as skin diseases due to heat risks in the workers (Berumen-Rodríguez et al. 2021; Hamid et al. 2022; Jahan et al. 2016; Kownacki et al. 2018; Raza et al. 2021a, b; Raza et al. 2021a, b; Rupakheti et al. 2018; Sanjel et al. 2016 2017; Sinaga et al. 2022; Thomas et al. 2014; Thygerson et al. 2019). Not only respiratory illnesses but also various other health issues like human reproductivity are also affected due to inhalation of harmful gases (Jahan et al. 2016). Access to essential services like healthcare, education, and electricity are practically nonexistent for brick kiln workers living in remote areas (Baishya et al. 2019; Pawar 2022; Pellenq et al. 2022; Roshania et al. 2023; Siddaiah et al. 2018). It is crucial to address these health concerns and improve the working conditions of brick kiln workers, including fair wages (Kubasiewicz et al. 2022). Governments, nongovernmental organizations, and brick kiln owners need to work collaboratively to ensure that workers' health and rights are protected and upheld.

5 Conclusion

In this chapter, a comprehensive review of mapping techniques for brick kilns and their environmental aspects has been done. While brick kilns continue to be vital for construction industries, they also face challenges related to environmental sustainability, labor practices, and social welfare. This review highlights the crucial role of technology in understanding and addressing the complex challenges posed by these kilns. The integration of advanced tools such as aerial remote sensing, satellite imagery, machine learning and deep learning, coupled with GIS techniques enables comprehensive and real-time data collection, facilitating accurate mapping of brick kilns. As the world seeks to strike a balance between economic growth and ecological responsibility, understanding the significance and impact of brick kilns becomes crucial for promoting sustainable development and ensuring a resilient built environment for future generations. Adopting sustainable brick-making technologies, promoting efficient resource use, and implementing proper environmental regulations are vital steps to mitigate the negative impacts of brick kilns on land, soil, water, and forests. Adopting modern kiln designs, such as zigzag firing or vertical shaft brick kilns, which are more energy-efficient and produce fewer emissions will promote environmental sustainability.

References

About the Global Brick Industry | Habla Zig Zag Kilns. (n.d.). Retrieved January 4, 2023, from https://hablakilns.com/the-brick-industry/the-brick-market/

Achakzai, K., Khalid, S., Adrees, M., Bibi, A., Ali, S., Nawaz, R., & Rizwan, M. (2017). Air pollution tolerance index of plants around brick kilns in Rawalpindi, Pakistan. *Journal of Environmental Management*. https://doi.org/10.1016/j.jenvman.2016.12.072

Adrees, M., Ibrahim, M., Ibrahim, M., Ibrahim, M., Shah, A. M., Abbas, F., Saleem, F., Saleem, F., Rizwan, M., Hina, S., Jabeen, F., & Ali, S. (2016). Gaseous pollutants from brick kiln industry decreased the growth, photosynthesis, and yield of wheat (Triticum aestivum L.). *Environmental Monitoring and Assessment*. https://doi.org/10.1007/s10661-016-5273-8

Ahmad, M., Ahmad, M., Ahmad, M., Ahmad, M., Ahmad, M., van den Berg, L. J. L., Shah, H. U., Masood, T., Büker, P., Emberson, L., & Ashmore, M. (2012). Hydrogen fluoride damage to vegetation from peri-urban brick kilns in Asia: a growing but unrecognised problem? *Environmental Pollution*. https://doi.org/10.1016/j.envpol.2011.11.017

Akinshipe, O., & Kornelius, G. (2018). Quantification of atmospheric emissions and energy metrics from simulated clamp kiln technology in the clay brick industry. *Environmental Pollution*. https://doi.org/10.1016/j.envpol.2018.01.074

Aniyikaiye, T. E., Edokpayi, J. N., Odiyo, J. O., Piketh, S., & Piketh, S. J. (2021). Traditional Brick Making, Environmental and Socio-Economic Impacts: A Case Study of Vhembe District, South Africa. *Sustainability*. https://doi.org/10.3390/su131910659

Arif, M., Arif, M., Arif, M., Kumar, R., Kumar, R., Kumar, R., Eric, Z., Zusman, E., & Gourav, P. (2018). Ambient black carbon, PM2.5 and PM10 at Patna: Influence of anthropogenic emissions and brick kilns. *Science of The Total Environment*. https://doi.org/10.1016/j.scitotenv.2017.12.227

Baishya, B., Satpathy, A., Nayak, R., & Mohanty, R. (2019). Oral hygiene status, oral hygiene practices and periodontal health of brick kiln workers of Odisha. *Journal of Indian Society of Periodontology*. https://doi.org/10.4103/jisp.jisp_383_18

Bashir, Z., Amjad, M., Raza, S. F., Ahmad, S., Abdollahian, M., & Farooq, M. (2023). Investigating the Impact of Shifting the Brick Kiln Industry from Conventional to Zigzag Technology for a Sustainable Environment. *Sustainability*. https://doi.org/10.3390/su15108291

Beard, J. D., Thygerson, S. M., Olivares, A., Tadje, J. E., Willis, S., & Johnston, J. D. (2022). Gaseous Air Pollutants and Respirable Crystalline Silica Inside and Outside Homes at Brick Kilns in Bhaktapur, Kathmandu Valley, Nepal. *International Journal of Environmental Research and Public Health*. https://doi.org/10.3390/ijerph191912431

Berumen-Rodríguez, A. A., León-Martínez, L. D. de, Zamora-Mendoza, B. N., Orta-Arellanos, H., Saldaña-Villanueva, K., Barrera-López, V., Gómez-Gómez, A., Pérez-Vázquez, F. J., Díaz-Barriga, F., & Flores-Ramírez, R. (2021). Evaluation of respiratory function and biomarkers of exposure to mixtures of pollutants in brick-kilns workers from a marginalized urban area in Mexico. *Environmental Science and Pollution Research*. https://doi.org/10.1007/s11356-021-15375-3

Bhanarkar, A. D., Gajghate, D. G., & Hasan, M. Z. (2002). Assessment of Air Pollution from Small Scale Industry. *Environmental Monitoring and Assessment*. https://doi.org/10.1023/a:1020636930033

Boyd, D. S., Jackson, B., Wardlaw, J., Foody, G. M., Marsh, S., Bales, K., & Bales, K. (2018). Slavery from Space: Demonstrating the role for satellite remote sensing to inform evidence-based action related to UN SDG number 8. *Isprs Journal of Photogrammetry and Remote Sensing*. https://doi.org/10.1016/j.isprsjprs.2018.02.012

Boyd, D. S., Perrat, B., Li, X., Jackson, B., Landman, T., Ling, F., Bales, K., Choi-Fitzpatrick, A., Goulding, J., Marsh, S., & Foody, G. M. (2021). Informing action for United Nations SDG target 8.7 and interdependent SDGs: Examining modern slavery from space. *Humanities and Social Sciences Communications*, *8*(1). https://doi.org/10.1057/s41599-021-00792-z

Bruce, C. W., Corral, A. Y., & Lara, A. (2007). Development of Cleaner-Burning Brick Kilns in Ciudad Juarez, Chihuahua, Mexico. *Journal of The Air & Waste Management Association*. https://doi.org/10.3155/1047-3289.57.4.444

Bukhari, S., McElligott, A., McElligott, A. G., Parkes, R. S. V, & Parkes, R. (2021). Quantifying the Impact of Mounted Load Carrying on Equids: A Review. *Animal*. https://doi.org/10.3390/ani11051333

Chen, Y., Chen, Y., Du, W., Zhuo, S., Liu, W., Liu, Y., Shen, G., Wu, S.-P., Li, J., Li, J., Zhou, B., Wang, G., Zeng, E. Y., Zeng, E. Y., Cheng, H.,

Cheng, H., Liu, W., & Tao, S. (2017). Stack and fugitive emissions of major air pollutants from typical brick kilns in China. *Environmental Pollution*. https://doi.org/10.1016/j.envpol.2017.02.022

Chowdhury, N., & Rasid, Md. M. (2020). Assessment of Soil Fertility and Crop Nutrient Status in Agricultural Soils Near a Brick Kiln Cluster. *The Journal of Agricultural Science*. https://doi.org/10.5539/jas.v13n1p122

Chowdhury, N., & Rasid, Md. M. (2021). Evaluation of brick kiln operation impact on soil microbial biomass and enzyme activity. *Soil Science*. https://doi.org/10.37501/soilsa/132232

Clark, P. J., & Evans, F. C. (1954). Distance to Nearest Neighbor as a Measure of Spatial Relationships in Populations. *Ecology*, *35*(4), 445–453. https://doi.org/10.2307/1931034

Climate and Clean Air Coalition. (2015). *Mitigating Black Carbon and Other Pollutants From Brick Production*. 3.

DeWitt, J. D., Chirico, P. G., Alessi, M. A., & Boston, K. M. (2021). Remote Sensing Inventory and Geospatial Analysis of Brick Kilns and Clay Quarrying in Kabul, Afghanistan. *Minerals*. https://doi.org/10.3390/min11030296

Environmental Impact of Brick Kilns | Habla Zig Zag Kilns. (n.d.). Retrieved March 29, 2023, from https://hablakilns.com/environment/environmental-impact/

Farhat, S. F., McLean, A., Mahmoud, H. F. F., & Mahmoud, H. F. F. (2020). Welfare Assessment and Identification of the Associated Risk Factors Compromising the Welfare of Working Donkeys (Equus asinus) in Egyptian Brick Kilns. *Animal*. https://doi.org/10.3390/ani10091611

Foody, G. M., Ling, F., Boyd, D. S., Li, X., Li, X., Li, X., & Wardlaw, J. (2019). Earth Observation and Machine Learning to Meet Sustainable Development Goal 8.7: Mapping Sites Associated with Slavery from Space. *Remote Sensing*. https://doi.org/10.3390/rs11030266

Gahlot, N., Rana, K., & Singh, K. (2020). Workplace environment assessment of brick kiln units. *The Pharma Innovation Journal*. null

Ghosh, S., & Biswas, S. (2022). Spatiotemporal expansion of brick kilns along the downstream course of Ichamati River and its impact on the channel mobility in the lower Gangetic plain of West Bengal, India. *Arabian Journal of Geosciences*. https://doi.org/10.1007/s12517-022-10177-4

Glaciers of the Himalayas World Bank Report. (2021). Retrieved June 15, 2023, from https://www.worldbank.org/en/news/press-release/2021/06/01/to-slow-himalayan-glacier-melt-curbing-air-pollution-is-key

Goal 8 | Department of Economic and Social Affairs. (n.d.). Retrieved July 30, 2023, from https://sdgs.un.org/goals/goal8

Guttikunda, S. K., Begum, B. A., Winebrake, J. J., & Wadud, Z. (2013). Particulate pollution from brick kiln clusters in the Greater Dhaka region, Bangladesh. *Air Quality, Atmosphere & Health*. https://doi.org/10.1007/s11869-012-0187-2

Guttikunda, S. K., & Calori, G. (2013). A GIS based emissions inventory at 1 km × 1 km spatial resolution for air pollution analysis in Delhi, India. *Atmospheric Environment*, *67*, 101–111. https://doi.org/10.1016/j.atmosenv.2012.10.040

Haack, B., & Khatiwada, G. (2007). Rice and Bricks: Environmental Issues and Mapping of the Unusual Crop Rotation Pattern in the Kathmandu Valley, Nepal. *Environmental Management*. https://doi.org/10.1007/s00267-006-0167-0

Hamid, A., Riaz, A., Noor, F., & Mazhar, I. (2022). Assessment and mapping of total suspended particulate and soil quality around brick kilns and occupational health issues among brick kilns workers in Pakistan. *Environmental Science and Pollution Research*. https://doi.org/10.1007/s11356-022-22428-8

Hammond, M. D. P. (1977). Brick Kilns: An Illustrated Survey. *10. 1179/Iar.1977.1.2.171*, 1(2), 171–192. https://doi.org/10.1179/IAR.1977.1.2.171

Haque, S. E., Shahriar, M. M., Nahar, N., & Haque, Md. S. (2022). Impact of brick kiln emissions on soil quality: A case study of Ashulia brick kiln cluster, Bangladesh. *Environmental Challenges*. https://doi.org/10.1016/j.envc.2022.100640

Hassan, M. M., Juhasz, L., & Southworth, J. (2019). Mapping Time-Space Brickfield Development Dynamics in Peri-Urban Area of Dhaka, Bangladesh. *ISPRS International Journal of Geo-Information*. https://doi.org/10.3390/ijgi8100447

Hossain, M. A., Hossain, M. A., Zahid, A. M., Arifunnahar, M., Siddique, M. N. A., & Siddique, M. N. A. (2019). Effect of brick kiln on arable land degradation, environmental pollution and consequences on livelihood of Bangladesh. *Journal of Science Technology and Environment Informatics*. https://doi.org/10.18801/jstei.060219.50

Imaduddin, S., Khan, Y. A., Mirza, K., & Bhadra, B. K. (2023). Detection of Brick Kilns Using Multi-Spectral Bands of Sentinel-2 Imagery. *Artificial Intelligence and Symbolic Computation*. https://doi.org/10.1109/aisc56616.2023.10085085

Jahan, S., Falah, S., Ullah, H., Ullah, A., Ullah, A., & Rauf, N. (2016). Antioxidant enzymes status and reproductive health of adult male workers exposed to brick kiln pollutants in Pakistan. *Environmental Science and Pollution Research*. https://doi.org/10.1007/s11356-016-6454-2

Joshi, S. K., I, D., & Dudani, I. (2008). Environmental health effects of brick kilns in Kathmandu valley. *Kathmandu University Medical Journal*. null

Khan, M. W., Ali, Y., Felice, F. De, Salman, A., Petrillo, A., & Petrillo, A. (2019). Impact of brick kilns industry on environment and human health in Pakistan. *Science of The Total Environment*. https://doi.org/10.1016/j.scitotenv.2019.04.369

Komal, N., uz Zaman, Q., Yasin, G., Nazir, S., Ashraf, K., Waqas, M., Ahmad, M., Batool, A., Talib, I., & Chen, Y. (2022). Carbon Storage Potential of Agroforestry System near Brick Kilns in Irrigated Agro-Ecosystem. *Agriculture*. https://doi.org/10.3390/agriculture12020295

Kownacki, K. L., Lundgren-Kownacki, K., Kjellberg, S. M., Gooch, P., Dabaieh, M., Anandh, L., & Venugopal, V. (2018). Climate change-induced heat risks for migrant populations working at brick kilns in India: a transdisciplinary approach. *International Journal of Biometeorology*. https://doi.org/10.1007/s00484-017-1476-0

Kubasiewicz, L. M., Watson, T., Norris, S. L., Chamberlain, N., Nye, C., Perumal, R. K., Saroja, R., Raw, Z., & Burden, F. A. (2022). One welfare: Linking poverty, equid ownership and equid welfare in the brick kilns of India – ERRATUM. *Animal Welfare*. https://doi.org/10.1017/awf.2023.14

Kumar, S., Zhao, M., Zhang, H., Rahman, Md. A., Rahman, A., Luo, C., & Rahman, M. M. (2021). Distribution, contamination status and source of trace elements in the soil around brick kilns. *Chemosphere*. https://doi.org/10.1016/j.chemosphere.2020.127882

Larmar, S. A., O'Leary, P., Chui, C. H.-K., Benfer, K. A., Zug, S., Zug, S., & Jordan, L. P. (2017). Hazardous child labor in Nepal: the case of brick kilns. *Child Abuse & Neglect*. https://doi.org/10.1016/j.chiabu.2017.08.011

Lee, J., Brooks, N. R., Tajwar, F., Burke, M., Ermon, S., Lobell, D. B., Biswas, D., & Luby, S. P. (2021). Scalable deep learning to identify brick kilns and aid regulatory capacity. *Proceedings of the National Academy of Sciences of the United States of America*, *118*(17). https://doi.org/10.1073/pnas.2018863118

Li, X., Li, X., Li, X., Foody, G. M., Boyd, D. S., & Ling, F. (2019). Aging brick kilns in the Asian brick belt using a long time series of Landsat sensor data to inform the study of modern day slavery. *IEEE International Geoscience and Remote Sensing Symposium*. https://doi.org/10.1109/igarss.2019.8898981

Lone, F. A., & Afroza, B. (2008). Impact assessment of brick kiln emissions on the foliage of some vegetable crops. *Journal of Phytological Research*.null

Mahboob, S., Hussain, B., Naqvi, S. A. R., Naqvi, S. A. A., Anwar, S., Shah, S. A. R., Shah, S. A. R., Shah, S. A. R., ul Hassan, R. H., & Shah, A. A. (2021). Zig-zag technology adoption behavior among brick kiln owners in Pakistan. *Environmental Science and Pollution Research*. https://doi.org/10.1007/s11356-021-13837-2

Maheshwari, H., & Jain, K. (2017). Carbon Footprint of Bricks Production in Fixed Chimney Bull's Trench Kilns in India. *Indian Journal of Science and Technology, 10*(16), 1–11. https://doi.org/10.17485/ijst/2017/v10i16/112396

Misra, P., Imasu, R., Hayashida, S., Arbain, A. A., Avtar, R., & Takeuchi, W. (2020). Mapping Brick Kilns to Support Environmental Impact Studies around Delhi Using Sentinel-2. *ISPRS International Journal of Geo-Information*. https://doi.org/10.3390/ijgi9090544

Motalib, M. A., Lasco, R. D., Lasco, R. D., Lasco, R. D., Pacardo, E. P., Rebancos, C. M., & Dizon, J. T. (2015). Health Impact Of Air Pollution On Dhaka City By Different Technologies Brick Kilns. *International Journal of Technology Enhancements and Emerging Engineering Research*. null

Nasir, M., Nasir, M., Nasir, M., Rehman, F. U., Kishwar, S., Bashir, S., & Adil, M. T. (2021). Air pollution and child health: the impact of brick kiln pollution on children's cognitive abilities and physical health in Pakistan. *Environment, Development and Sustainability*. https://doi.org/10.1007/s10668-021-01229-y

Nath, A. J., Lal, R., & Das, A. K. (2018). Fired Bricks: CO2 Emission and Food Insecurity. *Global Challenges (Hoboken, NJ)*. https://doi.org/10.1002/gch2.201700115

Nazir, K. I. (2018). Study of Workplace Environment and Occupational Hazards of Brick Kiln Industry and Its Impact on Female Workers. *Null*. null

Nazir, U., Khurshid, N., Bhimra, M. A., & Taj, M. (2019). Tiny-Inception-ResNet-v2: Using Deep Learning for Eliminating Bonded Labors of Brick Kilns in South Asia. *CVPR Workshops*. https://www.semanticscholar.org/paper/c79d960466516b1ec5ea436688a4c826a318bd0e

Nazir, U., Mian, U. K., Sohail, M. U., Taj, M., & Uppal, M. (2020). Kiln-Net: A Gated Neural Network for Detection of Brick Kilns in South Asia. *IEEE Journal of Selected Topics in Applied Earth Observations and Remote Sensing, 13*, 3251–3262. https://doi.org/10.1109/JSTARS.2020.3001980

Nazir, U., Taj, M., Uppal, M., & Khalid, S. (2023). Mitigating climate and health impact of small-scale kiln industry using multi-spectral classifier and deep learning. *ArXiv.Org*. https://doi.org/10.48550/arxiv.2303.11654

Öztaş, S. K., & Yüksek, İ. (2022). LIFE CYCLE ENVIRONMENTAL ANALYSIS OF BRICK PRODUCTION: TURKEY AS A CASE STUDY. *Journal of Green Building*. https://doi.org/10.3992/jgb.17.2.125

Pangtey, B. S., Kumar, S., Kumar, S., Kumar, S., Kumar, S., Bihari, V., Bihari, N, M., Mathur, N., Rastogi, S., Rastogi, S. K., Sk, R., Rastogi, S. K., Srivastava, A. K., Srivastava, A. K., Srivastava, A. K., & Srivastava, A. K. (2004). An environmental profile of brick kilns in Lucknow. *Journal of Environmental Science & Engineering*. null

Parvez, M. A., Rana, I. A., Nawaz, A., & Arshad, H. S. H. (2023). The impact of brick kilns on environment and society: a bibliometric and thematic review. *Environmental Science and Pollution Research International*. https://doi.org/10.1007/s11356-023-26011-7

Paul, A., Bandyopadhyay, S., & Raj, U. (2022). Brick kiln detection in remote sensing imagery using deep neural network and change analysis. *Spatial Information Research*. https://doi.org/10.1007/s41324-022-00458-1

Pawar, R. S. (2022). Socio-economic Condition of Brick Kiln workers in Ahmednagar District, Maharashtra. *International Journal of Health Sciences (IJHS)*. https://doi.org/10.53730/ijhs.v6ns1.5758

Pellenq, C., Lima, L., & Gunn, S. (2022). Education, Age and Gender: Critical Factors in Determining Interventions for Child Brick Workers in Pakistan and Afghanistan. *International Journal of Environmental Research and Public Health*. https://doi.org/10.3390/ijerph19116797

Pervaiz, S., Enviornemntal Protection Agency Punjab Lahore, P., Akram, M. A. N., Khan, F. Z., Javid, K., & Zahid, Y. (2021). Brick Sector and Air Quality: An Integrated Assessment towards 2020 Challenge of Environment Development. *Environment and Natural Resources Journal*. https://doi.org/10.32526/ennrj/19/2020203

PERVAIZ, S., KHAN, F. Z., JAVID, K., ALTAF, A., ASLAM, F., TAHIR, M. A. A., TALIB, B., ARIF, H., & HAYAT, S. (2022). DEVELOPMENT OF AIR QUALITY AND BRICK KILNS DURING THE ONSET OF COVID-19: AN ANALYSIS. *Biological & Clinical Sciences Research Journal*. https://doi.org/10.54112/bcsrj.v2022i1.122

Purakayastha, S. (2021). Geo-Spatial Distribution and Comparative Assessment of Brick Kiln Industry: Appraisal on Kulpi Block in Diamond Harbour Sub-Division, South 24 Parganas, West Bengal. *International Journal for Research in Applied Science and Engineering Technology*. https://doi.org/10.22214/ijraset.2021.38112

Raza, A., Raza, A., Ali, Z., & Ali, Z. (2021). Assessment of obstructive and restrictive patterns of lung function among the workers of brick kilns. *International Journal of Occupational Safety and Ergonomics*. https://doi.org/10.1080/10803548.2021.1992151

Raza, A., Raza, A., Raza, A., Ali, Z., & Ali, Z. (2021). Impact of Air Pollution Generated by Brick Kilns on the Pulmonary Health of Workers. *Journal of Health and Pollution*. https://doi.org/10.5696/2156-9614-11.31.210906

Redmon, J., & Farhadi, A. (n.d.). *YOLOv3: An Incremental Improvement*. Retrieved July 30, 2023, from https://pjreddie.com/yolo/.

Ren, S., He, K., Girshick, R., & Sun, J. (2017). Faster R-CNN: Towards Real-Time Object Detection with Region Proposal Networks. *IEEE Transactions on Pattern Analysis and Machine Intelligence, 39*(6), 1137–1149. https://doi.org/10.1109/TPAMI.2016.2577031

Ripley, B. D. (1977). Modelling Spatial Patterns. *Journal of the Royal Statistical Society: Series B (Methodological), 39*(2), 172–192. https://doi.org/10.1111/J.2517-6161.1977.TB01615.X

Rodrigues, J. B., Sullivan, R., Judge, A., Norris, S. L., & Burden, F. (2020). Quantifying poor working equid welfare in Nepalese brick kilns using a welfare assessment tool. *Veterinary Record*. https://doi.org/10.1136/vr.106135

Roshania, R. P., Giri, R., Cunningham, S. A., Young, M. F., Webb-Girard, A., Das, A., Mala, G. S., Srikantiah, S., Mahapatra, T., & Ramakrishnan, U. (2022). Early life migration and undernutrition among circular migrant children: An observational study in the brick kilns of Bihar, India. *Journal of Global Health*. https://doi.org/10.7189/jogh.12.04008

Roshania, R. P., Webb-Girard, A., Das, A., Giri, R., Mala, G., Srikantiah, S., Young, M., Mahapatra, T., & Ramakrishnan, U. (2023). Food Environments, Food Security, and Household Food Availability of Circular Migrant Families: A Mixed-Methods Study Among Brick Kiln Laborers in Bihar, India. *Food and Nutrition Bulletin*. https://doi.org/10.1177/03795721231152057

Rupakheti, D., Pradhan, P. M. S., & Basel, P. (2018). Occupational Safety and Health Vulnerability among Brick Factory Workers in Dhading District, Nepal. *Annals of Global Health*. https://doi.org/10.29024/aogh.2313

Saha, M. K., Sarkar, R. R., Ahmed, S. J., Sheikh, A. H., & Mostafa, G. (2021a). Impacts of brick kiln emission on agricultural soil around brick kiln areas. *Null*. https://doi.org/10.3126/njes.v9i1.34918

Saha, M. K., Sarkar, R. R., Ahmed, S. J., Sheikh, A. H., & Mostafa, M. G. (2021b). Impacts of brick kiln emission on agricultural soil around brick kiln areas. *Nepal Journal of Environmental Science, 9*(1), 1–10. https://doi.org/10.3126/njes.v9i1.34918

Sanjel, S., Khanal, S. N., Thygerson, S. M., Carter, W. S., Johnston, J. D., & Joshi, S. K. (2017). Respiratory symptoms and illnesses related to the concentration of airborne particulate matter among brick kiln workers in Kathmandu valley, Nepal. *Annals of Occupational and Environmental Medicine*. https://doi.org/10.1186/s40557-017-0165-0

Sanjel, S., Khanal, S. N., Thygerson, S. M., K, K., Khanal, K., Zd, P., S, T., & Joshi, S. K. (2016). Airborne Particulate Matter and Health Condition in Brick Kiln Workers in Kathmandu Valley, Nepal. *Kathmandu University Medical Journal*. null

Schmidt, C. W. (2013). Modernizing artisanal brick kilns: a global need. *Environmental Health Perspectives*. https://doi.org/10.1289/ehp.121-a242

Seay, B., Adetona, A. M., Sadoff, N., Sarofim, M. C., & Kolian, M. (2021). Impact of South Asian brick kiln emission mitigation strategies on select pollutants and near-term Arctic temperature responses. *Environmental Research Communications*. https://doi.org/10.1088/2515-7620/ac0a66

Siddaiah, A., Kant, S., Haldar, P., Rai, S. K., Rai, S. K., Rai, S. K., & Misra, P. (2018). Maternal health care access among migrant women labourers in the selected brick kilns of district Faridabad, Haryana: mixed method study on equity and access. *International Journal for Equity in Health*. https://doi.org/10.1186/s12939-018-0886-x

Simonyan, K., & Zisserman, A. (2015). *VERY DEEP CONVOLUTIONAL NETWORKS FOR LARGE-SCALE IMAGE RECOGNITION*. http://www.robots.ox.ac.uk/

Sinaga, H. T., Achmad, N., Manalu, M., Perangin-Angin, S., & Doloksaribu, L. G. (2022). Length Attainment and Stunting Prevalence of Children aged 0–24 Months Living in the Area of Clay Brick Kilns in Pagar Merbau Sub District, Indonesia. *Open Access Macedonian Journal of Medical Sciences*. https://doi.org/10.3889/oamjms.2022.7969

Singh, A. L., & Asgher, Md. S. (2005). Impact of brick kilns on land use/landcover changes around Aligarh city, India. *Habitat International*. https://doi.org/10.1016/j.habitatint.2004.04.010

Skinder, B. M., Pandit, A. K., P, A. K., Sheikh, A. Q., & Ganai, B. A. (2014). Brick kilns: Cause of Atmospheric Pollution. *Journal of Pollution Effects and Control*. https://doi.org/10.4172/2375-4397.1000112

Skinder, B. M., Sheikh, A. Q., Pandit, A. K., & Ganai, B. A. (2014). Brick kiln emissions and its environmental impact: A Review. *Journal of Ecology and the Natural Environment*. https://doi.org/10.5897/jene2013.0423

Skinder, B. M., Sheikh, A. Q., Pandit, A. K., Ganai, B. A., & Kuchy, A. H. (2015). Effect of brick kiln emissions on commonly used vegetables of Kashmir Valley. *Food Science and Nutrition*. https://doi.org/10.1002/fsn3.252

Suresh, R., Kumar, S., Mahtta, R., & Sharma, S. (2016). Emission Factors for Continuous Fixed Chimney Bull Trench Brick Kiln (FCBTK) in India. *Null*. null

Tahir, R., Imran, M. S., Minhas, S., Sabahat, N., Ilyas, S. H. W., & Gadi, H. R. (2021). Brick Kiln Detection and Localization using Deep Learning Techniques. *2021 International Conference on Artificial Intelligence, ICAI 2021*, 37–43. https://doi.org/10.1109/ICAI52203.2021.9445267

Thomas, B., Charles, N., Watson, B., Chandrasekaran, V., Kumar, R. S., Dhanalakshmi, A., Wares, F., & Swaminathan, S. (2014). Prevalence of chest symptoms amongst brick kiln migrant workers and care seeking behaviour: a study from South India. *Journal of Public Health*. https://doi.org/10.1093/pubmed/fdu104

Thygerson, S. M., Beard, J. D., House, M. J., Smith, R. L., Smith, R., Burbidge, H. C., Andrus, K. N., Weber, F. X., Weber, F. X., Chartier, R., & Johnston, J. D. (2019). Air-quality assessment of on-site brick-kiln worker housing in Bhaktapur, Nepal: Chemical speciation of indoor and outdoor PM2.5 pollution. *International Journal of Environmental Research and Public Health*. https://doi.org/10.3390/ijerph16214114

Ullah, M. S., Ullah, S., Javed, M. F., Ullah, R., Akbar, T. A., Ullah, W., Baig, S. A., Aziz, M., Mohamed, A., & Sajjad, R. U. (2022). Assessment and Impacts of Air Pollution from Brick Kilns on Public Health in Northern Pakistan. *Atmosphere*. https://doi.org/10.3390/atmos13081231

Watson, T. L., Kubasiewicz, L. M., Chamberlain, N., Nye, C., Raw, Z., & Burden, F. (2020). Cultural "Blind Spots," Social Influence and the Welfare of Working Donkeys in Brick Kilns in Northern India. *Frontiers in Veterinary Science*. https://doi.org/10.3389/fvets.2020.00214

Weyant, C. L., Athalye, V., Ragavan, S., Rajarathnam, U., Lalchandani, D., Maithel, S., Baum, E., & Bond, T. C. (2014). Emissions from South Asian Brick Production. *Environmental Science & Technology*. https://doi.org/10.1021/es500186g

Yadav, S. K. (2003). Remote Sensing Based Management of Degraded Soil Due to Brick Industry for Sustainable Development - A Case Study. *Null*. https://doi.org/10.1080/09709274.2003.11905651

Assessment of Hydro-Meteorological Variables in Gandak River Basin with an Integrated Approach for Climate Resilient Sustainable Water Resource Management

Rina Kumari, Ujjawal Kumar, Nang Swatie Manpoong, and Virendra Padhya

1 Introduction

Indo-Gangetic basin (IGB) is one of the world's largest quaternary alluvial sedimentations and constitutes one of the largest foreland basins (Sinha and Friend 1994). Ganga River alone makes up the largest river system in the nation constituting a catchment area of almost 86.1 m ha (Mukherjee et al. 2015). IGB sustains thriving agriculture making it the world's largest connected agricultural area, providing food to approximately one billion people (Biemans et al. 2019). Derived from the sediment incoming from the Himalayas and reorganized by the Indus, Ganges and Brahmaputra rivers, the Indo-Gangetic plains in Pakistan, northern India, Southern Nepal, and Bangladesh have low topographic relief aquifers, unconsolidated/fissured alluvium and shallow groundwaters (Bonsor et al. 2017). In India, it spreads to approximately 43.7 m ha with eight agro- ecological region and fourteen agro-ecological subregions. Covering 13% of the country, it supports 40% of the population by producing 50% of the total crops (Pal et al. 2009). The IGB regions not only provide food security to the resident population but also to South Asia (Singh and Sontakke 2002).

Ingo-Gangetic basin lies across a large area with various groundwater recharge processes and interaction of surface water (Lapworth et al. 2021). Groundwater is an essential natural resource in the IGB providing a vast majority of the population with drinking water, for irrigation and industrial purposes and supports riverine ecosystem via baseflow (MacDonald et al. 2016; Mukherjee et al. 2018). In recent decades, the dependency of food production has shifted from surface water and rain-fed agriculture to groundwater irrigation mainly due to the erratic rainfall patterns in these regions (Gadgil et al. 1999; Shah 2009). IGB relies on mountain water (Glacier-fed) resources along with groundwater particularly for irrigation when rainfall is scare though climate change potentially may weaken these processes and could strongly affect the crop production in one of the world's largest food hubs (Biemans et al. 2019). In addition, human activities such as agricultural leave behind contaminants in the environment eventually leading to leaching of contaminants in the water resources changing the physicochemical environmental characteristics. Hence, investigation and management of these water resources for its sustainable development and for the riverine ecosystem require have come into highlight.

1.1 Water Resources in Indo-Gangetic Basin

Bihar state in India, an agrarian state has abundant groundwater and surface water resources with water table varying from shallow in the north-eastern to deep toward southern regions of the state. Snow and glacier melt along with monsoon rainfall are the two major sources of water in the state. With reservoirs, canals, and advanced development, the state has become an intense food-production system (Biemans et al. 2019). According to the Agricultural Department, Government of Bihar (https://state.bihar.gov.in/krishi/CitizenHome.html), almost 24.84% (2011–12) of the state's gross domestic production (GDP) is derived from agriculture with 77% of workforce. Moreover, in the state almost 80% of the water demands are met by groundwater extraction (Sinha et al. 2018). Moreover, from 1985 to 2021, there has been an increase in the net irrigated area from 2808 thousand hectares to 3083 thousand hectares (Sharma and Kumari 2024). Despite the availability of water, the state's primary challenge is providing potable, uninterrupted and sanitary

R. Kumari (✉) · U. Kumar · N. S. Manpoong
School of Environment and Sustainable Development, Central University of Gujarat, Gandhinagar, Gujarat, India
e-mail: rina.sesd@cug.ac.in

V. Padhya
Geosciences Division, Physical Research Laboratory, Ahmedabad, Gujarat, India

D. Kumar et al. (eds.), *Data Analytics and Remote Sensing-Based Analysis for Sustainable Development*, Advances in Science, Technology & Innovation, https://doi.org/10.1007/978-3-031-98095-4_8

water to a total population of 104.1 million (Mishra 2009; Saha and Sahu 2014).

The hydrogeology of the state is unconsolidated/fissured formation with almost all the regions of north and some regions of south of alluvial landforms. The potential of the aquifers in Bihar are gradually decreasing toward the south with some sub-Tarai having autonomous flow.

Hydrometeorologically, it has a sub-humid climate. Most of the rainfall occurs during the month of June to September (Southwest Monsoon) with normal annual rainfall of 1200 mm with intense rainfall in the north which subsequently decreases toward the south (CGWB 2022). As the north and the central region of the state receives more rainfall, the yield of the aquifer can go up to 250m^3/hr to 20–50 m^3/hr in the marginal southern area (Saha and Sahu 2014). The total replenishable groundwater potential of the state is stated to be 33.35 BCM and the annual groundwater extraction is estimated to be 13.5 BCM (Sharma and Kumari 2024). This monsoon rain also is the main source of recharge accounting for 60.17% of the total annual groundwater recharge in the state and the rest 36.39% coming from different sources such as agricultural return flow, canal seepage, water bodies, and water conservation structures. In the state, artificial recharge of groundwater is being promoted within 1655 km^2 area whereas, the annual surface water recharged was estimated to be approximately 574 MCM (CGWB 2022).

Based on the Groundwater Extraction Methodology—2015 (GEC-2015 Methodology), the Central Groundwater Board, 2022 reported that out of the 535 assessment units in the state 8 (1.49%) were over-exploited, 12 (2.24%) were critical, 46% (8.60%), and 469 (87.66%) were safe. Furthermore, in the quality assessment, it was reported that 141 units (26.35%) were affected by high fluoride contamination and 64 units (11.96%) were affected by high arsenic contamination in the groundwater.

Based on the above background, the present study has been carried out in the Gandak River Basin to assess the water resources and hydro-metrological variation in the Basin.

Gandak River basin (GRB) overlays the Indo-Gangetic basin aquifer in India. Gandak River Basin has an exceptional topography with glacier snowfields and mountains in Nepal and alluvial plains in India (Chaubey et al. 2019). GRB supports irrigation, fisheries, riverine, and forest ecosystems with vast biodiversity (Choudhary et al. 2012). In the India-Nepal border, a large canal network was built at the Shivalik-Himalaya hills (Lapworth et al. 2021). This canal sustains irrigation in the basin, mostly in the dry-season although at times suffers from waterlogging in the command area (Chowdary et al. 2008; Lapworth et al. 2021).

It is home to large populations of critically endangered species such as Gharial crocodile (*Gavialis gangeticus*) as well as river dolphin (*Platanista gangetica*) (Choudhary et al. 2012). In the end of the GRB, the Valmiki Tiger Reserve and conservation reserve is situated as an initiation of Gandak Biodiversity Plan (Lapworth et al. 2021).

The unique topography of the GRB results in various hydrological processes and extreme precipitation events in this river basin. Climate variability also influences significant change in hydrometeorology of the river basin (Chaubey et al. 2019). Along with the change in hydrometeorology, it adversely affects water resources (quality and quantity), irrigation, terrestrial ecosystem, and biodiversity. Change in temperature could shift the crop growing seasons affecting the food security and could lead to severe water security and flooding (Singh et al. 2014). In addition, urbanization in the basin has alters the natural ecosystem including the soil, vegetation cover, hydrological cycle, land use/land cover, and surface runoff (Agarwal et al. 2023; Whitford et al. 2001). A study conducted by Nath et al. 2017 invigilated the spatiotemporal characteristics of drought on the Upper Middle Gangetic Plain (UMGP) and its effects on agriculture. It concluded relative dryness of the area and inadequate moisture, making the region drought prone in the recent decades with occurrence frequency of 40–45%. There was also a gradual decline of trend from 2000 of the total food produced by this UMGP region.

Management of water resources as well as for the well-being of the river flow, spatiotemporal understanding of river hydrology, the groundwater recharge and connectivity of surface water are very essential. Furthermore, with increased living standards, freshwater availability both quality and quantity has made it necessary for consistent and extensive hydrological research on water for sustaining the vast population.

To aid with the delineation of the provenance of ground and surface water, stable isotopes of hydrogen and oxygen become ideal tracing tools for tracing water in the hydrological cycle as water molecules have distinguishable isotopic signature during its transition from one reservoir to another (Gupta and Deshpande 2005). Isotope hydrology has been proven to be very important as it can be used in various hydrological investigations such as groundwater-surface interaction, genesis of freshwater, efficacy of artificial recharge, dynamics of water bodies (Jeelani et al. 2012; Pandey et al. 2023; Maurya et al. 2011). Various studies have been also done on river basin using stables isotopes of water molecules (Kumar et al. 2019; Fan 2016; Xie 2022; Rina et al. 2013).

In recent years, Bihar has witnessed a change in precipitation. Temporal and spatial variation of precipitation in Bihar leads to a drought-flood syndrome eventually leading to water crisis (Agrawal et al. 2021). Precipitation along with soil moisture are primary parameters for understanding the basin in terms of their hydrological responses for water resources management (Chaubey et al. 2019). Changes in the soil, water as well as vegetation of an area are interconnected to evapotranspiration, surface runoff, flux in sensible heat and eventual reduction of water and air quality. The fluxes in vegetation in an urban agglomerate consequently affect the energy demand and development of urban heat island as they influence the environment via absorption and reflection of radiation and regulate latent and sensible heat exchange (Huang et al. 1987; Yue et al. 2007; Gallo et al. 1993). Land surface temperature depicts the temperature of the earth's surface (Taloor et al. 2008; Niclos et al. 2018). It indicates the surface water-heat exchange phenomenon, giving extensive insight on environmental change (Deng et al. 2018).

Various studies have been conducted using LST for interpretation of evapotranspiration, climate variability, vegetation monitoring, hydrologic cycle, urban heat islands, and sensible heat indices (Taloor et al. 2008; Mannstein 1987; Seller et al. 1998; Van Leeuwen et al. 2011; Kumar et al. 2020) (35; 39; 40; 41; 42; 43).

Land surface temperature affects vegetation health status and overall quality, and it has a great influence on soil moisture which in turn affects crop productivity. They are inversely proportionate to each other (Sun and Kafatos 2007). Various vegetative indices such as Normalized Difference Vegetative Index (NDVI), Greenness Index (GI), Normalized Difference Moisture Index (NDMI), and Normalized Difference Built-up Index (NDBI) can be used for estimating the relation between land surface temperature (LST) and vegetation (Dasak et al. 2023; Deng et al. 2018).

With the advancement remote sensing and Geographical Information Systems (GIS) have proven to be useful for the interpretation of the urban vegetation and thermal variations over an urban ecosystem at all spatial and temporal scales (Weng 2003; Yue et al. 2007). For assessment of LST, various series of satellite like Landsat, MODIS, Terra and Aqua with high resolutions are used (Ghosh et al. 2022; Zhu et al. 2013).

Together with the help of quantitative physical data from visible and near-infrared sensor and GIS, various spatial and temporal insight can be obtained to classify the land surface temperature and the vegetative indices (Yue et al. 2007).

The specific objective of this study in the Gandak River Basin is (1) to assessment groundwater and surface water for drinking purpose, (2) to delineate the origin of ground and surface water using stable isotopes, and (3) to study the spatial and temporal variation in LST and vegetative indices.

2 Study Area

The present study (Fig. 1) has been carried out in Gandak river basin which is a major tributary of Ganga in India supplying third highest flow to Ganga River (MacDonald et al. 2015). Originating from the Tibetan plateau at Dhaulagiri in Tibet at an altitude of 7620 m above MSL, Gandak river flows across Nepal to enter India in the north-western region of Bihar. The river is flowing in southeast direction through following districts of Bihar namely, Champaran, Muzaffarpur, Gopalganj, Siwan, Saren and Vaishali, and it enters in Utter Pradesh near Gorakhpur and Deoria district for 335 km finally joining River Ganga near Patna (https://www.fmiscwrdbihar.gov.in/fmis/riverbasin.aspx; Lapworth et al. 2021) (Fig. 2).

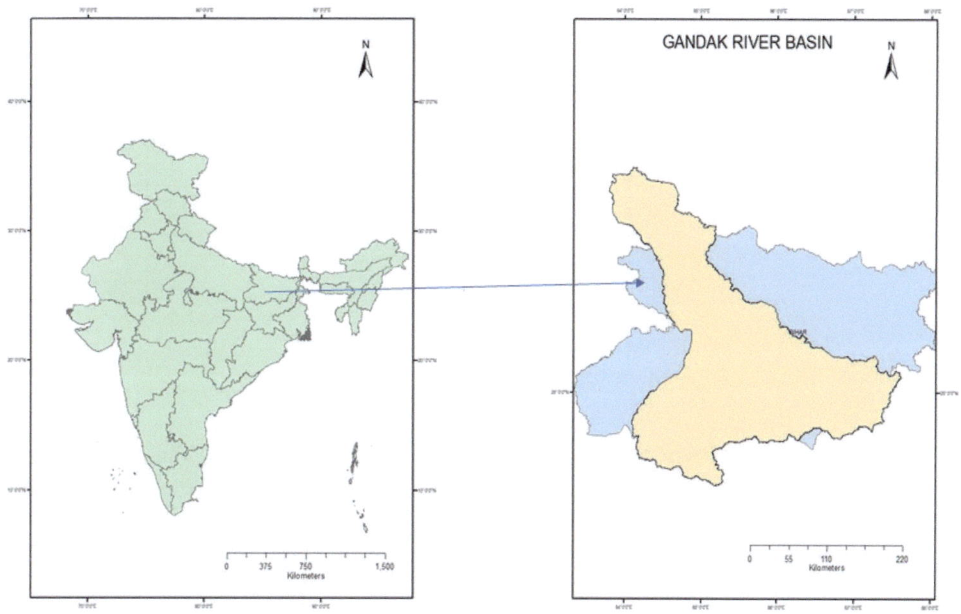

Fig. 1 Location of Gandak River Basin in Bihar

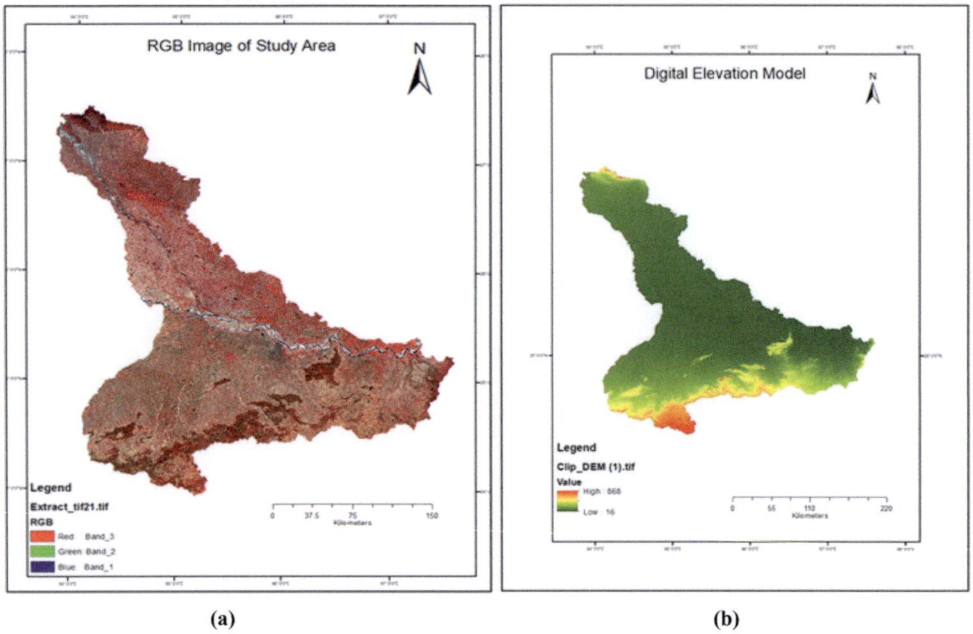

Fig. 2 **a** RGB and **b** DEM of the Basin

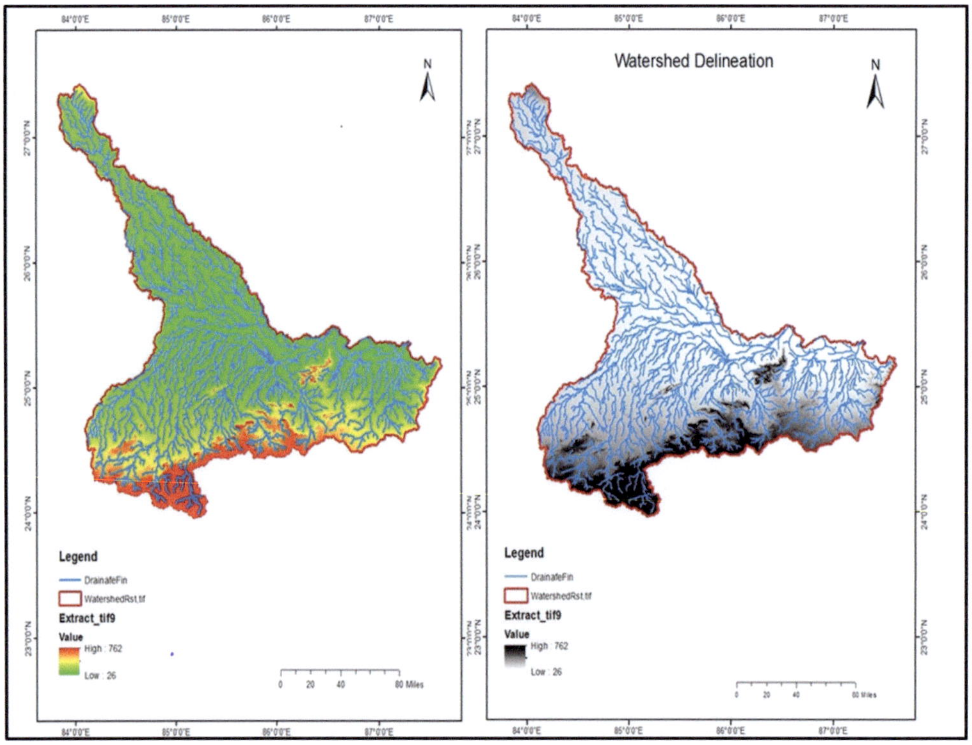

Fig. 3 Watershed delineation of the study area

2.1 Hydrometeorology and Hydrogeology of the River Basin

The Himalayan glaciers and southwestern monsoon govern the Ganga River Basin. The presences of peaks of Langtang, Machhapuchhre, Dhawalagiri, and Annapurna Mountain lead to fluctuation of the weather resulting in subtropical or mesothermal climate. Gandak river basin during summer (March to May) is said to have hot and humid, monsoon (June to September) mild comfortable, and winter (October and February) mild and cold climate. According to the Flood Management Improvement Support System, Bihar (https://fmiscwrdbihar.gov.in/), the catchment receives an annual average rainfall of an estimated 1125 mm. The total drainage area was 40,553 km^2 with surface water resources to be 51,998 MCM. Figure 3 represents the Watershed Delineation and Stream Flow Channel.

GRB has a dynamic geological formation having Precambrian rocks, Cretaceous-Tertiary igneous and metamorphic rock, and Quaternary sediments. It consists of aggraded flood plains with variations in flow regimes since the Quaternary period and having persisted in the present. Consequently, leading to the formation of a variety of erosional and depositional landforms. In the central portion of Gandak river, alluvial sediments are rich in silt, sand, clay, and kankar (carbonate nodule) (Sinha and Sarkar 2009; Densmore et al. 2016). The deposited thickness of sediment is highly variable, its thickness varies from kilometer in the northern river Ganga plain to 50 m in Jharkhand.

Holocene and Pleistocene alluvial formation cover the study area. Investigations using stable isotopes inferred the major source of groundwater recharge in the shallow Holocene (0-40 m bgl) and Pleistocene aquifer system (>40 m bgl) in the GBR is the local rainfall mainly during the Monsoon season and partially from local water bodies such as river, canal, agricultural river flow (Lapworth et al. 2021).

The eastern canal branch of the Gandak river supply water to Bihar with almost 50% leakage loss further enhancing the groundwater recharge in that upper catchment (Jha and Prasad 2002). There is extensive use of groundwater in the northern part of the catchment (O'Keeffe et al. 2020).

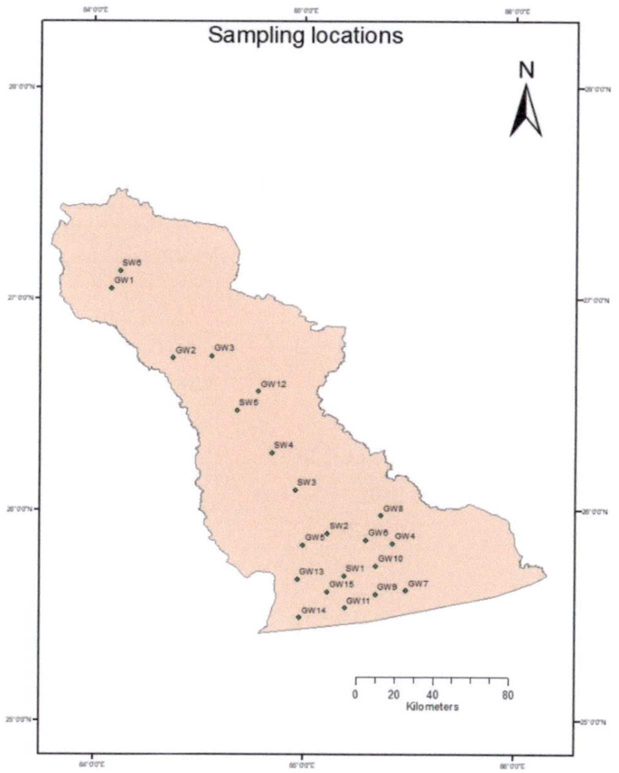

Fig. 4 Water samples sites in Gandak River Basin

3 Methods and Methodology

3.1 Sampling and Analysis

For the collection of samples, extensive field survey was conducted based on satellite image and topographic sheet. Field coordinates were recorded by Garmin GPS (Garmin GPS, model-Oregon 650). Samples were collected based on spectral signature of vegetation on satellite images, as well as accessibility to river. A total of 6 surface water samples and 15 groundwater samples were collected from the aquifers and banks adjacent to Gandak river in district Muzaffarpur, Gopalganj, Siwan, Saren, and Vaishali of Bihar and district Gorakhpur and Deoria of Uttar Pradesh. Samples sites are plotted in Fig. 4.

A total of 11 water quality parameters and water stable isotopes ($\delta^{18}O$ and δD) were studied for assessment of water quality, genesis and geochemical evolution of the groundwater and surface water samples. Groundwater samples were mostly collected from borewells, handpumps, and dug wells from nearly same depth. Before sample collection from borewells and handpumps, the bottle was rinsed two–three times with the samples, and from dug wells, the water was kept running for at least five minutes for obtaining stable temperature and minimizing influence of iron pipes. Whereas surface water samples were mostly collected from small streams, tributaries, and open channels.

For storage of the samples, two separate high-density polypropylene bottles (Tarson 125, and 250 ml) each were taken for analysis of anions and stable isotopes. For quality assurance and control, these bottles were soaked in 1:1 HCl solution for 24 h, washed with milliQ water and finally air dried. The unacidified bottles (250 ml and 125 ml) containing samples were used for the analysis of anions and stable isotopes. Bottles were labeled and tightly sealed after sampling to avoid hassle and any chances of reaction with the atmosphere. American Public Health Association (APHA 2012) standard norms were followed to ensure data accuracy and uniformity in cases of collecting the samples, handling, preservation, and analysis. Certain parameters such as temperature, pH, electric conductivity (EC), and total dissolved solids (TDS) were measured on site using electrodes (Hanna, HI 98311). For further analysis, the sample bottles were kept in Styrofoam boxes, taken to laboratory, and preserved in 4 °C. Moreover, for the removal of potential contaminants in the samples they were filtered using 0.22 μm Millipore paper. For analysis of stable isotopes, a sealed aliquot was sent to IWIN, Physical Research Laboratory, Ahmedabad.

Table 1 shows the various physicochemical parameters and stable isotopes along with methods used for estimation. Analysis of anions, namely carbonate and bicarbonate, was done using the titrimetric method as soon as after sampling.

Table 1 Physicochemical parameters of surface and groundwater in mg/L unit except pH, temperature (°C), and EC (μS/cm)

Parameters	Sampling water types				WHO limit
	Surface water		Groundwater		
	Range	Average ± SD	Range	Average ± SD	
Temperature	25.3–31.1	27.9 ± 1.9	25.7–33.1	29.0 ± 1.8	–
pH	6.4–8.3	7.7 ± 0.8	7.1–8.1	7.4 ± 0.3	6.6–8.5
EC	309.0–485.0	358.8 ± 65.3	765.0–3155.0	1565.9 ± 792.8	400–1500
TDS	155.0–244.0	178.5 ± 33.9	383.0–1573.0	783.2 ± 395.1	500–2000
CO_3^-	0.0–12.0	3.0 ± 5.0	0.0–12.0	5.0 ± 3.9	10–500
HCO_3^-	61.0–140.3	93.5 ± 28.3	152.5–433.1	258.2 ± 66.3	30–400
Cl^-	355.0–1065.0	576.9 ± 249.4	355.0–16,507.5	4354.7 ± 4529.8	250–1000
SO_4^{2-}	12.7–67.6	44.50 ± 30.9	4.04–460.69	204.23 ± 154.36	200–400
NO_3^-	1.5–18.9	7.6 ± 6.7	1.9–52.07	15.6 ± 16.9	45 mg/L
PO_4^{3-}	0.0–0.1	0.0 ± 0.0	0.0–0.0	0.0 ± 0.0	0.3–0.304
Si^-	9.0–22.0	11.6 ± 5.1	5.3–13.6	11.4 ± 2.1	5–25

Fluoride was estimated using the SPADNS method, chloride was done using argentometric method, turbidimetric method was adopted for estimation of sulfate, phosphate was determined using ascorbic acid method, and lastly, dissolved silica was estimated using molybdo-silicate method.

Analysis of stable isotopes was done by the Delta V Plus Isotopic Ratio Mass Spectrometry (IRMS) in continuous flow mode with gas bench II. It is a standard equilibration method where samples (300 μl) are equilibrated, respectively, with CO_2 and O_2 to obtain $^{18}O/^{16}O$ and $^{2}H/^{1}H$ values which further is used for calculating $\delta\ ^{18}O$ and δ D denoted as ‰ (part per thousand or per million). The following formula for generation of $\delta\ ^{18}O$ and δ D–

$$\delta(\text{in‰}) = \left(\frac{R_{\text{Sample}}}{R_{\text{Standard}}} - 1\right) \times 1000 \quad (1)$$

An additional parameter, *d-excess* defined by Dansgaard (1964) was computed for characterization and kinetic effects related to evaporation (Pandey et al. 2023). It is calculated for each pair of calculating $\delta\ ^{18}O$ and δ D as –

$$d - \text{excess} = \delta\text{D} - 8 \times \delta^{18}O \quad (2)$$

The reproducibility of 0.1 ‰ for $\delta\ ^{18}O$ and 1.0 ‰ for δ D was estimated by repeating analysis of multiple aliquots of the secondary laboratory standards.

3.2 Data Sets Used and Software

The satellite data used for the study area, Gandak basin include Landsat 8 data using Operational Land Imager (OLI) and Thermal Infrared Sensor (TIRS) with 30 m spatial resolution was downloaded from USGS center (https://www.usgs.gov) with eleven bands for April and May 2023.

The software ArcGIS 10.3 was used for basic image processing, geoprocessing, basin demarcation, and obtaining watershed delineation (Ahmed et al. 2010). The methodology for generation of data has been shown in Fig. 5.

For the generation of digital elevation map, spatial analyst toolbox and Arc Hydro toolbox were used. It includes fill, flow direction and accumulation, defining watershed to finally obtaining watershed delineation.

Landsat data has several bands and panchromatic bands are used for better visibility and interpretation (Kshetri 2018) of vegetated land. False Color Composite was generated where the red, green, and blue values do not match the real red, green, and blue colors. In this FCC, a high degree of reflectance in the near-infrared (NIR) band, greenery appears in a variety of shades of red. As the rate of photosynthesis is high the red color of the visible spectrum gets absorbed and near-infrared bands are reflected which shows that the healthy vegetation absorbs redder color than the degraded vegetation.

Vegetation indices are very useful tools linking leaf area index and chlorophyll content giving insight to crop yield and net primary production, detecting vegetation type and LULC changes (Hussain et al. 2023). Normalized Difference Vegetation Index (NDVI) is commonly used for interpretation of health of vegetation (Taloor et al. 2021). For NDVI, red band (band 4) and near-infrared band (NIR) (band 5) were selected and applied under raster calculator. The following formula generates NDVI-

$$\text{NDVI} = \frac{(\text{NIR} - \text{Red})}{(\text{NIR} + \text{Red})} \quad (3)$$

(Kshetri 2018)

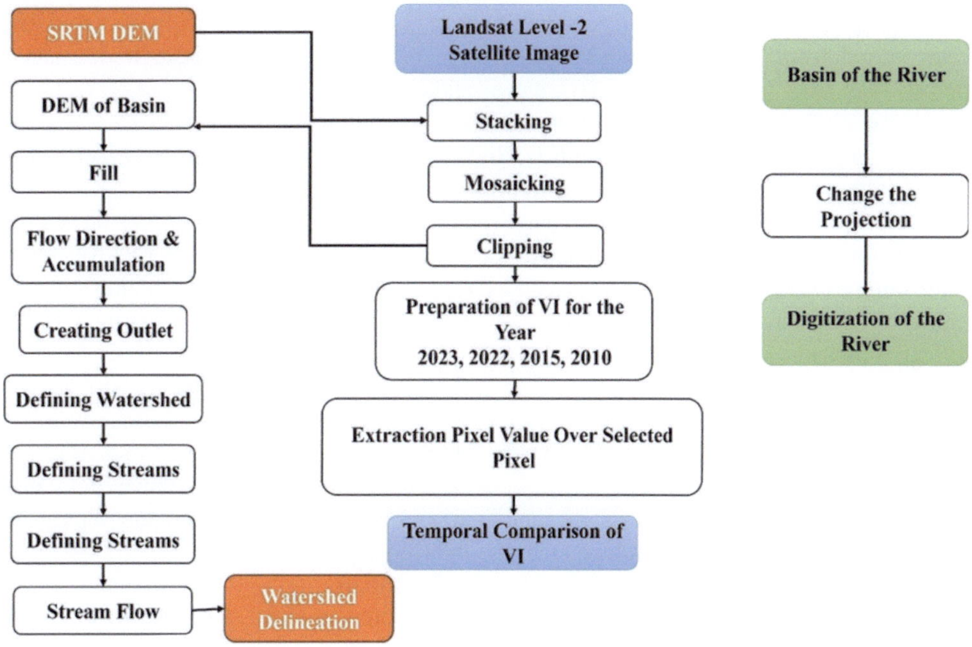

Fig. 5 Flow chart showing the methodology adopted for the study area

NDVI ranges from − 1 to + 1 where higher value infers healthy vegetation and greenery, and lower value infers unhealthy and spare vegetation.

Normalized Difference Moisture Index (NDMI) is commonly used for estimation of a crop's water stress level (Guo 1996). For NDMI, near-infrared band (NIR) (band 5) and shortwave infrared 1 band (SWIR) (band 6) are selected and applied under raster calculator. The following formula generates NDMI-

$$\text{NDMI} = \frac{(\text{NIR} - \text{SWIR})}{(\text{NIR} + \text{SWIR})} \quad (4)$$

(Taloor et al. 2021)

NDMI ranges from − 1 to + 1 where higher value infers good water moisture in the vegetation, and lower value infers bare to dry region. NDMI increases with increase in leaf layers (Gao 1996).

For spatial and temporal variation, NDVI and NDMI were generated for pre-monsoon and post-monsoon for the years 2010, 2015, 2022, and 2023.

For the estimation of land surface temperature, a well-defined calculation can be applied on Landsat images with 30 m spatial resolution (Hussain et al. 2023). At first, $L\lambda$ value (spectral radiance) is calculated using (Eq. 5).

$$L\lambda = \text{gain} \times \text{QCAL} \times \text{off set} \quad (5)$$

where, QCAL is a digital number (DN), a quantified calibrated pixel value.

off set is intercept of QCAL conversion factor.

then, $L\lambda$ is converted to temperature using Eq. (6)

$$T = \frac{K_2}{\text{In}\left(\frac{K_1}{L\lambda} + 1\right)} \quad (6)$$

where, K_2 and K_1 are 772.66 and 1321.07 for Landsat 8 (OLI/TIRS), and ln is natural logarithm.

Lastly, the temperature in Kelvin (K) is converted to Celsius (°C) using Eq. 7.

$$T(^{\circ}\text{C})m = T(K) - 273.15 \quad (7)$$

4 Results and Discussion

4.1 Physicochemical Parameter

The statistical description such as range, average, and standard deviation concentration of various physicochemical parameters along with the World Health Organization limits (2012) are given in Table 1.

In the surface water samples, the temperature ranged from 25.3 to 31.1 °C with an average of 27.9 °C, whereas in groundwater samples it ranged from 25.7 °C to 33.1 °C with an average of 29.0 °C. pH of sampled surface water samples ranged from 6.4 to 8.3 and groundwater ranged from 7.1 to 8.1.

Electrical conductivity of surface water ranged from 309.0 to 485.0 µS/cm with an average concentration ranged from µS/cm, whereas that of groundwater samples ranged from 765.0 to 3155.0 µS/cm with an average concentration of 1565.9 µS/cm.

In the present study, sulfate concentration of surface water varied from [12.7–67.6 mg/l with average concentration 44.5 mg/L] whereas sulfate concentration of the groundwater samples varied from [4.04–460.69 mg/L with average concentration of 204.23 mg/L]. High concentration of sulfate associated with either sodium or magnesium can cause serious gastrointestinal problems when consumed without proper treatment. Nitrate concentration in groundwater samples varied from [1.9–52.07 with average concentration of 15.6] whereas that of surface water ranged from [1.5–18.9 mg/l].

4.2 Isotope Systematics in the Study Area

Assessment of stable isotopes of oxygen and hydrogen helps track the movement of water molecules in the hydrological cycle, making it an important tool for investigation of its origin. Coupled with the ionic content of the water, stable isotope, it becomes a powerful tool to study the mixing between water masses of different origin, and is used to assess the genesis and evolution of water.

4.3 Isotopic Signature of Water in the Area

The isotopic composition of collected groundwater samples ($N = 15$) varies from [− 7.13‰ to − 0.61‰ with an average of − 5.5‰] for δ^{18}O; and for δD it varies from [− 48.95‰ to − 14.76‰ with an average of − 40.54‰]. However, the isotopic composition of collected surface water samples ($N = 6$) varied from [δ^{18}O = − 4.34 to − 9.50‰ with average − 7.99‰], and δD = [− 32.38 to − 63.10‰] with an average composition − 56.10‰.

Among these samples, most enriched composition is shown by groundwater [δ^{18}O = − 0.61‰] and most depleted composition [− 9.50‰] is shown by surface water at the highest elevation. Average isotopic composition in surface water were [− 7.99‰], whereas average isotopic composition of groundwater was 4.64‰. In general, the river waters have relatively enriched δ18O compared to the nearby groundwater (Deshpande et al. 2013); however, depleted isotopic composition in the month of May suggests Glacial melt in the river water. Also, there is a progressive enrichment of surface water from upstream to downstream.

4.4 d-excess in the Study Area

A parameter "*d-excess*" as $d = \delta D - 8{*}\delta$ 18O (‰) is said to assess the relative magnitude of kinetic fractionation in different water molecules. *d-excess* in the sampled groundwater ranged from [− 9.88 to 10.19‰] with an average 4.64‰. *d-excess* in the groundwater samples varied from − 9.88 to 10.99‰ with an average 4.64 ‰. Whereas minimum *d-excess* in surface water is 2.31 and maximum 13.14 with an average composition is 7.80. Higher *d-excess* in surface water as compared to groundwater suggests high kinetic fractionation due to prevailing high temperature.

4.5 Regression Line for δ^{18}O and δD

The relationship trend between δD and δ^{18}O was compared with the local meteoric water line (LMWL) calculated from precipitation at Patna. LMWL for Patna is given below:

$$\delta D = (7.76 \pm 0.66)\,\delta^{18}O + 3.02(\pm 7.43) \quad \left(n = 5,\ r^2 = 0.98\right)$$

The linear least-square regression between the δD and δ^{18}O for groundwater and surface water is plotted in Fig. 6a and that between the *d-excess* and δ^{18}O for groundwater and surface water is plotted in Fig. 6b. The regression line for δD vs δ^{18}O of groundwater samples has a slope of (5.08 ± 0.2) and intercept of (11.86 ± 0.59). The much lower slope and intercept values of regression line for groundwater as compared to LMWL suggests, groundwater has undergone significant evaporation before recharge (Gupta and Deshpande 2005). The regression line for δD vs δ^{18}O of surface water samples has a slope of (6.01 ± 0.51) and intercept of (8.07 ± 4.24) is also suggesting significant evaporation. Most of the samples are around the regression line, suggesting a common origin.

From Fig. 6b, d-*excess* vs δ^{18}O it can be depicted that surface water is having higher *d-excess* and depleted δ^{18}O in upper reaches of the river whereas in lower reaches enrich-

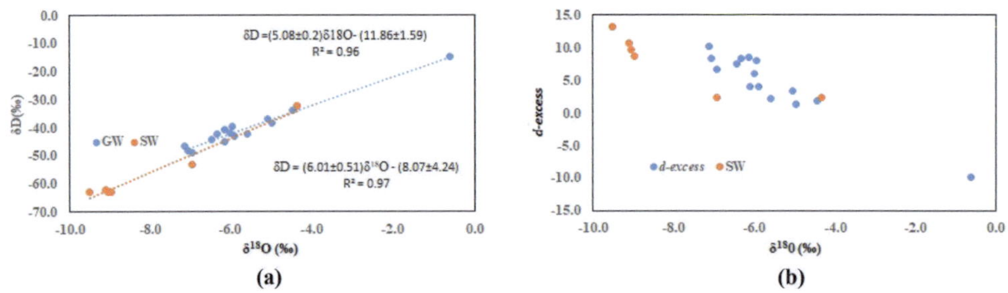

Fig. 6 **a** Regression line between the δD and δ^{18}O for groundwater and surface water; **b** *d-excess* and δ^{18}O for groundwater and surface water

ment in $\delta^{18}O$ is observed which suggests significant evaporation from upstream to downstream. Most of surface water samples in upstream is showing depleted $\delta^{18}O$ signature more than the precipitation suggesting significant contribution from glacier melt in summer. However, in case of groundwater, scatter plot of $\delta^{18}O$ vs *d-excess* suggests that enriched $\delta^{18}O$ and high d-excess have been observed in the study area. High *d-excess* in the study area suggests a prolonged process involving re-evaporation under low humidity and subsequent condensation with or without mixing with larger atmospheric reservoir causing high *"d-excess"* in resultant precipitation and groundwater (Deshpande et al. 2013). Although depleted isotopic signature in upstream suggests contribution of glacier melt, but regression line of $\delta^{18}O$–δD of surface water is below the LMWL of Patna.

So, we can say that in Gandak River both glacier melt and precipitation is sustaining the flow. Also, most groundwater is showing isotopic signature below precipitation, it suggests common origin. Any change in climate variables may impact the surface water hydrology and hence the groundwater hydrology.

4.6 Spatiotemporal Variation of Land Surface Temperature (LST), Vegetation Indices (NDVI), and Moisture Indices (NDMI) in the Study Area

Vegetation Health Status in the Basin

Normalized Difference Vegetation Index (NDVI) is used to measure the vegetation health status which in turn is associated with the other biophysical properties including vegetation cover, canopy, biomass, leaf area index. The main principle behind it is that healthy vegetation tends to reflect more amount of energy than unhealthy and sparse vegetation (Taloor et al. 2021).

In 2010, NDVI values during pre-monsoon ranged from − 0.735 to 0.599 whereas it ranged from − 0.219 to 0.530 during pre-monsoon of 2015 and for 2023 during pre-monsoon it ranged from − 0.149 to 0.502 (Fig. 8). However, during post-monsoon it ranged from − 0.650 to 0.740 in 2010, whereas in 2015 it ranged from − 0.149 to 0.669 and in 2022 post-monsoon it ranged from − 0.221 to 0.523. As compared to pre-monsoon, vegetation health status has increased in post-monsoon due to precipitation. Also, it can be seen that the healthy vegetation status in the basin has decreased from 2010 to 2015.

4.7 Spatial Variation of NDVI

NDVI values were compared at few selected points from upstream to downstream of the basin. It is observed at upstream, healthy vegetation status has reduced from 2010 to 2023 during pre-monsoon, whereas in midstream vegetation health has increased from 2010 to 2023; again in downstream vegetation health has shown to increase from 2010 to 2015 but poor vegetation status in 2023 (Fig. 7). Higher NDVI value in the midstream as compared to the upper part of the basin because of the availability of abundant of water due to presence of mainstream of river Ganga. In the upper part, less value of NDVI can be ascribed to presence of hills

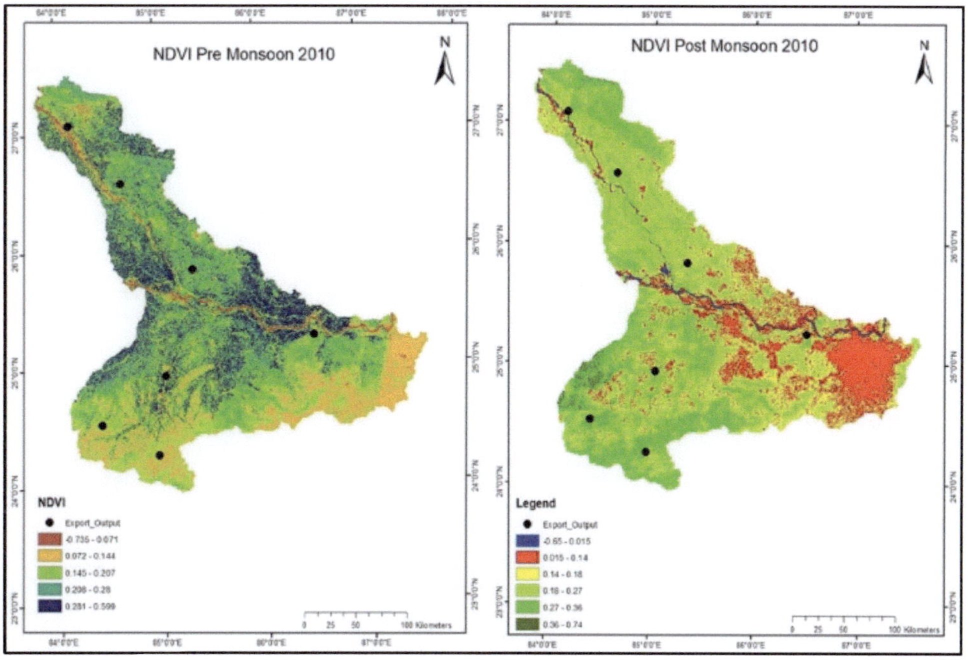

Fig. 7 Spatial and temporal variation in NDVI value for 2010 during both pre- and post-monsoon

and mountains where only some shrubs can adapt to the harsh conditions. Lastly, in the lower part of the basin, low NDVI value is due to absence of major perennial rivers, the seasonal river during the monsoon is governing the plants of nearby region.

5 Spatiotemporal Variation of Normalized Difference Moisture Index (NDMI)

Normalized Difference Moisture Index is the estimation of vegetation liquid water using remote sensing (Gao 1996). In simple words, it is the measurement of crop's moisture content, helping interpretate regions with water stress problems. Normalized Difference Moisture Index (NDMI) is a remote sensing-based vegetation index that describes the plant's water stress level.

The NDMI uses NIR and SWIR bands to assess moisture. The SWIR band reflects changes in both the vegetation water content and the spongy mesophyll structure in vegetation canopies, while the NIR reflectance is affected by leaf internal structure and leaf dry matter content but not by water content. The Normalized Difference Moisture Index can be used to monitor water stress in crop and agriculture field, which also helps in planning for tree harvest as well as in determining the combustibility level in fire prone areas.

The NDMI range is − 1 to + 1, where the lowest values (white to pale brown) indicate low vegetation water content, and the highest ones (in blue) correspond to high water content. In other words, a decrease in NDMI will indicate water stress in plant and field, whereas abnormally high NDMI is reflecting waterlogging condition.

The results for NDMI for 2010 during pre-monsoon ranged from − 0.658 to 0.366 whereas during post-monsoon it ranged from − 0.631 to 0.338 (Fig. 8), for 2015 during pre-monsoon it ranged from − 0.556 to 0.322 whereas post-monsoon it ranged from − 0.507 to 0.359, and for 2022 post-monsoon it ranged from − 0.614 to 0.462 and for 2023 pre-monsoon it ranged from − 0.610 to 0.342 (Fig. 9). NDMI for 2022 post-monsoon shows highest moisture content (Figs. 10 and 11).

From Figs. 8 and 9, it can be depicted that NDMI value for post-monsoon is higher than pre-monsoon. Monsoon rain nourishes the soil and overall reduces water stress and enhancing the moisture content in post-monsoon. High variation in NDMI for the two seasons also hints at the contribution of precipitation on the vegetation and water recharge in the study area.

NDMI for 2010 and 2023 during pre-monsoon varied from − 0.658 to 0.366 and − 0.610 to 0.342, respectively. Furthermore, NDMI for 2010 and 2022 during post-monsoon varied from − 0.631 to 0.338 and − 0.614 to 0.462, respectively. In both scenarios, the range for NDMI for 2010 is narrower than that of 2023.

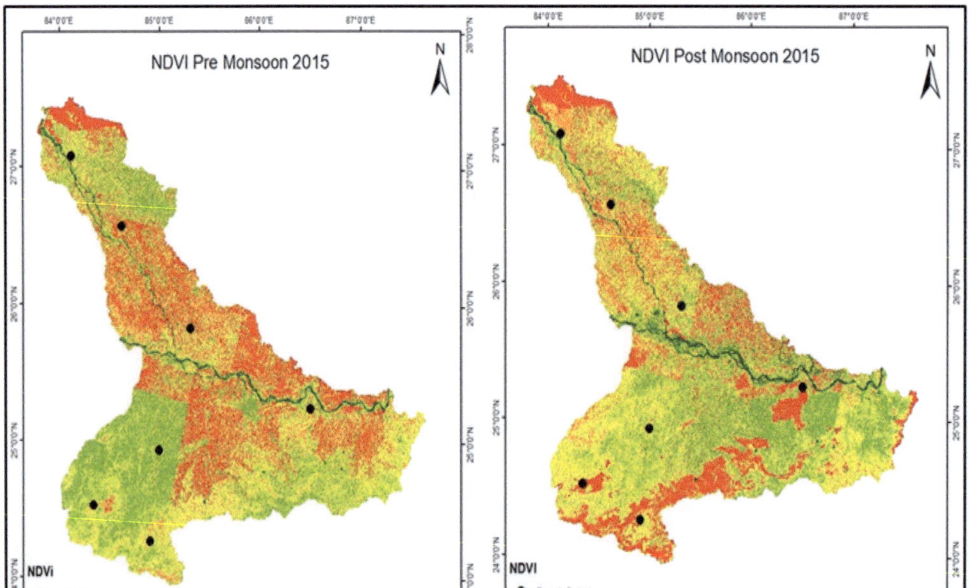

Fig. 8 Spatial and temporal variation of NDVI in 2015 during both pre- and post-monsoon

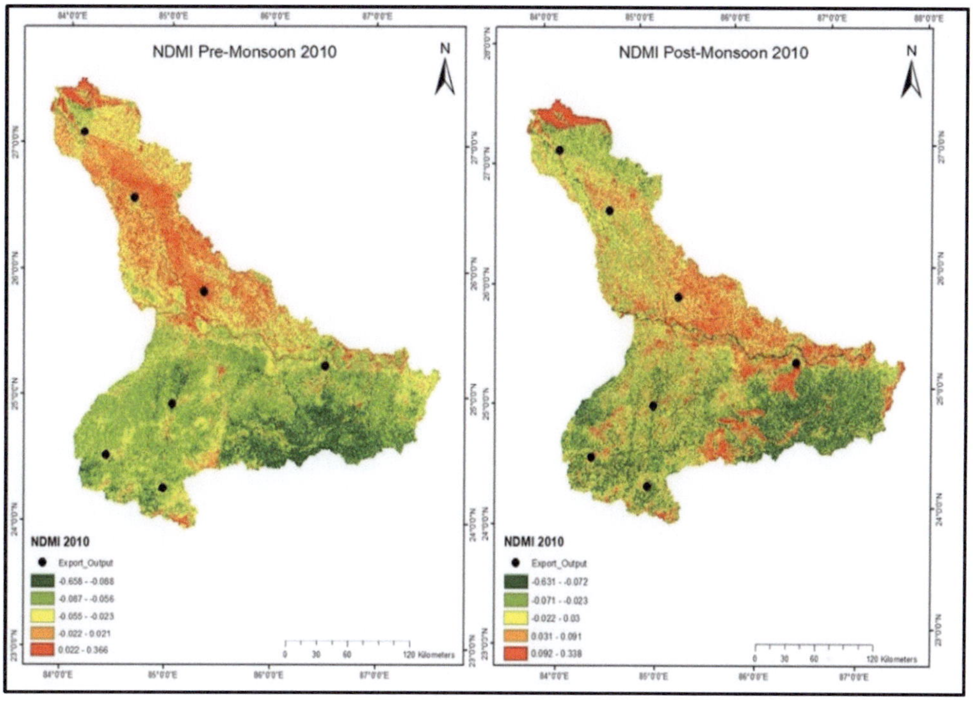

Fig. 9 Spatial and temporal variation of NDMI value for year 2010 (both pre- and post-monsoon)

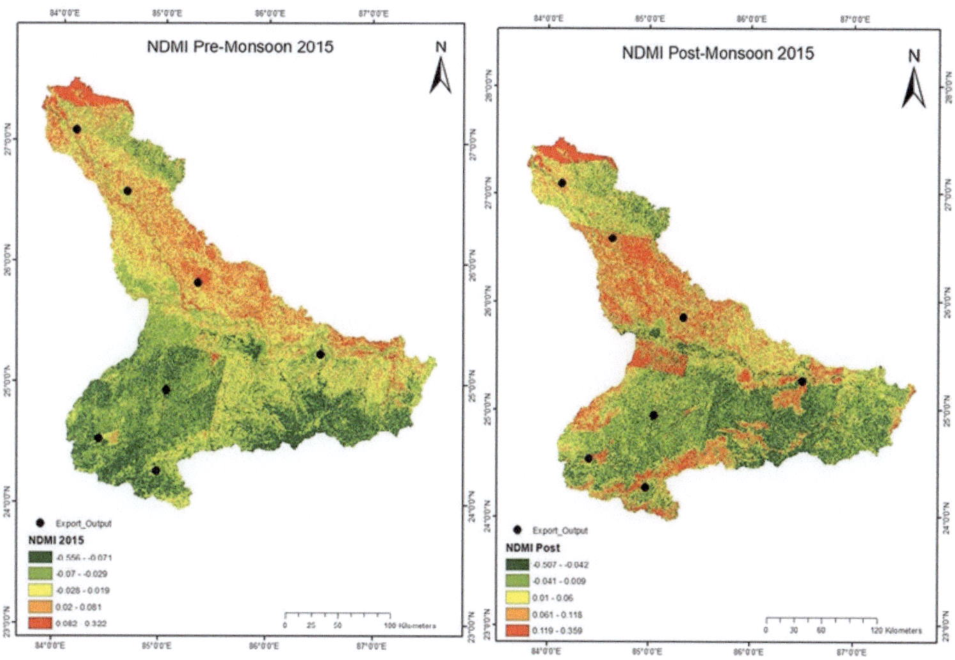

Fig. 10 Spatial and temporal variation of NDMI value for year 2015 (both pre- and post-monsoon)

5.1 Spatial Variation of NDMI

From upstream to downstream, at seven locations were selected to compare variation of NDMI from 2010 to 2022. From the extracted value it can be observed that the NDMI at upstream is low, while in the midstream it is higher in both seasons and again in the downstream the moisture is low. The graph plotted shows elevated value of NDMI at 3 locations in the study area.

In the midstream, River Ganga meets Gandak river near Patna, due to this confluence moisture content in this area is high. The lowest value of moisture is observed in location 7

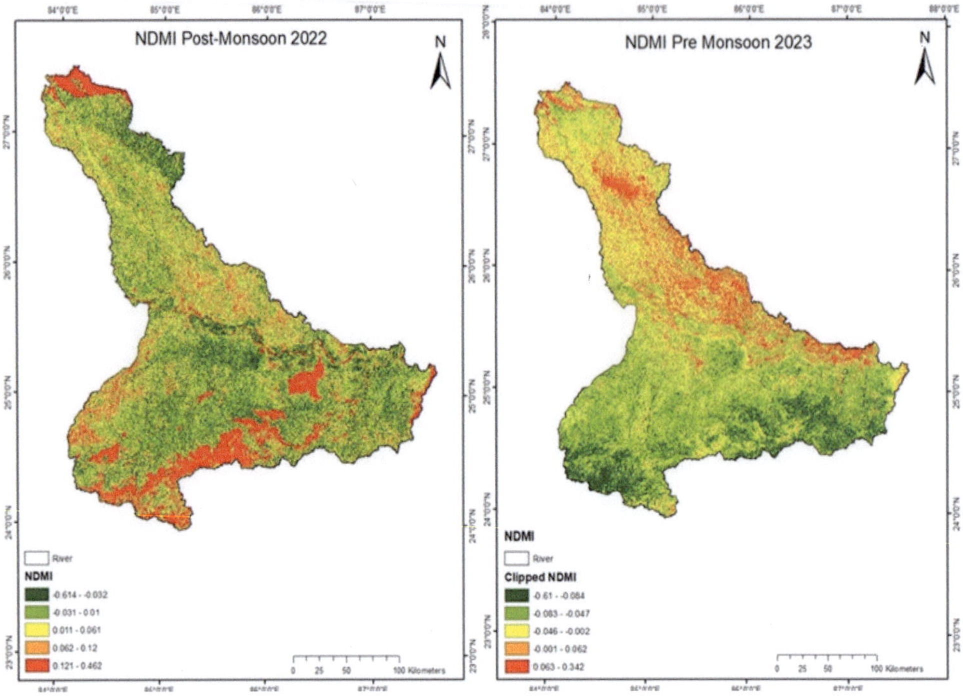

Fig. 11 Spatial and temporal variation of NDMI value for year 2022 (post-monsoon) and 2023 (pre-monsoon)

(downstream) where there is absence of any river. In the lower part of the basin, due to absence of any major perennial rivers, the seasonal river during the monsoon maybe governing the plants of nearby region.

5.2 Relation Between NDVI and NDMI

Vegetation growth depends on moisture available in the region. Fig. 13 illustrates the variation of NDVI and NDMI values for some selected location in the study area. It can be depicted that both in 2010 and 2015 vegetation health and cover are said to increase (increase in NDVI) with the increase in availability of water (NDMI) for the plant's growth.

5.3 Variation of Land Surface Temperature in the Study Area

Vegetation cover and water on the surface acts as a buffer to climate change. Having said that decrease in vegetation cover primarily causes shift in microclimate, altering the regions land surface temperature (Ibrahim 2016; Ghosh et al. 2022). Furthermore, anthropogenic activities such as high energy usage and increased land use/land cover exponentially increase land surface temperature (Ghosh et al. 2022).

Using Eqs. 5, 6, and 7, land surface temperature for post-monsoon, 2022, and pre-monsoon, 2023 has been calculated in Table 2. Furthermore, map has been generated from the same for assessment of spatial and temporal variation in land surface temperature of the study area (Fig. 12).

Table 2 Land surface temperature and NDVI for post-monsoon 2022 and pre-monsoon 2023

LST_2022	NDVI post-2022	LST 2023	NDVI pre-2023
29.189	0.132	42.328	0.108
29.389	0.118	40.711	0.123
28.636	0.253	44.640	0.289
27.579	0.356	41.590	0.116
31.895	0.187	52.674	0.179
25.506	0.277	47.889	0.133
28.174	0.269	47.852	0.153

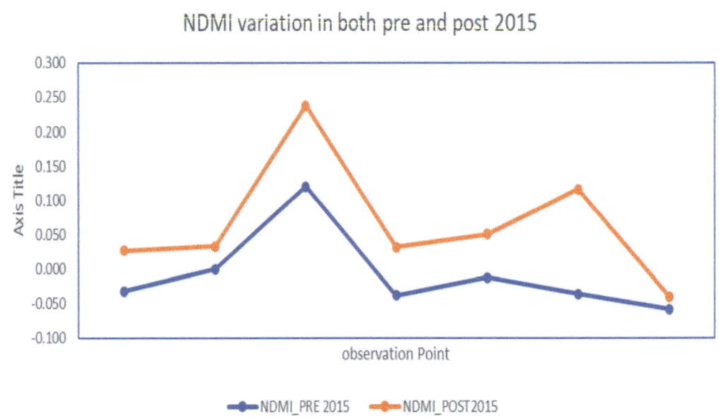

Fig. 12 NDMI variation at selected locations

Overall, pre-monsoon shows higher value (27.617 to 54.574 °C) than post-monsoon (17.411 to 37.153 °C) (Fig. 8). This can be due to the monsoon rain in the monsoon season cooling the surface. Downstream of GRB has higher LST,

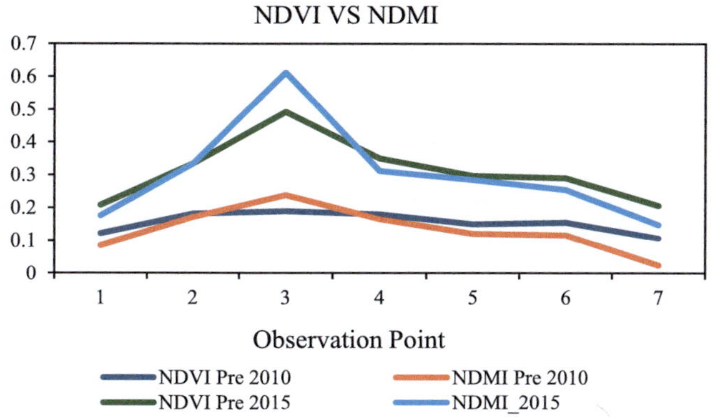

Fig. 13 Relation between NDVI and NDMI for year 2010 and 2015

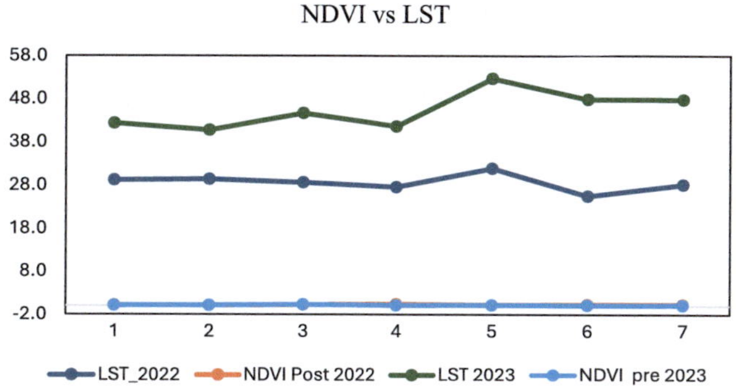

Fig. 15 NDVI versus LST

which can be due to lack of any water body on the surface in the area (Ibrahim 2016). According to Sandholt et al. (2002), regions having higher land surface temperature represent bare soil (edge effect) and regions with lower vegetation represent healthy vegetation. Figure 12, suggest lower LST for regions with vegetation. Many factors influence LST such as climate, land use/land cover, soil moisture, and vegetation cover and health. Vegetation cover is one of the most factor impacting variation and distribution of LST over a region (Ghosh et al. 2022) (Fig. 14).

5.4 Land Surface Temperature and NDVI

A major factor controlling LST can be evapotranspiration as it influences the energy balance of the surface. Low evapotranspiration indicates presences of higher energy for sensible heat (Sandholt et al. 2002). Elevated evapotranspiration indicates decline of the overall health and cover of vegetation (Ghosh et al. 2022). In order to understand the rise in temperature with vegetation cover, Fig. 13 was plotted for the estimation of relationship between LST and NDVI (Fig. 15).

Fig. 14 Spatial and temporal variation in land surface temperature of the study area

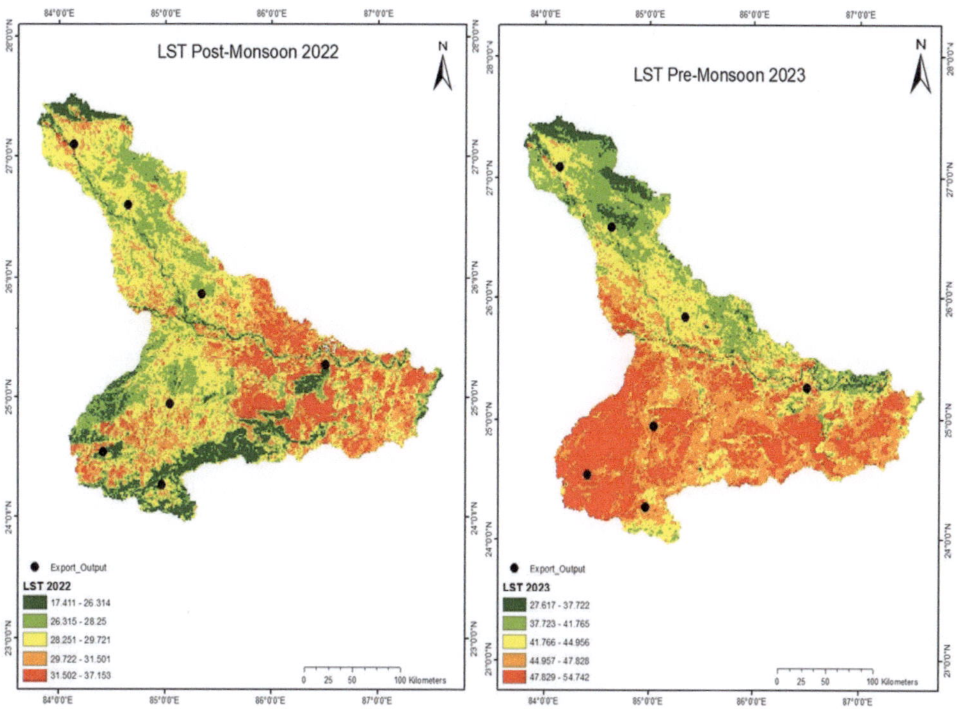

In Fig. 14, for 2022 during post-monsoon, the highest mean LST was 31.9 °C with mean NDVI as 0.19, and lowest mean LST was 25.5 °C with mean NDVI as 0.28. Similarly, for 2023 during pre-monsoon, the highest mean LST was 52.7 °C with mean NDVI as 0.18, and lowest mean LST was 40.7 °C with mean NDVI as 0.12. Overall, post-monsoon has lower LST value due to monsoon rain in comparison with pre-monsoon. Furthermore, the values conclude that higher the NDVI values, lower the LST. This suggests that green cover weakens the effect of urban heat and it suggests inverse relation the plot between NDVI and LST.

6 Concluding Remarks

In Global water cycle, rivers are important component, and it plays an important role in geochemical cycling of nutrient and heavy metals and thus it supports the ecosystem of that region. Worldwide anthropogenic activity is affecting the dissolved and particulate load of the river which is affecting surface water and groundwater chemistry. River systems integrate various land surface hydrologic processes, regulating water quality, food security, enhancing urban water supply and influencing diffuse water pollution. Thus, assessment of surface and groundwater quality of the river basin is important to manage the sustainable water resources in the area. In the present study, hydrochemical and isotopic analysis has been done, which suggests deterioration of water quality at few locations. Isotopic analysis suggests the admixture of precipitation water and glacier melt in surface water. LST from Earth observation data suggests there is an increase in temperature in the basin. It was also observed that as compared to upper and lower part, moisture and vegetation health status was good in middle part due to confluence of river Ganga with the Gandak River Basin. However, detail and extensive study is required to quantity water pollution as well as impact of climate change on water resources as Gandak River Basin is very important with respect to food security in BIHAR.

Acknowledgements Corresponding author Dr. Rina Kumari sincerely acknowledges IWIN laboratory of PRL for providing stable isotopes ($\delta^{18}O$ and δD) analysis facility.

References

Agarwal, D. S., & Bharat, A. (2023). Nature-based solutions for flood–drought mitigation using a composite framework: a case-based approach. *Journal of Water and Climate Change, 14*(3), 778–795.

Agrawal, A. K., Samanta, S. K., Roy, I., & Shukla, R. R. (2021, September). Groundwater Potential of Bihar: An Insight into its Sustainable Development and Management. In *Conference GSI* (pp. 90–103).

Ahmed SA, Chandrashekarappa KN, Raj SK et al (2010) Evaluation of morphometric parameters derived from ASTER and SRTM DEM –a study on Bandihole sub-watershed basin in Karnataka. J Indian Soc Remote Sens 38(2):227–238

Basak, D., Bose, A., Roy, S., & Chowdhury, I. R. (2023). Understanding the forest cover dynamics and its health status using GIS-based analytical hierarchy process: a study from Alipurduar district, West Bengal, India. In *Water, land, and forest susceptibility and sustainability* (pp. 475–508). Elsevier.

Biemans, H., Siderius, C., Lutz, A. F., Nepal, S., Ahmad, B., Hassan, T.,... & Immerzeel, W. W. (2019). Importance of snow and glacier meltwater for agriculture on the Indo-Gangetic Plain. *Nature Sustainability, 2*(7), 594–601.

Bonsor, H. C., MacDonald, A. M., Ahmed, K. M., Burgess, W. G., Basharat, M., Calow, R. C.,... & Zahid, A. (2017). Hydrogeological typologies of the Indo-Gangetic basin alluvial aquifer, South Asia. *Hydrogeology Journal, 25*(5), 1377.

Integrated assessment of extreme events and hydrological responses of Indo-Nepal Gandak River Basin Pawan K. Chaubey1 · Prashant K. Srivastava1 · Akhilesh Gupta2 · R. K. Mall.

CHEMICAL QUALITY OF GROUND WATERIN HYDROGRAPH NETWORK STATIONS OF BIHAR STATE. https://www.cgwb.gov.in/cgwbpnm/public/uploads/documents/170800000811006116630file.pdf

Choudhary, S., Dey, S., Dey, S., Sagar, V., Nair, T., Kelkar, N., 2012. River dolphin distribution in regulated river systems: implications for dry-season flow regimes in the Gangetic basin: river dolphin distribution in regulated river systems. Aquatic Conserv.: Mar. Freshw. Ecosyst. 22 (1), 11–25. https://doi.org/10.1002/aqc.1240.

Chowdary, V.M., Chandran, R.V., Neeti, N., Bothale, R.V., Srivastava, Y.K., Ingle, P., Ramakrishnan, D., Dutta, D., Jeyaram, A., Sharma, J.R., Singh, R., 2008. Assessment of surface and sub-surface water-logged areas in irrigation command areas of Bihar state using remote sensing and GIS. Agric. Water Manag. 95 (7), 754–766. https://doi.org/10.1016/j.agwat.2008.02.009.

Dansgaard, W., 1964. Stable isotopes in precipitation. Tellus 16, 436–438.

Deng, Y., Wang, S., Bai, X., Tian, Y., Wu, L., Xiao, J.,... & Qian, Q. (2018). Relationship among land surface temperature and LUCC, NDVI in typical karst area. *Scientific reports, 8*(1), 641.

R.D. Deshpande, A.S. Maurya, B. Kumar, A. Sarkar, S.K. Gupta (2013). Kinetic fractionation of water isotopes during liquid condensation under super-saturated condition. *Geochimica et Cosmochimica Acta. 100*, 60–72

Fan, Y., Chen, Y., He, Q., Li, W., & Wang, Y. (2016). Isotopic characterization of river waters and water source identification in an inland river, central Asia. *Water, 8*(7), 286.

Gadgil, S., Abrol, Y.P., Rao, S.P., 1999. On growth and fluctuation of Indian foodgrain production. Curr. Sci. 76 (4), 548–556.

GALLO, K.P., MCNAB, A.L., KARL, T.R., BROWN, J.F., HOOD, J.J. and TARPLEY, J.D., 1993, The use of a vegetation index for assessment of the urban heat island effect. International Journal of Remote Sensing, 14, pp. 2223–2230.

Gao, B.C., 1996. Ndwi - a normalized difference water index for remote sensing of vegetation liquid water from space. Rem. Sens. Environ. 58, 257–266.

Ghosh, S., Kumar, D., & Kumari, R. (2022). Assessing spatiotemporal dynamics of land surface temperature and satellite-derived indices for new town development and suburbanization planning. *Urban Governance, 2*(1), 144-156.

Gupta, S.K. and Deshpande, R.D. (2004) Water for India in 2050: First Order Assessment of Available Options. Current Science, 86, 1216–1224.

Gupta, S.K. and Deshpande, R.D. (2005) Groundwater isotopic investigations in India: What has been learned? Current Science, 89(5), 10 SEPTEMBER 2005

Hu, W., Lu, A., & Jia, S. (2013). Estimation of daily maximum and minimum air temperature using MODIS land surface temperature products. Remote Sensing of Environment, 130, 62–73. https://doi.org/10.1016/J.RSE.2012.10.034

Huang, Y. J., Ritschard, R., Bull, J., & Chang, L. (1987). Climate indicators for estimating residential heating and cooling loads.

Hussain, S., Raza, A., Abdo, H. G., Mubeen, M., Tariq, A., Nasim, W.,... & Al Dughairi, A. A. (2023). Relation of land surface temperature with different vegetation indices using multi-temporal remote sensing data in Sahiwal region, Pakistan. *Geoscience Letters*, *10*(1), 33.

Ibrahim, G. R. F. (2017). Urban land use land cover changes and their effect on land surface temperature: Case study using Dohuk City in the Kurdistan region of Iraq. Climate, 5(1). https://doi.org/10.3390/cli5010013

Jeelani, G., Feddema, J. J., van der Veen, C. J., & Stearns, L. (2012). Role of snow and glacier melt in controlling river hydrology in Liddar watershed (western Himalaya) under current and future climate. *Water Resources Research*, *48*(12).

Jha, S.N., Prasad, R.R., 2002. History of Gandak project. Water Resources Department, Government of Bihar, Patna, p. 430.

Jin Q, Wang C (2018) The greening of Northwest Indian subcontinent and reduction of dust abundance resulting from Indian summer monsoon revival. Sci Rep 8:4573

F.N. Kogan Operational space technology for global vegetation assessment Bull. Am. Meteorol. Soc. (9) (2001), pp. 1949–1964.

Kshetri, T. (2018). Ndvi, ndbi & ndwi calculation using landsat 7, 8. *GeoWorld*, *2*, 32-34.

Kumar, D., & Shekhar, S. (2015). Statistical analysis of land surface temperature–vegetation indexes relationship through thermal remote sensing. *Ecotoxicology and environmental safety*, *121*, 39–44.

Kumar, A., Sanyal, P., & Agrawal, S. (2019). Spatial distribution of δ18O values of water in the Ganga river basin: Insight into the hydrological processes. *Journal of hydrology*, *571*, 225–234.

D. Kumar, A.K. Singh, A.K. Taloor, D.S. Singh Recessional pattern of thelu and swetvarn glaciers between 1968 and 2019, bhagirathi basin, garhwal himalaya, India Quat. Int. (2020). https://doi.org/10.1016/j.quaint.2020.05.017

Lapworth, D. J., Dochartaigh, B. Ó., Nair, T., O'Keeffe, J., Krishan, G., MacDonald, A. M.,... & Jackson, C. R. (2021). Characterising groundwater-surface water connectivity in the lower Gandak catchment, a barrage regulated biodiversity hotspot in the mid-Gangetic basin. *Journal of Hydrology*, *594*, 125923.

MacDonald, A, Bonsor, H., Ahmed, K. et al. Groundwater quality and depletion in the Indo-Gangetic Basin mapped from in situ observations. *Nature Geosci, 9*, 762–766 (2016). https://doi.org/10.1038/ngeo2791.

MacDonald, A.M., Bonsor, H.C., Taylor, R., Shamsudduha, M., Burgess, W.G., Ahmed, K. M., Mukherjee, A., Zahid, A., Lapworth, D., Gopal, K., Rao, M.S., Moench, M., Bricker, S., Yadav, S.K., Satyal, Y., Smith, L., Dixit, A., Bell, R., van Steenbergen, F., Basharat, M., Gohar, M.S., Tucker, J., Calow, R.S., Maurice, L., 2015. Groundwater resources in the Indo-Gangetic Basin: resilience to climate change and abstraction. Nottingham, UK, British Geological Survey, 51pp. (OR/15/047) http://nora.nerc.ac.uk/id/eprint/511898/.

H. Mannstein Surface energy budget, surface temperature and thermal inertia Remote Sensing Applications in Meteorology and Climatology, Springer, Dordrecht (1987), pp. 391–410.

Maurya, A. S., Shah, M., Deshpande, R. D., Bhardwaj, R. M., Prasad, A., & Gupta, S. K. (2011). Hydrograph separation and precipitation source identification using stable water isotopes and conductivity: River Ganga at Himalayan foothills. *Hydrological Processes*, *25*(10), 1521–1530.

Mishra, D. S. (2009). Safe drinking water status in the state of Bihar, India: Challenges ahead.

Mukherjee, A., Saha, D., Harvey, C.F., Taylor, R.G., Ahmed, K.M., Bhanja, S.N., 2015. Groundwater systems of the Indian Sub-Continent. J. Hydrol.: Reg. Stud. 4, 1–14. https://doi.org/10.1016/j.ejrh.2015.03.005.

Mukherjee, A., Bhanja, S.N., Wada, Y., 2018. Groundwater depletion causing reduction of baseflow triggering Ganges river summer drying. Sci. Rep. 8 (1), 1–9.

Nath, R., Nath, D., Li, Q., Chen, W., & Cui, X. (2017). Impact of drought on agriculture in the Indo-Gangetic Plain, India. *Advances in Atmospheric Sciences*, *34*, 335–346.

R. Niclòs, J.A. Valiente, M.J. Barberà, M.J. Estrela, J.M. Galve, V. Caselles Preliminary results on the retrieval of land surface temperature from MSG- SEVIRI data in Eastern Spain EUMETSAT 2009: Proceedings of Meteorological Satellite Conference (2009), pp.

O'Keeffe, J., Moulds, S., Scheidegger, J.M., Jackson, C.R., Nair, T., Mijic, A., 2020. Isolating the impacts of anthropogenic water use within the hydrological regime of north India. Earth Surf. Process. Landforms 45 (5), 1217–1228. https://doi.org/10.1002/esp.4799

Pal, D. K., Bhattacharyya, T., Srivastava, P., Chandran, P., & Ray, S. K. (2009). Soils of the Indo-Gangetic Plains: their historical perspective and management. *Current Science*, 1193–1202.

Pandey, A., Padhya, V., Chakra, S., Ganguly, A., & Deshpande, R. D. (2023). Groundwater recharge in Central India and its spatio-temporal variation: Insights and implications from oxygen and hydrogen isotopes. *Journal of Hydrology*, *617*, 129040.

Rina, K., Datta, P. S., Singh, C. K., & Mukherjee, S. (2013). Isotopes and ion chemistry to identify salinization of coastal aquifers of Sabarmati River Basin. *Current Science*, 335–344.

Saha, D., & Sahu, S. (2014). Ground water resource availability and its development potential for enhancing assured irrigation in Bihar state located in middle Ganga plain (January).

Sandholt, I., Rasmussen, K., & Andersen, J. (2002). A simple interpretation of the surface temperature/vegetation index space for assessment of surface moisture status. *Remote Sensing of environment*, *79*(2–3).

Sharma, Y., & Kumari, M. (2024). A Study on Groundwater Irrigation Expansion in Bihar, India. *Asian Journal of Agricultural Extension, Economics & Sociology*, *42*(5), 53–60.

Singh, N., & Sontakke, N. A. (2002). On climatic fluctuations and environmental changes of the Indo-Gangetic plains, India. *Climatic Change*, *52*, 287–313.

Singh, C. S., Anima, K., Kumar, B., Gautam, E., & Kumari, E. A. (2014). Environmental challenge due to climate change in Bihar, Developing state of India. *Journal of Natural Sciences Research*, *4*(13), 21–28.

Sinha, S.K., 2018. The Gharials of the Gandak: a conservation success story. Down to Earth. Sep 25, 2018. https://www.downtoearth.org.in/blog/wildlife-biodiversity/the-gharials-of-the-gandak-a-conservation-success-story-61723.

Sinha, R., Friend, P.F., 1994. River systems and their sediment flux, Indo-Gangetic plains, Northern Bihar, India. Sedimentology 41 (4), 825–845.

Sinha, R., Sarkar, S., 2009. Climate-induced variability in the Late Pleistocene-Holocene f luvial and fluvio-deltaic successions in the Ganga plains, India: a synthesis. Geomorphology 113 (3–4), 173–188.

Sun, D., & Kafatos, M. (2007). Note on the NDVI-LST relationship and the use of temperature-related drought indices over North America. Geophysical Research Letters, 34(24). https://doi.org/10.1029/2007GL031485

Taloor, A. K., Manhas, D. S., & Kothyari, G. C. (2021). Retrieval of land surface temperature, normalized difference moisture index, normalized difference water index of the Ravi basin using Landsat data. *Applied Computing and Geosciences*, *9*, 100051.

J.A. Voogt, T.R. Oke Thermal remote sensing of urban climates Rem. Sens. Environ., 86 (3) (2003), pp. 370–384

WENG, Q., 2003, Fractal analysis of satellite-detected urban heat island effect. Photogrammetric Engineering and Remote Sensing, 69, pp. 555–565.

Whitford V., EnnosA. R. & Handley J. F.2001 'City form and natural process' - Indicators for the ecological performance of urban areas and their application to Merseyside, UK. Landscape and Urban Planning 57 (2), 91–103

Xie, C., Zhao, L., Eastoe, C. J., Wang, N., & Dong, X. (2022). An isotope study of the Shule River Basin, Northwest China: Sources and groundwater residence time, sulfate sources and climate change. *Journal of Hydrology*, *612*, 128043.

Yue, W; Xu, J; Tan, W; Xu, L (2007). The relationship between land surface temperature and NDVI with remote sensing: application to Shanghai Landsat 7 ETM+ data. International Journal of Remote Sensing, 28(15), 3205–3226. https://doi.org/10.1080/01431160500306906

Remote Sensing of Active Tectonics and Geomorphic Signatures for Sustainable Development

Gurav Chandrakant and Md Babar

Abbreviations

AF	Basin asymmetry factor
CFB	Continental flood basalt
DVP	Deccan Volcanic Province
GIAT	Geomorphic indices of active tectonic
HI	Hypsometric integral analysis
KLZ	Kurduwadi lineament zone
PGC	Peninsular gneissic complex
RB	Bifurcation ratio
Re	Basin elongation ratio
RLP	River longitudinal profile
SEDVP	South East Deccan Volcanic Province
SL	Stream length gradient ratio
SoI	Survey of India
T	Transverse topographic symmetry
TD	Trellis drainage
W	Watersheds
WBEDCSZ	Western boundary East Dharwar Craton strike-slip zone

1 Introduction

Most of the study area is covered by the Deccan Trap, which were formed due to sub-horizontally stratified basaltic lava flows. Seismic activity has been recorded since 1993 in the region surrounding the Lendi river, particularly in the Nanded and Latur districts of Maharashtra. In addition to the seismicity, other geological evidences also supports the idea that this terrain is tectonically active (Sukhija et al. 2006; Babar et al. 2012; Chunchekar 2012). Identification of the surficial expressions of ancient faults, which are present within and below the Deccan Trap, is very difficult to identify. Hence, most of the researchers have focused their research on the soft sediments study. Some of them are Peshwa and Kale (1996), Sukhija et al. (2006), Babar et al. (2012), and Kaplay et al. (2017). The investigation of the brittle sheared basaltic dykes from Sambhajinagar (Aurangabad) region reveals that the trend of dyke is similar to Narmada Tapti lineaments (Babar et al. 2017). The micro-seismic activities are recorded in Nanded city since November 2006 (Kaplay et al. 2017). In Kinwat area, Kaplay et al. (2019) have studied northwest–southeast strike-slip faults and lineaments and found that this area will be future location of intra-plate seismicity in central India.

The northeast–southwest trending strike-slip faults have also reported in the basement granite from Diglur area by Babar et al. (2019). Phani et al. (2017) conducted both megascopic and microscopic sharing studies of the granitic basement in Manar river adjoining to the Lendi river. The authors Kaplay et al. (2017) and Babar et al. (2019) named this region as "South East Deccan Volcanic Province (SEDVP)." Granitoid belt exposed at Diglur town (eastern part of Lendi river) belongs to "Western boundary East Dharwar Craton strike-slip zone (WBEDCSZ)" (Babar et al. 2019). Kaplay et al. (2013, 2017) reported deformative structures such as brittle fracturing and displacement of pipe amygdales (minor faults), antiform folds with associated brittle/brittle–ductile faults dipping toward the east, ramps with flat planes dipping toward the east, ductile shear of basaltic veins, and Quaternary deposits deformation in SEDVP.

Many researchers have focusing research activity in and around Terna River after the Killari earthquake, which was recorded in 1993. The study reveals that asthenosphere beneath the Latur region lies at a much shallower depths

G. Chandrakant (✉)
Department of Geology, Rajaram College, Kolhapur, Maharashtra 416004, India
e-mail: chandrakantgurav123@gmail.com

M. Babar
Department of Geology, Dnyanopasak College, Parbhani, Maharashtra 431401, India

D. Kumar et al. (eds.), *Data Analytics and Remote Sensing-Based Analysis for Sustainable Development*, Advances in Science, Technology & Innovation, https://doi.org/10.1007/978-3-031-98095-4_9

of around ~100 ± 10 km compared to 175 km below the adjacent western Dharwar Craton, and ~ 250 to 300 km in other global stable shields (Polet and Anderson 1995; Agrawal and Pandey 1999; Artemieva and Mooney 2001). Asthenosphere below Latur might also be contributing some way, to the continuous buildup of localized stress along weak zones due to differential isothermal rise and large lateral temperature differences (Agrawal and Pandey 1999). Another geochemical study reveals that up to 338 m of crystalline basement has been consisting up of basaltic composition, is consists of CO_2, Cl, FeO, and CaO-rich, high-density (2.82 g/cm^3) - high velocity (avg. Vp: 6.20 km/s), and shows amphibolite-granulite facies rocks. This study pointed out that the nucleation of the Killari earthquake may be related to the regional metasomatism (Pande et al. 2009). Block faulting reported by Venkatarao and Prasad (1995) near Latur and Kurduwadi area based on geophysical study. The evidence like lift over of historical civilization, thick alluvium in the river channel, and abandoned river channels point toward continuous seismicity occurs in the Terna river basin (Sukhija et al. 2006; Babar et al. 2012). Peshwa and Kale (1996) found that NW–SE trending Kurduwadi Lineament Zone (KLZ) that enters Deccan Volcanic Province (DVP) near Gulbarga, is of Precambrian origin, and it also reflects tectonic evidence in Tertiary to Quaternary period. They also observed that the river like Sina, Terna, and Dhod along the KLZ indicates strong lithological structural control in the channel development. Rajendran et al. (1996) argued that the 1993 Latur event is part of a series of events, for which they have found geologic evidence in trenches made across the 1993 rupture area. Sukhija et al. (2006) observed paleoseismic evidence like faults and liquefaction features along the Terna River, two near Ter, and one each at Halki and Shivoor villages. Similarly lineaments, topographic sinuosity index, river longitudinal profile, radiometric dating of charcoal samples collected from folded and faulted Quaternary sedimentary formation, etc., pointed that seismic activity might have occurred in Terna River between 120 AD and AD 1671 (Babar et al. 2012). Rajendran (1997) and Rajendran and Rajendran (1999) carried out radiocarbon dating (from flexure, warps, buckle folds, and vertical offsets) and used archeological evidence at Ter suggesting that the area was experienced earthquakes occurs around 350 to 450 AD.

Quantitative physiographic method can be helpful for evolution and nature of streams (Horton 1945; Abrahams 1984). Hence, geomorphic evidence indication tectonic activities need to be associated with seismicity by considering factors such as drainage pattern and the presence of soft sediment deformation in alluvial and colluvium deposits (Chetty and Rao 1994). Geomorphic indicators can reveal relationship between basin morphology and tectonics (Keller and Pinter 1996). The geomorphic indices of active tectonic (GIAT) parameters are considered important for studying the tectonic nature on a large scale. This parameter study has been carried out in Maharashtra, by Kale and Shejwalkar (2008), Herlekar and Sukhtankar (2011), and Gurav et al. (2020). Kale and Shejwalkar (2008) reported low tectonic activity in Western Ghats using the GIAT and found that it does not help in detecting the subtle imprints of the tectonic activity, even if the DVP has undergone significant uplift in recent time. They also pointed out that tectonics may leave their faint imprints on the present-day landscape or river. In the coastal tract extending from Dabhol Creek Jaigarh Creek the nature and spatial distribution of geomorphic features along the coast indicates that area is under the neotectonic activity (Herlekar and Sukhtankar 2011). Singh (2008a, b) and Singh and Jain (2009) also find tectonic characteristics in Mohnad ridges, NW Himalaya, India. Other examples of tectonic study in India are Sonai river basin by Dutta and Sarma (2013) and Daaki fault of Northeast India by Sharma and Sarma (2017). Gurav et al. (2020) using GIAT parameters pointed some structural control in the Dhamani River. All the reports show that GIAT parameters can be useful to study the tectonic nature of the basin. The study highlights the following objectives:

- To analyse GIAT parameters such as bifurcation ratio (Rb), basin elongation ratio (Re), hypsometric integral analysis (HI), basin asymmetry factor (AF), transverse topographic symmetry (T), stream length gradient ratio (SL), river longitudinal profile (RLP), lineaments, and the abnormal behavior of streams with the main river for study the basin tectonic evidences.
- To understand of neotectonic activity in the form of lineament, knick points along the stream bed, and deposition character of Quaternary sediments.
- To find the GIAT parameters relations with the watersheds in Deccan Trap, and granite terrain.

2 Study Area

Lendi River originates in Balaghat hill range on the eastern slope. And it confluences with Manjra river near Tamlur village in Diglur tehsil in Nanded district of Maharashtra, India. The area is bounded by latitude 18°17′20″N–18°41′26″N and longitude 76°53′23″E–77°45′9″E and covering 2536.26 km^2 area (Fig. 1). The study area has a semiarid climate and receives rainfall primarily from southwest monsoon between June to September. In this region, the average maximum temperature reaches 42 °C, while the average minimum temerature stands 10 °C. The area registers its highest elevation at 688 meters above mean sea level (msl), observed at the basin divide southwest of Rohina village in Udgir tehsil (Latur district). This site is situated along the Tiru river, a significant tributary of the

Lendi river. In contrast, the lowest elevation of 343 meters (msl) is noted near Tamlur village in Diglur tehsil in Nanded district, Maharashtra, India. After its origin, Lendi River takes a trend northwest up to Udgir tehsil, then, it takes a turn to north-east up to Mukarmabad village, and then it flows from west to east. Most of the rivers originated over the Peninsular Indian Plateau and flow eastward, indicating that the region is easterly tilted, which shows Cenozoic morphotectonic evolution (Dole et al. 2022).

3 Geology

Lendi River spreds ~81.648% by Deccan Volcanic Basaltic (DVB) rock, and granite and granite gneissic is ~18.31%. Geologically, the entire study area is characterized by three types of rock formations: Quaternary deposits of Pleistocene to Holocene age, Deccan Volcanic Province (DVP) with upper (late) Cretaceous to lower (early) Eocene age, and Peninsular Gneissic Complex (PGC) of Precambrian age as documented by Gurav in 2019 and illustrated in Fig. 3. The alluvial deposits found throughout the region comprise a mixture of pebbles, cobbles, sand, silt, and clay derived from weathered granite and basaltic rock fragments, extensively studied along the river. This geological composition signifies a younger stage in the stratigraphic succession of the area, as detailed in Wesanekar's 1994 study and presented in Table 1. The sediment thickness along the Lendi river ranges from 2.0 m to 8.0 m. These sediments were deposited over the DVP. The DVP features horizontally disposed continental flood basalt (CFB), resulting from fissure type of lava eruption, as described by Watts and Cox in 1989. This geological formation spans across the Indian states of Maharashtra, Karnataka, Gujarat, and Madhya Pradesh states in India. Commonly known as the Deccan Trap, the DVP ranks among the largest CFB provinces in Earths history, encompassing an extensive area of 500,000 km^2. It predominantly comprises horizontal to sub-horizontal basaltic lava flows intruded by mafic dykes, as outlined by Dole et al. in 2022. The passage is closely linked to the terminal Cretaceous mass extinction, as indicated by Kale et al. (2020) and Krishnamurthy (2020). The study area exhibits predominantly "aa" type (compact) horizontal lava flows, with occasional occurrences of vesicular amygdaloidal basalt (compound pahoehoe) formations in southern, central, and eastern Maharashtra. The lava flows are intermittently separated by red bole beds (Tachylitic bands) which were utilized by Chunchekar in 2012 for lava flow marking in Terna River. In Godbole et al. (1996) and Gurav (2019) assigned the basalt flows in the Lendi river to the Ajanta formation, which is considered the stratigraphic equivalent of the Upper Ratangad formation in the Wai subgroup of western Maharashtra. The eastern part of the basin is covered by Precambrian crystalline rocks commonly known as Mysore plateau (Dole et al. 2022). In the southern part of Nanded district, Maharashtra, Proterozoic granitoid rocks (2074–2385 Ma) exposed, marking the northeastern extension of PGC in Eastern Dharwar Craton (EDC). Deformation in granite has been extensively reviewed by various researchers, including Banerjee et al. (1993, 2008), Wesanekar (1994), Wesanekar and Patil (2000), Phani et al. (2017), Babar et al. (2019).

Table 1 Stratigraphic succession of the study area

Stratigraphic age	Lithology
Pleistocene to holocene	Clay mixed with silt, sand, and gravel
Lower cretaceous to upper eocene	Deccan trap basalt lava flows
Proterozoic to achaean	Quartz-feldspar-epidotic veins
	Dolerite dykes
	Younger pegmatites
	Older pegmatites and aplites
	Pink granite
	Gray granites, granodiorite
	Migmatites and granitic gneiss
	Basic enclaves (diorite, etc.)

Source Wesanekar (1994)

4 Methodology

In this research, ArcGIS software was used to analyze morphotectonic parameters for the Lendi river and its 22 watersheds. In this study, nine commonly used GIAT parameters were used, which is represented in Table 2. These parameters were applied for Lendi River and its watersheds. The result of the each and every watershed is classified based on geological formation. The watershed numbers 01, 02, 03, 04, 05, 06, 16, 17, 18, 19, 20, 21, and 22 are occupied by basaltic rocks, the watershed numbers 07, 08, 10, and 15 are occupied by both basalt and granite, and the watershed numbers 09, 11, 12, 13, and 14 are consisting of only granitic formation. In short methodology adapted for present work is also represented in flowchart (Fig. 2). For this study, Survey of India (SoI) topographic maps number 56 B/14 (E43K14), 56 B/15 (E43K15), 56 F/3 (E43L3), 56 F/6 (E43L6), 56 F/7 (E43L7), 56 F/10 (E43L10), 56 F/11 (E43L11), and 56 F/14 (E43L2) on the scale 1:50000 edited in 2011 have been used. These topographic maps were used for base map preparation (Fig. 1), drainage map (Fig. 3), basin asymmetry and transverse topographic symmetry map (Fig. 4), and data manipulation. The drainages in the toposheets were digitized using Strahler (1964) method of stream ordering, and onscreen digitization technique was adapted. For these, each first order, second

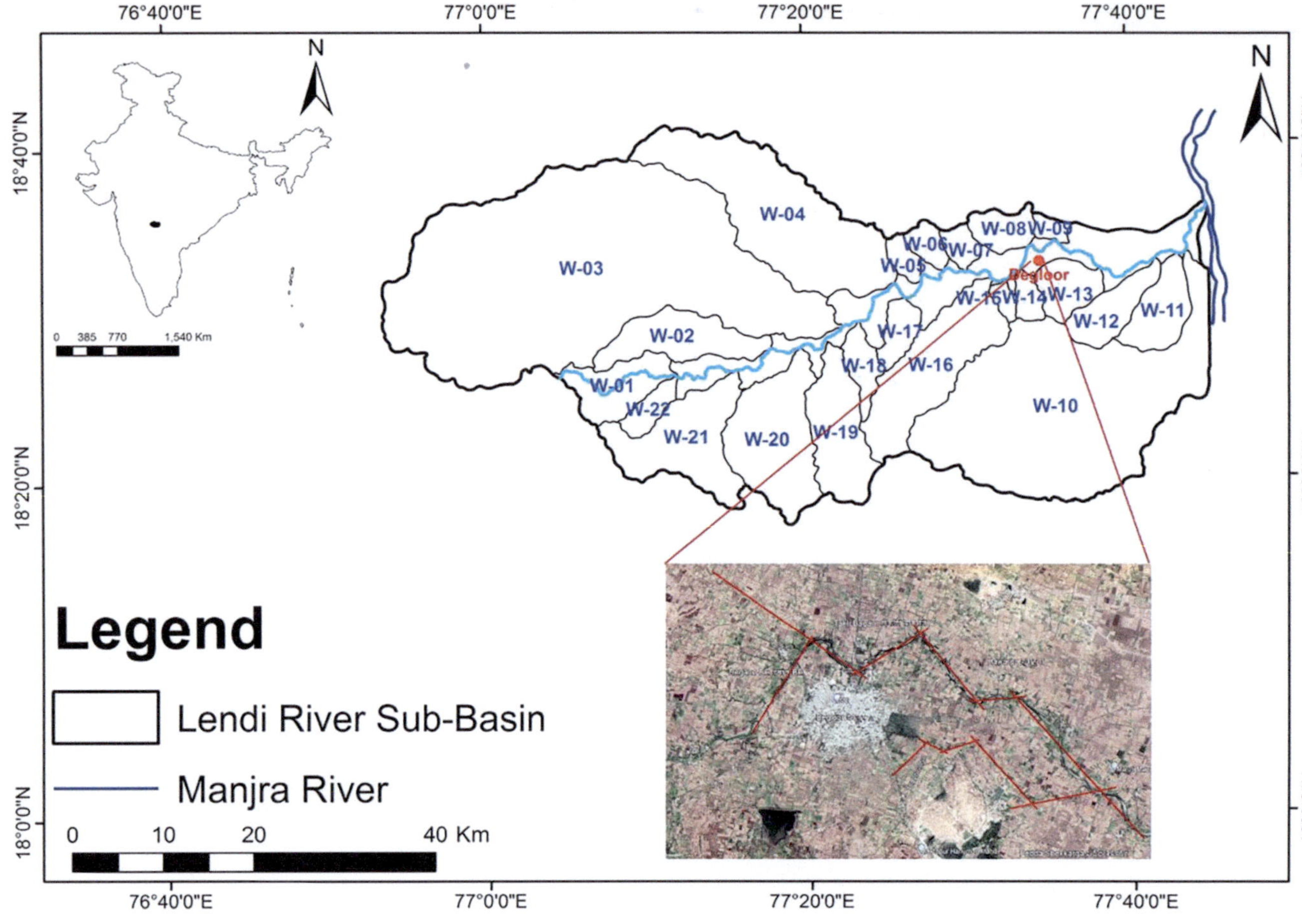

Fig. 1 Location map of Lendi River, and google map shows sites of structurally controlled lineament, site **a** shows waterfall with gorge and strike-slip fault site (Babar et al. 2019), site **b** shows tilted sediments near Malkapur, and site **c** shows cross-bedding structure observed near Begam Takali village. All the sites are located near Degloor town

order … and so on was digitized in different color as line. Four types of drainage pattern were observed in the Lendi river sub-basins, these are shown in Fig. 3: dendritic drainage (DD), sub-parallel drainage (SPD), radial drainage (RD), and trellis drainage (TD). The pattern of any drainage system reflects geology and its structures, geomorphic history, and basin slope (Thornbury 1968). The dendritic drainage (DD) pattern is characterized by irregular branching of tributary streams in all directions, while trellis drainage (TD) gives primary evidence of structural controls of most stream courses. Radial drainage (RD) pattern, also known as centrifugal pattern, is formed by the streams which diverge in all directions from a central higher point. The abnormal behavior of streams with the main river is also observed, and it is shown in Fig. 3 L1–L6.

The basin asymmetry has been mapped using the stream order. To map the basin asymmetry line, first- to seventh-order stream lines have been chosen for mark the line, and then they were used to study the area of the both sides in the Lendi River. Using the formula shown in Table 2, the area of both the sides of the Lendi River is calculated, and basin asymmetry map are shown in Fig. 4. Similarly, water divider line (Basin boundary) has been used to mark the transverse topographic symmetry (T) line. For this, both the sides (north and south boundary) of Lendi River, have calculated the length between them and then divided it into two equal parts, i.e. middle points. Using the same method, nearly, twenty-one points were marked in the basin, these points were then connected to create a central line, referred to as the transverse topographic symmetry line. For calculating the T value, the formula from Table 2 is used, and the results are displayed in Fig. 4. For visualization of the obtained results, maps classifying watersheds were prepared shown in Fig. 5a, b, c, and d and Fig. 6a, b. The examination of lineaments involved analyzing the straight course of the streams, as illustrated in Fig. 6c (rose diagram). This lineament rose diagram then correlated with the strike-slip fault rose diagram shown in Fig. 6d which was previously studied by Babar et al. in 2019.

For the accuracy of the results, contours in the study area were manually digitized at a 20 m interval. For this, topographic maps have been used as a base map. This 20-m contour interval has been further used for hypsometric integral

Table 2 GIAT parameters used for Lendi River

Symbols	Formula	References
Rb	Rb = Nu/Nu+1 where Rb is the bifurcation ratio, and Nu is the total number of streams of the next higher order	Strahler (1964)
Re	Re = $(2\sqrt{A}:\sqrt{\pi})/LB$ where, A = Basin area, L = Basin length	Bull and McFadden (1977)
HI	% Hypsometry integral method	Strahler (1964), Cannon (1976)
AF	AF = 100(Ar/At) where Ar is the right-hand side area of drainage looking downstream and At = Total area	Keller and Pinter (1996)
T	T = Da/Dd where Da is the distance from middle of basin of active channel, and Dd is the distance from basin midline to basin divide	Cox (1994)
SL	SL index = (H1-H2)/(ln L2-L1) where H1 and H2 are elevation of each end, and L1 and L2 are the distances	Hack 1973
RLP	Relationship between height and distance	Hack 1973
Lineaments	Linear surficial features especially straight course of the river	Kaplay et al. (2016)
ABS	Abnormal behavior of the streams such as 90° turn with the main stream	Babar et al. (2012)

Basin elongation ratio (Re), hypsometric integral analysis (HI), Basin Asymmetry (AF), transverse topographic symmetry (T), stream gradient length ratio (SL), river longitudinal profile (RLP) and $\pi = 3.14$

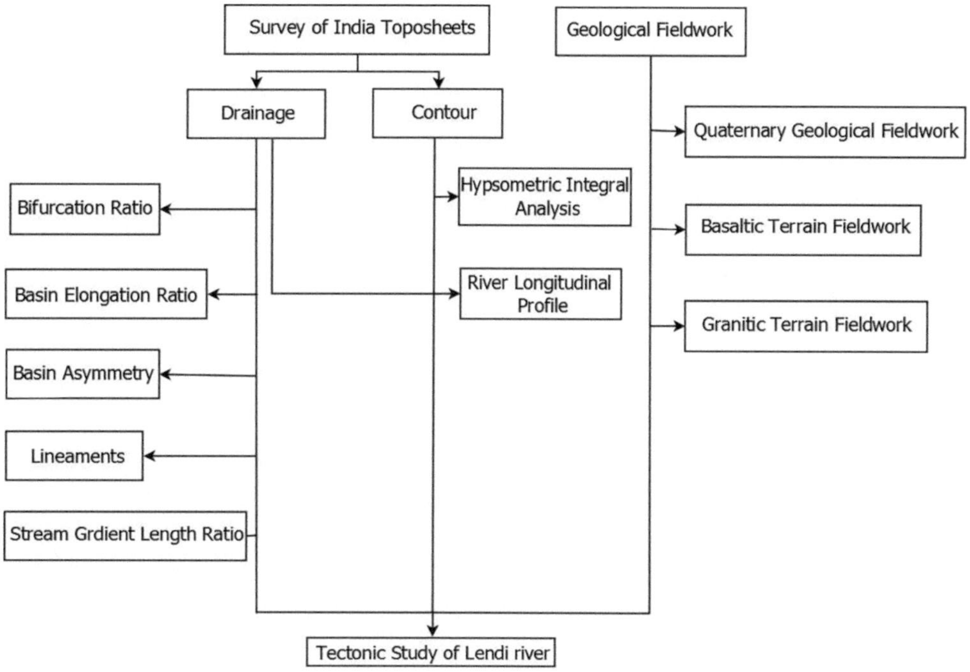

Fig. 2 Detail methodology adapted flowchart of the present work

analysis, i.e., present erosional status (Fig. 7), and preparation of longitudinal profile of the Lendi River as well as its watersheds (Fig. 8). To study the erosional condition, a 20-m contour covered area has been used for percentage hypsometric integral analysis method, which is previously successfully adapted by Gurav and Babar (2018) for Gharni River. The maps have been prepared for studying the youth, mature, and old stage of the river/watersheds. For this, a/A against h/H graphs have been plotted using Microsoft Excel. The percentage of the HI has been calculated by using the covered area below the HI obtained line against total area in the square box. The elevations against length of the main stream data were entered into Microsoft Excel, and it was used for preparation of longitudinal profiles. Same method was also implemented to calculate the percentage area in the longitudinal profile. The GIAT results have been tabulated in Table 3.

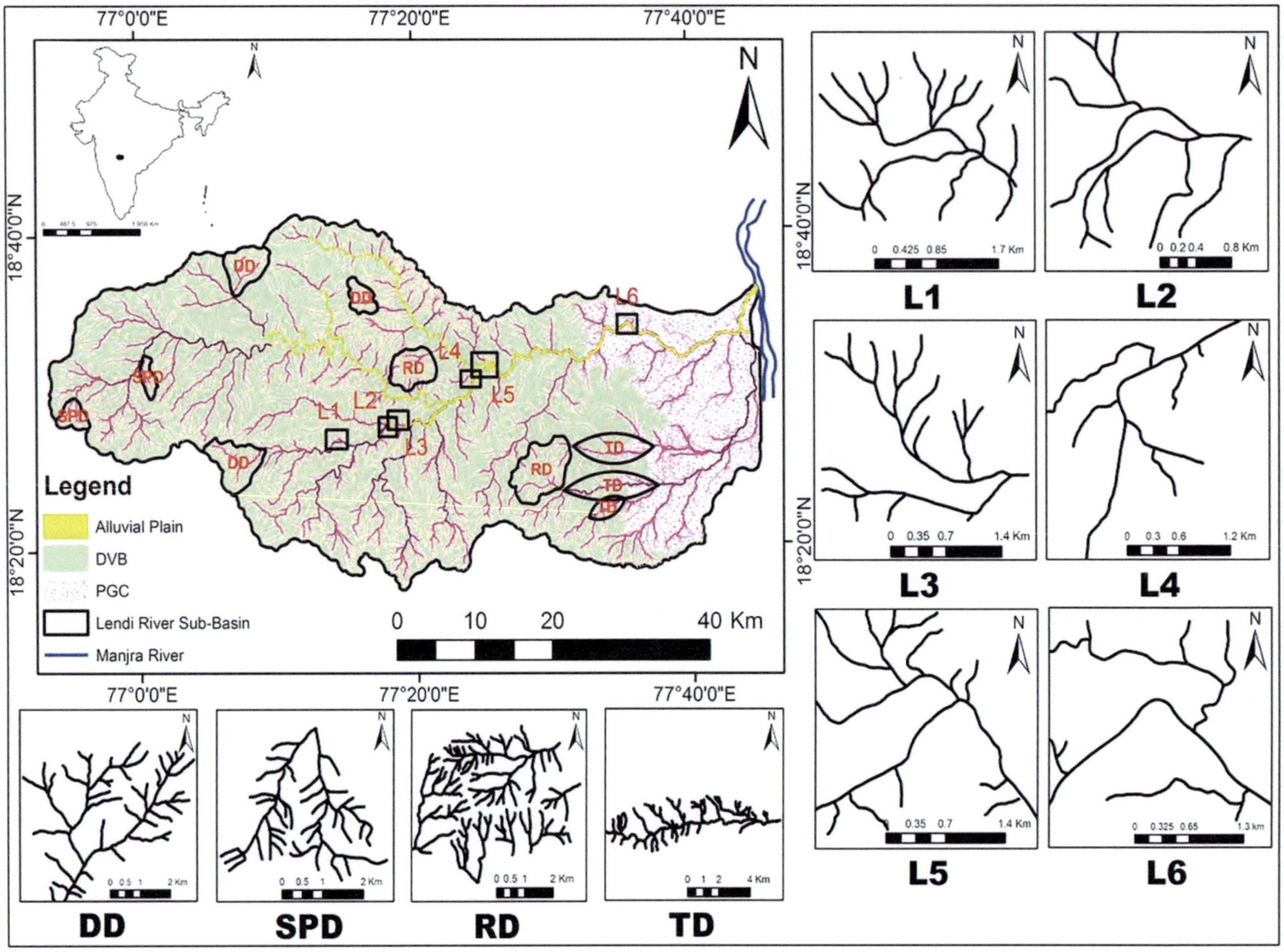

Fig. 3 Drainage map of Lendi River. Sites L1– L6 are the magnifying view locations of anomalous behavior of streams with Lendi River and dendritic drainage (DD), sub-parallel drainage (SPD), radial drainage (RD), and trellis drainage (TD)

5 Results

5.1 Geomorphic Indices of Active Tectonics (GIAT)

The efficacy of the GIAT parameters, including basin bifurcation ratio (Rb), basin elongation ratio (Re), hypsometric integral (HI), basin asymmetry factor (AF), transvers topographic symmetry (T), stream gradient length ratio (SL), river longitudinal profile (RLP), lineaments, and abnormal behavior of streams with the main river have been demonstrated as powerful reconnaissance tools for evaluate the connection between tectonics and basin morphology (Bull and McFadden 1977; Keller and Pinter 1996).

5.2 Bifurcation Ratio (Rb)

In Strahler (1964) defined Rb as the "ratio of number of streams of lower order (u) to the number of streams of higher orders (u+1)." In the Lendi River sub-basin Rb value have ranges from 2.40 to 5.00, an average is 4.16 (Gurav and Babar 2021). The average bifurcation ratio of watersheds in Lendi River have been tabulated in Table 3. For present study, Rb value of watersheds has classified into three categories as 2.92–3.00, 3.01–5.00 and 5.01–6.58 for flat, mountain and structurally controlled basin respectively. Based on this, watersheds classification map is prepared and it has shown in Fig. 5a. Based on this it is observed that Lendi River is flowing in mountainous region but watershed number W-01, W-02 and W-16 shows structurally controlled drainages.

5.3 Elongation Ratio (Re)

In 1956, Schumm defined it as "the ratio between the diameter of a circle of the same area as the drainage basin and the maximum length of the basin". The value of Re obtained from Lendi River sub-basin is 0.63, which indicates a semi-elongated basin (Gurav and Babar 2021). Bull and McFadden

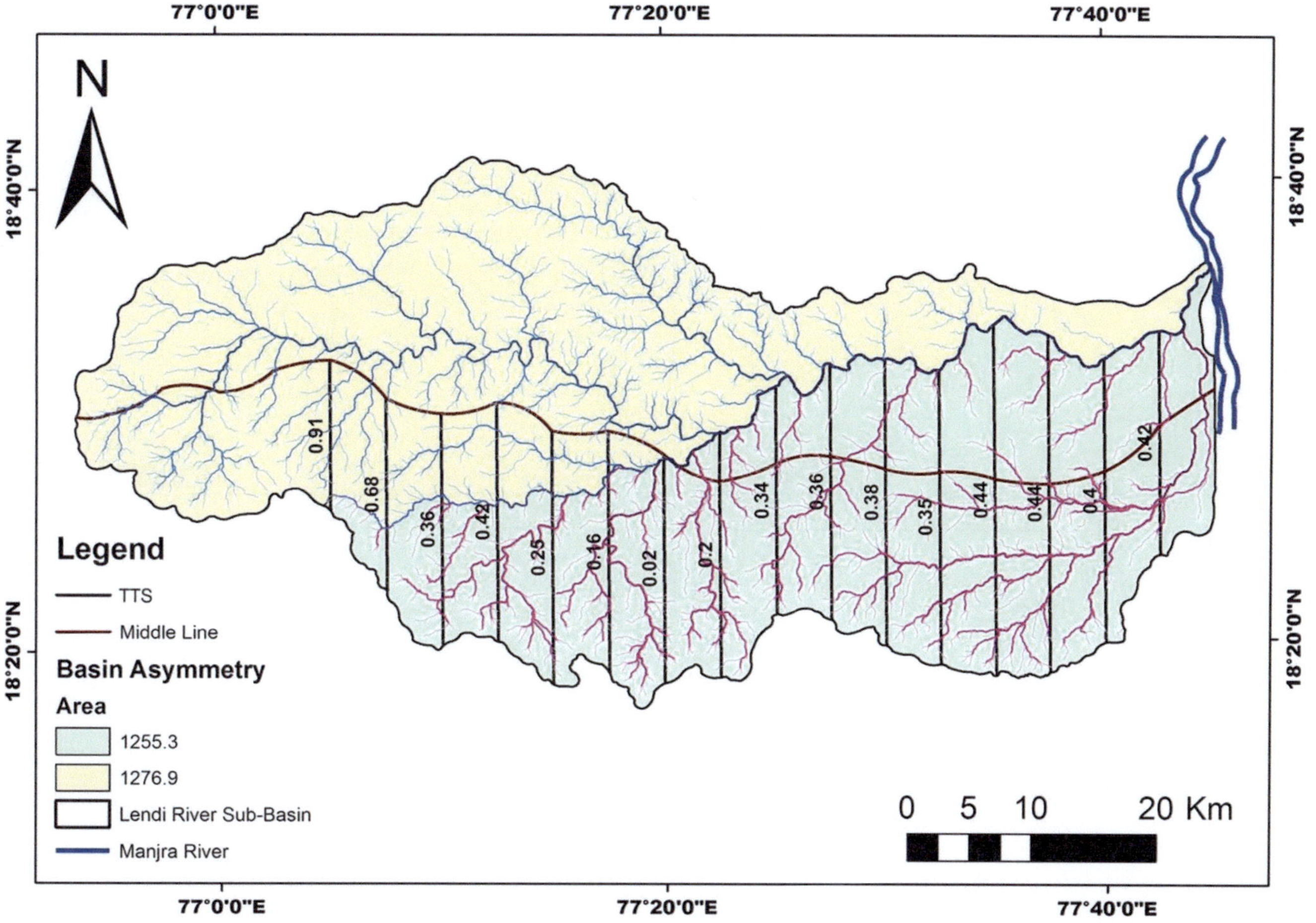

Fig. 4 Basin asymmetry (AF) and transverse topographic symmetry (TTS) map of Lendi River

(1977) categorized Re values into three groups: < 0.50, 0.51–0.75, and > 0.76 for tectonically active, slightly active, and inactive setting, respectively. This classification has been effectively applied by Cuong and Zuchiewicz (2001) in their study of the Lo river fault near Tam Dao in North Vietnam and Gurav et al. (2020) in their investigation of the Dhamani River in Kolhapur district, Maharashtra, India. The watersheds based on above classification are shown in Table 3, and its map is shown in Fig. 5b.

5.4 Hypsometric Integral Analysis

The erosional condition of a watershed is determined by factors such as the climate and geology and tectonic setting, as well as shape, and size (area) of the watershed or sub-basin or basin (Gurav and Babar 2018). The hypsometric integral (HI) curves will give indications of the "cycle of erosion" (Strahler 1956; Ritter et al. 2002). The cycle of erosion divided into the three stages by Strahler in 1956 as 0.00 < 0.30 or 30.00%, 0.3–0.60 or 30.01–60.01%, and > 0.60 or 60.01% for old or monadnock, mature (fully stabilized), and youth (tectonically active), respectively (Strahler 1956). The Lendi River's HI value obtained is 37.41%, signifying that the basin is in the late mature stage of erosional development. The Lendi River's HI graph (Fig. 7) shows that the area below the curve line is under youth, mature, and old stages. The youth stage covers an area of 500.14 km^2 area (19.73%); the mature stage, 1070.34 km^2 (42.27%); and the old stage, 961.78 km^2 (37.98%). The area was estimated on the basis of graph showing hypsometric integral. To study the erosional condition of the watershed within the Lendi river Strahler (1956) proposed a classification that has been used, and it is also represented in Fig. 5c.

5.5 Basin Asymmetry Factor (AF)

The AF is formulated to identify tectonic tilting at drainage basin scale or on large scale, as explained by Hare and Gardnerin 1985 and Keller and Pinter in 1996. According to Hare and Gardner (1985), AF values close to 50 signify a stable region for the stream network, while values lower or higher than 50 suggest the occurrence of tectonic tilting within the basin. AF value of Lendi River sub-basin is very close to the

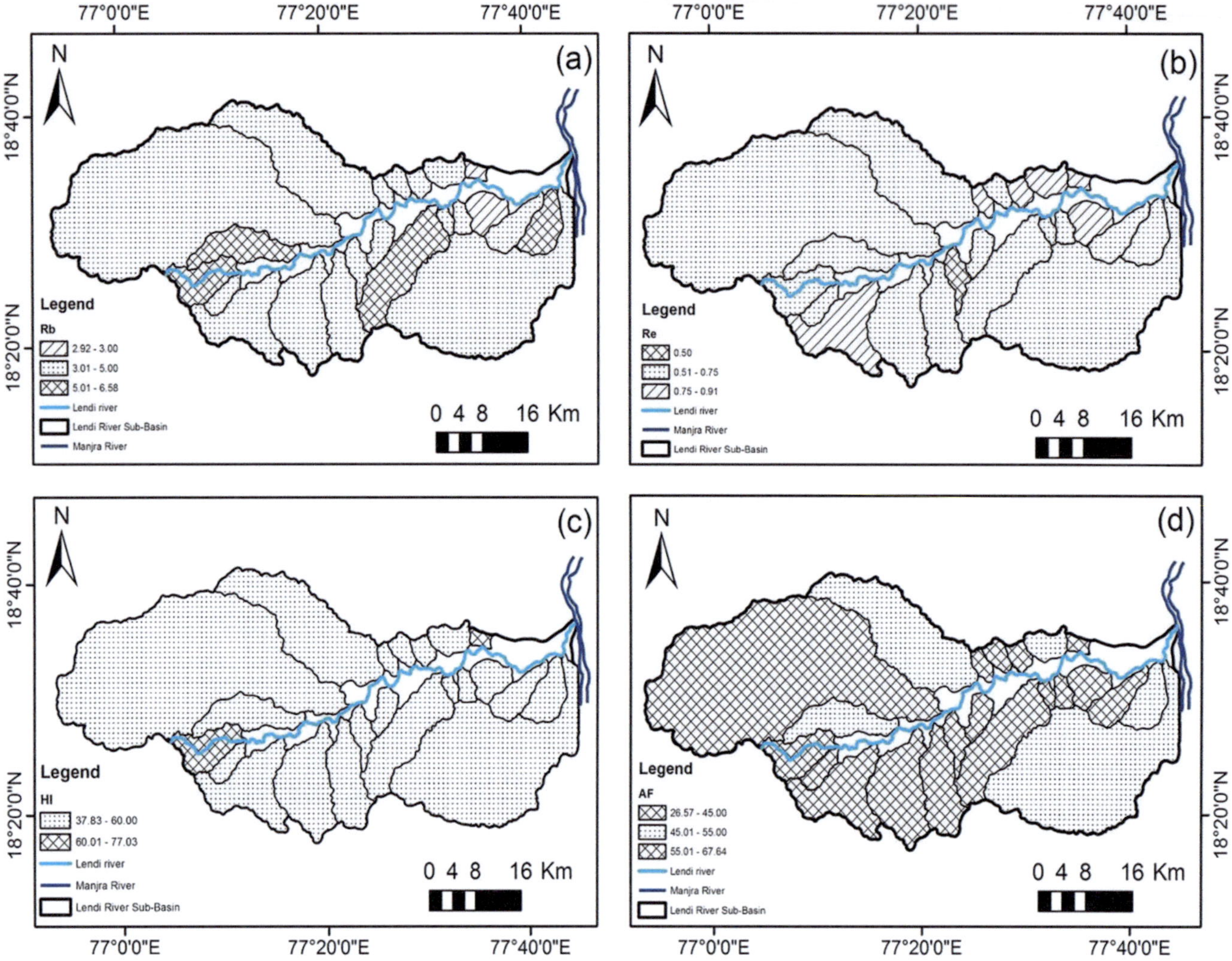

Fig. 5 Classification maps of the watersheds, **a** bifurcation ratio, **b** basin elongation ratio, **c** hypsometric integral analysis, and **d** basin asymmetry map

50.00%, i.e., 49.57%, which indicates that basin is symmetrical and not tectonically tilted (Fig. 4). It has been calculated using the equation, which is shown in Table 1. By using basin asymmetry, Lendi basin is divided into two parts i.e. north block (1277.00 km^2) and south block (1255.26 km^2). The individual morphometric analysis of both blocks has been calculated. The northern block shows Dd is 1.96 km/km^2, Fs is 2.48 $streams/km^2$, Rt is 12.98 km^{-1}, and HI value is 47.40. And, southern block shows Dd is 1.83 km/km^2, Fs is 2.14 $streams/km^2$, Rt is 12.05 km^{-1}, and HI value is 37.85. In this study, AF values between 45 and 55 are considered as near-normal and tectonically stable, while values from 26.57 to 45.00 and 55.01 to 67.63 were assumed to be tectonically tilted watersheds. Based on this, watershed classification is done, and it is shown in Table 3 and Fig. 5d.

5.6 Transvers Topographic Symmetry (T)

It indicates shifting of the stream channel, tectonic tilting of the basin, and nature of the valley as noted by Hare and Gardner (1985). It is a vector quantity ranging from 0 to 1, representing a perfectly asymmetric basin or one that is tilted, respectively (Keller and Pinter 1996). The T values for the Lendi River sub-basin range between 0.02 and 0.91 (Fig. 4), with an average of 0.38. The average T value of watersheds has been used to classify each watershed in three categories: stable (0.09–0.20), slightly active (0.21–0.40), and tectonically active (0.41–0.51). Based on this classification, watershed numbers 06, 07, 08, and 10 show stable condition, watershed numbers 01, 02, 03, 04, 05, 11, 12, 13, 14, 15, 16, 17, 18, 19, 20, 21, and 22 are slightly active, and watershed number 09 is tectonically active (Fig. 6a).

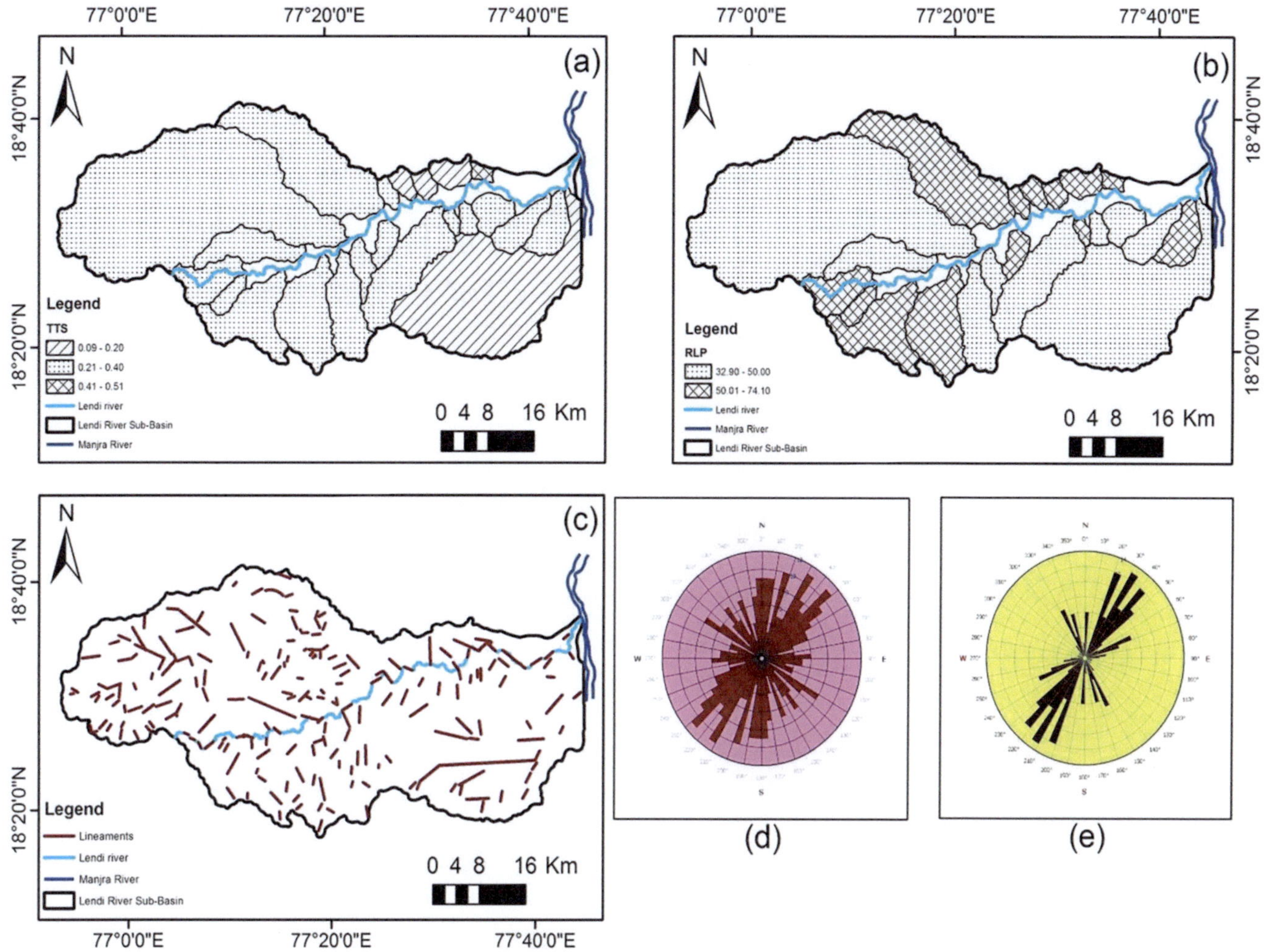

Fig. 6 Classification maps of the watersheds, **a** transverse topographic symmetry, **b** river longitudinal profile, **c** lineaments, **d** rose diagrams, and **e** trend of strike-slip faults near Degloor (Babar et al. 2019)

5.7 River Longitudinal Profile (RLP)

The river longitudinal profile has been plotted from the total drop in elevation from the source to the mouth, and horizontal distance between two points is measured along the channel and is shown in Fig. 8. The concave-up shape of the river longitudinal profile may indicate instable state of channel due to tectonic, climatic, or rock-type perturbation (Hack 1957). The longitudinal profile ratio of Lendi River sub-basin is 26.3. It is used to study the rapid tectonic deformation analysis by Keller and Pinter (1996). Based on the ratio values, two categories were established: > 50.01 as tectonically active (concave) and < 50.00 as inactive or stable (convex). Based on this, it is found that the watershed nos. 01, 04, 05, 06, 07, 08, 09, 11, 15, 17, 20, 21, and 22 are the tectonically active. The watershed nos. 02, 03, 10, 12, 13, 14, 16, 18, and 19 are inactive (Fig. 6b).

5.8 Stream Gradient Length Ratio (SL Index)

The Stream gradient length ratio (SL) initially proposed by Hack (1973), serves the purpose of identifying factors that controls the drainage network. These factors encompass geological setting, alterations in local or base level, tributary confluence, and human interference. Variations in SL index ratio reflect the downstream variations in the discharge and stream power; lithological or tectonic controls on channel slope, and the anomalous uplift within a landscape. This ratio provides a means to comprehend the complex interplay of these factors and their influence on the evolution of drainage networks and landscapes. The SL index of Lendi River is 6.73. The data regarding the SL index values of watershed of Lendi River are tabulated in Table 4. According to Seeber and Gornitz (1983), SL index value less than 2 means there are

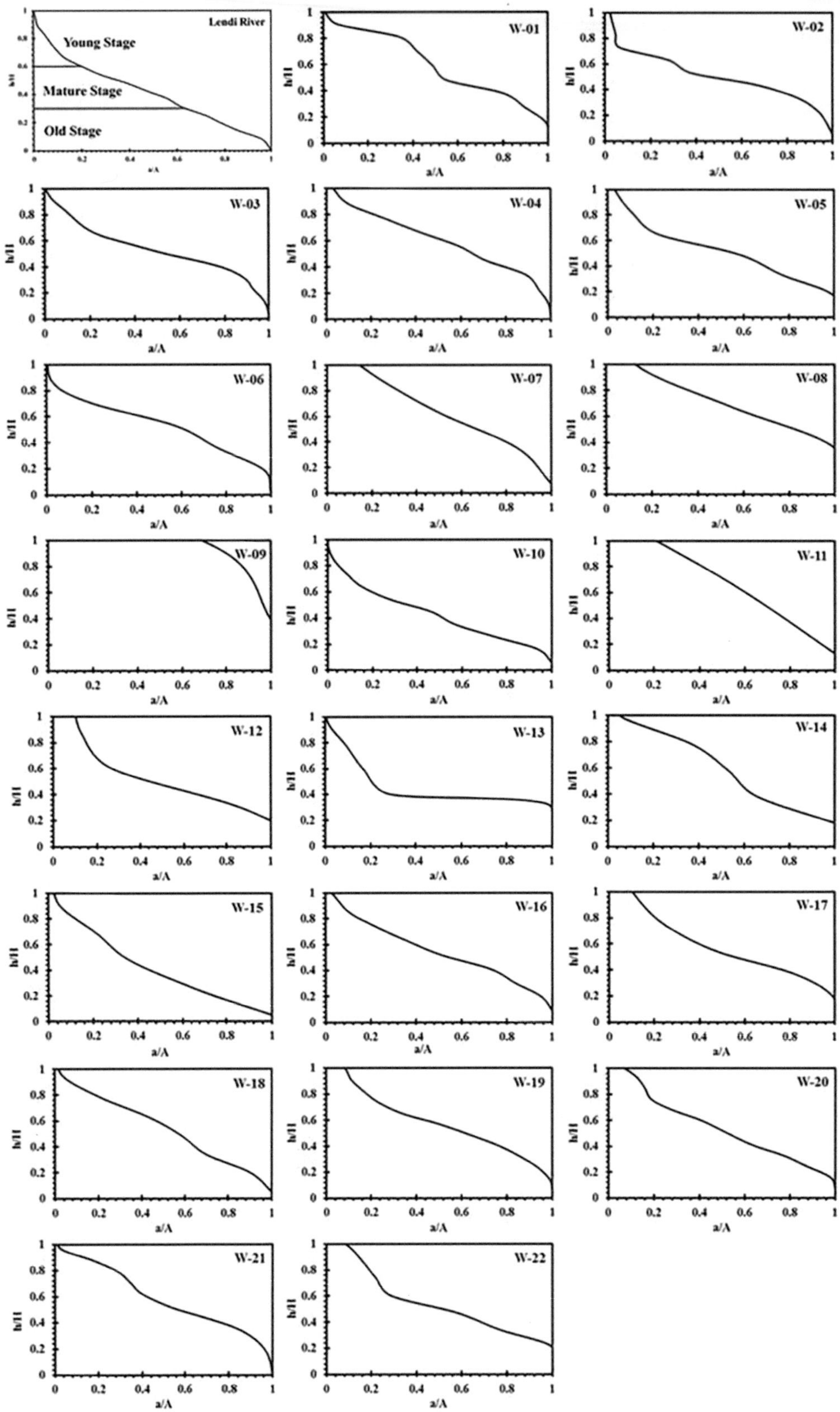

Fig. 7 Hypsometric integral graphs Lendi River and its watersheds

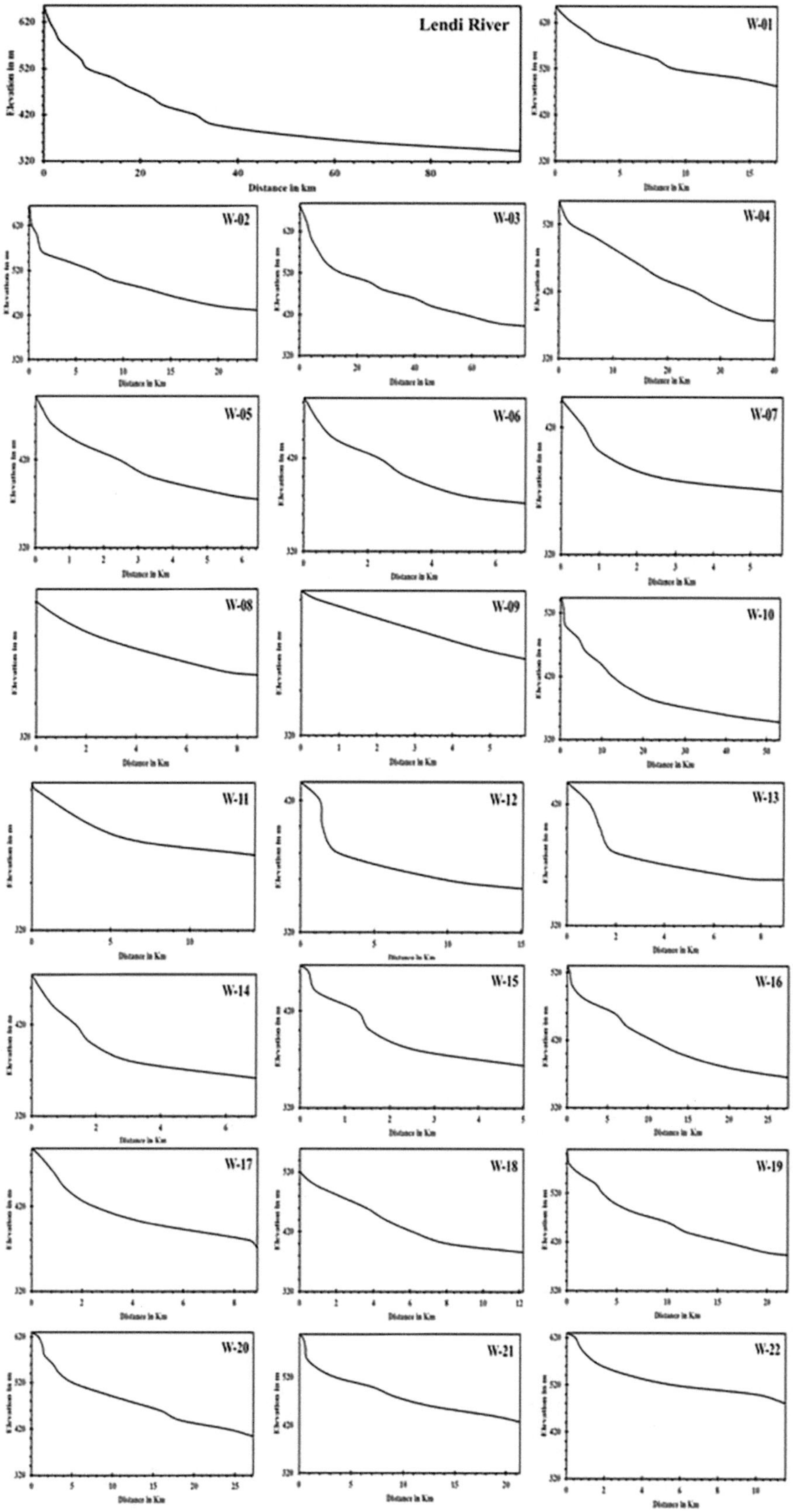

Fig. 8 Longitudinal profiles of Lendi River and its watersheds

Table 3 GIAT analysis of Lendi River watersheds

W- Id	A	Lb	u	Rb	Re	AF	T	HI	RLP
Basaltic formation									
W-01	58.0	11.7	4	5.02	0.73	59.9	0.32	59.6	66.0
W-02	74.2	18.9	4	5.29	0.52	51.9	0.22	42.3	49.2
W-03	752.6	48.2	6	4.28	0.64	40.5	0.28	53.2	40.9
W-04	231.6	31.1	6	3.78	0.55	51.4	0.29	50.8	52.4
W-05	12.9	5.35	4	3.50	0.76	56.4	0.31	43.8	53.0
W-06	15.9	6.01	4	3.22	0.75	41.8	0.12	44.8	52.2
W-16	130.8	22.7	4	6.58	0.57	55.9	0.24	55.0	45.1
W-17	23.1	8.31	4	3.63	0.65	63.4	0.24	49.3	54.0
W-18	24.7	11.1	4	3.53	0.51	67.6	0.24	45.1	47.1
W-19	86.6	18.2	5	4.13	0.58	40.1	0.24	48.4	48.3
W-20	125.7	19.5	6	3.35	0.65	40.0	0.26	54.3	51.3
W-21	113.8	16.9	5	3.88	0.76	36.1	0.34	58.7	55.5
W-22	19.8	8.36	4	4.12	0.60	43.8	0.30	54.7	67.7
Avg.	**–**	**–**	**5**	**4.20**	**0.60**	**49.9**	**0.30**	**50.8**	**52.5**
Granite and basalt formation									
W-07	14.7	5.15	3	4.84	0.84	63.7	0.19	54.4	53.4
W-08	28.1	6.55	4	3.83	0.91	48.3	0.17	52.0	59.8
W-10	445.6	34.4	6	3.75	0.69	54.0	0.09	42.9	32.9
W-15	8.50	4.61	3	3.63	0.71	38.8	0.37	41.9	50.5
Avg.	**–**	**–**	**4**	**4.00**	**0.80**	**51.2**	**0.20**	**47.8**	**49.2**
Granitic formation									
W-09	9.20	4.43	3	3.00	0.77	31.0	0.51	77.0	74.1
W-11	49.0	11.5	3	5.29	0.69	49.9	0.22	57.6	64.8
W-12	42.8	12.8	4	3.33	0.58	40.2	0.29	55.6	44.7
W-13	35.9	7.52	4	2.92	0.90	26.6	0.39	37.8	47.4
W-14	14.0	5.66	3	3.80	0.75	42.3	0.21	49.8	45.6
Avg.	**–**	**–**	**3**	**3.67**	**0.74**	**38.0**	**0.32**	**55.6**	**55.3**

Watershed (W), elongation ratio (Re), Basin asymmetry (AF), transverse topographic symmetry (T), stream gradient length ratio (SL), hypsometric integral analysis (HI) and Longitudinal profile of watersheds (LPW)

no anomalies, the value obtained between 2 and 10 is second-order anomaly, and the value equal to or exceeding 10 is first-order anomaly. Based on this classification, it is found that the Lendi River and its 22 watersheds shown second-order anomaly.

5.9 Lineament Analysis

In this work, a first- and second-order stream lineament was not considered because most of them are formed due to regional slope of the area. There are 208 lineaments marked. Out of these, 207 are straight, and one is concave curved. It is found that less than 2.30 km length lineaments are 79.32% and its length is 57.56% and 2.30–16.43 km length lineaments are only 20.68%, but its length is 42.44% of all the lineaments. All the lineament orientations are plotted in rose diagram and found that the most of the lineament direction is NE–SW (Fig. 6d). The second classification was carried out based on orientation direction of the lineaments. For this, every 10° azimuth range of N90° and N270° has been determined. Accordingly, total numbers of lineaments and number percentage in each 10° interval have been computed, and also length measurements and length percentage have been presented in Table 5. There are 207 lineaments studied in this classification. The total number of lineaments in between N0° and N90° is 122. The average number is 13.6 with number of frequency (N%) of 58.9%. Similarly, total number of lineaments between N270° and N360° is 85, and the average is 9.40 with number of frequency (N%) 41.1%. The total length of lineaments between N0° and N90° is 217.9 km, and the average is 24.2 km with length of frequency (L%) 56.6%. Similarly, total length of lineaments between N270° and N360° is 167.4 km, and the average is 18.6 km with length

Table 4 Stream gradient length ratio of Lendi River watershed

W-Id	δH total (m)	δL (m)	L (m)	SL section = (δh/δL)*L	SL total = δh/log(L)	SL index = SL section/SL total
Basaltic formation						
W-01	173	13,130	17,100	225.31	40.87	5.51
W-02	242	19,500	24,050	298.46	55.24	5.40
W-03	294	45,990	78,420	501.32	60.07	8.35
W-04	178	31,830	40,020	223.80	38.68	5.78
W-05	117	3400	6460	222.30	30.71	7.24
W-06	113	5840	6930	134.09	29.42	4.58
W-16	165	22,450	27,800	204.32	37.13	5.50
W-17	118	8300	8900	126.53	29.88	4.24
W-18	143	11,050	12,200	157.88	34.99	4.51
W-19	216	18,080	22,180	264.98	49.70	5.33
W-20	225	19,500	27,190	313.73	50.74	6.18
W-21	182	19,790	21,340	281.65	42.12	6.69
W-22	148	8030	11,630	214.35	36.40	5.89
Average						**5.78**
Granite and basalt formation						
W-07	74	5240	5830	82.33	19.65	4.19
W-08	51	7300	8790	61.41	12.93	4.75
W-10	196	37,040	53,180	281.41	41.48	6.78
W-15	103	4750	5010	108.64	27.83	3.90
Average						**4.91**
Granitic formation						
W-09	30	3400	5920	52.23	7.95	6.57
W-11	33	11,310	14,060	41.02	7.96	5.15
W-12	81	12,650	15,110	96.75	19.38	4.99
W-13	80	7400	8920	96.43	20.25	4.76
W-14	113	5850	6920	133.67	29.43	4.54
Average						**5.20**

Watersheds (W), Relief of the watershed in m (δH), Length of the watershed in m (δL), stream length in m (L)

of frequency (L%) 43.5%. It is observed that 98 lineaments are oriented in N60° direction with number of frequency (N%) 47.3%, and the total length of the lineaments is about 157.7 km with length of frequency percentage (L%) 40.9%.

5.10 Abnormal Behavior of Streams with the Main River

The abnormal behavior of streams with the main river can illustrate the structural control on development of the drainage basin. It includes sharp bend in streams, sudden right angle turn of the streams, stream flowing parallel to the major river for considerable distances, convergence and divergence of streams, etc. The streams marked with rectangles at location (L) 1–6 (Fig. 3) and its magnifying view (Fig. 3 L1–L6). It is found that the streams are flowing parallel to Lendi River for considerable distance in downstream direction instead of joining at shorter distance and then join to the main river indicating the structural control over drainage. Anomalous or sudden changes in the river's drainage pattern such as sharp bends, linear courses, convergence or divergence, deflection drainages can be inferred to indicate sub-surface faults (Chetty and Rao 1994).

5.11 River Bed and Nature of Quaternary Sediments

The Quaternary uplifts have controlled the development of the river network in the Lendi river. Such evidence was studied with the help of kink points, pothole development, waterfall formation, river incision, and gorge formation. Knickpoints, notable as landforms, arise as significant disruptions

Table 5 Classification based on orientation of the lineaments

Sr. no.	Range	N	N%	L in km	L%
1	0–10	19	9.2	26.4	6.9
2	11–20	14	6.8	20.1	5.2
3	21–30	13	6.3	20.0	5.2
4	31–40	19	9.2	34.0	8.8
5	41–50	19	9.2	28.7	7.5
6	51–60	14	6.4	28.5	7.4
7	61–70	09	4.4	17.6	4.6
8	71–80	08	3.9	10.9	2.8
9	81–90	07	3.4	31.7	8.2
Total		**122**	–	**217.9**	–
Average		**13.6**	**58.9**	**24.2**	**56.6**
10	270–280	14	6.8	22.0	5.7
11	281–290	06	2.3	21.0	5.5
12	291–300	05	2.4	9.31	2.4
13	301–310	10	4.8	29.2	7.6
14	311–320	08	3.9	9.73	2.5
15	321–330	12	5.8	26.4	6.9
16	331–340	09	4.4	11.4	3.0
17	341–350	07	3.4	13.3	3.5
18	351–360	14	6.8	25.0	6.5
Total		**85**	–	**167.4**	–
Average		**9.4**	**41.1**	**18.6**	**43.5**
Total lineaments		**207**	–	**385.3**	–

or breaks in a river longitudinal profile. These locations serve as sites where greater energy is dissipated, typically utilized for cutting through bedrock via erosion and transporting sediments from upstream. The presence of erosional features, including potholes, grooves, waterfalls, and gorges along the Lendi river near Diglur, indicates that the maximum energy has been directed toward active channel erosion. This effect is particularly pronounced during monsoon floods when flow depth and velocity peak (Fig. 9).

Potholes, which are circular or oval-shaped has been formed because of the river erosion, abrasion, and carving of the bed rock. The potholes are considered to be significant for channel incision and are accountable for the characteristic presence in river channels (Sengupta and Kale 2011). The actual work of abrasion is done either by hard grains of sand or by large pebbles called "grinders" suspended by the rapidly flowing water in the bottom of the potholes that were agitated constantly by the current swirling around inside (Douglas 1975). During the fieldwork, we observed small 20–30 cm length and depth is only 15–25 cm initial stage potholes on the top of the granite rock near Degloor (Fig. 9a). The second geomorphotectonic feature which is observed in Lendi River is waterfall. Hydraulic jumps at the base of a waterfall can create enormous powers to erode the bed rock causing the formation of waterfall. Often over time, the waterfall will recede back to form a canyon or gorge downstream as it recedes upstream, and it will carve deeper into the ridge above it. Such type of gorges is found in the Lendi River bed near Diglur town (Fig. 9b).

In swiftly flowing streams with steep gradients, erosion primarily targets the bottom of the stream channel. In case of steeply sloping streams, the lateral erosion on the sides of the valleys is not much when compared to the streams flowing on low and gentle slopes. Because of active lateral erosion, streams flowing over gentle slopes develop sinuous or meandering courses. It is common to find meandering courses over floodplains and delta plains where stream gradients are very gentle. But very deep and wide meanders can also be found cut in hard rocks. Such meanders are called incised or entrenched meanders. Meanders develop over original gentle surfaces in the initial stages of development of streams, and the same meanders get deeply rooted into the rocks usually due to erosion or slow, continued uplift of the land over which they originated. These valleys widen and deepen over time and can be found as deep gorges and canyons in hard rock areas. Same type of gorge has also developed near Diglur in Lendi River (Fig. 9c). They give an indication on the status of original land surfaces over which streams have developed might have locally uplifted.

Fig. 9 Field photographs: **a** potholes, **b** river incision and gorge, **c** waterfall formation near Degloor, **d** tilting of the coarse sediments near Malkapur, and **e** cross-bedding structure developed in coarse sediments near Begam Takali about 6.3 m above present river bed

Peshwa and Kale (1996) observed that the fluvial deposits in the upper and middle reaches of the river basins attest to either differential block subsidence along the channels or uplift of the flank, so as to accommodate the sediment fills. River channel sediments are the good sites to study the paleoseismicity as well as tectonic (Sukhija et al. 2006). In general, tilting of the sediments in the river channel is due to abrupt change in velocity of water or due to river rejuvenation process. In this present study, the cross-bedding structure which is generally developed due to change in wave motion of the water is in general observed along the river channel of younger alluvial plain. In Lendi River at Begam Takali village, it is observed 6.3 m above the present river channel of the Lendi River (Fig. 9d, e). Sukhija et al. (2006) observed the features in deep trenches (at Thar) could be surficial manifestation of deep-seated disturbances in the region.

6 Discussions

The Lendi River sub-basin is third major tributary of river Manjra, which is the tributary of westerly flowing river Godavari. So it is a part of Godavari river system in peninsular India. The Lendi River covered total area of 2532.26 km^2, and it has been classified as seventh-ordered stream (Gurav and Babar 2021). For present study, nine important GIAT parameters such as bifurcation ratio (Rb), basin elongation ratio (Re), hypsometric integral (HI), basin asymmetry factor (AF), transverse topographic symmetry (T), stream gradient length ratio (SL), river longitudinal profile (RLP), lineaments, and abnormal behavior of streams with the main river have taken in to consideration. The watershed average Rb value ranges from 2.92 to 6.58. Watershed numbers 01, 02, 11, and 16 lie in basaltic terrine and show Rb > 5, which indicates the geological structures have control on the drainage development. Lendi River Re value is 0.63 which indicates that basin is semi-elongated in shape and slightly tectonically active (Gurav and Babar 2021). The watershed Re value ranges from 0.55 to 0.91. It indicates watersheds are highly elongated to nearly circular in shape (Schumm 1956; Kaplay et al. 2017). The hypsometric curves show three distinct features such as the stage of the basin development and geomorphic and tectonic evolution of drainage basins (Schumm 1956; Strahler 1956, 1964). The hypsometric integral value of whole Lendi River sub-basin is 37.41, which indicates basin is at present in the late mature stage of river development. The watershed numbers 02, 03, 04, 05, 06, 07, 8, 10, 11, 12, 13, 14, 15, 16, 17, 18, 19, 20, 21, and 22 show mature stage/slightly active, while watershed numbers 01 and 09 show younger stage/tectonically active (Fig. 5c). The small watershed is less affected to erosion (Gurav and Babar 2018). The influence of tectonics on the drainage pattern is also reflected by the basin asymmetry (Kaplay et al. 2017). The AF is the areal morphometric variable that deciphers the presence or absence of regional tilt in the basin (Keller and Pinter 1996). In the Lendi River sub-basin, AF value is very close to the 50.00%, i. e., 49.57% (Fig. 4); this indication of basin is not tectonically tilted. The AF values of watershed numbers 02, 04, 08, and 10 are between 26.57 and 45.00, and watershed numbers 01, 05, 07, 16, 17 and 18 are between 55.01 and 67.63 indicating that these watersheds are slight to moderately tectonically tilted (Fig. 5d). SL index is first proposed by Hack (1973) to facilitate the identification of factors that control the drainage network such as occurrence of resistant rock, tectonic activity, change in the local base level, tributary confluence, and human interference. The average stream length gradient index value of Lendi River is 6.73. The SL index for watersheds is between 3.90 and 8.35 (Table 4). In this study, we observed second-order anomaly exists all over the basin. In Lendi River, nearly 207 lineaments are marked and classified based on length and orientation (Table 5). It is found that maximum lineaments are trending in NE–SW direction (Fig. 6d).

The smaller depressions occurring on the river bed cause formation of eddies, which, along with it the sediments spins, lead to erosion through whirling motion, resulting in the formation of the potholes. These depressions get further eroded, excavated, and broadened to form archetypal potholes. Further, these potholes progressively expand and merge ultimately to produce internal channels and develop in deep gorges. The modest relationship between the rock mass strength and the morphological properties of potholes suggests that this factor may not be the most critical factor determining the presence and morphology of potholes in this area of more or less uniform lithology and high-energy expenditure. Erosional disequilibrium between river systems has been commonly recognized as one of the causes of knick point formation and upstream migration (Hack 1973). The alternate clarification is that the formation of waterfall at Degloor perhaps may be an example of local or block uplift. In these locations, Babar et al. (2019) have identify strike-slip faults in granite terrain. Here, it is appropriate to remark that the southern limit of the bedrock-meander-dominant area coincides with the NW–SE trending regional lineament and shear zone (Babar et al. 2019). Rose diagram of strike-slip faults is shown in Fig. 6e.

Kale and Shejwalkar (2008) pointed out some well-known knick points, with scablands, inner channels, waterfalls, and potholes, which occur on the Godavari river near Gangapur and Puntamba, on the Pravara river at Randha, on the Mandvi river near Chilewadi, on the Kas river near Bote, on the Mula river near Mandwe, on the Indrayani river at Shelarwadi, on the Kukdi river near Nighoj, and on the Shivaganga river at Baneshwar. Such prominent breaks in the long profiles have been generally associated with geological or structural control or zones of uplift. It is one of the facts that finding a relation between knick points and faults/lineaments or lithology is very essential to understand the recent tectonic activity. In Fig. 1, Google Map shows the three sites (a, b, and c) of geological fieldwork evidence which was studied during the fieldwork. In Fig. 1, site "a" shows the waterfall as well as river incision and gorge. Its field photograph is shown in Fig. 9b, c.

In this place, Lendi River also takes NE–SW trend with abnormal behavior of streams with the main river. Site "b" shows tilted sediments near Malkapur village (Fig. 9d). Similarly, site "c" near Begam Takali observed cross-bedding structure at 6.3 m above the present river channel (Fig. 9e). Here also river takes the NE–SW trend. It is found that the trend of strike-slip fault and lineaments near the Degloor is found to be similar to NE–SW. All these three pieces of site evidence prove that there is moderate to strong structural control on the river development near Degloor town in Nanded district, Maharashtra, in Lendi River in Granitic basement which is activated in Pre-Cambrian as well as Quaternary time.

7 Conclusions

Lendi River sub-basin is elongated and occurs in the late mature stage of erosional development. The break in slope was studied using the river longitudinal profile. Similarly, uplifted areas within the river have been studied with the help of waterfalls and a gorge with sediments tilting and cross-bedded structure. Investigation of the topographical maps of the river indicates that the valleys in Deccan Trap do not reveal any uplift, but the valley in the PGC shows the evidence of Quaternary uplift and structurally controlled lineaments and drainage of the river bed. It is observed that all the morphologic features imply that the existing valleys are the product of usual fluvial erosion. A waterfalls with gorge has formed near Diglur village, where Lendi River takes NE–SW trend. This is accompanied by lineament-controlled abnormal behavior of streams in relation to the main river. Similarly, the presence of strike-slip faults at Diglur, tilting of coarse sediments at Malkapur, and cross-bedding structure observed above 6.3 m height from present river channel near Begam Takli village (both are close to Degloor town) point toward Degloor area in Lendi River, which is a seismic zone and is activated in Precambrian as well as Quaternary time. However, further investigations on the geophysical survey, a detailed tectonic and seismic study, need to be done in the future.

Authors' Contributions Gurav Chandrakant performed data collection, investigation, and formal analysis and was responsible for conceptualization and writing of the original draft.

Md. Babar provided supervision, assisted with original draft, prepared the visualization, and conducted reviewing and editing.

Acknowledgements The authors are grateful to Head of Geology department Prof. Dr. Pathare Sanjay. V. Rajaram College, Kolhapur Maharashtra, India for his keen interest and consistent support during the present research work.

References

Abrahams AD (1984) Channel Networks: A Geomorphological Perspective. Water Resour Res 20:161–188. https://doi.org/10.1029/WR020i002p00161

Agrawal PK, Pandey OP (1999) Relevance of hot underlying asthenosphere to the occurrence of Latur earthquake and Indian peninsular shield seismicity. J Geodyn 28:303–316. https://doi.org/10.1016/S0264-3707(99)00002-2

Artemieva IM, Mooney WD (2001) Thermal thickness and evolution of Precambrian lithosphere: A global study. J Geophys Res 106:16387–16414. https://doi.org/10.1029/2000JB900439

Babar M, Chunchekar R, Yadava MG, Ghute B (2012) Quaternary Geology and Geomorphology of Terna River Basin in West Central India. E&G Quaternary Sci J 61:156–167. https://doi.org/10.3285/eg.61.2.04

Babar Md, Kaplay RD, Mukherjee S, et al (2019) NE–SW Strike-Slip Fault in the Granitoid from the Margin of the South East Dharwar Craton, Degloor, Nanded District, Maharashtra, India. In: Mukherjee S (ed) Tectonics and Structural Geology: Indian Context. Springer International Publishing, Cham, pp. 115–134

Babar Md, Kaplay RD, Mukherjee S, Kulkarni PS (2017) Evidence of the deformation of dykes from the Central Deccan Volcanic Province, Aurangabad, Maharashtra, India. Geological Society, London, Special Publications 445:337–353. https://doi.org/10.1144/SP445.13

Babar Md, Kaplay RD, Panaskar DB (2000) Structural controls on drainage pattern in the Terna river catchment, Central west India. Indian Journal of Geomorphology 5:126–133

Banerjee R, Jain SK, Shivkumar K (2008) Geochemistry and Petrogenesis of Uraninite bearing Granitoid and Radioactive Phosphatic Cherty Cataclasite of Thadisaoli area, Nanded district, Maharashtra. Memo Geol Soc India 73:55–84

Banerjee R, Krishna V, Pandey BK, Parthasarathy TN (1993) Rb-Sr Geochronology of the Radioactive granites of Nanded area Maharashtra, India. Jour Atonic Min Science 1:111–117

Bull WB, McFadden LD (1977) Tectonic Geomorphology North and South of the Garlock Fault, California. In: Doehring, D.O., Ed., Geomorphology in Arid Regions: A Proceedings Volume of the 8th Annual Geomorphology Symposium, State University of New York, Binghamton

Cannon PJ (1976). Generation of explicit parameters for quantitative geomorphic study of the Mill Creek drainage basin, Oklahome Geology Note, vol. 36(1), pp. 3-16.

Chetty TRK, Rao MN (1994) Latur earthquake of September 30, 1993, surface deformation and lineament pattern, Latur earthquake,(ed. H.K. Gupta). Geol Soc of India, Mem 35:65–74

Chunchekar RV (2012) Quaternary Geology and Geomorphology of Terna river basin, Eastern Maharashtra. Swami Ramanand Teerth Marathwada University, Nanded. http://hdl.handle.net/10603/217203

Cox RT (1994). Analysis of drainage basin asymmetry as a rapid technique to identify areas of possible Quaternary tilt-block tectonics: an example from the Mississippi embayment. Geological Society of America Bulletin, vol. 106, pp. 571-581.

Cuong NQ, Zuchiewicz WA (2001) Morphotectonic properties of the Lo River Fault near Tam Dao in North Vietnam. Nat Hazards Earth Syst Sci 1:15–22. https://doi.org/10.5194/nhess-1-15-2001

Dole G, Das S, Kale VS (2022) Tectonic framework of geomorphic evolution of the Deccan Volcanic Province, India. Earth-Science Reviews 228:103988. https://doi.org/10.1016/j.earscirev.2022.103988

Douglas EC (1975) On the Interpretation of Potholes. Creation Research Society Quarterly 12(1):25–31

Dutta N, Sarma JN (2013) Morphotectonic analysis of Sonai River Basin, Assam, NE India. Global Research Analysis 2:2

Gurav C (2019) Quaternary stratigraphy and geomorphology of Lendi River Sub-Basin in West-Central Maharashtra. Swami Ramanand Teerth Marathwada University, Nanded. http://hdl.handle.net/10603/313727

Gurav C, Babar Md (2021) Morphometric Analysis of Lendi River Basin Using Geographical Information System (GIS) Techniques. In: Pawar PM, Balasubramaniam R, Ronge BP, et al. (eds) Techno-Societal 2020. Springer International Publishing, Cham, pp. 37–46

Gurav C, Babar Md (2018) Hypsometric analysis of Gharni river sub-basin of manjra river, Maharashtra, India-using geographical information system (GIS) techniques. Journal of Applied Geochemistry 20:447–454

Gurav C, Babar Md, Jagdale A (2020) Morphotectonics Study of Dhamani River Basin in Kolhapur District, Maharashtra, India. Bulletin of Pure & Applied Sciences- Geology 39f:131–149. https://doi.org/10.5958/2320-3234.2020.00013.X

Hack JT (1957) Studies of Longitudinal Stream Profiles in Virginia and Maryland. USGS Profpaper

Hack JT (1973) Stream-profile analysis and stream-gradient index. Journal of Research of the U.S. Geological Survey 1:421429

Hare PW, Gardner (1985) Geomorphic Indicators of Vertical Neotectonism along Converging Plate Margins, Nicoya Peninsula Costa Rica. In: Morisawa, M. and Hack, J.T., Eds., Tectonic GeomorphologyProceedings of the 15th Annual Binghamton Geomorphology Symposium, Allen and Unwin, Boston

Herlekar MA, Sukhtankar RK (2011) Morphotectonic Studies along the Part of Maharashtra Coast, India. International Journal of Earth Sciences and Engineering 04:24

Horton RE (1945) Erosional Development of Streams and Their Drainage Basins; Hydrophysical Approach to Quantitative Morphology. Geol Soc America Bull 56:275. https://doi.org/10.1130/0016-7606(1945)56[275:EDOSAT]2.0.CO;2

Kale VS, MakarBodas, Chatterjee P, Pande K (2020) Emplacement history and evolution of the Deccan Volcanic Province, India. Episodes 43:278–299. https://doi.org/10.18814/epiiugs/2020/020016

Kale VS, Shejwalkar N (2008) Uplift along the western margin of the Deccan Basalt Province: Is there any geomorphometric evidence? J Earth Syst Sci 117:959–971. https://doi.org/10.1007/s12040-008-0081-3

Kaplay RD, Babar Md, Mukherjee S, et al (2019) Structural Features of Kinwat Peninsular Gneissic Complex Along the Western Margin of Eastern Dharwar Craton, India. Arab J Sci Eng 44:6509–6523. https://doi.org/10.1007/s13369-019-03948-x

Kaplay RD, Babar MD, Mukherjee S, Vijay Kumar T (2017) Morphotectonic expression of geological structures in the eastern part of the South East Deccan Volcanic Province (around Nanded, Maharashtra, India). Geological Society, London, Special Publications 445:317–335. https://doi.org/10.1144/SP445.12

Kaplay RD, Vijay Kumar T, Sawant R (2013) Field evidence for deformation in Deccan Traps in microseismically active Nanded area, Maharashtra. Current Science 105:1051–1052

Keller EA, Pinter N (1996) Active tectonics: Earthquakes Uplift and Landscapes. Prentice Hall, New Jersey

Krishnamurthy P (2020) The Deccan Volcanic Province (DVP), India: A Review: Part 1: Areal extent and distribution, compositional diversity, flow types and sequences, stratigraphic correlations, dyke swarms and sills, petrography and mineralogy. J Geol Soc India 96:9–35. https://doi.org/10.1007/s12594-020-1501-5

O"Leary DW, Friedman JD, Pohn HA (1976) Lineament, linear, lineation: Some proposed new standards for old terms. Geol Soc America Bull 87:1463. https://doi.org/10.1130/0016-7606(1976)87<1463:LLLSPN>2.0.CO;2

Pandey OP, Agrawal PK (1999) Lithospheric Mantle Deformation beneath the Indian Cratons. The Journal of Geology 107:683–692. https://doi.org/10.1086/314373

Peshwa VV, Kale VS (1996) Neotectonics of the Deccan Traps Province: Focus on the Kurduwadi Lineament. Jour of Geophysics XVIII:77–86

Phani RC, Ningam N, Srinivas M, Cyeint Limited, Plot11, Software Units Layout, Infocity, Madhapur, Hyderabad-500 081. (2017) Field Geology and Petrography of Granitoid rocks in southeastern part of Nanded district, Maharashtra State. IJTSRD Volume 1:988–996. https://doi.org/10.31142/ijtsrd4710

Polet J, Anderson DL (1995) Depth extent of craton as inferred from tomographic studiesGeology 23:205–208

Rajendran CP (1997) Deformational features in the river bluffs at Ther, Osmanabad district, Maharashtra: Evidence for an ancient earthquake. Current Science Association, Vol 72, No 10, pp 750–755

Rajendran CP, Rajendran K (1999) Geological investigation at Killari and Ther, Central India and implications for paleoseismicity in the shield region. Tectonophysics. https://doi.org/10.1016/S0040-1951(99)00086-4

Rajendran CP, Rajendran K, Unnikrishnm R, John B (1996) Paleoseismic indicators of the rupture zone of the 1993 Killari (Latur) central India, earthquake. Current Science, Issue 70, pp 385–390

Ritter DF, Kochel RC, Miller JR (2002) Process geomorphology. McGraw Hill, Boston, pp 593

Schumm SA (1956) Evolution of Drainage Systems and Slopes in Badlands at Perth Amboy, New Jersey. Geol Soc America Bull 67:597. https://doi.org/10.1130/0016-7606(1956)67[597:EODSAS]2.0.CO;2.

Seeber L, Gornitz V (1983) River profiles along the Himalayan arc as indicators of active tectonics. Tectonophysics 92:335–367. https://doi.org/10.1016/0040-1951(83)90201-9

Sengupta S, Kale VS (2011) Evaluation of the role of rock properties in the development of potholes: A case study of the Indrayani knickpoint, Maharashtra, J Earth Syst Sci 120:157–165. https://doi.org/10.1007/s12040-011-0005-5

Sharma S, Sarma JN (2017) Application of drainage basin morphotectonic analysis for assessment of tectonic activities over two regional structures of the northeast India. Geol Soc India 89:271–280. https://doi.org/10.1007/s12594-017-0599-6

Singh T (2008a) Hypsometric analysis of watersheds developed on actively deforming Mohand anticlinal ridge, NW Himalaya. Geocarto International 23:417–427. https://doi.org/10.1080/10106040801965821

Singh T (2008b). Tectonic implications of geomorphometric characterization of watersheds using spatial correlation: Mohand Ridge, NW Himalaya, India. zfg 52:489–501. https://doi.org/10.1127/0372-8854/2008/0052-0489

Singh T, Jain V (2009) Tectonic constraints on watershed development on frontal ridges: Mohand Ridge, NW Himalaya, India. Geomorphology 106:231–241. https://doi.org/10.1016/j.geomorph.2008.11.001

Strahler AN (1956) Quantitative Slope Analysis. Geol Soc America Bull 67:571. https://doi.org/10.1130/0016-7606(1956)67[571:QSA]2.0.CO;2

Strahler A (1964) Quantitative Geomorphology of Drainage Basins and Channel Networks. In: Handbook of applied hydrology, McGraw Hill Book, New York, pp 439–476

Sukhija BS, Lakshmi BV, Rao MN (2006) Widespread Geologic Evidence of a large Paleoseismic event near the Meizoseismal Area of the 1993 Latur Earthquake, Deccan Shield, India. Journal of Indian Geophysical Union 10:1–14

Venkatarao K, Prasad V (1995) Magnetostratigraphy of Deccan traps along Kurduwadi-Latur traverse, Maharashtra. Geol Surv India, Spl Pub 27:99–108

Watts AB, Cox KG (1989) The Deccan Traps: an interpretation in terms of progressive lithospheric flexure in response to a migrating load. Earth and Planetary Science Letters 93:85–97. https://doi.org/10.1016/0012-821X(89)90186-6

Wesanekar P (1994) Geological, Geochronological and Fluid Inclusion Thermocryometric studies of Granite rocks around Degloor, Nanded dristrict Maharashtra. Shivaji University, Kolhapur. http://hdl.handle.net/10603/142191

Wesanekar P, Patil RR (2000) Rb-Sr dating of pink Granites of Deglur, Nanded district, Maharashtra, India. In: Proc. Nat. Sem. “Tectono-magmatism, Geochemistry and Metamorphism of Precambrian Terrains”, pp 181–187

Exploring the Relationship Between Women's Health and WASH Conditions in Slum Areas Towards Sustainable Development

Renu Dhupper and Suhasini Jindal

1 Introduction

India's economy is currently one of the fastest expanding in the world and is becoming more industrial and service-oriented. With 1.423 billion people, India may have already surpassed China as the world's most populous country, according to WPR, an independent group that focuses on census and demographics (Joshi 2011; Tooley and Dixon 2007). In the upcoming years, India, a country with a population that is half under 30 years old, is predicted to have the world's fastest-growing major economy (Ma et al. 2021). India is becoming more prominent on the international stage. It is seen as having a substantial claim to being the software capital of the world, the business process outsourcing for multinational corporations, and the centre for cutting-edge research in information technology and even biotechnology. This is supported by its brain power, which includes five lakh engineers, 2.5 lakh doctors and 75 lakh graduates (Chowdhury et al. 2022; Gurkov et al. 2021).

However, the predicted increase in India's urban population by 2030 is expected to be the largest of any nation at 404 million people. According to the World Urbanization Prospects Report, urban areas, particularly those in lower-middle-income nations like India, would face more obstacles to sustainable development (Li et al. 2019; Manoli et al. 2019; Pramanik 2024; Restrepo et al. 2021). There is a backlog in the implementation of policies, infrastructure and allowing for additional urban growth throughout the urban systems (Egal and Forster 2020; Goonetilleke et al. 2007; Pechsiri et al. 2023). According to a report by McKinsey & Company (2010), urban gridlock may result if Indian cities stick with their current strategies. Metropolitan cities offer a wealth of chances for employment and a living. This forces individuals from rural areas around the nation to migrate to these cities in search of employment and a means of livelihood and settle in squatter settlements called slums. This has put a strain on how people live in these cities (Shekhar 2021; Wang et al. 2019).

Slums are the outcome of rapid urban growth without the necessary development of urban infrastructure (Atahigwa et al. 2020; Iddi et al. 2020). These are run-down neighbourhoods in any city with horrendously subpar housing and living conditions. Slums range from huge, unplanned squatter colonies without legal recognition or rights to densely populated, filthy tenements in the heart of cities. Although they have existed since the beginning of time, they have proliferated greatly over the last several decades. Geographically, a slum is an urban area with a very disparate geographical, structural and socioeconomic urban environment. Figure 1 shows the typical view of a slum area.

It is a densely populated region with at least 300 people, or roughly 60–70 homes in shoddily constructed cramped tenements, in an unhygienic environment typically with poor infrastructure and lacking basic drinking water services (Kumpel et al. 2016; Pille and Säumel 2021). Slums typically have deteriorated and infirm housing structures, poor ventilation, severe overcrowding, improper street alignment, dim lighting, a lack of access to safe drinking water, flooding during rainstorms, a lack of toilet facilities and a lack of essential physical and social services (Hanks et al. 2021; Lundy and Wade 2011; Nöjd et al. 2020; Wang et al. 2024). The living conditions in slums are frequently unsanitary, go against all rules for planned urban growth, and have a significant role in hastening the spread of many air and waterborne diseases. Greater Mumbai, Kolkata, Delhi, Chennai and Hyderabad are the five largest cities in India, and together they account for more than 64% of all people living in slums in metropolitan areas (Martinopoulos et al. 2019; Realini et al. 2021; Shimomura and Ferreira 2018). Therefore, the megacities are where most people who live in slums reside.

R. Dhupper (✉) · S. Jindal
Amity Institute of Environmental Sciences, Amity University Uttar Pradesh, Noida, Uttar Pradesh, India
e-mail: rdhupper@amity.edu

D. Kumar et al. (eds.), *Data Analytics and Remote Sensing-Based Analysis for Sustainable Development*, Advances in Science, Technology & Innovation, https://doi.org/10.1007/978-3-031-98095-4_10

Fig. 1 View of a slum area

Slums unsuited for human habitation are home to every sixth urban Indian. Slums are so prevalent that they can be found in 65% of Indian municipalities. Nearly four out of ten people who live in slums do not have access to treated water, and six out of ten are located near contaminated sewers (Karpf and Krebs 2005; Zhao et al. 2020). Nine states have higher slum populations than the national average. With 36.1% of its urban population living in slums, Andhra Pradesh leads the list. The other states are: Jammu & Kashmir, Sikkim, Madhya Pradesh, Maharashtra, Odisha, West Bengal and Haryana. In India, 35% of slum residents lack access to treated drinking water (Garcia-Suarez et al. 2019; Heberer et al. 2002; Molinos-Senante and Sala-Garrido 2017).

A significant portion of slum families in Odisha lack access to treated tap water (64.9%) and are either without a drainage connection or connected to an open drain (90.6%). Out of 1, 37, 49,424 slum families, 11, 92,428 utilise untreated water for drinking purposes. Nearly 60% of these are concentrated in just five states: *West Bengal, Tamil Nadu, Andhra Pradesh, Madhya Pradesh and Andhra Pradesh.* 44% of the 63% of Indian slum households without a functioning drainage system have open drains, while close to 19% of the families have no drainage connection (Fig. 2).

While Swachh Bharat Abhiyaan (Clean India Mission) is in its second year and India's economy has been expanding for decades, the country's 360 million poorest inhabitants continue to live in some of the worst conditions possible. The slums have evolved into an essential and negative aspect of our nation, which we cannot be proud of. It is so ironic that lakhs of people live in Delhi's slums (and all of India's slums) while their labour makes the lives of the country's wealthier residents simpler and more comfortable while forcing them to endure the worst living conditions. They are living in a state of indignity and infection since they cannot even access a fundamental necessity like working toilets.

They require special consideration in terms of public health, family planning and reproductive health programmes because to their low socioeconomic situation, low level of education and awareness, deteriorated physical structure, high fertility and death rates (Buechley et al. 1972; Penfold et al. 2021; Sheikhtaheri et al. 2021; Verhoeff and Menzel 2011). Slum expansion at a rapid rate has led to widespread environmental deterioration throughout the city. The growth of slums in metropolitan areas is currently one of the main difficulties that urban planners throughout the world must deal with. Slum expansion, on the other hand, has been an unstoppable issue in metropolitan areas (By and Process n.d.; Pratomo et al. 2016; Shekhar 2014; Wang et al. 2019). The main causes of slum development are quick and exclusive urbanisation trends sparked by an increase in rural-to-urban migration. There are following types of slums:

(a) **Notified slums**—All areas in a town or city that have been designated as "slums" by the state, the UT administration, or local government according to any statute, including a "slum act," are considered NOTIFIED SLUM.
(b) **Recognised slums**—All locations that have been designated as "slums" by the state, UT administration, local government, housing and slum boards; nevertheless, they are not required to be legally recognised as such by law.
(c) **Identified slums**—There should be 60–70 houses in a small region with at least 300 inhabitants. It is made up of

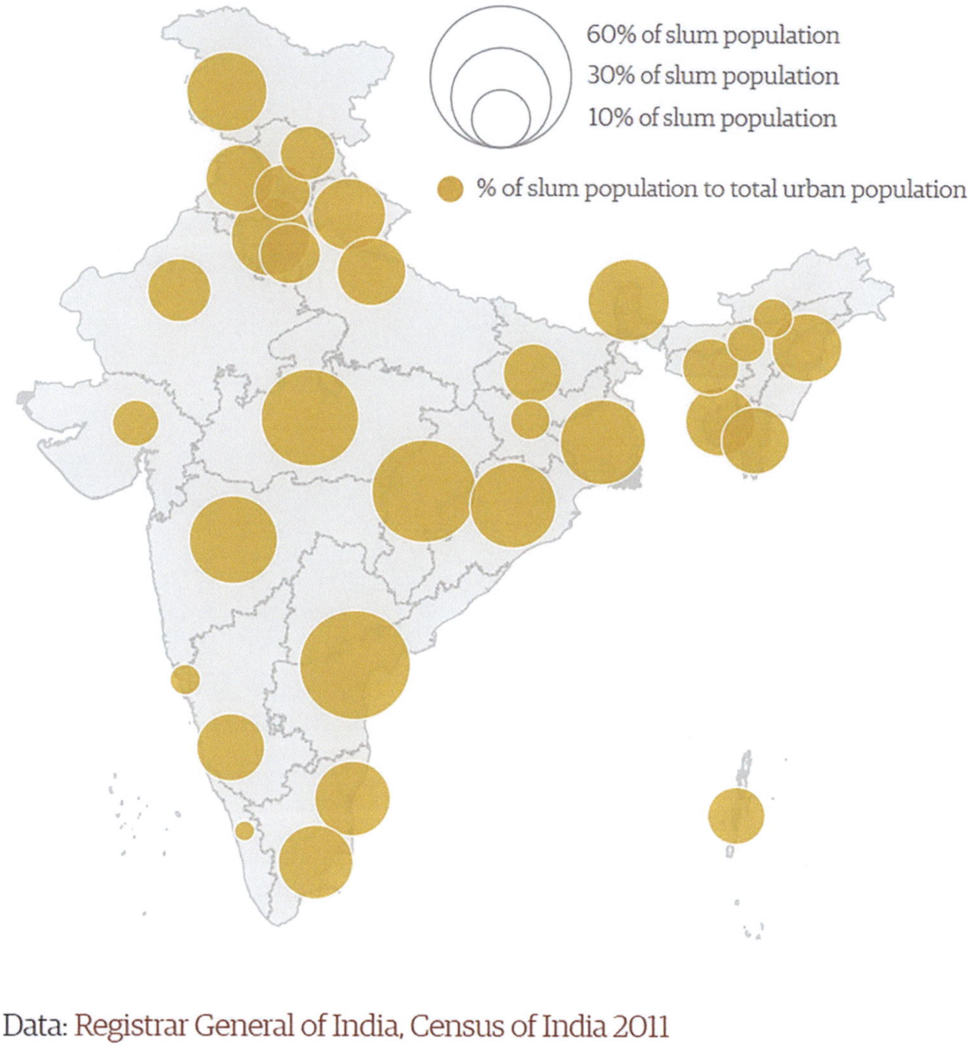

Fig. 2 Exhibit for some towns and cities reporting slums by state and the proportion of slum dwellers

poorly constructed, crowded tents, filthy circumstances, typically with poor infrastructure, and a lack of essential services including sanitary and drinking water facilities.

Causes of Slum Formation

(a) **Decentralisation**: When wealthy and middle-class residents migrate out to the outskirts of a town or city, the urban poor are abandoned in the congested centre of the community.
(b) **The economic conditions**: (i) The most significant and pervasive aspect of slums is poverty. (ii) The affected individuals may be forced to continue living in ill conditions due to unemployment and the fast growing population. Slums are thought to be a direct product of industrialisation, which in its early phases never bothered to provide housing for labourers.
(c) **Lack of livelihood resources and employment opportunities in rural areas**: Due to the extremely low employment level possibilities in rural areas, migration from rural areas to urban centres is encouraged.
(d) **Migrants**: Squatters are those who occupy land without a permit illegally. Slums could develop as a result of squatter communities. In major cities, the daily migrant flow increases. A third of the people reside in squalid huts with no public services.
(e) **Low wages**: Low-wage workers make up a significant portion of the workforce in our industrialised cities. Many of them work in temporary positions. They lack the resources to maintain decent housing, services and facilities. They are so compelled to live in slums.
(f) **Improper use of land**: Slums may arise if land that is suitable for a specific use, such as residential and commercial, is used for a different one.

Fig. 3 Living conditions of people living in slum area

(g) **Powers and local authority**: Slums may arise if local authorities do not have enough power to oversee urban growth.
(h) **Lack of zoning**: Slums may arise if the town is not separated into proper zones or if land usage and development are permitted to occur at random.
(i) **Repairs and maintenance**: If inexpensive homes built for rent collection are not adequately maintained, slum-favouring conditions may emerge.
(j) **Ineffective land reforms**: Ineffective land reforms, there is a long-standing custom for landowners to employ landless people as labourers. The wage labourers in the industrial sector draw the landless labourers, who then migrate away from their traditional or familial occupations and towards the industries and urban centres.

The inadequate availability to basic sanitation services and hygiene in these regions makes the escalating strains put on by these movements blatantly clear. India appears to be set for a serious sanitation crisis as 53.1% of its households lack a toilet, according to Census 2011 data. According to the WHO/UNICEF Joint Monitoring Programme on sanitation for the Millennium Development Goals, 59% of Indians still lack access to toilets and practise open defecation. Most of these people also reside in rural regions. The UN Sustainable Development Goal (SDG)-6.2 urges "access to sufficient and equitable sanitation and hygiene for all and to prevent open defecation, paying special attention to the needs of women and girls and those in vulnerable situations" to solve this pressing issue (Fig. 3).

Women's health, dignity and human rights are disproportionately impacted by poor urban sanitation. Despite being regarded as a fundamental human right, access to adequate water and sanitation is a problem for millions of urban poor women. Given that many women living in urban informal settlements contribute disproportionately to economic and community activities, the health of women is frequently correlated with the health of children and communities in general.

When mothers in slums have to wait a long time to use the toilet, they may have to leave their children unattended or put off doing household chores. Women's health is at risk because they are more likely to develop reproductive tract infections because of unhygienic public restrooms and latrines. The necessity for proper sanitation is even more pressing for women who are menstruating. In addition, women are more likely to contract diseases linked to contact with human excreta because females are typically in charge of disposing of human waste when sanitation is not provided.

For slum dwellers, especially women, health is a major economic concern. Sickness brought on by an adverse physical environment necessitates medical attention, which reduces workdays and causes financial loss. The inability to invest in a clean environment is a result of economic loss. The vicious loop keeps going. The health of those living in slums is negatively impacted by bad housing conditions, an overcrowded environment, poor sanitation, occupational risks, group rivalries and conflicts, stressful situations, a lack of open space for women's enjoyment, etc. A sombre picture emerges from an analysis of the health state of women and children. Globally, impoverished women—and particularly those in developing nations—are more likely to die or get sick.

Sanitation in India's local communities has substantially benefited from international assistance. Following the launch of UNICEF's WASH (Water, sanitation and hygiene) campaign in more than 100 developing nations, the private sector has agreed to contribute more than USD 5 million to help solve the problem. According to reports, 25,000 individuals have stopped using open defecation, and 175,000 people have access to clean water. The Asian Development Bank, however, projects that it will cost 4.7 million USD and 3.3 million USD, respectively, to install family toilets in rural and urban regions, respectively. More money must be committed to sanitary development, and enforcement of public infrastructure project completion is also necessary. The World Bank estimates that in low- and middle-income nations, 1.2 trillion USD are spent on infrastructure projects, including the building of toilets connected to an underground sewer system, yet 30% of those projects are left unfinished. Infrastructure for sanitisation needs constant upkeep but is frequently not properly supervised. More people turn to open defecation because of latrine overflow or malfunction. To reduce the number of people who practise open defecation and subsequently improve the quality of sanitation, it is necessary to address the incompleteness of infrastructure projects and uneven maintenance.

As many beliefs and predispositions held towards toilets and sanitation are misconstrued, particularly when gender and femininity are involved, the sociological implications of open defecation have not been thoroughly examined. India's capital, Delhi, is regarded as one of the country's most urbanised states. This nation's capital offers a wealth of chances for employment and a living. This forces individuals from rural areas around the nation to migrate to this state in search of employment and a means of livelihood. This has put a strain on how people live in this state. Delhi City's estimated population for 2023 is 1,52,17,000 people. There are 3,32,022 slums in the entire city of Delhi, home to 16, 17,239 people. This represents about 14.66% of Delhi's whole population. The growth of squatter settlements in Delhi's slums is the subject of this study. The primary focus of this study will be on living conditions, health and sanitary conditions. It will also focus on security, health and sanitation issues, as well as how they affect the settlements' female populations negatively.

This field study aims to investigate the physical and social conditions of health, sanitation and hygiene of informal settlements in Delhi-NCR considering the growing concerns among experts to achieve SDG 6.2 by 2030 and in light of the Swachh Bharat Mission, an initiative launched in 2014 to eradicate open defecation (OD) in India.

2 Methodology

2.1 Study Design and Period

It is designed to evaluate the wash conditions in two Delhi slums, a community-based cross-sectional study employing a structured questionnaire was carried out in April and May 2023.

2.2 Study Area

Delhi, the nation's capital, was selected as the study region. 1,52,17,000 people are anticipated to live in Delhi City in 2023. In the entire city of Delhi, there are 3,32,022 slums, housing 16, 17,239 people. This is roughly 14.66% of Delhi's total population. One of the slums for the study, Indira Camp in Kalyan Puri, was chosen, with an estimated area of 8000 sqm and approximately 250 homes in Part 1. Another slum, Harijan Camp, Mandawali, New Delhi was also chosen as a study area with an estimated 150 houses.

2.3 Database and Methodology

A total of 200 households were interviewed for an in-depth study. Interviews were conducted with the female members of the household. Their consent to participate in the study was obtained. To better understand the difficulties and problems associated with women's access to sanitation facilities and its impact on their health, the study required that a greater percentage of respondents be women. The survey was conducted during the daytime when most of the male members were found to be out at work. All the responders were over the age of 18. The interviewer went to the next home if the members of the first household refused to be interviewed.

Data was collected by the researcher well acquainted with both Hindi and English. Data was obtained using a semi-structured socioeconomic questionnaire. The interview was conducted in Hindi, which is the most spoken language in Delhi. To confirm the accuracy of the responses, a few field observations were taken and recorded following each household interview. The observations were accompanied by photographic evidence, which included images of houses, lanes, community taps, and public toilets as well as video evidence of some individual interviews. To make sure the meaning of the questions was not changed, the surveys were translated into the native language and then back into English.

After each interview, specific first-hand observations were conducted in the field to add to the data gathered and to guarantee accuracy. Transect walks were conducted across the chosen slum colonies, paying close attention to the sanitation infrastructure present there. People were approached at random during these walks and questioned about the state of sanitation in the neighbourhood. Common medical conditions, regularity of solid waste collection, the distance between the house and the community toilet, and the condition of the community toilet were among the inquiries made.

3 Results

Socioeconomic conditions encompass a broader range of factors that are connected to an individual's or a family's income, education and occupation. Since current socioeconomic situations have a direct impact on sanitation, it is best to understand the latter in the context of the former. The respondents of the survey were females aged between 25 and 70. The majority of slum residents were discovered to be engaged in a variety of tiny businesses and jobs in the secondary sector, such as scrap collection and selling, domestic help, vegetable vending, daily wage workers, masonry and rickshaw pulling. A small fraction of people were employed on a salary in jobs like driving.

Indira Camp, Chota Hasanpur Village, New Delhi

The majority of the respondents (55.1%) were living in houses with 1 room, followed by 32.6% in 2 rooms and 12.4% in more than 2 rooms. 100% of respondents had pucca houses. Out of all the households, 51.7% houses had 4–6 members, while 33.7% had more than 6 members and 14.6% had less than 4 members. Out of 70 households, 54% had 0–2 females in the house and 46% had more than 2 females in the house. 77.6% of houses had 0–2 children, 21.2% had 2–4 and 1.2% had more than 4 children. 58.4% of the respondents had just 1 earning member in the house, 38.2% had 2 and 3.4% had more than 2 earning members. 48.3% of the households have monthly income of less than Rs. 10,000, 28.1% have Rs. 10,000 and 23.6% have more than Rs. 10,000 monthly income.

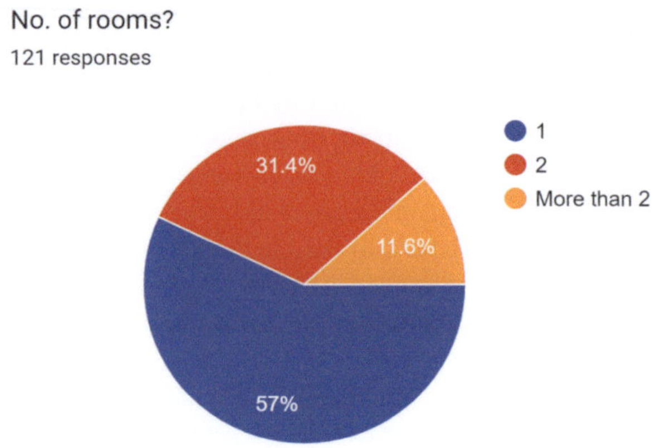

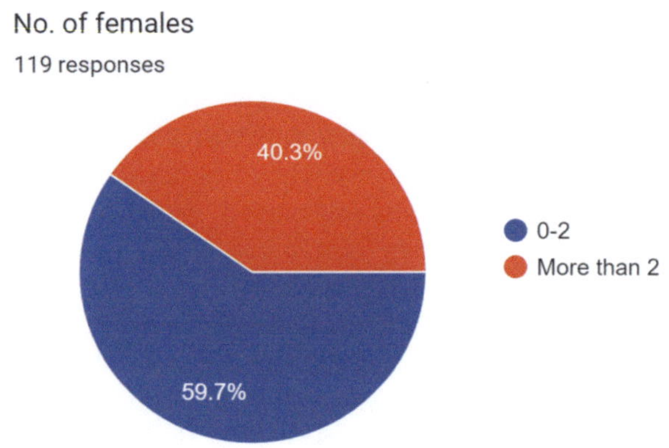

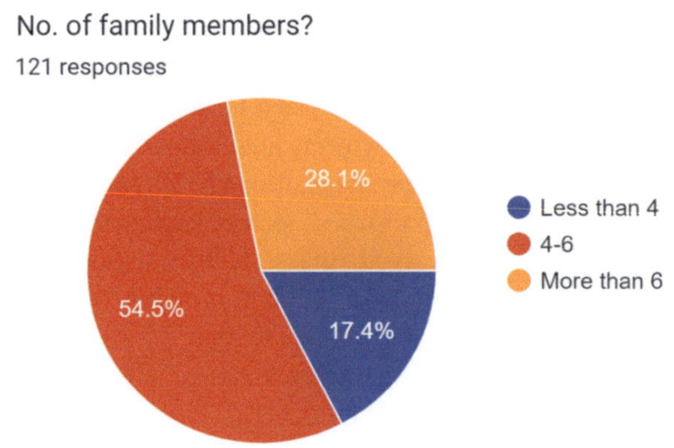

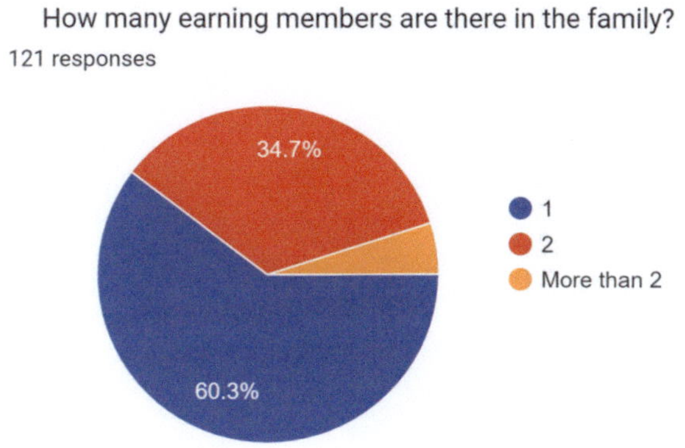

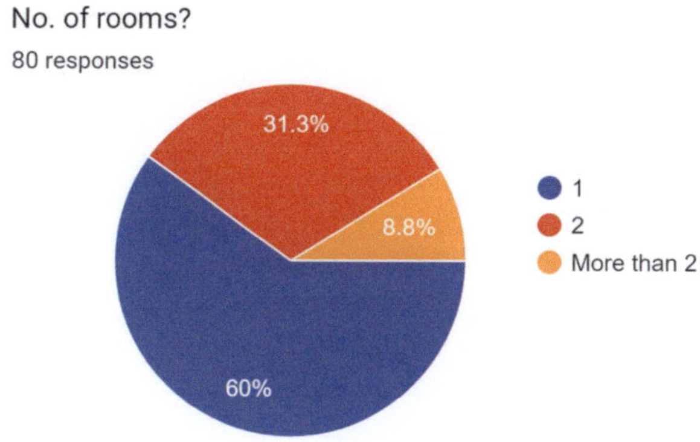

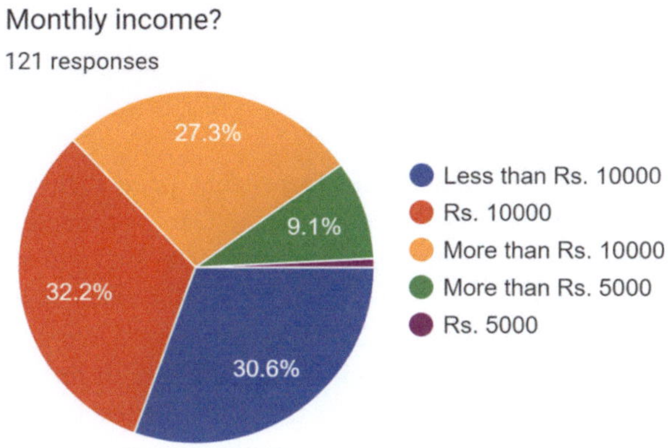

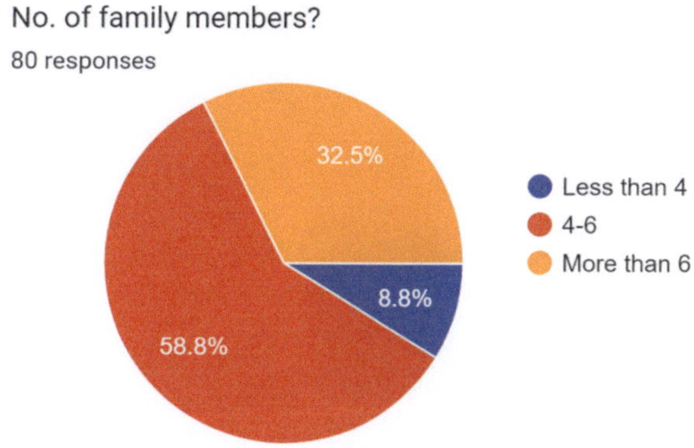

Harijan Camp, Mandawali, New Delhi

The majority of the respondents (60%) were living in houses with 1 room, followed by 31.3% in 2 rooms and 8.8% in more than 2 rooms. 100% of respondents had pucca houses. Out of all the households, 58.8% of houses had 4–6 members, while 32.5% had more than 6 members and 8.8% had less than 4 members. Out of 80 households, 65% had 0–2 females in the house and 35% had more than 2 females in the house. 52.6% of houses had 0–2 children, 44.9% had 2–4 and 2.6% had more than 4 children. 46.9% of the respondents had just 1 earning member in the house, 48.1% had 2 and 4.9% had more than 2 earning members. 45% of the households have monthly income of Rs. 10,000, 31.3% have less than Rs. 10,000 and 23.8% have more than Rs. 10,000 monthly income.

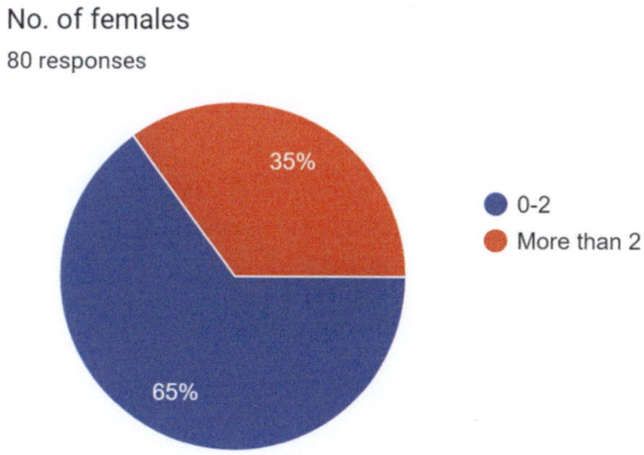

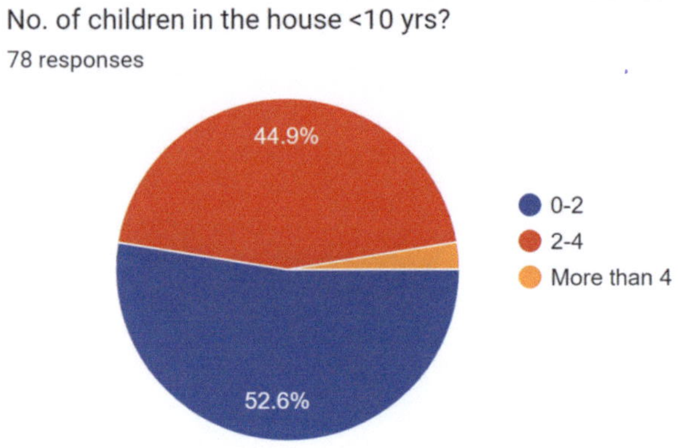

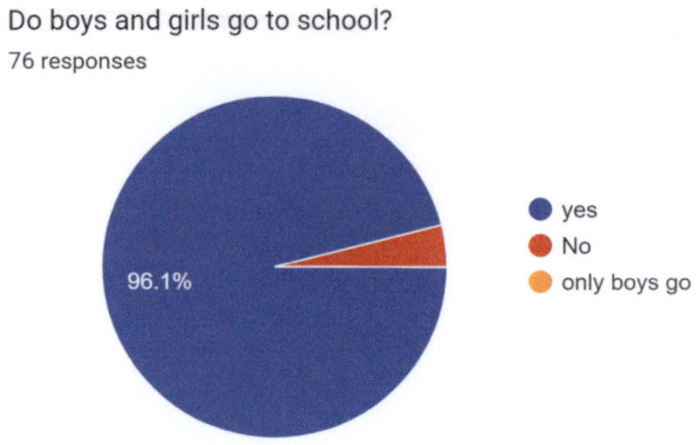

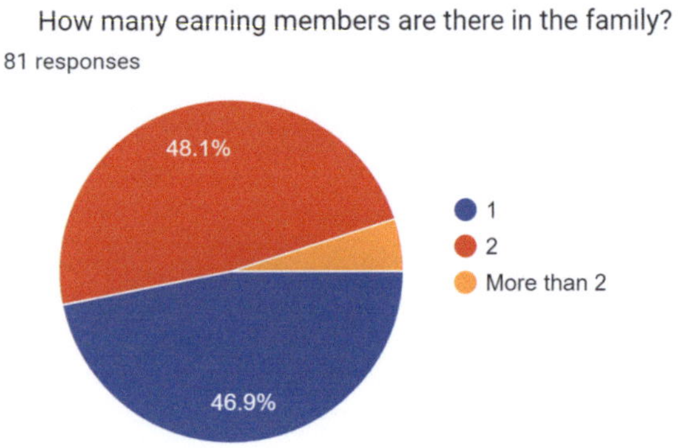

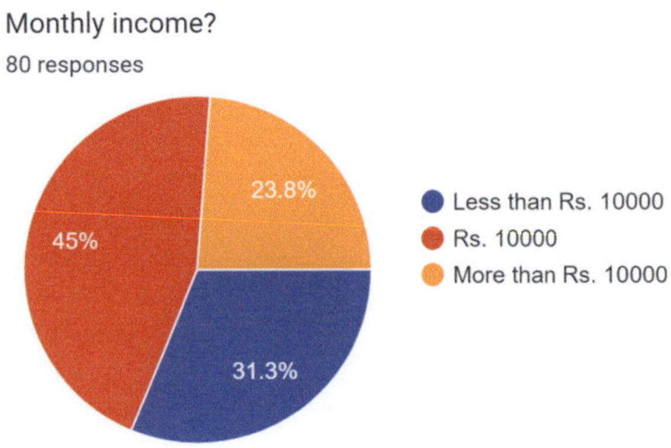

3.1 Water Access

Indira Camp

97.8% of the respondents are dependent upon community taps to fulfil their water requirements. One community tap is located at the starting of every lane in the area and about 10–15 households living in one lane depend on one single tap to meet their water needs. 2.2% of the households living in the same lane have taps near their houses because of their hard efforts to get a pipeline in their lane. The survey found out that the water in the community taps comes two times during the day, one hour in the morning and one hour in the evening, although no fixed timings. 94.4% of women agreed that the responsibility of collecting water lies with them, 2.2% agreed for girls, 2.2% for boys and 1.1% for men undertake any water treatment, 79.8% refused to do anything while 20.2% boiled water for drinking according to 58.4% percent respondents, the water seemed to be potable, although they inferred that they did not have any other option irrespective of its portability. 34.8% of respondents found the water to be unfit for consumption, let alone drinking.

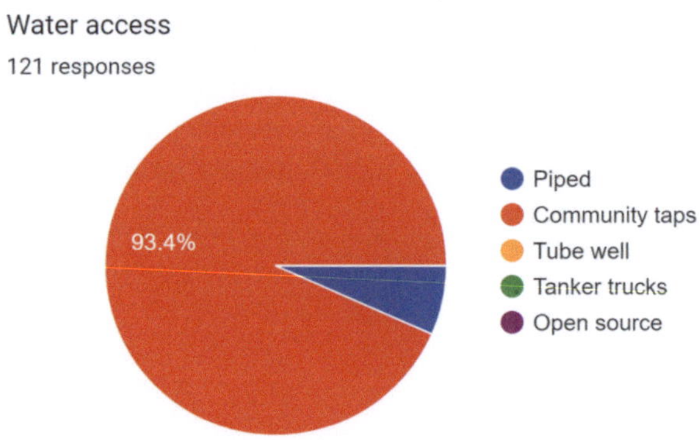

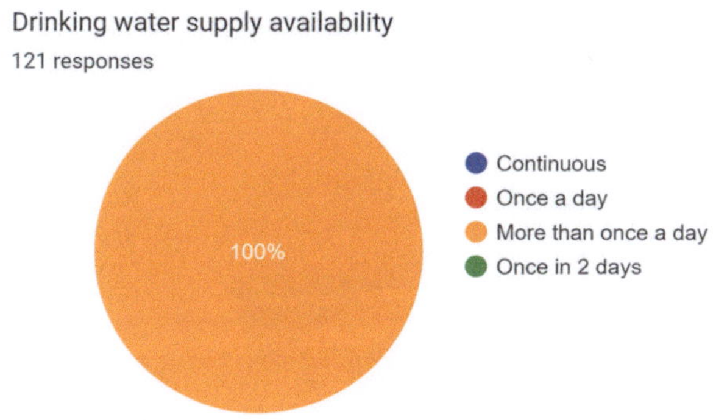

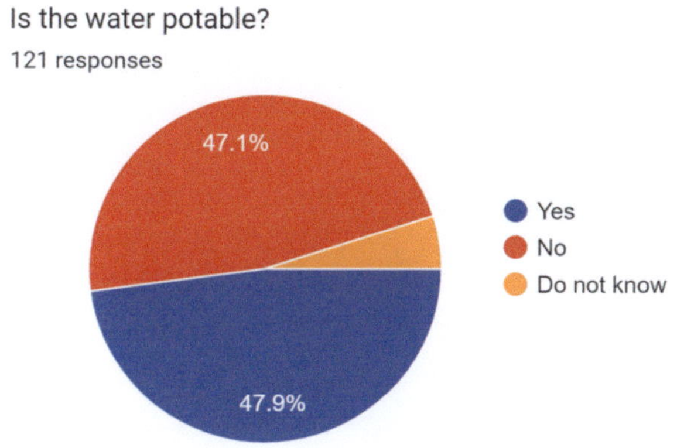

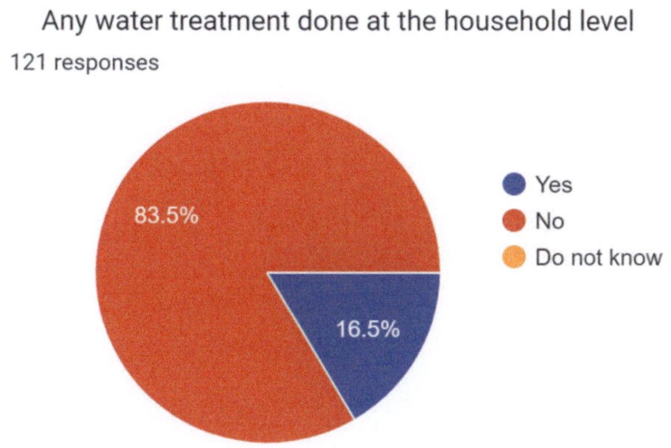

Harijan Camp

98.8% of the respondents are dependent upon community taps to fulfil their water requirements. One community tap is located in the entire area which sustains approximately 90–100 households to meet their water needs. 1.3% of the households that is just one household in the area living in the same lane have taps in their house which they got installed by paying a hefty price. The survey found out that the water in the community tap comes abruptly during the day, with no fixed timings. 100% of women agreed that the responsibility of collecting water lies with them. According to 66.3% per cent respondents, the water seemed to be unsafe for drinking, although they inferred that they did not have any other option irrespective of its portability. 31.3% of respondents found the water to be fit for consumption. When asked if they undertake any water treatment, 100% of respondents refused to do any treatment of water before consumption.

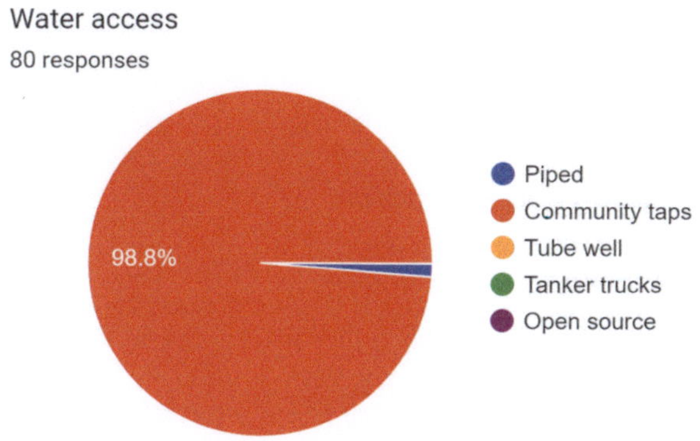

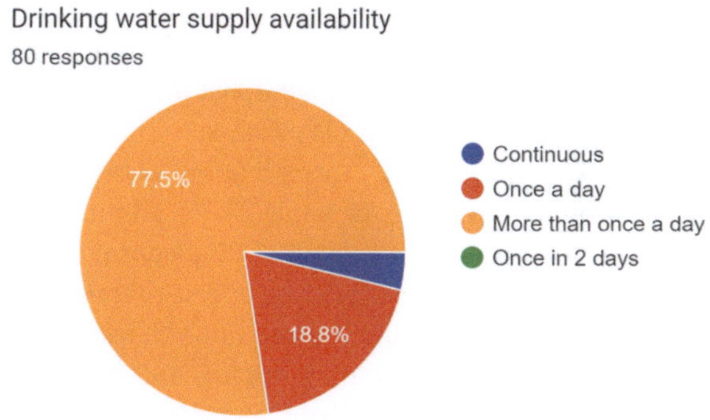

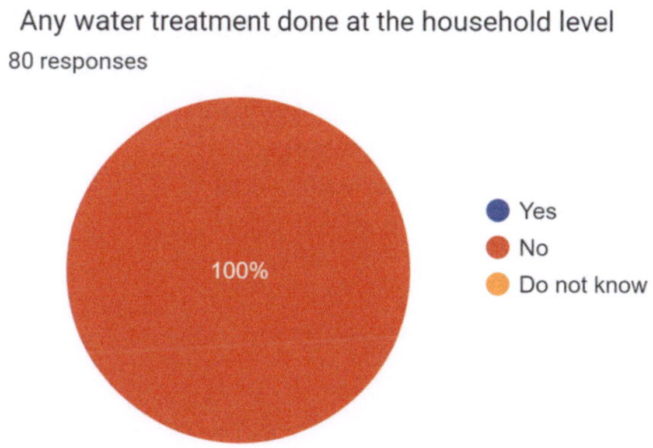

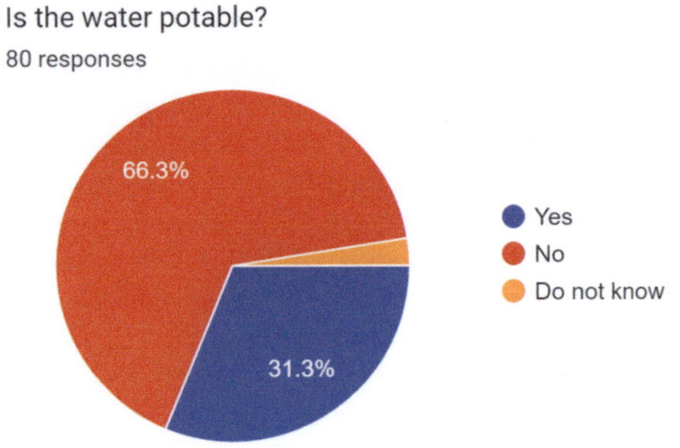

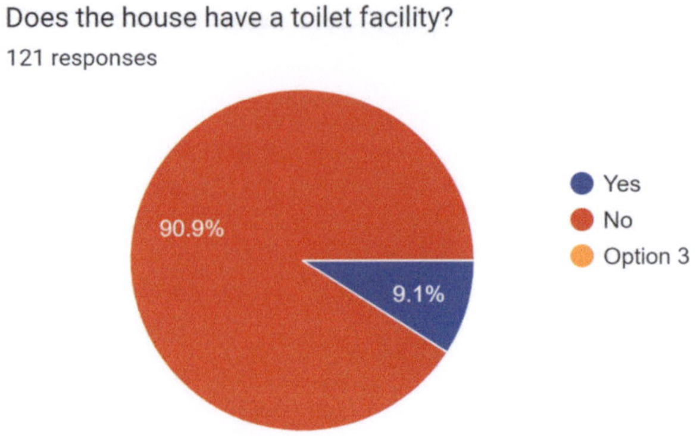

3.2 Toilet and Sanitation

Indira Camp

The survey enquired whether they had a toilet facility in their house or not. 89.9% did not have a toilet in their house, while 10.1% just had a bathing facility. 100% of the respondents depended upon the community toilet situated at the entry of the slum. When enquired about the cleanliness of the community toilet, 95.5% found the condition of the toilet to be extremely poor, with broken seats, tiles, ill-ventilation, obnoxious smell, undisposed sanitary napkins, filthy toilets, despite a sweeper present. 4.5% found that the toilet needs a lot of improvement in terms of cleanliness.

When asked about their menstruation habits, 72.7% admitted using sanitary napkins while 27.3% admitted using cloth. They were also asked about their disposal methods to which all the respondents acknowledged practising safe disposal in newspapers and polythene into the municipality garbage vans. The survey also attempted to assess the hygienic conditions of the slum by enquiring about the sanitation of the lanes and the sewers in the lanes. 61.8% of respondents revealed that the sweeper does not visit their slums often, just once in two or three months. 27% agreed that the sweeper visits their lanes while 11.2% agreed that the sweeper does not visit their lane at all. The majority of the population revealed that they clean the lanes and the sewers. Even if the sweeper comes, he does not clean properly, rather makes a mess in their lanes, which ultimately results in slum dwellers cleaning themselves.

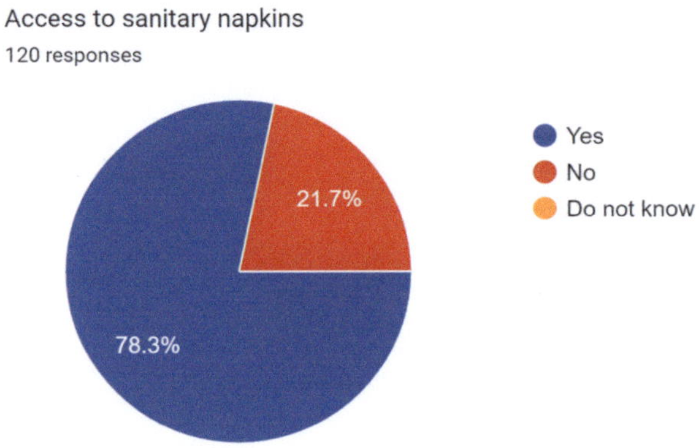

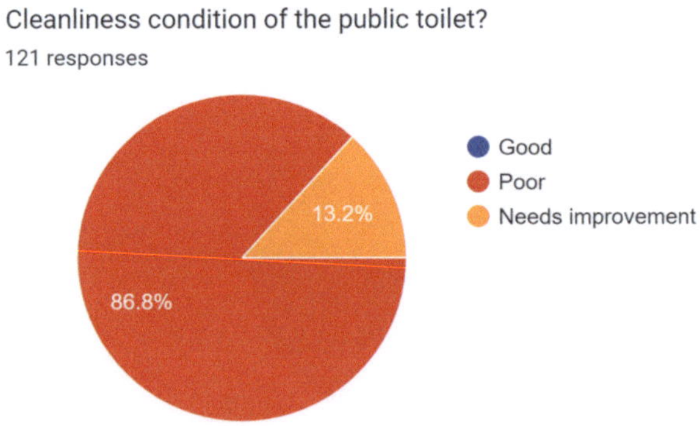

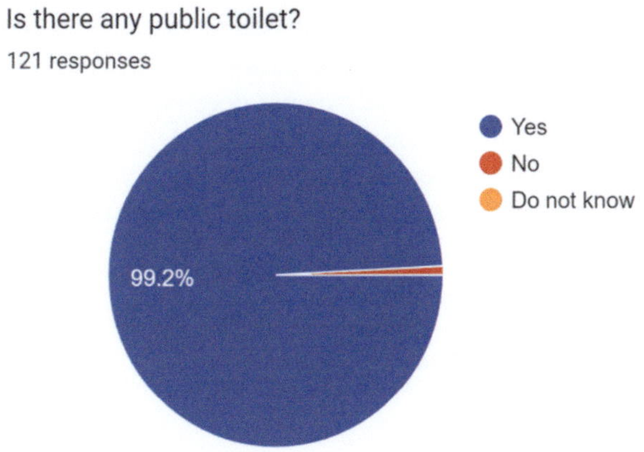

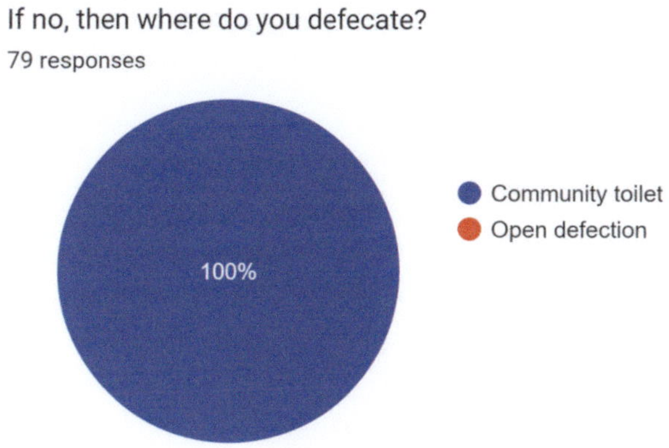

Harijan Camp

The survey enquired whether they had a toilet facility in their house or not. 97.5% did not have a toilet in their house, while 2.5% just had a bathing facility. 100% of the respondents depended upon the community toilet situated at the entry of the slum. When enquired about the cleanliness of the community toilet, 75% found the condition of the toilet to be extremely poor, with broken seats, tiles, ill-ventilation, obnoxious smell, undisposed sanitary napkins and filthy toilets, despite a sweeper present. 25% found that the toilet needs a lot of improvement in terms of cleanliness. When asked about their menstruation habits, 96.3% admitted using sanitary napkins while 3.7% admitted using cloth. They were also asked about their disposal methods to which all the respondents acknowledged practising safe disposal of newspapers and polythene into the municipality garbage bins.

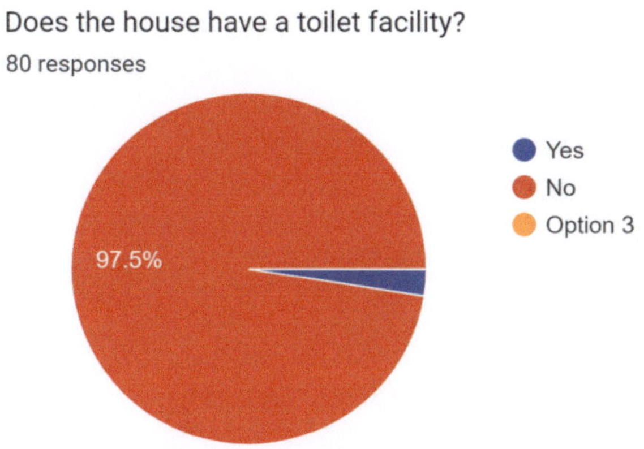

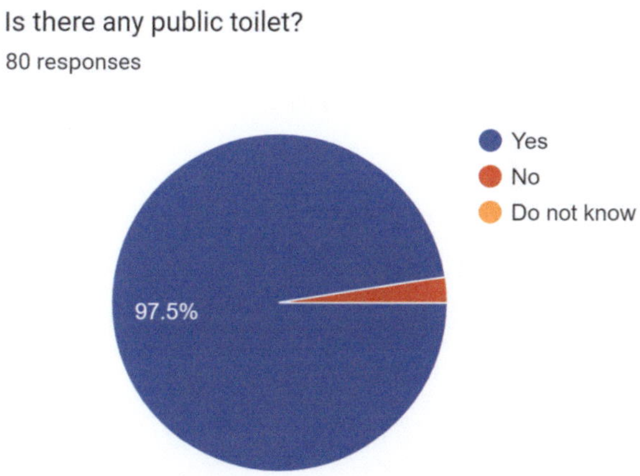

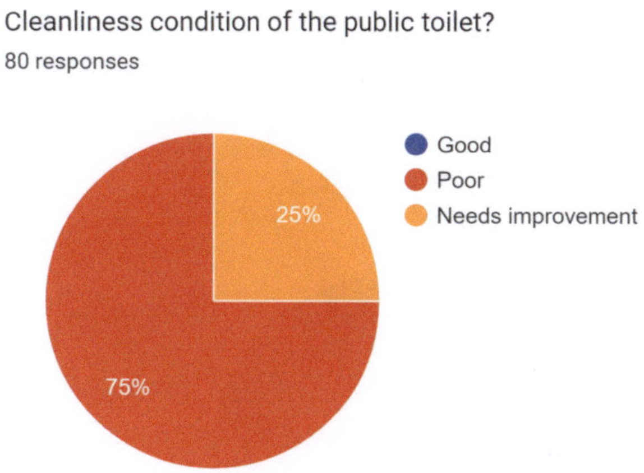

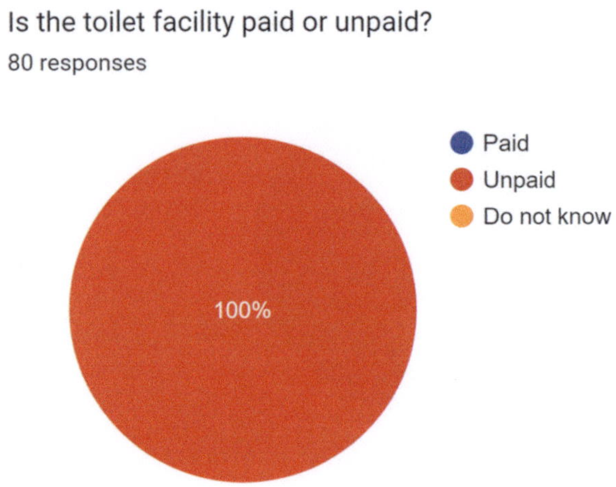

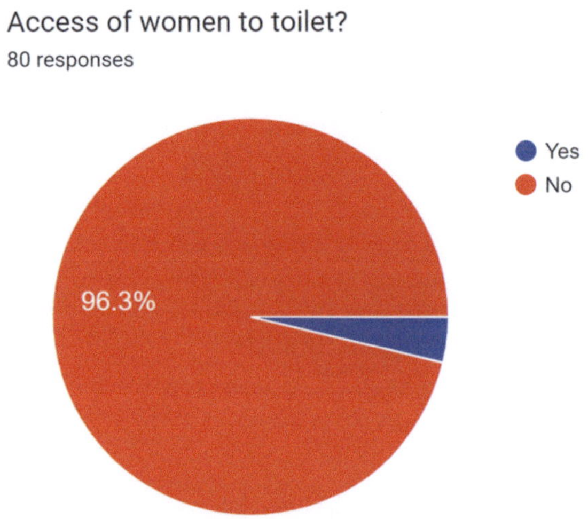

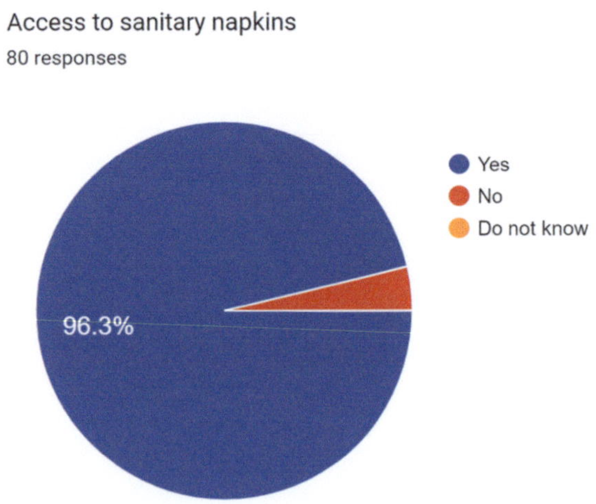

The survey also attempted to assess the hygienic conditions of the slum by enquiring about the sanitation of the lanes and the sewers in the lanes. 53.8% of respondents revealed that the sweeper does not visit their slums often, just once in two or three months. 26.3% agreed that the sweeper visits their lanes while 20% agreed that the sweeper does not visit their lane at all. The majority of the population revealed that they clean the lanes and the sewers. Even if the sweeper comes, he does not clean properly, rather makes a mess in their lanes, which ultimately results in slum dwellers cleaning themselves.

3.3 Waste Disposal

Indira Camp

The respondents were also asked about waste disposal and 95.5% admitted that municipality vans coming to fetch the waste, while 4.5% indulged in open dumping as they believed that the waste collector did not come inside the slum, as a result, they often miss the van.

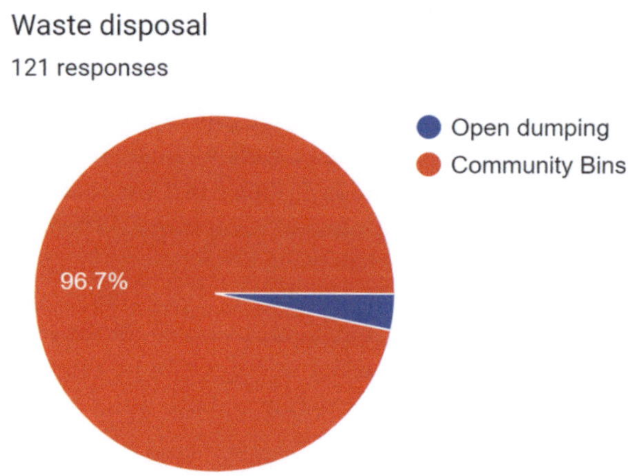

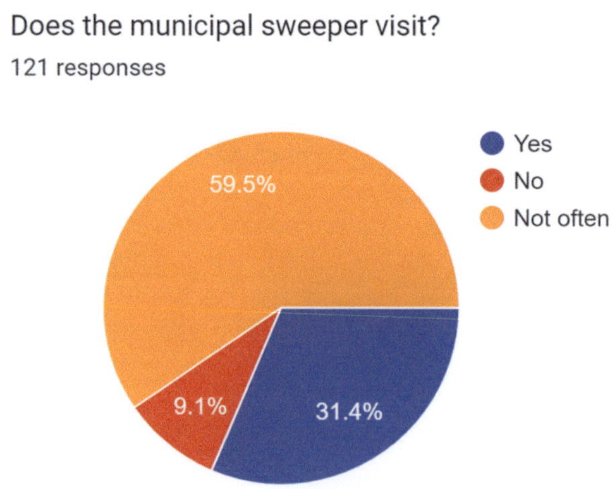

Harijan Camp

The respondents were also asked about waste disposal and 100% admitted that municipality vans coming to fetch the waste, although they revealed that nobody comes inside the slum to collect the waste. Also, the bins are not cleared on time which results in the overflow and scattering of waste all around. Also, a disposal ground was seen just behind the back of the jhuggis, with pigs scavenging over the garbage.

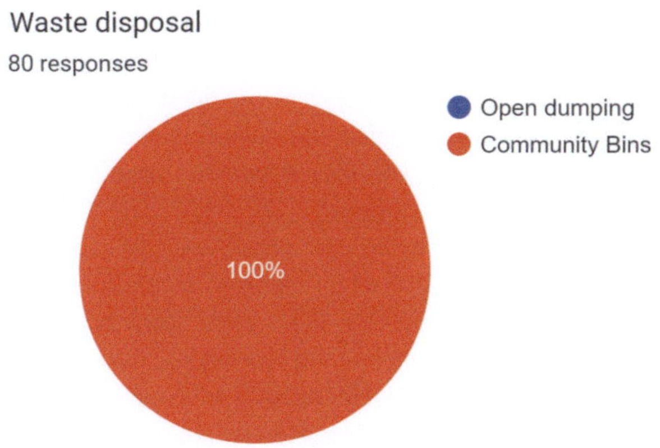

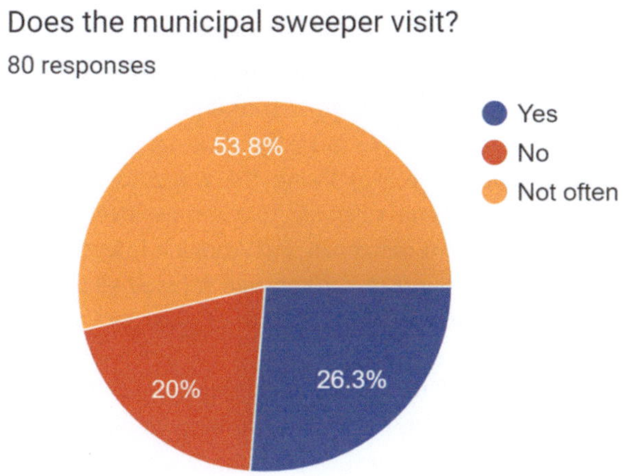

3.4 Healthcare

Indira Camp

It was found that women and children were the most affected due to unsanitary conditions of the water, toilets and lanes which often results in frequent illness. When asked about the approximate monthly expenditure on healthcare, 57.3% revealed it to be much more than Rs. 2000, 29.2% spent approximately Rs. 2000, 10.1% spent approximately Rs. 1000 and 3.4% spent less than Rs. 500. They were also asked if there is any dispensary in the area, to which 100% agreed on having one, out of which 57.3% respondents go to dispensary for day to day small illnesses, 38.2% go to private clinics and hospitals and 3.4% go to government hospitals.

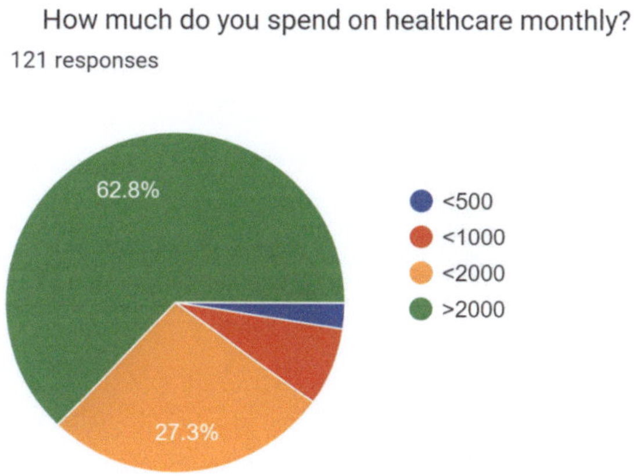

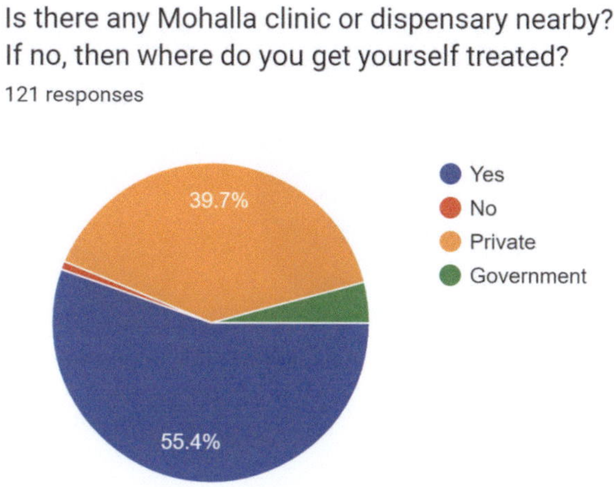

Harijan Camp

It was found that women and children were the most affected due to unsanitary conditions of the water, toilets and lanes which often results in frequent illness. When asked about the approximate monthly expenditure on healthcare, 87.3% revealed it to be much more than Rs. 2000, and 12.7% spent approximately Rs. 2000. They were also asked if there is any dispensary in the area, to which 100% agreed on having one, out of which 41.3% respondents go to private clinics and hospitals, 40% go to dispensary for day to day small illnesses and 18.8% go to government hospitals.

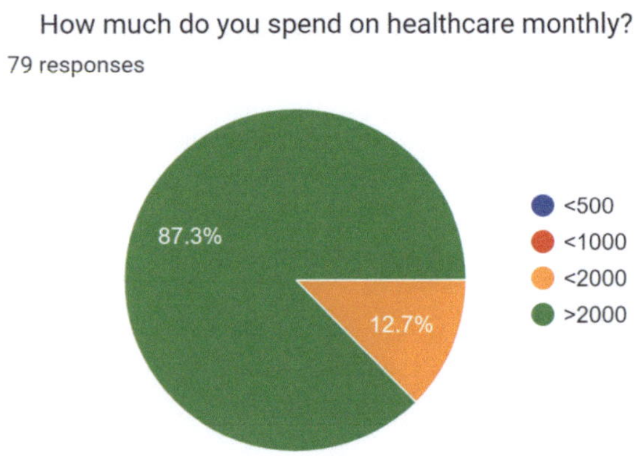

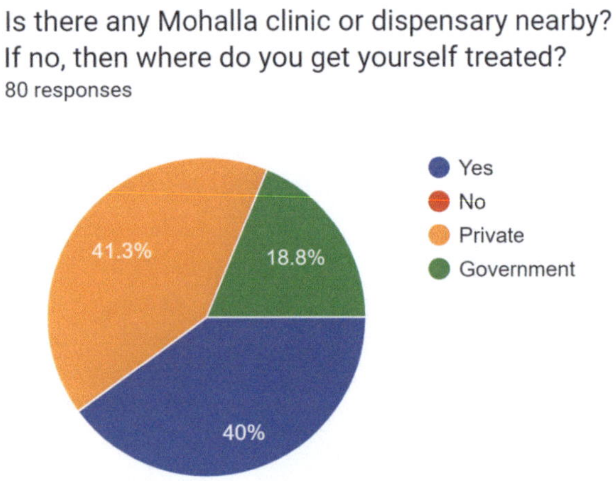

4 Discussion

4.1 Water Access, Sanitation, Waste Disposal and Women's Health

People's health is significantly impacted by water quality. This relates to both the drinkable water quality and the quality of other domestic uses. The respondents' judgements of the water's cleanliness were used to determine the quality of the water. To evaluate the quality of the water, no chemical analysis was done. Although 60% of the respondents said that the water they were getting seemed fine on the outside, they emphasised the fact that they had no other option than to use that water to meet their daily requirements due to their economic conditions. 30% of respondents found the water to be highly unfit for drinking. They revealed that in the window of 1 h that they receive water, they get dirty water in the first 20 min, followed by clean water thereafter. That is when they start collecting the water. Some respondents claimed that during summers and monsoons, they often spot insects coming out of the water taps. This is one of the factors which results in frequent fevers, stomach infections, typhoid, diarrhoea and dehydration in women and children in this slum. 10–15 households are dependent on just one community tap outside their lane, which often results in daily quarrels and violence among the residents. A large chunk of respondents refused to treat water at their houses due to unawareness, expensive cooking fuel and wastage of time, while 23% admitted to boiling water before drinking.

On the contrary, people in Harijan Camp, the majority of the people that is (66.3%) found the water to be unfit for drinking. One resident revealed that almost every child in the area suffers from diarrhoea and stomach infections daily due to unfit drinking water. Everyone refused to boil water before consumption owing to a lack of time and expensive LPG cylinders.

When it comes to toilets, the majority of the respondents did not have a toilet facility in their houses and were completely dependent on the community toilet located near the entry of the slum. Almost every respondent agreed to the extremely poor condition of the community toilet with broken seats, windows and tiles, poor ventilation, broken doors, soiled seats, obnoxious odour and most importantly used sanitary napkins lying on the floor of the toilet. The women revealed that despite a sweeper present in the toilet, the cleanliness of the toilet is extremely poor. To keep public toilets functional and usable, regular solid waste disposal and cleaning of the drains are essential. Toilets would be a perfect breeding ground for infectious bacteria if they were not regularly cleaned and used for solid waste collection. This is resulting in several illnesses in women of Indira Camp like vaginal infections, urinary tract infections, fevers and fungal infections that occur frequently.

Most of the women expressed their plight during menstruation because of inefficient food and medications due to their economic conditions. It was found that many women in Indira camp contract problems like extreme weakness, tiredness, low blood pressure, fatigue and dizziness very frequently. 30% of women did not have access to sanitary napkins and used cloth instead because of monetary concerns. Using a cloth during menstruation is one of the major causes of vaginal infections.

In the Harijan camp, the toilets were found to be ill-maintained by the majority of respondents. Most of the women expressed their anger over the condition of the toilets. One of the residents revealed that most of the women in the area suffer from urinary tract infections, but are hesitant to admit it. This is major because of the unhygienic condition of the community toilets. She inquired about the fate of the survey and requested for the toilets to be rebuilt with proper facilities and regular cleaning.

Another important aspect related to health is the sanitation of the lanes and sewers in the area. The women of Indira camp were unhappy about the sweeper not visiting their lanes frequently. Even if he does, he would not clean the drains properly, instead makes a complete mess in the lane, which is ultimately cleaned by the women themselves. These drains are the breeding grounds for disease-causing pathogens, mosquitos and flies, in addition to emitting obnoxious smells.

4.2 Household Income and Women's Health

The health of women living in these communities can be significantly impacted by their household income. It was reported that 76% of the households had a monthly income of </= Rs. 10,000 supporting an average of 4–6 members per family and the majority living in a single room. They stated that all the money is spent either on rent, school fees, healthcare, electricity, etc. leaving almost nothing in hand. Women stated that taking care of sick family members takes up a significant amount of household time and money. Also, it was noted that 57% of the households spent a lot more than Rs. 2000 monthly on healthcare. 57% of respondents visited the local dispensary for their treatment, although they expressed their dissatisfaction over the treatment they received there. Therefore, most of the residents prefer going to private clinics to get themselves treated by taking loans. Women expressed their grief over being deeply in debt to pay for their treatment in private clinics due to ineffective treatment and medications at the dispensary or government hospitals.

Many women were reported to be using cloth instead of sanitary napkins during menstruation because of their inability to buy. Although the majority of women (70%), reported to be using sanitary napkins. Some women also claimed to get free sanitary napkins from their daughter's school, which proved to be one of the significant developments in the area of menstrual hygiene and health.

5 Recommendations

5.1 Infrastructure Modifications for Water Distribution

Slums' water supply infrastructure has to be upgraded using a multifaceted strategy. The following are some methods the government can employ to improve the slums' water supply infrastructure:

(a) **Create a comprehensive water supply plan**: To update the water supply infrastructure in slums, the government should create a comprehensive water supply plan that details the necessary measures. The requirements of slum dwellers should be given top priority in this strategy, which should also include financing, implementation and monitoring measures. A detailed study of the current water supply infrastructure, the needs of slum residents, and the available resources are necessary to develop a complete water supply strategy for slums. Here is a rough outline for creating an elaborate water supply strategy for slums:

(b) **Develop a complete water supply plan**: Develop a complete water supply plan for slums by conducting a needs assessment to determine the infrastructure already in place, the demand for water, and the unique requirements of slum dwellers. Additionally, any current problems with water quality and potential sources of water supply should be identified via this study.

(c) **Determine potential water sources**: such as groundwater, surface water, or rainfall harvesting, under the needs assessment. The environmental impact of exploiting these sources and any applicable regulations should also be taken into account in this assessment.

(d) **Create a water supply system**: Create a system for supplying water that includes the locations of treatment centres, distribution networks and water sources that have been located. This infrastructure should be created to accommodate the demand for water and offer slum dwellers dependable, safe access to clean water.

(e) **Create a finance strategy**: Create a financing strategy that details the funding requirements, building, operation and maintenance costs for the water delivery system. This strategy should specify prospective funding options including federal grants, private investments, or foreign help.

(f) **Implement the water supply system**: Implement the water supply system including building new infrastructure, restoring existing infrastructure, setting up water treatment facilities and putting distribution networks in place. The places that are most in need of its implementation should be given priority and carried out in stages.

(g) **Monitoring and Evaluation**: Once the water supply system starts functioning, it should be monitored and evaluated for performance, taking into account the quantity, quality and accessibility of the water. To find any problems and make the required changes to the system, this monitoring should be carried out frequently.

(h) **Create a maintenance strategy**: Create a maintenance strategy that covers regular upkeep and repairs of the water supply infrastructure. This plan should also have a backup strategy for handling any unanticipated problems and urgent repairs.

(i) **Community involvement**: Include locals in the design and construction of the water supply system. This involvement can take the form of organising community committees, giving training and education, and encouraging community involvement in the upkeep of the water supply infrastructure.

(j) **Build new water treatment facilities**: To ensure that inhabitants have access to clean and safe drinking water, the government should invest in building new water treatment facilities in slums. These facilities can be made to treat a variety of water sources, including surface water and groundwater.

(k) **Enhance existing water treatment facilities**: To increase the quality and quantity of water supplied to slums, the government should enhance existing water treatment facilities. This may entail upgrading filtration systems, replacing dated pipes and upgrading water storage facilities.

(l) **Encourage rainwater collection**: To supplement the water supply in slums, the government should encourage rainwater collection. This can involve promoting the installation of rainwater collection systems in homes and offering training and education on rainwater harvesting methods.

(m) **Address water theft**: Water theft in slums needs to be addressed because it can cause water shortages and poor water quality. Adding security measures and installing water meters are some of the ways to prevent unauthorised access to water delivery networks.

(n) **Involve communities in water supply project planning and execution**: The government should involve communities in water supply project planning and execution in slums. This may entail organising community committees, giving local leaders training and developing their skills.

5.2 Improving Access to Sanitation Facilities in Slums

Government, public and private organisations, as well as residents, must all work together to improve access to sanitation facilities in slums. Here are some possible approaches:

(a) **Create a comprehensive sanitation plan**: To make it easier for people to access sanitary services in slums, the government should create a comprehensive sanitation plan. The requirements of slum dwellers should be given top priority in this strategy, which should also include financing, implementation and monitoring measures.

(b) **Public toilets**: To increase access to sanitation facilities, the government should provide public toilets in key areas across the slum. All inhabitants, including those with disabilities, should have access to these restrooms, which should be kept up well. Although the slum that was surveyed had one public toilet, but had extremely ill conditions due to the whole community using that very toilet along with irregular and improper cleaning.

(c) **Encourage the building and use of domestic toilets in slums**: Both public and private organisations may encourage the construction and use of domestic toilets. This can involve giving households who construct and maintain toilets financial incentives or subsidies as well as educating people on the value of good sanitation practices.

(d) **Address privacy and safety concerns**: To attract more slum dwellers, particularly women and girls, to public restrooms, the government and private organisations should address privacy and safety concerns. It may also entail establishing separate restrooms for men and women in addition to installing lighting, locks and other safety elements.

(e) **Encourage waste management**: To lower the danger of disease transmission in slums, the government and private organisations can encourage waste management practices. This can entail offering trash pickup services and advocating for proper disposal methods. The slum under survey had municipality vans coming to fetch the trash, but irregular timings forced the slum dwellers to practice open dumping.

(f) **Include locals in the planning and execution of sanitation projects**: Locals should be included in the planning and execution of sanitation projects in slums by the government and private organisations.
(g) **Educate the public on good hygiene practices**: Public and private organisations should educate the public on good hygiene practices, emphasising the value of hand washing and secure waste disposal. The requirements and cultural customs of those living in slums should be taken into account when designing this education.

5.3 Develop an Effective Menstrual Hygiene Plan

(a) **Identify the existing menstrual hygiene practises and facilities**: Identify the existing menstrual hygiene practises and facilities, the accessibility and affordability of menstrual hygiene products, and the particular requirements of slum dwellers by conducting a needs assessment as the first stage in creating a menstrual hygiene plan.
(b) **Build partnerships**: Work with local groups, governmental organisations and other stakeholder groups to create alliances that can aid in carrying out and maintaining the menstrual hygiene strategy.
(c) **Give access to menstrual products**: Establish a system that makes it possible for people to get affordable, high-quality menstrual hygiene supplies including reusable cloth pads, menstrual cups and sanitary pads. To raise awareness of menstruation hygiene and the significance of using safe and sanitary menstrual products, this system should incorporate outreach and teaching.
(d) **Establishing facilities for safe and hygienic menstrual management**: Establishing facilities for safe and hygienic menstrual management, such as clean, private restrooms with running water, areas for washing and drying reusable menstrual products, and facilities for disposing of used menstrual products, is important.
(e) **Education and training**: Educate and instruct slum dwellers on period hygiene management, including safe and hygienic menstrual management practices, the proper use of menstrual products, and the significance of menstrual hygiene for overall health.
(f) **Community participation**: Community participation is important for the development and execution of the menstrual hygiene strategy, particularly for women and girls. Women's organisations can be established, girls and women can receive leadership training, and men and boys can participate in the promotion of menstrual hygiene.
(g) **Monitoring and evaluation**: Monitor and assess the plan's performance, taking into account factors such as community involvement, the availability and grade of menstrual hygiene facilities and goods, and the effect on menstrual hygiene practises and health outcomes.
(h) **Develop a long-term strategy**: Create a sustainable strategy with a budget for continuous upkeep and administration of the menstrual hygiene products and facilities, as well as a schedule for routine monitoring and evaluation to guarantee the plan's effectiveness over time.

5.4 Increasing Waste Management Infrastructure

Developing waste management infrastructure in slums necessitates a customised strategy that considers the particular difficulties slum populations encounter. The following are some ways the government can improve the slums' garbage disposal facilities:

(a) **Improve garbage collection services**: Increasing the frequency of garbage collection, providing separate bins for recyclable material, and putting in place a system for the segregation of waste at the source are all ways that the government might improve waste collection services.
(b) **Decentralised waste management systems**: The government can create decentralised waste management systems in slums. This may entail the construction of small-scale waste treatment facilities that can process garbage locally, eliminating the need to transfer waste to facilities outside the slum.
(c) **Advocate waste reduction and recycling**: To raise the public's understanding of the advantages of waste reduction and recycling, the government can launch awareness-raising campaigns and educational initiatives. By offering financial incentives to those who reduce their trash or recycle their resources, the government can help encourage waste reduction and recycling.
(d) **Enhance accessibility to garbage disposal facilities**: The government should make waste disposal facilities in slum areas more accessible. Construction of garbage transfer stations or the placement of skip bins or other waste disposal containers at key slum locations may be necessary to accomplish this.
(e) **Enforce laws**: The government has the authority to impose laws governing trash disposal in slum regions. This may entail keeping an eye on how garbage is disposed of, enforcing rules governing the operation of waste treatment facilities, and levying fines or other penalties for violations of waste management laws.

6 Conclusion

Women in these communities face significant difficulties and sufferings due to inadequate water, sanitation and hygiene facilities, which have an impact on their health, safety and general well-being. The lack of readily available, safe water sources is one of the biggest problems that women in slums must deal with. Women find it challenging to meet their basic needs, including drinking, cooking and maintaining personal cleanliness, due to the limited supply of water. Furthermore, the difficulties experienced by women in slums are made worse by the lack of adequate sanitary facilities. Their safety, privacy and dignity are at risk because of inadequate or shared lavatory facilities. The danger of illnesses and infections rises when women are compelled to use public restrooms or other dirty, unhealthy facilities.

It is difficult for women to manage their period cycles with dignity and comfort due to the lack of adequate menstrual hygiene management facilities. Women's health issues are additionally exacerbated by the unhygienic living conditions in slums. Diseases can spread more easily in areas with inadequate waste management, bad drainage and crowded living conditions. Women are more susceptible to the health concerns associated with unsanitary settings, especially those who are in charge of childcare and home duties. Additionally, women face social and economic repercussions as a result of the substandard wash facilities in slums. They are unable to pursue school, employment, or other income-generating activities because of the time and effort required to collect water, manage sanitation requirements and deal with health risks.

Promoting gender equality, strengthening women's health and raising their standard of living all depend on addressing the harsh conditions in slums. It necessitates a multifaceted strategy that emphasises menstrual hygiene management, building and maintaining safe and private toilet facilities and increasing access to clean water sources. For societal norms and behaviours relating to hygiene practices to change, community involvement and awareness programmes are crucial. Enhancing wash facilities in slums is essential for attaining sustainable development as well as basic human rights. We can build a more inclusive and equitable society where women may flourish and participate by prioritising their needs and removing the barriers they encounter in accessing water, sanitation and hygiene services, thus helping in the progress of the city and the country.

References

Atahigwa, C., Kadengye, D. T., Iddi, S., Abrams, S., Van Rie, A., & for the NUHDSS. (2020). Trends and determinants of health facility childbirth service utilization among mothers in urban slums of Nairobi, Kenya. *Global Epidemiology, 2.* https://doi.org/10.1016/j.gloepi.2020.100029

Buechley, R. W., Van Bruggen, J., & Truppi, L. E. (1972). Heat island = death island? *Environmental Research, 5*(1), 85–92. https://doi.org/10.1016/0013-9351(72)90022-9

By, S., & Process, S. (n.d.). *Slum Free City Plan of Action.*

Chowdhury, M. A., Nowreen, S., Tarin, N. J., Hasan, M. R., Zzaman, R. U., & Amatullah, N. I. (2022). WASH and MHM experiences of disabled females living in Dhaka slums of Bangladesh. *Journal of Water Sanitation and Hygiene for Development, 12*(10), 683–697. https://doi.org/10.2166/washdev.2022.060

Egal, F., & Forster, T. (2020). Biodiversity, food systems and urban-rural linkages. In *Biodiversity, Food and Nutrition: A New Agenda for Sustainable Food Systems.* Taylor and Francis. https://doi.org/10.4324/9780429030574-6

Garcia-Suarez, T., Kulak, M., King, H., Chatterton, J., Gupta, A., & Saksena, S. (2019). Life cycle assessment of three safe drinking-water options in india: Boiledwater, bottledwater, andwater purified with a domestic reverse-osmosis device. *Sustainability (Switzerland), 11*(22). https://doi.org/10.3390/su11226233

Goonetilleke, A., Carroll, S., Dawes, L., & Hargreaves, M. (2007). Water quality profile of an urbanising catchment. *ASABE - Individual and Small Community Sewage Systems XI, Proceedings of the 11th National Symposium.* https://www.scopus.com/inward/record.uri?eid=2-s2.0-56349144283&partnerID=40&md5=113ef92278854b4c22f8c12150bf3148

Gurkov, I., Kokorina, A., Saidov, Z., & Balaeva, O. (2021). Rituals of opening ceremonies of overseas manufacturing facilities of multinational corporations in Russia. *Thunderbird International Business Review, 63*(2), 145–157. https://doi.org/10.1002/tie.22110

Hanks, L., Line, N. D., & Zhang, L. (2021). Expanding the Methodological Approach to the Social Servicescape: Moving From Measurement to Manipulation. *Cornell Hospitality Quarterly, 62*(1), 157–168. https://doi.org/10.1177/1938965520949213

Heberer, Th., Reddersen, K., & Mechlinski, A. (2002). From municipal sewage to drinking water: Fate and removal of pharmaceutical residues in the aquatic environment in urban areas. *Water Science and Technology, 46*(3), 81–88. https://doi.org/10.2166/wst.2002.0060

Iddi, S., Kadengye, D. T., Kiwuwa-Muyingo, S., Mutua, M. K., Asiki, G., & for the NUHDSS. (2020). Associated factors of pregnancy loss in two urban slums of Nairobi: A generalized estimation equations approach. *Global Epidemiology, 2.* https://doi.org/10.1016/j.gloepi.2020.100030

J. ., Pratomo;H., Kuffer;J., Martinez;D., K. (2016). Uncertainties in Analysing the Transferability of the Generic Slum Ontology. *Geoforum, 1.*

Joshi, V. (2011). *Census of India 2011.*

Karpf, C., & Krebs, P. (2005). Application of a leakage model to assess exfiltration from sewers. *Water Science and Technology, 52*(5), 225–231. https://doi.org/10.2166/wst.2005.0137

Kumpel, E., Albert, J., Peletz, R., De Waal, D., Hirn, M., Danilenko, A., Uhl, V., Daw, A., & Khush, R. (2016). Urban water services in Fragile States: An analysis of drinking water sources and quality in port Harcourt, Nigeria, and Monrovia, Liberia. *American Journal of Tropical Medicine and Hygiene*, *95*(1), 229–238. https://doi.org/10.4269/ajtmh.15-0766

Li, M., Shan, R., Hernandez, M., Mallampalli, V., & Patiño-Echeverri, D. (2019). Effects of population, urbanization, household size, and income on electric appliance adoption in the Chinese residential sector towards 2050. *Applied Energy*, *236*, 293–306. https://doi.org/10.1016/j.apenergy.2018.11.088

Lundy, L., & Wade, R. (2011). Integrating sciences to sustain urban ecosystem services. *Progress in Physical Geography*, *35*(5), 653–669. https://doi.org/10.1177/0309133311422464

Ma, J., Wang, J., & Shen, X. (2021). Economic policy uncertainty and green economy efficiency: power or resistance?—Empirical evidence from Chinese major urban agglomerations. *Economic Research-Ekonomska Istrazivanja*. https://doi.org/10.1080/1331677X.2021.1931911

Manoli, G., Fatichi, S., Schläpfer, M., Yu, K., Crowther, T. W., Meili, N., Burlando, P., Katul, G. G., & Bou-Zeid, E. (2019). Magnitude of urban heat islands largely explained by climate and population. *Nature*, *573*(7772), 55–60. https://doi.org/10.1038/s41586-019-1512-9

Martinopoulos, G., Alexandru, A., & Papakostas, K. T. (2019). Mapping temperature variation and degree-days in metropolitan areas with publicly available sensors. *Urban Climate*, *28*. https://doi.org/10.1016/j.uclim.2019.100464

Molinos-Senante, M., & Sala-Garrido, R. (2017). Energy intensity of treating drinking water: Understanding the influence of factors. *Applied Energy*, *202*, 275–281. https://doi.org/10.1016/j.apenergy.2017.05.100

Nöjd, S., Trischler, J. W., Otterbring, T., Andersson, P. K., & Wästlund, E. (2020). Bridging the valuescape with digital technology: A mixed methods study on customers' value creation process in the physical retail space. *Journal of Retailing and Consumer Services*, *56*. https://doi.org/10.1016/j.jretconser.2020.102161

Pechsiri, J. S., Thomas, J.-B. E., Bahraoui, N. E., Fernandez, F. G. A., Chaouki, J., Chidami, S., Tinoco, R. R., Martin, J. P., Gomez, C., Combe, M., & Gröndahl, F. (2023). Comparative life cycle assessment of conventional and novel microalgae production systems and environmental impact mitigation in urban-industrial symbiosis. *Science of the Total Environment*, *854*. https://doi.org/10.1016/j.scitotenv.2022.158445

Penfold, R. B., Johnson, E., Shortreed, S. M., Ziebell, R. A., Lynch, F. L., Clarke, G. N., Coleman, K. J., Waitzfelder, B. E., Beck, A. L., Rossom, R. C., Ahmedani, B. K., & Simon, G. E. (2021). Predicting suicide attempts and suicide deaths among adolescents following outpatient visits. *Journal of Affective Disorders*, *294*, 39–47. https://doi.org/10.1016/j.jad.2021.06.057

Pille, L., & Säumel, I. (2021). The water-sensitive city meets biodiversity: Habitat services of rain water management measures in highly urbanized landscapes. *Ecology and Society*, *26*(2). https://doi.org/10.5751/ES-12386-260223

Pramanik, S. (2024). AI's function in sustainable development's renewable energy planning. In *Next Generation Materials for Sustainable Engineering*. IGI Global. https://doi.org/10.4018/979-8-3693-1306-0.ch016

Realini, A., Borgarello, M., Viani, S., Maggiore, S., Nsangwe-Businge, C., & Caruso, C. (2021). Estimating the potential of ride sharing in urban areas: The Milan metropolitan area case study. *Journal of Sustainable Development of Energy, Water and Environment Systems*, *9*(3). https://doi.org/10.13044/j.sdewes.d8.0362

Restrepo, L., Aguilar, J., Toro, M., & Suescún, E. (2021). A sustainable-development approach for self-adaptive cyber–physical system's life cycle: A systematic mapping study. *Journal of Systems and Software*, *180*. https://doi.org/10.1016/j.jss.2021.111010

Sheikhtaheri, A., Zarkesh, M. R., Moradi, R., & Kermani, F. (2021). Prediction of neonatal deaths in NICUs: development and validation of machine learning models. *BMC Medical Informatics and Decision Making*, *21*(1). https://doi.org/10.1186/s12911-021-01497-8

Shekhar, S. (2014). Improving the Slum Planning Through Geospatial Decision Support System. *ISPRS - International Archives of the Photogrammetry, Remote Sensing and Spatial Information Sciences*, *XL–2*(October), 99–105. https://doi.org/10.5194/isprsarchives-XL-2-99-2014

Shekhar, S. (2021). Slums in India. *Urban Book Series*, 21–43. https://doi.org/10.1007/978-3-030-72292-0_2

Shimomura, A. R. P., & Ferreira, A. T. (2018). LCZ in metropolitan regions: Surface temperature in urban and rural areas. In N. E., F. S., & R. C. (Eds.), *PLEA 2018 - Smart and Healthy within the Two-Degree Limit: Proceedings of the 34th International Conference on Passive and Low Energy Architecture* (Vol. 2, pp. 790–795). School of Architecture, The Chinese University of Hong Kong. https://www.scopus.com/inward/record.uri?eid=2-s2.0-85086584652&partnerID=40&md5=d33ecf73952382f3b4ed77d797e76535

Tooley, J., & Dixon, P. (2007). Private schooling for low-income families: A census and comparative survey in East Delhi, India. *International Journal of Educational Development*, *27*(2), 205–219. https://doi.org/10.1016/j.ijedudev.2006.08.002

Verhoeff, A., & Menzel, H. (2011). Social capital to bridge the valley of death, simulating critical incidents in innovation. *International Journal of Entrepreneurship and Small Business*, *14*(1), 149–169. https://doi.org/10.1504/IJESB.2011.042167

Wang, J., Kuffer, M., & Pfeffer, K. (2019). The role of spatial heterogeneity in detecting urban slums. *Computers, Environment and Urban Systems*, *73*, 95–107. https://doi.org/10.1016/j.compenvurbsys.2018.08.007

Wang, Y., Gong, J., & Zhu, Y. (2024). Integrating social-ecological system into watershed ecosystem services management: A case study of the Jialing River Basin, China. *Ecological Indicators*, *160*. https://doi.org/10.1016/j.ecolind.2024.111781

Zhao, Z., Yin, H., Xu, Z., Peng, J., & Yu, Z. (2020). Pin-pointing groundwater infiltration into urban sewers using chemical tracer in conjunction with physically based optimization model. *Water Research*, *175*. https://doi.org/10.1016/j.watres.2020.115689

Science of Data Analytics and Role of Microclimate in Crop Disease Prediction for Sustainable Development

Pritimoy Sanyal

1 Introduction

The role of agriculture in the economy of developed and developing countries is very important. However, climate change in a country has a great adverse effect on the agriculture of that country (Nalley 2016; Bonman 1991; Koutroubas 2009). A branch of science known as Phenology has been developed to study the dependence of crops on changes in climatic factors. Climate factors that have an important effect on agriculture include humidity, temperature, amount of carbon dioxide, precipitation, amount of rainfall, amount of sunlight etc. (Lynch 2021; Guerrero 2022). For example, low precipitation can be partially balanced out by irrigation, and too much rainfall can be drained away. But if low precipitation is present in a region for a long time, then the amount of soil water in that region is bound to decrease. Yield production in a region depends on the annual rainfall amount time of precipitation and soil composition of that region. For example, heavy rainfall in a short period reduces the soil's water-holding capacity, but well-rooted, humus-rich soil can hold water quickly. Similarly, crop production in a region depends on a particular range of temperatures in that region. For example, 10–25 degrees centigrade is ideal for wheat production. If it is present for longer or longer than this range temperature, it creates cold and heat stress respectively on the crop which reduces the yield of that crop (Karasi 2020). The annual mean, maximum, and minimum temperature rise from 1901 to 2019 were 0.61 °C/100 year, 1.0 °C/100 year, and 0.22 °C/100 year, respectively. In India, the change in weather has a great impact on the production of kharif crops. Again, if the temperature drops too much in winter, it has a great impact on the quality of Rabi crops. The intensity of sunlight controls the yield and quality of crops. For example, excessive sunlight reduces the rate of photosynthesis of plants, which affects their growth rate. The amount of biomass a plant produces in a unit of time is called its growth rate. That is, if the intensity of sunlight increases the temperature, the rate of photosynthesis decreases, and the production of biomass decreases, thus its growth rate decreases. So, in the case of wheat, mild winters and relatively cool summers increase yield production. The amount of carbon dioxide in the air affects yield production and its quality. For example, when the amount of carbon dioxide in the air increases, the amount of protein and nitrogen in wheat, rice, and soybeans decreases. Although in the case of these C3 plants, if the amount of carbon dioxide in the air increases, the rate of photosynthesis increases, i.e., carbon fertilization increases, but in the case of C4 plants such as sugarcane or maize, it creates adversity. It should be remembered that these C4 plants control $1/4^{th}$ of all crops by value. Again, when the amount of carbon dioxide in the air increases, the temperature increases and the humidity increases, the attack of weeds, pests, and fungi increases, which greatly reduces the yield production. Thus, it is understood that climate factors directly and indirectly control yield production as well as they control agricultural growth also (CCSP 2008; IPCC 2014).

But climate provides statistical information of a relatively large area revealing the spatial and temporal variation of the mean values of its components. As a result, the regional variation of these elements is manifested in relatively controlled areas, especially in mountainous areas coastal areas or islands, i.e., in the case of topographically dynamic zones. For example, when air moves over a mountain, the pressure decreases, causing it to release moisture as rain or snow. But when this air goes down the mountain, it gets compressed and heated, so the lower part of the mountain becomes drier and hotter. This change in topography leads to changes in the vegetation of the upper or lower part of the hill which is expressed through its variation or density. So, there is a succession of different types of plants at different heights of the hill. Generally, climate influences over a wide area. However, it

P. Sanyal (✉)
Technology Cell, Maulana Abul Kalam Azad University of Technology, Kolkata, West Bengal, India
e-mail: pritimoy.sanyal@makautwb.ac.in

D. Kumar et al. (eds.), *Data Analytics and Remote Sensing-Based Analysis for Sustainable Development*, Advances in Science, Technology & Innovation, https://doi.org/10.1007/978-3-031-98095-4_11

has now been found that changes in the climate of a particular or small region have a considerable effect on the production of food grains in that place (Landsberg 1997). This regional variation of climate is called microclimate. A microclimatic region extends from a few metres to a few kilometres. Mainly the microclimate of a particular region depends on factors such as temperature, humidity, wind speed, soil moisture, amount of carbon dioxide in the air, etc. It has been observed that the changes in all these elements or any element greatly affect the food grain production as well as the agricultural growth rate of the region (Mislan 2008). For example, soil moisture is an important microclimate factor, because when soil moisture is 60%, it not only controls soil and air temperature but also rainfall. Again, if the amount of moisture in the soil is optimum, the microorganisms present in the soil can work, which increases the fertility of the soil, resulting in proper plant growth and yield production. Soil temperature also affects agricultural growth. As the soil temperature increases or decreases too much, it affects the growth of plants as well as affects the activities of microorganisms present in the soil. Again, soil temperature controls evapotranspiration. Soil temperature depends on the colour of the soil, i.e., the darker the soil, the more sunlight and incoming radiation it absorbs and heats up. On the other hand, light-coloured soil reflects most of the sunlight or radiation, so the temperature of that soil does not increase much, it remains cold. The air temperature of a particular region depends on the vegetation of that region. As the surface above the vegetation is dense, i.e., high leaf density, it absorbs maximum radiant energy and absorbs most of the momentum of wind speed. This results in an inefficient exchange of temperature, water vapour, and carbon dioxide between the leaf surface and the adjacent air. As a result, the lower part of the vegetation has a lower temperature than the upper part, which increases the agricultural growth of that region. The amount of sunlight a vegetation surface absorbs is called albedo. Since a cropland reflects more sunlight into space than a forest, the albedo of a cropland is higher than that of a forest, which helps reduce the temperature of the surrounding air. An increase in wind speed around the plant removes the warm air trapped on the plant, which increases the transpiration rate, thereby increasing the rate of water uptake by the plant from the soil. Increased air circulation reduces excess moisture and regulates carbon dioxide levels, thereby accelerating plant growth and reducing disease attack rates as wet conditions and increased temperatures accelerate pest attacks. As a result, it appears that the elements of the climate of a particular region, i.e., microclimatic region play an important role in yield production as well as controlling the agricultural growth of that region (Debangshi 2021).

However, no method has yet been discovered to predict the possible outbreak of diseases on food crops due to the change of these elements of microclimate. It is very important to know in advance what is the probability of the manifestation of a disease, especially when changing the elements of the microclimate. No research work has been done on it yet. Some diseases cannot be detected with the naked eye. Such as rice plant blast disease, one of the most devastating diseases of rice which are very difficult to detect early. This disease depends on various elements of microclimate. For example, if there is little air movement during the night, if the amount of humidity in the air increases rapidly, or if the temperature at night is between 22 and 32 °C, then the probability of this disease is greatly increased.

In this research, we have collected weather information such as temperature, humidity, air pressure, cloud coverage, and several different elements present in the air for the last year from 5 blocks of each of the 20 districts, i.e., a total of 100 blocks of West Bengal, one of the most important agricultural states of India. These elements are collected from all those block levels at fixed time intervals through API from different weather satellites like NOAA sent by NASA. Microclimatic clustering of regions is done by analysing how this large amount of data is changing across these regions with the machine learning (ML) method. An AI-based model has been trained on how weather factors are responsible mainly for blast disease in a particular region. Executing the model on microclimatic clusters predicts the likelihood of the disease occurring somewhere in those regions. The entire application has been developed as an Android app named "MCIS-Smriti" for common farmers to use on their mobiles, which can be downloaded from the Play Store for free. If necessary, it is possible to give weather forecasting in regional language along with disease warnings through this app as well as early warning of natural disasters up to the block level. Important issues related to this research are discussed in the next section before entering the Materials and Methods section of this case study.

2 Review of Important Issues Related to the Case Study

In the previous section, we have seen that microclimate factors in a particular region play an important role in causing crop disease as well as influencing the crop productivity of that region. We considered the following components of microclimate in our case study: Air temperature, air humidity, wind speed, UV Index, and PM 2.5. We studied the role of these elements and their changes in crop disease and whether changes in these elements could provide early warning of crop disease. We will discuss the role of these elements of microclimate in the occurrence of blast disease, one of the most dangerous diseases of rice plants, and whether early warning of this disease can be given depending on the changes in the number of microclimatic elements:

2.1 Rice Blast Disease and Its Importance

Rice blast is one of the most dreaded rice diseases caused by the fungus *Magnaporthe oryzae*. Due to its worldwide spread and destructiveness, this disease is considered one of the most common diseases of rice. Almost all parts of the rice plant are affected by this disease (Sanyal 2006; Sanyal 2011). The pathogen infects and produces lesions on the following parts of the rice plant: leaf (leaf blast), leaf collar (collar blast), culm, culm nodes, panicle neck node (neck rot), and panicle (panicle blast). Lesions or spots are mainly seen when rice plants are first affected by this disease. With the naked eye, these spots are indistinguishable from brown spot disease, another rice disease. If this spot is not identified properly in the beginning, the disease spreads very quickly to acre after acre of land, and if the environment is favourable for the disease, total crop failure occurs in a very short period. The shape of these spots is elliptical in which the central part is grey-white and the periphery of the spot is red-brown. Mature lesions are approximately 0.4 to 0.7 inch long and 0.1 to 0.2 inch wide. The worldwide spread of this disease disrupts about 30% of rice production every year, which is equivalent to the food intake of 60 million people. The yield loss can be prevented by chemical application, but uncontrolled use of chemicals can change the pH value of the soil, which prevents the soil from absorbing water. There are several methods for the automatic detection of rice blast disease using machine learning and image processing algorithms where features are extracted from healthy rice leaves and blast disease-affected leaves to train the machine (Sanyal 2008). However, no predictive model has been developed yet to predict the probability of this disease attacking rice plants. It has been found that changes in the microclimate components of a particular region can accelerate the development of blast disease in that region (Fig. 1).

2.1.1 Role of Temperature for Occurring Blast Disease

Temperature plays an important role in spore germination and lesion development of rice blast disease pathogens. Because of this role of temperature, the disease is more prevalent in tropical and subtropical regions of the world. Studies have shown that spore germination and growth of rice blast pathogen is favoured at 27 °C, but temperatures between 22 and 32 °C favour the spread of the disease (Rajput 2017). Therefore, if the temperature increases as a result of climate change or global warming, this disease will spread very quickly. For example, growing rabi rice in the southern region of India is in the temperature range of approximately 18 degrees to 24 °C. So, if this temperature increases slightly, then the possibility of blast disease increases rapidly. The increase in the lesion is proportional to the increase in temperature. It has been observed that if the temperature of 25–32 °C is present for 12 h, the size of the lesions increases during the first 8 days and then the rate of growth decreases (Silvino 1994).

2.1.2 Role of Humidity for Occurring Blast Disease

An increase in relative humidity increases the probability of blast disease. It has been found that blast pathogen spore germination is rapidly initiated when the relative humidity is 72–95%. However, when the ambient temperature is 28 °C and the relative humidity is more than 92%, the blast infection increases rapidly. Again, if the relative humidity and temperature increase, and if it is prolonged, the rate of blast infection increases rapidly. For example, temperatures between 19 and 29 °C particularly in the range of 23–26 °C, during more than 16 h when the relative humidity was above 90%, are highly favourable for blast disease development and propagation (Choong 1998). However, in the case of a panicle blast, the increase in relative humidity does not have any effect, 65% relative humidity produces spores quickly (Silvino 1994).

Fig. 1 **a** Leaf blast lesion at initial stage, **b** Blast lesions at matured stage **c** Panicle Blast and **d** Sheath Blast

2.1.3 Role of Wind Speed for Occurring Blast Disease

The life cycle of the blast disease pathogen begins by shedding conidia on rice leaves. These microscopic conidia are firmly attached to the surface of hydrophobic rice leaves. Larger numbers of these conidia are produced from mature lesions when there is little or no air circulation at night, and when air circulation is low, dew deposition is greater on rice leaves which discharges large amounts of conidium. It has been observed that if the wind speed is 12.25 KMH at night with rainfall of 83 mm/day and relative humidity of 90%, the most favourable conditions for the proliferation of conidia are considered. Again, if the wind speed is 15 KMH and if there is heavy rain, then the immature and mature conidia fall from the leaves of the infected rice plant due to the wind speed and spread to the healthy rice plant parts (Silvino 1994; Berger 1997).

2.1.4 Role of UV for Occurring Blast Disease

Studies have shown that UV radiation stimulates plant defence systems to ward off various fungal attacks. However, after sporulation of the blast spore, the germ tube formation is enhanced in the 366–340 nm range of UV radiation. If UV radiation is prolonged, it activates rice's defence mechanism against blast pathogens and the growth rate and sporulation of pathogens are inhibited (Katsantonis 2017).

Therefore, it can be concluded from the above discussion that microclimate factors such as temperature, humidity, wind speed, and UV radiation have important effects on rice blast disease, and if the changes in these factors are monitored at regular intervals depending on the region, the possibility of developing blast disease can be predicted. We have included in our study the state of West Bengal, one of the important agricultural states of India, to study the regular monitoring of microclimate factors and whether they can lead to the emergence of blast disease.

2.2 West Bengal, One of the Important Agricultural States of India

In the eastern part of India lies the state of West Bengal which occupies 2.7% of India's total area but has 8% of India's total population. The state is primarily agrarian with a wide range of crops cultivated for its geographical diversity. The state ranks first in India in the production of paddy and vegetables. Blast disease is one of the most common diseases of rice plants in the rice-producing districts of West Bengal. For example, in three districts of West Bengal, namely Purulia, Bankura, and Birbhum which are mono-cultivated and 80% of the total area of the districts are under paddy cultivation, rice production is disrupted due to this blast disease. This agrarian state has faced many famines in the past where crores of people died of starvation. The famine of 1770 and the famine of 1943 are among them. One of the reasons for the famine of 1943 was the loss of crop production over large areas due to the attack of brown spot disease on rice plants (Fig. 2).

The climate of a particular region affects the crop production of that region. These climatically dependent regional units are called agro-climatic zones. Depending on the

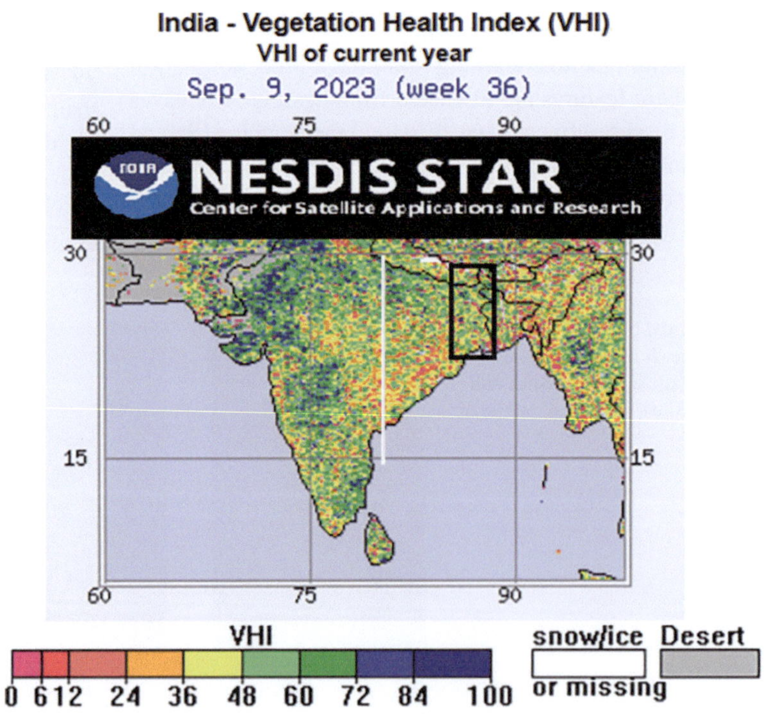

Fig. 2 Satellite imagery regularly monitors microclimate elements to reveal the climate characteristics of a region and its influence on vegetation. Thermal stress affects the vegetation health (VH) of a region. Time series of Vegetation Condition Index (VCI) and Temperature Condition Index (TCI) shows that in combination moisture-stress and thermal stress determine whether or not a region will experience drought or how severe it may be, which also affects the Vegetation Health Index (VHI). The figure shows the satellite imagery of these elements on the map of India where the location of West Bengal is shown as a black box. *Source* https://www.star.nesdis.noaa.gov/smcd/emb/vci/VH/vh_browse.php

geographical diversity of West Bengal, three agro-climatic zones can be seen, namely: (1) The Terai and Teesta alluvial region of north Bengal: This region includes Cooch Behar, Jalpaiguri, Alipurduar and North Dinajpur districts, (2) The Gangetic alluvial region in the southeast: includes East Bardhhaman, Hooghly, Howrah, Kolkata, Nadia, Murshidabad and North 24 Paraganas districts, and (3) The lateritic red a gravely undulating region in the west which includes Purulia, Bankura, Birbhum, West Bardhhaman, and West Medinipur districts (Dutta 2018). We have considered all 20 districts of West Bengal in our study and we have considered five blocks from each district. That is, in this case study, the microclimate data of a total of 100 blocks of West Bengal have been considered. The rest of this paper is discussed as follows: Part 3 depicts the literature survey, Part 4 explains the details of how we conducted the case study, and Part 5 discusses the experiments and their results. Part 6 is the conclusion part of the paper.

3 Literature Survey

The high prevalence and high losses of yields due to rice blast disease, research is being done all over the world to predict this disease. Research is underway on how weather factors affect the origin and spread of this disease. Crop production is particularly disrupted due to global warming. William R. Cline (2008) has shown in his research that this production disruption is greater in countries that are closer to the equator because the temperature here is closer to the crop tolerance level. The increase in temperature is the cause of various crop diseases. Laxman Singh Rajput et al. (2017) showed in their study that an increase in temperature changes the pathogenicity of blast pathogen. Especially the growth and sporulation of the blast pathogen is favourable at 27 °C but when the temperature increases to 32 °C, or if the temperature decreases to 22 °C, then the pathogenicity of this pathogen decreases. Shafaullah, Muhammad et al. (2011) have done almost similar work where they have shown that not only temperature but also relative humidity and rainfall play an important role in blast severity. Y. Lia, W. Uddina and J. E. Kaminski showed in their study that relative humidity helps in the colonization of blast pathogens (Li 2014). Blast disease can be predicted if changes in these weather elements are known in advance. Silvino B. Calvero Jr., developed predictive models in his master's degree thesis through statistical analysis of meteorological parameters (Silvino 1994). In the 1970s and 1980s, a group of Japanese scientists attempted to develop a simulation model using computer hardware and software to predict blast disease (Fukuoka 1975, 1984; Hashimoto et al. 1982; Oota 1982; Takai et al. 1985; Ishiguro 1986). Koizumi and Kato developed a dynamic simulation model for blast disease prediction (Koizumi 1991). Several such blast disease prediction models like EPIBLA, BLASTSIM.2, EPIBLAST, BLASTSIM, DYMEX, CLIMEX, BLASTMUL, EPIRICE, etc., are notable (Katsantonis 2017).

However, artificial intelligence and machine learning (AI & ML) are now becoming important, so they are being used for blast disease prediction. Li-Wei Liu used this ML in their research to develop a rice blast prediction model where the highest and lowest temperature, average relative humidity, soil temperature, and solar energy are used as input to the model. Basically, in this model multilayer perceptron (MLP), support vector machine (SVM), Elman recurrent neural network (Elman RNN), and probabilistic neural network (PNN) are primarily used, and sensitivity analysis (SA) is used to assess the importance of these weather elements for blast disease occurrence (Liu 2021). Saharat Tabonglek et al. developed a model using the finite difference method (FDM) to predict the severity of rice blast disease (Tabonglek 2022). Alvin R. Malicdem and Proceso L. Fernandez used principal component analysis (PCA) for their dimensionality reduction by synchronizing weather data obtained from two government agencies in their model. They determined which elements of the weather are important for blast disease. Based on this, ANN and SVM binary classifiers and regression models have been formulated (Malicdem 2015).

However, all these above-mentioned studies have been based on the meteorological elements of a relatively large area. But Rukhsana et al. (2021) have shown in their research that elements of the weather in a very small area called microclimate also play an important role in the origin and spread of rice blast disease. But the sad thing is that regular observation of how this microclimate is changing in a particular region, its analysis, and predicting whether this change has any effect on rice blast disease in that region or forecast whether changes in these factors can cause blast disease in that region—these important works have not yet been done. In our case study, research work has been done for developing a ML-based application called "MCIS-Smriti" where regular monitoring of microclimate elements of 100 blocks of West Bengal, an important agricultural state of India, is done automatically and predict whether blast disease is likely to occur in that region or not.

4 Materials and Methods

The MCIS-Smriti application module has two broad submodules, namely (1) Live Weather Information Module and (2) Weather Prediction Module. The Live Weather Information Module or LWIM presents the latest information on microclimate components up to the block level. The crop disease prediction section is present in this module along with the language process and speech generation section. On the other hand, the weather predictive module can predict block-level weather for the next seven days by analysing the block-level microclimate elements over a couple of days. Currently, MCIS-Smriti can provide this information in one of the most widely used languages of the world, i.e., Bengali as useful information for the general public.

Three weather stations have been configured, and each one has been assigned a task to make the entire application work smoothly. The load balancer is designed so that these three weather stations can work in coordination with each other. The names of these three AI and ML-based weather stations are Brahma Weather Station (BWS), Bishnu Weather Station (BiWS), and Maheswar Weather Station (MWS). The load balancer is named Pawandoot Load Balancer (PLB), these three weather stations and load balancer are integrated with the "Smriti" grid. These weather stations and load balancers are configured on four dedicated GPUs that are connected to the dedicated university cloud (MAKAUT-Cloud) through this "Smriti" grid. MCIS-Smriti application is built by Python framework, and the app is developed by Android Studio. The backend database is created by PostgreSQL.

4.1 Live Weather Information Module (LWIM)

Live Weather Information Module or LWIM has six main sections: (1) Live Weather Fetch Section, (2) Live Weather Execute Section, (3) GIS-based processing and plotting section, (4) Crop Disease Prediction Section, (5) Human Health Hazard Section, and (6) Language Process and Speech Generation Section. The specific functions of each section are summarized below:

4.1.1 Live Weather Fetch Section (LWFS)

Meteorological satellites are currently used to analyse weather or climate data. This type of satellite is of two types, namely polar orbiting (covering the entire Earth asynchronously) and geostationary (hovering over the same spot on the equator). Currently, there are many geostationary satellites of different countries, among them GOES series, Elektro-L No.1, MTSAT-2, INSAT series, etc., are notable. Among the polar satellites, the NOAA series, Metop series, Meteor, and RESURS series, etc., are notable. All these weather satellites have different types of sensors through which they scan the Earth, taking measurements of reflected light and infrared temperatures. These measurements are digitized and sent to various ground stations on Earth. LWFS of MCIS-Smriti collects all this information through API and stores it in the database.

4.1.2 Live Weather Execution Section (LWES)

A total of 8 times of weather data are collected through API at three-hour intervals in a day. LWES's noise reduction system removes unacceptable information from this huge amount of data and stores it in the database. Temperature, humidity, wind speed, amount of clouds, amount of different gases present in the air, UV information, and PM 2.5 are considered for future processing.

4.1.3 GIS-Based Processing and Plotting Section

At present, microclimate data of only 100 blocks of West Bengal have been considered. The predefined GIS data of these 100 blocks are mapped with LWES data received through API, and this mapping information is plotted on the map of West Bengal. The blocks are marked in red colour on the map of West Bengal where the weather data of the region are integrated which is available from LWES. As a result, when this interface is accessed from the end-user level, clicking on this red spot hits the trigger which fires the weather information, so the end-user can see the latest weather information of the area in that interface.

4.1.4 Crop Disease Prediction Section

It has been discussed earlier that changes in climatic factors promote rice blast disease. Generally, maximum, and minimum temperature, relative humidity, amount of rainfall, intensity of sunlight, soil temperature—all these factors—affect blast disease. Weather factors play an important role, especially during blast spore sporulation and blast lesion growth. In our case study, we used weather satellite data such as temperature, humidity, wind speed, and UV intensity—these four parameters—in a blast disease prediction model. Our main objective in this case study is to collect block-level microclimate data at fixed intervals from different weather satellites and develop a blast event forecasting model by applying these data to ANN.

Data Collection

Generally, favourable conditions for the spread of blast disease pathogens are created when a certain range of temperature, humidity, wind speed, and UV intensity is present at a given time for a given interval of days. In this case study, 8 times data were captured at a daily frequency of 3 h for a fixed period and a total of 2420 microclimate elements were recorded during the last 12 months. Since the maximum extent of blast disease lesions occurs in 7–10 days under favourable weather conditions and sporulation occurs in 7–14 days, the ANN model in this case study uses P10 (before 10 days of rice blast occurred) and P20 (10 days to 20 days after the initiation of rice blast) as input variable periods.

Model Selection for Prediction of Blast Disease

ANN is considered in this case study as an ML technique. MLP, the most widely used model among ANNs, is primarily used as a blast prediction model. Classic MLP is made up of three layers, namely the input layer, hidden layer, and output layer. A hidden layer can have multiple neuron layers where the output of one layer serves as the input of the next layer. A single hidden layer is considered in this case study. MLP is a particularly noteworthy thing in that a backpropagation or backward propagation algorithm is incorporated here that is designed to test for errors working back from output nodes to input nodes. As a result, the accuracy of the prediction level increases. In this case study, NeuroSolution 7.1 software from NeuroDimension, Inc. is used to generate MLP where the data normalization method described by Liu (2021) is followed to avoid overwhelming errors in case of extreme values.

Database of Rice Blast Disease

The rice blast disease database was obtained from the Rice Research Station located in Hooghly district of West Bengal. The data were collected from 2007 to 2009. There are a total of 110 rice blast datasets of which 80% are used for training the classifier, and the remaining 20% are used for testing purposes.

4.1.5 Human Health Hazard Section

The work of this section is to warn about possible health hazards by analysing changes in microclimate elements obtained from weather satellites. For example, studies have shown that floating particles with a diameter of less than 2.5 μm (PM2.5) present in the air can cause adverse effects on the human body such as increased overall mortality, as well as cause-specific mortality, cardiovascular and pulmonary diseases, asthma, and lung cancer. Satellite-derived PM2.5 values are regularly stored and processed to predict potential hazards in the Bengali language.

4.1.6 Language Process and Speech Generation Section

Discrete meteorological data received from satellites are converted into corresponding English sentences while keeping the meaning unchanged. Then using encoder–decoder recurrent neural network each English sentence is translated into Bengali language keeping the meaning unchanged. LWIM's speech generation system converts human accents to speech using AI-based text-to-speech (TTS) assistive technology. In this case study, Google TTS is used for this purpose.

4.2 Weather Prediction Module (WPM)

The application of ARIMA in weather prediction is already established (Shivhare 2019). This ARIMA has been done in WPM to predict block-level weather in a regular manner. The data of microclimate elements at the block level of the last twelve months have been used in this model. In this case, two things have been observed, firstly climate records for a particular location can exhibit great variations from an average level within a few months, and secondly, time series of climate variables, especially in the case of extreme indices, skewed distributions have been observed.

In this study, twelve months of data (September 2022 to August 2023) are mainly divided into three parts, and each part is applied to a specific task, i.e., (1) September 2022 to December 2022 data which is used for training the forecasting model, (2) January 2023 to May 2023 data which is used as monitoring set, and (3) June 2023 to August 2023 data which is used to validate this model. This ARIMA model is designed based on ARIMA (p,d,q) where p is partial autocorrelation, d is the lagged difference between current and previous values, and the value obtained from autocorrelation is q. Basically in this case study, ARIMA(2,1,3) is used for temperature, ARIMA(2,1,2) is used for humidity, and ARIMA(3,1,2) is used for wind speed. The root means square values obtained for these temperature, wind speed, and humidity are 0.084, 0.091, and 0.094, respectively, which proves the acceptance of the forecasting model.

5 Result and Discussion

The MCIS-Smriti application is currently live in two modes: web interface (http://sattrack.makautwb.net) and Android app (available at Android Play Store freely in the name of MCIS-Smriti). Currently, nearly 100 people are using this app regularly on their mobiles and the feedback received is quite satisfactory.

Schematically, this interface is divided into five parts: (1) Information at a Glance: where the weather forecast from WPM is given and Blast Disease information is given at block level from Crop Disease Prediction System (2) Live weather dashboard: where maximum and minimum temperature, maximum and minimum humidity, maximum and minimum wind speed, maximum and minimum PM 2.5, maximum and minimum UV index and rainfall data at block level are given, (3) Real-time microclimate data at block level is given on interactive dashboard where the blocks of West

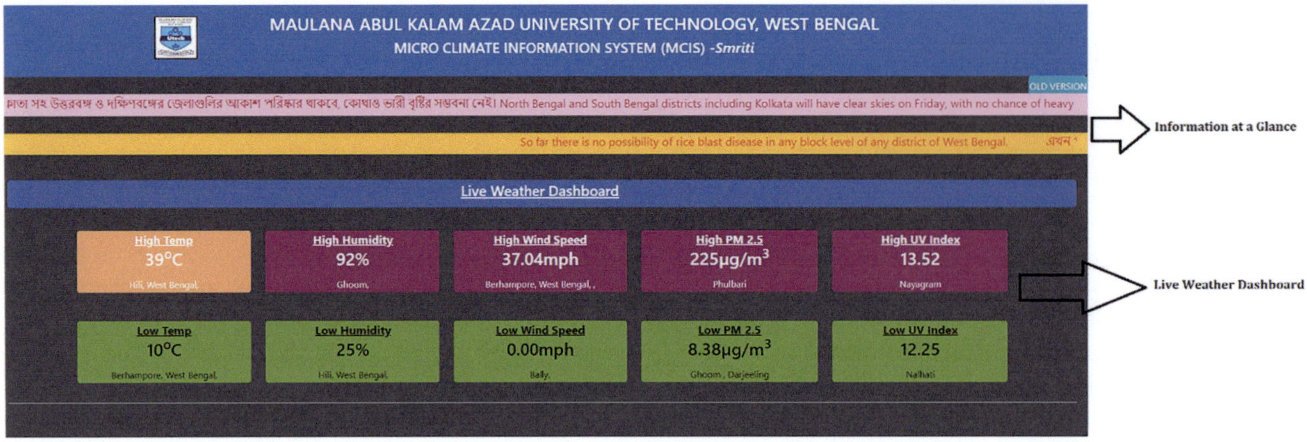

Fig. 3 Web Interface of MCIS-Smriti showing information briefly and live weather dashboard sections

Fig. 4 GIS-based processing and plotting section of MCIS-Smriti application

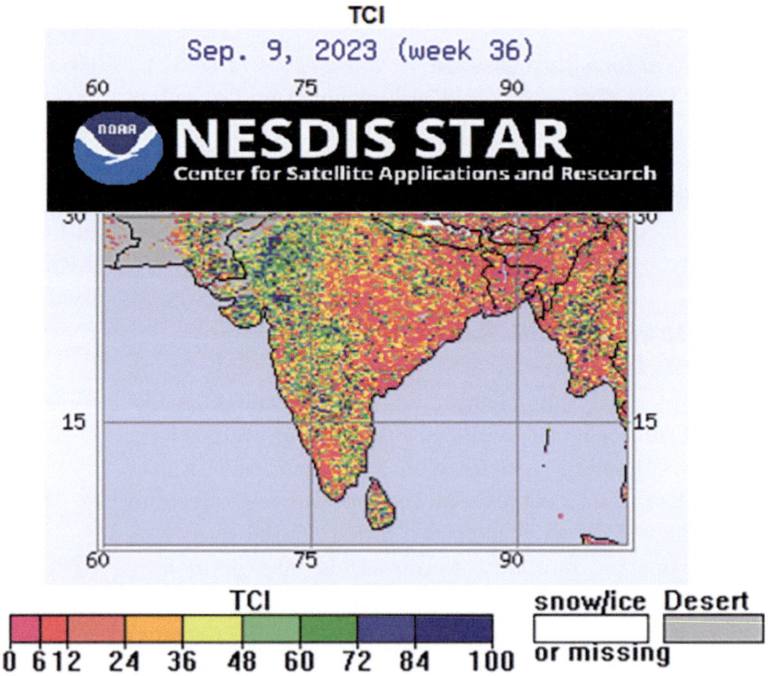

Bengal are mapped as red spots with which the information from LWES is integrated, so clicking on the specific red spot gives the microclimate information of that block (4) District wise dashboard section with three buttons for each of the 100 blocks of the 20 districts of West Bengal. Three audio files are integrated with three buttons at each block, such as a live audio button where the microclimate information of that block can be heard in the Bengali language, a forecast button explains the weather forecast for the next seven days of that block in Bengali language and the health button explains how the microclimate elements in that block area can affect the human body and also provide the information on how to protect against it in Bengali language. Besides that, it also predicts whether there is a possibility of rice blast disease or blast pathogen sporulation at that block level in the Bengali language. 5) In the data analytics visual section, the recent trends of microclimate elements at the block level of West Bengal are shown in graphical mode, besides that, it also plots more than 2.5 magnitude seismic activity across the world in the last 7 days. Next, a schematic diagram of the web interface of MCIS-Smriti is depicted (Figs. 3, 4 and 5).

As mentioned earlier, in this case study, block level microclimate data of the last 12 months from September 2022 to August 2023 have been considered although MCIS-Smriti is collecting this information regularly to date and will continue to do so in the future. In the case of West Bengal, paddy is cultivated thrice a year as the temperature for sowing rice is favourable almost throughout the year. The period from planting to harvesting of pre-Kharif rice was available when this case study was conducted. Besides, matured Ravi rice

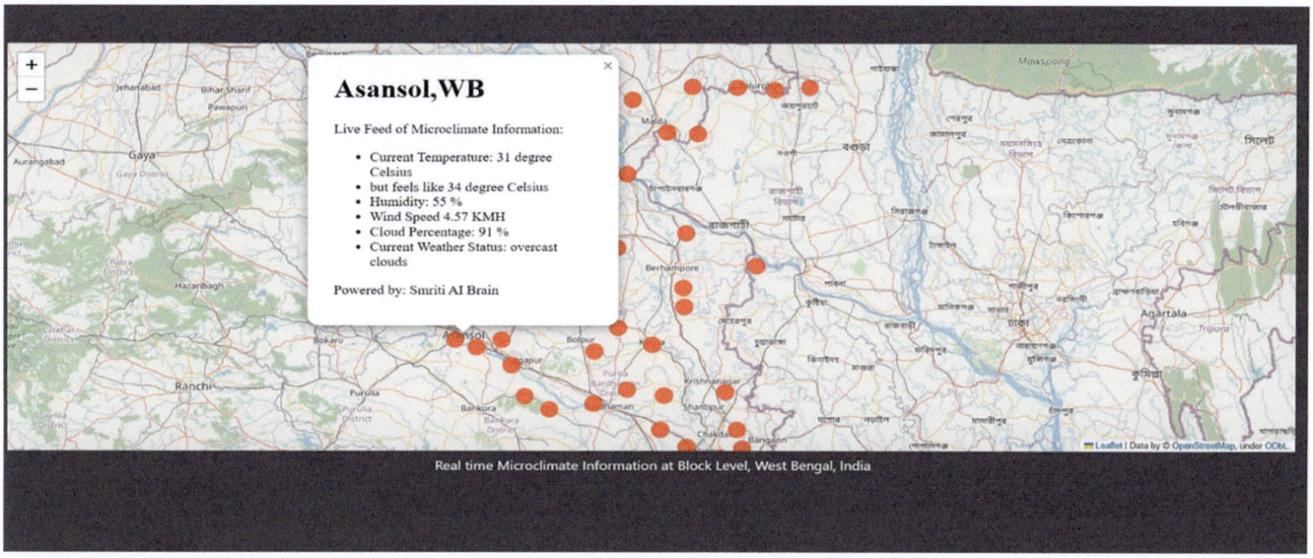

Fig. 5 Audio information of microclimate at block level and data analytics section of MCIS-Smriti

is also included in this case study. The average temperature during this period is considered in the case study at the block level. Based on that the blocks are divided into four clusters.

(a) Blocks whose average temperature is below 16 °C are marked with label A.
(b) Blocks with an average temperature of 17–25 °C are marked as label B.
(c) Blocks with an average temperature of 26–32 °C are marked with label C.
(d) Blocks whose average temperature is more than 32 °C are marked with label D.

Based on this, the following clusters are formed:

MCIS-Smriti is currently capable of seven-day temperature and humidity block-level predictions. Till now this ML model has made this prediction with 91% accuracy. Here are the temperature predictions and their real values plotted for four blocks of East Burdwan district (Fig. 6):

According to the data obtained from the MLP-based model used in this case study, the rate of blast disease spread is inversely proportional to the rate of increase in temperature but proportional to the rate of increase in relative humidity. A region with a temperature of 22–28 °C with a relative humidity of more than 92% if present for two days at night is very favourable for the spread of blast disease and its sporulation. Considering this fact, it is possible to predict the probability of blast disease spread at the block level. Based on this fact

Name of the cluster	Average temperature of this cluster (°C)	The total number of blocks belonging to this cluster	Blocks belong to agro-climatic zone(s)
Cluster-A	Below 16	04	The terai and Teesta alluvial region of North Bengal
Cluster-B	17–25	14	The terai and Teesta alluvial region of North Bengal
Cluster- C	26–32	54	The Gangetic alluvial region in the southeast and part of the lateritic red & gravely Undulating region in the west
Cluster- D	More than 32	28	The lateritic red & gravely Undulating region in the west and the part of Gangetic alluvial region in the southeast

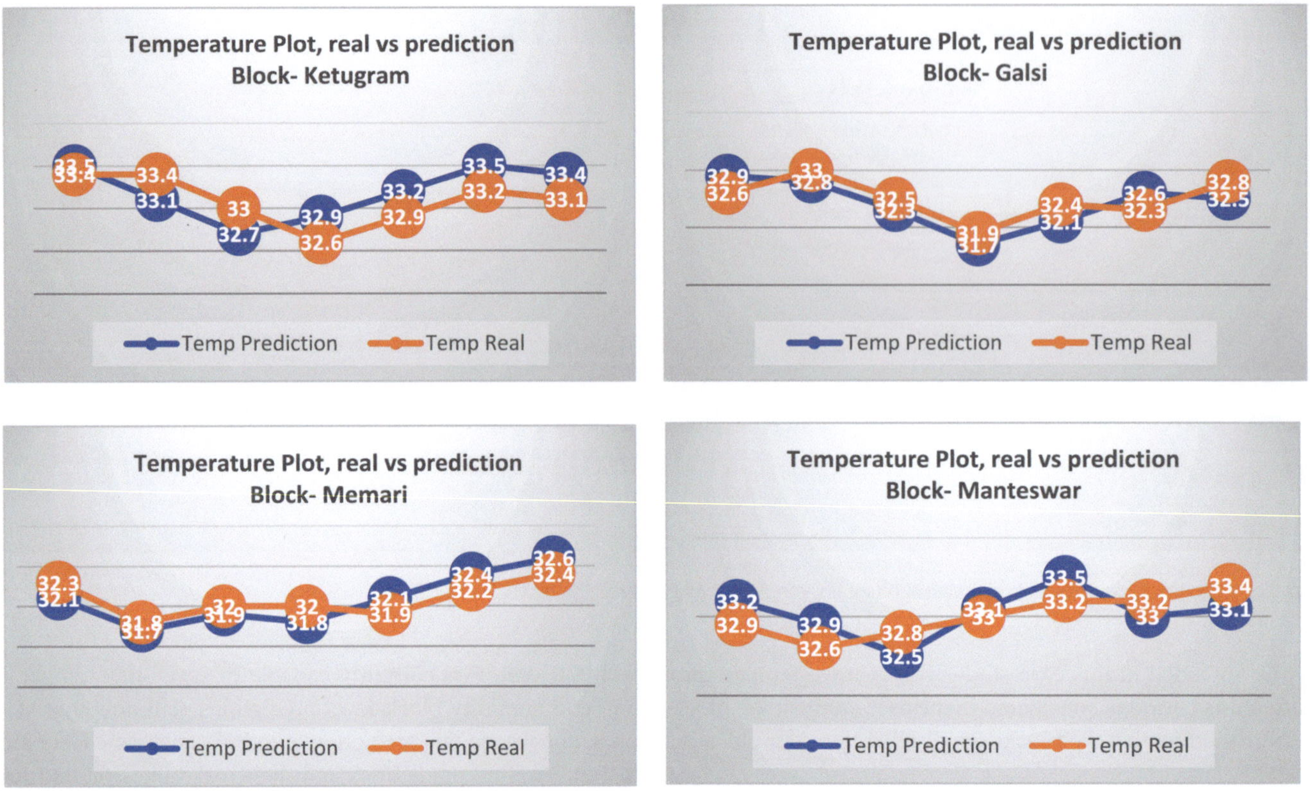

Fig. 6 MCIS-Smriti can predict seven-day microclimate elements at the block level. Here the predicted values and real values of temperature for four blocks are plotted

and by analysing the microclimate data of 100 blocks, the blocks can be divided into three categories in terms of blast disease occurrence, namely less sensitive, sensitive, and very sensitive. MCIS-Smriti publishes this blast disease information for public information through its web interface and app (Fig. 7).

Plotting the temperature and humidity of the 100 blocks leads to the following conclusions:

(a) Thirty-five percent of the 100 blocks of West Bengal are less sensitive to blast disease because in this case the relative humidity is high, but the temperature is between 10 and 20 °C, in another case the temperature is more than 35 °C but the relative humidity is less than 60%, both cases are not favourable for the manifestation and spread of blast disease.
(b) Twenty-nine percent of the blocks are sensitive to this disease, because in this case, the temperature is between 30 °C and the relative humidity is between 80%, whereas if the temperature decreases, a little and the relative humidity increases, which will be favourable for the spread of blast disease.
(c) Thirty-six percent of the blocks are more sensitive to this disease, because in this case, the temperature is between 25 and 30 °C and the relative humidity is more than 90%, which is favourable for the development as well as the spread of blast disease.

It has been observed that in Dhupguri, Dinhata, Falakata, Haldibari, Jaigaon, Jalpaiguri, Kalchini, Nagarakata, and

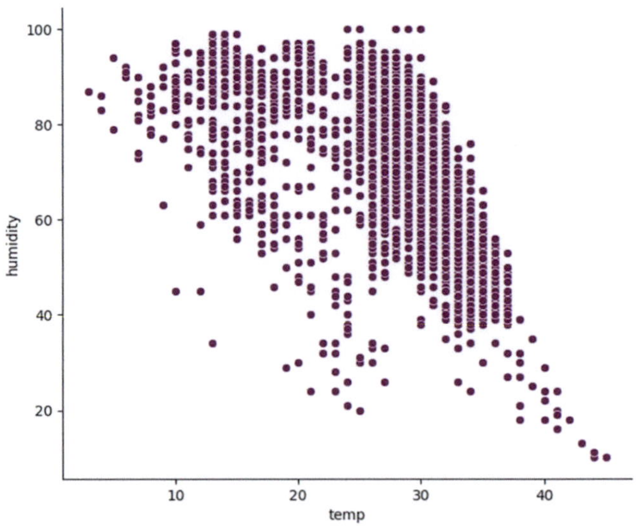

Fig. 7 Twelve months of temperature and relative humidity data of 100 blocks are plotted to illustrate the relationship between them

Tufanganj blocks in North Bengal, Barasat, Basirhat, Habra,and Hasnabad blocks in North and South 24 Parganas, Haringhata, Kalyani, and Krishnanagar blocks in Nadia and Nalhati, Sainthia, Labpur, and Jamuria blocks of in western part of West Bengal are most susceptible to rice blast disease. On the other hand, some blocks of West and East Burdwan, Hooghly, Murshidabad, Medinipur and Maldah like Beldanga, Belpahari, Bhatar, Bolpur, Glossi, Jalangi, Khanakul, Ketugram, Monteshwar, Memari, Pandua, Singur, and Tarkeshwar blocks are sensitive to blast disease because of changes in temperature or humidity can accelerate the onset of this blast disease.

However, an important point is that this case study has only been done for Ravi's cultivation data at a particular time. India is an agricultural country where rice is grown three times a year. Therefore, similar case studies should be done for the remaining two types of cultivation. Moreover, out of the existing 341 blocks in West Bengal, 100 blocks have been included in this case study. Case studies can be done by increasing this number in future.

6 Conclusion

Rice blast disease is one of the most destructive diseases of rice which greatly reduces yield production. Its wide presence and rapid spread all over the world have made this disease even more dangerous. Hence, yield production loss can be reduced if there is a method to predict this disease. Changes in weather factors especially temperature, relative humidity, and rainfall are conducive to the development of this disease. Hence, regular monitoring of these elements of weather and their analysis can be considered an important input in blast disease forecasting models. Based on this, multiple prediction models are being developed. All these models work on meteorological elements over a large area. Changes in these elements of weather in a limited area, called microclimate, influence blast disease in that area. Therefore, regular monitoring and analysis of these microclimate elements can play an important role in developing blast disease prediction models. This case study is carried out on West Bengal, one of the important agricultural states of India, where microclimate data of 100 blocks have been regularly collected, monitored, and analysed, and based on these results, the forecast of blast disease can be done in that region.

To make this machine learning-based application suitable for general use, the web interface has been developed, and to make it usable on mobile, the Android app has been developed in a regional language, which is available on the Play Store. This application collects meteorological data from satellites at the block level, analyses it, and predicts whether conditions are favourable for blast disease. Although temperature, relative humidity, UV intensity, and PM 2.5 have been considered as microclimate factors in this case study, it has been observed that the intensity of blast disease at any block level depends on the amount of temperature and relative humidity there and in that case the UV intensity and PM 2.5 does not play a significant role. This type of ML-based application at the microclimate-based block level is unique and is currently live successfully. In the coming days, voluminous microclimate data obtained through this application will be analysed, predictions on other diseases of cash crops will be given as well, and the number of 100 blocks will be increased so that it can reach many people who are directly or indirectly dependent on agriculture. As a result, this AI-ML-based application "MCIS-Smriti" will be able to reduce yield loss from crop disease to a large extent.

References

Berger R D, Bergamin F A, Amorim L, Lesion expansion as an epidemic component, Phytopathology, Vol: 87, 1997, pp- 1005–1013

Bonman JM, Estrada BA, Kim CK, Lee EJ. Assessment of blast disease and yield loss in susceptible and partially resistant rice cultivars in two irrigated lowland environments. Plant Dis. 1991; 75(5): 462–466

Cline R W, Global Warming and Agriculture, Finance & Development, International Monetary Fund, Vol: 45, March 2008

CCSP (2008) The Effects of Climate Change on Agriculture, Land Resources, Water Resources, and Biodiversity in the United States. A Report by the U.S. Climate Change Science Program and the Subcommittee on Global Change Research. Backlund, P., A. Janetos, D. Schimel, J. Hatfield, K. Boote, P. Fay, L. Hahn, C. Izaurralde, B.A. Kimball, T. Mader, J. Morgan, D. Ort, W. Polley, A. Thomson, D. Wolfe, M. Ryan, S. Archer, R. Birdsey, C. Dahm, L. Heath, J. Hicke, D. Hollinger, T. Huxman, G. Okin, R. Oren, J. Randerson, W. Schlesinger, D. Lettenmaier, D. Major, L. Poff, S. Running, L. Hansen, D. Inouye, B.P. Kelly, L Meyerson, B. Peterson, and R. Shaw. U.S. Environmental Protection Agency, Washington, DC, USA.

Debangshi U, Crop Microclimate Modification to Address Climate Change, International Journal of Research and Review (ijrrjournal.com) Vol.8; Issue: 9; September 2021, pp- 384–395

Dutta K, The Comparative Study of Three Agro-Climatic Zones in West Bengal, India, International Journal of Creative Research Thoughts (IJCRT), Vol: 6(1), 2018, pp- 1348–1364

Fukuoka Agricultural Experiment Station,. Report of special study on development of using method of electronic computer for crop disease forecasting – rice blast disease [In Japanese], 1975, 153 pp.

Guerrero, S., et al. (2022), "The impacts of agricultural trade and support policy reform on climate change adaptation and environmental performance: A model-based analysis", *OECD Food, Agriculture and Fisheries Papers*, No. 180, OECD Publishing, Paris, https://doi.org/10.1787/520dd70d-en.

Hashimoto A., K. Matsumoto and T. Adachi, Some trials for forecasting rice leaf blast epidemics by application of the computer simulation. Annual Report of the Society of Plant Protection of North Japan 33, 1982, pp- 12–14.

Hashimoto A., K. Hirano and K. Matsumoto. Studies on the forecasting of rice leaf blast development by application of the computer simulation. Special Bulletin of the Fukushima Prefecture Agricultural Experiment Station 2, 1984, pp- 75–78.

Ishiguro K., Composing a simulation model for epidemics of rice panicle blast [in Japanese]. Nogyo Gijutu 41, 1986, pp- 491–495.

IPCC (2014) Romero-Lankao, P., J.B. Smith, D.J. Davidson, N.S. Diffenbaugh, P.L. Kinney, P. Kirshen, P. Kovacs, and L. Villers Ruiz, 2014: North America. In: Climate Change 2014: Impacts, Adaptation, and Vulnerability. Part B: Regional Aspects. Contribution of Working Group II to the Fifth Assessment Report of the Intergovernmental Panel on Climate Change [Barros, V.R., C.B. Field, D.J. Dokken, M.D. Mastrandrea, K.J. Mach, T.E. Bilir, M. Chatterjee, K.L. Ebi, Y.O. Estrada, R.C. Genova, B. Girma, E.S. Kissel, A.N. Levy, S. MacCracken, P.R. Mastrandrea, and L.L. White (eds.)]. Cambridge University Press, Cambridge, United Kingdom and New York, NY, USA, pp. 1439–1498.

Karasi B, Mills, Laurence V. Madden and Pierce A. Paul, Quantifying the Effects of Temperature and Relative Humidity on the Development of Wheat Blast Incited by the Lolium Pathotype of Magnaporthe oryzae, Plant Disease, Volume 104, No. 10, 2020, pp-2622–2633

Katsantonis D, Kadoglidou K, Dramalis C, Puigdollers P, Rice blast forecasting models and their practical value: a review, Phytopathologia Mediterranea, Vol 56 (2), 2017, pp- 187–216

Kim Choong-Hoe, Mackenzie, D R and Rush M C, Field testing a computerized forecasting system for rice blast disease, Phytopathology, Vol:78(7), 1998, 931–934

Koizumi S. and H. Kato, Dynamic simulation of blast epidemics using a multiple canopy spore dispersal model. In: Rice Blast Modeling and Forecasting: Proceedings of the International Rice Research Conference, Suweon, Korea. Edited by International Rice Research Institute, Manila, Philippines, 1991 pp- 75–88.

Koutroubas SD, Katsantonis D, Ntanos DA, Lupotto E. Blast disease influence on agronomic and quality traits of rice varieties under Mediterranean conditions. Turk. J. Agric. 2009; 33(5): 487–494. http://journals.tubitak.gov.tr/agriculture/abstract.htm?id=10484

Landsberg J.J, Gower S T, Canopy Architecture and Microclimate, Applications of Physiological Ecology to Forest Management, Academic Press, pp- 51–88, 1997

Li Y, Uddin W, Kaminski, Effects of Relative Humidity on Infection, Colonization and Conidiation of Magnaporthe orzyae, Plant Pathology, Vol: 63, 2014, pp- 590–597

Liu, L.-W.; Hsieh, S.-H.; Lin, S.-J.;Wang, Y.-M.; Lin,W.-S. Rice Blast (Magnaporthe oryzae) Occurrence Prediction and the Key Factor Sensitivity Analysis by Machine Learning. Agronomy **2021**, 11, 771. https://doi.org/10.3390/agronomy11040771

Lynch J, Cain M, Frame D and Pierrehumbert R (2021) Agriculture's Contribution to Climate Change and Role in Mitigation Is Distinct From Predominantly Fossil CO2-Emitting Sectors. Front. Sustain. Food Syst. 4:518039. https://doi.org/10.3389/fsufs.2020.518039 466.

Malicdem A, Fernandez P, Rice Blast Disease Forecasting for Northern Philippines, WSEAS Transactions on Information Science and Applications, Vol: 12, 2015, pp- 120-129

Mislan K.A.S., Helmuth B., Microclimate, Encyclopedia of Ecology, Academic Press, pp-2389–2393, 2008

Nalley L, Tsiboe F, Durand-Morat A, Shew, A, Thoma G (2016) Economic and nvironmental Impact of Rice Blast Pathogen (*Magnaporthe oryzae*) Alleviation in the United States. PLoS ONE 11(12): e0167295. https://doi.org/10.1371/journal.

Oota K., The forecasting of rice blast disease [in Japanese]. Kongetu No Nooyaku 26, 1982, 12–19.

Rajput L S, Sharma T, Madhusudhan, Sinha P, Effect of Temperature on Growth and Sporulation of Rice Leaf Blast Pathogen *Magnaporthe oryzae,* International Journal of Current Microbiology and Applied Science, Vol: 6 (3), 2017, pp- 394–401

Rukhsana, Alam A, Mandal I, Impact of Microclimate on Agriculture in India: Transformation and Adaptation, Agriculture, Food and Nutrition Security, 2021, pp- 41–59

Sanyal P, Bandyopadhyay S K, Application of intelligent techniques towards improvement of crop productivity, International Journal of Engineering Science and Technology (IJEST), Vol: 3 (1), 2011

Sanyal P, Bhattacharya U, Parui S K, Bandyopadhyay S K, Color Texture Analysis of Rice Leaves for Detection of Blast Disease, 20th CSI Conference Proceedings, 2006, pp- 45–48

Sanyal P, Patel S C, Pattern recognition method to detect two diseases in rice plants, The Imaging Science Journal, Vol: 56 (6), 2008, pp-319–325

Shafaullah, Khan A M, Khan A N, Mahmood, Effect of Epidemiological Factors on the Incidence of Paddy Blast (Pyricularia oryzae) Disease, Pakistan Journal of Phytopathology, Vol: 23(2):108, 2011, pp- 108–111

Shivhare N, Rahul A K, Dwivedi S B, ARIMA based daily weather forecasting tool: A case study for Varanasi, MAUSAM, Vol: 70, 2019, pp- 133–140

Silvino B C Jr., Developing Models to Predict Favourable Environments for Rice Blast, Thesis submitted to Oregon State University, June, 1994

Tabonglek S, Khan A, Humphries W U, An extension of mathematical model for severity of rice blast disease, AIMS Mathematics, Vol: 8(1), 2022, pp- 2419–2434

Takai A., K. Kanai, M. Kabayashi, K. Murata and R. Komori, Studies on the forecasting of rice blast by system dynamic model. Forecast for number of lesions and infection area of rice leaf blast by application of computer simulation. Bulletin of the Ibaraki Agricultural Experiment Station 25, 1985, pp- 97–131.

Mitigating Climate Change Effects on Food and Nutrition Security for a Sustainable Future

Rosy Bansal, Gorakh Singh Teja, and Monika Hans

1 Introduction

Climate change presents significant challenges to global food and nutrition security, threatening vulnerable populations and sustainable development. Rising temperatures, altered precipitation patterns, and extreme weather events disrupt agricultural productivity, jeopardizing food systems. These challenges necessitate immediate action to mitigate adverse effects through climate-resilient practices, sustainable resource management, and emission reduction strategies. Ensuring equitable access to nutritious food requires integrated policies addressing environmental, social, and economic factors.

Despite the significant progress that has been made in the last few decades to end hunger, as of 2015, almost 800 million people still experience chronic undernourishment. 161 million children under the age of five are thought to be stunted. 500 million people are obese at the same time. The vital micronutrients required for two billion people to lead healthy lifestyles are not available to them. By 2050, food production would have to expand by 60% to meet the rising demand brought on by population growth and dietary changes, according to the FAO. The availability of our food is threatened by climate change's devastating effects. At the same time, a significant contributor to the crisis is food production itself. Farmers throughout the world are directly experiencing this. Rising temperatures, copious rainfall, violent storms, and protracted droughts are all effects of climate change (Mahendra Dev 2012).

Due to these harsh weather conditions, many crops fail. This affects not only the farmers themselves but also the food they produce and, in turn, the safety of the world's food supply.

When everyone, at all times, has physical and financial access to enough, safe, and nutritious food that satisfies their dietary needs and food choices for an active and healthy life, that is when food security exists (World Food Summit 1996). According to this concept, food security has four components: availability, accessibility (both physically and monetarily), usage (how it is absorbed and assimilated by the human body), and stability of these three components.

Collaborative global efforts, innovation, and knowledge sharing are essential to safeguard food security and build resilience against climate change impacts. This article explores solutions to ensure a sustainable future through targeted mitigation and adaptation strategies. Food sector responsibility: climate is affecting our food sector can be explained by three major factors explained below:

(a) **Monoculture**: Farmers exclusively cultivate one crop on one piece of land under our present, prevalent kind of agriculture, which depletes the soil and has a severe impact on biodiversity such as the sizable Indonesian palm oil plantations or the vast American maize fields. Monoculture depletes the soil, but good soil is essential for absorbing CO_2. Hazardous substances make this issue worse. Additionally, bees and other important pollinators flee monocultures. Our future food supply would be severely threatened without the assistance of these tenacious creatures.

(b) **Greenhouse gas emission:** More than a third of the world's greenhouse gas emissions are caused by the food industry. The production, processing, packaging, and transportation of food—up to and including consumption—are all included in this. We are aware that cultivating rice and producing beef both contribute considerable amounts of CO_2 emissions. However, moving

R. Bansal (✉) · G. S. Teja
GSSDGS Khalsa College Patiala, Patiala 147001, Punjab, India
e-mail: rosydeepak@gmail.com

M. Hans
Government PG College for Women Gandhi Nagar, Jammu, Jammu and Kashmir, India

D. Kumar et al. (eds.), *Data Analytics and Remote Sensing-Based Analysis for Sustainable Development*, Advances in Science, Technology & Innovation, https://doi.org/10.1007/978-3-031-98095-4_12

food by air or water is also incredibly polluting. Last but not least, garbage: in addition to the greenhouse gases released during waste processing, plastic litter also seriously harms the ecosystem.

(c) **Forests are cut down to make way for agriculture:** Around 80% of all forests that are being cleared globally must make way for arable land. For instance, the Amazon is losing trees at a startling rate to make way for soy fields. The ironic irony is that since woods absorb CO_2, they are crucial in the fight against climate change.

It is critical to acknowledge that agriculture contributes significantly to climate change when thinking about the intricate relationships between food and nutrition security and climate change. Agri-food systems currently contribute roughly 30% of the world's greenhouse gas emissions (GHGs), with livestock playing a significant role. The production of food alone will exceed the world targets total GHG emissions if current trends continue. To reduce the expected increase in agriculture's impact on climate change and relieve pressure on other natural resources, such as water, a mix of actions is required. Reducing food loss and waste, enhancing agronomic practices, and altering dietary consumption are examples of mitigation methods. It is important to pick mitigation strategies that do not negatively impact food and nutrition security themselves.

The food system and agriculture are crucial to combating global climate change. Combining demand-side initiatives like changing dietary preferences and reducing food loss and waste with supply side measures like efficient production, transport, and processing lowers GHG emissions and strengthens the resilience of the food system (high confidence). Large-scale land-based adaptation and mitigation methods can be implemented with the help of these combined approaches, all without endangering food security due to heightened competition for agricultural land and rising food prices. Negative consequences would include a rise in the number of malnourished individuals and negative effects on smallholder farmers if food system reforms in farm management, supply chains, and demand were not coupled (medium evidence, high agreement). All that is required to remedy these impacts are transitions.

In many locations, climate change causes a great deal of uncertainty regarding future water supplies. It will have an impact on precipitation, runoff, snowmelt, hydrological systems, water quality, temperature, and groundwater recharge. The increased water scarcity brought on by climate change will pose a significant obstacle to climate adaptation in many parts of the world. The salinity of surface and groundwater in coastal areas will change as a result of sea level rise (Bhuyan et al. 2020) (Fig. 1).

Extreme event frequency and intensity are predicted to change as a result of climate change. Extreme weather

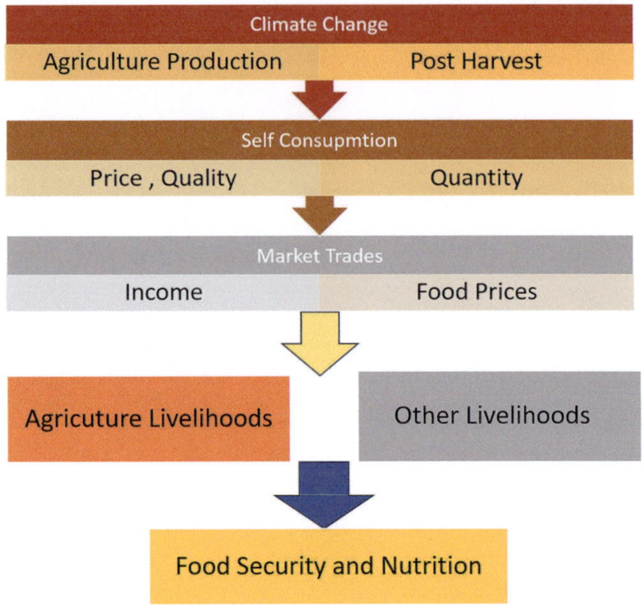

Fig. 1 Schematic representation of the climate change impacts on food security and nutrition

already has significant effects on agriculture. FAO's recent analysis of 78 post-disaster needs assessments in 48 developing countries spanning the 2003–2013 period shows that 25% of all economic losses and damages inflicted by medium- and large-scale climate hazards such as droughts, floods, and storms in developing countries affect the agriculture sectors.

2 Climate Change's Impacts on Food and Nutrition Security

On the one hand, rising CO_2 levels are beneficial for crop growth, but on the other hand, CO_2 emissions are causing frequent climatic changes, such as extreme heat, severe weather, and droughts, which pose a serious threat to crops in high demand like wheat and maize. According to some predictions, global yields could decrease by up to 30% by 2050 if successful adaptation is not made. The worst effects of these consequences are likely to fall on nations already struggling with concerns like violence, pollution, deforestation, and other problems. The two billion people who already go hungry would be worst hit, especially smallholder farmers and those people living in poverty (Tacoli et al. 2013). Despite decades of international commitment, the rates of hunger and food insecurity are alarmingly high. Approximately 750 million people worldwide experienced extreme food insecurity in 2019, according to the State of Food Security and Nutrition in the World report. Undernourishment and food insecurity are on the rise, and climatic shocks are a major contributor. If quick action is not taken, climate change will increase food costs, decrease food availability, and encourage instability and violence as a result of competition for water and arable land.

Due to the majority of its population being impoverished, urban India not only contributes significantly to the world's greenhouse gas emissions but also suffers from the effects of climate change. As was already noted, urban food insecurity would be significantly impacted by climate change. Therefore, there is a pressing need to address urban food insecurity. Both the access and absorption components of urban food insecurity must be included in the strategy for addressing it. The implementation of efficient public distribution systems is required to increase access to wholesome food. It is important to make an effort to learn from states with successful public distribution systems and superior nutritional outcomes, like Tamil Nadu. Living conditions in urban informal settlements must be improved to promote food absorption. One step in the right direction may be the Swachh Bharat Mission, which seeks to build 10.4 million private toilets, and 0.5 million public restrooms, and implement scientific solid waste management in 4041 municipalities.

3 Climate Change's Effects on Food and Nutrition Security

There is growing evidence that global climate change has adverse effects on human health, including direct and indirect effects that are mediated through ecosystems and socioeconomic systems. The effects of climate change on food systems are a significant component of the overall repercussions on human and planetary health, and they are already being felt in the EU. The European Environment Agency has conducted a thorough investigation that shows how changes in temperature, precipitation, and weather extremes are already having an impact on crop yields and livestock output across the EU. In general, climate change is anticipated to make northern Europe more suitable for growing cereal crops, while reducing crop yield in significant portions of southern Europe, even though the growing season there may move into the winter as partial compensation. According to future impact modeling, there could be significant yield decreases in central and southern Europe depending on how severe climate change is. Additional effects on crop quality and quantity may result from rising atmospheric CO_2 concentrations. In the EU, climate change also has a severe impact on vegetable and fruit crops, with further ramifications for nutrition and health. Although the negative effects on fisheries in the European region may be smaller than elsewhere, there will be uncertainty and local variation as a result of fish stocks shifting from tropical to temperate latitudes, and ocean acidification will lower shellfish productivity. Everyone will be impacted by climate change, although some environments and demographic groups are more vulnerable than others. Certain areas of Europe, such as the Arctic and the Mediterranean, may be particularly vulnerable in terms of food and nutrition security. Changing dietary consumption patterns would promote public health in societies that already consume excessive amounts of calories and food from animal sources, as well as aid in mitigating climate change and reconciling current goals with the needs of future generations. What is feasible by changing existing Western dietary intakes to environmentally friendly patterns shows that advantages in all-cause mortality risk were generally inversely correlated with the extent of animal-based food restriction and reduction in environmental footprints. However, greater focus must be placed on comprehending what a sustainable, healthy diet entails and how to protect low-income, vulnerable populations that may already be suffering from micronutrient deficiencies that would be made worse by a levy on animal products. How to manage the effects of changes in food production and consumption on rural development and agricultural communities. Although encouraging sustainable and healthy food choices through food product labeling is one way to do so, there is a chance that entrenched interests will exploit consumers and confuse them.

4 Undernutrition and Climate

According to FAO/WFP (2012), undernutrition is still one of the most important socioeconomic and health issues in the world. Undernutrition has tremendous human and socioeconomic costs that disproportionately affect the poorest people, particularly women and children. The millions of people worldwide who were undernourished as children confront numerous difficulties as they mature (Simon et al. 2011). When they are young, they face an increased risk of disease and death, struggle in school, and frequently are not able to fully contribute to the social and economic growth of their families, communities, and countries. Due to several factors, including the increasing frequency and severity of climate hazards, declining agricultural yields and reduced production in vulnerable regions, rising health and sanitation risks, increasing water scarcity, and intensifying conflicts over limited resources, which would result in new humanitarian crises and increased displacement, climate change threatens to exacerbate existing threats to food security and livelihoods (IPCC 2007). All of the factors that determine food security—availability, access, stability, and utilization—are predicted to be impacted by climate change. Changes in agricultural production and changes in arable land both have an impact on the total availability of food (Hall 2019). The ability of poor households to get food could be impacted by changes in food production, along with other factors. Reduced water availability and quality in some regions could lead to a rise in hygienic issues like diarrhea, which when combined with altered patterns of vector-borne diseases, has the potential to worsen malnutrition and impair the use of food. The stability of the food supply as well as people's livelihoods are both impacted by extreme weather events. Climate change-related increases in extreme weather, such as drought and floods, will worsen this trend and will have a detrimental effect on livelihoods dependent on climate-sensitive activities like rain-fed agriculture and livestock raising. Food security is defined as "all people, at all times, having physical and economic access to sufficient, safe and nutritious food that fits their dietary needs and food preferences for an active and healthy life." by the World Food Summit. Four major components form the foundation of food and nutrition security: Food that is sufficient and of high quality that is provided through domestic production or imports, with a focus on local and national altitudes. Access is all about giving people and households the means to buy the right foods and maintain a healthy diet. Income and/or food poverty issues have a close relationship with food access. Utilization: appropriate use of food based on education, eating habits, availability of enough water, and sanitation. Stability: both food access and availability are impacted. This pillar emphasizes the significance of always having access to food, both now and in the future as well as during the month and year. Because of unexpected shocks, a household or individual must be at risk of not having access to food. The two billion people who already go hungry would be worst hit, especially smallholder farmers and those people living in poverty. Despite decades of international commitment, the rates of hunger and food insecurity are alarmingly high. Undernourishment and food insecurity are on the rise, and climatic shocks are a major contributor. If quick action is not taken, climate change will increase food costs, decrease food availability, and encourage instability and violence as a result of competition for water and arable land. The Indian Council of Agricultural Research (ICAR), which is part of the Ministry of Agriculture and Farmers Welfare, has launched a flagship network project to address the challenges of maintaining domestic food production in the face of climate change. To increase food grain production in the face of climate change, ICAR has created resilient varieties of several crops that are tolerant to climatic shocks (Kumar et al. 2009).

5 Conclusion

Mitigating climate change impacts on food and nutrition security is crucial for sustainability. Strengthening climate-adaptive agricultural practices, reducing emissions, and enhancing global cooperation can protect vulnerable populations. Integrated policies and innovations are essential for equitable access to food resources, ensuring resilience and fostering a sustainable future for generations to come. The magnitude of India's food security concerns has grown as a result of climate change. Despite the complexity of the relationship between food security and climate change, the majority of research only considers one aspect of food security, namely, food availability. The Sustainable Development Goals (SDGs) are set goals with the aims of eradicating hunger, achieving food security, and enhancing nutrition. Because the country's relatively high rates of economic growth have not resulted in a decrease in hunger and undernutrition, India continues to place a high priority on food security when it comes to development. While significant portions of the Indian population experience severe malnutrition, rising wealth and increased urbanization are quickly shifting the food basket away from cereals and toward high-value agricultural products like fish and meat. Low productivity is one of the greatest problems facing Indian agriculture. In comparison with affluent nations like North America (6671 kg per ha), East Asia and the Pacific (5184 kg per ha), and the Euro area, India's cereal yields are significantly lower (5855.4 kg per ha) (Source: World Bank Database). The Sustainable Development Goals (SDGs) are set goals with the aims of eradicating hunger, achieving food security, and enhancing nutrition. Since the country's relatively high rates of economic growth have not resulted in a decrease in hunger and undernutrition, India continues to place a high priority on food security when it comes to development. Between 1990 and 1991 and 2013 and 2014, India's gross domestic product at factor cost and per capita income increased by 7% and 5% year, respectively. However, the prevalence of undernutrition has only slightly decreased from 210.1 million in 1990 to 194.6 million in 2014, thus India has not succeeded in half the number of people who experience hunger, as stated in the Millennium Development Goal. The Global Hunger Index classifies about 12 Indian states as having an "alarming" level of hunger. According to the National Family Health Survey 2015–16, states like Bihar (43.9%), Madhya Pradesh (42.8%), and Andhra Pradesh have a disproportionately high percentage of children under the age of five who are underweight (31.9%) (Mittal 2006).

An integrated assessment of the impact of climate change on India's food security is required to design climate-resilient plans and implement appropriate policy responses. Less research has been done so far on how climate change may affect aspects of food security outside production. Wherever possible, research efforts should be focused on determining the effect of climate change on undernutrition and food absorption.

References

Bhuyan, B., Sahoo, B. K., and Suar, D. (2020). "Food insecurity dynamics in India: a synthetic panel approach. Soc. Sci. Human". Open 2:100029. doi: https://doi.org/10.1016/j.ssaho.2020.100029

FAO (2006). Food Security; Policy Brief. Food and Agriculture Organization of the United Nations; Rome

Food and Agriculture Organisation (FAO), (2016). "Climate change and food security: risks and responses", http://www.fao.org/3/a-i5188e.pdf

Hall, C. (2019). "The impact of climate change on food and nutrition security in Africa and potential adaptation options for the future: a case study of Malawi" (Doctoral dissertation, University of Dundee)

Impact of climate change on agriculture. https://pib.gov.in/PressReleasePage.aspx?PRID=1884236

Intergovernmental panel for climate change, Newsroom Post: The evidence is clear: the time for action is now. We can halve emissions by 2030. — IPCC

IPCC (2007) Fourth Assessment Report. Cambridge: Cambridge University Press. ISDR, UNDP and IUCN (2009) Making Disaster Risk Reduction Gender-Sensitive. Geneva: UN

Praduman Kumar, P. K. Joshi and Pratap S. Birthal, (2009) "Demand Projections for Foodgrains in India", *Agricultural Economics Research Review*, 22: 237

S. Mahendra Dev, (2012) "Small Farmers in India: Challenges and Opportunities", Working Paper, WP-2012-014, Indira Gandhi Institute of Development Research, Mumbai, http://www.igidr.ac.in/pdf/publication/WP-2012-014.pdf

Ministry of Agriculture and Farmers welfare, Press Release: https://pib.gov.in/PressReleasePage.aspx?PRID=1696468

Surbhi Mittal, "Structural Shift in Demand for Food: Projections for 2020", Working Paper 184, Indian Council for Research on International Economic Research, New Delhi, August 2006. http://www.icrier.org/pdf/WP_184.pdf

Simon J. Llyod, R. Sari Kovats and Zaid Chalabi, (2011): "Climate Change, Crop Yields, and Undernutrition: Development of a Model to Quantify the Impact of Climate Scenarios on Child Undernutrition", *Environmental Health Perspectives*, 119 1817, http://www.ncbi.nlm.nih.gov/pmc/articles/PMC3261974/

Cecilia Tacoli et al., (2013). "Urban poverty, food security and climate change", Working Paper 37. International Institute for Environment and Development, London, http://pubs.iied.org/10623IIED.html

UNICEF, Press Release: One billion children at 'extremely high risk' of the impacts of the climate crisis – UNICEF

WFP (2009). Emergency Food Security Assessment Handbook. Second edition. World Food Programme; Rome

WHO (2013). "Diet, Nutrition and the Prevention of Chronic Diseases. Report of a joint WHO/FAO expert consultation". Technical report series 916. Geneva: World Health Organization

Challenges and Future Trends of Data Analytics and Remote Sensing Based Analysis for Sustainable Development

Deepak Kumar, Nick P. Bassill, and Khushboo Rani

1 Introduction

The necessity for effective and sustainable management of natural resources has become more crucial due to growing environmental challenges such as climate change, resource depletion, and biodiversity loss (David et al. 2024; Jehanzaib and Kim 2020). Data analytics with remote sensing skills have emerged as powerful tools in addressing these challenges, providing valuable insights into the health of ecosystems, monitoring environmental changes, and optimizing resource use (Hamud et al. 2021). Remote sensing involves collecting data from satellites, drones, and sensors and offers the ability to monitor vast areas and track changes in real time, while data analytics processes this large-scale data to identify patterns, trends, and actionable insights. These technologies are critical in progressing sustainable development in areas such as agriculture, water management, disaster response, and urban planning. They enable informed decision-making, improve resource efficiency, and support the achievement of global sustainable development goals (SDGs) (Shrestha et al. 2021; Zarbakhshnia et al. 2020). So far, despite their immense potential, several challenges limit the widespread adoption and effectiveness of data analytics and remote sensing in sustainable development. These challenges include issues related to data quality and accuracy, high costs, data integration, and the skills gap in both developing and developed countries. The future trends of data analytics and remote sensing offer promising solutions to these challenges. Advancements in satellite technology, cloud computing, and machine learning are improving the accessibility, accuracy, and scalability of remote sensing data (Parente et al. 2019; Waleed and Sajjad 2022). Open-source platforms and reduced costs are expected to make these tools more available to governments, organizations, and local communities, particularly in developing regions (Reygadas et al. 2021; Vu et al. 2021). They will increasingly play a central role in creating more resilient and sustainable systems across the globe, as these technologies continue to evolve. This will explore the challenges faced in the application of data analytics and remote sensing for sustainable development, as well as the future trends to address these obstacles and enhance their role in promoting global sustainability.

D. Kumar (✉)
Atmospheric Sciences Group, Department of Geosciences, College of Arts and Sciences, Texas Tech University, Lubbock, Texas, USA
e-mail: ; deepakdeo2003@gmail.com

D. Kumar · N. P. Bassill
Center of Excellence in Weather and Climate Analytics, Atmospheric Sciences Research Center (ASRC), University at Albany, State University of New York (SUNY), Albany, New York, USA

N. P. Bassill
State Weather Risk Communication Center (SWRCC), Atmospheric Sciences Research Center (ASRC), University at Albany, State University of New York (SUNY), Albany, New York, USA

K. Rani
Bradley Department of Electrical & Computer Engineering, Virginia Tech, Blacksburg, Virginia, USA

2 Developments in Data Analytics and Remote Sensing for Sustainable Development

These technologies are primarily used for military and scientific purposes but now become indispensable tools for managing environmental, agricultural, and urban systems to address critical issues such as climate change, resource depletion, and environmental degradation (Cooney et al. 2021; Hu, 2020; Sobocká et al. 2021). Recent advancements in data analytics and remote sensing have significantly enhanced our ability to monitor, manage, and preserve natural resources, contributing directly to the pursuit of sustainable development.

D. Kumar et al. (eds.), *Data Analytics and Remote Sensing-Based Analysis for Sustainable Development*, Advances in Science, Technology & Innovation, https://doi.org/10.1007/978-3-031-98095-4_13

2.1 Advancements in Remote Sensing Technologies

Remote sensing has seen remarkable progress over the last few years with the innovations in satellite, drone, and sensor technology. These advancements have resulted in the ability to collect high-resolution data at a global scale, making it possible to monitor vast areas in near real time. The key developments in remote sensing technologies include:

(a) **Improved Satellite Imagery**: The launch of new satellites with higher resolution capabilities has made it easier to capture detailed information about land use, vegetation, and urban development. Satellites like Sentinel-2 and Landsat 8 provide data that can be used to monitor deforestation, desertification, water quality, and crop health at unprecedented spatial and temporal scales.
(b) **Miniaturization and Cost Reduction**: The development of smaller, more affordable satellites (often referred to as CubeSats) and unmanned aerial vehicles (UAVs), or drones, has significantly lowered the costs of obtaining high-quality remote sensing data. This has democratized access to critical environmental data, allowing smaller organizations, local governments, and even individuals to use these technologies for sustainable development efforts.
(c) **Advances in Sensor Technology**: New sensors like LiDAR (Light Detection and Ranging) and hyperspectral sensors offer more precise measurements and greater versatility in capturing environmental data. LiDAR is highly effective in mapping forests and measuring vegetation structure, while hyperspectral imaging can detect subtle changes in plant health, soil conditions, and water quality.

2.2 Progress in Data Analytics for Sustainable Development

Data analytics is evolving in tandem with remote sensing to enhance the ability to process and analyse vast amounts of environmental data. The key developments include:

(a) **Big Data and Cloud Computing**: The growth of big data analytics harmonized with the development of cloud computing platforms, has enabled more efficient processing, storage, and analysis of the massive amounts of data generated by remote sensing systems (Amani et al. 2020; Parente et al. 2019). Cloud-based platforms allow for the aggregation of diverse datasets (e.g. satellite imagery, sensor data, weather data) into accessible repositories for stakeholders to access and analyse environmental data without the need for expensive infrastructure (Mahdianpari et al. 2020).
(b) **Machine Learning and AI**: AI-based systems can also improve the accuracy of predictive models, supporting decision-making processes for sustainable resource management and disaster response. Developments in artificial intelligence (AI) and machine learning are empowering more sophisticated analyses of remote sensing data. Machine learning algorithms automatically identify patterns and anomalies in environmental data like detecting changes in land cover, predicting crop yields, or identifying illegal activities like deforestation or poaching (Maity et al. 2024; Toland, 2018).
(c) **Data Integration and Interoperability**: The recent developments have improved the ability to integrate and interoperate data from diverse sources, including satellites, drones, ground-based sensors, and social media. This integration allows for more comprehensive analyses and better decision-making to combine geospatial data with real-time monitoring systems and crowdsourced data and provide a richer, more accurate picture of the environment (Lunetta et al. 1991; Shi and Meng 2006).
(d) **Real-time Data Analytics**: Real-time data analytics is important for disaster management for timely data to save lives and reduce damage. The availability of real-time data—facilitated by advancements in satellite technology and the Internet of Things (IoT) has transformed environmental monitoring. Real-time data analytics allows for immediate responses to environmental changes like detecting forest fires, monitoring air quality, or tracking floods (Egaji et al. 2021; Kumar et al. 2021).

2.3 Impact on Sustainable Development Goals (SDGs)

The integration of remote sensing and data analytics is playing an essential role in achieving the UN Sustainable Development Goals (SDGs). These technologies help monitor progress in areas such as:

(a) **SDG 2 (Zero Hunger)**: Remote sensing with data analytics is used to monitor crop health, predict yields, and optimize irrigation practices, all of which contribute to improved food security.
(b) **SDG 6 (Clean Water and Sanitation)**: Remote sensing is employed to track water bodies, assess water quality, and monitor changes in water availability, aiding in sustainable water resource management.

(c) **SDG 13 (Climate Action)**: Satellite imagery and climate models assist in monitoring greenhouse gas emissions, tracking temperature changes, and assessing climate vulnerability, helping countries mitigate and adapt to climate change.

2.4 Challenges in Advancing These Developments

The developments in data analytics and remote sensing have created significant opportunities for sustainable development but there are still several challenges to overcome:

(a) **Access to Data**: High-resolution satellite imagery is still often behind paywalls, and data access can be restricted due to national security concerns, commercial interests, or the high cost of data licencing.
(b) **Data Quality and Accuracy**: Inaccurate data result in poor decision-making and hinder efforts to address sustainable development challenges. Remote sensing data in regions with frequent cloud cover or extreme weather conditions may still suffer from low accuracy or resolution.
(c) **Integration and Interoperability**: Integrating data from various sources like satellite imagery, sensors, and historical databases is a complex task. The different formats, scales, and resolutions of the data make it difficult to create a cohesive, accurate analysis that can guide decision-making.
(d) **Skills Gap**: Building capacity and providing training is crucial for empowering local stakeholders to leverage these technologies effectively. There is a significant gap in technical expertise in developing countries, where professionals with the skills to interpret remote sensing data and apply advanced analytics are in short supply.

The developments in data analytics and remote sensing are fundamentally transforming how we approach sustainable development. These technologies provide new opportunities for monitoring environmental changes, optimizing resource use, and supporting decision-making across various sectors. Nevertheless, addressing challenges such as data access, integration, and skill gaps is essential for ensuring their widespread adoption and maximizing their potential (Hamud et al. 2021; Pham et al. 2024). As these tools continue to evolve, they will play an increasingly central role in fostering sustainable practices and achieving global development goals.

2.5 Timeline of the Key Developments (in Data Analytics and Remote Sensing)

1980s—Early Remote Sensing Advancements

- **Satellite Launches**: The launch of Landsat satellites in the 1970s and 1980s provided crucial Earth observation data, laying the groundwork for long-term environmental monitoring.
- **Initial Remote Sensing Applications**: Remote sensing data was primarily used in scientific research, focusing on land cover mapping, vegetation studies, and natural resource management.

1990s—Growing Integration of Geospatial Data

- **Introduction of Geographic Information Systems (GIS)**: The integration of GIS with remote sensing data began makes it easier to analyse and visualize geospatial data for resource management.
- **First Applications in Sustainable Development**: Remote sensing started being used for monitoring environmental factors like deforestation, urbanization, and water resource management.

2000s—High-Resolution Satellite Imagery and Data Analytics Begin

- **High-Resolution Satellites**: The launch of satellites like IKONOS (1999) and QuickBird (2001) provided high-resolution imagery with 1-m spatial resolution, marking a leap forward in environmental monitoring.
- **Early Data Analytics Integration**: Data analytics started to play a role in interpreting large volumes of remote sensing data with manual and basic techniques.

2010s—Big Data, Cloud Computing, and AI Integration

- **Advancements in Satellite Technology**: The launch of Sentinel-1 and Sentinel-2 satellites by the European Space Agency (ESA) (2014–2017) provided frequent and high-quality Earth observation data, focusing on vegetation, land use, and disaster monitoring.
- **Cloud-Based Platforms**: The development of cloud computing platforms like Google Earth Engine (launched in 2010) allowed for faster processing and analysis of satellite data, making it accessible to a global community of researchers and developers.

- **Machine Learning and AI**: AI and machine learning algorithms were integrated into remote sensing data analytics to automate image classification, change detection, and predictive modelling, helping to improve decision-making for sustainable development.

2020s—Democratization and Real-Time Monitoring

- **Miniaturization of Satellites (CubeSats and Drones)**: The introduction of CubeSats and widespread use of drones enabled more affordable and frequent collection of high-resolution data. Platforms like Planet Labs (launched in 2016) began providing daily satellite imagery at a global scale.
- **Real-Time Data Analytics**: Remote sensing combined with IoT (Internet of Things) sensors and real-time data analytics to monitor environmental conditions comprising the air quality, water resources, and agricultural conditions, enabling immediate action during disasters and crises.
- **Enhanced Accessibility through Open Data**: Initiatives such as the Open Data for Development (OD4D) network and NASA's Earth Observing System Data and Information System (EOSDIS) provide open access to remote sensing data, increasing availability for developing countries and local organizations.

2025 and Beyond—Future Trends and Technological Integration

- **AI and Deep Learning Expansion**: Ongoing developments in AI will enable more accurate and advanced predictive models, allowing for better forecasting of environmental changes and their impacts on ecosystems and societies.
- **Collaborative Global Platforms**: Global platforms will further integrate data from satellites, drones, ground sensors, and social media, enhancing cooperation among international organizations, governments, and local communities.
- **Sustainable Development Goal (SDG) Tracking**: Data analytics and remote sensing will become more integrated into tracking progress towards the UN SDGs, using real-time data to assess resource use, monitor climate change, and ensure sustainable practices are adopted globally.
- **Blockchain and Data Integrity**: Innovations like blockchain could improve the transparency and security of remote sensing data, ensuring that it is verifiable and trustworthy for decision-making in sustainable development initiatives.

This timeline highlights the key advancements in remote sensing and data analytics for sustainable development over the decades. These fields will play an increasingly crucial role in achieving a more sustainable and resilient future with technology.

3 Contribution of Data Analytics and Remote Sensing for Sustainable Development

Data analytics and remote sensing play an instrumental role in advancing sustainable development to offer invaluable tools to monitor, manage, and preserve natural resources, ecosystems, and human societies (Onifade et al. 2023; Zhang and Zhao 2024). The integration of these technologies has enabled the creation of more informed policies, resource management strategies, and disaster response mechanisms, all of which contribute to a more sustainable and resilient world.

(a) **Monitoring and Managing Natural Resources**

Satellites and drones provide real-time, high-resolution imagery that can be used to track environmental conditions (Alotaibi and Nassif 2024; Coppin et al. 2004). Remote sensing contributes to sustainable development by monitoring natural resources at a large scale such as:

- **Deforestation**: Tools like the Global Forest Watch rely on satellite data to monitor forest loss in real time, enabling governments and NGOs to take prompt action. Remote sensing allows for the continuous monitoring of forest cover to identify areas at risk of deforestation and support conservation efforts.
- **Water Resources**: Remote sensing is used to track changes in water bodies, monitor droughts, floods, and water quality, and assess groundwater levels. This helps ensure that water resources are managed sustainably, particularly in areas facing water scarcity.
- **Soil and Agriculture**: Remote sensing offers crucial data for assessing soil conditions and health, optimizing irrigation practices, and monitoring crop yields. This contributes to more sustainable agricultural practices, reducing water waste and improving food security.

(b) **Climate Change Monitoring and Mitigation**

Data analytics and remote sensing are critical in tracking and understanding the effects of climate change. Remote sensing satellites measure various environmental parameters like as temperature, sea-level rise, glacier melting, and carbon emissions to provide detailed insights into climate change impacts

(Ahmad et al. 2019; Rana and Ilina 2021). The data generated from these sensors, when analysed through data analytics, helps policymakers and scientists:

- **Predict and mitigate climate-related risks: It** offers timely information for disaster preparedness and early warning systems such as extreme weather events (e.g. hurricanes, floods, and wildfires).
- **Track carbon emissions**: Remote sensing technologies coupled with advanced data analytics estimate greenhouse gas emissions from deforestation, industrial activities, and land use changes, helping to inform national and global climate policies.

(c) **Enhancing Disaster Risk Management**

Data analytics and remote sensing have been crucial in enhancing disaster management strategies to capture real-time data from disaster-prone regions for monitoring and assessing disaster risks, such as floods, wildfires, earthquakes, and hurricanes (Carrão et al. 2018; Erian et al. 2014). The key contributions include:

- **Disaster Preparedness and Early Warning**: Remote sensing allows for the tracking of natural hazards to provide valuable information for disaster preparedness and risk management. Flood-prone areas are identified with satellite data to anticipate flooding events, while drought and fire risks are assessed with temperature and vegetation analysis.
- **Damage Assessment and Response**: After a disaster, remote sensing data is crucial in assessing the extent of damage. Drones and satellites provide detailed imagery to help emergency responders identify affected areas, prioritize relief efforts, and allocate resources effectively.

(d) **Supporting Sustainable Urban Development**

Remote sensing along with data analytics are essential for sustainable urban planning (Rosales Carreón and Worrell 2018; Zhao et al. 2008). These technologies provide insights into land use, urban sprawl, infrastructure, and human settlement patterns and these contributions include:

- **Monitoring Urban Growth**: This information is critical for managing sustainable urban growth and minimizing environmental degradation. Remote sensing data helps to monitor urban expansion and the impact of **urbanization** on natural habitats and agricultural land.
- **Urban Heat Island Effect**: Remote sensing and data analytics are used to study the urban heat island effect, where cities experience higher temperatures than surrounding rural areas. More sustainable environments are identified with urban heat hotspots to help cities plan for cooler spaces with the introduction of green spaces, energy-efficient infrastructure, and better urban designs.

(e) **Advancing Sustainable Agriculture**

Data analytics with remote sensing contribute significantly to sustainable agriculture with the improvement of resource efficiency to reduce environmental impacts, and increase productivity (Efremova et al. 2023; Hüttel and Mußhoff 2024; Lal and Gupta 2009). Remote sensing data is used to:

- **Monitor crop health**: Remote sensing technologies like multispectral and hyperspectral sensors detect changes in crop conditions, allowing farmers to take proactive measures to address issues like pest infestation, water stress, or nutrient deficiency.
- **Precision Agriculture**: Data analytics enables precision farming with the integration of remote sensing data with local weather, soil, and crop data. This allows farmers to optimize irrigation, fertilization, and pesticide use, reducing waste and environmental impact while maximizing yields.

(f) **Supporting Biodiversity Conservation**

Data analytics and remote sensing are invaluable tools for monitoring biodiversity and ecosystem health (Gopal and Chauhan 2006; Raj et al. 2024; Wang et al. 2010). Remote sensing can track:

- **Habitat loss**: Monitoring changes in land use and forest cover helps to identify critical habitats at risk due to human activities like agriculture or urbanization. This information supports conservation efforts by highlighting areas in need of protection.
- **Wildlife Tracking**: Remote sensing coupled with data analytics is used to track wildlife populations and migration patterns. Satellite tags and GPS collars provide real-time data on animal movements, helping to inform conservation strategies for endangered species.

(g) **Facilitating Policy Development and Global Cooperation**

The data collected through remote sensing and analysed via data analytics contribute to the creation of informed policies for sustainable development (Zhang and Zhao 2024; Zhao et al. 2008). Governments and international organizations use this data to:

- **Track progress on Sustainable Development Goals (SDGs)**: Remote sensing and data analytics provide measurable indicators to support the tracking of SDG progress. Satellite data can be used to monitor poverty, hunger, clean energy, and **climate action**, to facilitate evidence-based policymaking.
- **Global Collaboration**: Data-sharing platforms such as Google Earth Engine and NASA Earth Observing System provide open access to satellite data, enabling global collaboration between governments, NGOs, and research organizations. This collaborative approach allows for coordinated actions to address global challenges such as climate change, deforestation, and biodiversity loss.

Data analytics and remote sensing for sustainable development contribute to resource management, climate monitoring, disaster risk management and biodiversity conservation. The real-time large-scale data and advanced analytical tools empower governments, organizations, and local communities to make informed, proactive decisions (Park et al. 2016; Solecki and Rosenzweig 2020). These technologies continue to advance, and their role in achieving a sustainable and resilient future will become increasingly indispensable.

4 Challenges for Sustainable Development with Data Analytics and Remote Sensing

Sustainable development tries to stabilize environmental health, social equity, and economic growth for present and future generations. Data analytics and remote sensing are essential tools for monitoring, analysing, and managing resources to ensure that development is sustainable (Deilami et al. 2022; Erian et al. 2014; Khosravi et al. 2019; Reiche et al. 2016). Even though this application faces numerous challenges some of the key obstacles include:

(a) **Data Quality and Accuracy**

Remote sensing data often comes with inherent limitations such as cloud cover, atmospheric interference, or sensor calibration issues. These factors can impact the quality and accuracy of data, which is critical for informed decision-making. Though inaccurate data can lead to wrong conclusions, affecting critical decisions related to environmental monitoring, agricultural practices, urban planning, and more. This undermines the effectiveness of sustainability initiatives, especially in areas like deforestation, water management, and biodiversity monitoring.

(b) **Data Integration and Interoperability**

Sustainable development decisions require integrating diverse data sources (e.g. satellite imagery, ground sensor data, social media, and economic statistics). These datasets often vary in format, scale, and resolution, making integration complex. The inability to effectively combine and analyse different data types can lead to incomplete or fragmented insights, preventing a holistic understanding of sustainability challenges like climate change, land use changes, or population growth.

(c) **Limited Accessibility to Data**

High-resolution satellite data and advanced analytics tools may be excessively expensive, particularly for developing countries. Limited access to remote sensing data hampers decision-making in regions that need these tools the most. This lack of data can slow down progress in addressing issues like poverty, resource depletion, and environmental degradation. Besides, political or security concerns restrict access to crucial environmental data in some regions.

(d) **High Costs of Technology and Infrastructure**

The powerful data processing tools and infrastructure needed to handle and analyse vast amounts of data from remote sensing technologies comprising satellite imaging, drones, and sensors, require significant investment. This financial constraint restricts the ability to monitor sustainability and implement data-driven solutions. The costs of these technologies are a significant barrier for many governments or organizations with limited resources.

(e) **Technical Expertise and Skill Gaps**

Decision-makers may struggle to interpret complex data, leading to less effective sustainability strategies. Data analytics and remote sensing tools involve expertise in geospatial sciences, data processing, machine learning, and other technical fields. Many regions including developing ones may not have enough trained professionals to leverage these technologies. The insights generated from data analytics and remote sensing are underutilized due to the proper technical expertise.

(f) **Ethical and Privacy Concerns**

The use of remote sensing data can raise ethical concerns when combined with other data sources such as social media. These concerns include privacy violations or misuse of data in ways that could harm vulnerable communities. Public distrust or resistance to data-driven solutions may hinder the adoption of remote sensing technologies in regions with strong privacy regulations. It could also limit data sharing across borders or among organizations.

(g) **Data Overload and Complexity**

Remote sensing systems produce massive amounts of data and processing and analysing of these data in real time can beat existing systems and decision-makers. The sheer volume and complexity of data can make it difficult to extract actionable insights. Data excess can lead to slower decision-making or missed opportunities. This complexity can devastate organizations or governments with delayed implementation of sustainable development initiatives.

(h) **Timeliness of Data**

While remote sensing data is invaluable for monitoring changes over time (e.g. deforestation, land use changes), there is often a delay between data collection, processing, and analysis. In cases such as natural disasters or urgent environmental events, real-time data becomes crucial. Delays in obtaining and analysing critical data may limit the ability to respond quickly to urgent sustainability issues like climate events, natural resource depletion, or pollution crises.

(i) **Geospatial Challenges**

Remote sensing data, especially satellite images, must be accurately georeferenced. Variations in the geographical area or differences in terrain can complicate the interpretation of spatial data. The localized environmental factors may not always be captured at a broad scale. Poor geospatial accuracy or misalignment between different datasets will lead to incorrect interpretations of land use, vegetation cover, or ecosystem health, which in turn could affect decision-making and policy development.

(j) **Political and Governance Barriers**

Some governments may restrict access to certain environmental data due to national security concerns or political motivations. The political climate may make it difficult to implement effective data-sharing mechanisms or enforce policies based on data insights. Political barriers to data access and collaboration can prevent the international sharing of data or hinder national-level environmental monitoring programmes. These barriers reduce the effectiveness of regional or global efforts to address sustainability challenges such as climate change, cross-border pollution, or shared resource management.

(k) **Scalability and Adaptability of Solutions**

Data resolution, processing power, and the ability to adapt solutions to local contexts may vary. Solutions that work well in one area or sector may not be easily replicated in others, leading to inefficiencies when trying to apply these technologies to global challenges such as poverty reduction, sustainable agriculture, or urban sustainability. While data analytics and remote sensing may work well for specific, localized problems, scaling these solutions to larger regions or broader sustainability challenges is difficult.

(l) **Public Awareness and Engagement**

The complexity of the data and the specialized nature of the technology often make it difficult for the general public to understand or act upon the insights generated. A lack of public awareness can lead to apathy or resistance to sustainability measures. Without clear communication of the value and benefits of data-driven approaches, public participation in sustainability efforts may be limited. Thus, a comprehensive public engagement is crucial for remote sensing and data analytics to contribute to sustainable development.

Despite the vast potential, these tools face significant challenges in supporting sustainable development. Remote sensing and data analytics can become powerful enablers for achieving sustainable development goals (SDGs) by addressing these challenges (Chirwa and Amodu 2021; Deilami et al. 2022; Riahi et al. 2024; Zabihi et al. 2023). Overcoming ethical, political, and scalability barriers is critical to ensuring that data-driven solutions tackle sustainability challenges. It is required to address the issues related to data quality, accessibility, cost, and technical capacity to unlock their full potential.

5 Effects of Data Analytics and Remote Sensing on Sustainable Development

The integration of data analytics and remote sensing has profound effects on sustainable development to offer a more informed and efficient approach to addressing global challenges. Remote sensing technologies like satellites and drones provide detailed real-time data on environmental conditions, land use, climate change, and biodiversity (Mbow et al. 2014; Tewari et al. 2019; Wei et al. 2021). This consortium allows for better monitoring and management of natural resources, aiding in sustainable agriculture, water management, and urban planning. Satellite imagery allows the monitoring of deforestation and environmental degradation, facilitating timely interventions to protect ecosystems. Remote sensing supports to improve crop yields for minimizing water and pesticide usage, leading to more sustainable food production (Alberti et al. 2022; Efremova et al. 2023; Moazeni et al. 2021; Taylor et al. 2007). These knowhows play a critical role in disaster management to offer real-time data during natural disasters, enabling quicker responses and recovery efforts. They assess the impact of climate change and enable adaptive strategies to mitigate its effects like rising sea levels and extreme weather events. It facilitates more efficient resource use to enhance environmental protection to promote informed decision-making, data analytics and remote sensing in achieving the goals of sustainable development, contributing to a more resilient and equitable future. Their efficacy depends on overcoming challenges like data accessibility, integration, and capacity-building in developing regions (Ackerman et al. 2014; Pramanik, 2024; Restrepo et al. 2021). These tools provide a comprehensive understanding of natural systems, human impacts, and trends that can help address critical challenges in sustainable development. The full potential depends on overcoming certain barriers, including data quality, access, and capacity-building. The major effects include:

(a) **Environmental Monitoring and Protection**

The technology satellites and drones enable continuous, real-time monitoring of the environment to capture enormous amounts of land use, vegetation cover, deforestation, water quality, and biodiversity. This information when combined with data analytics advances actionable for tracking environmental changes and understanding their reasons. Governments and organizations quickly detect environmental issues such as illegal deforestation, soil degradation, or pollution hotspots for timely interventions, more effective conservation efforts, and better protection of critical ecosystems to protect biodiversity.

(b) **Sustainable Resource Management**

Data analytics provides the analysis of large datasets created by remote sensing to optimize the use of natural resources. The analysis of satellite data on soil health, water availability, and crop conditions improves land management, water conservation, and agricultural practices. This leads to more efficient and sustainable use of resources like water, energy, and land, helping reduce waste and mitigate over-exploitation. Precision agriculture uses satellite imagery to assess soil moisture and crop health to guide farmers to make data-driven decisions on irrigation and fertilizer for reducing water waste and environmental pollution.

(c) **Climate Change Monitoring and Adaptation**

It plays a crucial role in monitoring the effects of climate change including sea-level rise, temperature fluctuations, melting ice caps, and extreme weather events. These technologies capture long-term trends and provide insights into the local and global impact of climate change to process this information to generate climate models, predict future scenarios, and develop adaptive strategies for vulnerable regions. Policymakers use this data to prioritize climate adaptation efforts and take proactive measures to reduce climate risks. It monitors the melting of polar ice caps and sea-level rise to enable countries to develop policies for coastal protection and flood management in low-lying areas.

(d) **Disaster Risk Management and Response**

In the case of floods, satellite imagery tracks water levels and flood extents in real time to allow emergency responders to assess damage, allocate resources, and provide assistance to affected communities. It offers real-time information during natural disasters like hurricanes, floods, and wildfires. Satellite images with ground-based sensors help monitor the progression of disasters to assess their impact on infrastructure, ecosystems, and communities. The integration of these data sources with analytics, governments, and humanitarian organizations improves their preparedness, response times, and recovery efforts. Early warning systems rely on satellite data to predict floods or storms and help evacuate at-risk populations.

(e) **Biodiversity and Ecosystem Conservation**

Remote sensing helps monitor biodiversity to track species habitats, vegetation changes, and ecosystem health. It provides insights into the effects of human activities like urbanization or agriculture, on biodiversity. This information can guide conservation efforts and the creation of protected areas. The identification of critical habitats and areas of high biodiversity to ensure the preservation of ecosystems in essential services like carbon sequestration, water filtration, and pollination. It is used to track forest cover loss and habitat destruction for organizations to intervene and protect areas crucial for maintaining biodiversity, tropical rainforests and coral reefs.

(f) **Urban Planning and Sustainable Cities**

Urban planners use satellite images to monitor the growth of cities, assess traffic patterns, and determine areas for green spaces or sustainable infrastructure (e.g. renewable energy installations) that should be incorporated to improve liveability and reduce carbon footprints. Cities use this data to make informed decisions on land development, public transportation, and waste management. Remote sensing monitors air quality, temperature, and green spaces to create healthier and more resilient urban environments. Remote sensing technologies provide data on urban sprawl, land use, infrastructure, and environmental conditions. These technologies help to design sustainable cities by optimizing space use, reducing pollution, and improving infrastructure with data analytics.

(g) **Water Resource Management**

Remote sensing monitors water bodies comprising rivers, lakes, and reservoirs to provide data on water levels, quality, and flow. Analytics evaluate water usage patterns, predict droughts, and evaluate the impact of climate change on water resources. This enables better water conservation practices, equitable distribution, and efficient use of water in agriculture, industry, and domestic consumption. The areas suffering from water scarcity, data-driven water management policies help to optimize water allocation and improve irrigation techniques. Satellite data monitor reservoir levels and water quality in drought-prone regions for the efficient management of available water resources and the development of irrigation systems to minimize water waste.

(h) **Sustainable Agriculture and Food Security**

Remote sensing offers critical information on crop health, soil conditions, and weather patterns. The data analytics improves planting schedules, irrigation practices, and the use of fertilizers and pesticides. This enables precision agriculture to maximize crop yields to minimize the environmental impact. Data-driven solutions predict food security risks to assess agricultural production, climate conditions, and market trends. Farmers use satellite imagery to monitor crop growth and detect early signs of diseases or pest infestations to apply targeted treatments and reduce the use of harmful chemicals.

(i) **Social Development and Poverty Reduction**

Combining remote sensing with socio-economic data for a comprehensive understanding of regional disparities, poverty, and access to basic services like healthcare, education, and sanitation. Data analytics identifies the high-risk areas and guides interventions for improving living standards. These are used to plan development projects, allocate resources effectively, and monitor progress towards poverty alleviation and other social goals. Satellite provides population density, infrastructure, and land use patterns and identifies areas for basic services (e.g. healthcare or clean water) and interventions.

The effects of data analytics and remote sensing on sustainable development for more informed decision-making and helping to address complex challenges across environmental, social, and economic dimensions. These technologies enable real-time monitoring, efficient resource management, and proactive interventions in areas such as climate change, disaster management, agriculture, urban planning, and biodiversity conservation. The full potential of these technologies is realized for widespread access to data, technical expertise, and the infrastructure.

6 Future Trends, Challenges and Opportunities

The future of sustainable development with data analytics and remote sensing supports the use of artificial intelligence (AI) and machine learning (ML) to analyse large datasets generated by satellites, drones, and IoT sensors. These technologies enable precise monitoring and forecasting of environmental changes, such as climate shifts, land degradation, and biodiversity loss, leading to more informed decision-making and proactive measures. The development of big data with cloud computing permits easier storage, sharing, and real-time analysis of data for global collaboration. Remote sensing tools are more affordable and accessible to enable the development of regions to tap into advanced technologies for environmental monitoring and resource management. However, challenges in data integration and standardization are often different platforms with variable accuracy and formats. Data privacy with security concerns collects sensitive information on land use or community well-being. The digital divide is another challenge in less-developed nations/regions to fully utilize these technologies. Sustainable agriculture benefits from precision farming techniques with the use of remote sensing reduces waste and improves food security.

Climate change mitigation is encouraged with the continuous monitoring of carbon emissions and deforestation, facilitating real-time policy interventions. Smart cities use data to enhance infrastructure, reduce energy consumption, and improve waste management. The future timeline for sustainable development with data analytics and remote sensing involves imagining how these technologies might evolve and impact global sustainability goals. The successful integration of data analytics and remote sensing into sustainable development requires overcoming these challenges to use the immense potential for a more resilient, equitable, and sustainable future. The possible timeline can be broken down for the future:

2025: **Initial Integration and Awareness**

- **Data Analytics**: Early-stage artificial intelligence (AI) and machine learning (ML) tools are adopted to interpret data from various sources. Governments and organizations use data analytics to monitor and analyse sustainability indicators like carbon emissions, energy use, and water consumption.
- **Remote Sensing**: Data from satellites like Landsat and new private satellites are more readily available for environmental analysis. Remote sensing technologies like satellite imagery are used for more widespread applications for land use, deforestation tracking, and urban sprawl monitoring.

- **International Collaboration**: Partnerships among governments, NGOs, and tech companies collect and analyse data on global sustainability challenges.

2030: **Expanding Data-Driven Decision Making**

- **Data Analytics**: Data from IoT sensors in cities, agriculture, and energy systems can be processed to optimize energy efficiency and reduce waste. The use of predictive analytics will become more sophisticated, allowing for better forecasting of environmental changes and resource needs.
- **Remote Sensing**: Remote sensing data can be integrated into decision-making processes, guiding climate change mitigation efforts and conservation projects. Real-time remote sensing tools and satellites will offer high-resolution data to monitor biodiversity, deforestation, and agricultural productivity.
- **Sustainable Development Goals (SDGs)**: Data visualization and interactive dashboards with common tools will present progress. Administrations can use data analytics to track progress towards SDGs like clean water, affordable energy, and sustainable cities.

2040: **Advanced AI and Autonomous Systems**

- **Data Analytics**: These insights can be used for autonomous systems in agriculture (e.g. precision farming), transportation (e.g. smart city infrastructure), and renewable energy (e.g. smart grids). Innovative AI models will process massive datasets to offer actionable insights for complex sustainability issues.
- **Remote Sensing**: Multi-spectral and hyperspectral sensors will detect subtle changes in ecosystems like soil health, water quality, and forest cover. Unconventional satellite collections and drones will provide near-continuous, real-time monitoring of global environmental trends.
- **Circular Economy**: Remote sensing will assist in trailing material flows at a global scale, helping identify inefficiencies. Data analytics will encourage industries to shift near a circular economy by optimizing supply chains, improving recycling, and minimizing waste.

2050: **Fully Integrated Sustainability Ecosystem**

- **Data Analytics**: AI and machine learning will automate analysis to enable real-time sustainability management and adaptive strategies for climate change. Data-driven decision-making is embedded in all levels of society, from local communities to international organizations.
- **Remote Sensing**: Climate change models are highly accurate, and mitigation strategies are implemented proactively. A global network of satellites and drones, combined with AI ensures sustainability indicators (e.g. land degradation, water stress, biodiversity loss) are continuously monitored and adjusted in real time.
- **Global Coordination**: Data analytics will facilitate global cooperation on climate change and sustainability. Countries will use public datasets to implement policies for equitable access to resources.

2060: **Climate Resilience and Ecological Restoration**

- **Data Analytics**: Predictive models for climate resilience will allow governments and businesses to plan for long-term sustainability. Resource management decisions will gradually move towards data-driven mode with real-time information from ecosystems, cities, and industries.
- **Remote Sensing**: Remote sensing will be used for large-scale ecological restoration projects. Satellite data will assist in monitoring restored areas to ensure ecosystem recovery. Monitoring biodiversity, soil health, and water quality will guide global reforestation and ocean conservation efforts.
- **Sustainability in Every Sector**: Data analytics and remote sensing will be fully integrated with every sector of the economy to ensure sustainability as the foundation for industries like agriculture, construction, transportation, and manufacturing.

2070 and Beyond: Earth's Resilient Future

- **Data Analytics**: Data analytics will be no longer just a tool and it may become a basis of governance. AI has automated decision-making for sustainability across all levels, and cities, industries, and nations will operate with sustainability as the core for their long-term survival and prosperity.
- **Remote Sensing**: The ultra-advanced remote sensing capabilities (e.g. quantum sensors) will track everything from urban heat islands to microclimates providing minute-level detail on environmental changes.
- **Global Equilibrium**: The world will adapt to a state where environmental sustainability, economic equity, and technological innovation harmoniously coexist. Data-driven policies will help to maintain this balance with humanity within the ecological limits.

The above-mentioned timeline illustrates data analytics and remote sensing technologies to transform global sustainability practices over the next few decades. The key to success will be increasing collaboration, continuous technology innovation, and the commitment to data-driven decision-making in achieving sustainability goals.

7 Summary

Over recent years, there have been significant developments in both fields, with advancements in satellite technology, machine learning, and cloud computing improving the accuracy, efficiency, and accessibility of data. The contribution of data analytics and remote sensing to sustainable development is quite substantial. Data analytics and remote sensing have become indispensable tools for driving sustainable development by enabling comprehensive monitoring and management of natural resources, ecosystems, and environmental changes. These technologies are essential for tracking environmental indicators such as deforestation, water quality, and climate change, as well as optimizing resource use in agriculture, energy, and water management. They support more informed decision-making, improve resource management, enhance disaster preparedness, and help achieve global sustainable development goals (SDGs). These technologies allow for the identification of trends, early warning of environmental hazards, and the optimization of practices in sectors like agriculture, urban planning, and conservation with real-time data. Remote sensing offers the advantage of monitoring large areas and hard-to-reach regions, while data analytics processes this data to generate actionable insights. The ability to analyse large datasets and generate predictive models supports better decision-making, more efficient use of resources, and improved disaster response, all of which are crucial for sustainable development. These technologies enhance environmental monitoring, resource optimization, and disaster resilience to improve the socio-economic conditions and contribute to environmental conservation. These technologies face challenges in the application in real time and issues such as data quality, high costs, data integration, and a shortage of skilled professionals delay the widespread adoption of remote sensing and data analytics, especially in developing countries. The lack of interoperability between different data sources and concerns over data privacy and ethical implications also pose significant barriers. Uncertainty in predictive models complicates long-term planning, making it difficult to predict environmental changes with high confidence.

Data analytics with remote sensing face challenges in data quality, costs, and accessibility but still they offer substantial opportunities for advancing sustainable development. Data analytics and remote sensing have the potential to drive more effective and equitable solutions to global sustainability challenges. The advancements in satellite technology, artificial intelligence (AI), and machine learning will continue to improve the accuracy and scalability of data analytics and remote sensing. The growing availability of open-source platforms and the decreasing cost of satellite imagery will increase the accessibility of these technologies, enabling a broader range of stakeholders, including small-scale farmers and local governments, to engage in sustainable practices. Also, cloud computing will provide more efficient data storage and processing, making large-scale analysis more feasible.

References

Ackerman, K., Conard, M., Culligan, P., Plunz, R., Sutto, M.-P., & Whittinghill, L. (2014). Sustainable food systems for future cities: The potential of urban agriculture. *Economic and Social Review*, *45*(2), 189–206.

Ahmad, F., Uddin, M. M., & Goparaju, L. (2019). Analysis of forest health and socioeconomic dimension in climate change scenario and its future impacts: remote sensing and GIS approach. *Spatial Information Research*, *27*(4), 385–397. https://doi.org/10.1007/s41324-019-00245-5

Alberti, M. A., Blanco, I., Vox, G., Scarascia-Mugnozza, G., Schettini, E., & da Silva, L. (2022). The challenge of urban food production and sustainable water use: Current situation and future perspectives of the urban agriculture in Brazil and Italy. Sustainable Cities and Society, 83. https://doi.org/10.1016/j.scs.2022.103961

Alotaibi, E., & Nassif, N. (2024). Artificial intelligence in environmental monitoring: in-depth analysis. *Discover Artificial Intelligence*, *4*(1). https://doi.org/10.1007/s44163-024-00198-1

Amani, M., Ghorbanian, A., Ahmadi, S. A., Kakooei, M., Moghimi, A., Mirmazloumi, S. M., Moghaddam, S. H. A., Mahdavi, S., Ghahremanloo, M., Parsian, S., Wu, Q., & Brisco, B. (2020). Google Earth Engine Cloud Computing Platform for Remote Sensing Big Data Applications: A Comprehensive Review. *IEEE Journal of Selected Topics in Applied Earth Observations and Remote Sensing*, *13*, 5326–5350. https://doi.org/10.1109/JSTARS.2020.3021052

Carrão, H., Naumann, G., & Barbosa, P. (2018). Global projections of drought hazard in a warming climate: a prime for disaster risk management. *Climate Dynamics*, *50*(5–6), 2137–2155. https://doi.org/10.1007/s00382-017-3740-8

Chirwa, D., & Amodu, N. (2021). Economic, Social and Cultural Rights, Sustainable Development Goals, and Duties of Corporations: Rejecting the False Dichotomies. *Business and Human Rights Journal*, *6*(1), 21–41. https://doi.org/10.1017/bhj.2020.34

Cooney, R., Wan, A. H. L., O'Donncha, F., & Clifford, E. (2021). Designing environmentally efficient aquafeeds through the use of multicriteria decision support tools. *Current Opinion in Environmental Science and Health*, *23*. https://doi.org/10.1016/j.coesh.2021.100276

Coppin, P., Jonckheere, I., Nackaerts, K., Muys, B., & Lambin, E. (2004). Review ArticleDigital change detection methods in ecosystem monitoring: a review. In *International Journal of Remote Sensing* (Vol. 25, Issue 9, pp. 1565–1596).

David, A. O., Ndambuki, J. M., Muloiwa, M., Kupolati, W. K., & Snyman, J. (2024). A Review of the Application of Artificial Intelligence in Climate Change-Induced Flooding—Susceptibility and Management Techniques. *CivilEng, 5*(4), 1185–1198. https://doi.org/10.3390/civileng5040058

Deilami, K., Shooshtarian, S., Rudner, J., Butt, A., & Amati, M. (2022). Resilience and Adaptation Strategies for Urban Heat at Regional, City and Local Scales. In *Disaster Risk Reduction for Resilience: Disaster Risk Management Strategies*. Springer International Publishing. https://doi.org/10.1007/978-3-030-72196-1_8

Efremova, N., Foley, J. C., Unagaev, A., & Karimi, R. (2023). AI for Sustainable Agriculture and Rangeland Monitoring. *Philosophical Studies Series, 152*, 399–422. https://doi.org/10.1007/978-3-031-21147-8_22

Egaji, O. A., Evans, G., Griffiths, M. G., & Islas, G. (2021). Real-time machine learning-based approach for pothole detection. *Expert Systems with Applications, 184*. https://doi.org/10.1016/j.eswa.2021.115562

Erian, W., Katlan, B., Khoday, K., & Murshed, Z. (2014). Social vulnerability to drought risks in the Arab region: Towards resilience-based development responses. *Proceedings of the 5th International Disaster and Risk Conference: Integrative Risk Management - The Role of Science, Technology and Practice, IDRC Davos 2014*, 237–241.

Gopal, B., & Chauhan, M. (2006). Biodiversity and its conservation in the Sundarban mangrove ecosystem. *Aquatic Sciences*. https://doi.org/10.1007/s00027-006-0868-8

Hamud, A. M., Shafri, H. Z. M., & Shaharum, N. S. N. (2021). Monitoring Urban Expansion and Land Use/Land Cover Changes in Banadir, Somalia Using Google Earth Engine (GEE). *IOP Conference Series: Earth and Environmental Science, 767*(1). https://doi.org/10.1088/1755-1315/767/1/012041

Hu, Z. (2020). When energy justice encounters authoritarian environmentalism: The case of clean heating energy transitions in rural China. *Energy Research and Social Science, 70*(April), 101771. https://doi.org/10.1016/j.erss.2020.101771

Hüttel, S., & Mußhoff, O. (2024). Sustainable Development of Agriculture: Promoting Biodiversity and Climate Neutrality; [Nachhaltige Entwicklung der Landwirtschaft: Förderung von Biodiversität und Klimaneutralität]. *Wirtschaftsdienst, 104*(3), 165–169. https://doi.org/10.2478/wd-2024-0048

Jehanzaib, M., & Kim, T.-W. (2020). Exploring the influence of climate change-induced drought propagation on wetlands. *Ecological Engineering, 149*. https://doi.org/10.1016/j.ecoleng.2020.105799

Khosravi, K., Melesse, A. M., Shahabi, H., Shirzadi, A., Chapi, K., & Hong, H. (2019). Flood susceptibility mapping at Ningdu catchment, China using bivariate and data mining techniques. *Extreme Hydrology and Climate Variability: Monitoring, Modelling, Adaptation and Mitigation*, 419–434. https://doi.org/10.1016/B978-0-12-815998-9.00033-6

Kumar, P., Jain, R., Kumar, D., & Rawat, S. (2021). Interdisciplinary Implementation of Supervised Real-Time Waste Management. *SN Computer Science, 2*(4). https://doi.org/10.1007/s42979-021-00651-3

Lal, C., & Gupta, V. (2009). Sustainable development of agriculture: Urban waste water a viable mean for irrigation. *Proceedings of World Environmental and Water Resources Congress 2009 - World Environmental and Water Resources Congress 2009: Great Rivers, 342*, 5772–5799. https://doi.org/10.1061/41036(342)585

Lunetta, R. S., Congalton, R. G., Fenstermaker, L. K., Jensen, J. R., McGwire, K. C., & Tinney, L. R. (1991). Remote sensing and geographic information system data integration: error sources and research issues. *Photogrammetric Engineering & Remote Sensing, 57*(6), 677–687.

Mahdianpari, M., Salehi, B., Mohammadimanesh, F., Brisco, B., Homayouni, S., Gill, E., DeLancey, E. R., & Bourgeau-Chavez, L. (2020). Big Data for a Big Country: The First Generation of Canadian Wetland Inventory Map at a Spatial Resolution of 10-m Using Sentinel-1 and Sentinel-2 Data on the Google Earth Engine Cloud Computing Platform [Mégadonnées pour un grand pays: La première carte. *Canadian Journal of Remote Sensing, 46*(1), 15–33. https://doi.org/10.1080/07038992.2019.1711366

Maity, R., Srivastava, A., Sarkar, S., & Khan, M. I. (2024). Revolutionizing the future of hydrological science: Impact of machine learning and deep learning amidst emerging explainable AI and transfer learning. *Applied Computing and Geosciences, 24*. https://doi.org/10.1016/j.acags.2024.100206

Mbow, C., van Noordwijk, M., Prabhu, R., & Simons, T. (2014). Knowledge gaps and research needs concerning agroforestry's contribution to Sustainable Development Goals in Africa. *Current Opinion in Environmental Sustainability, 6*(1), 162–170. https://doi.org/10.1016/j.cosust.2013.11.030

Moazeni, M., Maracy, M. R., Dehdashti, B., & Ebrahimi, A. (2021). Spatiotemporal analysis of COVID-19, air pollution, climate, and meteorological conditions in a metropolitan region of Iran. *Environmental Science and Pollution Research*. https://doi.org/10.1007/s11356-021-17535-x

Onifade, S. T., Haouas, I., & Alola, A. A. (2023). Do natural resources and economic components exhibit differential quantile environmental effects? *Natural Resources Forum, 47*(3), 355–374. https://doi.org/10.1111/1477-8947.12289

Parente, L., Taquary, E., Silva, A. P., Souza C., Jr., & Ferreira, L. (2019). Next generation mapping: Combining deep learning, cloud computing, and big remote sensing data. *Remote Sensing, 11*(23). https://doi.org/10.3390/rs11232881

Park, S., Im, J., Jang, E., & Rhee, J. (2016). Drought assessment and monitoring through blending of multi-sensor indices using machine learning approaches for different climate regions. *Agricultural and Forest Meteorology, 216*, 157–169. https://doi.org/10.1016/j.agrformet.2015.10.011

Pham, T. T., Dang, K. B., Giang, T. L., Hoang, T. H. N., Le, V. H., & Ha, H. N. (2024). Deep learning models for monitoring landscape changes in a UNESCO Global Geopark. *Journal of Environmental Management, 354*. https://doi.org/10.1016/j.jenvman.2024.120497

Pramanik, S. (2024). AI's function in sustainable development's renewable energy planning. In *Next Generation Materials for Sustainable Engineering*. IGI Global. https://doi.org/10.4018/979-8-3693-1306-0.ch016

Raj, A., Jhariya, M. K., Banerjee, A., Kittur, B. H., Bargali, S. S., Bargali, K., & Nema, S. (2024). Forest biodiversity conservation and restoration: Policies, plan, and approaches. In *Ecorestoration for Sustainability*. Wiley Blackwell. https://www.scopus.com/inward/record.uri?eid=2-s2.0-85168641849&partnerID=40&md5=1cd62195681d1d2aeebf5e3e9eced2a9

Rana, M. M. P., & Ilina, I. N. (2021). Climate change and migration impacts on cities: Lessons from Bangladesh. *Environmental Challenges, 5*(August), 100242. https://doi.org/10.1016/j.envc.2021.100242

Reiche, J., Lucas, R., Mitchell, A. L., Verbesselt, J., Hoekman, D. H., Haarpaintner, J., Kellndorfer, J. M., Rosenqvist, A., Lehmann, E. A., Woodcock, C. E., Seifert, F. M., & Herold, M. (2016). Combining satellite data for better tropical forest monitoring. In *Nature Climate Change*. https://doi.org/10.1038/nclimate2919

Restrepo, L., Aguilar, J., Toro, M., & Suescún, E. (2021). A sustainable-development approach for self-adaptive cyber–physical system's life cycle: A systematic mapping study. *Journal of Systems and Software, 180*. https://doi.org/10.1016/j.jss.2021.111010

Reygadas, Y., Spera, S., Galati, V., Salisbury, D. S., Silva, S., & Novoa, S. (2021). Mapping forest disturbances across the southwestern amazon: Tradeoffs between open-source, landsat-based algorithms. *Environmental Research Communications, 3*(9). https://doi.org/10.1088/2515-7620/ac2210

Riahi, M. R., Sadeghi, S. H., Mirnia, S. K., & Sadoddin, A. (2024). Coincidence of sustainable development indicators for the nekarood watershed with the United Nation's sustainable development goals. *Science of the Total Environment*, *917*. https://doi.org/10.1016/j.scitotenv.2024.170177

Rosales Carreón, J., & Worrell, E. (2018). Urban energy systems within the transition to sustainable development. A research agenda for urban metabolism. *Resources, Conservation and Recycling*, *132*, 258–266. https://doi.org/10.1016/j.resconrec.2017.08.004

Shi, W., & Meng, L. (2006). Integration and interoperability of multiple theme data on the basis of Google Earth. *Proceedings of SPIE - The International Society for Optical Engineering*, *6421*. https://doi.org/10.1117/12.713135

Shrestha, Y. R., Krishna, V., & von Krogh, G. (2021). Augmenting organizational decision-making with deep learning algorithms: Principles, promises, and challenges. *Journal of Business Research*, *123*(September 2020), 588–603. https://doi.org/10.1016/j.jbusres.2020.09.068

Sobocká, J., Saksa, M., Feranec, J., Szatmári, D., Holec, J., Bobáľová, H., & Rášová, A. (2021). Mapping of urban environmentally sensitive areas in Bratislava city. *Journal of Soils and Sediments*, *21*(5), 2059–2070. https://doi.org/10.1007/s11368-020-02682-4

Solecki, W., & Rosenzweig, C. (2020). Indicators and monitoring systems for urban climate resiliency. *Climatic Change*, *163*(4), 1815–1837. https://doi.org/10.1007/s10584-020-02947-4

Taylor, P., Way, J., Paris, J., Kasischke, E., Slaughter, C., Viereck, L., Imhoff, M., & Weber, J. (2007). International Journal of Remote Sensing The effect of changing environmental conditions on microwave signatures of forest ecosystems : preliminary results of the March 1988 Alaskan aircraft SAR experiment. *IEEE Transactions on Geoscience and Remote Sensing*, *December 2014*, 1119–1144. https://doi.org/10.1080/01431169008955084

Tewari, M., Yang, J., Kusaka, H., Salamanca, F., Watson, C., & Treinish, L. (2019). Interaction of urban heat islands and heat waves under current and future climate conditions and their mitigation using green and cool roofs in New York City and Phoenix, Arizona. *Environmental Research Letters*, *14*(3). https://doi.org/10.1088/1748-9326/aaf431

Toland, A. (2018). The learning machine and the spaceship in the garden. AI and the design of planetary "nature" [El autómata aprendiz y la nave espacial en el jardín. IA y el diseño de la naturaleza planetaria]. *RA Revista de Arquitectura*, *20*, 216–227+318 and 323. https://doi.org/10.15581/014.20.216-227

Vu, T. T., Vu, N. V. A., Phung, H. P., & Nguyen, L. D. (2021). Enhanced urban functional land use map with free and open-source data. *International Journal of Digital Earth*, *14*(11), 1744–1757. https://doi.org/10.1080/17538947.2021.1970262

Waleed, M., & Sajjad, M. (2022). Leveraging cloud-based computing and spatial modeling approaches for land surface temperature disparities in response to land cover change: Evidence from Pakistan. *Remote Sensing Applications: Society and Environment*, *25*. https://doi.org/10.1016/j.rsase.2021.100665

Wang, K., Franklin, S. E., Guo, X., & Cattet, M. (2010). Remote Sensing of Ecology, Biodiversity and Conservation: A Review from the Perspective of Remote Sensing Specialists. In *Sensors* (Vol. 10, Issue 11, pp. 9647–9667).

Wei, M., Zhang, Z., Long, T., He, G., & Wang, G. (2021). Monitoring Landsat Based Burned Area as an Indicator of Sustainable Development Goals. *Earth's Future*, *9*(6). https://doi.org/10.1029/2020EF001960

Zabihi, O., Siamaki, M., Gheibi, M., Akrami, M., & Hajiaghaei-Keshteli, M. (2023). A smart sustainable system for flood damage management with the application of artificial intelligence and multi-criteria decision-making computations. *International Journal of Disaster Risk Reduction*, *84*. https://doi.org/10.1016/j.ijdrr.2022.103470

Zarbakhshnia, N., Wu, Y., Govindan, K., & Soleimani, H. (2020). A novel hybrid multiple attribute decision-making approach for outsourcing sustainable reverse logistics. *Journal of Cleaner Production*, *242*. https://doi.org/10.1016/j.jclepro.2019.118461

Zhang, L., & Zhao, H. (2024). Sustainable development mechanism: The role of natural resources, remittance and policy uncertainty. *Resources Policy*, *90*. https://doi.org/10.1016/j.resourpol.2023.104621

Zhao, J., Liu, H., & Dong, R. (2008). Sustainable urban development: Policy framework for sustainable consumption and production. *International Journal of Sustainable Development and World Ecology*, *15*(4), 318–325. https://doi.org/10.3843/SusDev.15.4:5a

MIX
Papier aus verantwortungsvollen Quellen
Paper from responsible sources
FSC® C105338

If you have any concerns about our products,
you can contact us on
ProductSafety@springernature.com

In case Publisher is established outside the EU,
the EU authorized representative is:
Springer Nature Customer Service Center GmbH
Europaplatz 3, 69115 Heidelberg, Germany

Printed by Libri Plureos GmbH
in Hamburg, Germany

Stefano Boriani • Livio Presutti
Alessandro Gasbarrini • Francesco Mattioli
Editors

Atlas of Craniocervical Junction and Cervical Spine Surgery

Springer

Editors
Stefano Boriani
Bologna
Italy

Alessandro Gasbarrini
Bologna
Italy

Livio Presutti
Modena
Italy

Francesco Mattioli
University Hospital of Modena
Modena
Italy

ISBN 978-3-319-42735-5 ISBN 978-3-319-42737-9 (eBook)
DOI 10.1007/978-3-319-42737-9

Library of Congress Control Number: 2017939173

Printed on acid-free paper

This Springer imprint is published by Springer Nature
The registered company is Springer International Publishing AG
The registered company address is: Gewerbestrasse 11, 6330 Cham, Switzerland

Acknowledgment

The Authors are indebted and thank Carlo Piovani for his strenuous activity as anatomical artist and for imaging collection and elaboration. Thanks to him we could select the most attractive and demonstrating images and clarify the details of surgical anatomy.

Contents

Part I Anatomy

1 Anatomy of Craniocervical Junction 3
M. Alicandri-Ciufelli, M. Menichetti, M.P. Alberici, and L. Presutti

2 Anatomy of the Subaxial Cervical Spine 17
M. Girolami, R. Ghermandi, M. Ghirelli, A. Gasbarrini, and S. Boriani

3 Anatomy of the Neck 27
M. Bonali, D. Soloperto, E. Aggazzotti Cavazza, M. Ghirelli, and L. Presutti

Part II Planning

4 Interventional Radiology: Presurgical Selective Angiographic Embolization (SAE) in Hypervascular Cervical Spine Tumours 49
Luigi Simonetti, Carlotta Barbara, Salvatore Isceri, and Elena Mengozzi

5 Anesthesiological Management and Patient Positioning 63
M.R. Bacchin, M. Di Fiore, Y.E. Akman, M. Girolami, R. Ghermandi, A. Gasbarrini, and S. Boriani

6 Cervical Spine Instrumentation 69
M. Girolami, R. Ghermandi, A. Gasbarrini, Y.E. Akman, and S. Boriani

7 Tracheotomy Surgical Technique 87
M. Ghirelli, F. Mattioli, G. Molinari, I. Cena, and L. Presutti

8 Surgical Planning in Cervical Spine Oncologic Patients 95
S. Boriani, R. Ghermandi, M. Girolami, and A. Gasbarrini

Part III Surgical Approaches

9 Surgical Approaches to CCJ (Endoscopic Transnasal-Transoral-Transcervical and Robotic Transoral Approach) 105
F. Mattioli, G. Molteni, M. Bettini, E. Cigarini, and L. Presutti

10 Anterior and Lateral Approaches to Cervical Spine 129
M. Ghirelli, F. Mattioli, G. Molinari, and L. Presutti

11 Posterior Approach to Cervical Spine 175
R. Ghermandi, M. Girolami, A. Gasbarrini, and S. Boriani

12 Exemplificative Cases in Cervical Spine 185
A. Gasbarrini, M. Girolami, R. Ghermandi, Y.E. Akman, and S. Boriani

13 Complications of Cervical Spine Surgery 217
Gabriele Molteni, Marco Giuseppe Greco, and Pierre Guarino

Contributors

Elisa Aggazzotti Cavazza Otorhinolaryngology, Head and Neck Surgery Department, University Hospital of Modena, Modena, Italy

Yunus Emre Akman Orthopaedics and Traumatology Department, Metin Sabanci Baltalimani Bone Diseases Training and Research Hospital, Istanbul, Turkey

Maria Paola Alberici Head and Neck Surgery Department, University Hospital of Modena, Modena, Italy

Ciufelli Matteo Alicandri Head and Neck Surgery Department, University Hospital of Modena, Modena, Italy

Maria Renata Bacchin Rizzoli Orthopedic Institute, Bologna, Italy

Carlotta Barbara Emergency Interventional Unit, Ospedale Maggiore, Bologna, Italy

Margherita Bettini Head and Neck Surgery Department, University Hospital of Modena, Modena, Italy

Marco Bonali Otorhinolaryngology, Head and Neck Surgery Department, University Hospital of Modena, Modena, Italy

Stefano Boriani Oncologic and Degenerative Spine Surgery Department Rizzoli Orthopedic Institute, Bologna, Italy

Isida Cena Head and Neck Surgery Department, University Hospital of Modena, Modena, Italy

Elisa Cigarini Head and Neck Surgery Department, University Hospital of Modena, Modena, Italy

Maria Di Fiore Rizzoli Orthopedic Institute, Bologna, Italy

Alessandro Gasbarrini Oncologic and Degenerative Spine Surgery Department Rizzoli Orthopedic Institute, Bologna, Italy

Riccardo Ghermandi Oncologic and Degenerative Spine Surgery Department Rizzoli Orthopedic Institute, Bologna, Italy

Michael Ghirelli Otorhinolaryngology, Head and Neck Surgery Department, University Hospital of Modena, Modena, Italy

Marco Girolami Oncologic and Degenerative Spine Surgery Department Rizzoli Orthopedic Institute, Bologna, Italy

Marco Giuseppe Greco Ent Department University Hospital of Modena, Modena, Italy

Pierre Guarino Head and Neck Surgery Department, University Hospital of Modena, Modena, Italy

Salvatore Isceri Emergency Interventional Unit, Ospedale Maggiore, Bologna, Italy

Francesco Mattioli Head and Neck Surgery Department, University Hospital of Modena, Modena, Italy

Elena Mengozzi Emergency Interventional Unit, Ospedale Maggiore, Bologna, Italy

Marcella Menichetti Head and Neck Surgery Department, University Hospital of Modena, Modena, Italy

Giulia Molinari Head and Neck Surgery Department, University Hospital of Modena, Modena, Italy

Gabriele Molteni Department of Otolaryngology-Head and Neck Surgery, University Hospital of Modena, Modena, Italy

Livio Presutti Otorhinolaryngology, Head and Neck Surgery Department, Azienda Ospedaliera Universitaria Integrata, University Hospital of Verona, Modena, Italy

Luigi Simonetti Emergency Interventional Unit, Ospedale Maggiore, Bologna, Italy

D. Soloperto Otorhinolaryngology, Head and Neck Surgery Department, University Hospital of Verona, Verona, Italy

P. Zambito, MD ENT Department University Hospital of Verona, Modena, Italy

Part I

Anatomy

Anatomy of Craniocervical Junction

1

M. Alicandri-Ciufelli, M. Menichetti, M.P. Alberici, and L. Presutti

1.1 Osseous Anatomy

The craniocervical junction (CCJ) is an osteoligamentous complex between the occiput, atlas and axis, which provides both structural stability and movement [1].

1.1.1 Occipital Bone

The occipital bone extends from the clivus anteriorly to the lambdoid suture posteriorly, its embryologic origin being four primary cartilaginous centres laid down in the chondrocranium around the foramen magnum, and a fifth membranous element [2]. The superior nuchal line serves as a rough guide for the location of the transverse sinus, and the inion, found in the midline along this line, approximates the torcular herophili. The insertion of the semispinalis capitis may be the most accurate landmark for the confluence of the sinuses [3].

A fundamental anatomical part of the occipital bone is the foramen magnum, which has three parts:

1. The squamosal portion, located in the dorsal aspect of foramen magnum
2. The basal or clival portion located anterior to the foramen magnum
3. The condylar part that connects the squamosal and the clival parts [4]

The most posterior margin of the foramen magnum is called the opisthion. The most anterior midline of the foramen magnum is termed the basion. The sagittal diameter of the foramen magnum should be 35 ± 4 mm. The transverse diameter at the equator of foramen magnum is slightly less.

The condylar part includes the occipital condyles, which fall just at the level and anterior to the equator of the foramen magnum. The shape of these condyles positioned on either side of the foramen magnum allows the skull to articulate with the cervical spine, whilst the angles prevent excessive axial rotation at the craniocervical junction [5].

M. Alicandri-Ciufelli (✉) • M. Menichetti • M.P. Alberici
L. Presutti
Head and Neck Surgery Department, University Hospital of Modena, Modena, Italy
e-mail: matteo.alicandri@hotmail.it

S. Boriani et al. (eds.), *Atlas of Craniocervical Junction and Cervical Spine Surgery*, DOI 10.1007/978-3-319-42737-9_1

1.1.2 **Atlas** (Fig. 1.1)

The atlas, first cervical vertebra, has its origins in the fourth occipital and first cervical sclerotomes. It is unique among vertebrae in not having a body and is formed from three ossification sites: the anterior arch or centrum and two neural arches which fuse in later life to become a unified posterior arch, thereby completing the osseous ring which surrounds the spino-medullary junction [6]. An appreciation that this ring is incomplete in up to 5 % of patients is important if one is to avoid causing a durotomy or spinal cord injury when approaching the craniocervical junction posteriorly [7, 8].

The ring of the atlas consists approximately of one-fifth anterior arch, two-fifths posterior arch, with the remaining two-fifths being contributed by the lateral masses [9]. The longus colli muscles and the anterior longitudinal ligament, which contribute to anterolateral flexion and resistance to hyperextension of the cervical spine respectively, are attached to the anterior tubercle found in the midline on the anterior arch. Two important membranes also arise from this portion of the atlas: the anterior atlanto-occipital membrane, connecting the atlas to the occipital bone, and the anterior atlantoaxial ligament extending from the atlas to the axis immediately inferior. Atlantal lateral masses have both a superior articular facet and an inferior articular facet. These true synovial joints allow articulation with the occipital condyles and the axis, respectively. The atlanto-occipital joints' orientation at caudal angles of 129° from lateral to medial limits the rotation possible, compared with the atlantoaxial joint with a cranially biased angulation of between 130° and 135°, where much greater rotation is possible [10]. A posterior tubercle is found in the midline posteriorly providing attachment for the rectus capitis and the ligamentum nuchae. The posterior atlanto-occipital membrane extends from the superior border of the posterior arch of the atlas to the anterior surface of the rim of the foramen magnum [11, 12].

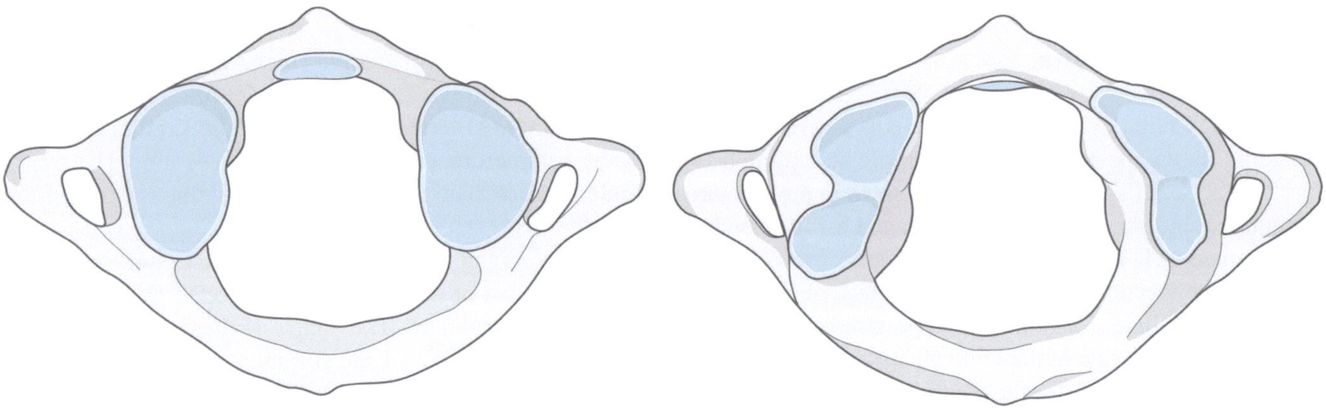

Fig. 1.1 C1 anatomy

1.1.3 **Axis** (Fig. 1.2)

The axis is the second cervical vertebra, and it is called the epistropheus (literally "to twist") because of its configuration that forms the pivot for the atlas and the head to rotate. The axis is formed from five ossification centres: one in the body, one in each vertebral arch and two in the odontoid process [13]. The odontoid process projects cephalad from its articulation with the axis body. On the ventral odontoid surface is an oval facet, which articulates with the dorsal surface of the anterior arch of the atlas. In the dorsal aspect of the dens, there is a transverse groove over which passes the transverse ligament of the atlas. The axis has a spinous process, which is large and deeply concave on its caudal border, that makes the axis the first bifid vertebra in the cervical spine [4].

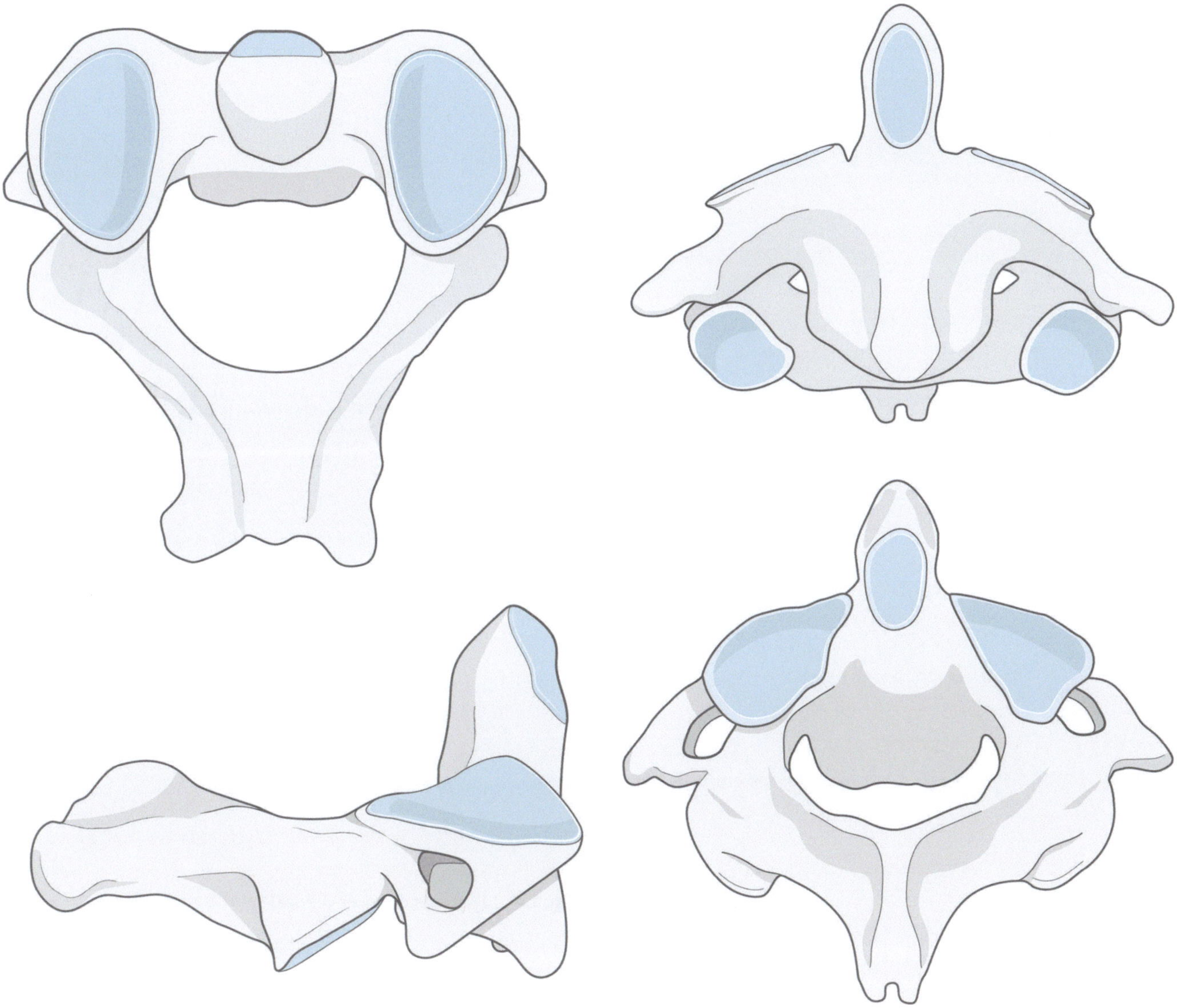

Fig. 1.2 C2 anatomy

1.2 Articular and Ligamentous Anatomy

The CCJ is composed of two major joints: the atlanto-occipital and the atlanto-axial. These joints are responsible for the majority of the movements of the cervical spine and operate on different biomechanical principles. The mechanical properties of the atlanto-occipital joint are primarily determined by bony structures, whereas the mechanical properties of the atlantoaxial joint are mainly determined by ligamentous structures [14, 15]. The prominent movements at the atlanto-occipital joint are flexion and extension. The primary movement at the atlantoaxial joint is axial rotation [16] (Figs. 1.3 and 1.4).

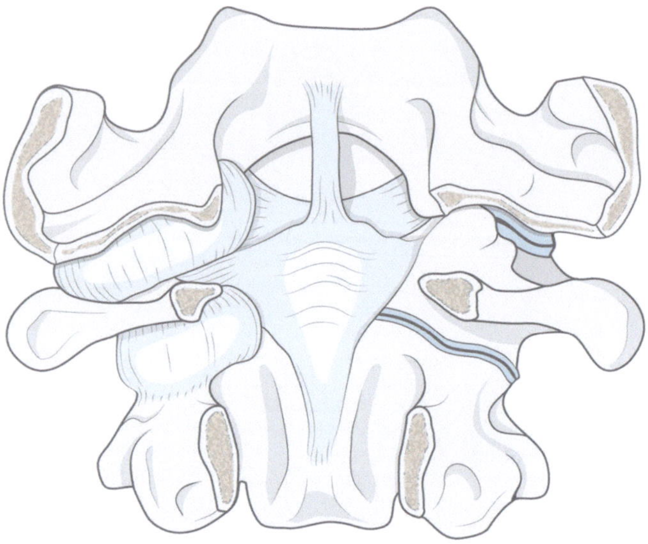

Fig. 1.3 Ligaments of upper cervical spine, posterior view

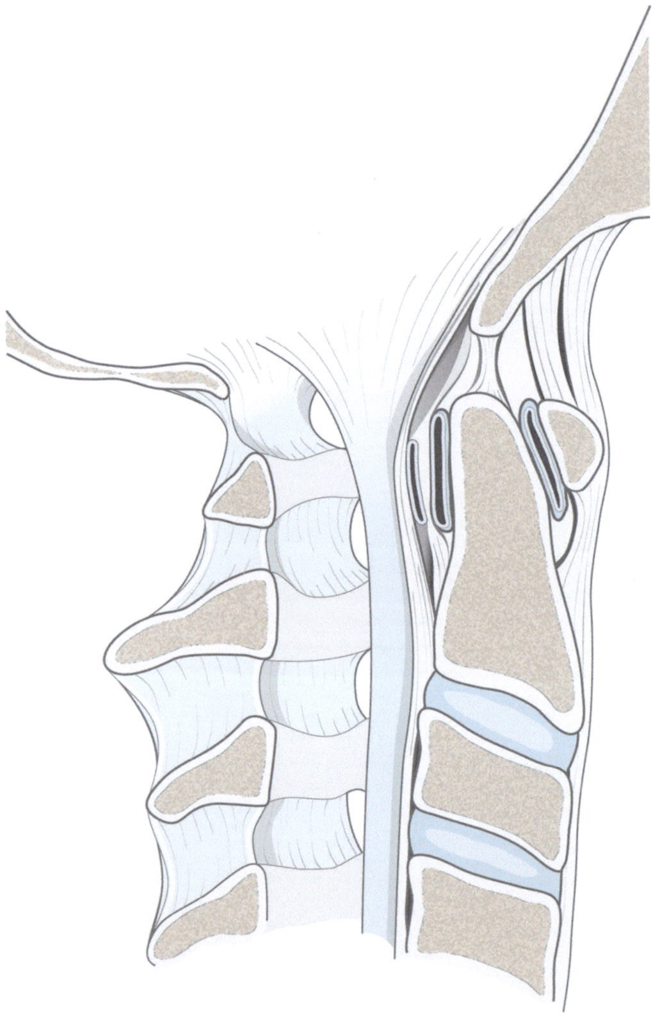

Fig. 1.4 Ligaments of upper cervical spine, sagittal view

1.2.1 Transverse Ligament

The transverse ligament of the atlas is the key component of the cruciform ligament and is one of the most important ligaments in the body. It is the largest, strongest and thickest craniocervical ligament (mean height/thickness 6–7 mm) [17]. The superior and inferior limbs of the cruciform ligament are extremely thin and offer no known craniocervical stability, whilst the transverse ligament maintains stability at the CCJ by locking the odontoid process anteriorly against the posterior aspect of the anterior arch of C-1, and it divides the ring of the atlas into two compartments: the anterior compartment houses the odontoid process, and the posterior compartment contains primarily the spinal cord and spinal accessory nerves. The transverse ligament runs posterior to the odontoid process of C-2 and attaches to the lateral tubercles of the atlas bilaterally. A synovial capsule is located between the odontoid process and the transverse ligament.

The ligament also has a smooth fibrocartilaginous surface to allow the odontoid process to glide against it [4]. The tectorial membrane, epidural fat and dura mater are located dorsal to the transverse ligament [17].

1.2.2 Alar Ligament

The alar ligament attaches the axis to the base of the skull, from the lateral aspects of the odontoid process to the anterolateral part of the foramen magnum and/or on the medial aspect of the occipital condyles [18–20].

1.2.3 Transverse Occipital Ligament

The transverse occipital ligament (TOL) is a small accessory ligament of the CCJ that is located posterosuperior to the alar ligaments and the odontoid process. It attaches to the inner aspect of the occipital condyles, posterosuperior to the alar ligament, superior to the transverse ligament, and extends horizontally across the foramen magnum [21].

Dvorak et al. [18] stated that the TOL is only present in about 10 % of the population, whereas Lang [1] identified the TOL in approximately 40 % of their specimens. The discrepancies in the occurrence of the TOL in specimens could be due to its proximity and similar morphology to the alar ligament that makes it difficult to distinguish the two easily.

1.2.4 Accessory Atlantoaxial Ligament

The accessory atlantoaxial ligament is an important but often ignored ligament that inserts medially into the dorsal surface of the axis and courses laterally and superiorly to insert on the lateral mass of the atlas, posterior to the transverse ligament [22, 23].

Tubbs et al. [24] suggested that this ligament could be more appropriately named the accessory atlantal-axial-occipital ligament, to underline its anatomical attachments.

1.2.5 Lateral Atlanto-Occipital Ligament

The lateral atlanto-occipital (LAO) ligament is another ligament of the CCJ that has been neglected in the literature.

It runs just lateral to the anterior atlanto-occipital membrane, attaching to the anterolateral aspect of the transverse process of the atlas and inserting onto the jugular process of the occipital bone [25, 26]. This ligament runs immediately posterior to the rectus capitis lateralis muscle and has fibres that extend in the opposite direction to the muscle (i.e. muscle runs lateral to medial and ligament runs medial to lateral).

1.2.6 Barkow Ligament

Barkow ligament has been rarely described. It is a horizontal band attaching onto the anteromedial aspect of the occipital condyles anterior to the attachment of the alar ligaments. This ligament is located just anterior to the superior aspect of the dens with fibres travelling anterior to the alar ligaments, but there is no attachment to these structures.

Barkow ligament, present in 92 % of studied cases [20], inserts anterior to the alar ligaments and is often adherent to the anterior atlanto-occipital membrane. Its primary function is thought to be in resisting extension of the atlanto-occipital joint, acting synergistically with the anterior atlanto-occipital membrane to achieve this.

1.2.7 Apical Ligament

The apical ligament, also known as the middle odontoid ligament or suspensory ligament, attaches the tip of the odontoid process to the basion. The ligament runs in the triangular area between the left and right alar ligaments known as the supraodontoid space (apical cave) [27] and travels just posterior to the alar ligaments and just anterior to the superior portion of the cruciform ligament.

1.2.8 Tectorial Membrane

The tectorial membrane is a thin structure at the CCJ that serves as the posterior border to the supraodontoid space [27]. It runs posterior to the cruciform ligament, and the accessory atlantoaxial ligament runs along its lateral border. The tectorial membrane is composed of 2–3 distinct layers that run the length of the ligament and then fuse together at the posterior longitudinal ligament.

The outermost layer is the widest and attaches as far laterally as the hypoglossal canals. The second layer is thicker and runs from the clivus to the body of the axis. A small bursa is often present between the two layers over the odontoid process. The third layer is the deepest and is discontinuous as it attaches to the clivus above and then becomes frayed in the area over the odontoid apex.

Nerves and vessels often run between the different layers of the tectorial membrane [27]. Descriptions of the tectorial membrane are insufficient and inconsistent regarding the anatomy and function.

1.2.9 Posterior Atlanto-Occipital Membrane

The posterior atlanto-occipital (PAO) membrane is a broad, thin ligament that attaches to the posterior arch of the atlas inferiorly to the posterior rim of the foramen magnum superiorly. It is continuous with the posterior atlantoaxial membrane and then the ligamentum flavum inferiorly [28]. This structure has been noted by several authors to extend laterally over the atlanto-occipital joint capsules [29]. The PAO membrane runs adjacent to the rectus capitis posterior minor muscle posteriorly and the spinal dura mater anteriorly. Several authors have noted connection or interdigitation of the PAO membrane with both the rectus capitis posterior minor muscle and the spinal dura mater [30–32].

1.2.10 Anterior Atlanto-Occipital Membrane

The anterior atlanto-occipital (AAO) membrane is a thin structure that attaches the anterior aspect of the atlas to the anterior rim of the foramen magnum [19, 20, 27–32]. It is located just posterior to the prevertebral muscles of the neck and anterior to Barkow ligament. Tubbs et al. [20] observed a connection of Barkow ligament to the midline onto the AAO membrane. The AAO membrane also serves as the anterior wall of the supraodontoid space, which houses the alar, apical and Barkow ligaments, as well as fat and veins [27].

1.2.11 Nuchal Ligament

The nuchal ligament is the cephalic extension of the supraspinous ligament, extending from the C-7 spinous process to the inion of the occipital bone. With the shorter spinous processes of the cervical vertebrae and the lordotic curve of the cervical spine, this ligament forms a midline septation dividing the posterior neck muscles on left and right sides. Moreover, some of these muscles attach medially to this structure.

1.3 Muscular Anatomy

The muscles of the craniocervical junction do not limit movements of the joints. It was felt that they only had a minor role in motion of the CVJ. Their principle function is one of initiating and maintaining movement of the craniocervical region [33]. They have been grouped into those that cause extension, flexion, abduction, adduction and rotation.

The muscles involved with the C1–C2 complex on the anterior aspect are the recti capitis anterior and lateralis, both stretching from the anterior aspect of the transverse process of C1, respectively, to the inferior surface of the basilar occipital bone. The longus capitis extends from the inferior surface of the basilar occipital bone and clivus to its attachment on the transverse processes of the third to sixth cervical vertebrae. The longus colli is placed anterior to the vertebral column and covered by the longus capitis in its superior aspect. It is divided in three parts and attaches to the transverse processes of the cervical and first thoracic vertebrae, as well as to the anterior aspect of the bodies of the first thoracic vertebrae.

On the posterior aspect, the short and thick bifid spinous process of C2 gives attachment to three muscles (Fig. 1.5):

- Semispinalis cervicis, extending from the second to fifth cervical spines to the transverse processes of the upper five or six thoracic vertebrae
- Inferior oblique, extending to the transverse process of C1 and forms the inferior limit of the suboccipital triangle
- Rectus capitis posterior major, extending to the inferior nuchal line forming the medial limit of the suboccipital triangle

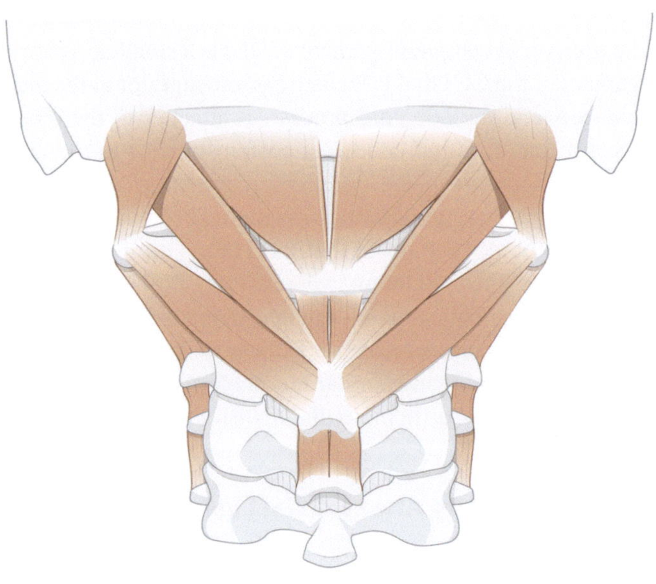

Fig. 1.5 Short neck muscles, posterior view

The lateral limit of the triangle is formed by superior oblique, which extends from the transverse process of C1 to the inferior nuchal line. Rectus capitis posterior minor converging from the inferior nuchal line to the posterior tubercle on the arch of C1 is the only muscle that attaches to the posterior arch of the atlas. Laterally the scalenus medius attaches to the transverse process of the axis and elevator scapulae to the transverse process of the atlas [34–35].

1.4 The Arterial System

The *vertebral artery* is the main vessel that supplies the cervical cord. There are two vertebral arteries (one for each side), and they take their origin from the first portion of the respective subclavian artery. They go upwards and laterally until they reach the foramen transversarium of the sixth cervical vertebra on each side (first segment, V1). The arteries then ascend passing through the foramina transversarium of each cervical vertebra until the axis (second segment, V2), and they continue laterally to reach the foramen transversarium of the atlas; at this point, the vertebral arteries turn posteriorly around the lateral masses of the atlas, perforate the posterior atlanto-occipital membrane and enter the foramen magnum (third segment, V3).

Afterwards, the two vertebral arteries come together to form the basilar artery (fourth and fifth segments, V4 and V5) (Fig. 1.6).

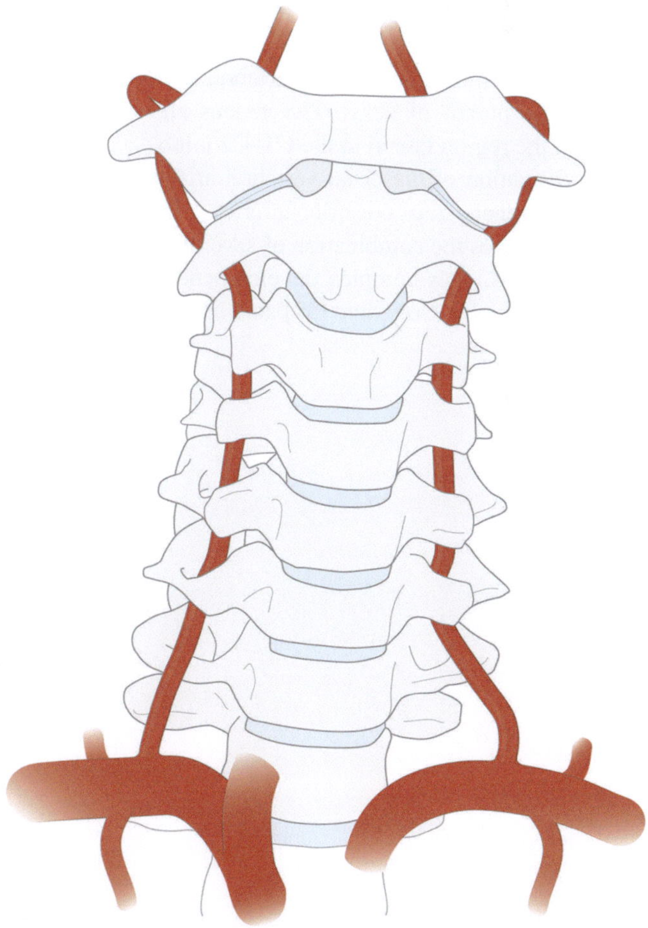

Fig. 1.6 Segments of vertebral artery

To sum up and assist in the description:

- The artery from its course from the C6 transverse foramina to the C2 transverse foramina is labelled as V.1
- The artery during its course from C2 transverse foramina to C1 transverse foramina is labelled as V2.
- The artery in its course from the transverse foramen of C1 to the point of its dural entry is labelled as V3 segment [36].

The left vertebral artery is dominant in 50 % of the subjects, the right vertebral artery is dominant in 25 % and in 25 % of cases right and left arteries have equal vessel diameters that contribute to flow in the basilar artery.

For a surgical approach, it's important to consider that running from their origin (the subclavian artery) to the foramen of C6, the vertebral arteries lie between the longus colli muscle and the anterior scalene muscle, and they are in relationship with the vertebral veins and the neurovascular bundle anteriorly and with the transverse process of C7 and ventral rami of 7th and 8th cervical nerves posteriorly.

During its entire course, the vertebral artery is covered with a large plexus of veins. The venous plexuses are the largest in the region lateral to the C1–C2 joint.

The formation of the distal vertebral artery (VA) and its principal branch, the posterior-inferior cerebellar artery (PICA), involves the combination of several embryonic vascular segments. This complex developmental anatomy was well described by Congdon [37] and Padget [38, 39] in human specimens.

1.4.1 V1 Segment

After its exit from the transverse foramen of the C3 vertebra, the V1 segment of the artery courses posterosuperiorly and forms a loop within the vertebral artery groove on the inferior surface of the superior articular facet and then exits from C2 transverse foramina. The total length of the segment ranges from 17.2 to 46.1 mm (average 23.4 mm). The distal part of the artery is intraosseous and cannot be seen until its bony unroofing. The artery courses medially and takes a reverse loop inside the vertebral artery bony foramina on the inferior surface of the superior facet of the C2 vertebra. The angle of the loop varies and ranges from 2° to 110° (average 86°) [36].

1.4.2 V2 Segment

The V2 exits from the C2 transverse foramen and takes an initial lateral bend and then traverses superiorly. The artery courses anterior to the two roots of the C2 ganglion. In an anatomical study by Ciaccola et al., the length of this segment of the artery ranged from 12.6 to 32.2 mm (average 15.7 mm). The distance of the lateral edge of the ganglion from the vertebral artery ranged from 5.1 to 11.1 mm (average 7.5 mm). The distance of artery from the lateral end of the dural tube ranged from 14.7 to 17.9 mm (average 15.3 mm). Two sets of branches arise from the vertebral artery in this segment, a relatively large muscular branch and a small artery traversing along the C2 ganglion into the spinal canal [36].

1.4.3 V3 Segment

After exiting C1 transverse process, the vertebral artery takes an approximately 90° posterior bend and turns medially to engage in the groove on the superior surface of the posterior arch of the atlas, where turning around the superior facet of the atlas, it bends anteriorly to enter the spinal canal. In an anatomical study by Ciaccola et al., the total length of this segment of the artery was 32.3–43.5 mm (average 35.7 mm). The C1 roots course posterior-inferior in relationship to the artery. The posterior-inferior cerebellar artery did not arise from the segment in any specimen. The distance between the most medial extension of the vertebral artery and the medial edge of the vertebral artery groove on the outer cortex of the posterior arch of the atlas ranged from 2.1 to 5.2 mm (average 4.24 mm) [40].

1.4.4 Anatomical Variation

The incidence of vertebral artery (VA) anomalies is about 3.5 %.

As regards the anomalous origins of the right vertebral artery, these conditions can be divided into three categories:

- VA originating directly from the aorta
- VA originating from the carotid arteries or from the brachiocephalic artery
- VA of duplicated origin

Also the anomalous origins of the left vertebral artery are divided into three categories:

- VA originating directly from the aorta
- VA originating from the left internal carotid artery
- VA of duplicated origin

Only two of the variants described have a higher frequency: the origin of the left vertebral artery directly from the aortic arch (between the left common carotid artery and left subclavian artery) and the right vertebral artery originating from the right common carotid, usually accompanied by an aberrant right subclavian artery.

There are also other important tips for the surgeon regarding the vertebral artery anatomical variants, during the approach to the deep cervical region:

- There is a risk to damage an abnormally long prevertebral VA segment when the surgeon is performing the incision of the muscles of the transverse spinal processes. In some cases, vertebral artery enters the spinal transverse foramina at other levels from C7 to C2, and not at C6; there can also be a symmetrical entrance in 85 % of cases and asymmetrical in 15 % (because the right VA enters at a lower level).
- Another important tip for the surgeon is that there can be anomalous carotid-VA anastomoses (Fig. 1.7).
- The third segment of the VA can have anomalies, and in particular (A) it doesn't pierce the dura at the atlanto-occipital membrane level, but it goes into the dura at the level of atlas or between atlas and axis and courses inside the spinal canal (C2 segmental type of VA); (B) it can have a fenestration. This condition is encountered when there are two branches of VA after the artery emerges from the C2 transverse foramina. One branch has a normal course, whilst the other pierces the dura between C1 and C2 and reaches the first branch inside the skull. (C) There can be a pseudofenestration, that is, a condition where the posterior-inferior cerebellar artery originates from the second or the third segment of the VA (instead of V4), pierces the dura at the atlantoaxial level and has an ascending course laterally to the cord (Fig. 1.8) [41–42].

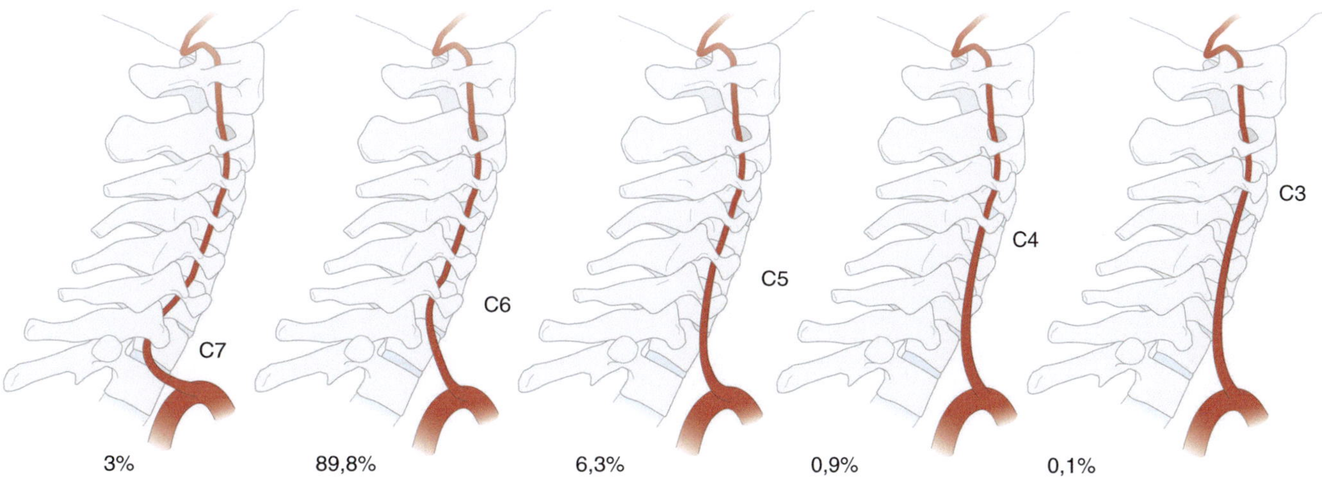

Fig. 1.7 Anatomy of vertebral artery and its variations

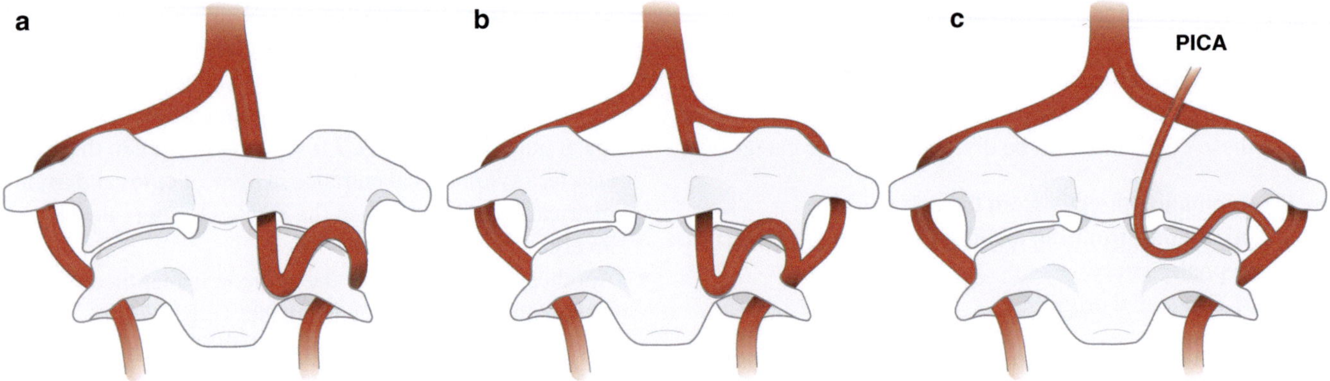

Fig. 1.8 Anatomy of vertebral artery and basilar artery at the level of C1–C2

1.5 The Venous System

Intracranial veins and venous sinuses converge to form major dural sinuses, the transverse sinus and the sigmoid sinus, which drain into extracranial veins. These major dural sinuses are connected by other venous structures to the skull base. These venous structures form a complex venous network that drains intracranial venous flow into extracranial veins at the craniocervical junction [43–44]. These venous structures are also known to have an important role as collateral pathways in cases of veno-occlusive disease.

The venous structures include the major dural sinuses and emissary veins that have a role as the main drainage route for cephalic venous blood flow. The emissary veins are also thought to have a function of redirecting blood outflow towards the vertebral venous system in the upright position [45].

In literature there are many radiological studies of the anatomy and/or variations of craniocervical venous structures using transcatheter angiography or MRI [46–48].

Tanoue et al. [49] (Fig. 1.9) studied the IPS, SCS, MS and ACV; the anatomical architectures of the LCV and PCV could be evaluated; the LCV was found to be absent in 27 sides (27 %), and the PCV was absent in 33 sides (33 %). The OS was depicted in only 18 cases, with 4 cases of prominently developed OS. There is an apparent discrepancy between these results and those of a previous anatomical study, which presented the frequency of absence of the OS as 35.5 % [50].

The IPS normally connects the cavernous sinus to the jugular bulb and passes along the petroclival fissure.

Most investigators have classified the IPS in the context of the patterns of the IPS-jugular bulb junction, for example, single or multiple connections, plexiform connection, no connection or connection with cervical venous plexus. The OS could be demonstrated in prominently developed cases and all drained into the right sigmoid sinus. According to a previous anatomical description, the number of OS running towards the right side is three times larger than the number of OS going to the left [51]. Most PCVs originated from the inferolateral aspect of the sigmoid sinus, but there were cases of a PCV originating from the ACC. In some cases, the PCV originated from the jugular bulb or, rarely, originated from the sigmoid sinus [44].

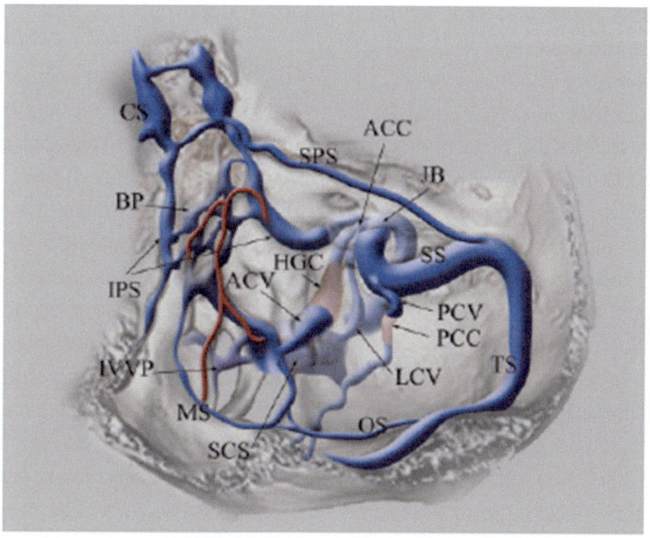

Fig. 1.9 Schematic drawing of the veins at the craniocervical junction. The inferior petrosal sinus (IPS) originates from the posterosuperior aspect of the cavernous sinus (CS), runs along the petroclival fissure and drains into the jugular bulb (JB). Basilar plexus (BP) lies on the clivus with connecting bilateral IPS. The anterior condylar vein (ACV) and lateral condylar vein (LCV) originate from the medial aspect of the JB, forming an anterior condylar confluence (ACC). ACV runs medially through the hypoglossal canal (HGC) and drains into the lateral part of the marginal sinus (MS). MS is contiguous to the medial part of the suboccipital cavernous sinus (SCS). LCV runs postero laterally and flows into SCS. The posterior condylar vein (PCV) originates from the sigmoid sinus (SS), runs through the posterior condylar canal (PCC) and flows into SCS. SCS lies under the occipital bone surrounding the horizontal portion of the vertebral artery. The occipital sinus (OS) originates from the torcular herophili, confluence of transverse sinus (TS) and straight sinus and drains into the posterior part of MS. MS is the round-shaped sinus surrounding the foramenmagnum. MS and the medial part of SCS are connected to the internal vertebral venous plexus (IVVP) [45]

1.6 Neural Tissue

Neural structures related to CVJ are:

1. Caudal portion of brainstem (medulla)
2. Cerebellum
3. Fourth ventricle
4. Rostral part of spinal cord
5. Lower cranial and upper cervical nerves

In cerebellum, only the tonsils, biventral lobules and the lower part of the vermis (nodule, uvula and pyramid) are related to CVJ. Biventral lobule is located above the lateral part of foramen magno and the tonsils lie above the posterior edge.

The afferents to the spinal nerves C1–C3 must include those originating in the upper cervical synovial joints, the upper cervical muscles, the C2–C3 disc, the craniovertebral ligaments and the dura mater of the upper spinal cord and posterior cranial fossa. Collectively, the upper three cervical levels are called the sinuvertebral nerves (SVNs) which could be responsible for mediating pain from the upper cervical spine and the posterior cranial fossa [52].

The SVN had been termed the recurrent branch of the spinal nerve, meningeal rami of spinal nerves and ramus meningeus [53]. However, most authors have adopted the term SVN now for the nerve that enters the intervertebral foramen to innervate the intraspinal structures.

All three SVNs, once formed in the intervertebral foramen, give off very tenuous branches to the dura mater and run in the epidural space anterior to the DRG. Both C2 and C3 SVNs, having entered the vertebral canal, pass upwards and continue to ascend by their main ascending branches over the next vertebra above, running parallel to the membrana tectoria in the vertebral canal and continuing upwards towards the ligaments of the CVJ.

The membrana tectoria is innervated by transverse branches of C2 and C3 SVN at their level of entry into the vertebral canal. In addition, the ascending branch of the C2 and C3 SVNs provides a second source of innervation to the membrane. They run parallel to the lateral edge of the membrana tectoria, supplying its dual laminated structure periodically along the way by tiny filaments to it. They also give off the very important tiny filaments to the overlying dura mater. The ascending and descending branches of the SVNs accompany the posterior ascending artery of the dens to the body and dens of the axis and are protected somewhat by overlying longitudinal venous sinuses and delicate neurovascular sheaths enclosing the nerves and the arteries as they ascend towards the dens where they form an apical arterial arcade between the loops of an apical venous plexus above and around the apex of the dens.

The cervical SVNs at C1–C2 and C2–C3 intervertebral levels have long ascending branches innervating structures at their level of entry and above in the vertebral canal and posterior cranial fossa. Descending branches from both SVNs could be identified and travel only one segment downwards from their level of entry into the vertebral canal.

C1 SVN, however, has a short course in the vertebral canal. The ascending branches are longer than the descending branches. The C1 SVN and its constituent roots are found between the posterior arch of the atlas below and the occipital bone above, immediately behind the atlanto-occipital joint [59].

C3 SVN innervates most of the craniovertebral structures medially, whereas the C2 SVN innervated the structures more laterally. The communication between the ascending and descending branches of C2 and C3 SVNs and between the ascending branch of C2 SVN and the hypoglossal canal is in keeping with the segmental arrangement of the innervation of the vertebral levels below. The hypoglossal canal is ranked developmentally as the next higher segmental level embryologically. Thus, the basiocciput receives at least part of its innervation from the upper SVNs. Interestingly, the communication between the ascending branch of the C2 SVN with the hypoglossal nerve confirms the long-held belief that the sensory communication does occur between cervical and cranial nerves.

Lower four cranial nerves are closely related to CVJ. Ninth and tenth cranial nerves arise from the medulla in the groove between the inferior olivary nucleus and the inferior cerebellar peduncle. They are separated by a dural sheath, which separates these nerves as they penetrate the dura to enter the jugular foramen. The accessory nerve is the only cranial nerve that passes through the FM.

References

1. Lang J (1986) Cervical region, osteology and articulations. Neuro Orthopedics 1:67–92
2. Shapiro R, Robinson F (1976) Embryogenesis of the human occipital bone. Am J Roentgenol 127:281–287
3. Martin MD et al (2010) Anatomic and biomechanical considerations of the craniovertebral junction. Neurosurgery 66(3S):2–6
4. Menezes AH, Traynelis VC (2008) Anatomy and biomechanics of normal craniovertebral junction (a) and biomechanics of stabilization (b). Childs Nerv Syst 24:1091–1100
5. Noble ER, Smoker WRK (1996) The forgotten condyle: the appearance, morphology, and classification of occipital condyle fractures. Am J Neuroradiol 17(3):507–513
6. Kim DM et al (2007) Surgery of the paediatric spine. Thieme Medical Publications, Stuttgardt. ISBN 978-3-13-141931-6
7. Torriani M, Lourenco J (2002) Agenesis of the posterior arch of the atlas. Revista do Hospital das Clínicas 57(2):73–76
8. Denaro L et al (ed) (2010) Pitfalls in cervical spine surgery: avoidance and management of complications. Springer Publications, Stuttgardt. ISBN 978-3-540-85019-9
9. Gray H (1918) Anatomy of the human body. http://www.bartleby.com/107/
10. Konig SA et al (2005) Anatomical data on the craniocervical junction and their correlation with degenerative changes in 30 cadaveric specimens. J Neurosurg Spine 3(5):379–385
11. Fitzgerald RH et al (2002) Orthopaedics. Mosby Inc., Stuttgardt. ISBN 0-323-01318-X
12. Murphy TM et al (2012). Rheumatoid arthritis-etiology, consequences and comorbidities. Elsevier.
13. Lustrin ES et al (2003) Paediatric cervical spine: normal anatomy, variants, and trauma. Radiographics 23(3):539–560
14. Steinmetz MP, Mroz TE, Benzel EC (2010) Craniovertebral junction: biomechanical considerations. Neurosurgery 66(3 Suppl): 7–12
15. Wolfla CE (2006) Anatomical, biomechanical, and practical considerations in posterior occipitocervical instrumentation. Spine J 6(6 Suppl):225S–232S
16. Tubbs RS et al (2011) Ligaments of the craniocervical junction. J Neurosurg Spine 14:697–709
17. Dickman CA, Mamourian A, Sonntag VK, Drayer BP (1991) Magnetic resonance imaging of the transverse atlantal ligament for the evaluation of atlantoaxial instability. J Neurosurg 75:221–227
18. Dvorak J, Schneider E, Saldinger P, Rahn B (1988) Biomechanics of the craniocervical region:the alar and transverse ligaments. J Orthop Res 6:452–461
19. Krakenes J, Kaale BR, Rorvik J, Gilhus NE (2001) MRI assessment of normal ligamentous structures in the craniovertebral junction. Neuroradiology 43:1089–1097
20. Tubbs RS, Dixon J, Loukas M, Shoja MM, Cohen-Gadol AA (2010) Ligament of Barkow of the craniocervical junction: its anatomy and potential clinical and functional significance. J Neurosurg Spine 12:619–622
21. Tubbs RS, Griessenauer CJ, McDaniel JG, Burns AM, Kumbla A, Cohen-Gadol AA (2010) The transverse occipital ligament: anatomy and potential functional significance. Neurosurgery 66(3 Suppl Operative):1–3
22. Schaeffer JP (ed) (1953) Morris' human anatomy a complete systematic treatise, 11th edn. The Blakiston Company, New York
23. Yuksel M, Heiserman JE, Sonntag VK (2006) Magnetic resonance imaging of the craniocervical junction at 3-T: observation of the accessory atlantoaxial ligaments. Neurosurgery 59:888–893
24. Tubbs RS, Salter EG, Oakes WJ (2004) The accessory atlantoaxial ligament. Neurosurgery 55:400–404
25. Pick TP, Howden R (eds) (1901) Gray's anatomy, descriptive and surgical. Lea Brothers, Philadelphia
26. Tubbs RS, Stetler W, Shoja MM, Loukas M, Hansasuta A, Liechty P et al (2007) The lateral atlanto-occipital ligament. Surg Radiol Anat 29:219–223
27. Haffajee MR, Thompson C, Govender S (2008) The supraodontoid space or "apical cave" at the craniocervical junction: a microdissection study. Clin Anat 21:405–415
28. Tubbs RS, Wellons JC III, Blount JP, Oakes WJ (2002) Posterior atlanto-occipital membrane for duraplasty. Technical note. J Neurosurg 97(2 Suppl):266–268
29. Zumpano MP, Hartwell S, Jagos CS (2006) Soft tissue connection between rectus capitus posterior minor and the posterior atlanto-occipital membrane: a cadaveric study. Clin Anat 19:522–527
30. Hack GD, Koritzer RT, Robinson WL, Hallgren RC, Greenman PE (1995) Anatomic relation between the rectus capitis posterior minor muscle and the dura mater. Spine 20:2484–2486
31. Thompson VP (1995) Anatomical research lives! Nat Med 1:297–298
32. Williams PL (ed) (1996) Gray's anatomy: the anatomical basis of medicine and surgery, 38th edn. Churchill Livingstone, London
33. VanGilder JC, Menezes AH, Dolan K (1987) The craniovertebral junction and its abnormalities. Futura, New York, pp 1–255
34. Menezes AH (2003) Developmental abnormalities of the cranio-cervical junction. In: Winn RH (ed) Youmans neurological surgery. Saunders, Orlando, p 3331–3345
35. Osenbach RK, Menezes AH (1992) Pediatric spinal cord and vertebral column injury. Neurosurgery 30:385–390
36. Cacciola F, Phalke U, Goel A (2004) Vertebral artery in relationship to C1–C2 vertebrae: an anatomical study. Neurol India 52:178–184
37. Congdon ED (1922) Transformation of the aortic-arch system during the development of the human embryo. Contrib Embryol 68:47–110
38. Padget DH (1948) The development of the cranial arteries in the human embryo. Contrib Embryol 212:205–261
39. Padget DH (1954) Designation of the embryonic intersegmental arteries in reference to the vertebral artery and subclavian stem. Anat Rec 119:349–356
40. Bruneau M, Cornelius JF, George B (2006) Antero-lateral approach to the V3 segment of the vertebral artery. Neurosurgery 58:29–35
41. Gaillard F (2008) Title of subordinate document. In: Anatomy and vertebral artery. Toshiba (Australia) Medical Division. Available via DIALOG. http://radiopaedia.org/articles/vertebral-artery. Accessed 26 Oct 2008
42. Duan S, He H, Lv S, Chen L (2010) Three-dimensional CT study on the anatomy of vertebral artery at atlantoaxial and intracranial segment. Surg Radiol Anat 32(1):39–44. doi: 10.1007/s00276-009-0552-5. Epub 2009 Aug 26
43. San Milla'n Rui'tz D, Gailloud P, Rufenacht DA, Delavelle J, Henry F, Fasel JH (2002) The craniocervical venous system in relation to cerebral venous drainage. AJNR Am J Neuroradiol 23:1500–1508
44. Tanoue S, Kiyosue H, Sagara Y, Hori Y, Okahara M, Kashiwagi J, Mori H (2010) Venous structures at the craniocervical junction: anatomical variations evaluated by multidetector row CT. Br J Radiol 83(994):831–840
45. Takahashi S, Sakuma I, Omachi K, Otani T, Tomura N, Watarai J et al (2005) Craniocervical venous anatomy around the suboccipital cavernous sinus: evaluation by MR imaging. Eur Radiol 15:1694–1700
46. Caruso RD, Rosenbaum AE, Chang JK, Joy SE (1999) Craniocervical junction venous anatomy on enhanced MR images: the suboccipital cavernous sinus. AJNR Am J Neuroradiol 20:1127–1131

47. Arnautovic' KI, al-Mefty O, Pait TG, Krisht AF, Husain MM (1997) The suboccipital cavernous sinus. J Neurosurg 86:252–262
48. Das AC, Hasan M (1970) The occipital sinus. J Neurosurg 33:307–311
49. Valdueza JM, von Mu¨ nster T, Hoffman O, Schreiber S, Einha¨upl KM (2000) Postural dependency of the cerebral venous outflow. Lancet 355:200–201
50. Hacker H (1974) Dural venous sinuses. In: Newton TM, Potts DG (eds) Radiology of the skull and brain, vol 2 Book 3 Veins. CV Mosby, Saint Louis. p. 1862–1872
51. Rennie C, Haffajee MR, Ebrahim MA (2013) The sinuvertebral nerves at the craniovertebral junction: a microdissection study. Clin Anat 26(3):357–366
52. Wiberg G (1949) Back pain in relation to the nerve supply of the intervertebral disc. Acta Orthop Scand 19:211–221
53. Kimmel DL (1961) Innervation of spinal dura mater and dura mater of the posterior cranial fossa. Neurology 11:800–809

Anatomy of the Subaxial Cervical Spine

2

M. Girolami, R. Ghermandi, M. Ghirelli, A. Gasbarrini, and S. Boriani

The cervical spine is a deep anatomical structure centrally located in the axial view of the neck and surrounded by a thick and powerful muscular compartment. It is contained by the deep layer of deep cervical fascia (Fig. 2.1).

This chapter describes the essential anatomy of subaxial cervical spine (C3-C7).

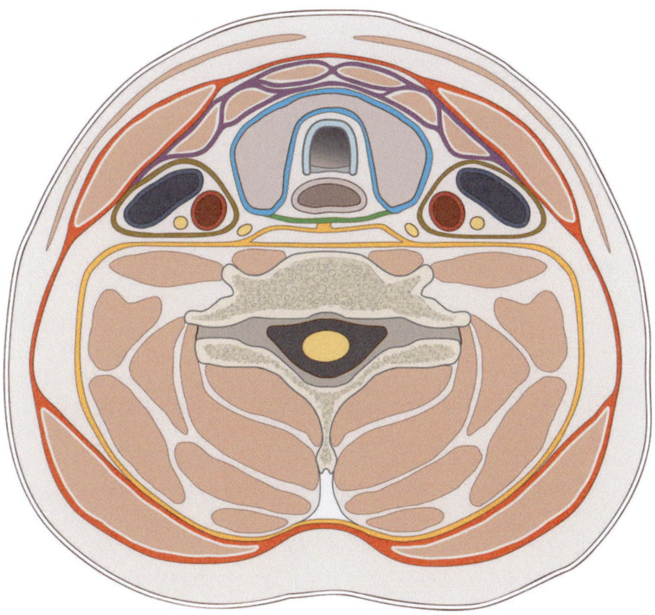

Fig. 2.1 Axial view of the subaxial cervical spine

M. Girolami (✉) • R. Ghermandi • A. Gasbarrini • S. Boriani
Oncologic and Degenerative Spine Surgery Department Rizzoli Orthopedic Institute, Bologna, Italy
e-mail: marco.girolami@yahoo.it

M. Ghirelli
Head and Neck Surgery Department University Hospital of Modena, Modena, Italy

2.1 Bony Anatomy

2.1.1 C3-C6

Vertebrae are composed by a vertebral body, anteriorly, and a neural arch, composed by pedicles and laminae, posteriorly (Fig. 2.2).

The typical lower cervical vertebra presents a small square vertebral body. A hook-shaped uncinate process, on each side of the superior endplate of vertebral bodies from third cervical to first thoracic, articulates with the corresponding uncus surface, on the inferior endplate of the vertebral body above, forming the uncovertebral joints (Luschka's joints).

The neural arch is formed by two pedicles and two laminae and forms the posterior border of the vertebral foramen which is triangular in shape, with apex directed posteriorly [1].

Extending posteriorly from where the two laminae meet is the spinous process which is bifid.

The transverse process extends laterally from the junction of the pedicle and the lamina on each side. Cervical transverse processes have anterior and posterior bars, which delimitate the foramen transversarium, and terminate laterally as corresponding tubercles (connected by a costal – or intertubercular – lamella).

The attachment of the dorsal bar to the pediculolaminar junction represents the morphological transverse process, while the attachment of the ventral bar to the vertebral body represents remnants of ancient coastal elements.

It is important to consider that the anterior tubercle of the transverse process of the 6th cervical vertebra is bulky, and it is an important surgical landmark; moreover, it has a close relationship with the common carotid artery, and so is also called the carotid tubercle (Chassaignac tubercle) [2].

S. Boriani et al. (eds.), *Atlas of Craniocervical Junction and Cervical Spine Surgery*, DOI 10.1007/978-3-319-42737-9_2

The junction between lamina and pedicle bulges laterally between the superior and inferior articular processes to form lateral masses on each side.

The superior articular facets, flat and ovoid, are directed superoposteriorly, whereas the corresponding inferior facets are directed mainly anteriorly, and lie nearer the coronal plane than the superior facets [3].

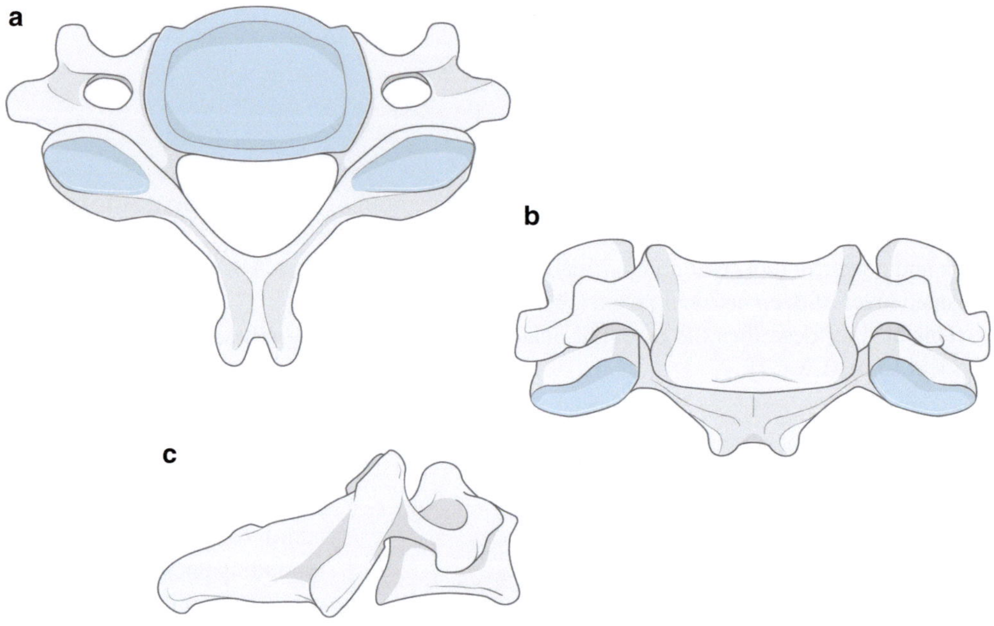

Fig. 2.2 Subaxial cervical vertebra: (**a**) superior, (**b**) anteroposterior, (**c**) lateral views

2.1.2 C7: Vertebra Prominens

C7 has some peculiar features that make it different from the other cervical vertebrae. Some of its characteristics are that of a transitional vertebra between the cervical and thoracic spine (Fig. 2.3).

The spinous process is not bifid; it is directed dorsally and downwards ending in a prominent tubercle for the attachment of the ligamentum nuchae.

The posterior tubercle of the transverse process is thicker than that in the other cervical vertebrae, and sometimes particularly prominent (cervical rib).

Articular surfaces are more vertical but still the superior, oriented dorsally, and the inferior, ventrally.

Another important consideration concerns the content and the shape of the foramen transversarium. In fact, it only contains the vertebral vein, and not the correspondent artery, as all other lower cervical vertebrae do. Moreover, it is often divided by a bony spicule, and its ventral burden is relatively thin and may be underdeveloped or partly deficient.

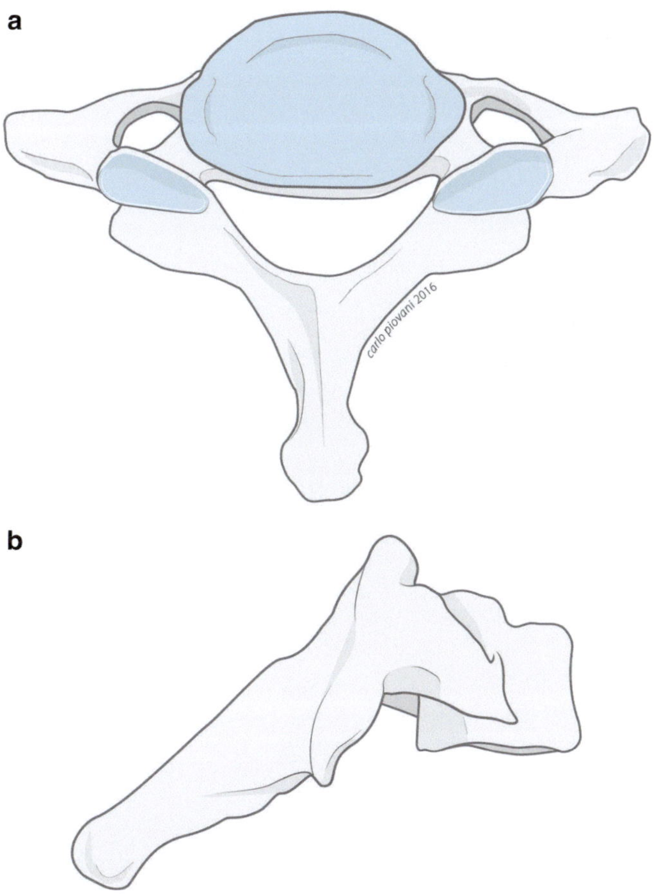

Fig. 2.3 C7 peculiar features: (**a**) superior and (**b**) lateral views

2.2 Joints and Ligamentous Anatomy

Joints within each motion segment are both anteriorly and posteriorly, between vertebral bodies and lateral masses, respectively.

Two adjacent vertebral bodies articulate within each other by means of three articulations: the two uncovertebral joints and a discosomatic joint. This latter is a synarthrosis since there is no joint cavity and the motion is obtained through the intervertebral disc located in between them. The intervertebral disc is a structure is formed by a central nucleus pulposus enclosed by the fibers of the annulus fibrosus in the peripheral part.

The uncovertebral joints are synovial joints formed by the uncinate process of the 3rd cervical to the 1st thoracic vertebra which articulates with the correspondent surface on the inferior part of the vertebral body above. These joints allow flexion-extension limiting lateral flexion.

Dorsally, two plain synovial joints formed by the inferior articular facet of a vertebra and the superior articular facet of the vertebra below. The orientation of the facet varies within the different regions of the spine and dictates the plane of motion at each relative level. In the cervical spine, superior articular facets are directed upwards and dorsally, whereas the corresponding inferior facets, downwards and ventrally. The capsules of these joints strongly contribute to segmental stability.

There is a system of ligaments (Fig. 2.4) connecting adjacent vertebrae, thus allowing segmental motion:

1. Anterior longitudinal ligament (ALL)
2. Posterior longitudinal ligament (PLL)
3. Ligamentum flavum
4. Interspinous ligaments
5. Ligamentum nuchae

2.2.1 Anterior Longitudinal Ligament

This fibrous structure extends on the ventral surface of the vertebral bodies from the base of the skull to the sacrum. It resists hyperextension.

2.2.2 Posterior Longitudinal Ligament

This ligament lies within the spinal canal on the dorsal surface of the vertebral bodies extending from the occiput to the sacrum. Caution is advised during dissection of the anterior epidural space, since this ligament might be in close relationship with the dural sac.

2.2.3 Ligamentum Flavum

This structure owes its name (lat. flavus, yellow) to the great prevalence of elastic fibers which determine its color. Ligamenta flava are coupled, one at each lamina (right and left) of the vertebra and might be separated by a central fissure. They extend from half the ventral surface of the lamina of a vertebra to the superior margin of the lamina of the vertebra below. Caution is advised during opening of the canal, since this ligament might be in close relationship with the dural sac.

2.2.4 Interspinous Ligaments

These ligaments connect two contiguous spinous processes and extend anteriorly up to reach the ligamentum flavum.

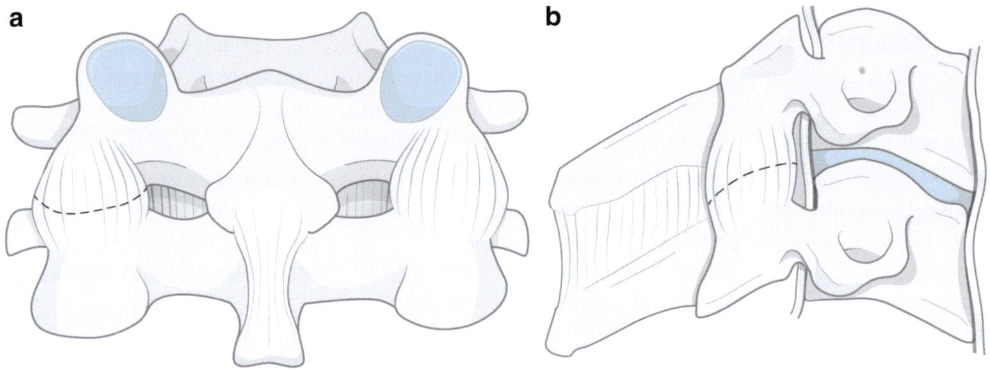

Fig. 2.4 (**a**) Posterior and (**b**) lateral views of a functional spinal unit with its posterior capsular and ligamentous structures (posterior ligamentous complex)

2.2.5 Ligamentum Nuchae

This structure is stretched from the external occipital protuberance to the spinous process of C7 (Fig. 2.5). It represents the homologous of the supraspinous ligament in the thoracolumbar spine. Its fibers deepen to reach the spinous processes of each cervical vertebra contributing to separate the right and left paravertebral muscle compartments [4].

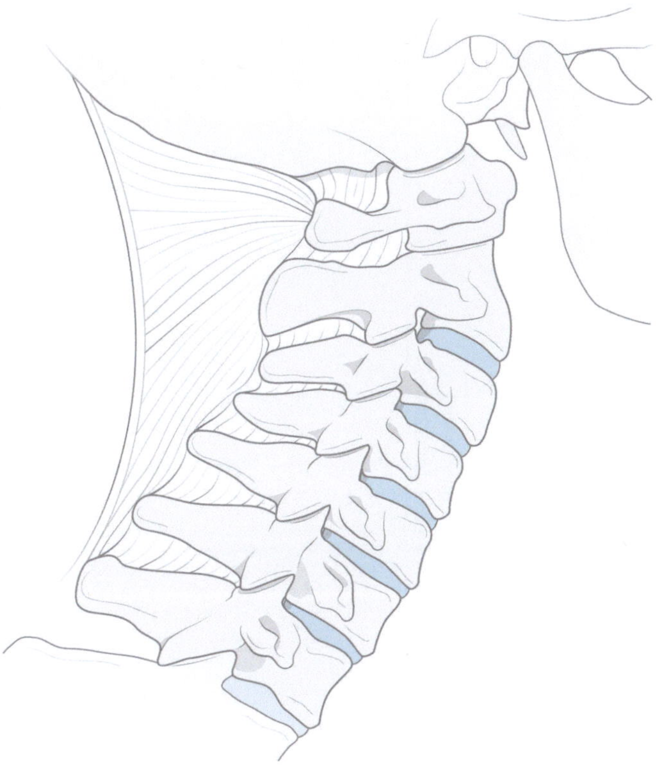

Fig. 2.5 Lateral view of the cervical spine and ligamentum nuchae

2.3 Muscular Anatomy

	Muscle		Origin	Insertion	Innervation	Action
Prevertebral	Longus colli	*Superior-oblique fibers*	Transverse processes from C3 to C5	Anterior arch of atlas	Branches of anterior rami of C2 to C8	Flexes the cervical spine Bends the spine to the ipsilateral side rotating it to the contralateral side
		Inferior-oblique fibers	Vertebral bodies from T1 to T3	Transverse processes from C5 to C7		
		Vertical fibers	Vertebral bodies from C5 to T3	Vertebral bodies from C2 to C4		
	Longus capitis		Transverse processes from C3 to C5	Basilar portion of the occipital bone	Ventral rami of C1 to C3	
	Rectus capitis anterior		Transverse process of the atlas	Basilar portion of the occipital bone	Ventral rami of C1 and C2	Flexes the head
	Rectus capitis lateralis			Jugular portion of the occipital bone		Bends the head to the ipsilateral side
	Anterior scalene		Transverse processes from C3 to C6	Scalene tubercle of 1st rib	Ventral rami of C4 to C6	Elevates the 1st rib
	Middle scalene		Transverse processes from C2 to C7	1st rib	Ventral rami of C5 to C8	Flexes the spine Bends the spine to the ipsilateral side rotating it to the contralateral side
	Posterior scalene		Transverse processes from C5 to C7	2nd rib	Ventral rami of C6 to C8	Extends the spine
Suboccipital (C0-C2)	Obliquus capitis superior		Transverse process of atlas (C1)	Occipital bone	Suboccipital nerve (dorsal rami of C1)	Extends head Lateral flexion of head
	Obliquus capitis inferior		Spinous process of the axis (C2)	Transverse process of the atlas (C1)		Rotates head to the ipsilateral side
	Rectus capitis posterior major			Inferior nuchal line (lateral portion) of the occipital bone		Extends head Rotates head to the ipsilateral side
	Rectus capitis posterior minor		Posterior arch of the atlas	Inferior nuchal line (medial portion) of the occipital bone		Extends head

Paravertebal	*Superficial layer*					
	Trapezius	*Descending part*	Superior nuchal line (medial portion), external occipital protuberance, and nuchal ligament	Lateral third of clavicle	Spinal accessory nerve (CN XI) ventral rami of and C3 and C4 nerves	Supports the weight of the arm
		Transverse part	Spinous processes from C7 to T3	Spine of scapula and acromion		Adducts scapula
		Ascending part	Spinous processes from T3 to T12	Scapula (medial margin)		Medially rotates and depresses the scapula
	Sternocleidomastoid		Manubrium of the sternum and clavicle	Superior nuchal line (anterior portion) and mastoid process of temporal bone	Spinal accessory nerve (CN XI) ventral rami of and C2 and C3 nerves	Flexes the head Rotates head to the contralateral side
	Rhomboideus minor		Nuchal ligament and C7 spinous processes	Scapula (medial margin)	Dorsal scapular nerve (C4-C5-C6)	Adducts scapula
	Levator scapulae		Transverse processes of C1 to C4	Vertebral border of scapula between medial angle and root of spine	C4-C5, and possibly branches from the dorsal scapular nerve	Lifts and adducts the scapula
	Intermediate layer					
	Serratus posterior superior muscle		Lower half of ligamentum nuchae and spinous processes from C6 to T2	Ribs from 2nd to 5th, beyond their angle	Posterior rami from C6 to T2	Elevates the ribs (aid deep inspiration)
	Splenius cervicalis		Spinous processes from T3 to T5	Transverse processes of C1 to C3	Posterior rami from C2 to C4	Extends the spine
	Splenius capitis		Lower half of ligamentum nuchae and spinous processes from C7 to T2	Superior nuchal line (lateral portion) and mastoid process of temporal bone	Lateral branches of posterior rami from C4 to C7	Extends the spine Bends head and neck to the ipsilateral side
	Deep layer					
	Erector spinae muscle	Ileocostalis cervicalis	Angle from 1st to 5th rib	Cervical transverse processes from C4 to C6	Posterior rami from C4 to C7	Extends the spine
		Longissimus cervicalis	Transverse processes from T1 to T6	Posterior tubercle of transverse processes from C2 to C5		
		Longissimus capitis	Transverse processes from C3 to T3	Mastoid process of temporal bone		
		Spinalis cervicalis	Spinous processes from C6 to T2	Spinous processes from C2 to C4		
		Spinalis capitis		Occipital bone, merging with semispinalis capitis		
	Trasversospinal group	Semispinalis cervicalis	Transverse processes from T1 to T6	Spinous processes from C2 to C5	Posterior rami from C4 to C7	Extends the spine Rotates spine to the contralateral side
		Semispinalis capitis	Transverse processes from C4 to T6	Occipital bone, merging with spinalis capitis		
		Multifidus	Transverse processes	Spinous processes of vertebrae above origin (1–4 segments)		
		Rotatores	Transverse processes	Spinous processes of the vertebra above		
	Interspinales cervicalis		Spinous processes	Spinous processes of the vertebra above	Posterior rami from segmental spinal nerve	Extends the spine
	Intertrasversarii cervicalis		Transverse processes	Transverse processes of the vertebra above		Bends head and neck to the ipsilateral side

2.4 The Arterial System

All the segments of the spinal cord receive blood supply from three spinal arteries, one anterior and two posterior. These arteries arise from the intracranial (V4) segment of the vertebral arteries before these merge into the basilar artery.

The anterior spinal artery extends downward in the anterior median fissure, on the anterior surface of the spinal cord. The two posterior spinal arteries go downward symmetrically along the posterior lateral groove of the spinal cord, just behind the dorsal roots of spinal nerves. These travel in the subarachnoid space and send branches into the spinal cord.

They form anastomoses (connections) via the anterior and posterior radicular arteries, branches of division of the medullary artery (from the spinal artery, branch of the segmental artery) that enters the dural sac along the exiting spinal nerve. These segmental connections origin at any level from the vertebral artery [5].

Two more vessels contribute to the arterial blood supply of the spinal region of the neck: the ascending cervical artery, branch of the inferior thyroid artery (thyrocervical trunk), and the deep cervical artery, branch of the costocervical trunk of the subclavian artery.

2.5 The Venous System

The venous system of the cervical spine can be subdivided into external and internal plexuses.

The first (Batson plexus) can be further subdivided into an anterior portion, covering the ventral surface of the vertebral bodies, and in a posterior portion, which is on the dorsal surface of the lamianae, lateral masses, and spinous processes.

The internal venous plexus is located in the epidural space and is further subdivided into an anterior portion and a posterior portion connected to the contralateral by the transverse plexus. This drains the blood from the vertebral body via the basivertebral veins and the blood from the spinal cord via the medullary veins.

The subarachnoid venous system is little different from the arterial: there is an anterior spinal vein but only one posterior spinal vein. These drain into the anterior and posterior radicular veins which follow the dorsal and ventral roots of the spinal nerve, and then into the intervertebral vein which leaves the epidural space through the foramina.

Blood drains then into the vertebral vein and the deep jugular vein that, ultimately, drains into the brachiocephalic trunk which forms the superior vena cava system.

2.6 The Cervical Spinal Nerves

Spinal nerves are formed by anterior and posterior roots: the first (motor fibers) originates from the anterior horn of the spinal cord and leaves it from the anterolateral sulcus, while fibers of the latter (sensory fibers) originate in the dorsal ganglion from pseudo-unipolar neurons. Its central axons enter the spinal cord at the posterolateral sulcus to reach the posterior horn of the spinal cord, and peripheral axons join the ventral root to form the spinal nerve.

Spinal nerves, eight pairs in the cervical spine, right after their course in the intervertebral foramen, subdivide into ventral and dorsal roots (Fig. 2.6).

The ventral ramus of each spinal nerve goes through the groove that is bounded anteriorly by the vertebral artery and anterior tubercle, posteriorly by the transverse process and posterior tubercle, medially by the vertebral body, and laterally by the facet joint.

The anterior rami of the 1st, 2nd, 3rd, and 4th cervical spinal nerves form the cervical plexus, while the 5th, 6th, 7th, 8th cervical, and 1st thoracic form the brachial plexus. Each cervical nerve receives sympathetic fibers via gray rami communicantes.

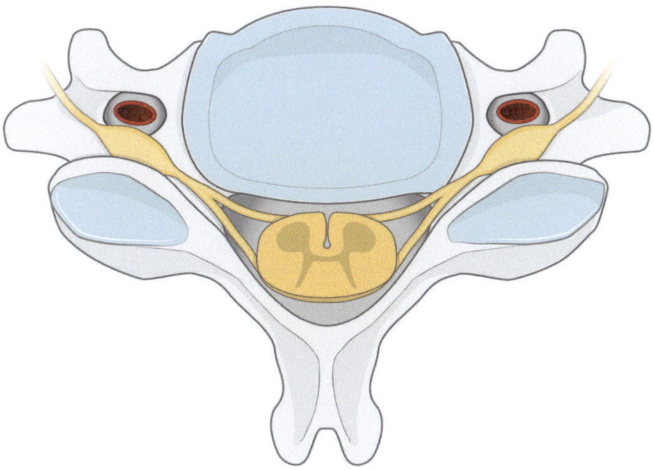

Fig. 2.6 Superior view of subaxial cervical vertebra with the spinal cord, nerves, and vertebral artery. Spinal nerves pass dorsal to the vertebral artery and subdivide into dorsal and ventral roots

Cervical sympathetic is composed by superior, middle, and inferior ganglia located on the ventral surface of the transverse processes and connected by a sympathetic trunk. They receive preganglionic fibers by a white rami communicantes at level of the 1st thoracic nerve and give postganglionic fibers with gray rami communicantes to the correspondent spinal nerve.

The dorsal roots are directed posteriorly to innervate posterior structures including joint capsules, muscles, and skin. The dorsal root of the 1st cervical spinal nerve (suboccipital nerve) is exceptionally composed of motor fibers for the suboccipital muscles.

The dorsal root of the 2nd cervical spinal nerve (greater occipital nerve of Arnold) is composed by a larger sensory branch, the greater occipital nerve for the skin of the scalp, and a thinner motor branch for the muscle inferior oblique, semispinalis, longissimus capitis, and trapezius.

The dorsal root of the 3rd cervical spinal nerve as well provides sensory to the scalp.

References

1. Norton NS (2016) Netter's head and neck anatomy for dentistry. Elsevier Health Sciences
2. Anastasi G (2007) Human Anatomy [Trattato di anatomia umana]. Edi. Ermes.
3. Bogart BI (2007) Elsevier's integrated anatomy and embryology. Mosby Elsevier.
4. Hoppenfeld S (2009) Surgical exposures in orthopaedics: the anatomic approach, 4th ed. Lippincott Williams & Wilkins.
5. Paulsen W (2011) Sobotta atlas of human anatomy: head, neck and neuroanatomy. Vol.3, 15th ed. Elsevier Urban & Fisher.

Anatomy of the Neck

3

M. Bonali, D. Soloperto, E. Aggazzotti Cavazza, M. Ghirelli, and L. Presutti

3.1 Topographic Anatomy

The cervical spine is a complex and vital region, and several critically important structures are in close proximity in a small area. The neck is limited superiorly by the pericranium cervical line that goes from the inferior portion of the mental symphysis to the inion, that corresponds to the point of confluence of dural sinuses. Inferiorly, the neck continues with the thorax and the upper limbs, by clavicles and shoulders.

The first cervical vertebra is located behind the angle of the mandible, and its transverse process is palpable between the angle of the mandible and the mastoid process. The hyoid bone is just anterior to the level of C3, the thyroid cartilage is anterior to the level of C4, and cricoid cartilage is opposite to the level of C6.

On transverse section, the neck appears to be roughly divided into two sections: a posterior or nuchal region (osteo-muscular) and an anterior or tracheal region (muscular-fascial). Usually the conventional dividing line extends from the transverse vertebral processes to the anterior edges of the trapezius muscle. Topographically, the anterior part of the neck is divided by the sternocleidomastoid muscle into an anterior and a posterior triangle. The description of the triangles will be the purpose of the specific section. The thoracic inlet, which is the anatomic region where the neck communicates with the mediastinum, corresponds to the superior aperture of the thorax. It is bounded by the spine posteriorly (at the level of T1), the cartilage of the first ribs laterally, and the sternal (jugular) notch anteriorly. The description of the anatomy is focused on structures and areas as they are approached during surgical procedures (Fig. 3.1).

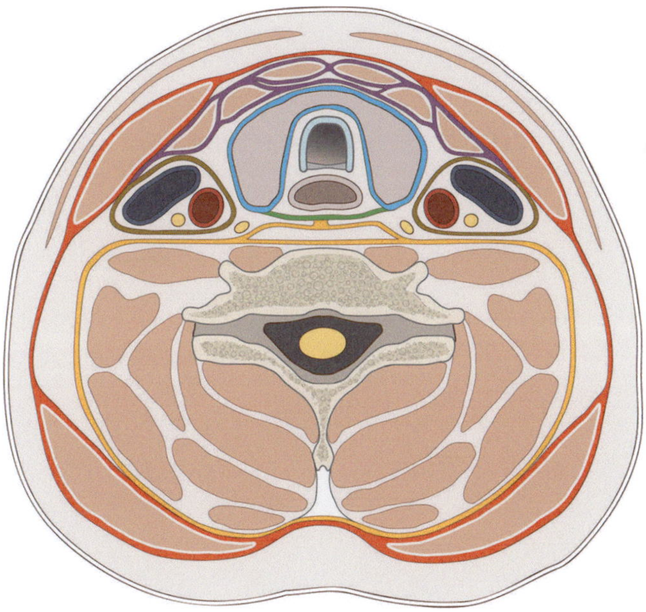

Fig. 3.1 Cross section of the main muscles and fascial layers of the neck

M. Bonali (✉) • E. Aggazzotti Cavazza • M. Ghirelli • L. Presutti
Otorhinolaryngology, Head and Neck Surgery Department, University Hospital of Modena, Modena, Italy
e-mail: bonamed1984@hotmail.it

D. Soloperto
Otorhinolaryngology, Head and Neck Surgery Department, University Hospital of Verona, Verona, Italy

S. Boriani et al. (eds.), *Atlas of Craniocervical Junction and Cervical Spine Surgery*, DOI 10.1007/978-3-319-42737-9_3

3.2 Muscles of the Neck

3.2.1 Platysma

Compared to all other muscles of the neck, which are skeletal, this is the only cutaneous. It is a wide and thin lamina in the subcutaneous tissue. It covers the frontal and lateral side of the neck and the lower face. This muscle is cut when a myocutaneous flap is created.

3.2.2 Sternocleidomastoid Muscle

It is a long and thick muscle, situated in the anterolateral region of the neck. It fits with one sternal, one mastoid, and one clavicular insertion. It covers the great vessels of the neck, and it is crossed superficially by the external jugular vein that goes behind the clavicle, near the frontal corner of the anterior triangle and the brachial plexus. At Erb's point, the great auricular nerve, which derives from the cervical plexus, obliquely crosses the surface of the muscle upward (Fig. 3.2).

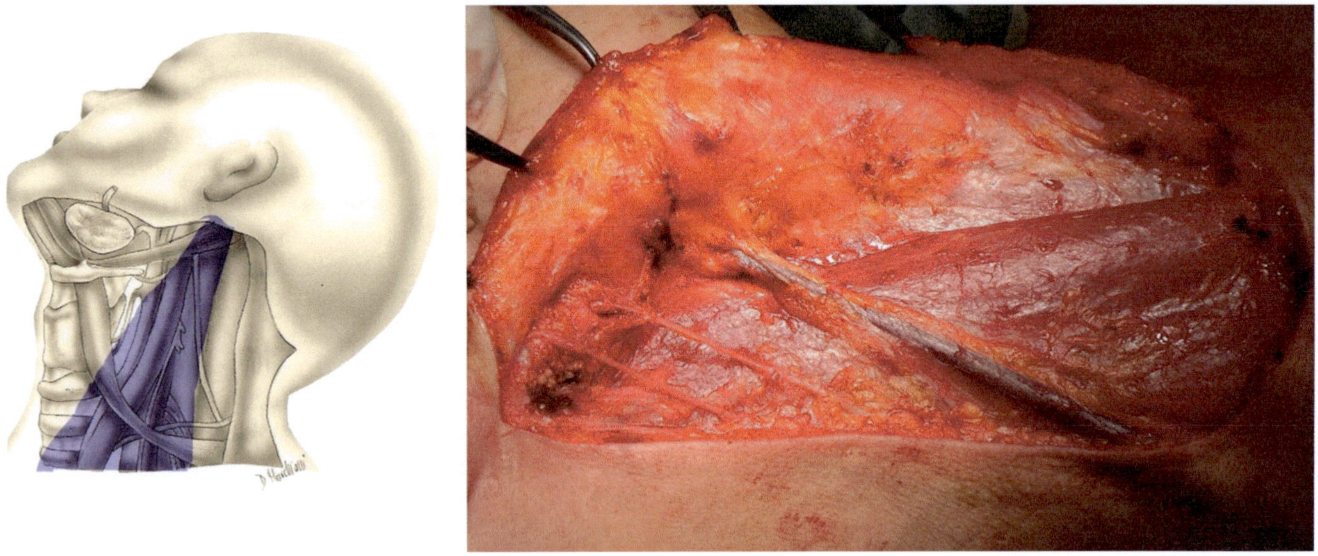

Fig. 3.2 The SCM on the right side and its relationship with the external jugular vein and the great auricular nerve (Courtesy of Prof. Daniele Marchioni, ENT Department, Verona, Italy)

3.2.3 Suprahyoid Muscles (Fig. 3.3)

The suprahyoid muscles are located in the submandibular region superiorly to the hyoid bone and connect the hyoid to the skull. They are the geniohyoid (deep layer), the mylohyoid, the stylohyoid (middle layer), and the digastric muscle (superficial layer).

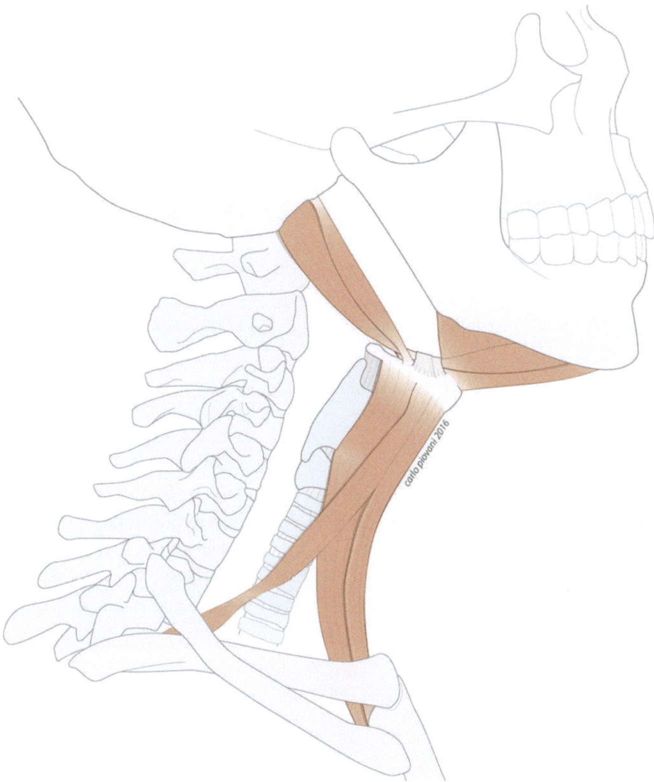

Fig. 3.3 Relationship between hyoid bone and muscles (suprahyoid and infrahyoid)

Muscle	Origin	Insertion	Innervation	Action
Digastric (anterior belly)	Inner side of the lower border of the mandible	Body of the hyoid, by an intermediate tendon continuous with the posterior belly	Mylohyoid nerve, branch of the trigeminal nerve (V3)	Elevates the hyoid. Helps depression and retraction of the mandible
Digastric (posterior belly)	Digastric notch of the temporal bone	Body of the hyoid, by an intermediate tendon continuous with the anterior belly	Facial nerve (CN VII)	Elevates the hyoid. Helps depression and retraction of the mandible
Stylohyoid	Styloid process	Body of the hyoid	Facial nerve (CN VII)	Helps to pull the hyoid bone up and backward during swallowing
Mylohyoid	Mylohyoid line of the mandible	Body of the hyoid and median raphe	Mylohyoid nerve, branch of the trigeminal nerve (V3)	Raises the hyoid and tongue during swallowing
Geniohyoid	Inferior genial tubercle	Body of the hyoid	Branch of C1 through the hypoglossal nerve	Elevates the hyoid bone and tongue

3.2.4 Infrahyoid Muscles ("Strap Muscles") (Fig. 3.3)

Infrahyoid muscles vary considerably in extent of their development. They connect the hyoid with the upper part of the thorax, shoulders, and larynx. They are the sternothyroid, thyrohyoid, sternohyoid, and omohyoid. They overlie the thyroid gland, the thyroid cartilage, the larynx, the trachea, and the esophagus.

Muscle	Origin	Insertion	Innervation	Action
Omohyoid (superior belly)	Body and greater cornu of the hyoid bone	Deep to the SCM muscle, connected to the inferior belly by a tendon	Descendens hypoglossi (branches of C2 and C3)	Depresses the hyoid
Omohyoid (inferior belly)	Upper border of scapula	Deep to the SCM muscle, connected to the superior belly by a tendon	Descending cervical nerve of cervical plexus	Depresses the hyoid
Sternohyoid	Posterior manubrium and medial third of the clavicle	Lower border of the hyoid	Ansa hypoglossi	Depresses the hyoid
Sternothyroid	Posterior manubrium and first costal cartilage	Oblique line of the thyroid cartilage	Ansa hypoglossi	Depresses the larynx and the thyroid cartilage
Thyrohyoid	Oblique line of the thyroid cartilage	Lower border of the body of the hyoid bone	Thyrohyoid branch of C1 through descendens hypoglossi	Depresses the larynx and hyoid bone, elevates the thyroid cartilage

3.2.5 **Omohyoid** (Fig. 3.4)

The omohyoid muscle is an important landmark in the neck because it divides the anterior and posterior cervical triangles into smaller triangles. Also known as "resident's friend," the omohyoid muscle is the anatomic separation of nodal levels III and IV. It has also been used as a reliable landmark in the supraclavicular region, for endoscopic exploration of the brachial plexus.

Having a common primordium with the sternohyoid muscle, the omohyoid muscle often varies in shape and insertions. It usually has two bellies (anterior and posterior) and extends obliquely from the hyoid bone to the scapula.

In addition to this, its intermediate tendon can have variable relationship with the internal jugular vein. It always covers the inferior part of internal jugular vein, but several abnormalities have been reported. The omohyoid is the most frequently absent muscle of the infrahyoid group. It may occur only on one side, or it may be found as a single-bellied muscle; one belly is absent more frequently than both bellies; the inferior belly may be doubled, with the second belly possibly arising from the coracoid process. When the inferior belly is absent, the superior belly arises from the clavicle, and the resulting muscle is named cleidohyoideus.

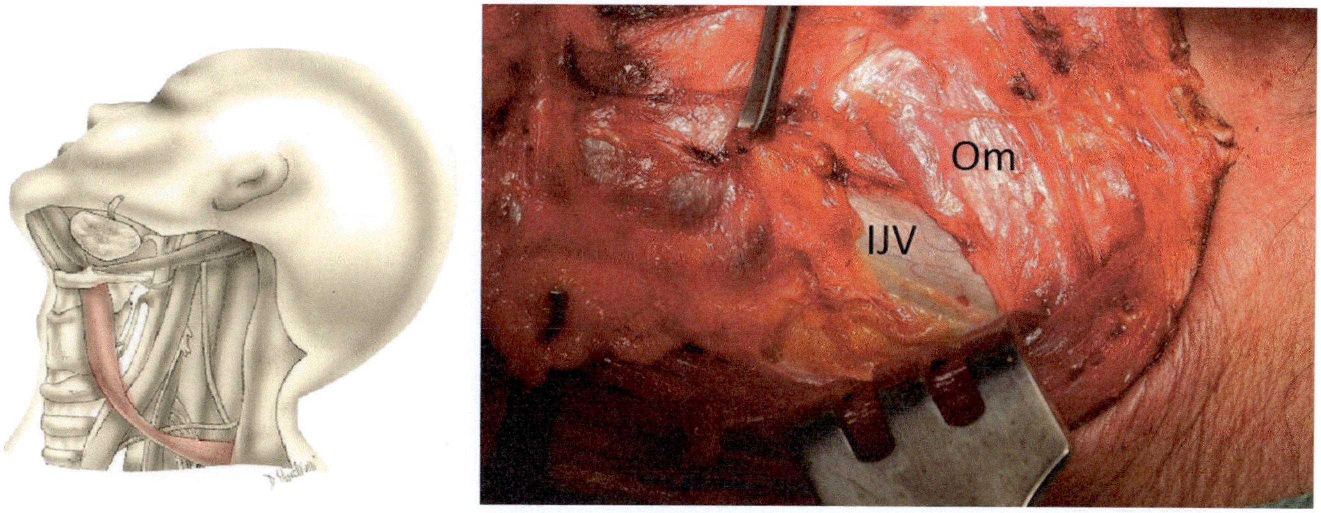

Fig. 3.4 The omohyoid muscle (Om) and its relationship with the internal jugular vein (IJV) (Courtesy of Prof. Daniele Marchioni, ENT Department, Verona, Italy)

3.3 Vascular Supply: Arteries and Veins (Fig. 3.5)

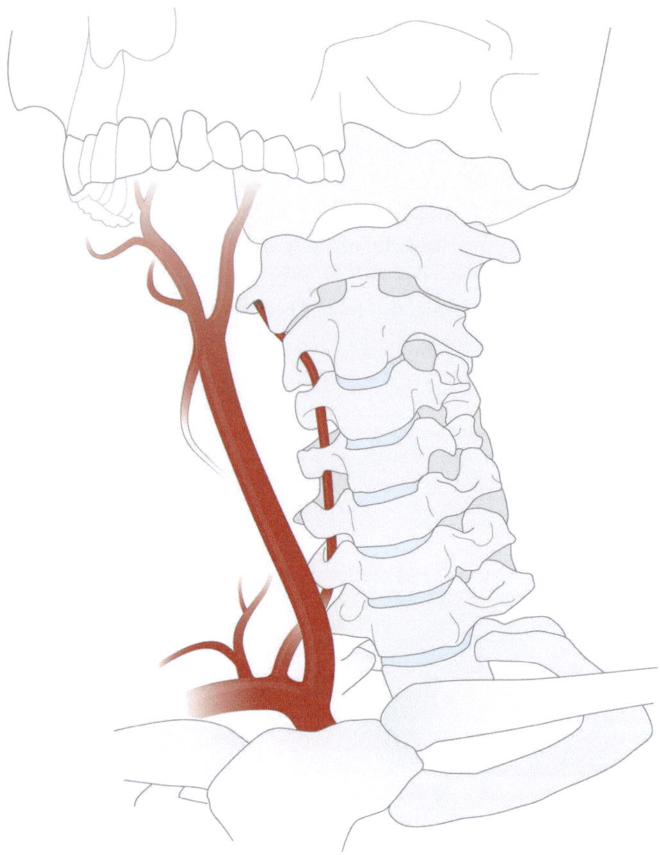

Fig. 3.5 Arterial blood supply of the neck (focus on carotid, subclavian and vertebral arteries)

3.3.1 Arteries

The major arteries of the neck are:

- *Subclavian – its branches involved in cervical spine approach are*:

Artery	Source	Description
Thyrocervical	A branch of 1st part of the subclavian a. along the anterior scalene muscle	Divides into four branches:
		Inferior thyroid
		Suprascapular
		Transverse cervical
		Ascending cervical artery
Costocervical	A branch of the first part of the right subclavian a. and the second part of the left subclavian a.	Divides into two branches:
		Deep cervical
		Supreme intercostal
Vertebral	The first part of subclavian a.	Described in Chapter 1

- *Common carotid – its branches involved in cervical spine approach are*:

Artery	Source	Description
Superior thyroid artery	The first branch of the external carotid artery from the bottom *Infrahyoid*	On its path to the thyroid gland, it passes inferiorly along the inferior constrictor m.
		The superior laryngeal a. arises from the superior thyroid a. and passes through the thyrohyoid membrane to supply the larynx
Lingual	The second branch of external carotid a. *Suprahyoid*	Passes superiorly and medially toward the greater cornu of the hyoid bone in an oblique fashion and makes a loop by passing anteriorly and inferiorly while traveling superficial to the middle constrictor m. While forming a loop, the artery is crossed superficially by the hypoglossal n.
		The lingual a. passes deep to the posterior belly of the digastric and stylohyoid mm. as it travels anteriorly.
		In this region, it gives rise to a hyoid branch that travels on the superior surface of the hyoid bone, supplying the muscles in the area
		It passes deep to the hyoglossus m. and travels anteriorly between the hyoglossus and genioglossus mm. to supply the tongue
Facial	Superiorly to the lingual artery *Suprahyoid*	Immediately passes superiorly deep to the posterior belly of the digastric and stylohyoid mm.
		Passes along the submandibular gland giving rise to the submental a., which supplies the gland
		Passes superiorly over the body of the mandible on the masseter m. with a tortuous pattern to supply the face
Ascending pharyngeal	Posterior portion of the external carotid artery, near the bifurcation of the common carotid a. *Infrahyoid*	The smallest branch of the external carotid artery
		Ascends superiorly between the lateral side of the pharynx and the internal carotid a. It has a series of branches:
		3–4 pharyngeal branches supply the superior and middle constrictor mm.
		The upper branch passes through the gap superior to the superior constrictor m.
		Gives rise to an inferior tympanic branch, which supplies the middle ear cavity, and to a posterior meningeal branch, which supplies the bones of the posterior cranial fossa and the dura mater.
Occipital	Lateral part of the external carotid artery *Suprahyoid*	Branches along the inferior margin of the posterior belly of the digastric and stylohyoid mm.
		The hypoglossal n. turns around the occipital a. from the posterior part of the vessel, traveling anteriorly, and then passes posteriorly along the mastoid process, making a groove on the bone
		Pierces the fascia that connects the attachment of the trapezius with the sternocleidomastoid m.
		Ascends in the connective tissue layer of the scalp, dividing into many branches
		Anastomoses with the posterior auricular and superficial temporal aa.
		The terminal part of the artery is accompanied by the greater occipital n.

3.3.2 Important Anatomical Variation of the Arteries of the Neck

3.3.2.1 Lusory Artery

It is the name of the most common embryologic anatomical anomaly of the aortic arch.

In this particular condition, there is an aberrant right subclavian artery and the brachiocervical trunk is not present. There are, indeed, four single arteries that arise from the aortic arch: (1) right common carotid artery, (2) left common carotid artery, (3) left subclavian artery, and (4) right subclavian artery, that originates in the extreme left position, distal to the other three blood vessels.

In this case the right subclavian artery goes toward the right usually passing behind the esophagus. When the artery eventually compresses the esophagus, the condition is referred to as "dysphagia lusoria" (Fig. 3.6).

An important anatomical condition related to the lusory artery is the right nonrecurrent laryngeal nerve that rises from the cervical vagus nerve but does not have the typical recurrent course.

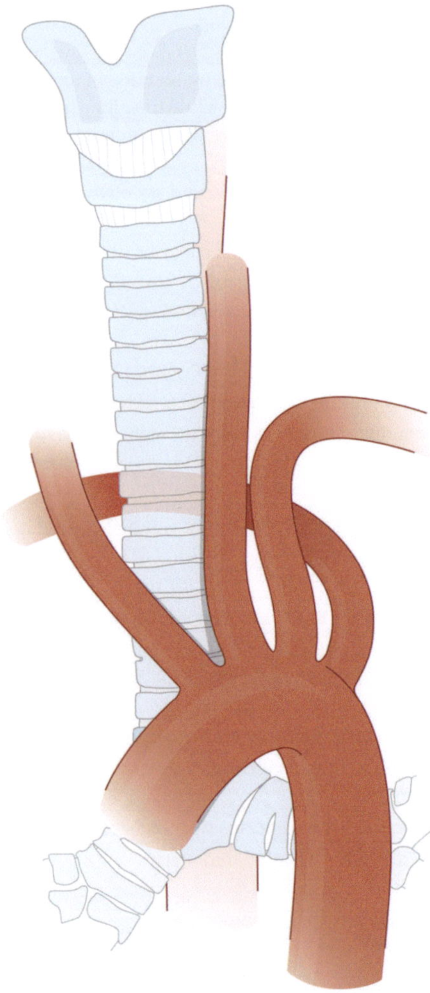

Fig. 3.6 Lusory artery. Right subclavian artery originates in the extreme left position and goes toward the right passing behind the esophagus

3.3.2.2 Veins (Fig. 3.7)

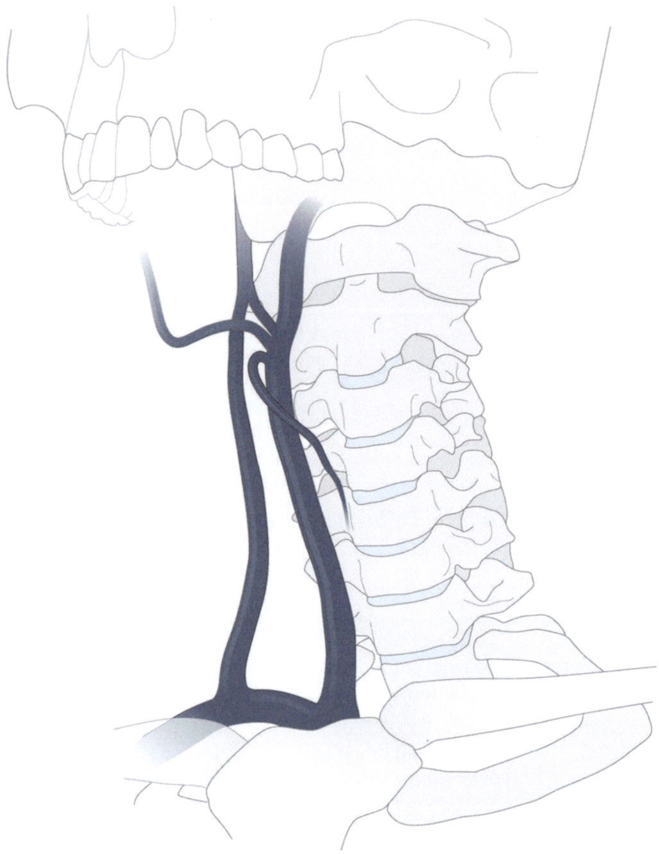

Fig. 3.7 Venous blood supply of the neck (focus on jugular veins)

Vein	Description
Anterior jugular	Arises by the joining of a series of superficial veins in the submental region
	Descends anteriorly to the sternocleidomastoid m. and passes deep to the muscle before draining into the external jugular or the subclavian v.
External jugular	Formed by the confluence of the posterior branch of the retromandibular and posterior auricular vv. in the parotid gland
	Lies deep to the platysma m. but superficial to the sternocleidomastoid m. as it descends vertically
	Passes into the posterior triangle of the neck, where it drains into the subclavian v. immediately lateral to the anterior scalene m.
Internal jugular	Continuous with the sigmoid sinus within the cranial cavity
	Begins at the base of the skull at a dilation called "superior bulb"
	Lies posterior to the internal carotid a. and the glossopharyngeal, vagus, and spinal accessory nn. as it initially descends
	Travels lateral to the internal carotid a. within the carotid sheath with the vagus n. posterior to the vessels
	Unites with the subclavian v. to form the brachiocephalic v. at the root of the neck
	Receives a series of branches
Facial	Passes inferiorly along the side of the nose, receiving the lateral nasal v.
	Continues in a posterior and inferior path across the angle of the mouth to the cheek, receiving the superior and inferior labial vv.
	While passing toward the mandible, the deep facial v. connects the facial v. to the pterygoid plexus
	In the submandibular triangle, the facial v. joins the anterior branch of the retromandibular to form the common facial v.
	Common facial v. drains into the internal jugular v.
Lingual	Passes with the lingual a., deep to the hyoglossus m., and ends in the internal jugular v.
	The vena comitans nervi hypoglossi (or accompanying v. of the hypoglossal n.) begins at the apex of the tongue and either joins the lingual v. or accompanies the hypoglossal n. and enters the common facial v., draining into the internal jugular v.
Thyroid veins	Superior: from thyro-lingual-facial trunk
	Medial: directly from the internal jugular vein
	Inferior: from brachiocephalic vein
Vertebral	Begins as a plexus in the suboccipital triangle and descends through the foramen transversarium of all of the cervical vertebrae before draining into the subclavian or, more commonly, the brachiocephalic v.

3.4 Nerves

3.4.1 Vagus

It is also known as cranial nerve X.

It branches from the medulla oblongata and passes through the jugular foramen together with the glossopharyngeal and spinal accessory nerves. It passes between the internal carotid artery and internal jugular vein.

A series of nerve branches from the vagus nerve as it passes from the base of the skull through the neck: auricular, pharyngeal, superior laryngeal, and recurrent laryngeal.

3.4.2 Superior Laryngeal Nerve

It travels inferiorly, posterior to the internal carotid and on the side of the pharynx, and divides into the:

- *Internal laryngeal nerve*
  – It passes inferiorly to the larynx through the thyrohyoid membrane along with the superior laryngeal vessels to distribute the fibers to the base of the tongue in the epiglottic region and to the mucous membranes of the larynx inferiorly as far as the false vocal folds and SVA fibers to the taste buds in the area (Figs. 3.8 and 3.9).
- *External laryngeal nerve*
  - It travels inferiorly along the inferior constrictor to supply the cricothyroid muscle and the inferior portion of the inferior constrictor (Figs. 3.10 and 3.11).

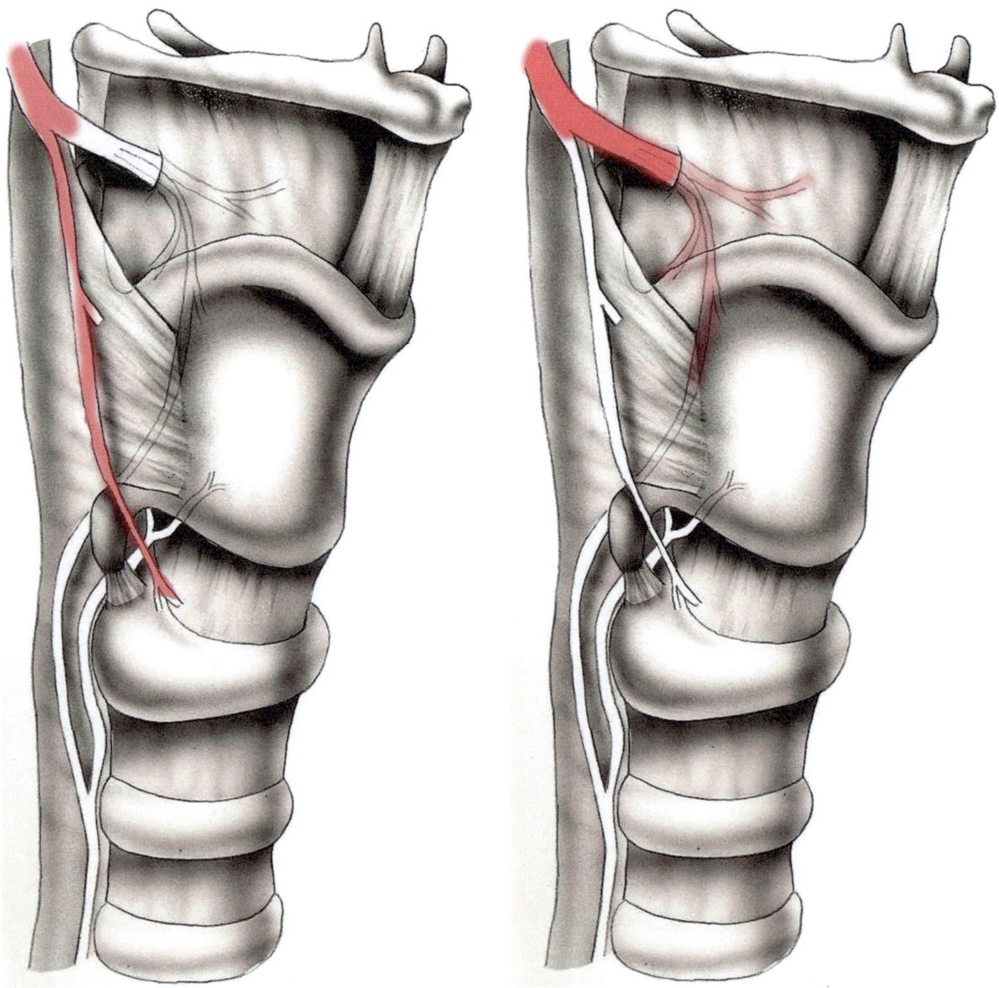

Figs. 3.8 and 3.9 Superior laryngeal nerve: internal branch and its relationship with the laryngeal artery (Courtesy of Prof. Daniele Marchioni, ENT Department, Verona, Italy)

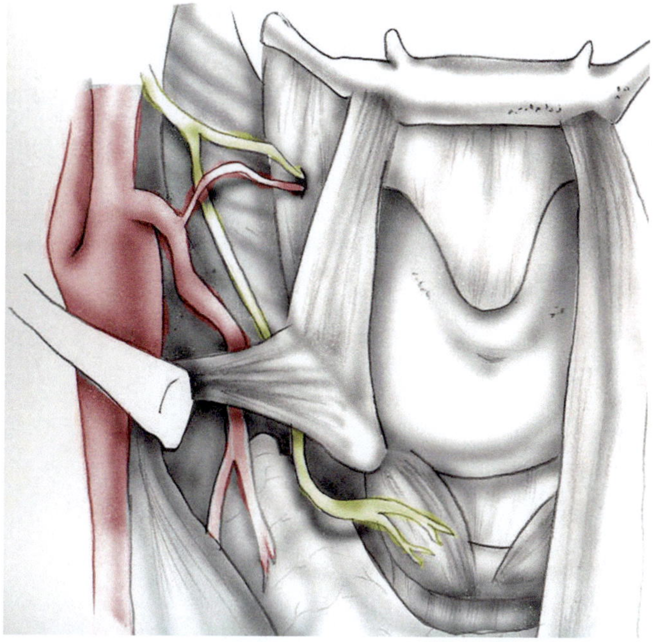

Fig. 3.10 External branch of superior laryngeal nerve (Courtesy of Prof. Daniele Marchioni, ENT Department, Verona, Italy)

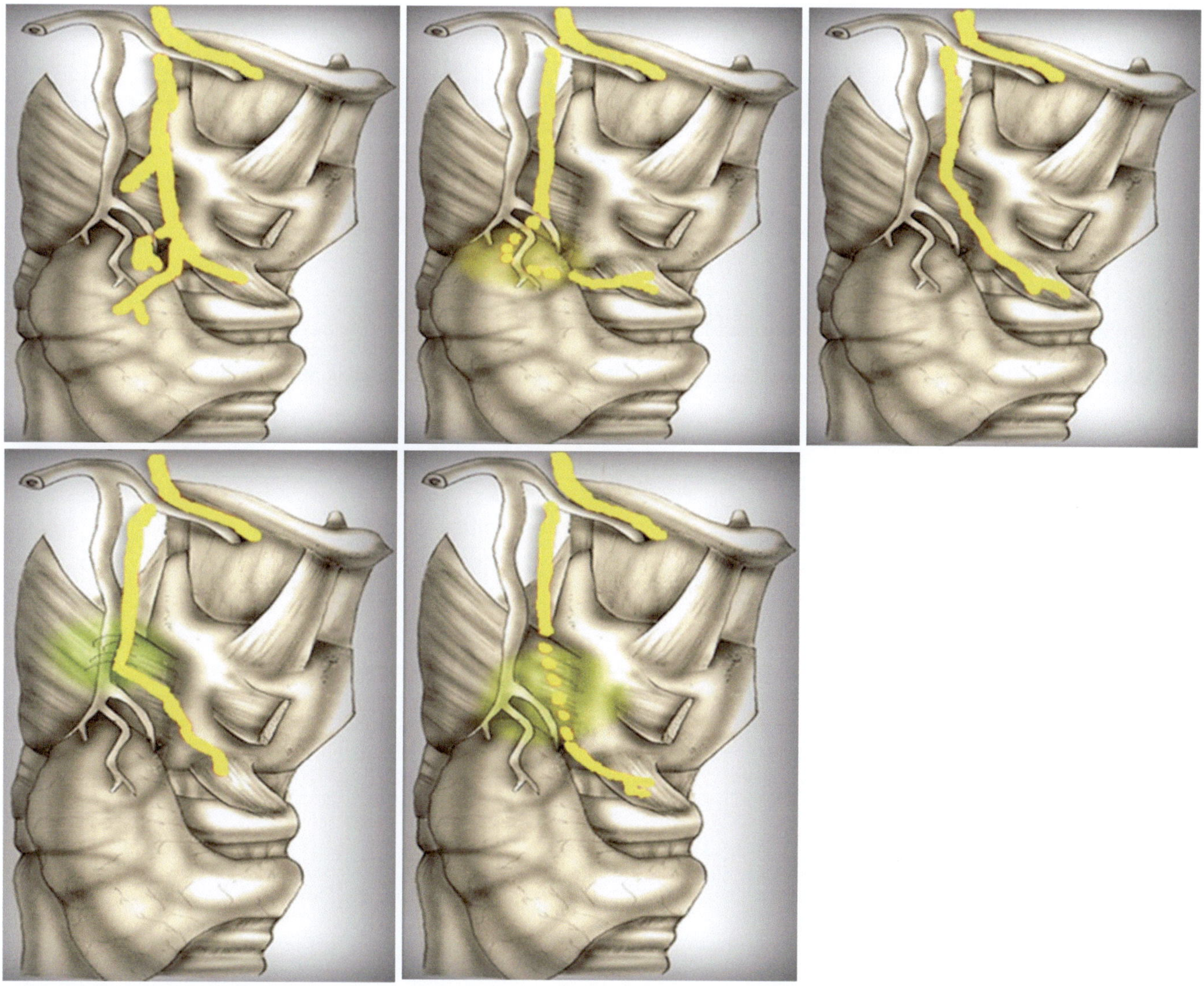

Fig. 3.11 Anatomical variants of external branch of the superior laryngeal nerve (Courtesy of Prof. Daniele Marchioni, ENT Department, Verona, Italy)

3.4.3 Inferior Laryngeal Nerve or Recurrent Laryngeal Nerve

On the left side the recurrent laryngeal nerve arises from the vagus at the level of the aortic arch, and, after having surrounded it, it assumes a vertical course toward the neck.

On the right side, the nerve is detached from the vagus nerve at the point where this crosses the subclavian artery, and then it surrounds the artery and ascends with an oblique course near the esophageal margin. It ascends on the lateral side of the trachea until reaching the pharynx, where it passes deep to the inferior constrictor muscle, innervating the mucous membranes below the false vocal folds and all of the intrinsic muscles of the larynx except from the cricothyroid (Figs. 3.12, 3.13 and 3.14).

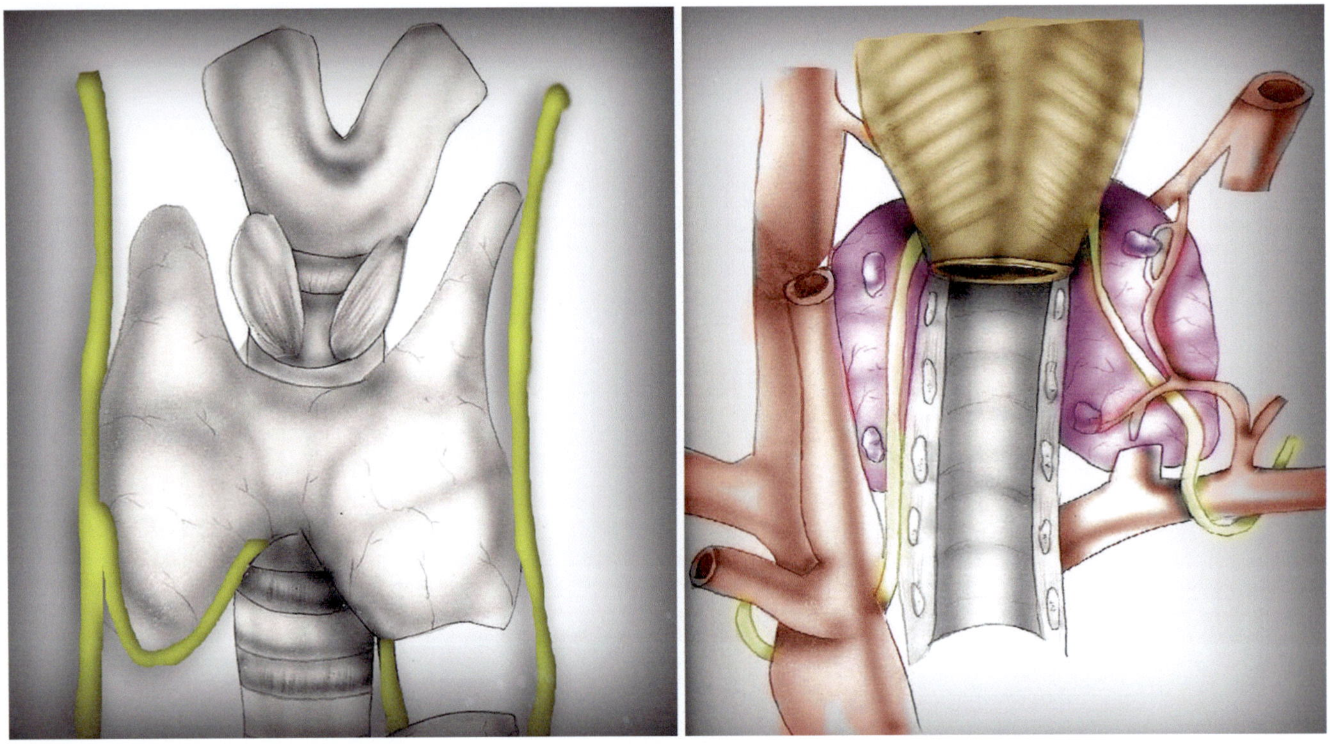

Fig. 3.12 Course of the laryngeal recurrent nerves (Courtesy of Prof. Daniele Marchioni, ENT Department, Verona, Italy)

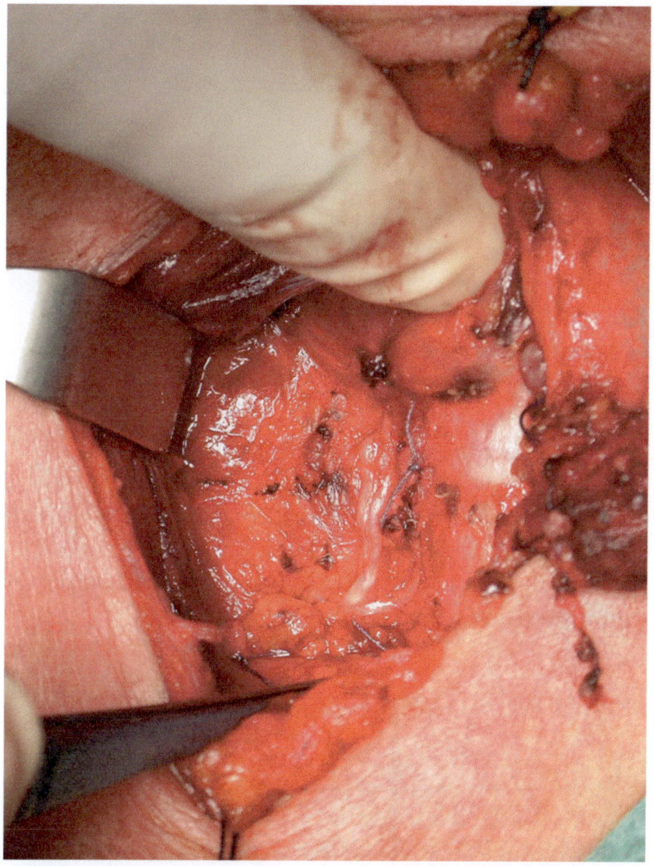

Fig. 3.13 Intraoperative photos of the RLN in the right side of the neck

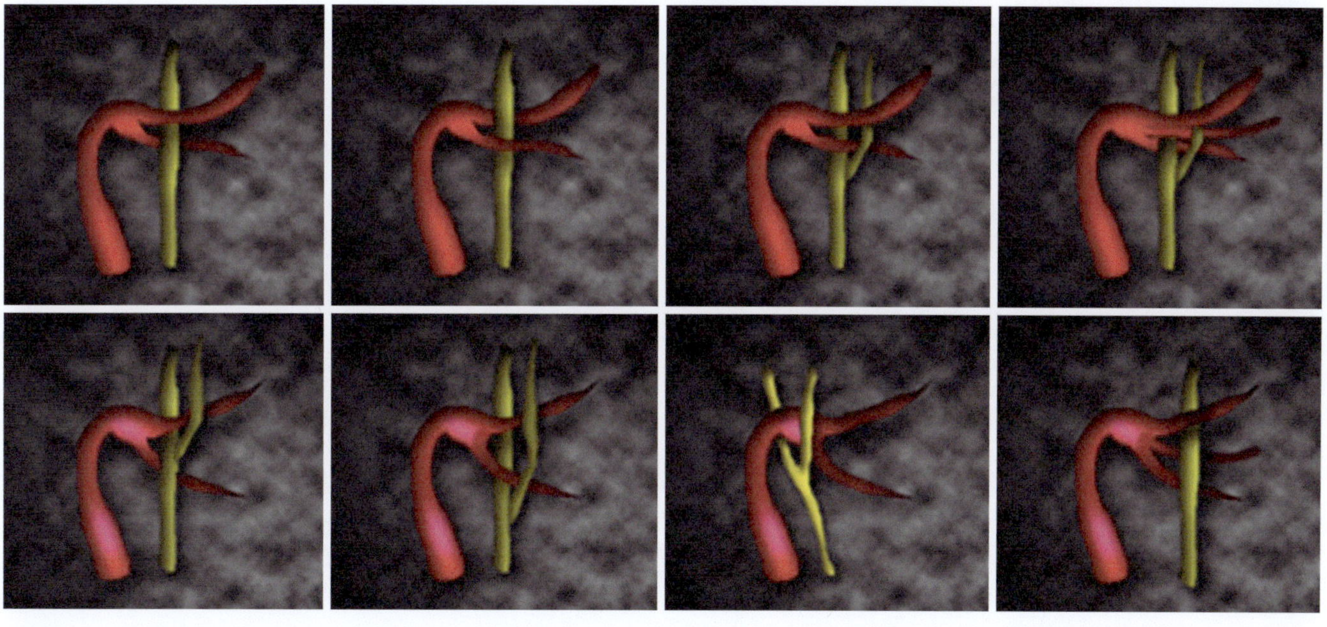

Fig. 3.14 Inferior laryngeal nerve: relationship with the inferior thyroid artery (Courtesy of Prof. Daniele Marchioni, ENT Department, Verona, Italy)

3.4.4 Nonrecurrent Inferior Laryngeal Nerve

This anatomical variation is associated with an anomaly in the embryological development. The fourth right aortic arch and the proximal dorsal aorta are occasionally obliterated, and the subclavian artery can originate from the right side of the aortic arch, being called "lusory" artery.

This anatomical variant occurs between 0.4 and 2.4 % of people on the right side, while it is extremely rare on the left side.

Many articles describe the course patterns of non-RLN:

Abboud and Aouad [1] reported three types of courses for non-RLNs (Fig. 3.15).

The course patterns of non-RLN are:

- *Type Ia* – the nerve has a straight course on the level with the upper thyroid pole.
- *Type Ib* – the most common, the nerve runs transversely at the level of the thyroid isthmus.
- *Type II* – the nerve curves downward, eventually reaching the lower pole of the thyroid gland.

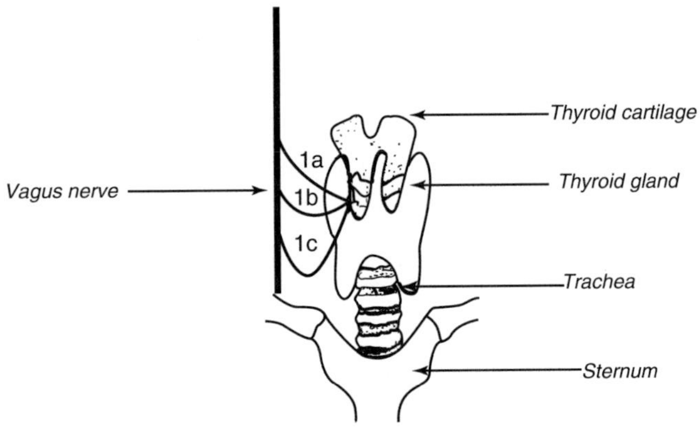

Fig. 3.15 Course patterns of non-RLN (from Abboud and Aouad [1])

In a case report by Yang et al. [2] the non-RLN courses are described in more detail, distinguishing four types (Fig. 3.16):

- In the descending type, the non-RLN descends after originating from the vagus nerve trunk.
- In the vertical type, the non-RLN runs vertically to the cricothyroid joint.
- In the ascending type, the non-RLN runs upward to the cricothyroid joint.
- In the V-shaped type, the non-RLN takes a downward direction and runs upward to the cricothyroid joint.

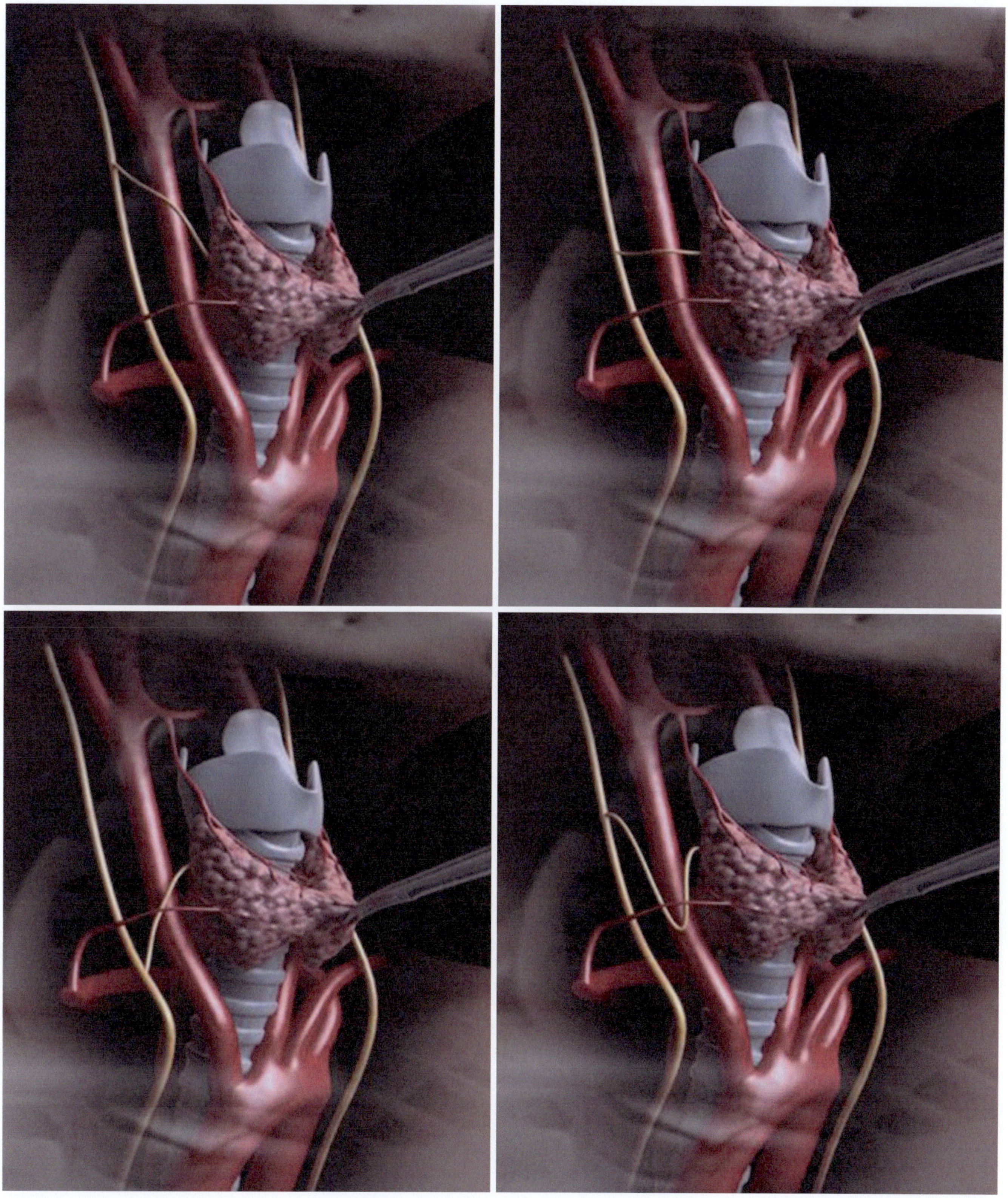

Fig. 3.16 Course patterns of non-RLN (from Yang et al. [2])

3.4.5 **Spinal Accessory Nerve** (Fig. 3.17)

It originates from the union of the vagal accessory nerve (parasympathetic fibers/visceral effector) and spinal accessory nerve (somatic motor); it exits from the posterior foramen lacerum and divides in two branches:

- *Internal or medial branch* participates in the larynx innervation.

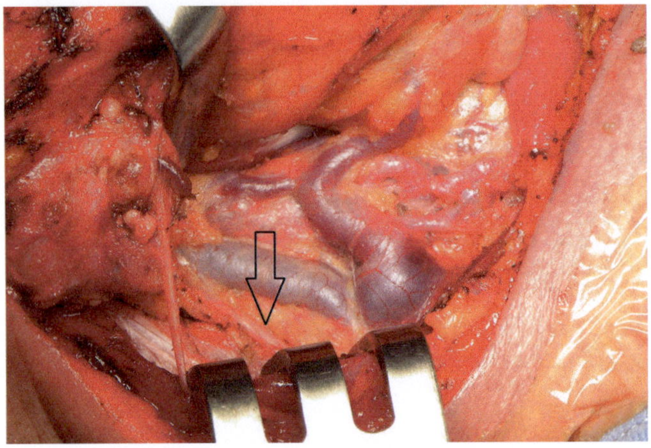

Fig. 3.17 Spinal accessory nerve on the right side (*arrow*)

- *The spinal portion* (*external or lateral branch*): passes anteriorly to the internal jugular vein, enters the sternocleidomastoid muscle (which it innervates), and exits in proximity to the posterior margin of the muscle. Running from the bottom to the top and from front to rear, the peripheral portion of the nerve enters the trapezius, which it innervates. The common trunk of the nerve usually runs above the internal jugular vein.

3.4.6 **Hypoglossum** (Fig. 3.18)

The hypoglossal nerve (HN), also known as cranial nerve XII, is the motor supply of the tongue. The HN can be divided into three main parts: cisternal, intracanalicular, and extracranial. When the HN emerges between the ICA and the internal jugular vein (IJV) on its course toward the tongue, inferior to the tendon of the posterior belly of the digastric muscle, it becomes more superficial. As the HN emerges from the hypoglossal canal, it is supplied by a vessel arising from the posterior neuromeningeal division of the ascending pharyngeal artery, the first branch of the posterior group of the ECA. When the HN exits through the external orifice of the hypoglossal canal, it descends medially to the IJV and posteromedially to the ICA. The descending part starts with the external exit of the hypoglossal canal, at which level the HN crosses medially to the IJV and posteromedially to the ICA. Then, the HN passes through laterally and descends beyond the posterior aspect of the vagal nerve and crosses the external carotid and lingual arteries below the tendon of the digastricus. It passes beneath the tendon of the digastricus, the stylohyoideus, and the mylohyoideus, lying between the last-named muscle and the hyoglossus, and communicates with the lingual nerve in the anterior border of the hyoglossus; it then continues forward in the fibers of the genioglossus as far as the tip of the tongue, distributing branches to its muscular substance. The descending ramus (*ramus descendens*; *descendens hypoglossi*), long and slender, quits the hypoglossal where it turns around the occipital artery and descends in front of or in the sheath of the carotid vessels; it gives a branch to the superior belly of the omohyoideus and then joins the communicantes cervicales from the second and third cervical nerves, just below the middle of the neck, to form a loop, the so-called ansa hypoglossi. From the convexity of this loop, some branches pass to supply the sternohyoideus, the sternothyroideus, and the inferior belly of the omohyoideus.

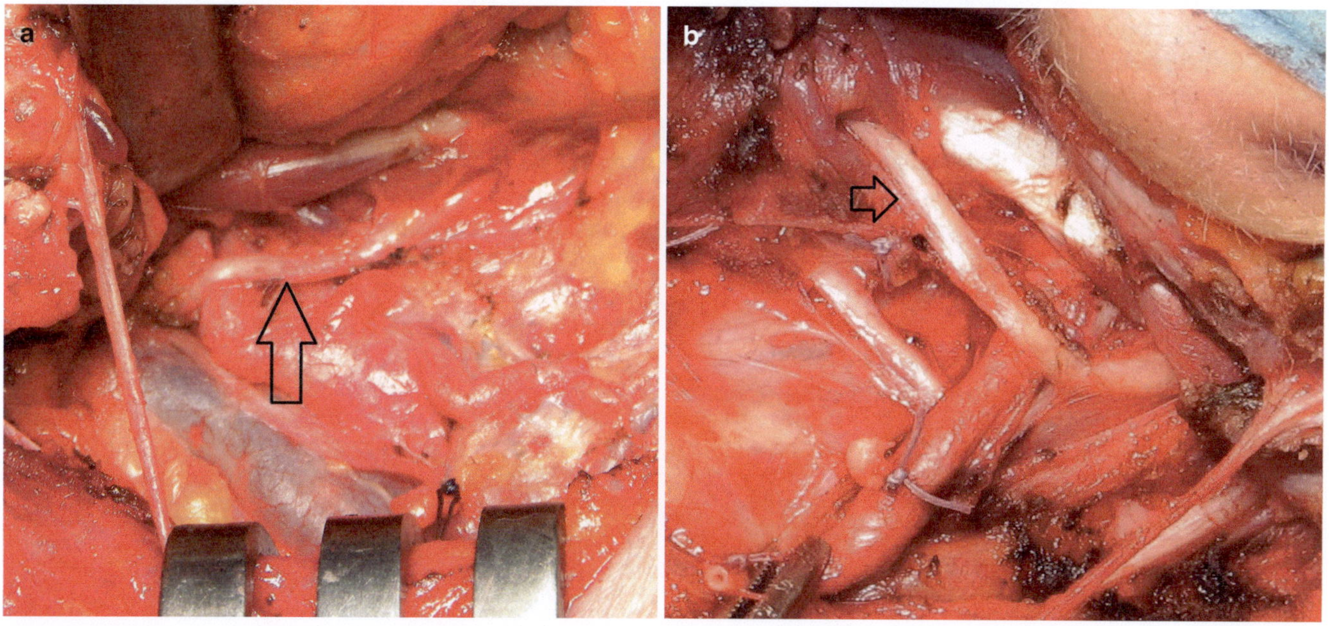

Fig. 3.18 HN on the *right side* (**a**) and on the *left side* (**b**)

3.4.7 Glossopharyngeal Nerve

The glossopharyngeal nerve exits the skull from the jugular foramen in close proximity to the vagus nerve and courses between the internal carotid artery and the internal jugular vein before passing between the stylopharyngeus and styloglossus muscles to enter the base of the tongue.

3.4.8 Cervical Plexus

It is formed by the anterior branches of the first four cervical spinal roots. It is located in front of the middle scalene muscle and the elevator scapulae muscles. All the branches are divided in an ascending and a descending branch that are connected to each other in loops.

- The superficial plexus branches are for the skin, while the deep ones for the muscles:
  - Superior ascending: C2 small occipital
  - C2–3 great auricular
  - C2–3 cutaneous transverse of the neck
  - Superior descending: C3–4 supra clavicular nerves
- Medial deep:
  - Anastomotic branches with con hypoglossal, vagus, and sympathetic trunk
  - Muscolar branches (rectus lateral of the head, rectus anterior of the head, long of the head, long of the neck, inferior root of the cervical loop, phrenic nerve)
- Lateral deep:
  - Anastomotic branches (accessory nerve)
  - Muscolar branches (SCM muscle, trapezius, elevator scapulae, middle scalene)

3.4.9 The Cervical Phrenic Nerve

It is the motor nerve of the diaphragm, and it also carries significant afferent fibers from the diaphragm, pericardium, pleura, and peritoneum. It originates in front of the lateral margin of the anterior scalene muscle most from C4, but also from C3 and C5. In its cervical course, it goes vertically downward in the thoracic region, covered by the prevertebral fascia, the SCM muscle, the inferior belly of the omohyoid muscle, the jugular vein, the transverse cervical artery, the suprascapular artery, and on the left, the thoracic duct. It is located anteriorly to the subclavian artery and posteriorly to the subclavian vein and enters the thorax crossing medially in front of the internal thoracic artery. The phrenic nerve ordinarily enters the thoracic cavity posteriorly to the subclavian vein, less commonly anteriorly to it, and rarely may penetrate it.

The right phrenic nerve is shorter and is separated from the subclavian artery by the anterior scalene muscle; the left phrenic nerve leaves the medial margin of the anterior scalene muscle passing over the subclavian artery and behind the thoracic duct. Prakash [4] describes a variation of the phrenic nerve that could be very close to the subclavian vein. In this dissection study, the authors describe an early communicating branch of phrenic nerve to the C5 root of the brachial plexus and just before entering the thorax, it is located in front of the subclavian vein instead of posterior or between vein and artery. Bigeleisen et al. [5] describe an accessory phrenic nerve in up to 75 % cadavers examined, it follows the phrenic nerve and joins it distally (Fig. 3.19).

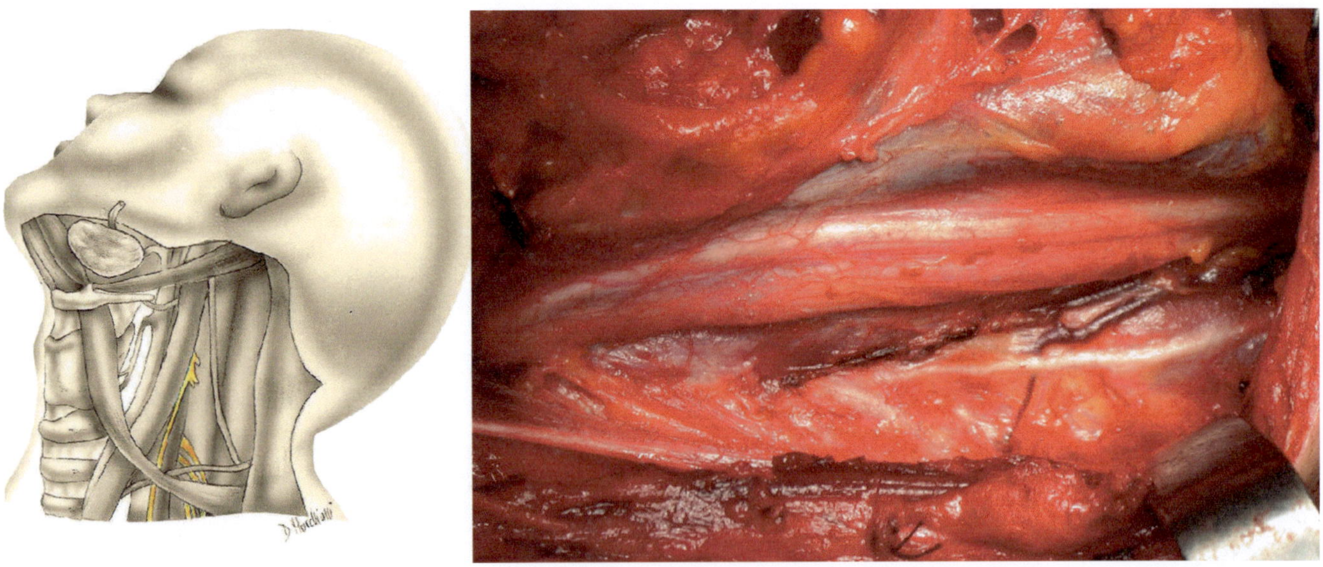

Fig. 3.19 Detail of the Cervical Phrenic Nerve (Courtesy of Prof. Daniele Marchioni, ENT Department, Verona, Italy)

3.4.10 Cervical Sympathetic Trunk

The sympathetic trunk is sometimes damaged during the anterior and anterolateral approach to the cervical spine, resulting in Horner's syndrome (ptosis, ipsilateral miosis, anhidrosis, and enophthalmos). The cervical sympathetic chain is located posteromedial to carotid sheath and runs over the longus muscles. It extends longitudinally from the longus capitis to the longus colli over the muscles and under the prevertebral fascia. From up to down, the lateral cervical muscles diverge laterally, whereas the cervical sympathetic trunk converges medially at C6. The superior ganglion is usually located on the level with C4. The location of the intermediate (middle) ganglion of the cervical sympathetic trunk has shown some variations.

3.5 Fascias of the Neck

Fascias of the neck have been variously described in the literature. Description terms are different.

Guidera et al. [3] made a clear review of the terminology of fascias description.

3.5.1 Superficial Cervical Fascia

The SCF of the neck contains the platysma and the superficial lymph nodes and is loosely arranged to facilitate neck movement. Cranial to the mandible, the SCF continues as the fascia that "invests" the muscles of facial expression and the occipitofrontalis in the scalp; it is referred to by different names as it progresses cranially. Interest in this layer was renewed with the arising popularity of facelift surgery and the definition of the superficial musculoaponeurotic system (SMAS). Starting from the vertex, the galea aponeurotica, occipitofrontalis, and orbicularis oculi muscles are in continuity with the temporoparietal fascia and, inferior to this, the SMAS over the zygomatic arch (although not all authorities agree on the latter). This layer, which incorporates the muscles of facial expression, is then continuous with the SCF and platysma in the neck.

3.5.2 Deep Cervical Fascia

It is divided into:

- *Superficial layer of deep cervical fascia (SLDCF)*
- *Middle layer of deep cervical fascia (MLDCF)*
- *Deep layer of deep cervical fascia (DLDCF)*

3.5.3 Superficial Layer of Deep Cervical Fascia

The superficial layer of deep cervical fascia (SLDCF) is usually described as "completely encircling the neck," although it has been suggested that it is incomplete between the sternocleidomastoid and trapezius muscles. A simplified "rule of twos" describes the SLDCF as enclosing two glands (submandibular and parotid), two muscles (sternocleidomastoid and trapezius), and two "spaces" (suprasternal space and the "subvaginal" space of the posterior triangle).

3.5.4 Middle Layer of Deep Cervical Fascia

This is described as having muscular and visceral divisions or simply a visceral part.

3.5.4.1 Muscular Layer

This term is usually used in conjunction with "visceral fascia" to describe that portion of the MLDCF that sheathes the strap muscles. Some authors subdivide it into sternohyoid/omohyoid and sternothyroid/thyrohyoid components, whereas others refer to it generally as the "strap fascia," or even part of the SLDCF. This layer runs between the bony attachments of these muscles.

3.5.4.2 Visceral Layer

This term has been used to describe all components of the MLDCF or only that part surrounding the larynx, pharynx, trachea, esophagus, and thyroid. The latter is generally agreed to blend inferiorly with the fibrous pericardium. However, there is disagreement about whether it extends superiorly only to the level of the hyoid bone or if a posterior continuation reaches the skull base.

3.5.4.3 Buccopharyngeal Fascia

The prefix "bucco" refers to the superior extension of this layer that is stated to continue over the buccinator muscle.

3.5.5 Deep Layer of Deep Cervical Fascia

This is consistently described as encompassing the vertebral column and paravertebral muscles, attaching to the transverse and spinous processes of the cervical vertebrae. As it passes laterally over the scalene muscles, it forms the floor of the posterior triangle. The term "prevertebral" is used to describe either the complete circumferential layer of fascia or just that part covering the prevertebral muscles anteriorly between the transverse processes. Laterally, it is described as being continuous with the axillary sheath and the suprapleural membrane (Sibson's fascia). Caudally, it is stated to extend to the coccyx or "fades away" in the thorax. The alar fascia is generally stated to be a division of the deep layer of deep cervical fascia (DLDCF) spanning between the transverse processes of the cervical vertebrae anterior to the prevertebral fascia and fusing laterally with the carotid sheath.

References

1. Abboud B, Aouad R (2004) Non-recurrent inferior laryngeal nerve in thyroid surgery: report of three cases and review of the literature. J Laryngol Otol 118(2):139–142
2. Hong KH, Park HT, Yang YS (2014) Characteristic travelling patterns of non-recurrent laryngeal nerves. J Laryngol Otol 128(6):534–539
3. Guidera AK et al (2014) Head and neck fascia and compartments: no space for spaces. Head Neck 36(7):1058–1068
4. Prakash, Prabhu LU et al (2007) A variation of the phrenic nerve: Case report and Review. Singapore Med. J. 48(12):1156–1157
5. Bigeleisen PE (2003) Anatomical variations of the phrenic nerve and its clinical implication for supraclavicular block. Br J Awaesth 91(6):916–917

Part II
Planning

4 Interventional Radiology: Presurgical Selective Angiographic Embolization (SAE) in Hypervascular Cervical Spine Tumours

Luigi Simonetti, Carlotta Barbara, Salvatore Isceri, and Elena Mengozzi

4.1 Introduction: Anatomy

4.1.1 Cervical Vascularization

The anatomical features of the vascular region of interest cannot be disregarded when performing interventional therapeutic procedures. Deep and detailed understanding of anatomy is an unavoidable necessity (Table 4.1).

The spinal cord and dura mater have a united and functionally indissociable vascularization. Characteristic of the intramedullary vessels of the spinal cord is to maintain their embryonal metameric disposition, which occasionally disappears in the arterial and venous extramedullary vessels.

The vascular structures feeding the rachis and spinal cord was first described by Adamkiewicz, Corbin, Crock and Lazorthes. There are however significant differences between vascularization at cervical level and vascularization in other portions of the rachis, where the metameric disposition is more evident. In the cervical region, the ascending branches of the right and left subclavian arteries feed the vertebromedullary region, from where the inferior thyroid and ascending cervical, vertebral and deep cervical arteries take their origin.

Below are listed the three cervical axes:

- Anterior or prevertebral, arising from the inferior thyroid and ascending cervical arteries (vertebral body and vertebral arch)
- Median or latero-vertebral, arising from the vertebral artery (anterior and posterior spinal arteries)
- Posterior, arising from the deep cervical artery (anterior and posterior spinal arteries)

The three axes anastomose with each other through horizontal intra-rachideal or paravertebral branches that provide blood supply to the vertebrae. At cervical level, the spinal axis receives most of the arterial vessels.

At high cervical level, up to C3, there are no anterior radiculomedullary arteries, and two anastomosing spinal arteries, arising from the vertebral ones, feed the anterior spinal axis.

At medium-low cervical level, the anterior axis is supplied blood by 2–4 anterior radiculomedullary arteries, randomly originating either on the right or on the left side from the vertebral or deep cervical arteries.

Two main radiculomedullary arteries are often present, one originating from the vertebral artery at the level of the foramen of conjugation (C5-C6 or C4-C5) and the other arising from the deep cervical artery and entering the spinal canal through the C7-T1 foramen. The anterior radiculomedullary arteries enter the spinal canal following the nervous root course.

Blood supply is provided to the posterior spinal axes by 4–6 slender posterior radiculomedullary arteries, arising from the vertebral artery, which runs between C3-C6. These small arteries penetrate into the spinal canal through the conjugation foramen, following the radicular course.

4.1.2 Anatomy of the Anterior Spinal Artery (ASA) at Cervical Level

The anterior spinal artery (ASA) course completely follows the median line of the spinal cord (Fig. 4.1).

It arises from two branches of the vertebral arteries in the superior cervical region, a little behind the origin of the basilar trunk. Both branches descend anteriorly to the bulb and unite forming the descending, median and anterior spinal trunks or the anterior spinal artery (ASA). The latter runs along the median anterior fissure, usually ending at the level of the fifth pair of the cervical nerves. At this level, 6–10 anterior radicular arteries, each of them feeding one descending and one ascending branch, form the anterior spinal trunk.

L. Simonetti (✉) • C. Barbara • S. Isceri • E. Mengozzi
Emergency Interventional Unit, Ospedale Maggiore – Bologna, Bologna, Italy
e-mail: lg.simonetti@gmail.com

S. Boriani et al. (eds.), *Atlas of Craniocervical Junction and Cervical Spine Surgery*, DOI 10.1007/978-3-319-42737-9_4

4.1.3 Anatomy of the Cervical Posterior Spinal Arteries

There are two cervical posterior spinal arteries, one located on the left-hand and the other on the right-hand side. They originate from the vertebral arteries and run posteriorly to the bulb and spinal cord, reaching the lateral aspect of the median posterior sulcus. Along their descending course, a variable number of posterior radicular arteries (10–23) reinforce them. Each spinal segment of the cervicodorsal region of the spinal cord averagely receives one or two tributaries.

4.1.4 Radicular Arteries or Lateral Spinal Arteries at Cervical Level

Even though of variable origin, most of the radicular arteries arise from the arteries running along the spinal roots to reach the vertebral canal. With segmental course, in the neck, they originate from the vertebral and cervical ascending arteries.

Table 4.1 Cervical vascularization

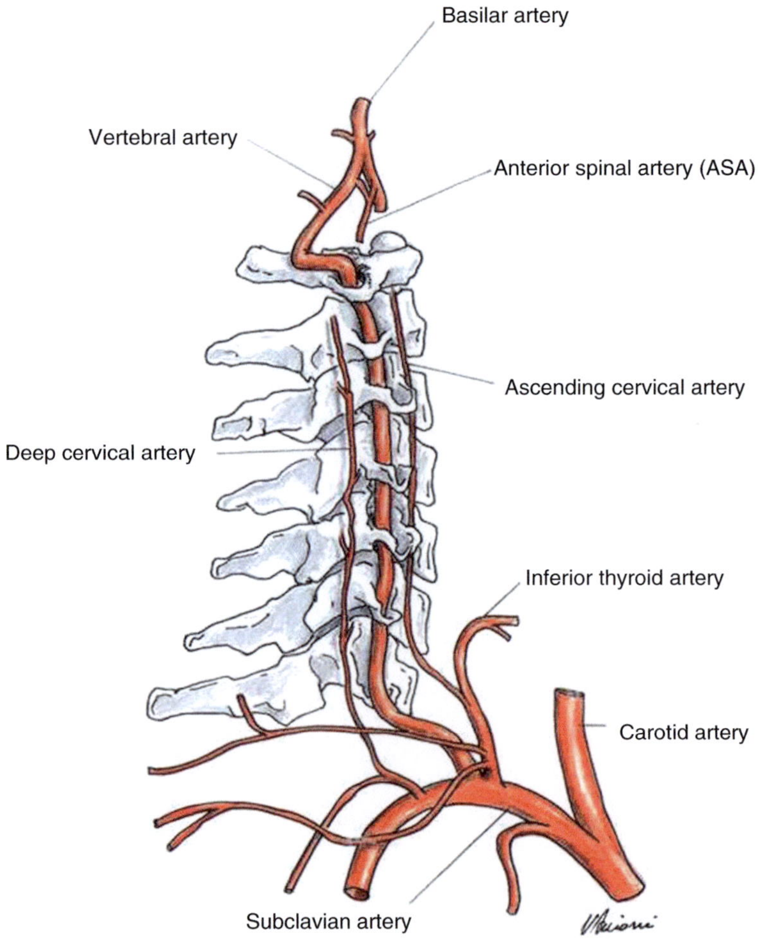

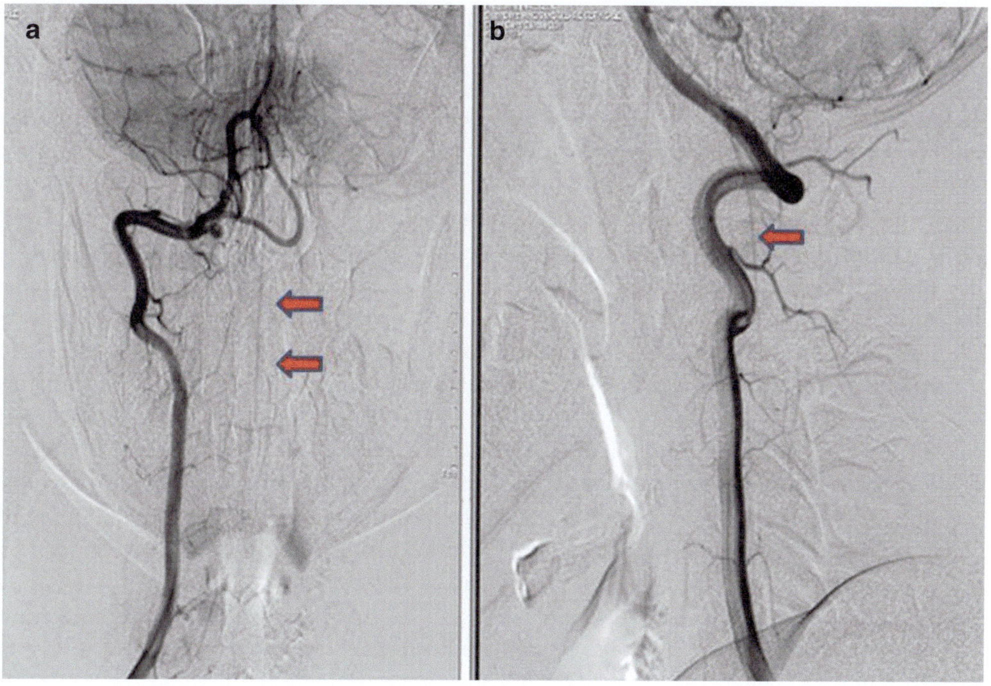

Fig. 4.1 Anterior spinal artery (ASA) (A) AP and (B) LL views of right vertebral artery angiography. ASA follows the median line of the spinal cord (arrows), arising from branches of the vertebral artery

4.2 Diagnostic Angiography of Spinal Cord: Technique

Diagnostic angiography prior to SAE is necessary for the following purposes:

(a) To accurately evaluate the pathological circle
(b) To identify possible presence of arteries ending in the spinal cord, which are not eligible for SAE (AKA, ASA, ASPL)
(c) For a safe embolization, to obtain stable selective catheterization, prior to selective micro-catheterization

We routinely employ 4 F introducers, with the exception of presurgical embolization of the vertebral arteries, where 6 F catheters are necessary. All diagnostic angiographic procedures of the vertebro-medullary tracts are performed using 4 F catheters (Glidecath Terumo®, Tokyo, Japan o Tempo Aqua Codman®, J&J, MA, USA).

Catheters with "curve" distal extremities are preferably used, because they guarantee a more stable catheterization, independently of vascular site and anatomical features of the region of interest. We use a curve catheter type "vertebral" for the study of the cervical district (vertebral, thyrocervical and costocervical arteries).

In case of cervical lesions, the vertebral arteries are examined at angiography on the two orthogonal planes. To study the vertebral arteries is necessary to identify possible feeders of the lesion, as well as presence of medullary links or anastomoses with regional arteries (ascending and deep cervical arteries). The study of thyrocervical and costocervical trunks usually necessitates use of the antero-posterior projections alone. The latero-lateral projections are fundamental for lesions of the inferior cervical tract (C5-C7), to distinguish the pathologic circle of origin of the thyrocervical trunk from the normal circle of the thyroid. In case of presurgical embolization of a vertebral artery, a bilateral examination of the internal carotid arteries is necessary, to identify possible presence of carotid-vertebral compensatory vessels at the level of the vertebral artery to be treated.

To summarize, diagnostic angiography must include the study of the following districts:

- Vertebral arteries
- Thyrocervical trunks
- Costocervical trunks
- Internal carotid artery (only in case of presurgical vertebral artery embolization)
- External carotid, bilaterally (lesions affecting C1-C2)

4.2.1 Technical Considerations on Super-Selective Catheterization

Catheterization is optimal if performed as super-selectively as possible within the arteries feeding the lesion of interest. Super-selective catheterization is carried out with 4 F catheters and coaxial catheterization technique (micro-guide, micro-catheter, carrier catheter). In our experience, we routinely use micro-catheters (Renegade®18 HI-FLO™) and micro-guides (Transcend EX .014; Stryker, USA). Both devices guarantee accurate navigation through vessels, albeit mobile and tortuous. Remodelling of micro-guide tip is often necessary to adapt it to the variable course of the arteries supplying blood to the lesion. The micro-catheters employed in our experience had a straight tip, and no remodelling was necessary. Road mapping provided by the carrier catheter is fundamental. Diagnostic angiography with 4 F catheters is necessary prior to injection of embolic agent and is performed through the micro-catheter. The injection is carried out at high pressure and with small syringes (3 ml). The preliminary angiography provides important morphological and haemodynamic information, since it allows:

- Visualization of the pathological circle
- Exclusion with reasonable certainty of presence of vessels ending in the spinal cord

Another advantage offered by injecting at high pressure through the micro-catheter is the visualization of possible metameric anastomoses, which in turn may help diffuse embolic agent to medullary feeders originating from the adjacent metameric levels.

4.2.2 Embolic Agents and Techniques

Embolic agents now available for SAE of tumours of the spine are listed below:

- Particles
- Acrylic glue (Glubran 2)
- Coils

4.2.2.1 Micro-particles

Particles or micro-particles are the most widely used material for embolization. They are of two different types:

- Stable
- Absorbable

Stable particles are non-absorbable tris-acryl gelatin microspheres, with calibre ranging from 50 to 1000 μ. In our experience, we employed Embosphere® Biosphere Medical, Roissy France, and Embozene® CeloNova BioSciences, San Antonio, Texas, USA.

Due to their flexibility, stable particles can be used for vessels with calibre inferior to their diameter. Stable particles do not significantly aggregate within the proximal tract of the main vessel. The particles mainly used in our experience ranged in diameter from 250 to 700 μ.

Stable particles are chosen for the following reasons:

- Particles with calibre inferior to 250 μ (from 50 to 200 μ, i.e. inferior or comparable to the diameter of vessels ending in the spinal cord) are not easily controlled during injection and can cause incidental embolization of very small vessels, not seen at angiographic examination.
- Particles with calibre exceeding 700 μ usually aggregate precociously and cause embolization at a quite proximal level, upstream of the target circle. We used these particles in a few selected cases where also the most proximal portion of the metameric arteries had to be closed.

Absorbable particles are polyvinyl alcohol (PVA) microspheres with irregular surface, ranging in size from 45 to 1180 μ.

The estimated absorption time is about 72 h. Recanalization of the occluded artery is brought about by proteolytic enzymes acting, after embolization, at variables times, depending on the size of the injected particles. The absorbable particles are preferably employed for visceral or skeletal posttraumatic embolization, where vessel recanalization is needed after haemostasis.

In our SAE experience, we employed stable and absorbable particles in the treatment of particularly highly vascularized lesions or when more proximal embolization of the pathologic artery was needed. In these cases, calibres ranging from 500 to 900 μ were preferably used.

PVA absorbable particles are similar to the stable ones as to their intralesional penetration capability. Without any doubt, their cost-effectiveness represents an advantage. There are, however, some limits, which are listed below:

- Precocious aggregation within the catheter, which makes their injection through the micro-catheter difficult. Injection is possible after dilution with contrast medium and has to be rapid and immediately followed by washout with physiological solution.
- Penetration capability of PVA particles is so high, as to increase the risk that a highly vascularized lesion become avascular, i.e. more fragile during surgery.

Injection of Micro-particles: Technique

After advancing the micro-catheter to the right position, an injection of micro-particles is performed. The micro-particles are preliminarily diluted with the proper quantity of iodate contrast medium (see producer instructions). Very low pressure is exerted during injection, in order to avoid possible reflux. It is suggested to follow the systolic frequency. Each injection of particles is followed by an injection of physiological solution to wash the catheter from the remaining particles. A subsequent injection of contrast medium is performed for control. This sequence of injections (particles, physiological solution and contrast medium) is repeated until complete embolization of the pathologic circle is obtained.

Super-selective micro-catheterization is not always possible. In most cases, this is due to the presence of a marked ostial stenosis of the metameric artery or severe tortuosity of its proximal tract. Both conditions do not allow advancement of either the micro-guide or the micro-catheter. In these cases, in the presence of a sufficient pathologic circle feeding the metameric one, the particles are injected through the carrier catheter. With the injection of embolic particles through the carrier catheter, reflux of micro-particles into the aorta may occur. For this reason, embolization through the carrier catheter is preferably performed only in the following cases:

- The position of the carrier catheter tip is sufficiently stable.
- No doubt subsists as to the presence of arteries ending to the spinal cord and originating from the metameric artery.

Under these conditions, the injection must be performed with the flow control technique.

Flow Control Technique

The ostium of the feeding artery to the pathologic circle is occluded pushing forward the carrier catheter tip. This allows for temporary obstruction of the anterograde flow in the artery and possible flow inversion within the distal arterial branches through the anastomoses with the adjacent metameric arteries. As a result, the embolic particles are pushed towards the lesion. At this point, through the continuous flow of physiological solution in the carrier catheter, the stable particles are injected via micro-catheter. It is sufficient that the micro-catheter tip occupies the metameric ostium. Owing to the concomitant absence of anterograde and inverted flow in the distal portion of the afferent artery, the particles will preferably spread in the main circle, i.e. the pathologic circle.

Haemodynamically speaking, the main flow of an artery feeding the pathologic circle is attracted by the lesion that "steals" most of the flow of that artery. Consequently, the injection of micro-particles via carrier catheter can result as reasonably safe, provided that the injection be performed with low pressure and the solution properly diluted. For the same mechanism, the injection via carrier catheter is to be considered as reasonably safe also in those cases not eligible for selective micro-catheterization.

To visualize possible presence of reflux, high image quality (obtained with mA increase) and maximum magnification are mandatory.

4.2.2.2 Acrylic Glue (Glubran 2®)

Glubran2® (GEM, Viareggio, Italy) is an acrylic glue introduced in Europe in 2002 as substitute of Histoacryl®. It has been initially employed in the neuroradiological settings for interventional treatments of dural arteriovenous fistulas (DAVF) and arteriovenous malformations (MAV). Previous studies report that Glubran2® penetrates into the pathological vessels causing occlusion and severe inflammatory responses of the vessel wall, with subsequent necrosis and endothelial damage, to a lesser degree with respect to Histoacryl®. The intravascular homogeneously adherent property of Glubran2® provides a good rendering of the angioarchitecture of the lesion as obtained at angiography with super-selective catheterization. Experimental studies have demonstrated that Glubran2® polymerization requires minor thermic degrees (exothermic reaction of about 45°) with respect to Histoacryl® and that its polymerization times are longer, owing to modification of Glubran2® through addition of methacryloxy-sulpholane monomer. In our experience, Glubran2® was employed in selected cases of lesions with "comb" distribution, such as vertebral bone angiomas or primitive or secondary neoplastic lesions, with similar vascular features. Care must be taken to prevent the catheter tip from being glued to the injection site. Therefore, the use of this embolic agent implies micro-catheter change after each injection. This is the reason why it is preferably employed in those cases where a low number of injections are needed.

Glubran2® Injection Technique

Prior to each Glubran2® injection, the micro-catheter must be accurately washed with glucose solution to remove blood and saline solution residuals. This is important, to avoid immediate glue polymerization within the catheter. Next, the glue is added with Lipiodol (Guerbet, Roissy, France), an oily, viscous mixture, composed of iodine combined with ethyl esters of fatty acids, through which the injectable solution becomes radiopaque. Different concentrations of Glubran2® and Lipiodol are used (1:1, 1:2; 1:3), depending on the desired fluidity, which, in turn, depends on the characteristics of the target region. Penetration into the lesion, in fact, depends on flow and dilution of Glubran2®. In our experience, the ratio was Glubran2/Lipiodol 1:2 and sometimes also 1:3. These ratios proved useful to facilitate intravascular penetration and to reach more distal vessels. The glue is uninterruptedly injected, while its penetration and deposition are monitored via subtraction fluoroscopy. When glue reflux appears into the feeding vessel, the injection is suspended and the employed catheter immediately withdrawn. Obstructing the metameric artery flow with the guide catheter may prevent too rapid glue polymerization and allow monitoring of glue injection into most of the distal vessels. In the majority of cases treated with super-selective micro-catheterization, glue injection follows the main flow course within the lesion. A good penetration of the glue into the vascularization of the lesion, even in the presence of collateral vessels with "comb" morphology, provides a good rendering of the angioarchitecture of the lesion as that obtained at angiography with super-selective catheterization and avoids too distal embolization, frequent when using small particles. Differently from particles, Glubran2® injection via catheter must be performed from a more stable and distal position, which is difficult to achieve in the presence of particularly tortuous vessels. Glubran2® injection may provoke severe pain in the patient. For this reason, neuroleptoanalgesia is induced to the patient a few minutes prior to injection of the embolic agent.

4.2.2.3 Coils

Coils are metallic endovascular embolic agents of variable size. Their choice depends on calibre and size of the artery to be occluded.

In our experience of SAE, coils are employed for vertebral artery occlusion in the presence of cervical tumours scheduled for surgical removal with en bloc technique. The aim is the complete occlusion of the artery along the whole tract from V2 to V3 (from the cervical intra-transversary segment to the cervico-occipital curve). When applicable (non-dominant vertebral artery, absence of macroscopic ASA feeders), coils with high filling coefficient, such as the detachable Penumbra Coil 400™ (Penumbra Inc., USA), are employed. In other cases, Target® Detachable Coils (Stryker Inc., USA) are employed.

Presurgical Occlusion of Vertebral Arteries: Technical Notes

Being one of the most delicate SAE procedures on vertebral lesions, presurgical occlusion of a vertebral artery through detachable coils requires high skill and proficiency.

Potential risks of the procedure are mainly related to the vertebrobasilar circle and its functional peculiarities.

Cerebral pan-angiography is recommended, including the study of vertebral and internal carotid arteries, bilaterally, to visualize possible asymmetry of vertebral arteries as well as presence of posterior communicating arteries.

The procedure is declined when marked asymmetry of the vertebral arteries is present and prevails in the artery to be occluded. In such cases, the occlusion of the dominant vertebral artery would carry a high risk.

When both vertebral arteries are significantly symmetrical, the only condition under which the procedure can be carried out safely is that the patient can haemodynamically and functionally tolerate the occlusion of a vertebral artery.

To establish this necessitates a tolerability test. Whenever an accurate tolerability test of the vertebral artery occlusion is not possible, as it frequently occurs during interventional procedures involving the intracranial carotid district, catheterization of the artery through a 6 F carrier catheter is suggested.

In most cases, the external calibre of a 6 F catheter (about 2.5 mm) causes marked reduction of the vertebral artery flow, which has a mean diameter of about 4 mm. If the patient does not sufficiently compensate contralaterally, severe vertigo will occur as a sign of vertebrobasilar insufficiency. In this case, the procedure must be suspended due to very high risk. Besides, this emphasizes the importance that the patient be awake and collaborating throughout the procedure.

In the absence of vertebrobasilar signs of insufficiency after catheterization of the vertebral artery with a 6 F carrier catheter, 5000 I.U. of heparin with low molecular weight are intravenously administered. Anticoagulant agents help prevent formation of blood clots in arteries with slow flow or after positioning of the first coil.

In our experience, we preferably employed a carrier catheter (Neuron 6 F 052) with micro-catheter PX 400™ dedicated to release of Penumbra Coil 400™ (Penumbra Inc., USA). The Neuron catheters have been chosen due to their soft distal portions. However, we also employed other catheters, such as Envoy 6 F Multipurpose D (Codman J&J, MA, USA).

The coils are retrogradely released, starting from the V3 tract. Possible presence of muscular branches at V3 can be useful to "hook up" the distal extremity of the first coil and avoid an incidental as well as dangerous intracranial migration. Multiple coils are subsequently released to create a solid skein retrogradely extending almost to reach V2. At the end of the procedure, a control angiography of the contralateral vertebral artery and internal carotid arteries is performed (Fig. 4.2).

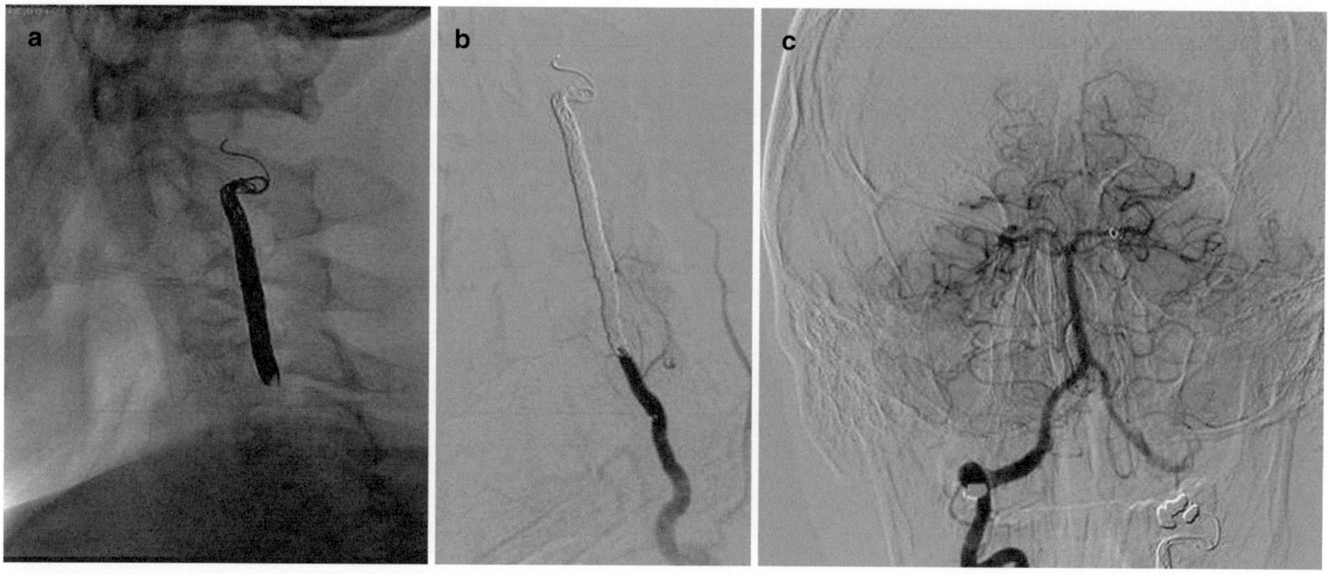

Fig. 4.2 Presurgical occlusion of left vertebral artery The coils are retrogradely released, starting from the V3 tract. Multiple coils are subsequently released to create a solid skein retrogradely extending almost to reach V2 (**a**, **b**). At the end of the procedure, a control angiography of the right vertebral artery (**c**) checks the retrograde injection of left posterior inferior cerebellar artery

4.3 Exemplificative Cases

Case 1
Cervical aneurysmal bone cyst at C3 level

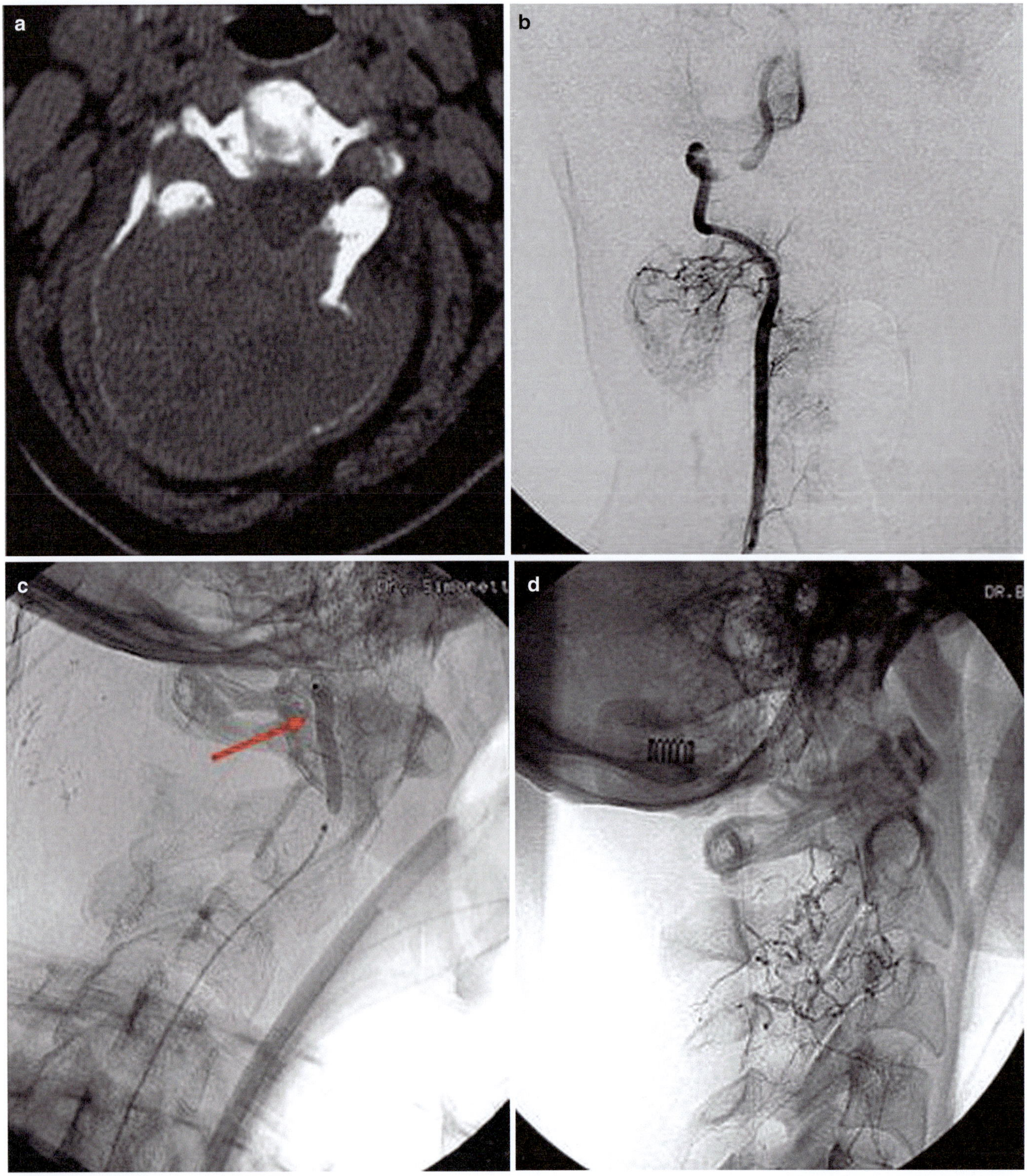

Fig. 4.3 CT scan (**a**) and angiography (**b**). After distal protection positioning balloon in the vertebral artery (*arrow* in **c**), we perform treatment with super-selective micro-catheterization. Glubran2 injection follows the main flow course within the lesion, with a good penetration of the glue into the vascularization of the lesion. Note the overlap of the final distribution of Glubran2 with pathological vascular network (**d**)

Case 2

Aggressive vertebral osteoangioma at C7 level

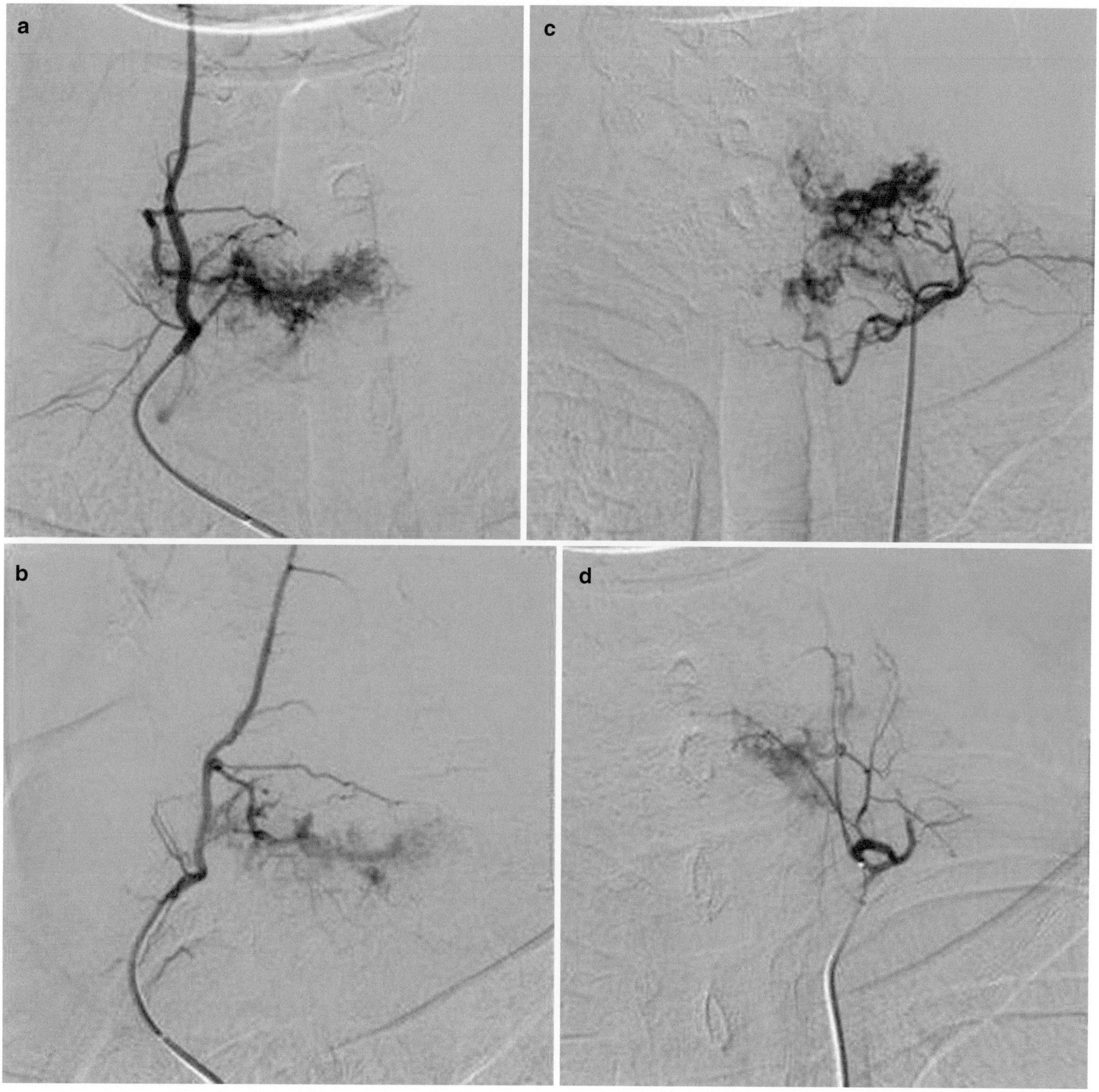

Fig. 4.4 Before (**a–c**) and after (**d**, **f**) super-selective embolization with stable microparticles

Case 3

Cervical aneurysmal bone cyst at C5 level in a 12-year-old girl. Right, drug-resistant, cervicobrachialgia.

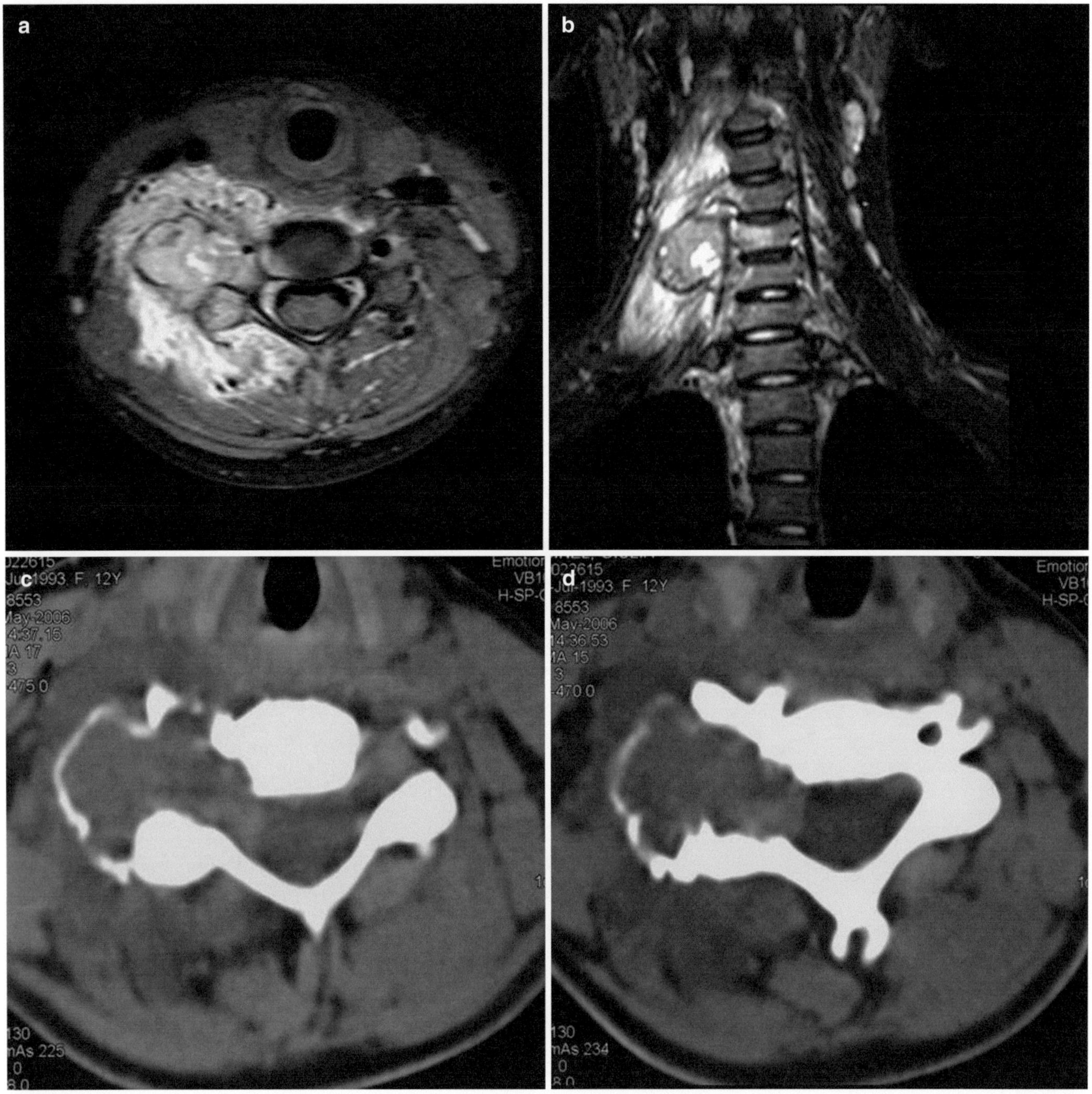

Fig. 4.5 MR scans, in axial (**a**) and coronal (**b**) planes, highlight the lesion and associated extensive inflammation of the surrounding soft tissues. CT scans (**c**, **d**) outline the bone changes of the vertebra. Right vertebral angiography (**e**) demonstrates the narrowing of the artery due to aneurysmal bone cyst and its pathological vessels. We performed the occlusion of vertebral artery: the coils are retrogradely released, starting from stenotic segment of artery (**f**). The pathological vessels disappeared. Then we proceeded with super-selective embolization with stable microparticles of pathological vascular network arising from deep cervical artery (**g**, **h**). CT scans three months (**i**, **j**) after embolization. CT scans at six months (**k**, **l**) after embolization. Progressive recalcification of osteolytic lesion. The pain disappeared two days after the procedure

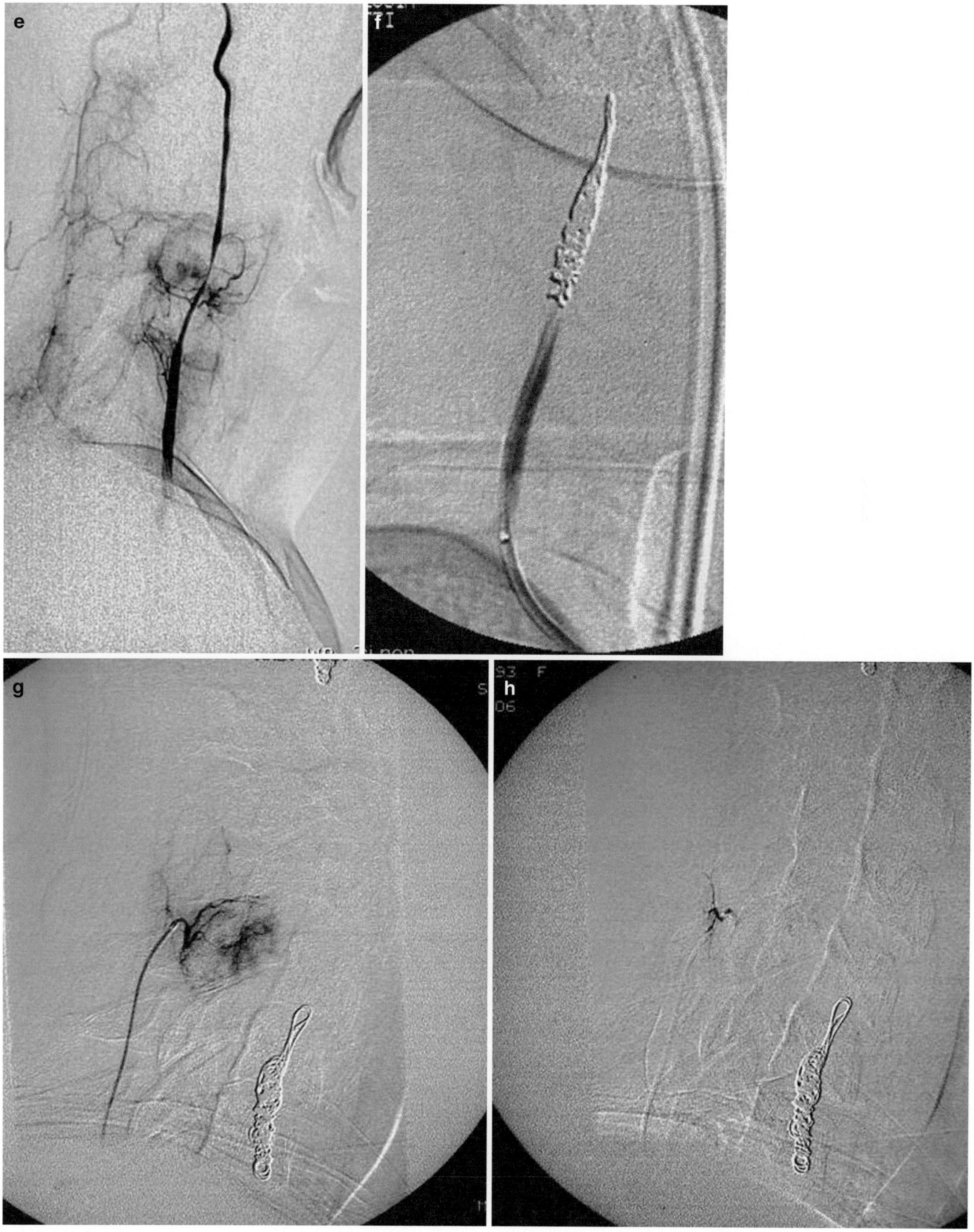

Fig. 4.5 (continued)

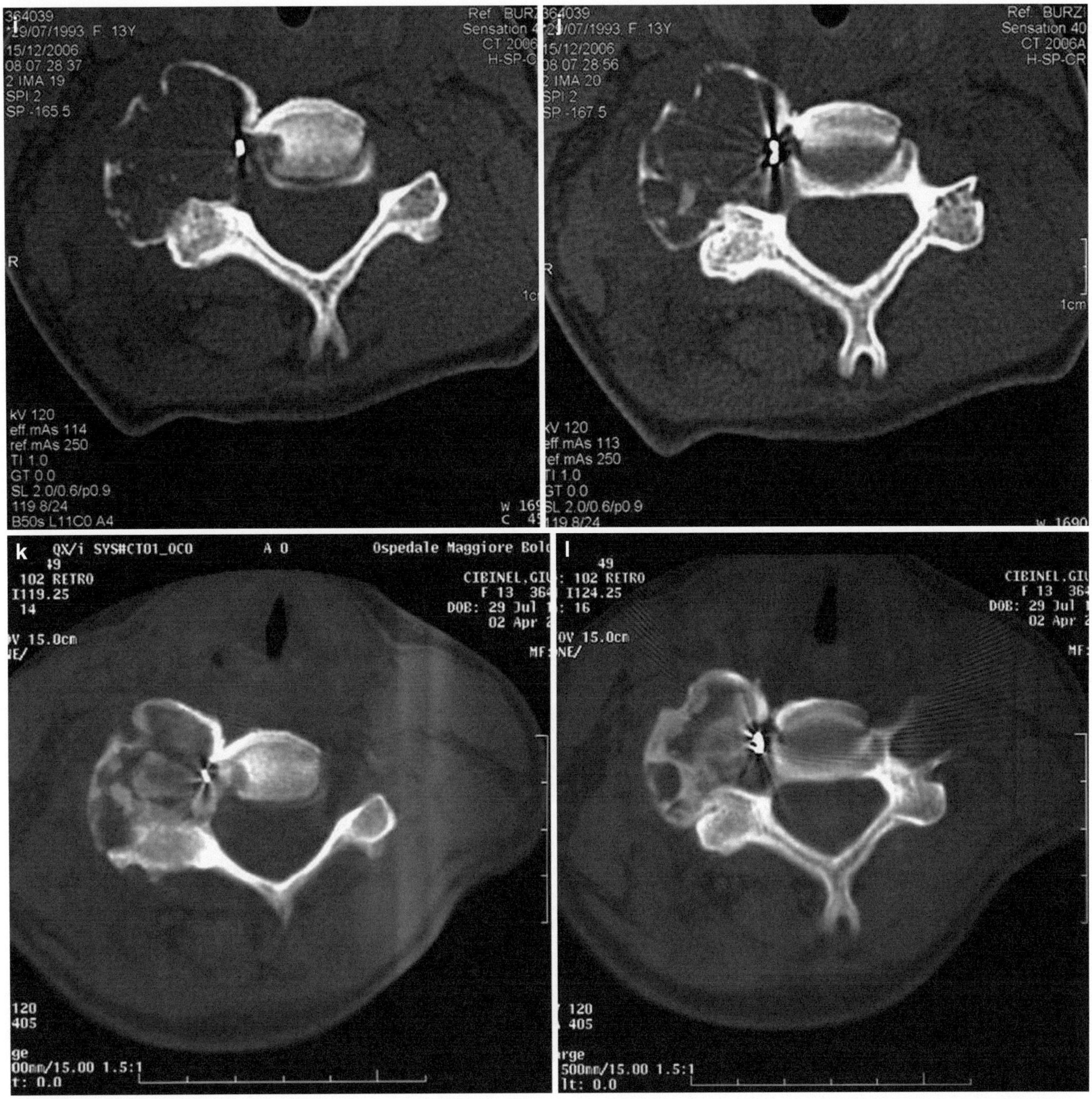

Fig. 4.5 (continued)

Selected References

1. Daniel H. Kim, Alexander R. Vaccaro, Curtis A. Dickman, et al (2005) Surgical anatomy and techniques to the spine. Thieme
2. Tomita K, Kawahara N, Murakami H, Demura S (2006) Total en bloc spondylectomy for spinal tumors: improvement of the technique and its associated basic background. J Orthop Sci 11(1):3–12
3. Broaddus WC, Grady MS, Delashaw JB, Ferguson RD, Jane JA (1990) Preoperative superselective arteriolar embolization: a new approach to enhance resectability of spinal tumors. J Neurosurgery 27(5):755–759
4. Guzman R, Dubach-Schwizer S, Heini P et al (2005) Preoperative transarterial embolization of vertebral metastases. Eur Spine J 14(3):263–268
5. M. Leonardi, L. Simonetti, R. Agati, M. Messia, F. De Santis, G. Dani(2001) Recent CT advances in spine imaging. NRJ Volume 14 (No. 2):207
6. Stafa A, Barbara C, Boriani** S, Simonetti L, Leonardi* M (2010) A little talk on Adamkiewicz's artery. Some practical considerations on the Pre-operative identification of this artery starting from a single team experience in Pre-surgical selective embolization of VascularizedSpinal lesions. Neuroradiol J 23:225–233
7. Simonetti L, Gasbarrini* A, Boriani* S, Leonardi M (2003) Presurgical selective arterial embolization of hypervascularized spinal tumours: embolization techniques and results in 290 patients(1998–2008). Interv Neuroradiol 9(4):339–349
8. Leonardi M, Barbara C, Simonetti L (2002) Glubran 2: a new acrylic glue for neuroradiological endovascular use experimental study on animal. Interv Neuroradiol 8:245–250
9. Leonardi M, Cenni P, Simonetti L et al (2003) Glubran 2®:a new acrylic glue for Neuroradiological Endovascular use. A complementary histological study. Interv Neuroradiol 9:249
10. Raffi L, Simonetti L, Cenni P, Bandiera S, Gasbarrini A, Boriani S, Leonardi M (2003) Presurgical embolization of spinal tumours using glubran 2 acrylic glue. Interv Neuroradiol 9(4):339–349
11. Vetter SC, Strecker EP, Ackermann LW, Harms J (1997) Preoperative embolization of cervical spine tumors. Cardiovasc Intervent Radiol 20(5):343–347
12. Marushima A, Matsumaru Y, Suzuki K et al (2009) Selective arterial embolization with n-butyl cyanoacrylate in the treatment of aneurysmal bone cyst of the cervical vertebra: a case report. Spine (Phila Pa 1976) 34(6):115

5 Anesthesiological Management and Patient Positioning

M.R. Bacchin, M. Di Fiore, Y.E. Akman, M. Girolami, R. Ghermandi, A. Gasbarrini, and S. Boriani

Prior to the surgery, reproduction of symptoms with extension of the neck should be assessed. Extension may help for better exposure and visualization. However, the severity of stenosis may limit the amount of extension that can be tolerated by the patient. In this case, a neutral neck position must be maintained till the end of the decompression procedure.

The head should be placed in slight hyperextension but not excessive traction since it causes tension on soft tissues. The combination of traction and hyperextension of the head will stretch the tissue planes, resulting to the separation and the dissection being harder and more dangerous as it is more difficult to distinguish structures in that case. All bony prominences must be appropriately padded. The extremities must be positioned in an appropriate position to avoid any peripheral neuropathy. The legs and thighs are tied to the table using straps and the feet are supported with a pillow. A roll is placed between the shoulder blades to allow the neck for extension. The shoulders are taped inferiorly as they are lowered, for better radiographic visualization of the lower cervical segments. Also, the arms can be extended with wrist bands to provide traction, but it should be kept in mind that excessive arm traction may cause stretching of the roots of the brachial plexus causing postoperative stupor of the brachial plexus. Excessive pressure by the wrist bands on wrists may cause median nerve injury. The iliac crest should be kept ready for an eventual autograft harvesting. The head is placed on a circular silicone pillow. The patient's eyes should be protected with appropriate material to avoid direct pressure around (Fig. 5.1). Especially in prone position, some intraocular increased pressure-related serious ophthalmologic complications may occur. These complications are corneal abrasion (the most common), blindness as a result of posterior ischemic optic neuropathy, and central retinal artery occlusion [1].

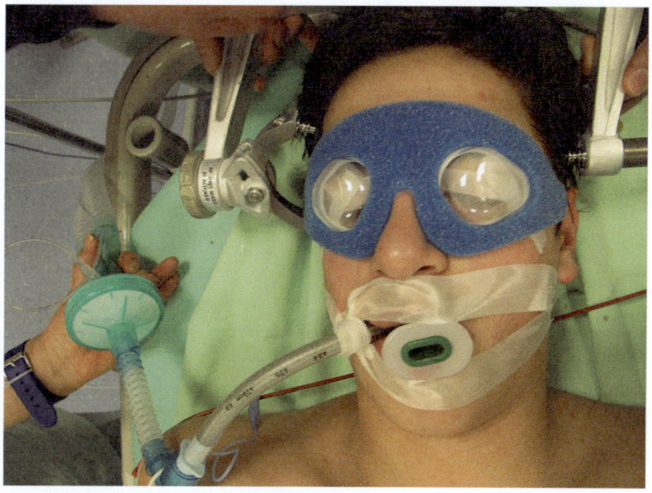

Fig. 5.1 The patient's eyes are protected to avoid direct pressure

The nasogastric and the endotracheal tubes must be carefully anchored to the mouth to avoid a possible displacement and must be kept away from the surgical area (Fig. 5.2). In some cases, a tracheostomy may be required (Fig. 5.3). Also the tubes must be easily accessible for the anesthesiologist for a fast replacement of the tubes during surgery. In such a case, the surgery must be immediately interrupted. The patient's position should be reassessed by the anesthesiologist as often as possible to be sure that any change has not occurred [2, 3].

For a more secure and fixed position, Mayfield traction tongs may be applied (Fig. 5.4). However, Mayfield traction should be used carefully in patients with conditions (such as tumors, osteoporosis, hemopathies, etc.) that cause fragility in the skull.

M.R. Bacchin (✉) • M. Di Fiore
Department of Anesthesiology, Perioperative and Pain Medicine, Rizzoli Orthopedic Institute, Bologna, Italy
e-mail: mariarenata.bacchin@ior.it

Y.E. Akman
Orthopaedics and Traumatology Department, Metin Sabanci Baltalimani Bone Diseases Training and Research Hospital, Istanbul, Turkey

M. Girolami • R. Ghermandi • A. Gasbarrini • S. Boriani
Oncologic and Degenerative Spine Surgery Department, Rizzoli Orthopedic Institute, Bologna, Italy

S. Boriani et al. (eds.), *Atlas of Craniocervical Junction and Cervical Spine Surgery*, DOI 10.1007/978-3-319-42737-9_5

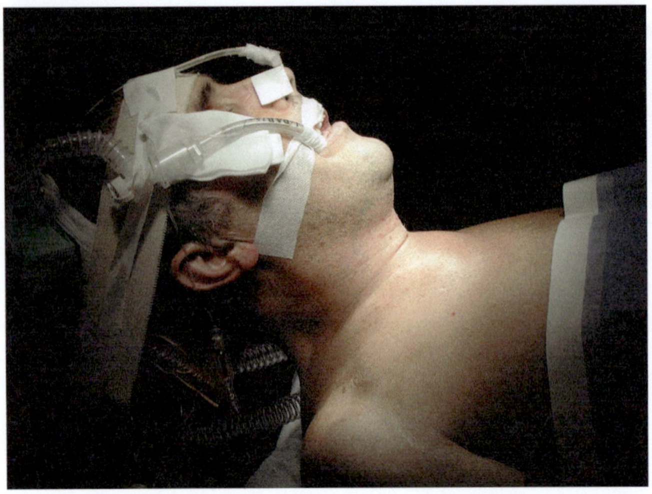

Fig. 5.2 Positioning of nasogastric and endotracheal tubes

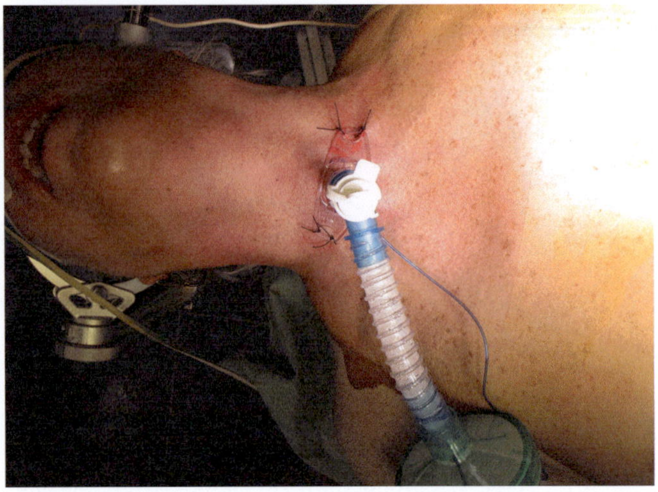

Fig. 5.3 Tracheostomy opened in a patient before surgery

Fig. 5.4 The patient is positioned supine using Mayfield tongs with appropriate support under the cervical spine

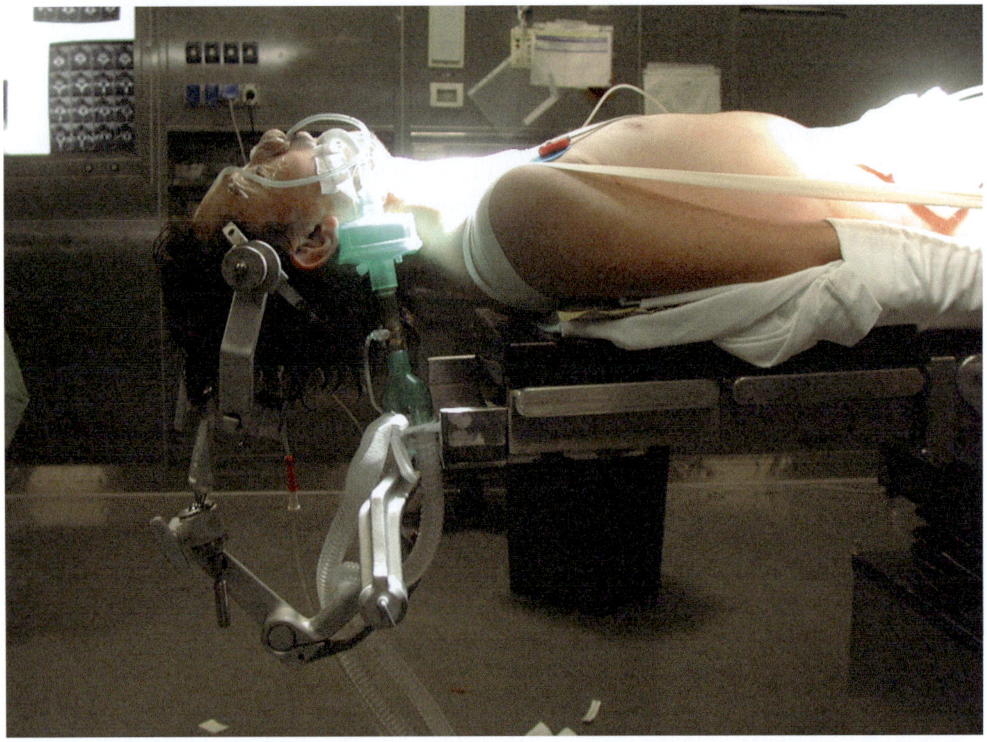

First, the skin is cleansed by an iodine or alcohol solution. The authors do not perform shaving before the application of the Mayfield pins. However, if the surgical area extends to the upper cervical segments, hair should be shaved at least 24 h before the procedure (Fig. 5.5). One pin is placed above the ear, as a reference (Fig. 5.6), and the other two pins are placed at the same level with it on the contralateral side. When placing the pins on this side, attention must be paid as the pin that is located on the temporal bone may cause temporal perforation. Mayfield tongs are squeezed together to make sure that the single contacts the skull. Then the screw is tightened until the pressure gauge shows between 60 and 80 lb (Fig. 5.7).

Because the surgical wound is above the heart level producing a subatmospheric pressure in the open veins, the risk of venous air embolism should be kept in mind. The fixating devices on the skull must be always removed in the supine position to avoid venous air embolism. During the entire

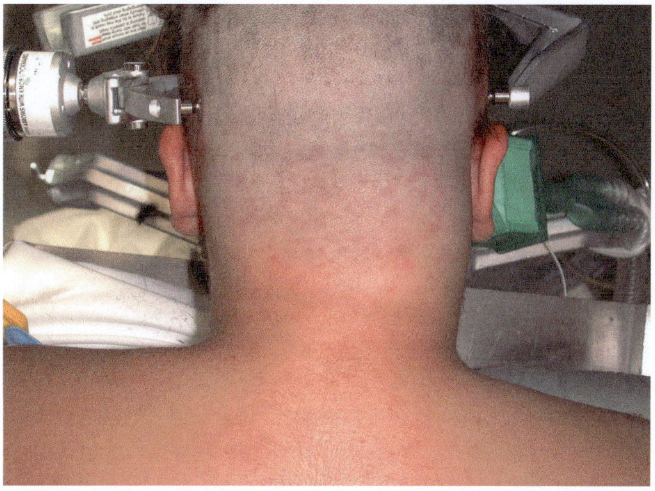

Fig. 5.5 For interventions in the upper cervical segments, the hair is shaved at least 24 h before surgery

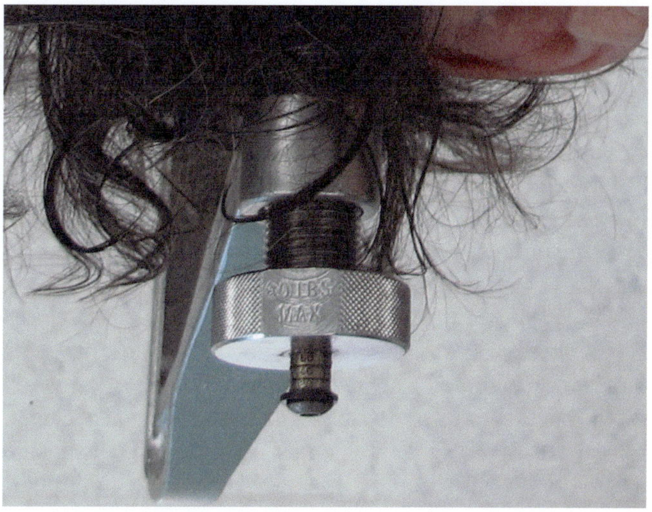

Fig. 5.7 Adequate pressure is applied in order to avoid perforations

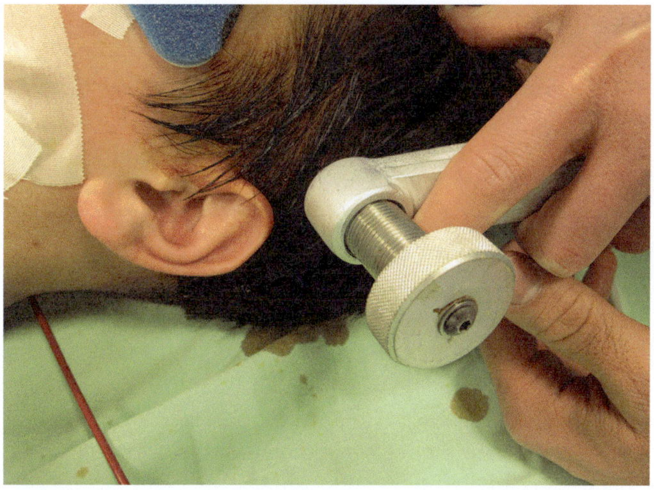

Fig. 5.6 The application of the first Mayfield pin above the ear

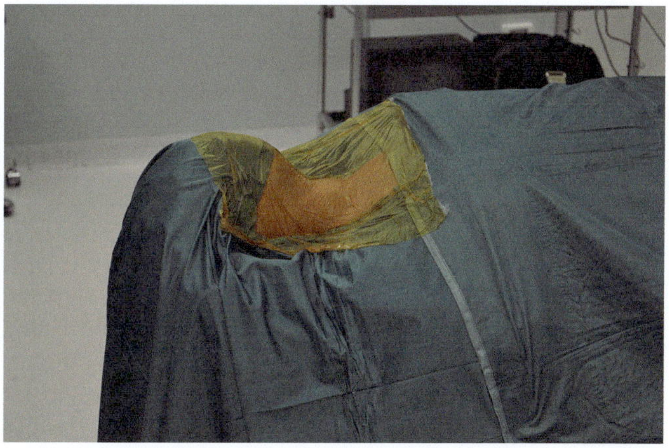

Fig. 5.8 Draping should be applied on the surgical area to be as wide as possible

operation procedure, the patient should be monitored for venous air embolism until all potential sites for air entry into the blood circulation have been closed [4]. Rigid support is essential for the neck (Fig. 5.3. The spinal processes should be well supported against pressure since considerable force may be applied on cervical spinal bodies (during graft insertion, screwing, etc.)

To determine the midline, sternal notch may be marked with a skin marker. A wide area as much as possible should be draped open, in order to let the surgeon orient for the overall position and the alignment of the patient (Fig. 5.8).

- *Anesthesia and its impact intraoperative neuromonitoring procedures*

Neuromonitoring protocols depend on the patient, surgeon, and institution and should be individualized accordingly. Both motor-evoked potentials (MEPs) and somatosensoryevoked potentials (SSEPs) may be used during surgical procedures. Communication with the anesthesia team is essential because inhaled agents and paralytic drugs must be avoided. Following the prepositioning, baseline values are obtained, then the patient's head and neck are extended, and the shoulders are taped inferiorly. MEPs and SSEPs are then retested to ensure no deviation from baseline [5, 6]. For neuromonitoring, neuromuscular blockade must be minimized or avoided. As a result, each stimulation will produce movement of limb and axial muscles. The amount of movement can be minimized by using a threshold-level stimulation protocol that is based on determining the lowest stimulus intensity that produces consistent muscle activation [7]. The variable that is used for monitoring for this technique is the change in threshold needed to elicit muscle activation. Even with this technique, however, it is necessary to warn the surgeon when a stimulus train is going to be delivered to minimize the risk of movement during a critical part of the surgery. In addition, MEP recording introduces constraints into choice of anesthetic agents (vide infra).

During the neuromonitoring, cerebral stimulation causes some other risks. The stimulation directly activates masseter muscles, and forceful contraction can produce tongue bite, tooth fracture, or even mandible fracture. These risks can be minimized or eliminated by proper use of bite blocks inside the mouth. MEP recording is considered to be contraindicated in patients with epilepsy, cortical lesions, skull defects, increased intracranial pressure, surgically implanted intracranial devices, cardiac pacemakers, or other implanted pumps. Actual experience with tceMEP monitoring has seen a very low incidence of complications [8].

The operating room is an electrical environment, and electrical interference and artifacts may deteriorate interpretation of the monitoring potentials. The depth of anesthesia, body temperature, and blood pressure may alter potentials. The monitoring personnel must be able to detect and understand the reasons of such variables in order to interpret them correctly. False interpretations may alert the surgeon in an undesirable way. It is not necessary or desirable to alert the surgeon to every change in the SSEP or tceMEP recordings, particularly if they are not physiologically meaningful. However, it is absolutely essential to warn the surgeon when a change in the potentials that may reflect neural injury is observed. Since there can be real-time changes in the potentials, any variation in the recordings must be sustained and reproducible [9].

The intraoperative neurophysiologic monitoring place requires additional demands on the anesthesiologist. In addition to conventional anesthesia for the surgical procedure, it is necessary to apply the anesthetic so that it helps maximize signal acquisition. The assessment of somatosensory- (SSEPs) and motor-evoked potentials (MEPs) are the most commonly used modalities to evaluate neural risk and help decision making during the surgical procedure. These have certain impacts on the anesthetic management.

tceMEPs are particularly sensitive to interference by anesthetic agents. tceMEPs are the more difficult to evaluate during operative procedures; anesthetic conditions that are optimized for tceMEPs will usually produce acceptable SSEPs. The anesthetic approach needs to be optimized for obtaining appropriate potentials and correct interpretations that can help in guiding the surgical procedure.

Most of the anesthetic agents depress evoked response amplitudes and increase latencies which cause intraoperative neurophysiologic monitoring to be more difficult. Notable exceptions include the intravenous anesthetic agents etomidate and ketamine which are the exceptions to increase SSEP and MEP amplitudes [9]. The infusion of etomidate causes adrenocortical suppression; thus, its use is limited to induction of anesthesia in potentially hemodynamically unstable patients [7, 10].

Significant temperature-related decreases in evoked potential amplitude may occur after the surgical exposure of the spine but before instrumentation and deformity correction. Hypothermia may lead to false-negative results [11]. As the esophageal temperature decreases, with hypothermia, a gradual increase occurs in the MEPs. An increase in stimulation threshold was also observed at lower temperatures [9]. In contrast, hyperthermia reduces the latency and increases the conduction velocity of evoked potentials. Above 42 °C, spinal SSEP amplitudes do not change, whereas cortical SSEPs and spinal MEPs deteriorate. Hypothermia increases latency and decreases conduction velocities. Below 28 °C, the spinal MEP amplitude decreases and the cortical SSEPs and spinal MEPs disappear. As hyperthermia and hypothermia cause significant changes in the latency of MEPs and SSEPs, it has been suggested that for a correct evaluation, evoked potential measurements should be performed in a range of 2–2.5 °C above or below the baseline temperature [12, 13].

Hypoxemia can lead to evoked potential deterioration before the other clinical parameters have been changed. Changes in carbon dioxide levels may cause alteration in spinal cord and cortical blood flow. The most apparent changes in cortical SEP occur when the carbon dioxide tension is extremely low, suggesting excessive vasoconstriction may produce ischemia ($PaCO_2$ 20 mmHg) [9]. This effect may be expected to produce some MEP changes. As hypocapnia may cause small SEP changes, and possibly MEP, baseline recordings should be obtained before the initiation of hyperventilation [14].

SSEP response may also change with hematocrit. An amplitude increase has been found with mild anemia, followed by an increase in latency at hematocrits of 10–15%. At hematocrits less than 10%, further latency changes and amplitude reductions are observed. No comparable studies have been performed with MEPs [9].

References

1. Stambough JL et al (2007) Ophthalmologic complications associated with prone positioning in spine surgery. J Am Acad Orthop Surg 15(3):156–165
2. Manfredini M et al (2000) Unilateral blindness as a complication of intraoperative positioning for cervical spinal surgery. J Spinal Disord 13(3):271–272
3. Yanagidate F, Dohi S (2003) Corneal abrasion after the wake-up test in spinal surgery. J Anesth 17(3):211–212
4. Grinberg F, Slaughter TF, McGrath BJ (1995) Probable venous air embolism associated with removal of the Mayfield skull clamp. Anesth Analg 80(5):1049–1050
5. Smith PN et al (2007) Intraoperative somatosensory evoked potential monitoring during anterior cervical discectomy and fusion in nonmyelopathic patients – a review of 1,039 cases. Spine J 7(1):83–87
6. Khan MH et al (2006) Intraoperative somatosensory evoked potential monitoring during cervical spine corpectomy surgery: experience with 508 cases. Spine (Phila Pa 1976) 31(4):E105–E113
7. Cohan P et al (2005) Acute secondary adrenal insufficiency after traumatic brain injury: a prospective study. Crit Care Med 33(10):2358–2366
8. MacDonald DB (2002) Safety of intraoperative transcranial electrical stimulation motor evoked potential monitoring. J Clin Neurophysiol 19(5):416–429
9. Pajewski TN, Arlet V, Phillips LH (2007) Current approach on spinal cord monitoring: the point of view of the neurologist, the anesthesiologist and the spine surgeon. Eur Spine J 16(Suppl 2):S115–S129
10. Wagner RL et al (1984) Inhibition of adrenal steroidogenesis by the anesthetic etomidate. N Engl J Med 310(22):1415–1421
11. Seyal M, Mull B (2002) Mechanisms of signal change during intraoperative somatosensory evoked potential monitoring of the spinal cord. J Clin Neurophysiol 19(5):409–415
12. Oro J, Haghighi SS (1992) Effects of altering core body temperature on somatosensory and motor evoked potentials in rats. Spine (Phila Pa 1976) 17(5):498–503
13. Sakamoto T et al (2003) The effect of hypothermia on myogenic motor-evoked potentials to electrical stimulation with a single pulse and a train of pulses under propofol/ketamine/fentanyl anesthesia in rabbits. Anesth Analg 96(6):1692–1697, table of contents
14. Gravenstein MA, Sasse F, Hogan K (1992) Effects of hypocapnia on canine spinal, subcortical, and cortical somatosensory-evoked potentials during isoflurane anesthesia. J Clin Monit 8(2): 126–130

6 Cervical Spine Instrumentation

M. Girolami, R. Ghermandi, A. Gasbarrini, Y.E. Akman, and S. Boriani

6.1 Anterior Instrumentation

6.1.1 Anterior Cervical Discectomy and Fusion (ACDF)

Anterior cervical discectomy and fusion (ACDF) is often used as a surgical treatment for cervical degenerative disorders (Fig. 6.1). Numerous long-term follow-up studies have shown that is effective in case of cervical myelopathy, relieving anterior compression on the spinal cord, or radiculopathy, decompressing nerve roots. This technique allows direct decompression of neural structures, restoration of disc height, and stabilization of the affected motion segments. The success of the procedure relies on achieving a solid fusion.

Anterior column (vertebral bodies and discs) is exposed through an anterior prevascular exposure (more often with transverse skin incision) and the level is identified using intraoperative fluoroscopic control.

The anterior longitudinal ligament (ALL) and the anterior portion of the annulus are incised, and discectomy is carried out in the usual manner. Vertebral bodies are prepared removing all the cartilaginous end plates with high-speed burr or curettes until bleeding bone.

By doing this, it is essential to identify the uncinate processes, which defines the lateral borders of a "safe zone" whose violation would put vertebral artery at risk.

Distraction to the disc space can be applied in order to increase the working space, thus, facilitating removal of the posterior half of the disc.

The two pins of the Caspar retractor are affixed slightly superior to the center of the vertebral bodies above and below (Fig. 6.2). Pin retractors also allow, after implant insertion, to put it under slight compression (Fig. 6.3).

Foraminotomy can be performed by removing the medial portion of the uncinate processes with a 1- or 2-mm Kerrison rongeur. By doing this, it is essential to remember the course of the exiting nerve root and to keep in mind that surgeons have the better visualization of the compression on the contralateral side to the one they are standing to.

Posterior longitudinal ligament (PLL) and posterior bony spurs can be removed allowing for complete anterior decompression of the cord.

Finally, the selected implant can be trialed and, after fluoroscopic control, implanted.

For a successful fusion, bony contact with the graft (or interbody device) must be maximized, so meticulous endplates preparation is essential.

Excessive intervertebral disc space distraction, as a result of a too high graft, may cause facet impingement and postoperative neck pain. Any remaining anterior osteophytes should be removed, and the implant should be placed approximately 4 mm anterior to the posterior margin of the vertebral body in order to prevent impingement of the spinal canal.

Finally, distraction should be released and position of the cage checked using fluoroscopy.

M. Girolami (✉) • R. Ghermandi • A. Gasbarrini • S. Boriani
Oncologic and Degenerative Spine Surgery Department Rizzoli Orthopedic Institute, Bologna, Italy
e-mail: marco.girolami@yahoo.it

Y.E. Akman
Orthopaedics and Traumatology Department, Metin Sabanci Baltalimani Bone Diseases Training and Research Hospital, Istanbul, Turkey

S. Boriani et al. (eds.), *Atlas of Craniocervical Junction and Cervical Spine Surgery*, DOI 10.1007/978-3-319-42737-9_6

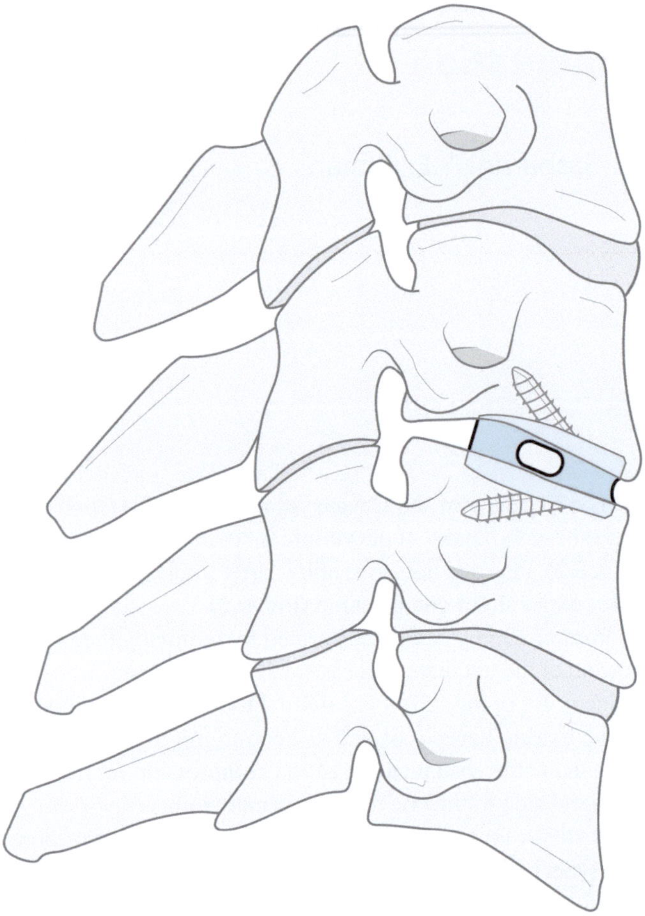

Fig. 6.1 ACDF with stand alone anterior cage

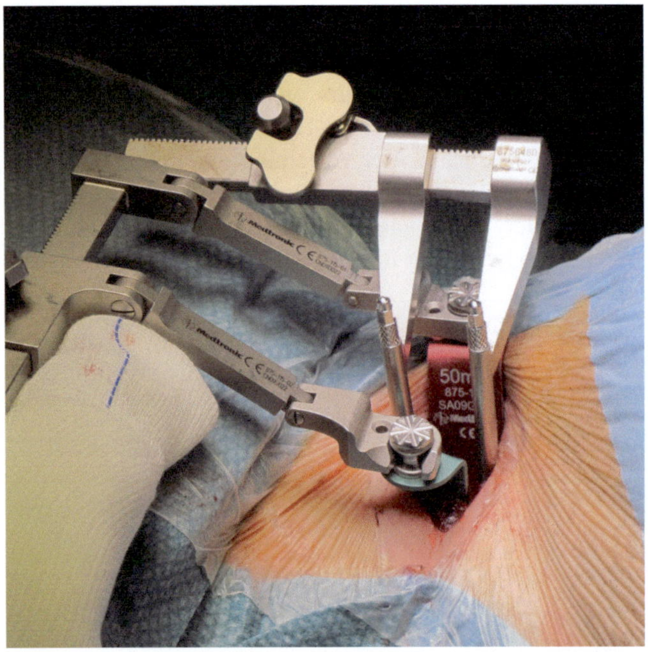

Fig. 6.2 Caspar pin retractors are applied in order to allow segmental distraction, easing discectomy and endplates preparation, and compression to provide stimulus for fusion

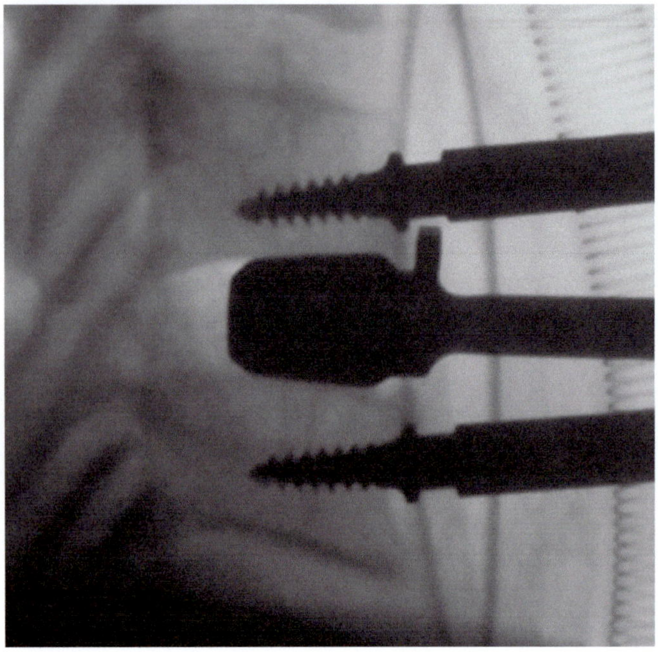

Fig. 6.3 Intraoperative x-ray showing Caspar pin distractors

6.1.2 Cervical Interbody: Cages and Grafting

The anterior approach to the cervical spine for discectomy and fusion by the insertion of an autologous iliac-crest tricortical bone graft was originally introduced by Robinson and Smith in 1955. In 1958, Cloward described a wide anterior cylindrical discectomy performed with a special reamer combined with anterior fusion by the insertion of autologous iliac bone graft of the same shape.

Nowadays, several implants are available on the market to perform anterior interbody fusion. Cages of different shapes and materials, which can be implanted with or without the need for additional anterior plating (Fig. 6.4). The cage size must be determined intraoperatively, under fluoroscopic control, using available trials. Use of allograft avoids the need for autograft harvesting, that produces significant morbidity to the donor site (i.e. iliac crest, fibula). Thus, authors prefer the use of bone bank allograft, but autograft can be easily obtained from the iliac crest too.

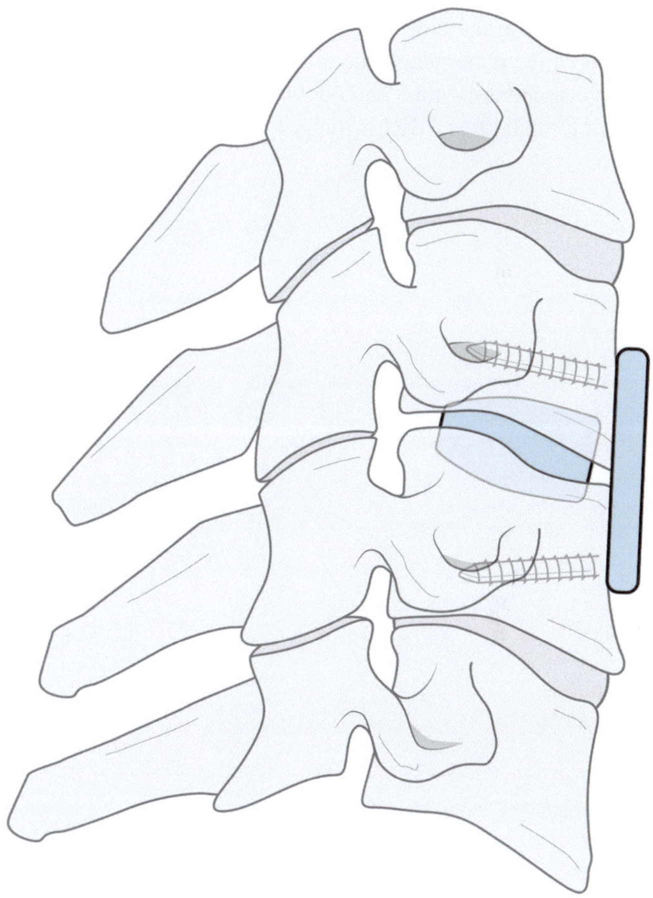

Fig. 6.4 ACDF with interbody grafting and anterior plate

6.1.3 Anterior Cervical Corpectomy and Fusion (ACCF)

ACCF has been proven to be an effective technique in case of ventral compression posterior to the vertebral body (i.e., trauma, degenerative disease, infection, tumor; Fig. 6.5).

Moreover, cervical kyphotic deformities (especially if fixed on flexion-extension films) generally require an anterior approach, since posterior-only decompression would not allow the spinal cord to migrate posteriorly and symptoms are likely to continue postoperatively.

Furthermore, an anterior approach resolves compression on the anterior spinal artery that supplies up to 75–80% of the anterior spinal cord. Accurate collection of the medical history of the patient and a detailed physical examination are always mandatory and should be verified with advanced imaging modalities (MRI is the gold standard).

Anterior column (vertebral bodies and discs) is exposed through an anterior prevascular exposure, and the level is identified using intraoperative fluoroscopic control. Surgical field must be large enough to include the disc above and below the targeted vertebral body.

Then, discectomies at the levels above and below are performed as previously described.

After the discectomies have been performed, two longitudinal grooves, parallel to the uncinate processes and as deep as the anterior two thirds of the vertebral body, are made using a high-speed burr.

The central bone between the grooves is removed with a rongeur and stored for grafting. Then, the posterior wall of the vertebral body is thinned using a high-speed burr and later removed with Kerrison rongeurs along with the PLL. Finally, the decompression is accomplished removing any eventual osteophyte left.

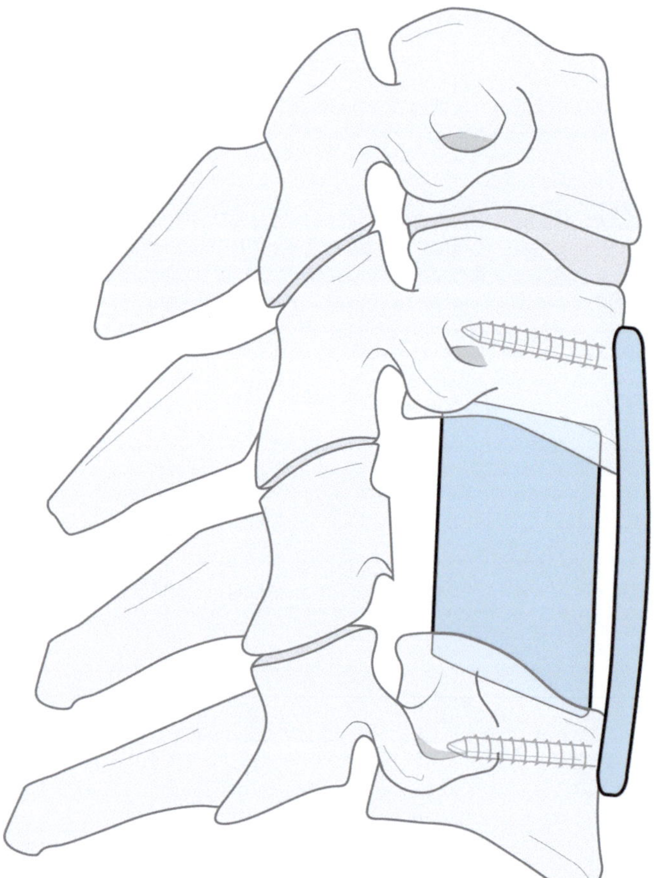

Fig. 6.5 Single-level ACCF, recontructed with strut graft and anterior plating

6.1.4 Reconstruction of the Anterior Column

Corpectomy defect can be reconstructed with various techniques, but proper size and appropriate positioning of the implant is cornerstone as the choice of reconstruction. After implant positioning, an anterior cervical plate is applied to stabilize the construct and prevent ventral migration.

Technical options for anterior column reconstruction:

- Strut bone grafts (allograft, autograft)
- PMMA (eventually armed with Steinman pins)
- Titanium mesh cages (Harms cages)
- Polyetheretherketone (PEEK) cages
- Carbon fiber cages
- Expandable cages

Variables to take into consideration when selecting the technique:

- Number of levels to be reconstructed
- Bone quality
- Further treatment planned (i.e., radiotherapy)
- Need to minimize artifacts (early detection of local recurrences)
- Fusion (secondary stability)

6.1.4.1 Graft Selection for Anterior Interbody Fusion: Autograft Versus Allograft?

- Donor site morbidity
- Single level vs. multilevel

6.1.5 Supplemental Posterior Fixation for Multilevel ACCF

Multilevel ACCF may lead to iatrogenic instability of the cervical spine, and therefore supplemental posterior fixation must be considered (Fig. 6.7). Generally, a supplemental posterior fixation is necessary when more than three levels are involved anteriorly.

Posterior fixation with respect to the sagittal balance of the cervical spine significantly decreases the risk of complications such as pseudarthrosis, graft failure, progressive kyphotic deformity, and plate dislodgement. Regardless of the number of levels involved, poor bone quality or an unstable anterior fixation should be always supported posteriorly.

In order to achieve secondary long-lasting stability with a solid fusion, decortication and grafting are recommended.

ACCF vs. double-level ACDF (Fig. 6.6)

ACCF	Double-level ACDF
Pros: Complete decompression	*Pros*: More stable construct
Cons: Blood loss Less stable	*Cons*: Less decompression

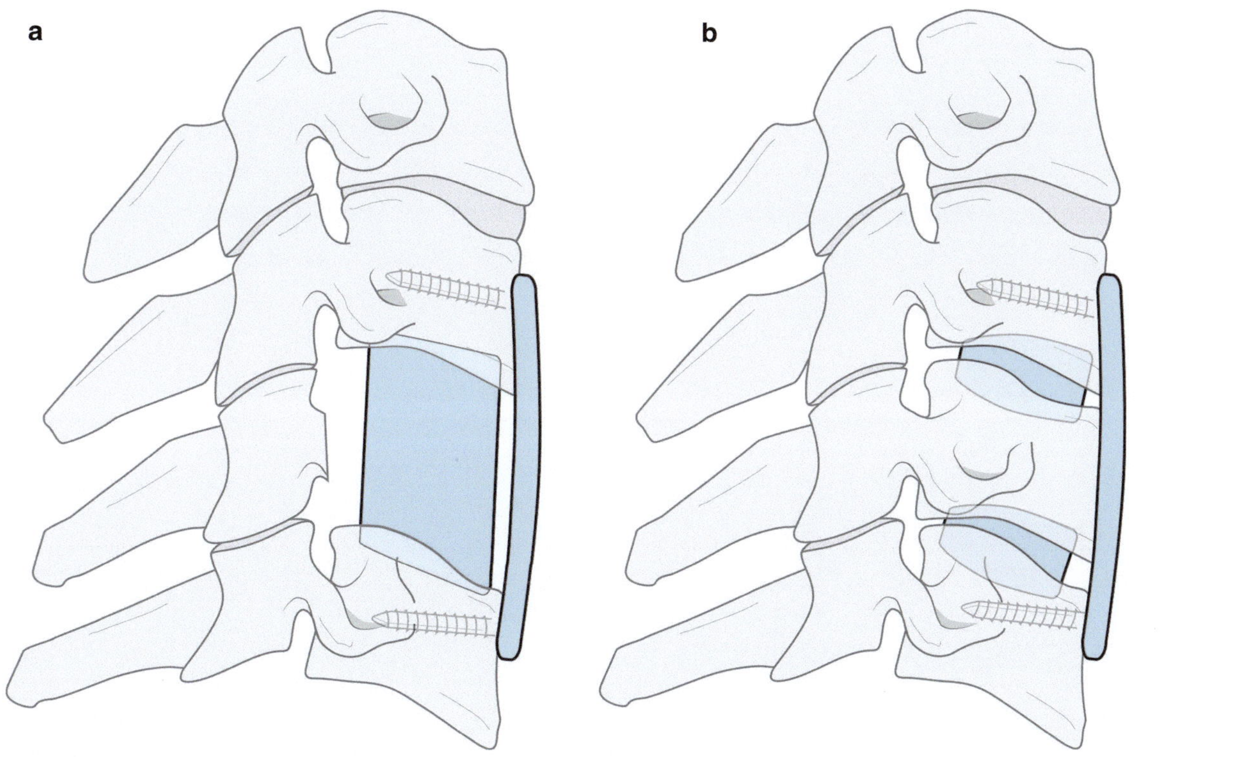

Fig. 6.6 (a) ACCF vs. (b) double-level ACDF

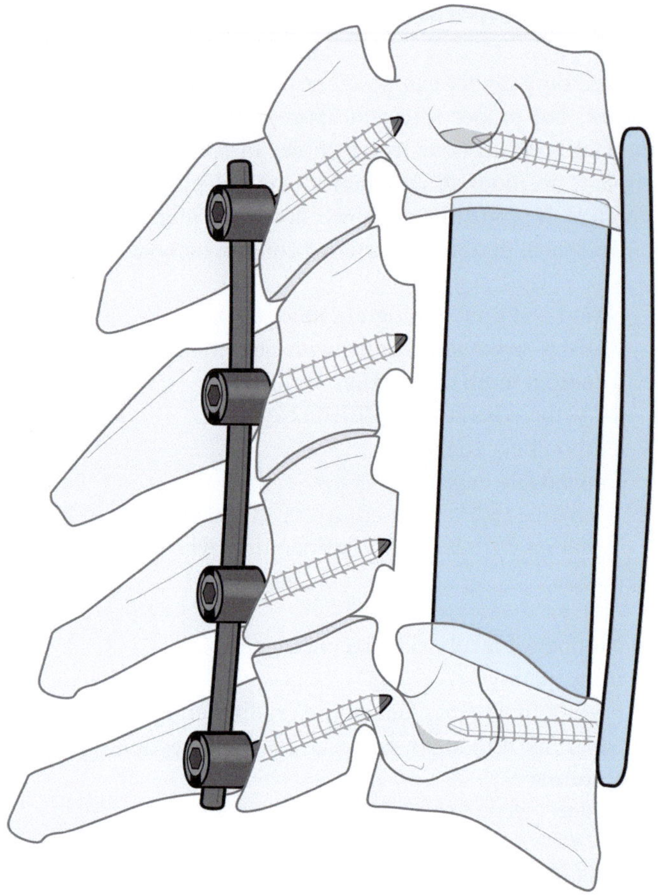

Fig. 6.7 Multi-level cervical corpectomy need to add posterior segmental instrumentation to enhance stability. Neverthless, long-term stability is obtained only if fusion occur

6.1.6 Anterior Cervical Plating

Anterior cervical plate has double biomechanical behavior depending on the applied forces; in fact, it acts as a tension band in extension and as a buttress plate in flexion.

Its use was initially meant to avoid graft dislodgement and/or subsidence.

First generation implants where nonconstrained load-sharing plates which required bicortical screw purchase in order to avoid excessive motion at the screw-plate interface and decrease screw loosening. Thus, engagement of the posterior vertebral cortex was technically challenging due to the risk of neurologic injury.

Then, constrained systems (static plates) that firmly lock the screws to the plate were developed. Securing the screws to the plates allows a more direct transfer of the applied forces from the spine to the plate and improved construct stiffness without the need for bicortical screw purchase.

Rigidly locked screw-plate systems theoretically expose the anterior column to potential stress shielding. Thus, dynamic semiconstrained plates were designed allowing for degree of load sharing between the plate and the spinal column (screw rotation, translation, or plate shortening).

Furthermore, in selected cases, specific materials need to be used (i.e. carbon-fiber plates) for oncologic patients undergoing adjuvant radiotherapy.

6.1.7 Motion Sparing Techniques: Cervical Disc Prosthesis

Cervical disc prosthesis gained popularity due to potential for the motion preservation and decreased rate of adjacent level degeneration (Fig. 6.8).

Despite these tempting promises, nowadays further studies still are needed to determine whether cervical disc replacement will have a lower rate of adjacent segment degenerative changes than anterior cervical discectomy and fusion. Different designs of prosthesis have been proposed, but none have been proved superior to the others, or to ACDF with regard to clinical outcomes at long-term follow-up.

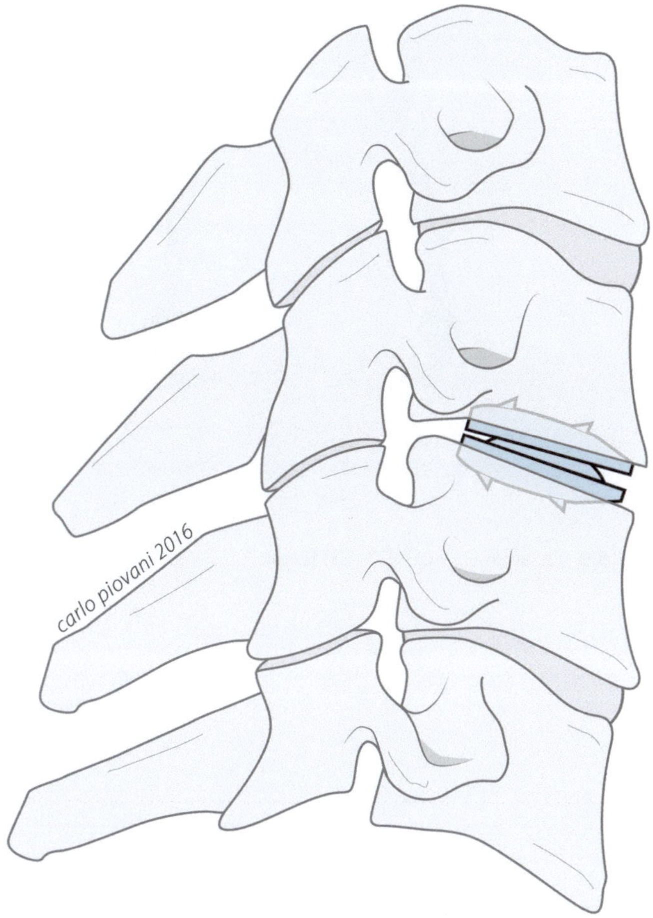

Fig. 6.8 Cervical disc prosthesis preserves motion while relieving symptoms at the operated segment

6.2 Posterior Instrumentation

6.2.1 Occipital Plate

Screws can be safely placed in the occiput, below the inion (to avoid damage to the sagittal intracranial sinus), but this requires a careful understanding of the occipital anatomy. The occipital cortex is thicker along the midline ridge, allowing for a strong bone purchase. The lateral cortex is thinner, only 3–7 mm, so screws must be placed bicortically and, if possible, with a trajectory convergent toward the midline in order to maximize screw length.

After drilling the hole, depth is checked, and when penetration of the distal cortex has occurred, they can be tapped bicortically and screws inserted.

Midline occipital plate must align well with the segmental screws placed below, but, if they do not, bilateral plates are preferred (if system allows).

Care must be taken in order to avoid injury to the cerebellum. Dural laceration with cerebrospinal fluid (CSF) leak is not uncommon and is dealt just by screw insertion.

Occipitocervical fixation (i.e., C0-C2, Fig. 6.9) can be achieved by combining occipital plating to the usual C1-C2 fixation techniques: poliaxial pedicle screw inserted across the C1-C2 joint (Magerl-Seeman technique) or segmental fixation with C1 lateral mass screws and C2 screws (either with pedicle, isthmic, or translaminar trajectory). In both cases, the construct is accomplished by connecting it to the occipital plate with bilateral rods.

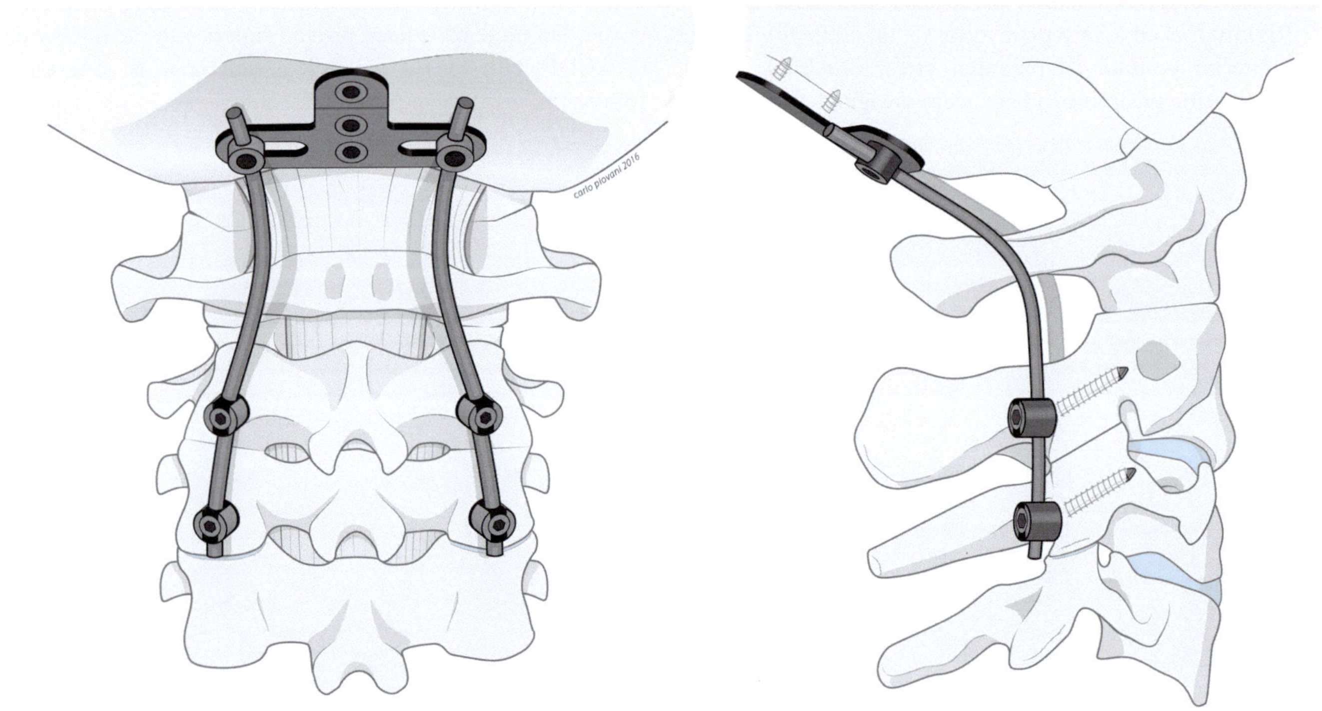

Fig. 6.9 Occipito-cervical (C0–C3) fusion

6.2.2 Segmental Screw Fixation

Segmental fixation in the cervical spine is achieved using various techniques: lateral mass (C1 and C3-C6) or pedicle screws (C2 and C7).

Fixation can be extended cranially up to the skull using occipital plates, and caudally down to the upper-thoracic spine using connectors or double-diameter rods. In alternative to screws' segmental fixation, hooks and wires were once used for fixation in the cervical spine.

6.2.2.1 C1 Lateral Mass Screws (Goel, Harms, and Melcher)

Entry point is just caudal to the posterior arch of C1, 3–4 mm lateral to the medial edge of the lateral mass (Fig. 6.10). The C2 spinal nerve and root ganglion (greater occipital nerve), which cross posterior to the C2 isthmus, must be mobilized distally. Some authors advocate its sacrifice, but in our experience, this is unnecessary. In fact, the C1 lateral mass screw is not fully threaded, in order to avoid irritating the C2 nerve.

Significant venous bleeding from the plexus can be expected during dissection, but can be controlled with Gelfoam and Surgicel and cottonoids.

Trajectory goes slightly medial (10°) and cephalald (10–20°). The tip of the screw should stop 3 mm before the anterior tubercle of C1 (as seen in the lateral fluoroscopy). The posterior arch can be used to properly orient the drill.

Although drilling must be bicortical, if too long screws are used, internal carotid artery and hypoglossal nerve are at risk (anterior to the C1 lateral mass).

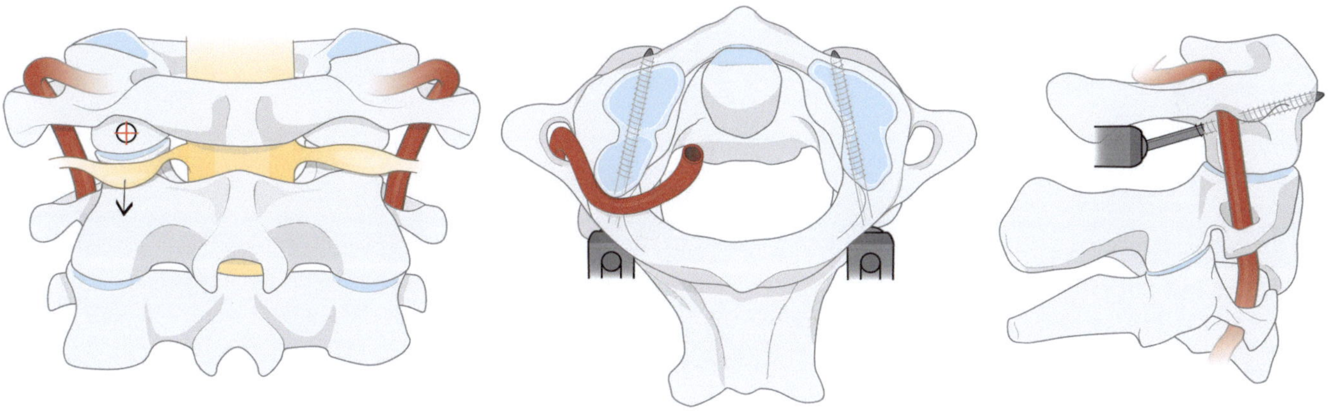

Fig. 6.10 C1 lateral mass screws positioning

6.2.2.2 C2 Pedicle Screws (Judet)

Entry point is identified determining the medial cortex of the isthmus and halfway between the upper and lower articular surfaces of C2 (Fig. 6.11).

Trajectory is oriented 20–30° cephalad and 15–25° medially as checked clinically by palpation of the medial wall of the isthmus of C2.

On its trajectory to the C2 body, the course of the vertebral artery is traversed, and therefore this structure is at risk for injury. Careful preoperative planning is needed because at least 8–10% of patients do not have anatomy that allows safe pedicle screw placement.

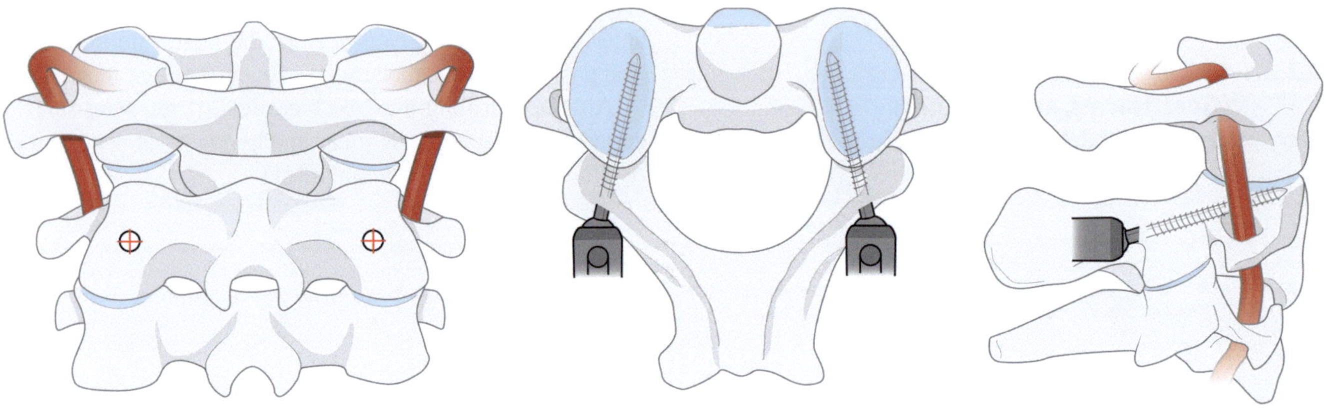

Fig. 6.11 C2 pedicle screws positioning

6.2.2.3 C2 Translaminar Screw Fixation (Wright)

This alternative technique (Fig. 6.12), described in 2004, was meant to further decrease the risk for vertebral artery injury. Generally, this method is used if anatomical constrains preclude C2 screws' placement with other trajectories.

Although, there are other risks related to injuries the canal content (dura or spinal cord), when properly done, these are small. Two screws are inserted through the base of the C2 spinous process and contained within the lamina on the contralateral side.

Translaminar screws can be connected to C1 lateral mass screws providing a biomechanical equivalent construct to the Harms technique for C1-C2 fixation.

This is not true if fixation includes the occiput. If it is extended below C2, rod contouring can be problematic because the translaminar screws are not aligned with the subaxial lateral mass screws.

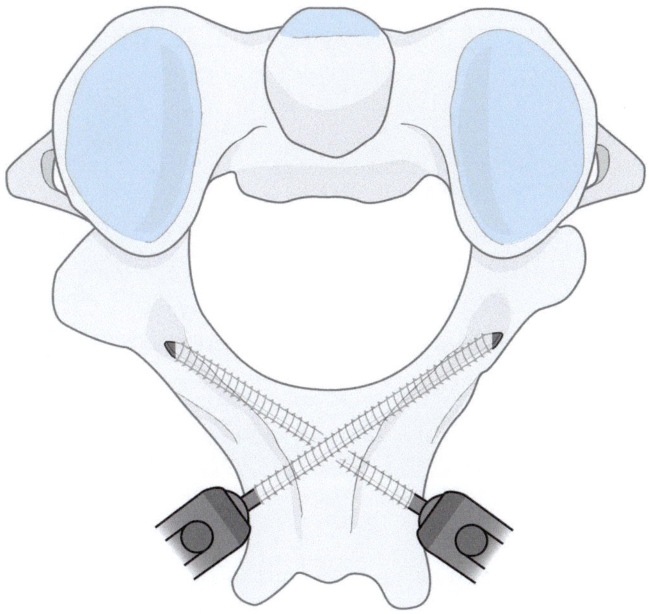

Fig. 6.12 Translaminar C2 screws

6.2.2.4 Subaxial Lateral Mass Screws (Magerl)

Lateral mass screws are placed by identifying the four boundaries of the lateral mass. Entry point is 1 mm medial and caudal to the geometric center of the lateral mass (Fig. 6.13). Penetrate the cortex using a high-speed burr. Trajectory is oriented laterally and cephalad until the far cortex (ideally parallel to the joint line, Fig. 6.14).

Appropriate length is measured and screw is placed bicortically (unicortical 14-mm screws have been shown to provide satisfactory fixation, thus, can be used if needed).

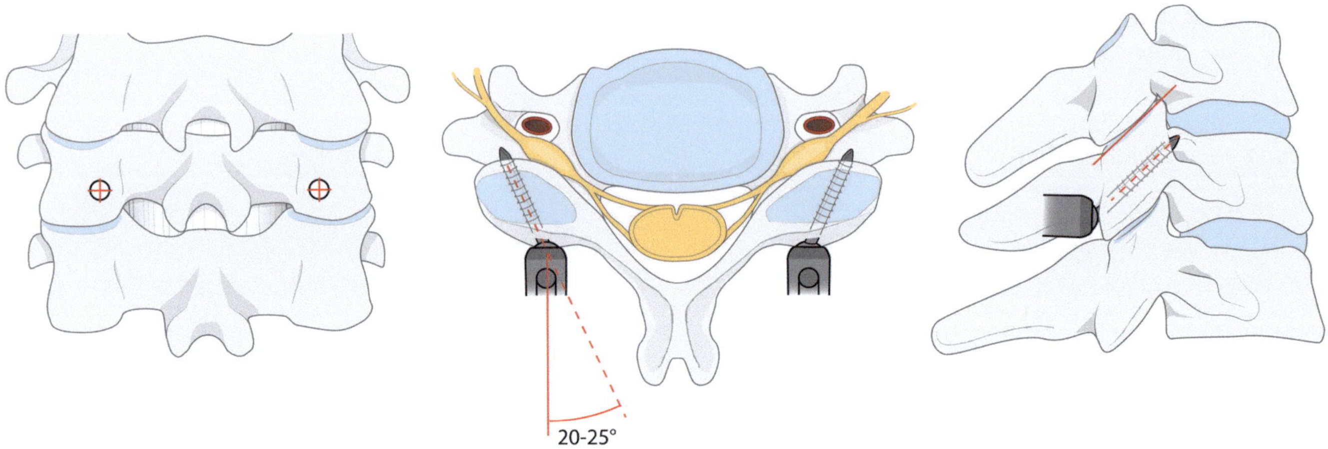

Fig. 6.13 Lateral mass screws positioning from C3 to C6 (Magerl technique)

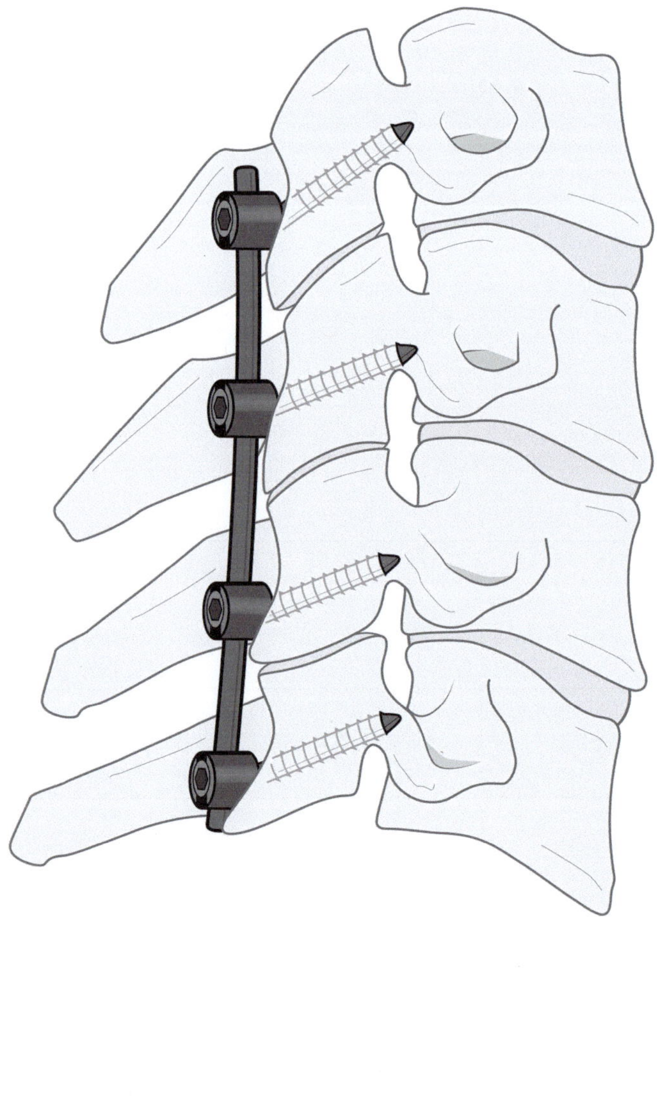

Fig. 6.14 Posterior instrumented fusion with lateral mass screw-rod contruct

6.2.2.5 C7 Transpedicular Screw

In case of small lateral masses, C7 can be fixed using transpedicular screws. Entry point is on the cranial part of the middle third of the lateral aspect of the lateral masses. Trajectory is 45°convergent and slightly caudal (Fig. 6.15).

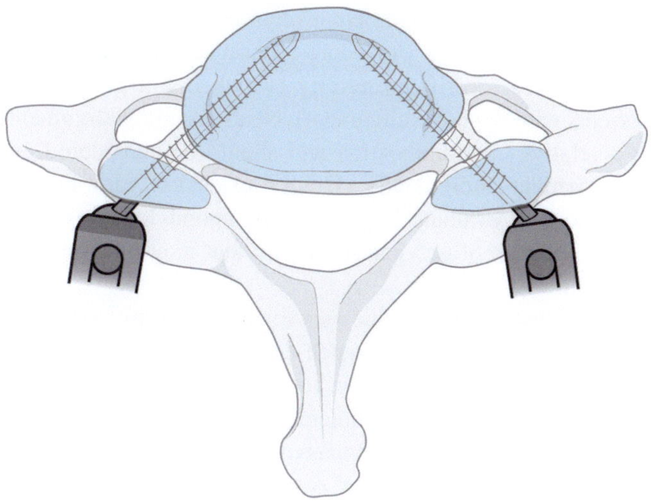

Fig. 6.15 C7 transpedicular screws positioning

6.2.3 Laminoplasty

Laminoplasty is indicated to relief multilevel compression on the spinal cord without causing iatrogenic instability (if properly performed) so supplemental fusion is not needed and cervical range of motion is preserved.

Lordotic sagittal alignment is required for posterior cervical canal expanding procedures, because the spinal cord must be free to drift posteriorly (which is not possible in patients with kyphotic alignment). Neutral sagittal alignment is a relative contraindication and should be considered on each patient individually. Flexion-extension radiographs may help to evaluate hypermobility, spondylolisthesis, or other evidence of cervical instability that may preclude the procedure.

The most popular techniques for laminoplasty are open-door laminoplasty and French-door laminoplasty. In the open-door technique, expansion of the spinal canal is achieved performing a complete opening at the level of the junction between the lateral mass and the lamina on one side, and a partial "greenstick-like" osteotomy on the contralateral side (which act as a hinge). The opening side of the lamina is chosen depending on symptoms.

In the French-door technique (Fig. 6.16), the spinous processes are splitted sagittally in the midline, and an incomplete hinge opening is made in both sides. The door is opened in the midline, and this creates a symmetric widening of the canal, which can be maintained using bone graft (resected laminae) or specifically designed plates.

6.2.3.1 Kurokawa Modification

In this modification of the French-door technique, the dorsal aspect of the spinous processes is cut in the midline using a high-speed drill, removed and used as grafts to hold open the laminoplasty.

6.2.3.2 Tomita Modification

In this modification, the spinous processes are split with a wire-saw. This has been termed the T-saw laminoplasty.

Foraminotomy may be indicated in addition to laminoplasty in the patients with concurrent symptomatic radiculopathy, or prophylactic, for prevention of postsurgical C5 palsy.

Performing the foraminotomy after laminoplasty on the hinge side puts the lamina at the risk of breaking. Care should be taken to preserve at least 50% of the facet joint to avoid instability.

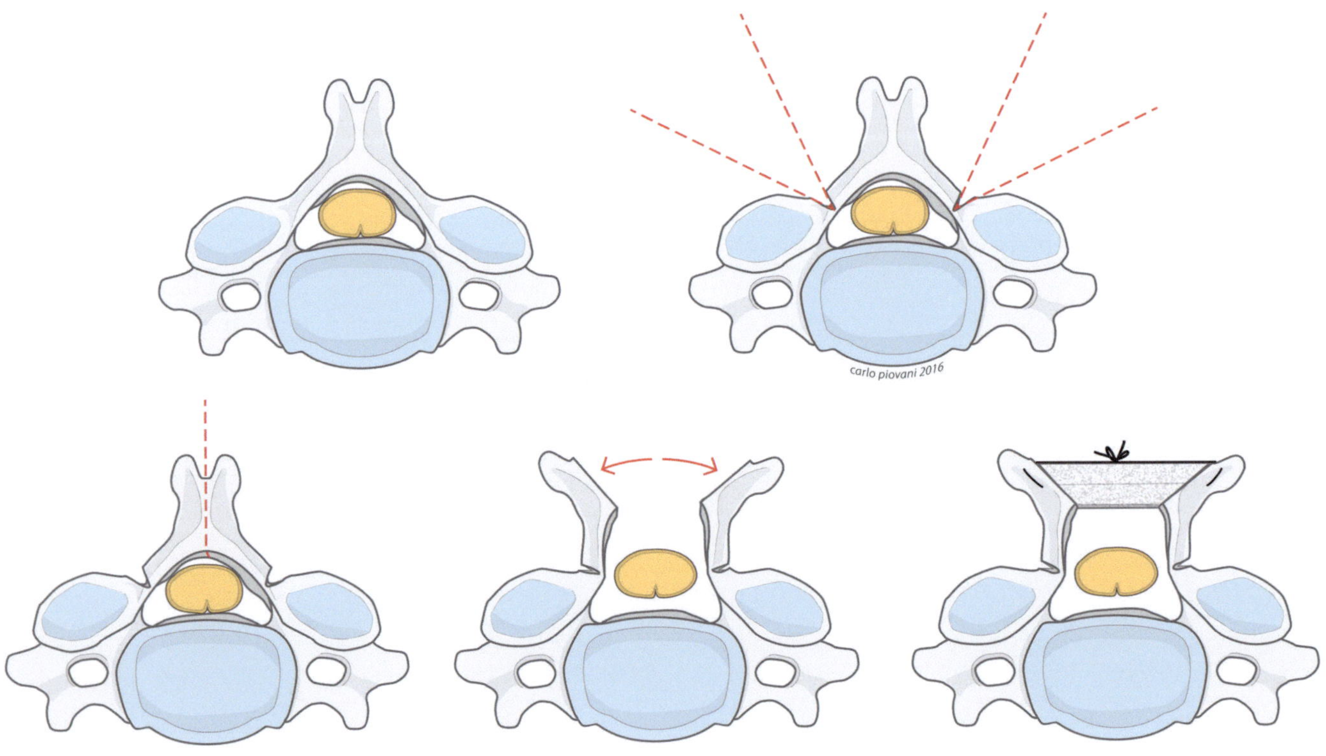

Fig. 6.16 Surgical steps for French-door laminoplasty

6.3 Focus on C1-C2 Fusion Techniques

6.3.1 Posterior C1-C2 Wiring (Gallie technique; Brooks and Jenkins technique)

Wiring techniques are less rigid and have proven efficacy since they have been used for many years.

Nowadays, they are rarely used as primary method of stabilization, but they remain useful for adjunctive stabilization and to maintain corticocancellous graft in position.

These techniques are relatively easy to perform, but require integrity of posterior arch of C1.

The Gallie technique (Fig. 6.17) is less resistant against rotational forces because fixation is at the midline, while the Brooks and Jenkins technique uses bilaterally located wires which provide more rotational and tensile stability. In fact, it creates a bone block between the arch of C1 and the C2 lamina.

Gallie technique is particularly resistant to flexion-extension solicitations, but is less effective in limiting translation, lateral bending, and rotation forces.

The wire is passed from the inferior aspect of C1 cranially, looped over the superior aspect of C1, and then carefully pulled caudal until it is distal enough to be looped over the spinous process of C2.

The arch of C1 and the lamina of C2 are decorticated and a rectangular corticocancellous bone graft is removed from the iliac crest and fashioned into a H configuration to snugly fit around the spinous process and over the lamina of C2. The two free ends of the wire, which are laterally placed, are then brought across the midline after the graft has been applied to the posterior surfaces of C1 and C2.

Violation of the canal exposes the spinal cord at risk of damage during the procedure (Brooks and Jenkins more than Gallie technique), especially if there is any degree of subluxation between vertebrae.

When used as a primary fusion method, halo vest immobilization is continued for 10–12 weeks.

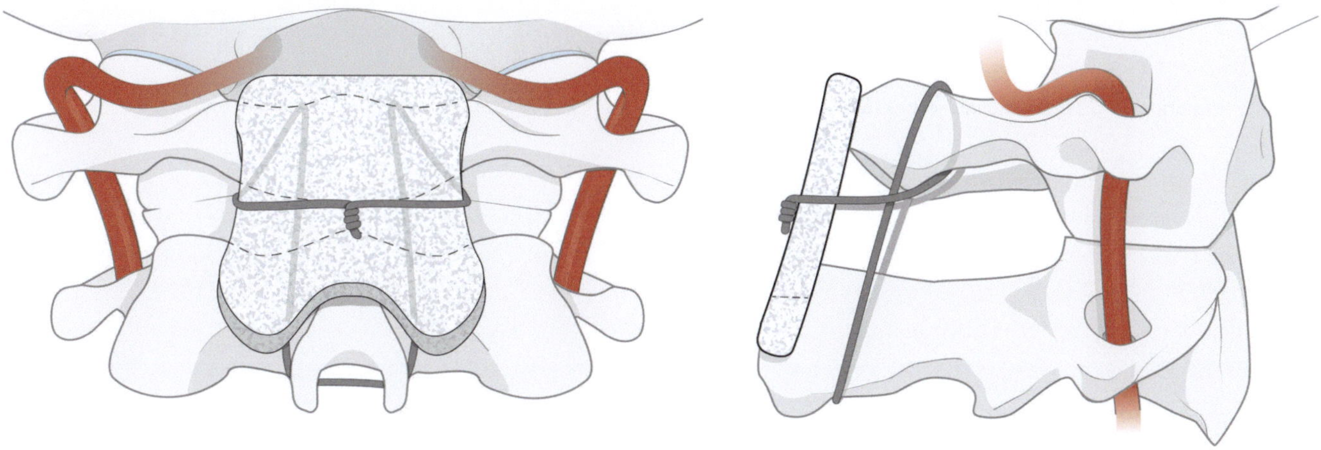

Fig. 6.17 Posterior C1-C2 wiring (Gallie technique)

6.3.2 Posterior C1-C2 Transarticular Screws (Magerl and Seeman)

More stable than wiring techniques (to whom may be easily combined) and does not necessitate integrity of the posterior arch. Nonetheless, it is technically difficult and requires perfect reduction of C1-C2 before screw insertion.

The atlanto-axial joints are exposed by opening the posterior capsule, thus, making the C1-C2 joint visible. The articular cartilage of the posterior half of the facet joint is removed with either a small chisel or a sharp curette, after which the joints are packed with cancellous bone and the screws are inserted.

The entry point of the screw is at the lower edge of the caudal articular process of C2, 3 mm above the C2-C3 joint line. The trajectory (Fig. 6.18) goes through the isthmus near to its posterior and medial surface in a strictly sagittal direction. It then enters the lateral mass of the atlas close to its posteroinferior edge. Anteriorly, the drill perforates the cortex of the lateral mass of C1. The internal carotid artery and the hypoglossal nerve, which typically lie less than 3 mm anterior to the C1 lateral mass, can be injured if the screw is too long. Drilling in horizontal direction must be avoided in order to protect the vertebral artery which runs upward anteriorly to the C1-C2 joint and could easily be damaged at the level of C2. A firm collar may be worn for 8–12 weeks, but patients are allowed to remove the collar for daily care. Aberrant trajectory of the vertebral artery may render it impossible.

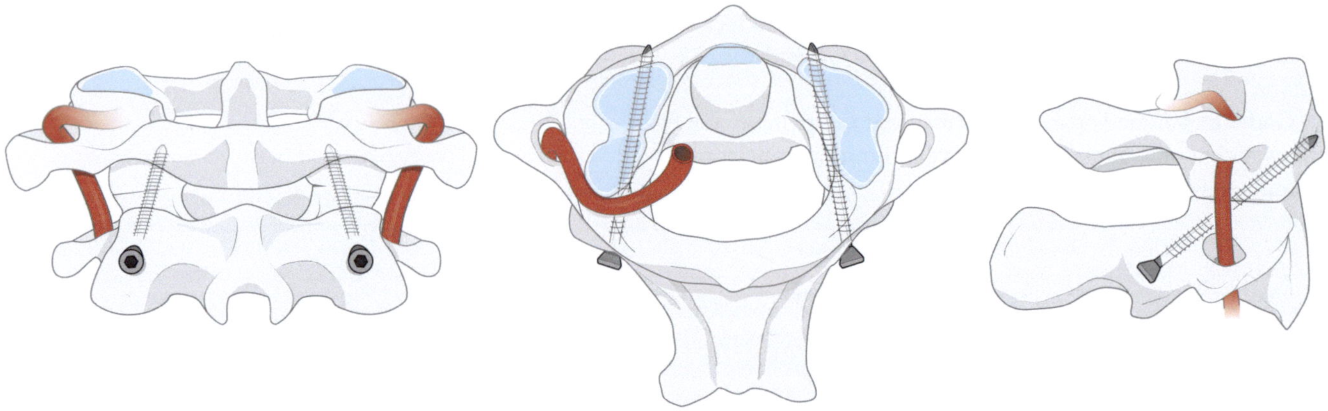

Fig. 6.18 Posterior C1-C2 transarticular screws (Magerl and Seeman)

6.3.3 Posterior C1-C2 Fusion Using Rod and Polyaxial Screw Construct (Harms Technique)

Fixation using screw rods technique (Fig. 6.19) provides the more rigid fixation method.

Postoperatively, a rigid cervical orthosis must be worn for 8–12 weeks, but patients are allowed to remove the collar for daily care. If additional posterior wiring has been used, a soft collar can be worn instead of a firm collar.

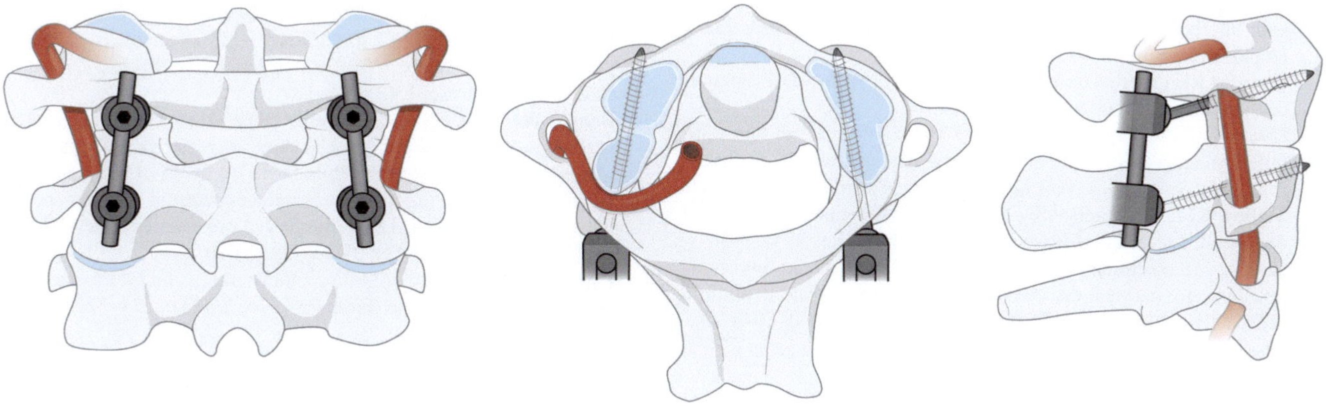

Fig. 6.19 Posterior C1-C2 stabilization using polyaxial screws (Harms technique)

References

1. Harms J, Melcher RP (2001) Posterior C1-C2 fusion with polyaxial screw and rod fixation. Spine (Phila Pa 1976) 26(22):2467–2471
2. Melcher RP, Puttlitz CM, Kleinstueck FS, Lotz JC, Harms J, Bradford DS (2002) Biomechanical testing of posterior atlantoaxial fixation techniques. Spine (Phila Pa 1976) 27(22):2435–2440
3. Puttlitz CM, Melcher RP, Kleinstueck FS, Harms J, Bradford DS, Lotz JC (2004) Stability analysis of craniovertebral junction fixation techniques. J Bone Joint Surg Am 86-A(3):561–568
4. Robinson RA, Smith G (1955) A nterolateral cervical disc removal and interbody fusion for cervical disc syndrome. Bull Johns Hopkins Hosp 96:223–224
5. Cloward RB (1958) The anterior approach for removal of ruptured cervical disks. J Neurosurg 15(6):602–617
6. Lowery GL, McDonough RF (1998) The significance of hardware failure in anterior cervical plate fixation. Patients with 2- to 7-year follow-up. Spine (Phila Pa 1976) 23(2):181–186. discussion 186–7
7. Malloy KM, Hilibrand AS (2002) Autograft versus allograft in degenerative cervical disease. Clin Orthop Relat Res 394:27–38
8. Bolesta MJ, Rechtine GR 2nd, Chrin AM (2000) Three- and four-level anterior cervical discectomy and fusion with plate fixation: a prospective study. Spine (Phila Pa 1976) 25(16):2040–2044. discussion 2045–6
9. Chou YC et al (2008) Efficacy of anterior cervical fusion: comparison of titanium cages, polyetheretherketone (PEEK) cages and autogenous bone grafts. J Clin Neurosci 15(11):1240–1245
10. Fransen P (2010) A simplified technique for anterior cervical discectomy and fusion using a screw-plate implanted over the Caspar distractor pins. Acta Orthop Belg 76(4):546–548
11. Hauerberg J et al (2008) Anterior cervical discectomy with or without fusion with ray titanium cage: a prospective randomized clinical study. Spine (Phila Pa 1976) 33(5):458–464
12. Emery SE, Fisher JR, Bohlman HH (1997) Three-level anterior cervical discectomy and fusion: radiographic and clinical results. Spine (Phila Pa 1976) 22(22):2622–2624. discussion 2625
13. Song KJ, Lee KB (2006) A preliminary study of the use of cage and plating for single-segment fusion in degenerative cervical spine disease. J Clin Neurosci 13(2):181–187
14. Beutler WJ, Sweeney CA, Connolly PJ (2001) Recurrent laryngeal nerve injury with anterior cervical spine surgery risk with laterality of surgical approach. Spine (Phila Pa 1976) 26(12):1337–1342
15. Jung A, Schramm J (2010) How to reduce recurrent laryngeal nerve palsy in anterior cervical spine surgery: a prospective observational study. Neurosurgery 67(1):10–15. discussion 15
16. Palit M et al (1999) Anterior discectomy and fusion for the management of neck pain. Spine (Phila Pa 1976) 24(21):2224–2228

Tracheotomy Surgical Technique

7

M. Ghirelli, F. Mattioli, G. Molinari, I. Cena, and L. Presutti

This procedure allows creating a reversible connection between the tracheal lumen and the skin.

Normally, tracheotomy is performed to guarantee safe ventilation after airways surgery or extensive neck surgery. Tracheotomy is mandatory in all surgical procedures at high risk of neck oedema or acute airway compromise.

In cervical spine surgery, the indications to tracheotomy are essentially correlated to avoid an upper airway obstruction (actual or potential). This event might occur in surgery involving:

- Multilevel approach to cervical spine from C3 to C6 (the retropharyngeal space is located at this level)
- Anterior and posterior combined approach
- Transmandibular approach to upper cervical spine with median lingual split

M. Ghirelli (✉) • F. Mattioli • G. Molinari • I. Cena • L. Presutti
Head and Neck Surgery Department, University Hospital of Modena, Modena, Italy
e-mail: michael.ghirelli@gmail.com

S. Boriani et al. (eds.), *Atlas of Craniocervical Junction and Cervical Spine Surgery*, DOI 10.1007/978-3-319-42737-9_7

7.1 Surgical Technique

The patient is placed on supine position, with the head extended using a shoulder roll.

Hyperextension of the neck is mandatory to obtain an optimal exposition of surgical landmarks.

7.1.1 Landmarks on the Skin

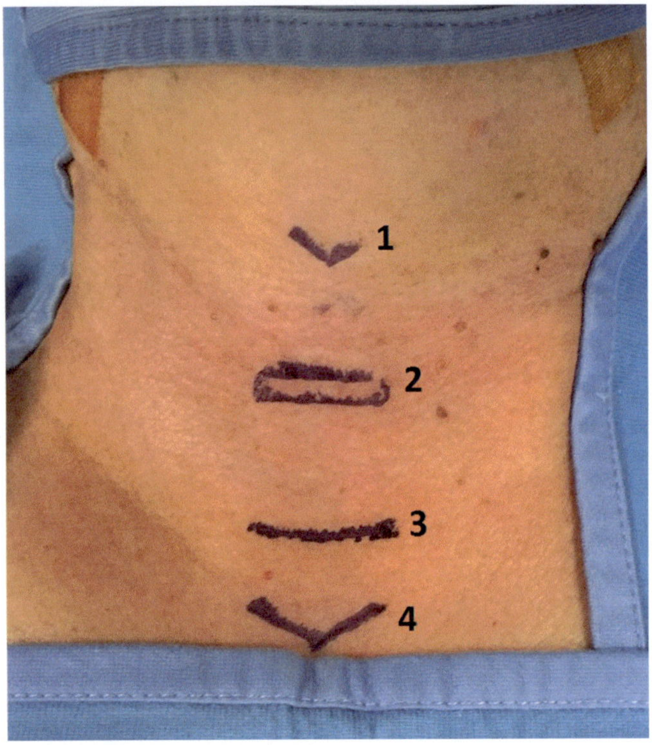

Fig. 7.1 Identification of skin landmarks before incision

The surgeon has to identify (Fig. 7.1):

1. *Notch of thyroid cartilage*
2. *Cricoid cartilage*
3. *Skin incision line*
4. *Suprasternal notch*

7.1.2 1st Surgical Step: Incision of the Skin

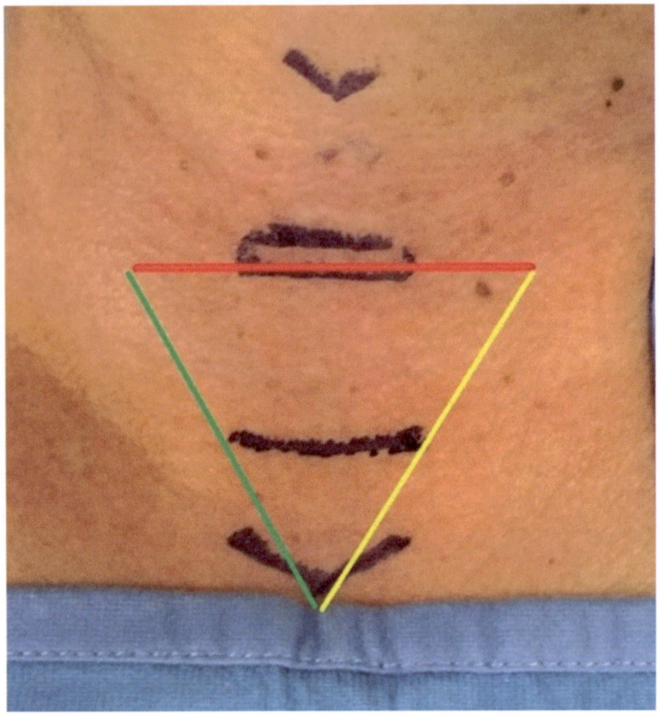

Fig. 7.2 Ideal triangle formed by cricoid cartilage, suprasternal notch and SCM

7.1.2.1 Anatomical Key Points (Fig. 7.2)

- Cricoid cartilage
- Suprasternal notch
- Sternocleidomastoid muscle (*SCM*)

7.1.2.2 Pearls and Potential Complications

The surgeon normally performs a 3–4 cm horizontal skin incision (Fig. 7.3). The incision is usually placed at least two fingerbreadths below the cricoid cartilage in an inverse triangular space formed by the cricoid cartilage (red line) at the base, the suprasternal notch at the apex and SCM in the lateral border (yellow and green line). The identification of the anterior jugular vein on the skin is useful to prevent its accidental section. The incision must involve the skin, subcutaneous tissue and platysma muscle until exposure of SLDCF (Fig. 7.4).

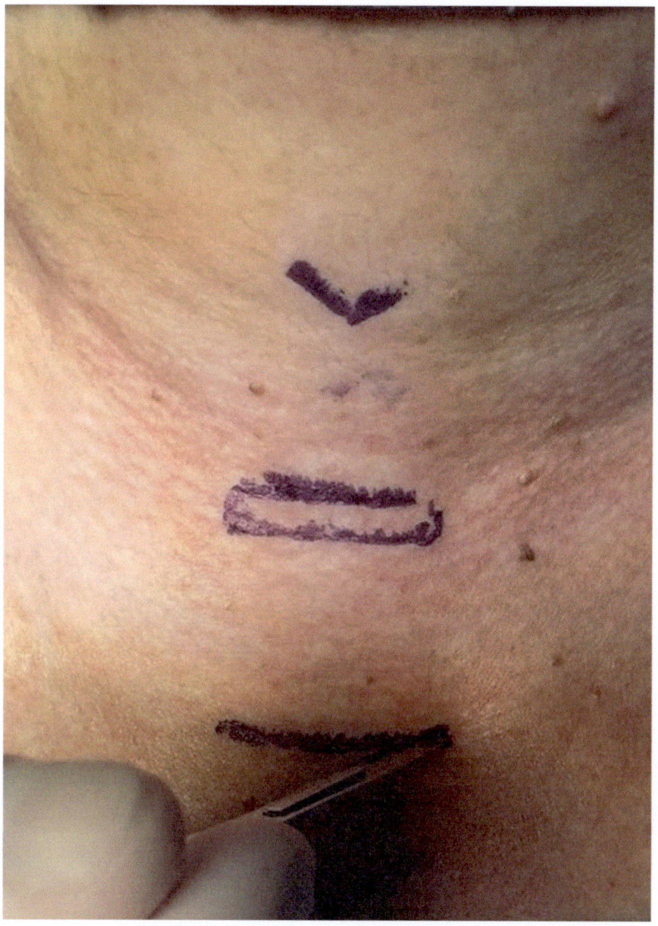

Fig. 7.3 Skin incision with 15 or 10 blade

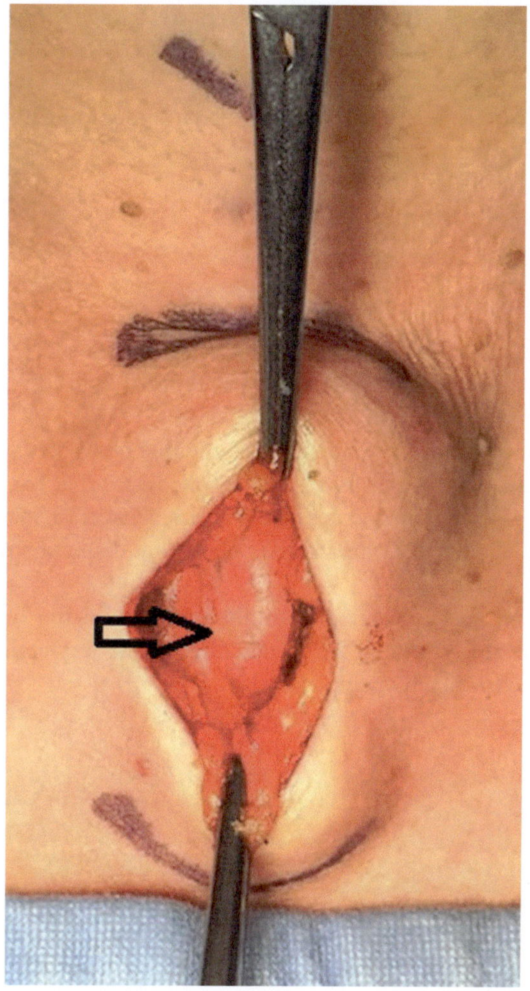

Fig. 7.4 The black arrow shows SLDCF

7.1.3 2nd Surgical Step: Incision of SLDCF and Identification of Median Line

The incision of SLDCF is performed in order to identify the so-called strap muscles.

7.1.3.1 Anatomical Key Points

- Sternohyoid muscles
- Sternothyroid muscles

7.1.3.2 Pearls and Potential Complications

The sternohyoid muscle is more superficial than the sternothyroid (Fig. 7.5).

The surgeon has to identify the midline between the strap muscles. This region is an aponeurotic space that has to be dissected. Principal complications are related to incorrect localization of the midline and consequent dissection lateral to the trachea.

In order to identify the midline more easily, the surgeon should always check the localization of the laryngeal axis and the trachea by palpation.

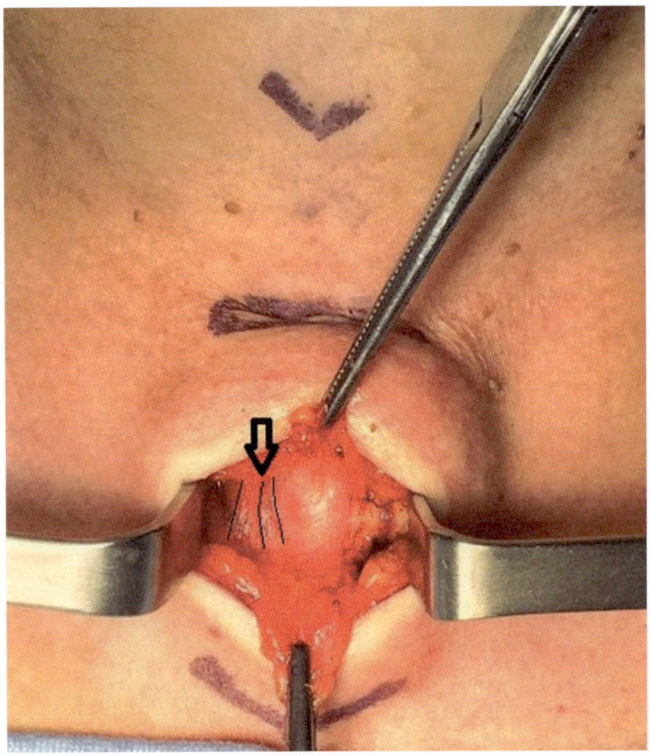

Fig. 7.5 Identification of sternohyoid and sternothyroid muscle on the right side (*black arrow*)

7.1.4 3rd Surgical Step: Incision of MLDCF and Identification of Thyroid Gland and Trachea

Below the strap muscles, along the midline, we can find the isthmus of the thyroid gland (Fig. 7.6).

If the isthmus covers a large portion of the trachea, it is necessary to ligate the thyroid vessels and perform isthmus dissection before tracheotomy procedure (transisthmus tracheotomy).

Normally, the tracheotomy is positioned below the isthmus.

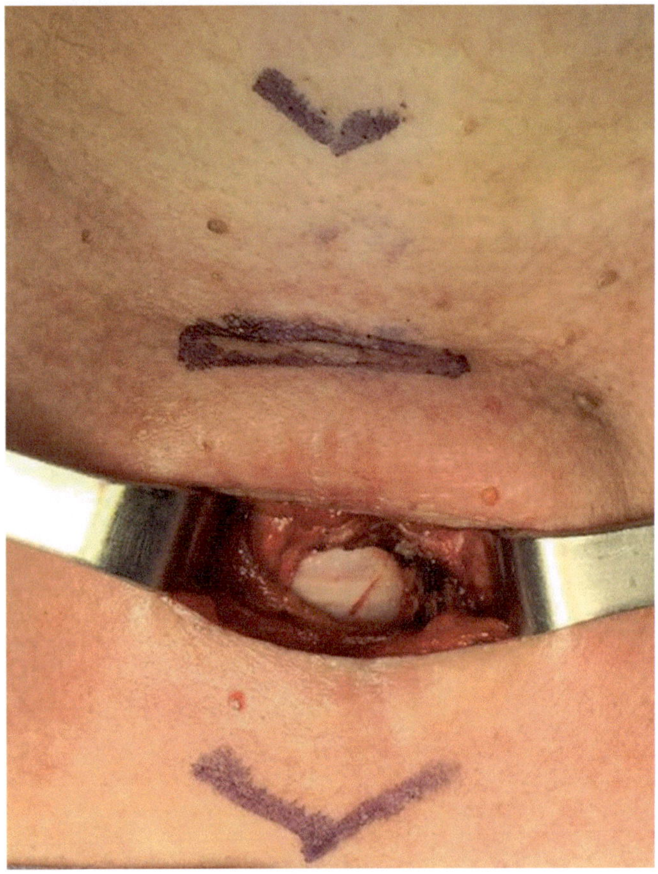

Fig. 7.6 Thyroid isthmus is not prominent (on each side strap muscles are visible)

7.1.5 4th Surgical Step: Incision of the Trachea

The cricoid should be palpated to establish the correct level of the tracheal incision.

The tracheotomy could be performed in two ways:

- By a linear incision between the second and third ring of the trachea
- By an inferiorly based flap, consisting of the anterior portion of a single tracheal ring being sutured to the inferior skin margin
- The choice between the two techniques depends on several factors (Table 7.1)

Table 7.1 How to choose tracheotomy technique

Liner incision	Inferiorly based flap
Normal position of the trachea	Deeper position of the trachea
Management of the tracheotomy requires the presence of an ENT surgeon or an anaesthesiologist	No-ENT or anaesthesiologist management of the tracheotomy required
First neck surgery/no neck scars	Previous neck surgery/neck scars
Thin neck	Thick and short neck
Absence of thyroid goitre	

7.1.5.1 Pearls and Possible Complications

Caustication of the incision line with bipolar forceps is useful to prevent bleeding and to dissect the fascia surrounding the trachea (Fig. 7.7).

The preliminary incision on the anterior wall of the trachea has to involve only the membranous part of it, in order to prevent perforation of the cuff of the anaesthesiology tube.

The monopolar blade is useful to perform this step (Fig. 7.8 a, b). Tracheotomy should be then completed with surgical scissors (Fig. 7.9).

After performance of complete tracheal incision, a Killian speculum is inserted to open the tracheal lumen safely (Fig. 7.10a).

A special curved endotracheal tube (Montandon endotracheal tube) is introduced in the gap to assure continuation of anaesthesia and later replaced by a tracheotomy tube at the end of the surgery (Fig. 7.10b).

The tracheotomy tube is sutured to the skin as an additional precaution to prevent accidental dislodgement of the tube. To avoid subcutaneous emphysema, pneumothorax and infection, the tracheotomy wound is never closed tightly around the tube.

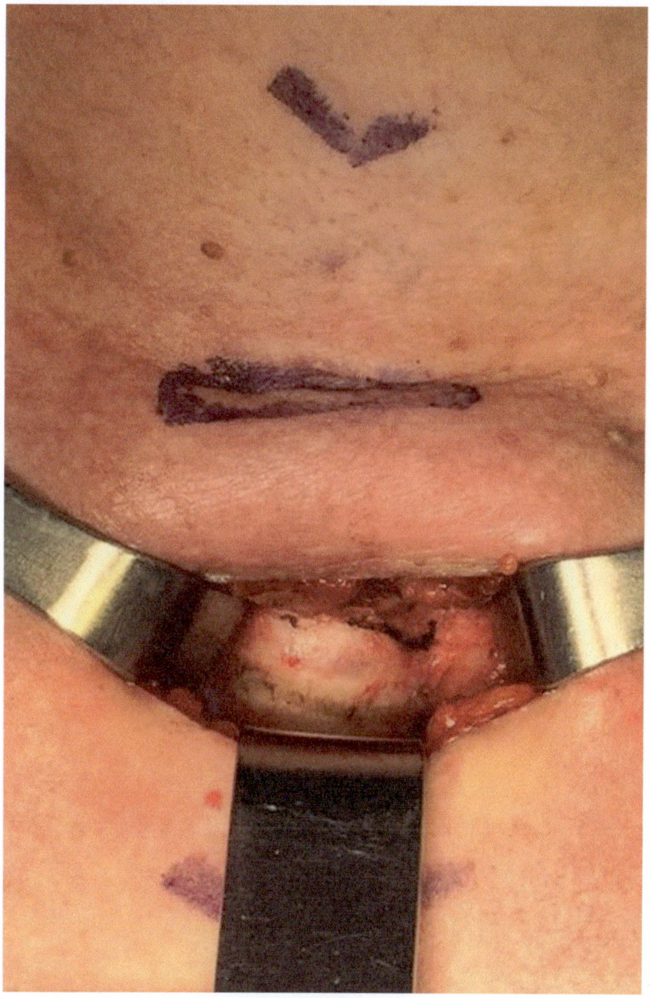

Fig. 7.7 Cauterization of tracheal incision line before actual incision

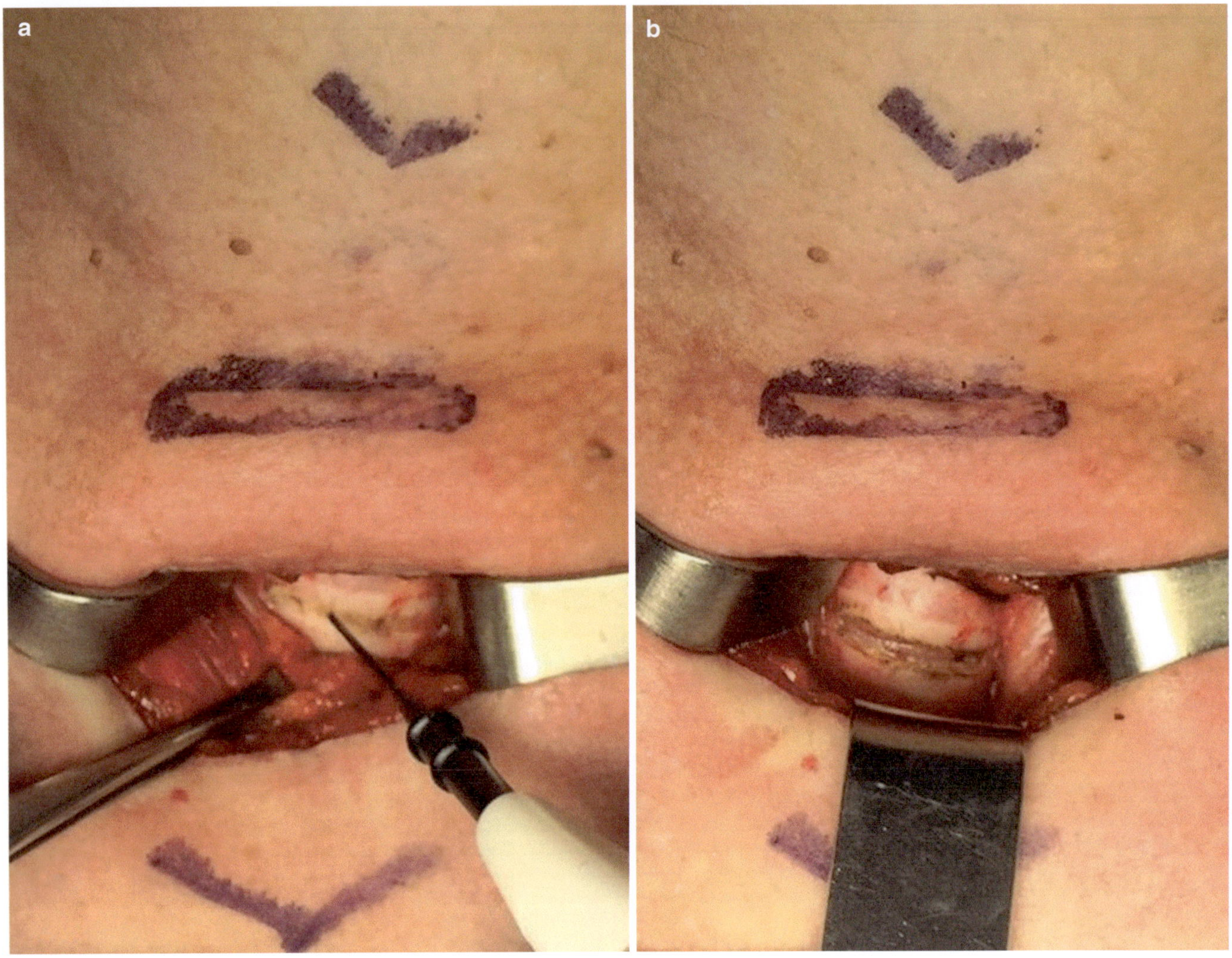

Fig. 7.8 (**a**) Use of fine monopolar blade (**b**) Incision of the membraneous part of the tracheal ring

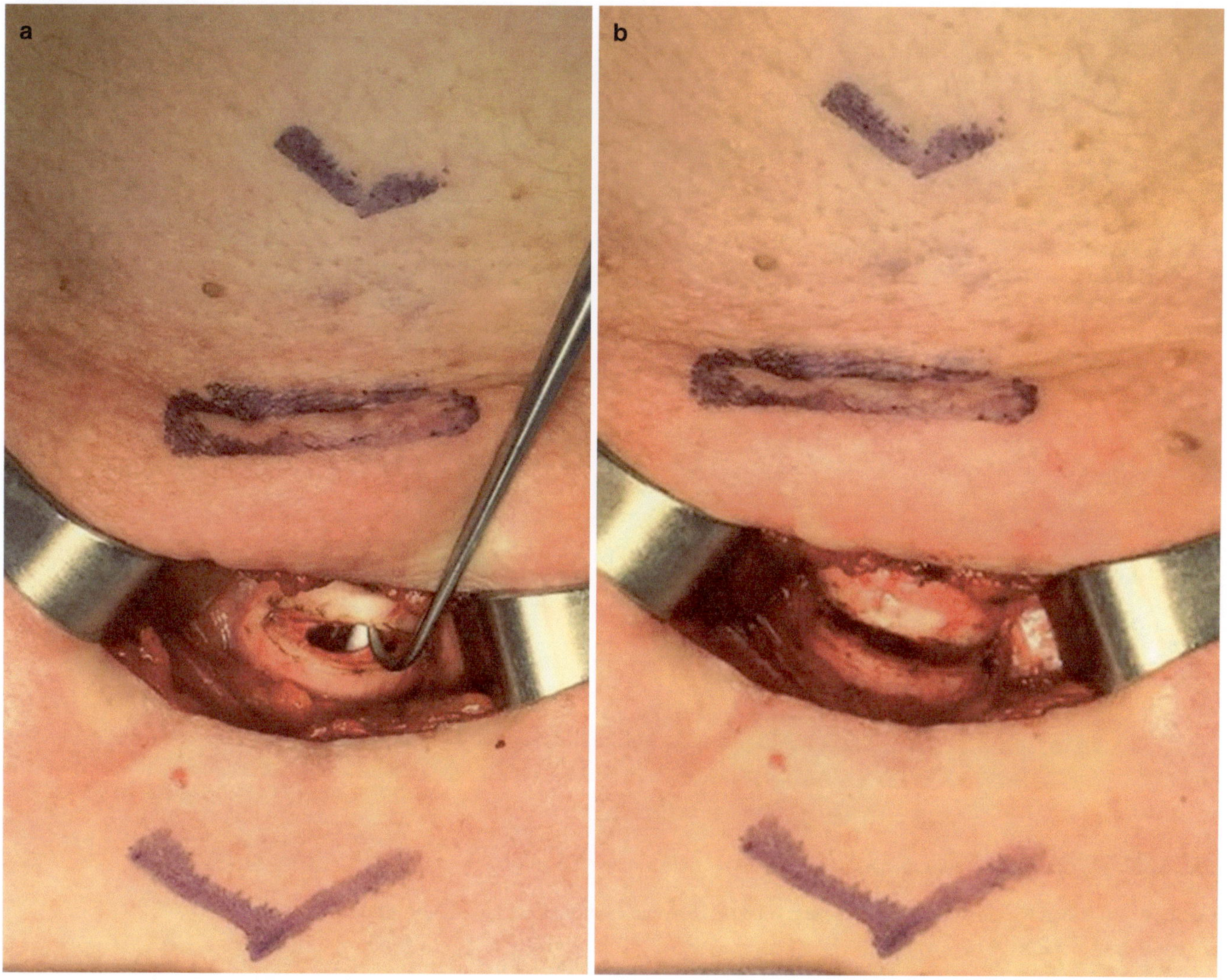

Fig. 7.9 (**a**) A hook could help separating tracheal rings; (**b**) Appearance of complete tracheal incision, after removal of the anesthesiology tube

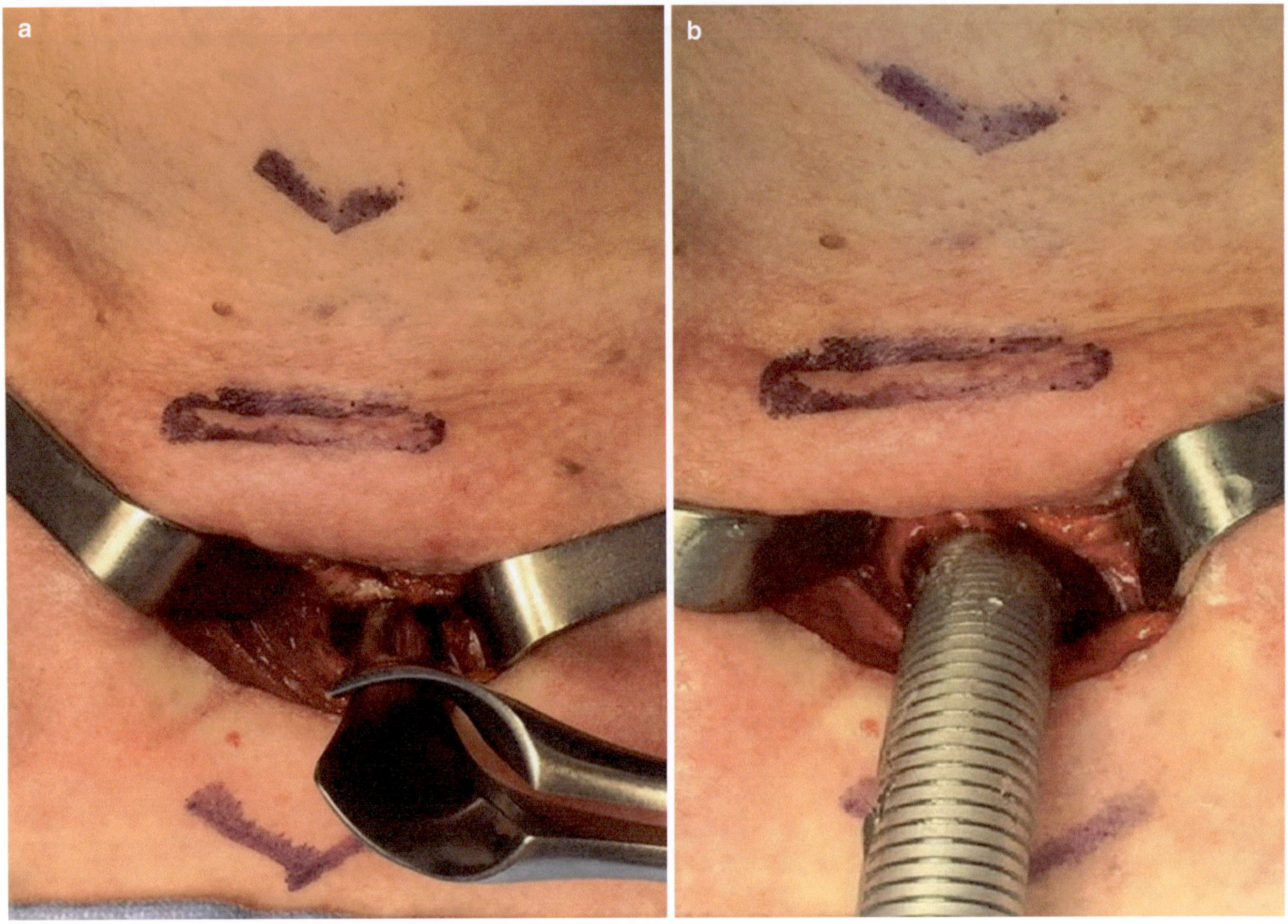

Fig. 7.10 (**a**) Use of Killian to enlarge the tracheal window; (**b**) Montandon endotracheal tube positioning

8 Surgical Planning in Cervical Spine Oncologic Patients

S. Boriani, R. Ghermandi, M. Girolami, and A. Gasbarrini

8.1 Epidemiology

Primary bone tumors of the spine are very rare [1]; they comprise only 10% or less of all bone tumors. In the United States, about 7,500 new cases are estimated per year. The overall world occurrence can be expected 2.5–8.5 cases per million inhabitants per year. Compared to primary spinal tumors, metastatic tumors are much more common in the spine: the spine is the most common skeletal region for secondary tumors. Basic knowledge about diagnosis and treatment strategy must be known for early recognition of a spine tumor. Further, diagnostic elements to differentiate primary from metastatic tumor are required.

8.2 Clinical Findings and Imaging

In the common population, back and neck pain related to activity are very frequent symptoms, particularly in the adults, and mostly related to disk prolapse, degenerative changes, and spondylolisthesis. These symptoms should be first approached with exercise, pain killer, and postural rehabilitation.

Only rarely a primary tumor can be suspected on symptoms and clinical findings. Pain is the most frequent onset symptom, but less common compared to metastases; cord compression and pathological fracture occur as well less frequently.

Latent lesions (most of hemangiomas, fibrous dysplasia, exostosis) are asymptomatic by definition, and diagnosis is achieved incidentally on imaging performed for other reasons.

Cases of incidental detection of chordoma are also reported. This malignant tumor is characterized by very slow growth; when arising in the sacrum, it can be diagnosed late, once already by large size.

Painful torticollis in an adolescent is strongly suggesting for osteoid osteoma or osteoblastoma located in the cervical spine, mostly not visible on standard radiograms, and technetium isotope bone scan helps to localize the lesion: CT scan – performed at the level of intense uptake – displays the pathognomonic image of osteoid osteoma, a small island of pathologic ossification surrounded by a lytic halo and frequently by a wide reactive bone formation.

Imaging studies (CT scan and MRI) can delimitate diagnostic hypothesis as some patterns are quite peculiar. Giant cell tumor and Ewing sarcoma are lytic conditions; most osteosarcomas are characterized by extensile aggressive pathologic bone formation with fuzzy borders. Multicameral balloon-like pattern with double density content is typical of aneurysmal bone cysts. Infiltrating erosions inside the cancellous bone of a vertebral body arising from the posterior wall is suggestive of chordoma. Soft tissue masses arising from the posterior elements with rounded calcifications are typical of peripheral chondrosarcoma.

Angiography shows pathologic vascularity, and selective arterial embolization has become an indispensable tool to reduce intraoperative bleeding.

Histological diagnosis should always be achieved by biopsy, but clinical, laboratory, and imaging studies are important to orientate diagnosis and select the biopsy technique.

8.3 Biopsy

The target of a biopsy is to obtain a specimen of the tumor which is representative of the lesion and large enough to allow histological and ultrastructural analysis as well as immunologic stains. The surgeon must be able to recognize the vital part of the tumor and to discard the necrotic or reactive part. Culture results may also confirm or rule out infection.

S. Boriani (✉) • R. Ghermandi • M. Girolami • A. Gasbarrini
Oncologic and Degenerative Spine Surgery Department, Rizzoli Orthopedic Institute, Bologna, Italy
e-mail: stefanoboriani@gmail.com

S. Boriani et al. (eds.), *Atlas of Craniocervical Junction and Cervical Spine Surgery*, DOI 10.1007/978-3-319-42737-9_8

Three traditional forms of spinal biopsy are available to the surgeon: incisional, by needle or trocar, and excisional.

A number of basic principles should be observed when performing incisional and excisional biopsy in order to prevent tumor contamination of the surrounding tissue, which is the major risk of biopsy. Transverse incisions and flaps should be avoided; the tumor should be approached in the most direct manner possible, but avoiding the anatomical interspaces (so-called *extracompartmental* spaces) commonly followed in orthopedic surgery. The approach should cross muscular structures.

If the planned definitive procedure is en bloc resection, the biopsy incision from the skin to the tumor mass should be placed so that it may be excised in a single bloc with the tumor and its margins. It is evident that in the spine – particularly in the cervical spine – these principles are extremely difficult and impossible to accomplish. When the tumor is growing in the posterior elements, it can be approached crossing the muscle, but when it is located in the vertebral body, this is impossible, and any open biopsy will necessarily contaminate one or more compartments.

Pathologic tissues should be handled carefully, hemostasis should be meticulous and a suture performed on all the anatomical layers. Bleeding from exposed bone or from uncauterized vessels and injured muscle will form a postoperative hematoma that may carry tumor cells beyond the margins of the intended excision and contaminate tissues far proximal or distal to the primary lesion. The margin of the soft-tissue mass is often the most helpful to biopsy, as central portions are frequently necrotic. The surgeon should take care not to crush or distort the specimen, so as to maintain its architecture.

Frozen section biopsy can be obtained during the excisional procedure when diagnosis is highly probable and must be just confirmed. This is frequently the case in cervical tumors.

Excisional biopsy can be considered only for conditions whose radiographic pattern is pathognomonic, for example, osteoid osteoma.

Trocar biopsies allow the minimal tumor contamination but are subject to sampling errors and provide small specimens for evaluation. The authors' preferred technique is biopsy by 12-gauge trocar performed under CT scan control. Trocar allows to remove carrots useful to study tumor histological architecture and to provide material for immunohistochemical studies. CT scan allows to reach even small lesions inside the vertebral body. In the cervical spine trocar biopsy can be performed by anterior approach prevascular or by posterior approach

Diagnosis should always be decided on histological diagnosis: biopsy is mandatory. Conversely some latent, asymptomatic lesions like hemangioma, fibrous dysplasia, and exostosis are characterized by such a pathognomonic imaging that biopsy is not immediately necessary. In this case however, it is wise to follow the evolution of the disease by sequential imaging each 3–4 months to be fully aware of the latency of the condition.

8.4 Oncological Staging

Enneking in 1980 [2] proposed a system to stage the biological behavior of tumors of bone and soft tissue. Later on, this system was applied also to the spine [3–6]

This system divides benign tumors into three stages (named S1, S2, and S3) and localized malignant tumors into four stages (named IA, IB, IIA, and IIB). Two further stages include metastatic high-grade intra- and extracompartmental malignant tumors (named IIIA and IIIB, respectively). This classification is based on clinical features, radiographic pattern, CT scan/MRI data, and histological findings. Each of these is related to overall prognosis and by necessity is very much linked to categories of surgical procedures based on the concept of "margin" [7].

8.4.1 Benign Tumors

The first stage of benign tumor (S1, latent, inactive) includes asymptomatic lesions, bordered by a true capsule, which is usually seen as a sclerotic rim on plain radiograms. These tumors do not grow or only grow very slowly. No treatment is required, unless palliative surgery is needed for decompression or stabilization. Benign tumors stage 2 (S2, active) grows slowly, causing mild symptoms. The tumor is bordered by a thin capsule and by a layer of reactive tissue, sometimes found on plain radiograms as an enlargement of the tumor outline. Bone scan is positive. An intralesional excision can be performed with a low rate of recurrence. The incidence of recurrences can be lowered further by local adjuvants (cryotherapy, embolization, radiation therapy). The third stage of benign tumors (S3, aggressive) includes rapidly growing benign tumors: the capsule is very thin, discontinued, or absent. The tumor invades neighboring compartments, and a wide reactive hypervascularized tissue (pseudocapsule) is often found, sometimes permeated by neoplastic digitations. Bone scan is highly positive, fuzzy limits are seen on plain X-rays, CT scan shows extracompartmental extension, and MRI clearly defines a pseudocapsule and its relationship to the neurological structures. Intralesional excision (curettage), even if augmented by radiation, can be associated with a significant rate of local recurrence. En bloc excision is the treatment of choice.

8.4.2 Malignant Tumors

Low grade malignant tumors are included in Stage I, subdivided into IA (the tumor remains inside the vertebra) and IB (tumor invades paravertebral compartments). No true capsule is associated with these lesions, but a thick pseudocapsule of reactive tissue permeated by small microscopic islands of tumor is seen. A resection performed along the pseudocapsule often leaves residual foci of active tumor; megavoltage radiation or proton-beam therapy can be added to reduce the risk of recurrence. The treatment of choice – if feasible – is a wide en bloc resection. High grade malignancies are defined as IIA and IIB. The neoplastic growth is so rapid that the host has no time to form a continuous reactive tissue. There is continuous seeding with neoplastic nodules (satellites). Moreover, these tumors can have neoplastic nodules at some distance from the main tumor mass (skip metastases). These malignancies are generally seen on plain radiograms as radiolucent and destructive and in many cases are associated with a pathologic fracture. Invasion of the epidural space is rapid in stage B, particularly in small cell tumors (Ewing sarcoma, lymphomas) characterized by semifluid tissue, being able to occupy the epidural space after infiltrating the cortical border of the vertebra.

The margin of the resection must be wide (it is not possible to achieve a "radical" margin, in the spine), and courses of radiation and chemotherapy (according to the tumor type) must be considered for the local control and for the avoidance of distant spread. Several times the anatomical constraints prevent from achieving an appropriate margin.

Stages IIIA and IIIB describe the same lesions as IIA and IIB, but are associated with distant metastasis.

8.5 Margins and Surgical Planning

As staging must be based on the natural history, so surgical planning must be based on what the procedure achieves in relation to the margin of the lesion. Stener was the first [8] to apply to the spine these concepts; starting from the early 1970s he made exhaustive reports of "enbloc resections" in the spine. Later on, Roy-Camille [9] and Tomita [10] popularized the techniques of "en bloc resection" in the thoracic spine. The application of these principles to the cervical spine is technically demanding. Till now, the technique of performing en bloc resection in the cervical spine has not been standardized.

Since the latest 1980s, Weinstein and Boriani tried to focus on a unified classification that could start to discuss envelopes of resection as Enneking had done for the extremities by retrospectively reviewing his cases and making the most important discovery about margins of resection.

This system – called WBB (Weinstein, Boriani, Biagini) Staging System [11] (Fig. 8.1) – has been subjected to several clinical evaluations and recently submitted to a study of reliability and validity by an international multidisciplinar group of spine tumor experts [12] resulting in a moderate interobserver reliability and substantial intraobserver reliability.

WBB Staging System. On the transverse plane, the vertebra is divided into 12 radiating zones (numbered 1–12 in a clockwise order) and into 5 layers (A to E from the prevertebral to the dural involvement). The longitudinal extent of the tumor is recorded by identifying the specific vertebrae involved. This system allowed for a more rational approach to the surgical planning, provided that all efforts are made to perform surgery along the required margins.

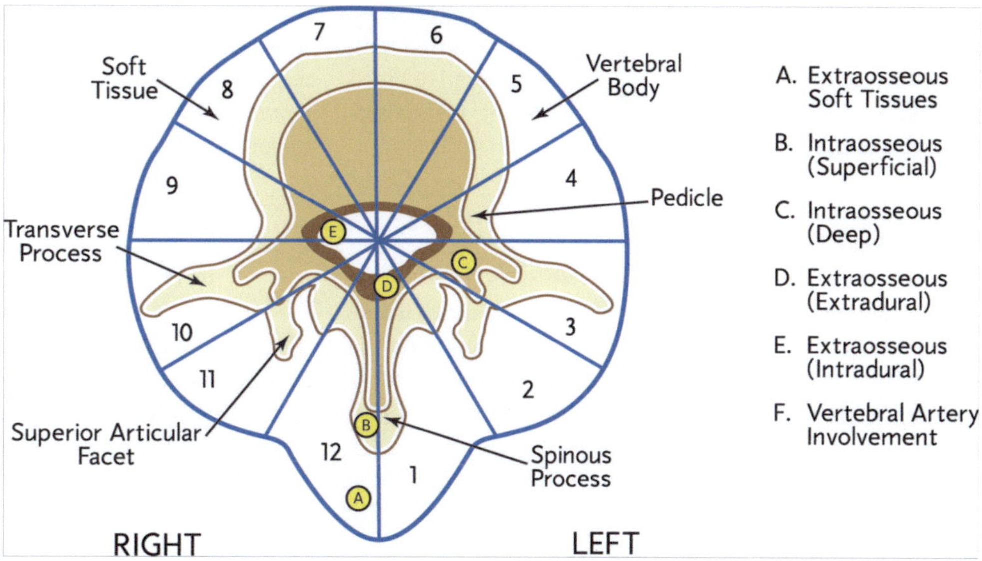

Fig. 8.1 WBB Surgical Staging System. (1) On the transverse plane, the vertebra is divided into 12 radiating zones (numbered 1–12 in a clockwise order) and into 5 layers (A to E from the prevertebral to the dural involvement)

8.6 Terminology

The application of a common terminology is mandatory to exchange informations and to compare the results among institutions. The term "vertebrectomy," as currently used by some, describes both an intralesional eggshell excision [13], and many authors use the term "radical" [14, 15] to mean any surgical margin that does not contain tumor.

Obviously, if the same terms used to describe en bloc resection are also used to describe gross total complete excision, it is difficult and confusing to compare results.

To stress the importance of a common language, the terms proposed by Enneking are recalled.

8.7 Intralesional Excision

Intralesional excision means a piecemeal removal of the tumor ("curettage," "gross total resection," "debulking"). It is an intralesional procedure; the tumor is violated, and any chance of obtaining tumor-free margins is lost. This procedure is appropriate for stage 2 benign tumors and for metastases. Anterior and or posterior approach according to the commonly known surgical procedures can be performed at any spine level. Combination of approaches must be considered to maximize the tumor removal (so-called extracapsular excision, i.e., excision of the tumor capsule). Microscopic and possible macroscopic tumor is however always left after these procedures, and local progression can be expected according to the growing potential of the disease. Local adjuvants can be associated. These become mandatory if intralesional surgery is performed in cases where en bloc is the treatment of choice, but it is not feasible due to surgical and anatomical constraints.

The ischemic effect on the tumor mass provoked by selective arterial embolization must be considered to substantially reduce bleeding, thus strongly reducing intraoperative bleeding and associated complications and allowing the surgeon to complete the excision.

8.8 En Bloc Resection

It is the attempt to remove the tumor as a single intact whole, fully encased by a cuff of healthy tissue (a circumferential margin of tumor-free normal tissue). It requires that the pathologist's confirms the resected tissue around the tumor mass (the so-called margin) is tumor free.

The effectiveness of the margin as a barrier depends on its quality and also on the aggressiveness of the tumor. The margin is defined as "wide" when the plan of dissection has been peripheral to the reactive zone through normal tissue. Quality of the margin and thickness are relevant. A fascial barrier represents a wide margin, whereas 1 cm of muscle or cancellous bone may be inadequate to represent an appropriate barrier. The major issue in the decision-making process of planning en bloc resection concerns the possible functional sacrifices related to the quality of the structure to be resected in a whole with the tumor to achieve a wide margin. *Marginal margin* means that the plane of dissection has been performed along the reactive zone, very close to the tumor and that satellites of tumor may be left. *Intralesional margin* means that the tumor periphery is violated, and the tumor specimen is not covered by healthy tissue. *Radical*: this term should be applied to the en bloc removal of the tumor together with the whole compartment of origin. An example would be a thigh amputation for a tumor arising in the tibia. Or a scapulectomy for a tumor arising and still contained in the scapula. Consequently a radical resection in the spine defines to resect in bloc a full vertebra provided the tumor is fully confined inside the vertebra, including the spinal cord above and below. Even in this dramatic case, if the tumor has grown into the epidural space, the term radical will not be appropriate, as the epidural space is to be considered an extracompartmental area extending from the skull to the sacrum.

8.9 Palliative Procedure

This term describes all the surgical procedures generally directed toward a functional purpose (decompress the thecal sac, stabilize the spine) without any effective surgical removal of the tumor. Partial piecemeal removal of the tumor can be required to release the compression on the cord. An example could be the fixation of a pathologic fracture due to a mechanical collapse of a latent lesion or the decompression and fixation of a myeloma or other radio/chemo-sensitive lesion, later to be submitted for curative purpose to such oncologic treatments.

8.9.1 Surgical Techniques: En Bloc Resections in the Cervical Spine

The techniques of en bloc resection for tumors of the vertebral body in the thoracic and lumbar spine are well known [8–10, 16], while only single case reports of en bloc resections of cervical spine tumors have been published till now [17–20]. WBB staging system can be helpful in standardizing the surgical planning of en bloc resection according to different tumor extension and different spine regions. The great variability of these two parameters dictates that the same surgical procedure cannot be performed in all cases and that surgical planning must be tailored on single case.

Under the guidance of WBB staging system, six different approaches or combinations of approaches have been proposed (ref), ending in a total of ten different types of surgery: only anterior; only posterior; anterior first, posterior second; posterior first, double anterior second and third; posterior first, combined anterior and posterior second; and anterior first, posterior second, combined anterior (contralateral) and posterior.

Three techniques of en bloc resection can be standardized in the cervical spine, considering three possibilities of tumor growing as criteria to achieve a tumor-free margin en bloc resection. In the author's opinion, different typology of tumor expansion will require violation of the tumor margin to save vital structures.

Single posterior approach is the obvious strategy to remove by en bloc resection a tumor arising in the posterior arch (Fig. 8.2). Criteria to achieve appropriate margins according to the WBB staging system (Fig. 8.1) include sectors 9 and 4 free from tumor. If the tumor grows in layer D, the margin will result intralesional during the release from the dura.

Three steps must be planned: the first provides the appropriate margin over the tumor posteriorly growing by resecting inside the posterior muscles which cover the tumor mass if it is expanding in layer A (I). The second step is a piecemeal excision of sectors 9 and 4 must be performed by the technique preferred: by high speed burr or ultrasound osteotome or alternatively an osteotomy can be performed by wire saw or chisel (II). Once a transverse laminotomy above and below is performed, the tumor is released from the dura, and the specimen is resected en bloc (III).

In cervical spine tumors partially occupying the vertebral body (not involving sectors 6 and 7, otherwise see Fig. 4) and the posterior arch (at least three sectors not involved) to perform en bloc resection aiming at achieving a tumor-free margin specimen, two approaches should be planned: anterior first and posterior second (Fig. 8.3).

Anterior approach is first performed to leave healthy tissue over the tumor growing in the lateral masses (I) and to perform a sagittal groove till the epidural space in the vertebral body (II). Diskectomies or transversal grooves in vertebral bodies are performed to define the upper and lower margins, including ligation of the vertebral artery.

The second stage is a posterior approach. Step III provides the appropriate margin over the tumor posteriorly growing by resecting inside the posterior muscles covering the tumor mass if it is expanding in layer A. Step IV is a piecemeal excision of the posterior arch not involved by the tumor. At least three sectors are required, starting from sector 4 or from sector 9. This allows to release of the dura from the tumor and section of the nerve root(s) involved by the tumor. Finally the specimen is removed – once the upper and lower diskectomies or osteotomies are finalized – by rotating around the dural sac (V). If the cervical tumor is particularly huge, extending over the midline, but not involving both vertebral arteries (Fig. 8.4), en bloc resection aiming to achieve a tumor-free margin can be planned, including three approaches: first posterior, second anterior contralateral to the tumor side, third anterior on the tumor side. A single wide anterior approach by transverse incision can be considered as well.

First stage is performed in prone position: the posterior arch not involved by the tumor is removed by piecemeal excision. At least three sectors are required, starting from sector 4 or from sector 9 (I) to allow the release of the dural sac and a safe manipulation. In case of tumor posteriorly growing, and invading layer A, an appropriate margin must be provided by resecting inside the posterior muscles covering the tumor mass (II). Then release the dura from the tumor (if the tumor grows in layer D, the margin will result intralesional) and section the nerve root(s) crossing the tumor. Second and third stages are performed in supine position. A sagittal grove is performed in the vertebral body not occupied by the tumor (III), till the vertebral artery, which must be saved as the other must be sacrificed as involved by the tumor. The anterior margin is provided by leaving healthy soft tissue over the tumor mass (IV). Diskectomies or transversal grooves in vertebral bodies are performed to define the upper and lower margins. The tumor is finally removed by the third approach (V), once the upper and lower diskectomies or osteotomies are finalized, including ligation of the vertebral artery.

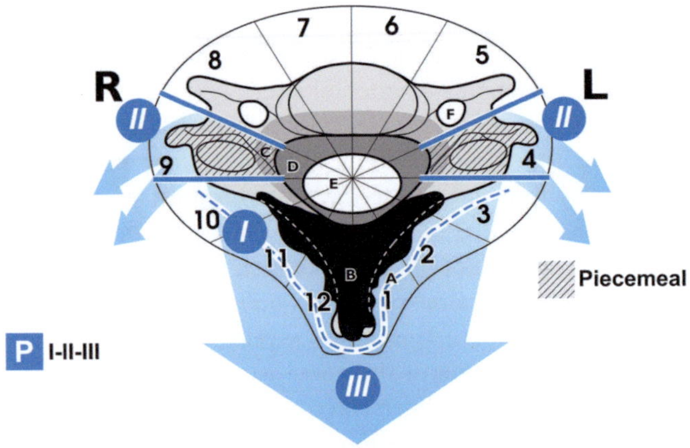

Fig. 8.2 Single posterior approach is the obvious strategy to remove by en bloc resection a tumor arising in the posterior arch. Criteria to achieve appropriate margins according to the WBB staging system (ref) include sectors 9 and 4 free from tumor. If the tumor grows in layer D, the margin will result intralesional during the release from the dura

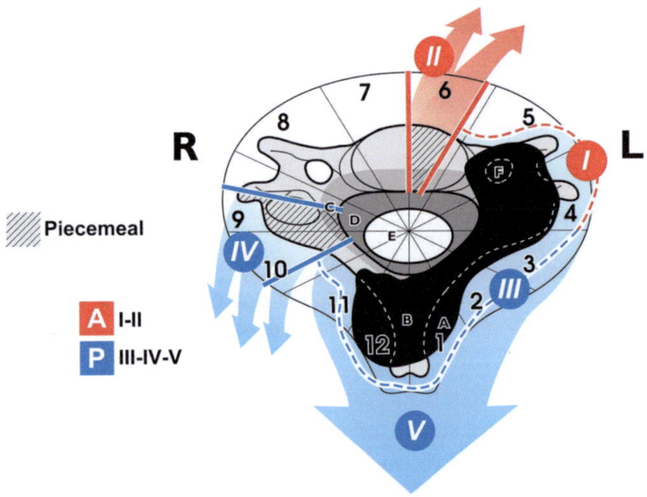

Fig. 8.3 In cervical spine tumors partially occupying the vertebral body (not involving sectors 6 and 7, otherwise see Fig. 4) and the posterior arch (at least three sectors not involved) to perform en bloc resection aiming at achieving a tumor-free margin specimen, two approaches should be planned: anterior first and posterior second

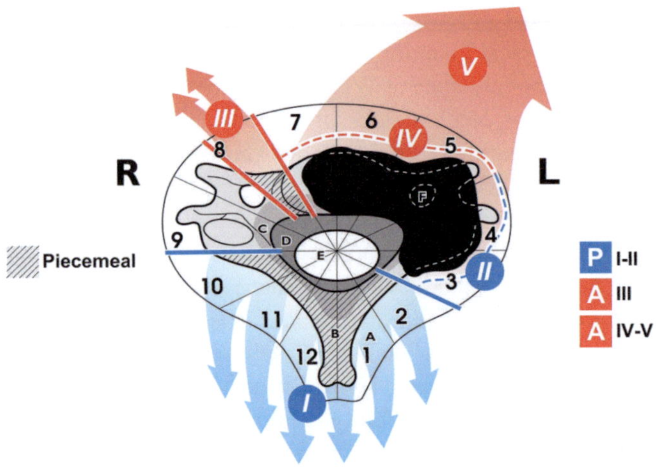

Fig. 8.4 If the cervical tumor is particularly huge, extending over the midline, but not involving both vertebral arteries, an en bloc resection aiming to achieve a tumor-free margin can be planned, including three approaches: first posterior, second anterior contralateral to the tumor side, third anterior on the tumor side. A single wide anterior approach by transverse incision can be considered as well

8.9.2 Morbidity and Complications

The morbidity of en bloc resections in the cervical spine is very high. First the decision-making process has to inquire about the relationship between the functional loss consequent to the sacrifice of structures included in the tumor compared to the expected better result in terms of prognosis and expected survival of en bloc resection. The role of margins in tumor surgery has been established since many years [21, 22]. In the spine, data are available in the literature in case of chordoma [23], chondrosarcoma, and osteosarcoma. While the evidence in case of Ewing sarcoma is not so relevant [24, 25]

The possible complications include all those related to anterior spine surgery [26, 27] with those of major posterior spine surgery [28] including all the anaesthesiological problems related to a long duration procedures [29]. Specific morbidity of spine tumor surgery has been recently analyzed [30] concluding that complications are mostly connected to the manipulation of important and vital structures, as required by the need of removing the tumor en bloc with an appropriate margin of healthy tissue, sometimes contiguous to vital structures. Incomplete intraoperative control of hemodynamics and too short posterior fixation and lack of anterior support heavily affect the morbidity. Previous surgery or radiation and double combined approach increase the risk of complications, as scar dissection will expose to lesions of neighboring vascular, nervous, visceral structures. Among late complications, infection is particularly threatening, due to the wide exposure and the reduced immunocompetence of these patients.

Mortality is not negligible (2.2 %). Radiation therapy did not result as a significant predictor of morbidity, but five of six patients who had a deep infection had been submitted to radiation therapy before surgery. The two cases of late dissection of the aorta occurred after RT (doses 42 and 44 Gy) was performed as an adjuvant after a vertebrectomy including anterior surgical approach and release of the aorta from the tumor[41].

Acknowledgments The author acknowledges the outstanding contribution of Carlo Piovani for data collection and original drawings.

References

1. Chi JH, Ali B, Patrick H, Tim W, Jean-Paul W, Gokaslan ZL (2008) Epidemiology and demographics for primary vertebral tumors. Neurosurg Clin N Am 19:1–4
2. Enneking WF, Spanier SS, Goodman MA (1980) A system for the surgical staging of musculoskeletal sarcoma. Clin Orthop Relat Res 153:106–120
3. Boriani S, Capanna R, Donati D, Levine A, Picci P, Savini R (1992) Osteoblastoma of the spine. Clin Orthop 278:37–45
4. Boriani S, Bandiera S, Biagini R, Bacchini P, Boriani L, Cappuccio M et al (2006) Chordoma of the mobile spine: fifty years of experience. Spine 31(4):493–503
5. Hart RA, Boriani S, Biagini R, Currier B, Weinstein JN (1997) A system for surgical staging and management of spine tumors. A clinical outcome study of giant cell tumors of the spine. Spine 22:1773–1782
6. Boriani S, De Iure F, Bandiera S, Campanacci L, Biagini R, Di Fiore M, Bandello L, Picci P, Bacchini P (2000) Chondrosarcoma of the mobile spine. Report on 22 cases. Spine 25:804–812
7. Simon MA (1998) Surgical margins. In: Simon MA, Springfield D (eds) Surgery for bone and soft tissue tumors. Lippincott-Raven, Philadelphia, pp 77–92
8. Stener B, Johnsen OE (1971) Complete removal of three vertebrae for giant-cell tumour. J Bone Joint Surg Br 53(2):278–287
9. Roy-Camille R, Montpierre H, Mazel C, Saillant G (1990) Technique de vertebrectomie totale lombaire. In: Roy-Camille R (ed) Rachis Dorsal et Lombaire Septieme Journees d'Orthopedie de la Pitié. Masson, Paris, pp 49–52
10. Tomita K, Kawahara N, Baba H, Tsuchiya H, Nagata S, Toribatake Y (1994) Total en bloc spondylectomy for solitary spinal metastases. Int Orthop 18(5):291–298
11. Boriani S, Weinstein JN, Biagini R (1997) Spine update. A surgical staging system for therapeutic planning of primary bone tumors of the spine. A contribution to a common terminology. Spine 22:1036–1044
12. Chan P, Boriani S, Fourney DR, Biagini R, Dekutoski MB, Fehlings MG et al (2009) An assessment of the reliability of the Enneking and Weinstein-Boriani-Biagini classifications for staging of primary spinal tumors by the Spine Oncology Study Group. Spine 34:384–391
13. Magerl F, Coscia MF (1988) Total posterior vertebrectomy of the thoracic or lumbar spine. Clin Orthop Relat Res 232:62–69
14. Fidler MW (1994) Radical resection of vertebral body tumours. A surgical technique used in ten cases. J Bone Joint Surg Br 76(5):765–772
15. Mazel C, Grunenwald D, Laudrin P, Marmorat JL (2003) Radical excision in the management of thoracic and cervicothoracic tumors involving the spine: results in a series of 36 cases. Spine 28(8):782–792
16. Boriani S (2000) Subtotal and total vertebrectomy for tumours. Surgical techniques in orthopedics and traumatology. Elsevier, Paris, Editions Scientifiques et Medicales
17. Fujita T, Kawahara N, Matsumoto T, Tomita K (1999) Chordoma in the cervical spine managed with en bloc excision. Spine 24(17):1848–1851
18. Rhines LD, Fourney DR, Siadati A, Suk I, Gokaslan ZL (2005) En bloc resection of multilevel cervical chordoma with C-2 involvement. Case report and description of operative technique. J Neurosurg Spine 2(2):199–205
19. Currier BL, Papagelopoulos PJ, Krauss WE, Unni KK, Yaszemski MJ (2007) Total en bloc spondylectomy of C5 vertebra for chordoma. Spine 32(9):E294–E299
20. Leitner Y, Shabat S, Boriani L, Boriani S (2007) En bloc resection of a C4 chordoma: surgical technique. Eur Spine J 16(12):2238–2242
21. Talac R, Yaszemski MJ, Currier BL et al (2002) Relationship between surgical margins and local recurrence in sarcomas of the spine. Clin Orthop 397:127–132
22. Bergh P, Gunterberg B, Meis-Kindblom JM, Kindblom LG (2001) Prognostic factors and outcome of pelvic, sacral, and spinal chondrosarcomas: a center-based study of 69 cases. Cancer 91(7):1201–1212
23. Boriani S, Saravanja D, Yamada Y, Varga Peter P, Biagini R, Fisher Charles G (2009) Challenges of local recurrence and cure in Low grade malignant tumors of the spine. Spine 34(22S):S48–S57

24. Boriani S, Amendola L, Corghi A, Cappuccio M, Bandiera S, Ferrari S, Picci R, Di Fiore M, Gasbarrini A (2011) Ewing's sarcoma of the mobile spine. Eur Rev Med Pharmacol Sci 15:831–839
25. Sciubba DM, Okuno SH, Dekutoski MB, Gokaslan ZL (2009) Ewing and osteogenic sarcoma: evidence for multidisciplinary management. Spine (Phila Pa 1976) 34(22 Suppl):S58–S68
26. Faciszewski T, Winter RB, Lonstein JE, Denis F, Johnson L (1995) The surgical and medical perioperative complications of anterior spinal fusion surgery in the thoracic, lumbar spine in adults. A review of 1223 procedures. Spine 20:1592–1599
27. McDonnell MF, Glassman SD, Dimar JR, Puno HMR, Johnson JB (1996) Perioperative complications of anterior procedures of the spine. J Bone Joint Surg Am 78:839–847
28. Smith JS, Saulle D, Chen CJ, Lenke LG, Polly DW Jr, Kasliwal MK, Broadstone PA, Glassman SD, Vaccaro AR, Ames CP, Shaffrey CI (2012) Rates and causes of mortality associated with spine surgery based on 108,419 procedures: a review of the Scoliosis Research Society Morbidity and Mortality Database. Spine (Phila Pa 1976) 37(23):1975–1982
29. Di Fiore M, Lari S, Boriani S, Formaro G, Perin S, Malferrari A, Zanoni A (1998) Major vertebral surgery: intra- and postoperative anaesthesia-related problems. Chir Org Mov 83:65–72
30. Boriani S, Bandiera S, Donthineni R, Amendola L, Cappuccio M, De Iure F, Gasbarrini A (2010) Morbidity of en bloc resections in the spine. Eur Spine J 19:231–241

Part III

Surgical Approaches

Surgical Approaches to CCJ (Endoscopic Transnasal-Transoral-Transcervical and Robotic Transoral Approach)

9

F. Mattioli, G. Molteni, M. Bettini, E. Cigarini, and L. Presutti

9.1 Introduction

This chapter is focused on the surgical approaches that are available for exposing lesions at the craniovertebral junction (CVJ). To reach the CVJ is necessary to cross anatomical regions that are the domains of several surgical subspecialties (neurosurgery, otolaryngology, and plastic/craniofacial surgery).

The surgeon should select an approach that allows the most extended operative field with the lowest rate of morbidity [12]. The location and extent of the lesion are the major determinants influencing selection of the appropriate cranial base approach.

When the skull base is considered in the axial plane, the CVJ can be approached from three directions:

- *Anteriorly*
- *Laterally*
- *Posteriorly*

Surgical approaches to the CVJ are classified (Fig. 9.1).

Anterior Approach

- *Transcranial* (transfrontal, orbitozygomatic, pterional)
- *Transnasal* (transsphenoidal, transmaxillary, endoscopic)
- *Transoral* (transpalatal, buccopharyngeal, translabiomandibular, translabioglossomandibular, endoscopic)
- *Transcervical*

Posterior Approach

- Suboccipital, retrosigmoid, extreme lateral

Lateral Approach

- Translabyrinthine, transcochlear, presigmoid, subtemporal transpetrosal, preauricular infratemporal, preauricular transtemporal

The anterior approaches are best suited for extradural lesions. Attacking intradural pathology through one of these approaches risks contamination with pharyngeal organisms, meningitis, and cerebrospinal fluid (CSF) leaks.

Consequently, intradural lesions located laterally and lesions involving the temporal bone are better treated through a lateral approach, while intradural posterior lesions are reached through a posterior approach [7, 12].

Details are shown in Fig. 9.2.

Our objective is to describe the anterior approaches performed by ENT and orthopedic surgical team, in particular, the minimally invasive endoscopic-assisted approaches.

Posterior and lateral approaches are not described in this chapter.

F. Mattioli (✉) • G. Molteni • M. Bettini • E. Cigarini • L. Presutti
Head and Neck Surgery Department,
University Hospital of Modena, Modena, Italy
e-mail: franz318@hotmail.com

S. Boriani et al. (eds.), *Atlas of Craniocervical Junction and Cervical Spine Surgery*, DOI 10.1007/978-3-319-42737-9_9

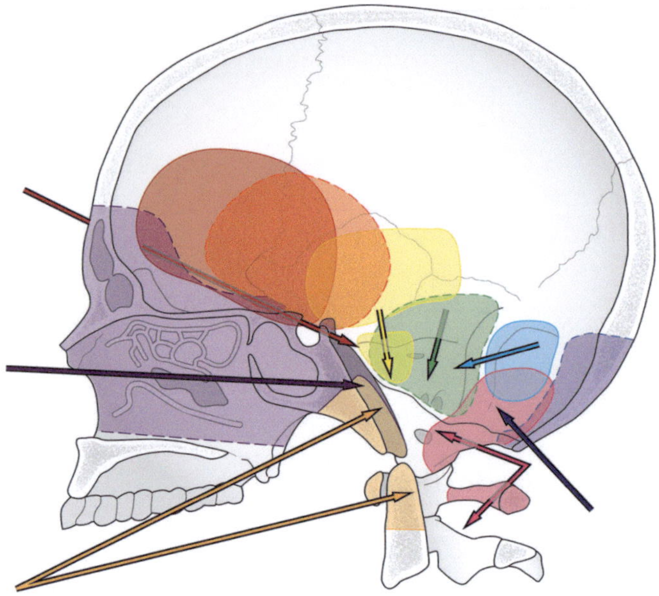

Fig. 9.1 Surgical approches to CVJ

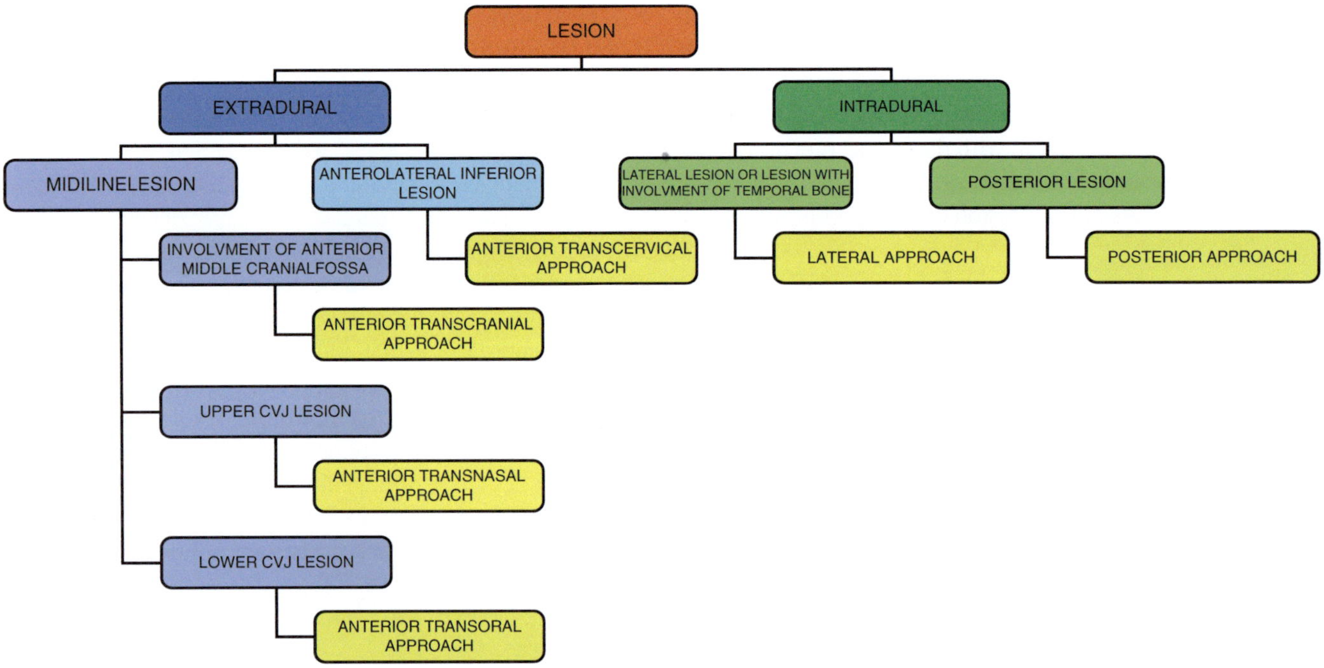

Fig. 9.2 Selection of the approach according to the site of the lesion

9.2 Anterior Approaches

CVJ is a complex region which can harbor neoplastic, degenerative, or inflammatory lesions that can cause compression of the cervicomedullary junction and craniovertebral instability. Treatment may require surgical decompression and subsequent craniocervical stabilization.

Several surgical approaches provide anterior access to the craniocervical junction and the upper cervical spinal including transnasal, transoral, and high transcervical approaches.

In particular, anterior or anterolaterally located CVJ lesions have traditionally been most difficult to resect with high morbidity and mortality resulting from anterior traditional approaches. Recently the development of endoscopic surgery allowed to perform decompression of the craniovertebral junction with endoscopic transnasal-transoral techniques, even in patients with challenging anatomic features [14].

Preoperatively, estimate the superior extent of the exposure by drawing an imaginary line in the plane of the hard palate toward the craniovertebral junction on a sagittal MRI. If the lesion is midline and is situated above this line, a transnasal approach may be suitable. If, however, the lesion is situated below the plane of the hard palate, a transoral approach alone may be sufficient. Alternatively, consider a Le Fort I maxillotomy with or without a palatal split if the lesion extends above and below the plane of the hard palate. If the lesion extends inferiorly beyond the line of sight of a standard transoral approach (roughly beyond the lower body of C2 to the C2–C3 disk space in most cases), a labiomandibular median glossotomy (transmandibular split) or cervical approach may be appropriate to gain more inferior exposure (Fig. 9.3) [11].

Transoral robotic surgery gets the opportunity to extend the approach along all upper cervical spine.

Prophylactic antibiotics are administered intraoperatively in transnasal and transoral approaches.

A high-resolution stereotactic computed tomography scan and MRI of the craniovertebral junction are necessary for preoperative planning. Moreover dynamic plain cervical radiographs in flexion and extension views to evaluate for preexisting instability at the craniovertebral junction are suggested. In some cases, the CVJ pathology has destabilized the occipitoatlantal or atlantoaxial joints, thus requiring a subsequent or prior stabilization procedure. If severe deformity or cranial settling is detected preoperatively, attempt reduction with cervical traction using Gardner-Wells tongs and mobilization in halo vest is recommended. Even if no instability is present, however, most cases will require stabilization after surgery because of postoperative iatrogenic instability [11]. Either occipitocervical or atlantoaxial stabilization can be performed postoperatively or simultaneously to the main procedure. Rarely ventral bone graft is placed intraoperatively into the bony defect because of high risk of infection, reabsorption, and dislodgement.

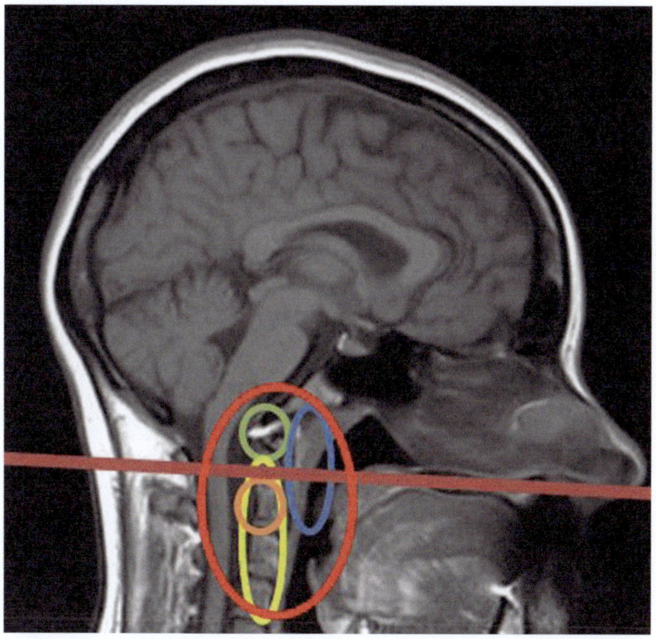

Fig. 9.3 Preoperative estimate of exposure and choice of the approach: lesions suitable for transnasal approach (*green*), for transoral approach (*orange*), for maxillotomy with or without palatal split (*blue*), for labiomandibular median glossotomy or cervical approach (*yellow*), and for transoral robotic surgery (*red*)

Take Home Massages

Advantages
- Low morbidity (endoscopic approaches)
- Direct extradural approach that does not require any brain retraction
- Low morbidity and mortality (traditional and extended approaches)

Disadvantages
- Risk of cerebrospinal fluid leak

Contraindications
- Intradural lesion (risk of meningitis)
- Lateral or posterior lesions

9.2.1 Radiological Planning of Anterior Approaches to Craniovertebral Junction [5]

It is possible to evaluate and compare the minimally invasive techniques to the anterior craniocervical junction using exclusive endoscopic approach or endoscopic-assisted approaches:

- *Endonasal (exclusive endoscopic)*
- *Transoral (endoscopic assisted)*
- *Transcervical (endoscopic assisted)*

9.2.1.1 Angles of Attack

The angles of attack were created with the use of a single entry point. The entry point selected allows the best "angle" to reach the surgical lesion (e.g., angled inferiorly as much as possible for the endonasal approach and anteriorly to posteriorly as much as possible for the transcervical approach).

9.2.1.2 Endonasal

For the endonasal approach, the entry point was defined as a point in the midline at the inferior edge of the nasal bone. The most superior access within the surgical field was arbitrarily defined as the point on the clivus at the base of the sella (the superior aspect of the clival recess). The most inferior access within the surgical field was then approximated by fashioning a straight line starting from the inferior midline of the nasal bone and ending at the lowest point on the odontoid or C2 while remaining tangential to, but not crossing, the hard palate. This approximates the most inferior point within the surgical field, but does not take into account the use of angled instruments that allow a more inferior reach of approximately 5 mm or an increase in the angle obtained by hard palate drilling.

9.2.1.3 Transoral

For the transoral approach, the entry point was defined as a point in the midline at the inferior edge of the incisors. The most superior access within the surgical field was defined as the point in the midline at the superior aspect of the lower third of the clivus. This was chosen based on the anatomic dissections as the point of superior access without splitting the soft palate. The most inferior access within the surgical field was chosen as the inferoposterior aspect of the body of C2.

9.2.1.4 Transcervical

For the transcervical approach, the entry point was defined as the midline on the skin at the C4–C5 cervical disk level. The most superior access within the surgical field was defined as the point in the midline at 1 cm above the basion. The most inferior access within the surgical field was the inferoposterior aspect of the body of C2.

The surgical goals of lower clival and odontoid (craniocervical) decompression were achieved using the endonasal and transoral approaches.

The transcervical approach was unable to achieve more than 1 cm of lower clival resection without undue retraction or was limited by the constraints of the mock chest on the angle of attack. Furthermore, it was difficult to maintain a midline dissection trajectory. This approach, like the endonasal and transoral approaches, did allow complete odontoid resection (Table 9.1).

The working angles and the overlapping working areas of each approach are illustrated in Fig. 9.4.

Table 9.1 Summary of the angles of attack, surgical working distance, and working area in each endoscopic approach

Approach	Distance to surgical target (mm)	Angle of attack, sagittal (degrees)	Working area (mm^2)
Transnasal	94	28	1305
Transoral	102	30	1406
Transcervical	100	15	743

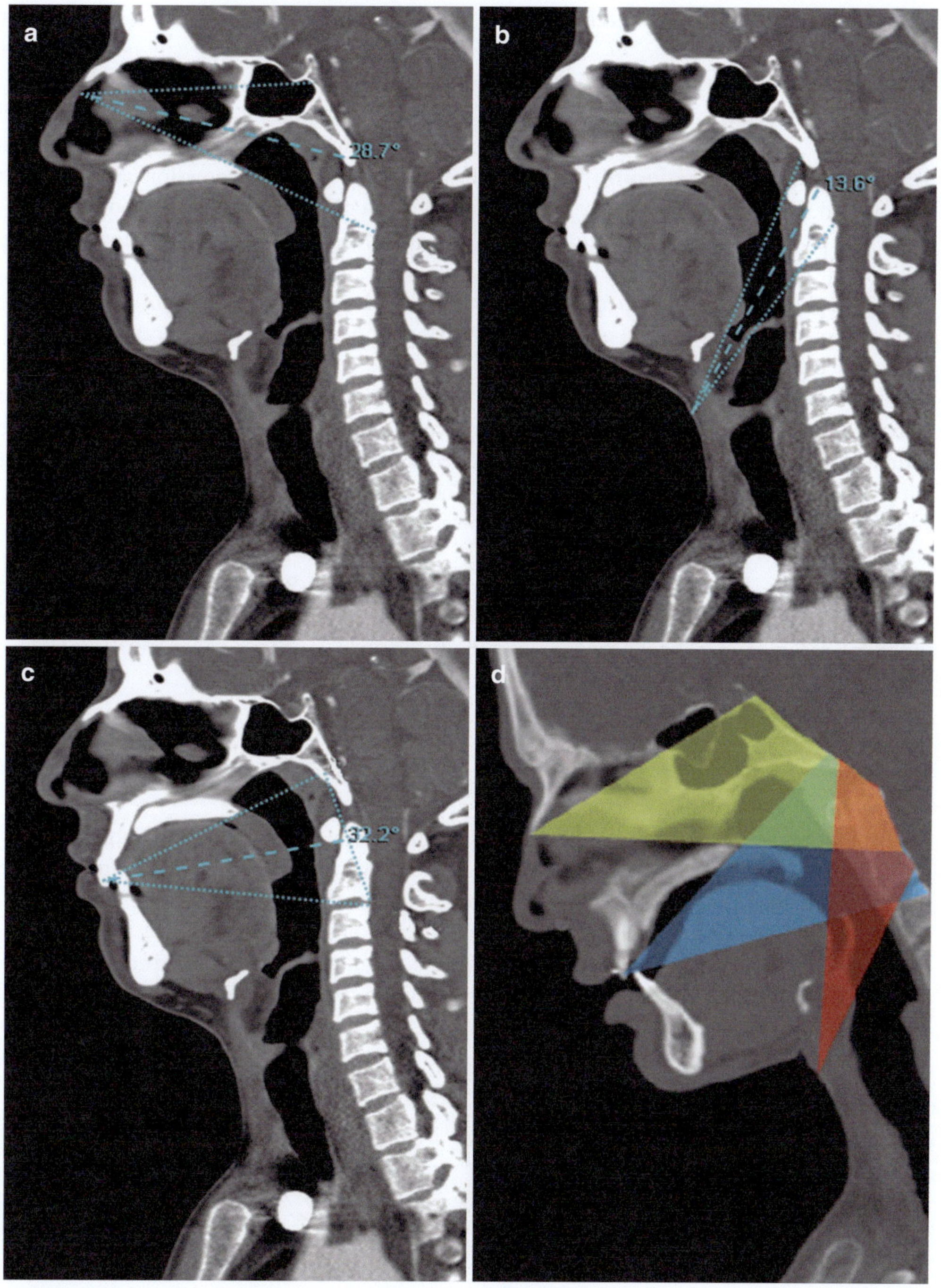

Fig. 9.4 Computed tomography scans demonstrating the surgical trajectory and angles for the endonasal approach (**a**), the transoral approach (**b**), and transcervical approach (**c**). (**d**) Common surgical area of the 3 approaches is represented by the overlapping illumination

9.3 Transnasal Approach

9.3.1 Standard Transnasal/Transsphenoidal Approach

9.3.1.1 Operative Positioning

The procedure is performed under general anesthesia with the patient positioned supine and intubated orally. Sometimes neuromonitoring with somatosensory-evoked potentials is used, and spinal traction is applied as needed to get the odontoid process into a more normal position. Neuronavigation is always used for surgical guidance.

The oropharynx is packed to prevent blood and secretions from the operative site from reaching the stomach. The patient is supine with the head slightly extended on a doughnut pad or fixated with a Mayfield head holder. The operating table is adjusted to provide a semirecumbent position for the patient.

9.3.1.2 Surgical Principles

In 2005 Kassam et al. [1] published the first report on endoscopic endonasal approach being used to perform odontoidectomy.

The endoscopic transnasal approach consisted of a bilateral approach to the nostril.

Kassam [1] described an expanded endonasal approach to perform resection of the odontoid process that consists in an elevation of a nasal septal flap that will be used for closure, a bilateral maxillary antrostomy, ethmoidectomy, middle turbinate resection posterior septectomy, and a wide sphenoidectomy to reach the CVJ. Actually not all patients require sphenoidectomy or middle turbinectomy. Depending on the location of the lesion, the sphenoid floor and/or clival bone were drilled if necessary. After cauterization the posterior nasopharyngeal mucosa was incised in the midline, and the prevertebral muscles were dissected vertically and elevated laterally to expose the anterior tubercle of the atlas. Decompression can be performed using a drill, curettes, and/or Kerrison rongeur [8, 13].

Neuronavigational devices can provide more accurate information and guidance. CSF leak and venous bleeding from the epidural venous plexus are common in this procedure. A fat graft is used to pack with fibrin glue which is used to pack the cavity. If CSF leak occurs, fat should be used to pack the entire sphenoid sinus, and fascia lata may also be necessary to enhance the closure. The graft is held in place with a plate or stent of nasal bone or cartilage. A Merocel sponge is packed in the nose bilaterally (Figs. 9.5, 9.6, and 9.7).

9.3.1.3 Postoperative Care

A period of postoperative lumbar drainage may be appropriate when a major CSF leak occurs.

Advantages

- Low morbidity

Disadvantages

- Limited access to the upper clivus

Contraindications

- Active nasopharyngeal infection

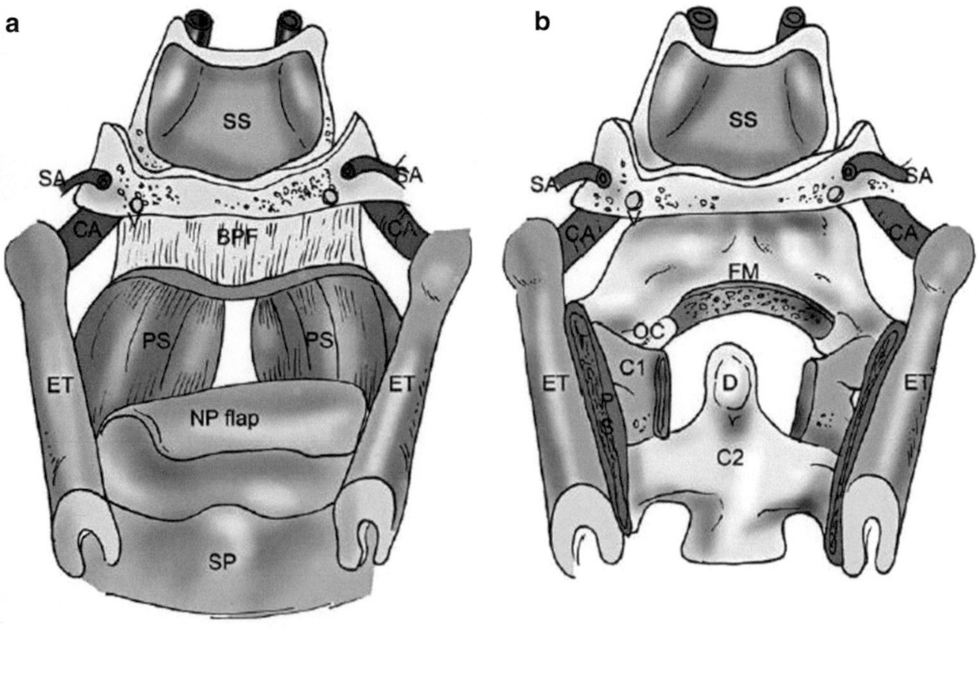

Fig. 9.5 From Presutti et al. eight descriptions of transnasal/transsphenoidal approach. (**a**) Creation of a U shape nasopharyngeal flap to identify the prevertebral structures (**b**) Exposure of odontoid process and anterior tubercle of the atlas

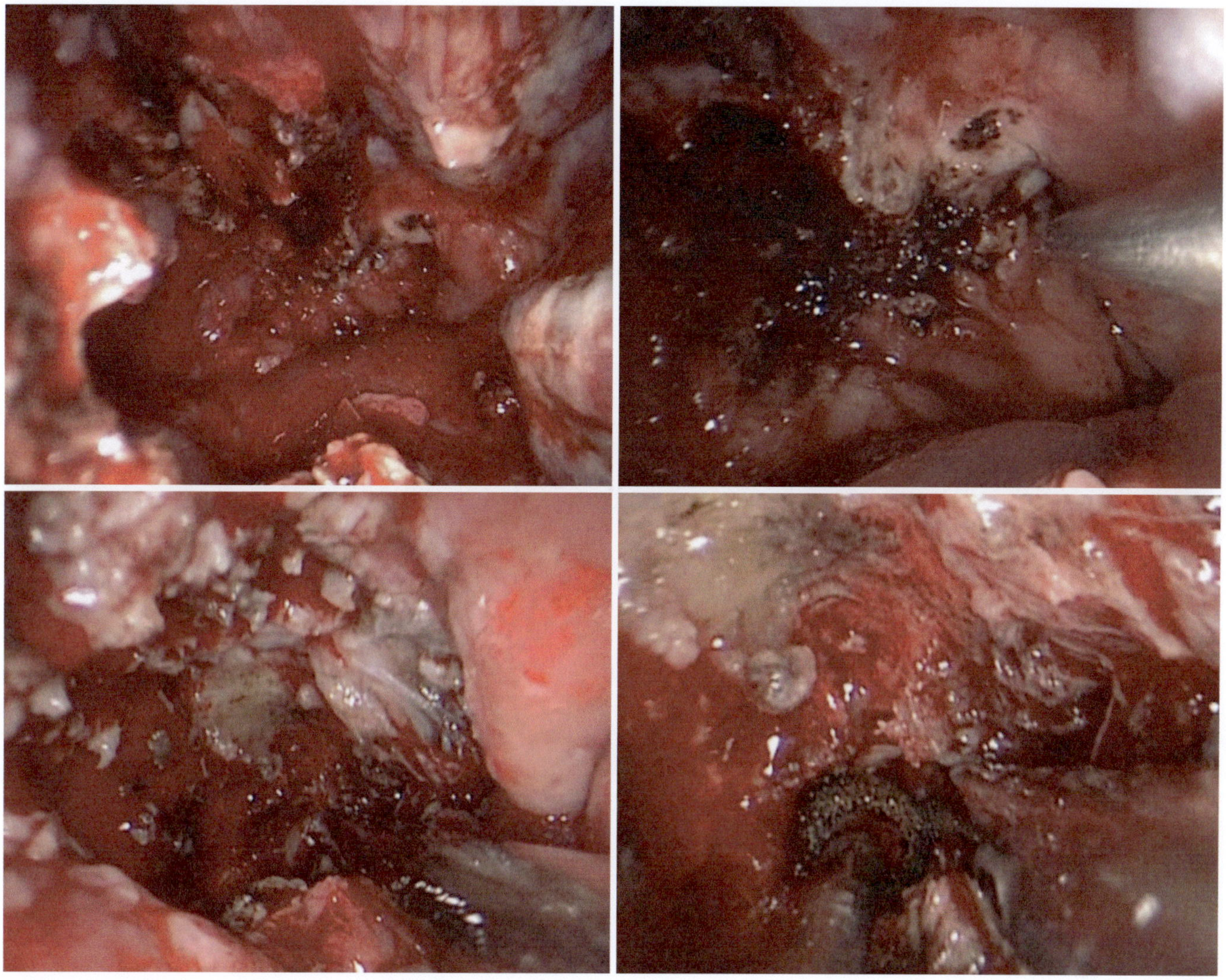

Fig. 9.6 Nasopharyngeal U-shaped flap to expose prevertebral plane

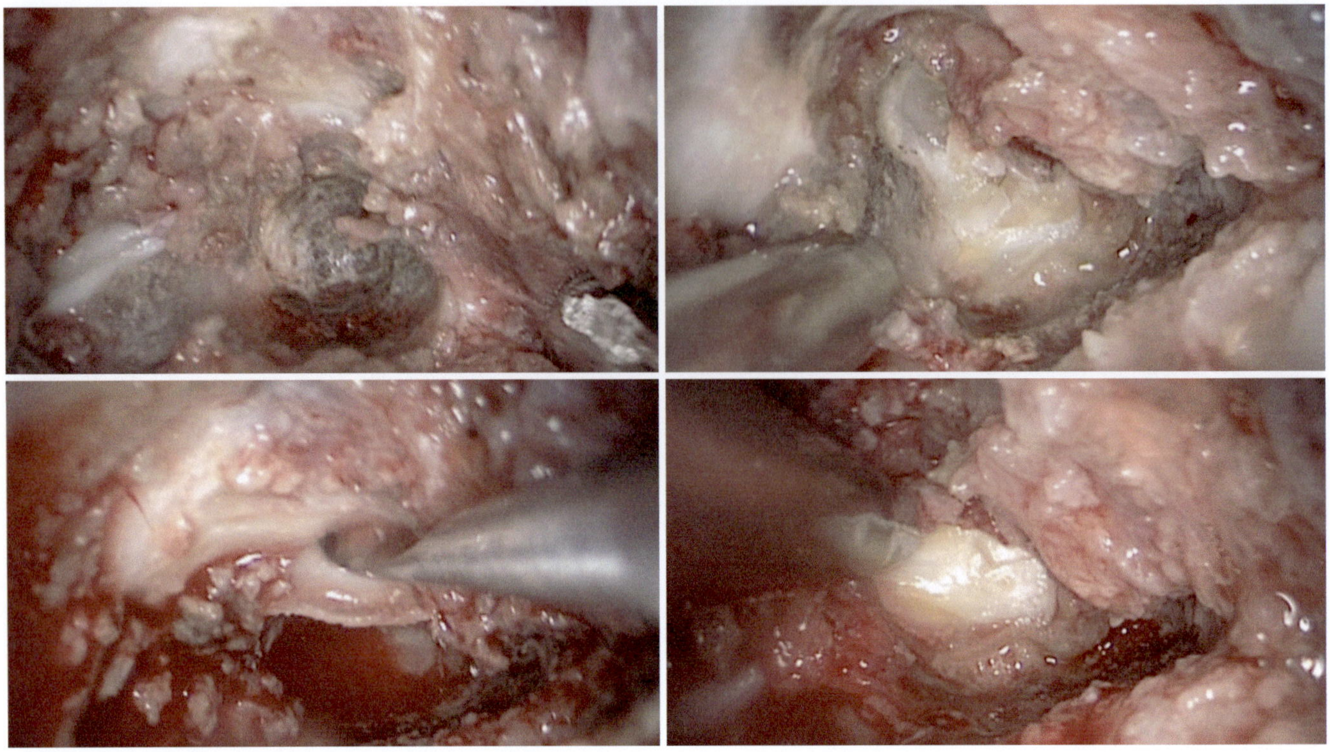

Fig. 9.7 Exposure of the C1 and C2 joints; dens removal

9.4 Transoral Approaches

The transoral approach provides direct midline access to the craniovertebral junction and is most suitable for extradural lesions, such as chordomas, chondrosarcomas, giant cell tumors, and rheumatoid or degenerative pannus [2, 9, 10, 13].

The corridor of exposure is limited by the extent to which the patient can open his or her mouth: the location of the hard palate relative to the CVJ limits superior exposure, whereas the mandible and base of the tongue limit the inferior exposure. To obtain adequate line of sight in the superior and inferior trajectories, the opening of the mouth should be at least 2.5–3 cm between the upper and lower teeth [11].

In most of cases, exposure can be obtained from the inferior clivus to C2 so the surgical procedure can be performed by simply retracting the soft palate and tongue without incising the soft-tissue structure of the mouth.

Extended transoral approaches involve additional incisions and facial osteotomies to mobilize structures that may be obstructing the surgeon's line of sight to the lesion. These approaches include transmaxillary (Le Fort I maxillotomy), transmaxillary with midline palatal split (extended "open-door" maxillotomy), or the frequently used transpalatal and median labiomandibular glossotomy (transmandibular split) (described in this chapter) (Fig. 9.8) [11, 13].

Advantages

- Access from the inferior clivus to the C2 vertebra
- Low morbidity (standard oral approach)

Disadvantages

- Constrained field due to restricted jaw opening or soft-tissue hypertrophy of the pharynx
- High morbidity (extended transoral approaches)

Contraindications

- Active nasopharyngeal infection
- Vascular structure (vertebral or basilar arteries) within or ventral to the lesion

Possible Complications

- Dysphagia
- Hypernasal speech
- Velopharyngeal incompetence
- CSF leak

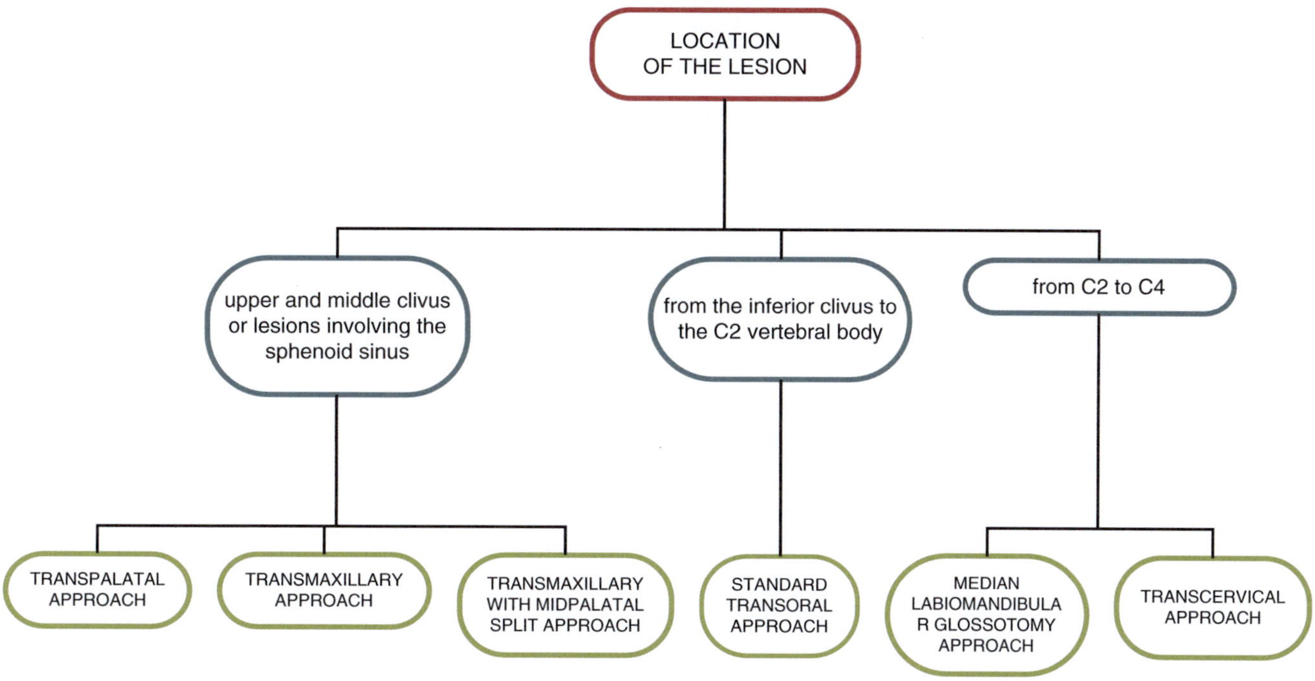

Fig. 9.8 Relationship between location of the lesion and type of the anterior approach

9.4.1 Median Labiomandibular Glossotomy Approach

Trotter [15] first described this approach in 1929 for resection midline tumors located at the base of the tongue and the posterior pharynx. It is also useful for removing lesions of the lower CVJ beyond C2 and C4 levels if they are not accessible through a standard transoral approach.

9.4.1.1 Operative Positioning

The procedure is performed under general anesthesia; preoperative tracheotomy is strongly recommended in this case.

9.4.1.2 Surgical Principles

A vertical curvilinear incision is made from the center of the lower lip to the prominence of the chin as showed in Fig. 9.9 and continued posteriorly in the midline to the hyoid bone. The mucosa membrane is split. Titanium plates and screws are pre-plated for later replacement. The mandible is cut in a step-cut fashion to facilitate accurate reapproximation. The tongue is split longitudinally through its central raphe and posteriorly toward the median glossoepiglottic fold. In this way the vascular supply and innervations from the hypoglossal nerves are preserved (Fig. 9.10). The halves of the mandible and tongue are retracted laterally, and the floor of the mouth is split with an incision between the submaxillary ducts anteriorly extended inferiorly to the level of the hyoid bone. The posterior pharyngeal wall can be opened to expose middle and lower clivus down to C3–C4 (Fig. 9.11a).

At the time of closure, tongue and mucosa are closed with absorbable sutures. The halves of mandible are reapproximated with preregistered plates and screws, and the skin is closed with nylon sutures (Fig. 9.11b, c, d).

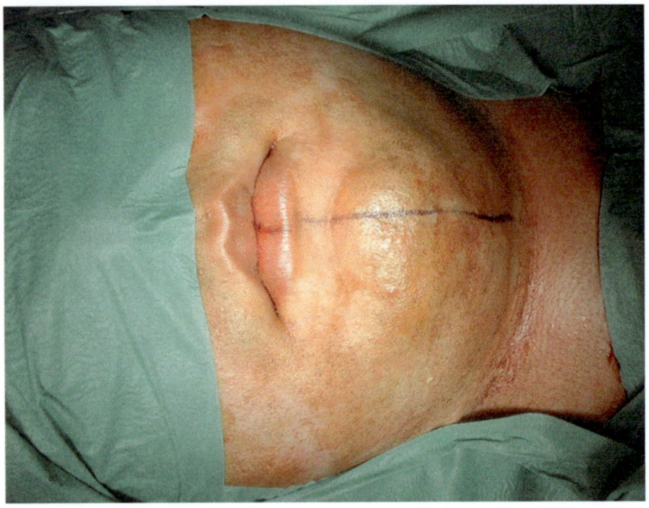

Fig. 9.9 Skin incision for median labiomandibular approach

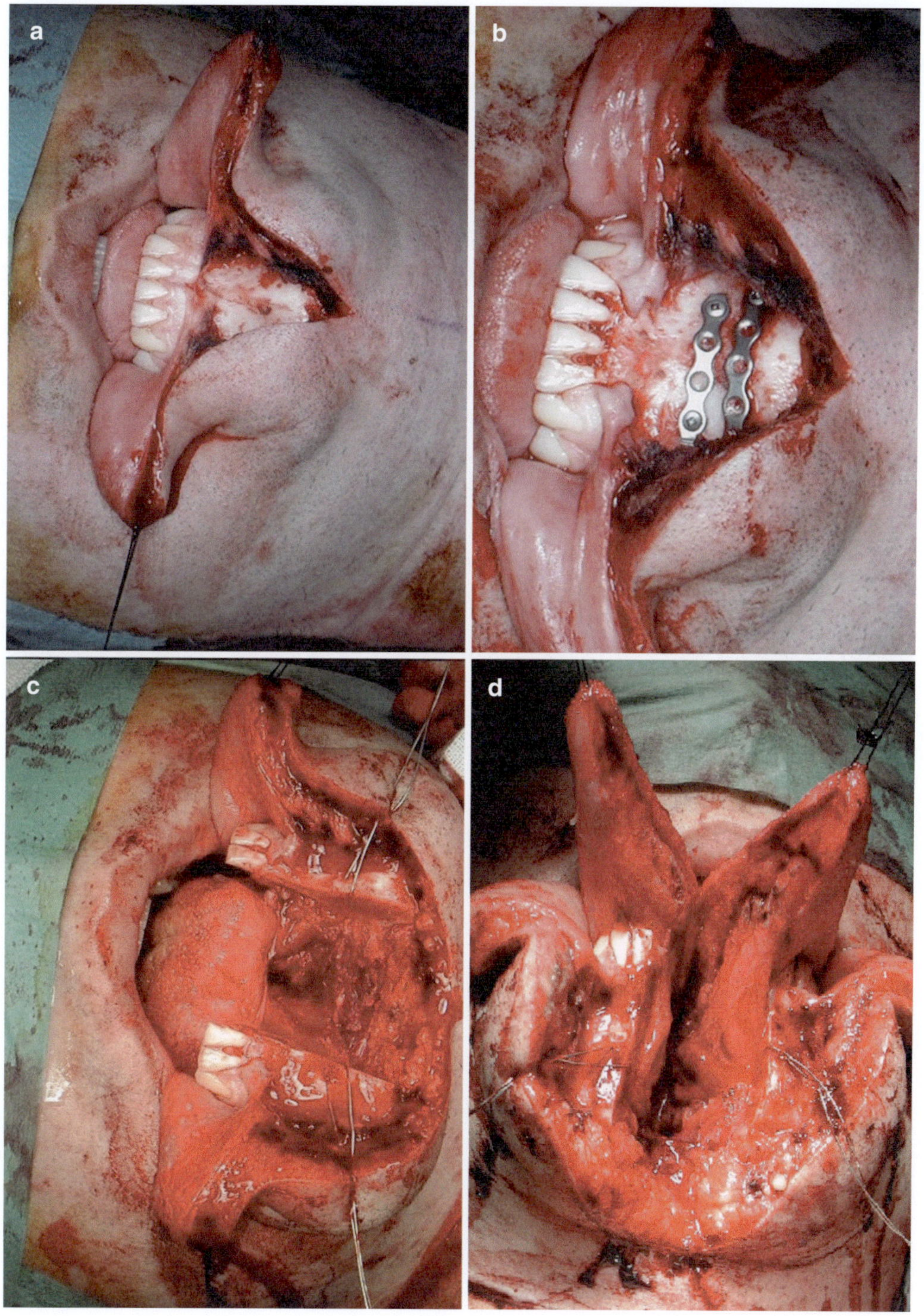

Fig. 9.10 Transmandibular approach to CVJ. (**a**) Labial split on the midline; (**b**) preplating before mandibular split. (**c**, **d**) Splitting of tongue at the midline in order to preserve main vascular and nervous structures

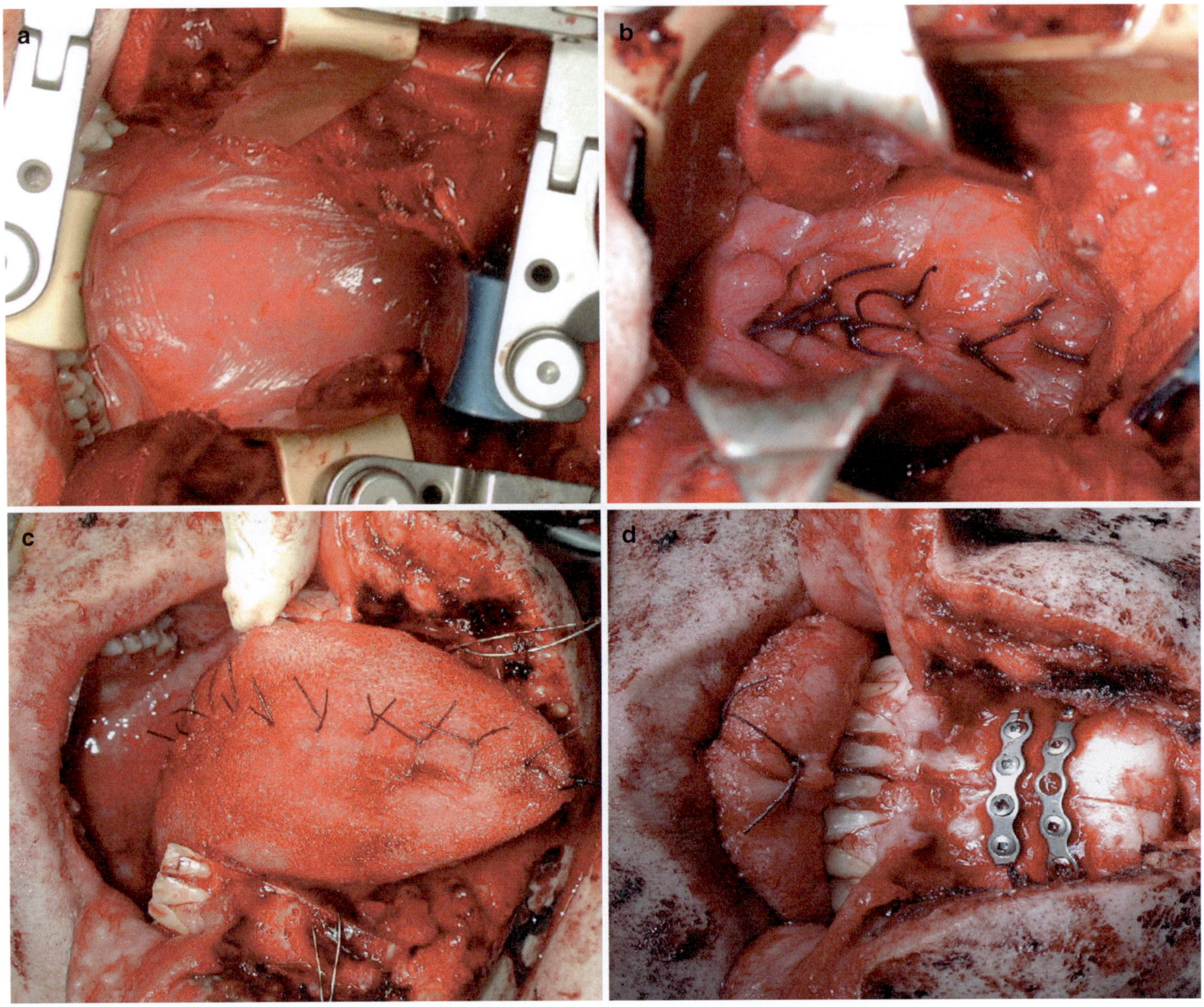

Fig. 9.11 (**a**) Exposition of posterior pharyngeal wall; (**b**–**d**) closure time

9.4.2 Endoscopic Transoral Approach

In 2002, Frempong-Boaudu reported for the first time seven cases of patients which underwent endoscopically assisted transoral CVJ surgery [3].

The endoscopic transoral approach is performed with soft palate retraction using one or two rubber catheters (Fig. 9.12a) tied to the uvula and pulled cranially through the nostrils. The oral cavity and tongue are retracted with a Spetzler-Sonntag oral retractor as in standard transoral approach. The endoscope is introduced into the oral cavity to visualize the posterior wall. A pharyngeal midline incision is created to expose the C1–C2 area (Fig. 9.12b–d). Microsurgical tools are used to remove the pathology and decompress CVJ. The use of endoscope allows a better exposure above the level of the palate because it could be navigated to look around the palate.

9.4.2.1 Surgical Principles

The surgeon locates the midline by palpating the tubercle of C1. Using a fine-tip monopolar cautery, he incises the posterior pharyngeal wall longitudinally in the midline over the region that is to be resected. Carry the incision progressively through the mucosa, the midline raphe between the pharyngeal muscles, and the anterior longitudinal ligament down to the bone.

The anterior longitudinal ligament is dissected subperiosteally till the C1 arch and the C2 body are identified. The longus colli and longus capitis muscles are mobilized laterally and held in place with tooth-bladed retractors. Lateral dissection may be limited to no more than 1.5 cm laterally because of the proximity of the internal carotid artery.

9.4.2.2 Removal of C1 Arch and Odontoidectomy (Endoscopic-Assisted Procedures)

The following section describes the removal of C1 arch and odontoid process in the absence of a tumor (e.g., rheumatoid pannus, basilar invagination, etc.). However odontoidectomy could be necessary to gain deeper access in case of extradural tumors.

The boundaries of the clivus, C1 arch, and the base of the odontoid process are identified; soft-tissue structures are detached with microsurgical tools.

Drills and Kerrison rongeur are used to resect the inferior portion of C1 arch preserving the superior rim as much as possible. Sometimes the anterior part if the C1 arch is removed completely. Apical and alar ligaments may be detached from the odontoid process before removing it to prevent upward retraction of the odontoid toward the clivus. The base of the odontoid process is transected anteriorly with drills and posteriorly with Kerrison rongeur, and the dens is pulled ventrally and caudally to remove it.

In case of pannus, transverse ligament, tectorial membrane, and any residual ligaments should be removed.

9.4.2.3 Extradural and Tumor Removal

After odontoidectomy dura should be identified removing transverse ligament and tectorial membrane. Soft-tissue extradural lesion can be excised in a piecemeal fashion using microsurgical tools, drills, curettes, and Kerrison rongeur. If tumors have involved the ligaments, they must be removed. If the dura mater has been violated by tumor, it must be resected without injuring the intradural structure such as the basilar artery, perforators, and brainstem [11].

If the dura mater has been violated, either by the tumor or by the surgical approach, the surgeon can reconstruct the dura mater with autologous or allograft fascia lata, fat, and fibrin glue in multiple layers. The placement of excessive fat may result in neural compression and should be avoided. A lumbar drain is inserted to reduce CSF pressure. Antibiotic cover for 1 week is necessary.

If the occipital condyle is damaged or drilled out because of tumor involved, the surgeon should be mindful of avoiding injury to the hypoglossal nerve into the hypoglossal canal located in the anterior third of the condyle.

The extent of decompression can be evaluated intraoperatively by injecting contrast into the epidural space and viewing this fluoroscopically.

Hemostasis is achieved and absorbable hemostatic material was placed. The mucosa, pharyngeal muscles, and ligaments are closed in a single [7] or double layer [11] with interrupted or running Vicryl suture to ensure a strong tissue closure.

9.4.2.4 Postoperative Care

In most of cases, patient is extubated immediately after the procedure unless surgeons are concerned about a difficulty airway.

A nasogastric feed tube ensures enteral feeding for the first week, and then patients are admitted for an oral soft diet.

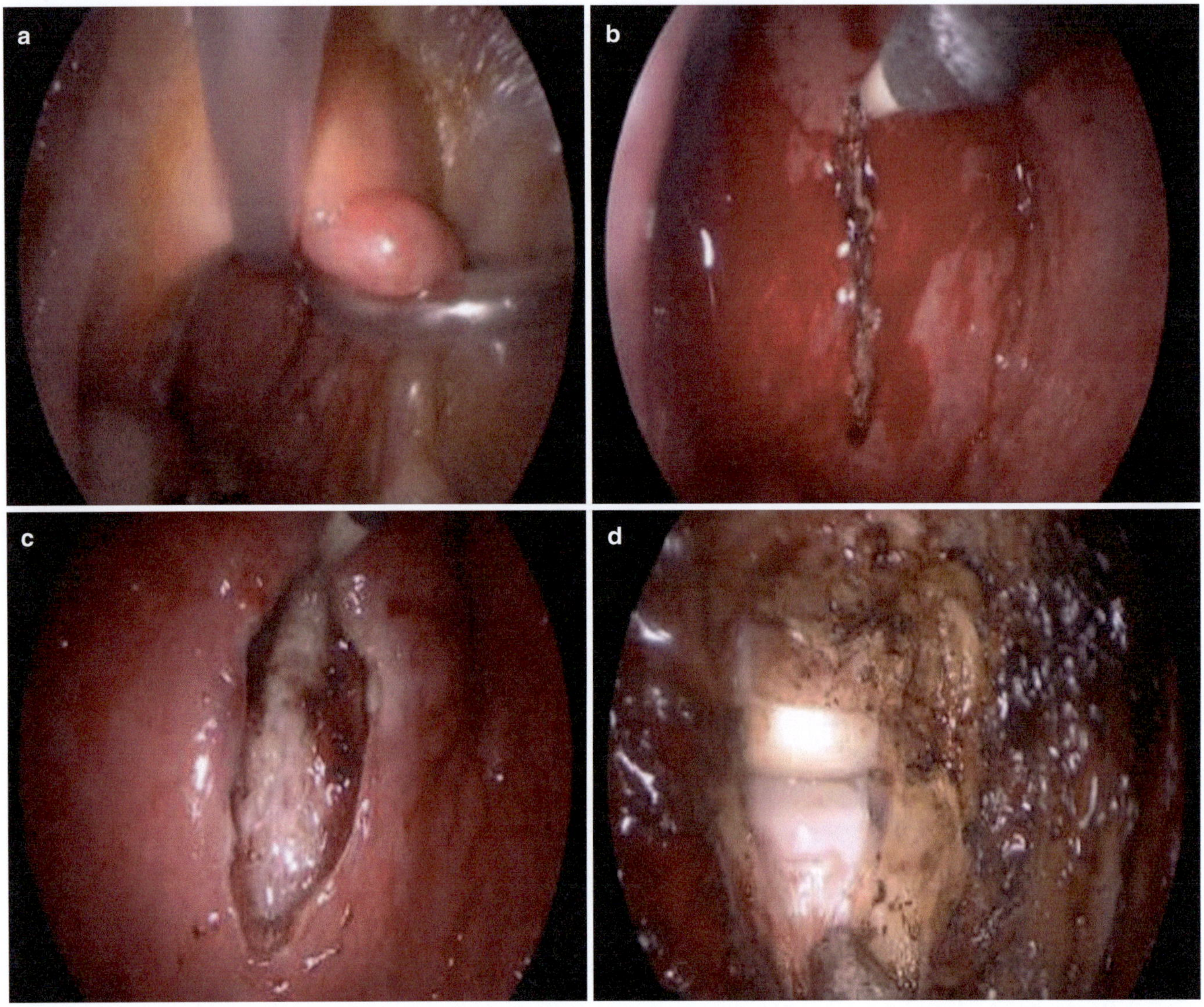

Fig. 9.12 Endoscopic transoral approach. (**a**–**c**) Steps from mucosa to prevertebral fascia; (**d**) evidence of odontoid malformations (os odontoideum)

9.4.3 Endoscopic Combined Transnasal-Transoral Approach

For this approach, the exposure is a combination of the abovementioned steps in both routes. Then the endoscope and surgical instruments are brought into surgical field alternatively through the nose and mouth in order to maximize the exposure with less dissection. Decompression is easier because visualization is gained from above and below the palate.

The most favorable feature gained via combined endoscopic approach is the possibility to reach out laterally, beyond the confines of the transnasal approach, through the oral cavity.

Limits, advantages, and disadvantages of endoscopic transnasal and transoral approaches are reported in Fig. 9.13 [4].

Table 2. Limits, Advantages, and Disadvantages of Approaches

	Transnasal	Transoral
Upper limit	Planum sphenoidale	Lower clivus
Lower limit	C2 body	C3 Body
Lateral limit	Medial pterygoid plates	Parapharyngeal carotid arteries
	Eustachian tube	
	Tubal elevation	
Width of exposure	Between tubal elevations	Between parapharyngeal carotid arteries
Average (range) mm	18.7 (15.1–22.5)	41.6 (36.0–47.0)
Working distance to odontoid tip	Shorter (3)	Longer (3)
Direction of approach to reach C1 and lower clivus	Straight ahead	Angled upward
Advantages	Greater rostral exposure	Wider approach unhampered by turbinates and septum
	Direct straight ahead access to clivus and C1	Greater caudal limit of exposure
	Shorter working distance	Easier maneuvering of instruments and suturing
	Avoids palatal, glossal, and mandibular split to reach mid-upper clivus	Reduced damage to eustachian tubes, petrous carotid artery, and vidian nerve
	Avoids tongue retraction and damage to teeth and reduces need for tracheotomy	
	Possibly reduced incidence of cerebrospinal fluid leakage and meningitis	
Disadvantages	Narrower operative corridor	Longer working distance
	Turbinates and septum limit exposure	Limited clival exposure
	Risks damage to eustachian tubes, petrous carotid artery, and vidian nerves	Incision bathed in saliva, with possibly increased meningitis
		Tongue swelling and necrosis
		Splitting soft and hard palates, tongue, or mandible in past to reach mid-upper clivus with hypermasal speech, nasal regurgitation, and velopharyngeal insufficiency
		Increased risk of need for tracheotomy

Fig. 9.13 Comparison between transnasal and transoral approach (from Askin Seker et al. [4])

9.4.3.1 Advantages and Disadvantages of Different Approaches

Endoscopic transoral approaches	Traditional transoral approaches
Better exposure in all directions	Limited exposure
Different angles of view	Limited angle of view
Two surgeons work in tandem	One surgeon works alone
Low morbidity	Low to high morbidity
Two-dimensional visualization	Three-dimensional visualization
Allows surgical procedure in people with challenging anatomic features	Jaw opening limits the working area
More difficult closure [6]	Easier closure

Extraoral approaches	Transoral approaches
Access limited to the clival region	Access allowed to the periclival region and subaxial areas
No risk of infection	Risk of infection

9.4.4 Transoral Robotic Approach

TORS approach allowed a minimally invasive surgical treatment with the intent to reduce morbidity and simplify the exposure if compared to the traditional transoral approach. The approach could be extended from lower clivus to C3.

The da Vinci Surgical System is used in this approach. Our experience started with cadaveric dissection to understand the main landmarks of the approach (Fig. 9.14).

9.4.4.1 Surgical Principles

The da Vinci System with three arms was used (a 5 mm Maryland forceps, a 5 mm spatula-tip monopolar cautery, and one endoscopic arm with integrated 30° camera for the tridimensional view). A bedside surgeon was necessary for suction and cauterization in case of bleeding; in the same surgery room, the first operator is seated at the console (Fig. 9.15).

After cadaveric dissections in a da Vinci laboratory, this approach was then used on patients.

The head were positioned supine without fixation. The patient underwent general anesthesia via orotracheal intubation.

Transoral exposure is obtained with a Feyh-Kastenbauer or Davis Meyer mouth gag.

Two nasogastric feeding tubes are inserted inside the nostrils, and a stitch is placed on the uvula to elevate soft palate to better expose the posterior wall of the oro- and rhinopharynx (Fig. 9.16).

Robotic transoral approach provided wide surgical field from nasopharynx and eustachian tubes through the hypopharynx.

An inferior-based U-shaped flap is performed using the da Vinci's monopolar cautery, through the mucosa, submucosa, and prevertebral muscles until vertebral periosteum of C1 and C2.

The flap is pushed down in the hypopharynx and the cervical spine or pathology is exposed.

The U-shaped mucosal flap based inferiorly resulted feasible and easy to perform and close compared to traditional approach. A U-shaped flap has minimal risk of dehiscence compared to midline linear incision because pharyngeal constrictor muscles movements are lower. Nonetheless the amount of saliva passing through the incision during each swallow is lower (Fig. 9.17).

At this moment surgical procedure switched to a "standard" transoral approach because no drill is actually available for the da Vinci arm. With this approach exposure is wide and goes from the clivus to C3–C4. In the picture, the below of C1 and base of C2 are shown after the soft tissue and anterior longitudinal ligament have been dissected with the robotic arms. The C1 arch was removed with the drill. Odontoid process is removed (Fig. 9.18).

At this moment great advantages from the use of the da Vinci system are shown. With TORS approach reconstruction of the mucosal flap was easily performed. A double continuous suture with V-Loc™ suture device is preferred compared to single stitches to avoid the need to tie knots, which may be difficult and time consuming up in the nasopharynx. Continuous suture of the mucosal flap is watertight and helps prevent dehiscence (Fig. 9.19).

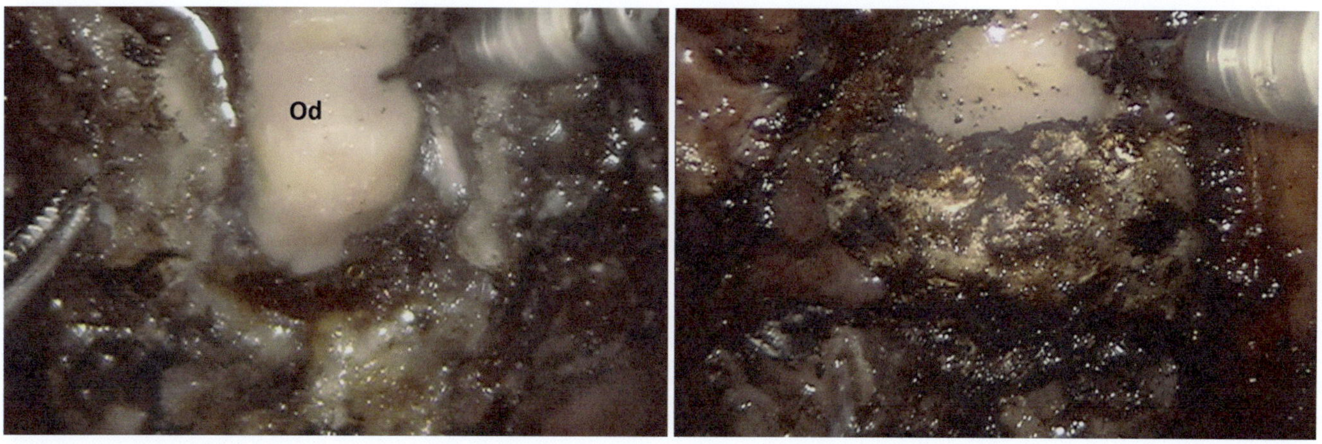

Fig. 9.14 *Od*: odontoid process dissection and its resection

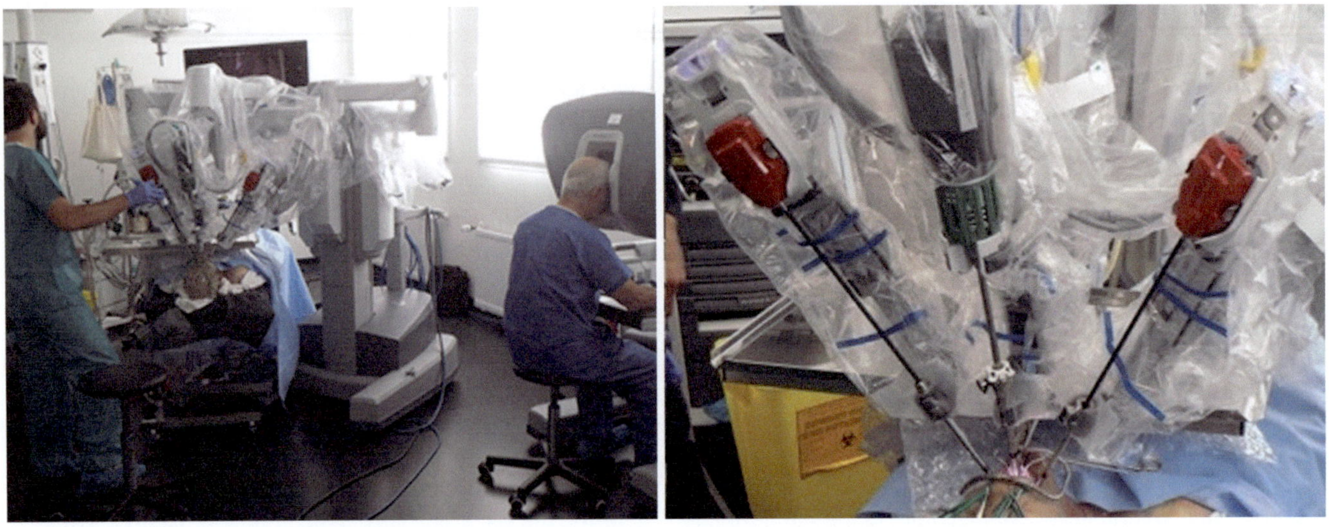

Fig. 9.15 The da Vinci system at cadaveric dissection lab

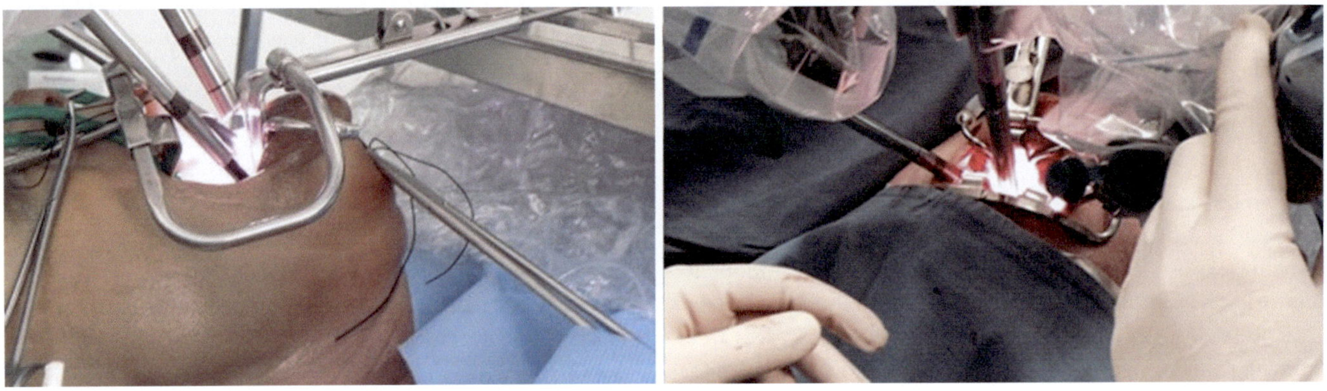

Fig. 9.16 Positioning of mouth gag and the da Vinci instrumentation

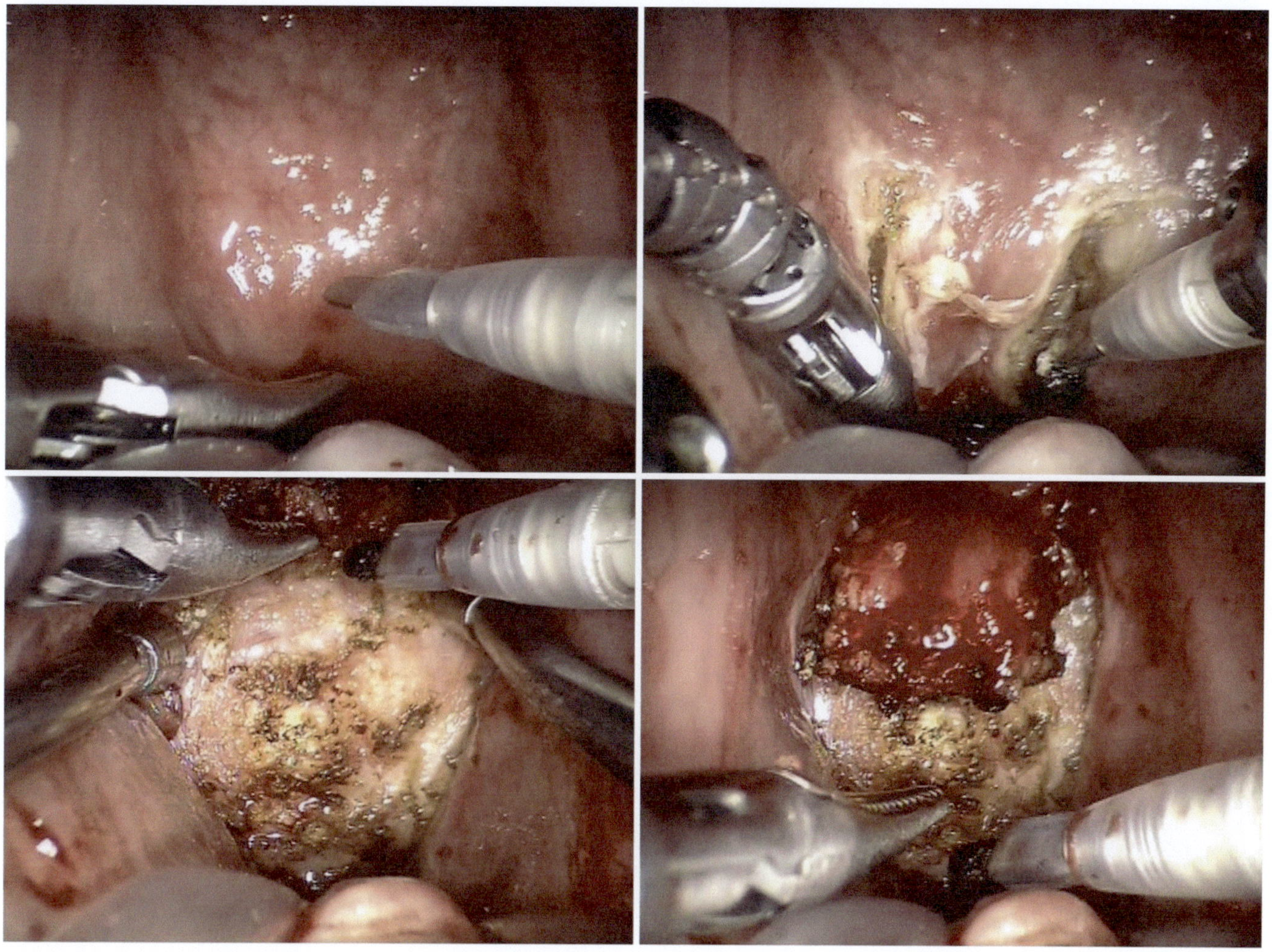

Fig. 9.17 U-shaped flap elevation and prevertebral fascia expose

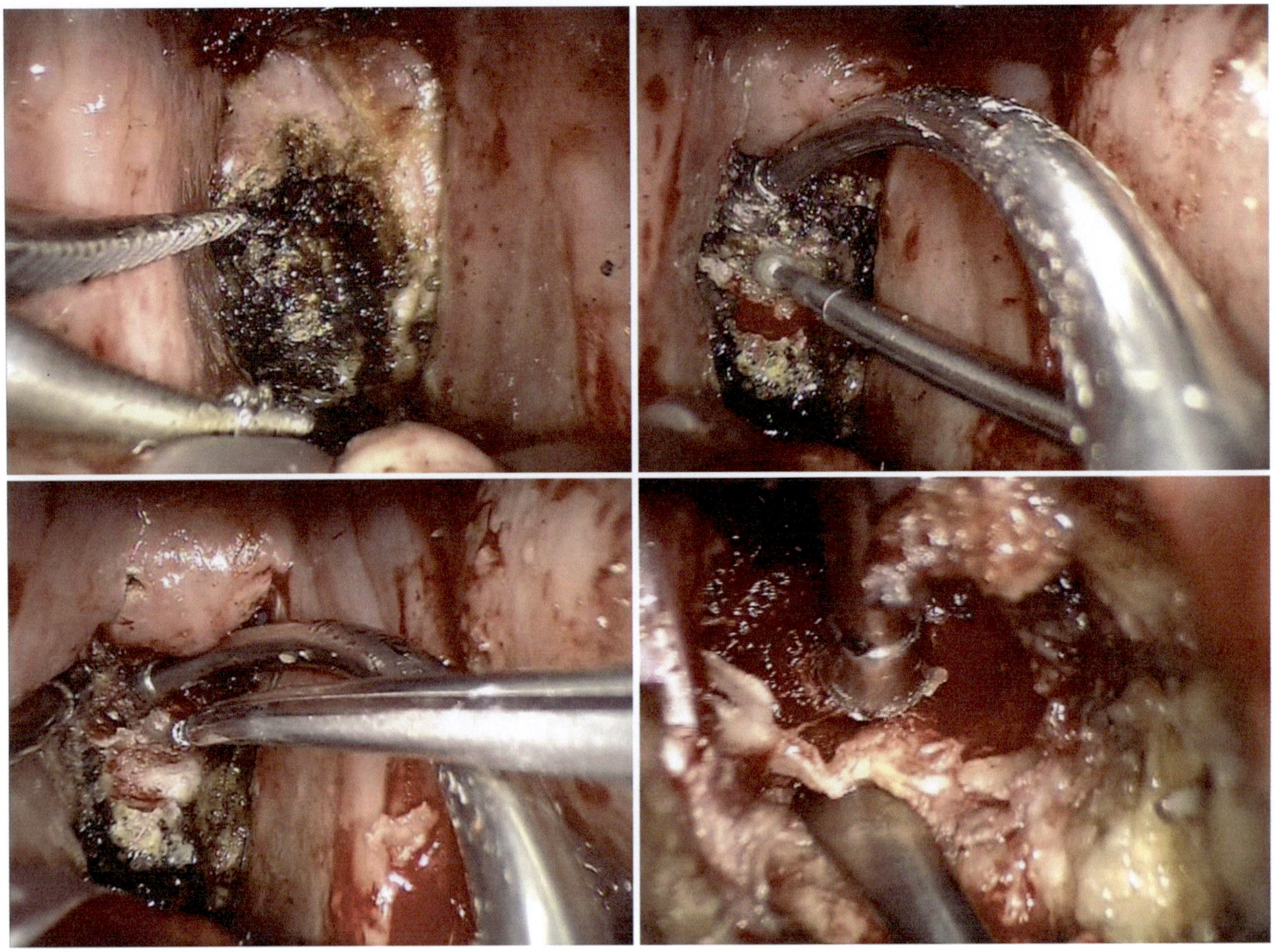

Fig. 9.18 CVJ dissection and dens drilling

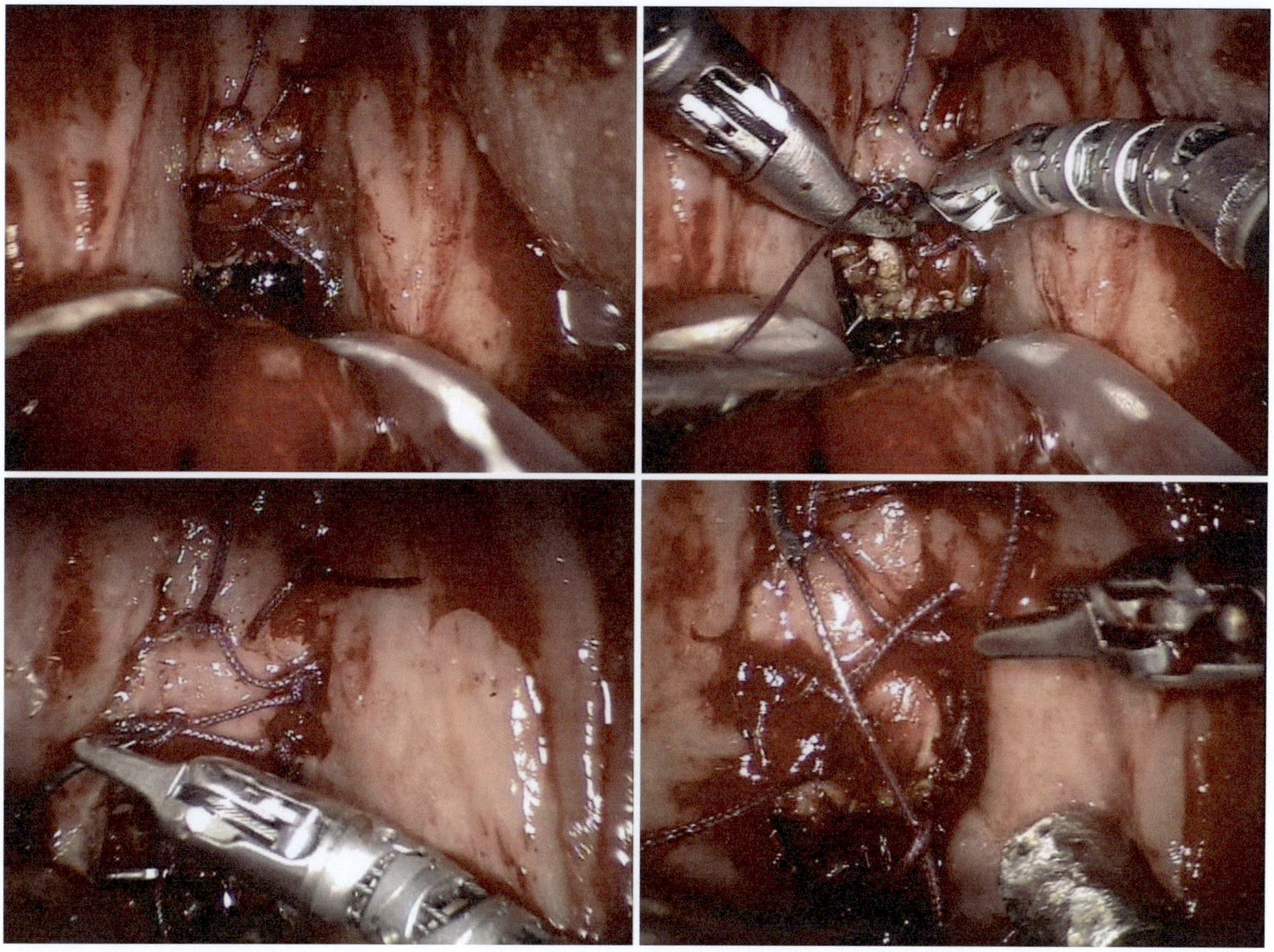

Fig. 9.19 Rhinopharynx suture

9.5 Transcervical Approaches

9.5.1 Standard Transcervical Approaches

This approach (Fig. 9.20) is well described in the anterior and lateral approach (Chap. 2.2.1).

9.5.2 Transcervical Endoscopic Approach

Intubation is performed via the nasotracheal route. The patient is positioned supine. Somatosensory-evoked responses and motor-evoked responses are monitored through the procedure. A shoulder roll is placed behind the patient and the neck is placed into gentle extension

With upper cervical spine approach, the spine is exposed rostrally to the level of the C1 tubercle. The longus colli muscles are dissected off the spine laterally to expose the ventral aspect of C2. Care must be taken at this point to avoid injury to the vertebral artery as they can lie ventral to C2. Neuronavigation system is very useful to recognize principal landmarks. A 0 or 30° 4 mm endoscope is attached to the endoscope arm or is used by the first surgeon. Retractor and suction are used by the second and third operators (Fig. 9.21).

The operation then proceeds using frameless stereotactic navigation and endoscopic visualization. The resection proceeds first between the posterior aspect of the anterior ring of C1 and the odontoid. Drilling proceeds rostrally until the tip of the odontoid is encountered. The resection then proceeds in a "top-down" fashion through the length of the odontoid until all bones have been removed. Resection of the odontoid in a top-down fashion avoids the problem of the odontoid tip becoming free-floating and difficult and dangerous to resect. The progress of the resection is continuously monitored by direct visualization through the endoscope and by using stereotactic neuronavigation (Fig. 9.2). The apical and transverse ligaments are not resected during resection of the odontoid, and they provide a protective barrier between the osseus resection and the dura mater. After the osseus resection has been completed, the ligaments (transverse, alar, and apical) and any pannus present are resected, exposing the underlying dura. After these resections, the cervical spine is unstable. In certain instances (especially in cases in which C1 has been assimilated into the occiput), the instability is between C1 and C2. In the majority of patients, however, there is instability from the occiput through C2. In the case of pure C1–C2 instability, an anterior arthrodesis can be achieved via the same approach through placement of bilateral anterior transarticular screw fixation and fusion across C1–C2 joints. If the anatomy is not suitable or if the instability is more extensive, then a second stage of OC fusion will be required (Fig. 9.22).

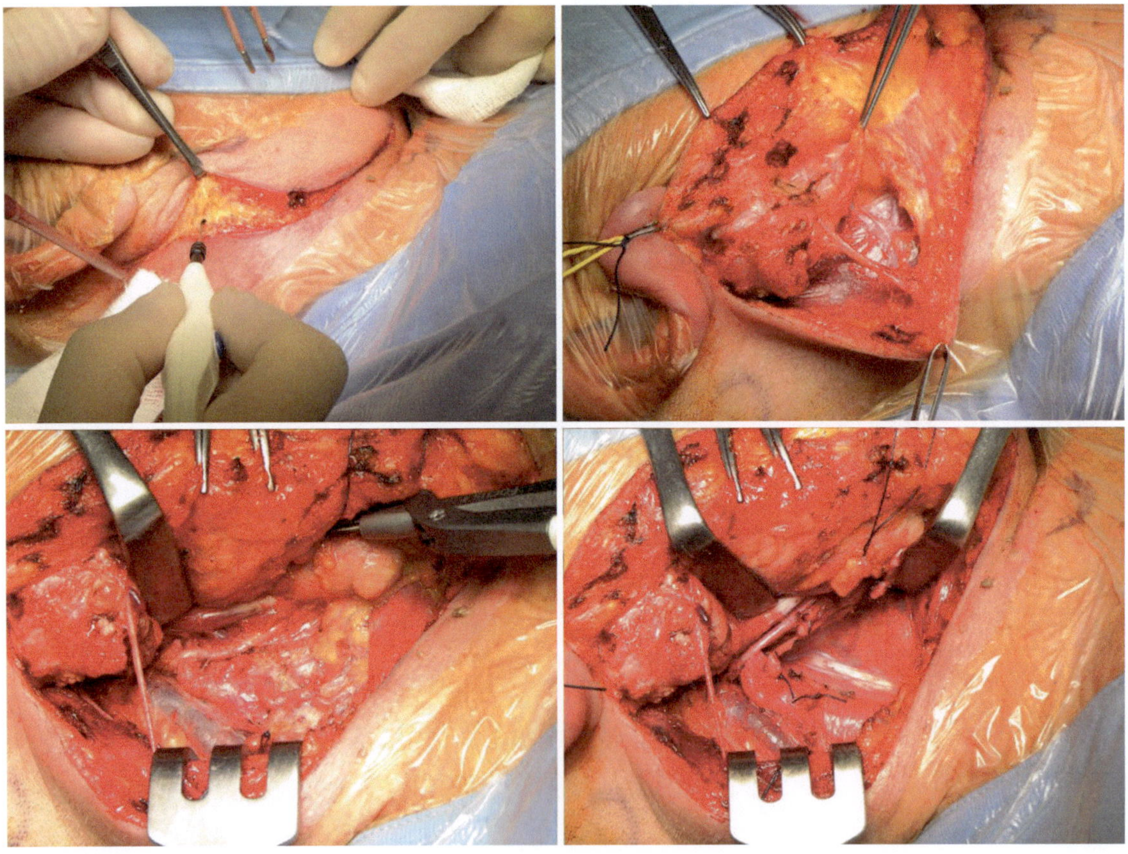

Fig. 9.20 Upper cervical spine approach on the right side

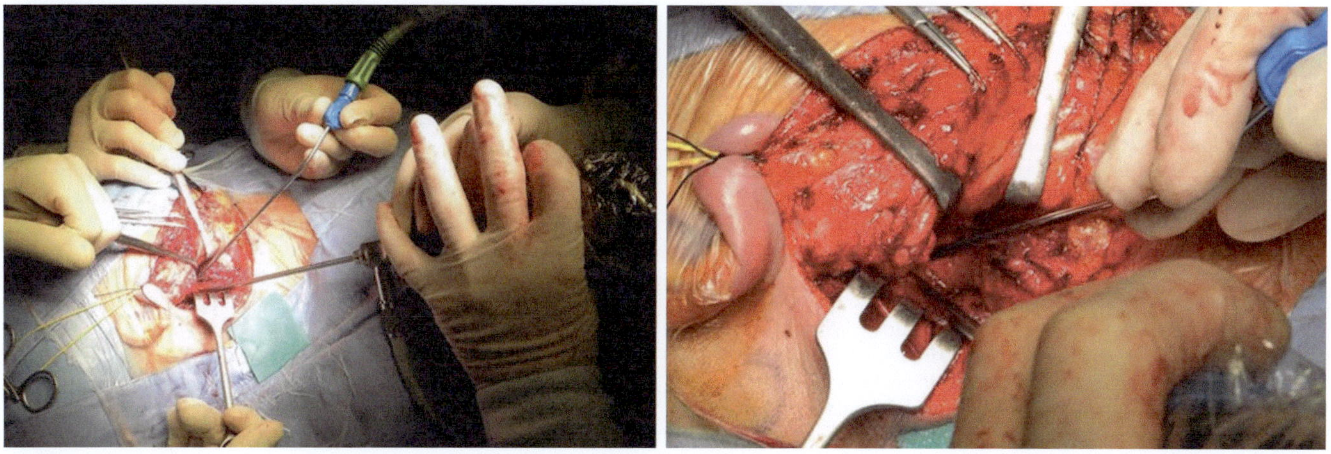

Fig. 9.21 Anterior approach to upper cervical spine on the *right side*. The surgeon with endoscope and two assistants with retractors

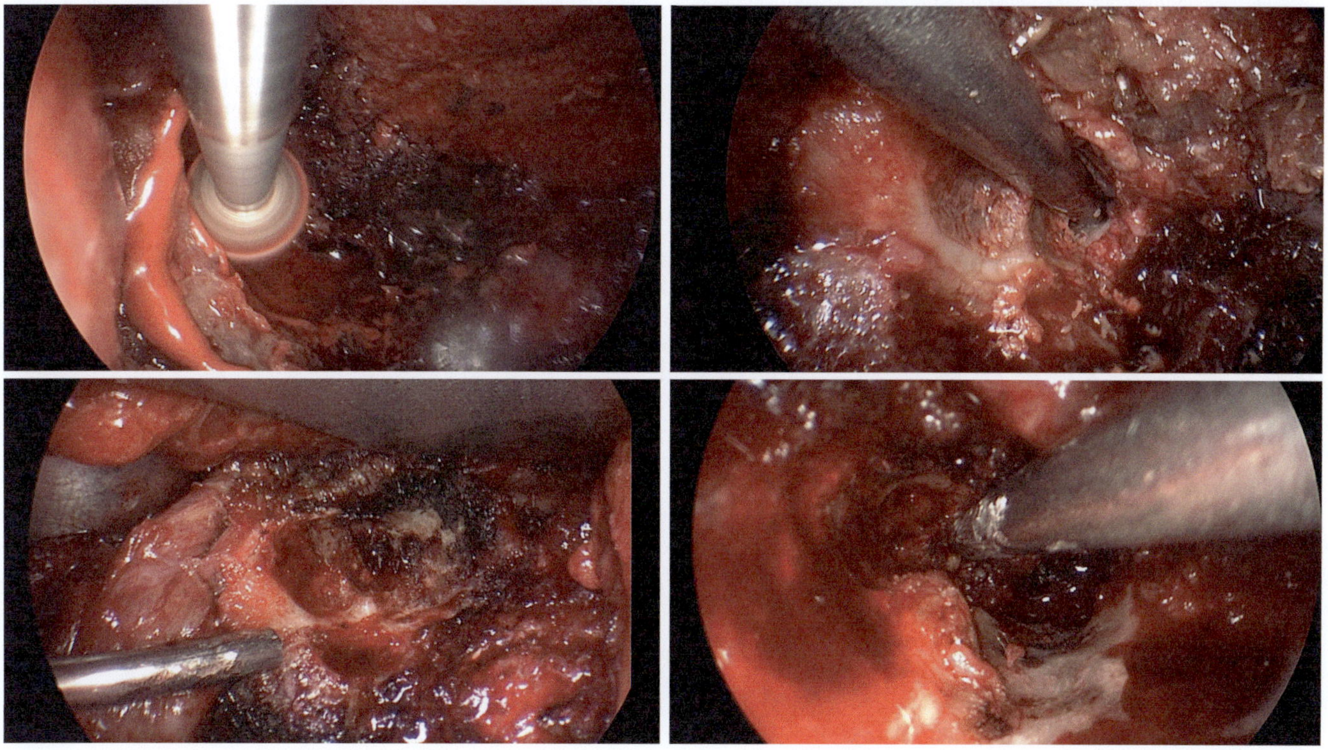

Fig. 9.22 Endoscopic view of drilling procedure on C2

References

1. Kassam AB (2005) The expanded endonasal approach: a fully endoscopic transnasal approach and resection of the odontoid process: technical case report. Neurosurgery 57(1 Suppl):E213
2. Ammirati, Bernardo (1999) Management of Skull Base Chordoma. Crit Rev Neurosurg: CR 9(2):63–69
3. Frempong-Boadu AK, Faunce WA, Fessler RG (2002) Endoscopically assisted transoral-transpharyngeal approach to the craniovertebral junction. Neurosurgery 51(5 Suppl):S60–S66
4. Seker A (2010) Comparison of endoscopic transnasal and transoral approaches to the craniovertebral junction. World Neurosurg 74(6):583–602. doi:10.1016/j.wneu.2010.06.033
5. Baird CJ, Conway JE, Sciubba DM, Prevedello DM, Quiñones-Hinojosa A, Kassam AB (2009) Radiographic and anatomic basis of endoscopic anterior craniocervical decompression: a comparison of endonasal, transoral, and transcervical approaches. Neurosurgery 65(6 Suppl):158–164
6. Dlouhy BJ (2015) Evolution of transoral approaches, endoscopic endonasal approaches, and reduction strategies for treatment of craniovertebral junction pathology: a treatment algorithm update. Neurosurg Focus 38(4):E8. doi:10.3171/2015.1.FOCUS14837
7. Dickman CA, Spetzler RF, Volker KH Sonntag (1998) Surgery of craniovertebral junction (1st edition) Thieme
8. Grammatica A, Bonali M, Ruscitti F, Marchioni D, Pinna G, Cunsolo EM, Presutti L (2011) Transnasal endoscopic removal of malformation of the odontoid process in a patient with type I Arnold-Chiari malformation: a case report. Acta Otorhinolaryngol Ital : Organo Ufficiale Della Società Italiana Di Otorinolaringologia E Chirurgia Cervico-Facciale 31(4): 248–252
9. Menezes H, VanGilder JC (1988) Transoral-transpharyngeal approach to the anterior craniocervical junction. Ten-year experience with 72 patients. J Neurosurg 69(6):895–903. doi:10.3171/jns.1988.69.6.0895
10. Menezes H, Traynelis VC, Gantz BJ (1994) Surgical approaches to the craniovertebral junction. Clin Neurosurg 41:187–203
11. Liu JK, Couldwell WT, Apfelbaum RI (2008) Transoral approach and extended modifications for lesions of the ventral foramen magnum and craniovertebral junction. Skull Base: Off J North Am Skull Base Soc 18(3):151–166. doi:10.1055/s-2007-994288
12. Bambakidis NC, Dickman CA (2012) Surgical of craniovertebral junction (second edition) Thieme
13. Shaha R (1993) Transoral-transpharyngeal approach to the upper cervical vertebrae. Am J Surg 166(4):336–340
14. Komotar RJ (2010) Approaches to anterior and anterolateral foramen magnum lesions: a critical review. J Craniovertebr Junction Spine 1(2):86–99. doi:10.4103/0974-8237.77672
15. Trotter W (1929) Operations for malignant disease of the pharynx. Br J Surg 16(63):485–495. doi:10.1002/bjs.1800166310

Anterior and Lateral Approaches to Cervical Spine

10

M. Ghirelli, F. Mattioli, G. Molinari, and L. Presutti

10.1 Anterior Approach to the Cervical Spine

Literature shows high morbidity rates after cervical spine surgery with anterior approach.

In 1958 Robinson and Smith [1] were the first to employ an anterior approach to the cervical spine. In the following years, this approach has been used for several types of spinal pathology, including tumors, infections, degenerative diseases, and traumas.

Some articles described different approaches to the cervical region [1–8].

The anterior cervical approach can be extended from the craniovertebral junction (CVJ) to the cervical thoracic junction (CTJ). Obviously, the extension of the approach is based on the pathology we have to treat. However, from mini-invasive to plane-by-plane extended dissection, it is fundamental to control all risk points, in particular neurovascular structures.

M. Ghirelli, MD (✉) • F. Mattioli, MD • G. Molinari, MD
L. Presutti, MD
Othorhinolaryngology- Head and Neck surgery Department, University Hospital of Modena, Modena, Italy
e-mail: michael.ghirelli@gmail.com

Our purpose is to analyze this surgical technique from anatomy to surgery, applying a precise method:

- *Description of the target*
- *Anatomical key points*
- *Pearls and potential complications*

This method is useful for both beginners and expert surgeons to perform safe procedures.

10.1.1 Preliminary Steps

The positioning of the patient has already been described in the specific chapter.

The choice of the side to approach depends on which cervical vertebral level is affected by the pathology and where it is specifically located.

Right-handed surgeons prefer right-sided approach, but to expose the vertebral bodies lower than C6, the standard approach is from the left side: this is done in order to reduce the risk of injury to the recurrent laryngeal nerve which ascends vertically within the tracheoesophageal groove on the left side, while the right homonymous nerve has an oblique direction. However, an approach from the right reduces the risk of iatrogenic esophageal injury, since the physiological curvature to the left of the esophageal tube promotes a better view of the prevertebral plane.

Certainly, the cervical region is one of most complex of the entire human body. To fully understand and to apply this surgical technique, it is necessary to keep in mind different landmarks, from the skin to the vertebral plane.

Surgeons have to identify on the skin: (Figs. 10.1, 10.2, 10.3, and 10.4)

S. Boriani et al. (eds.), *Atlas of Craniocervical Junction and Cervical Spine Surgery*, DOI 10.1007/978-3-319-42737-9_10

- Mastoid process
- Anterior border of the SCM
- Hyoid bone
- Cricoid cartilage
- Sternal notch
- Angle of the mandible
- Upper border of the clavicle

Some studies have defined a direct correlation between laringeal framework landmarks and cervical spine levels:

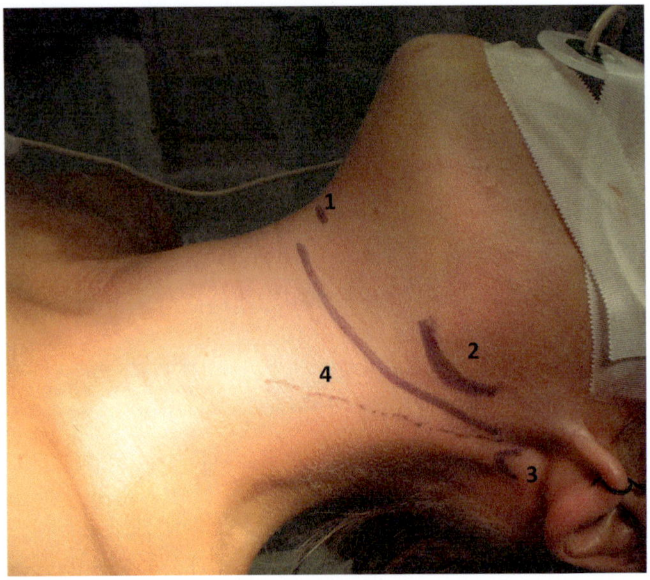

Fig. 10.1 Left side. Upper Cervical Spine (UCS) approach: (*1*) hyoid bone, (*2*) angle of the mandible, (*3*) mastoid tip, and (*4*) anterior border of sternocleidomastoid muscle

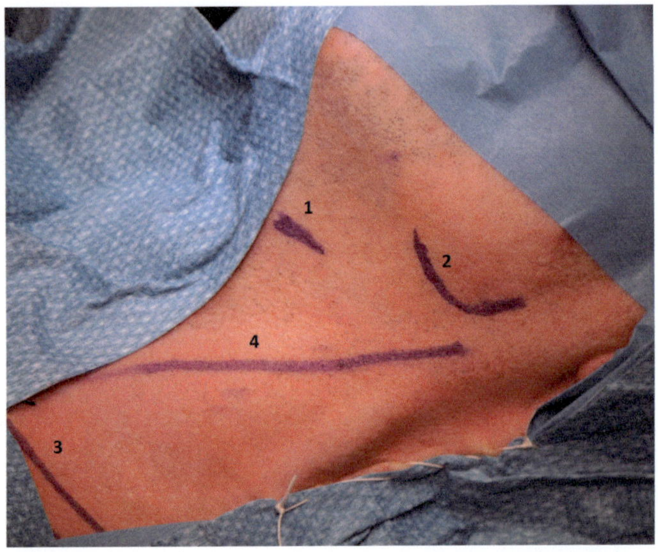

Fig. 10.3 Left side. Midcervical Spine (MCS) approach: (*1*) hyoid bone, (*2*) angle of the mandible, (*3*) clavicle, and (*4*) anterior border of sternocleidomastoid muscle

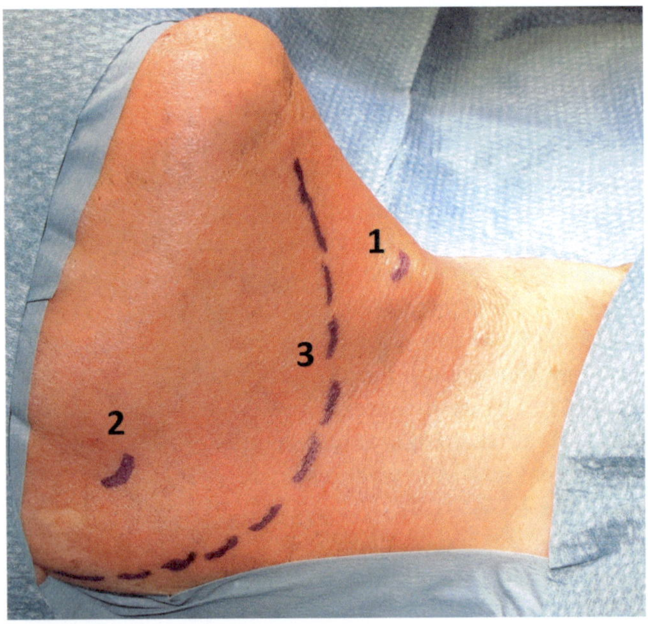

Fig. 10.2 Left side of UCS approach: (*1*) hyoid bone, (*2*) angle of the mandible, and (*3*) incision line

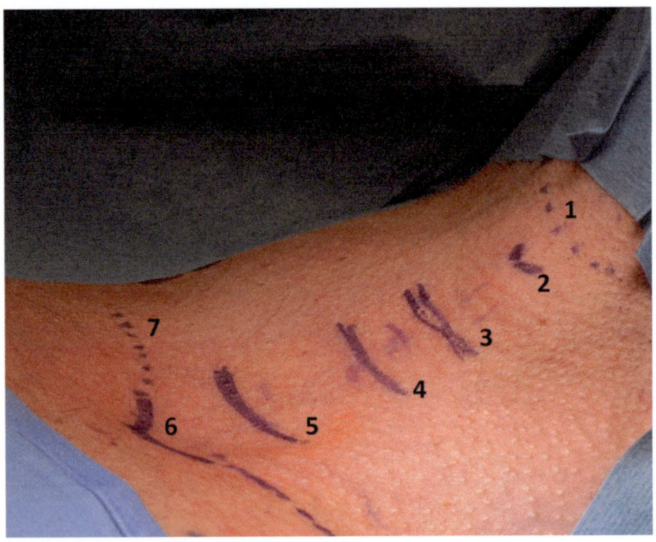

Fig. 10.4 Midline of neck: (*1*) hyoid bone, (*2*) notch of thyroid cartilage, (*3*) inferior border of thyroid cartilage, (*4*) cricoid cartilage, (*5*) incision line for tracheotomy, (*6*) suprasternal notch, and (*7*) clavicle

Landmark	Vertebral level
Hyoid bone	C3-C4
Thyroid cartilage	C4-C5
Cricoid ring	C6-C7
Carotid tubercle	C6

Four anatomical structures theoretically define three levels in which the cervical spine can be divided: (Scheme 10.1)

- From CVJ to C3: upper cervical spine (UCS)
- From C3 to C6: midcervical spine (MCS)
- From C6 to T2: lower cervical spine – cervical thoracic junction (CTJ)

The anterior approach to the cervical spine offers the possibility to extend the surgical field into these three levels, with a low percentage of morbidity.

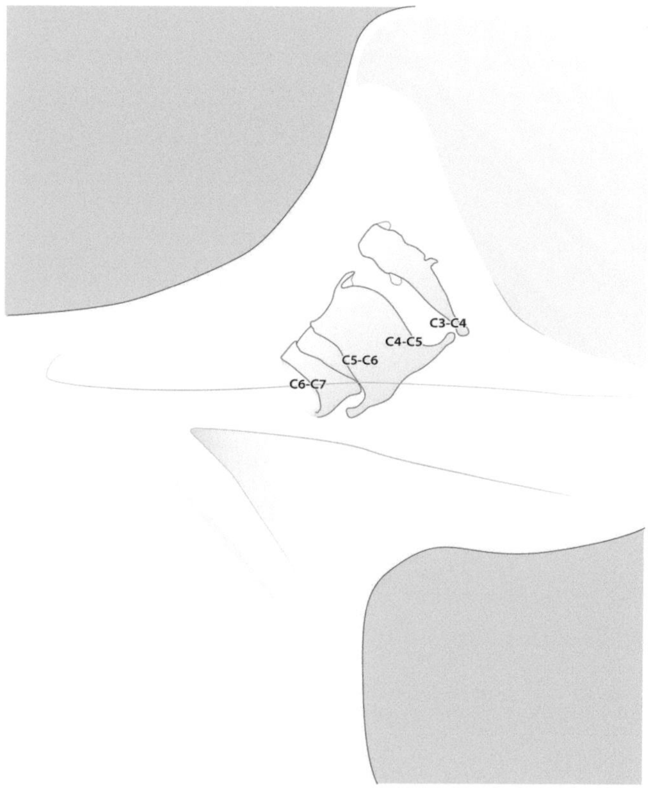

Scheme 10.1 Relationship between laryngeal framework and vertebras of cervical spine

10.1.2 **Upper Cervical Spine** (Scheme 10.2; Figs. 10.1, 10.2)

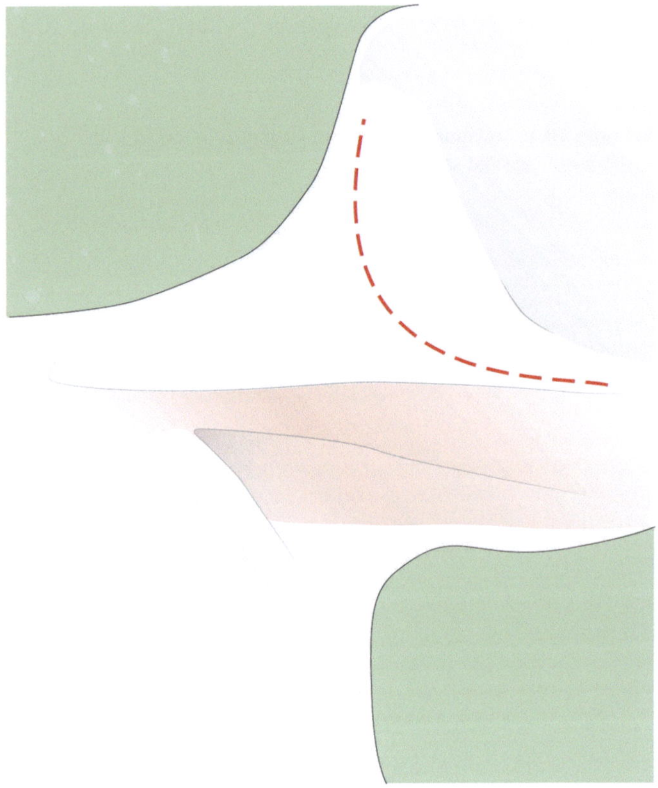

Scheme 10.2 Incision (Sebileau type) for upper cervical spine approach

10.1.2.1 First Surgical Step: Incision of the Skin

As defined by Sebileau, the incision is carried out in the submandibular region. The incision is placed at least two fingerbreadths below the edge of the mandible. Its length is about 7–8 cm, and it extends from the angle of the mandible to the midline of the neck near the hyoid bone. The previous identification of the external jugular vein on the skin is useful to prevent its accidental section. The incision must involve only the skin and the subcutaneous tissue, possibly not the platysma muscle (Scheme 10.3; Figs. 10.5, 10.6 and 10.7).

Scheme 10.3 Incision of subcutaneous tissue and platysma exposure

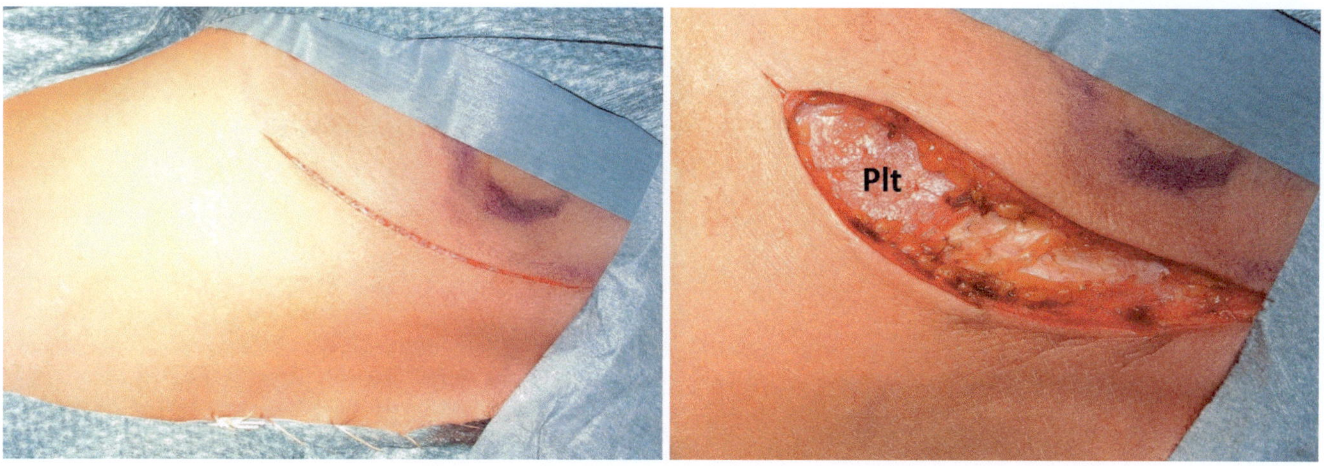

Figs. 10.5 and 10.6 Left side: incision in the submandibular region (Sebileau type) and exposure of the platysma (*Plt*)

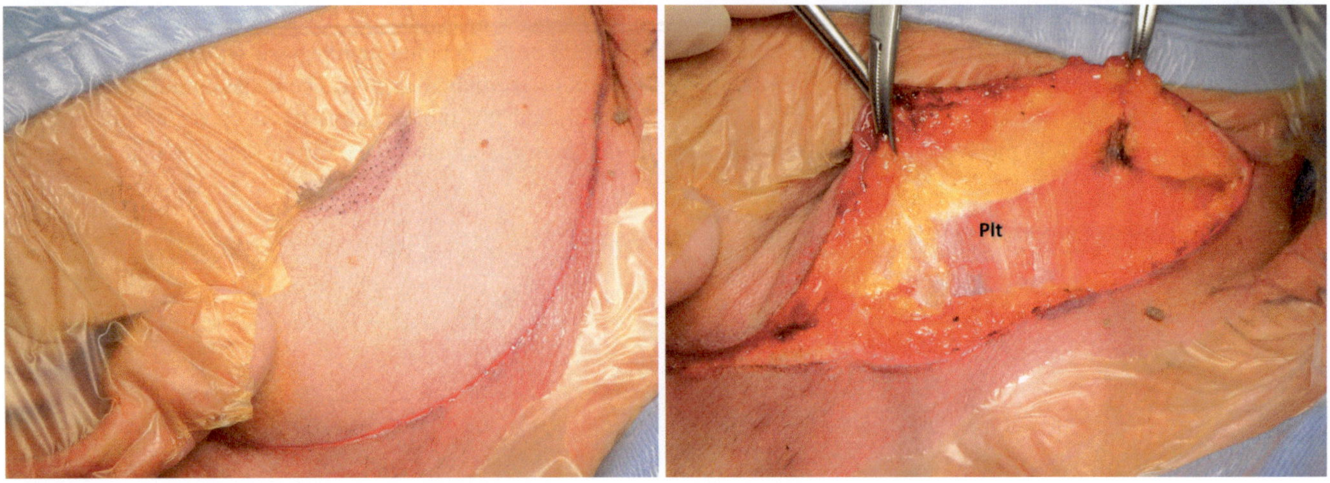

Fig. 10.7 Right side: incision in the submandibular region (Sebileau type) and expose of platysma (*Plt*)

10.1.2.2 Second Surgical Step: Dissection of the Subplatysmal Flap

Following the skin incision, a dissection of the *superficial cervical fascia* (*SCF*) – formed by the subcutaneous tissue and the platysma muscle – is performed. The flap is then elevated above the inferior border of the mandible to expose the *superficial layer of the deep cervical fascia* (*SLDCF*).

It is important to keep in mind that the marginal mandibular branch of the facial nerve passes close to the inferior margin of the corpus mandibulae in this dissection plane (Fig. 10.8).

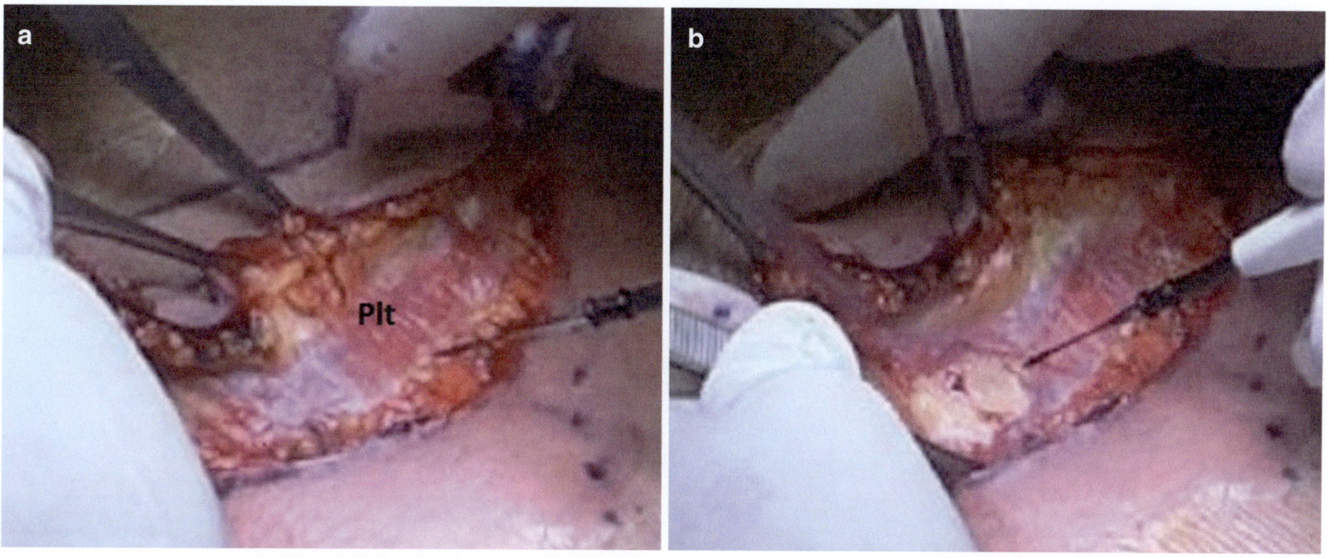

Fig. 10.8 Left side: (**a**) the platysma is exposed, after dissection of the subcutaneous tissue. (**b**) Incision of the platysma and, elevation of the subplatysmal flap, in order to expose the SLDCF

Anatomical Key Points (Figs. 10.9, 10.10, and 10.11; Scheme 10.4)

1. Sternocleidomastoid muscle (*SCM*)
2. External jugular vein (*EJV*)
3. Great auricular nerve (*GAN*)
4. Submandibular gland (*SG*)
5. Facial vein (*FV*)
6. Marginal mandibular branch of the facial nerve (Fig. 10.12)

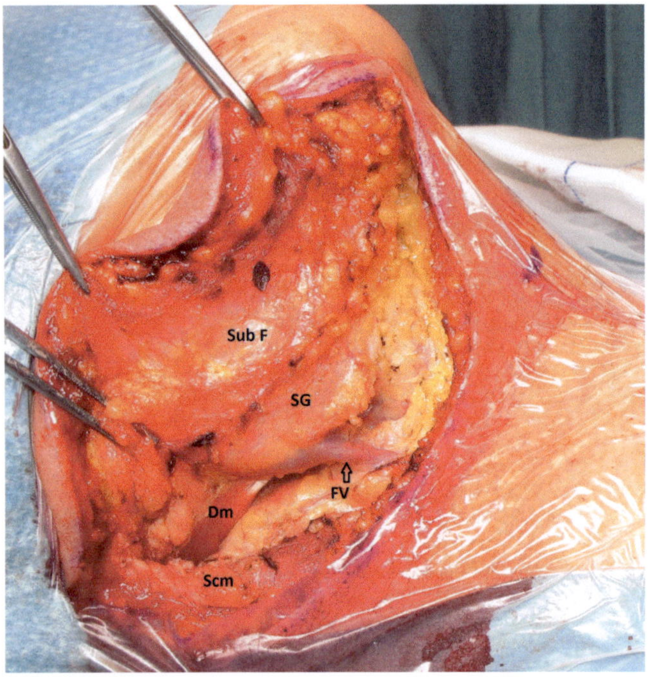

Fig. 10.9 Right side: subplatysmal flap (*Sub F*) elevation. Exposure of anatomical landmarks: digastric muscle (*Dm*), sternocleidomastoid muscle (*SCM*), facial vein (*FV*), and submandibular gland (*SG*)

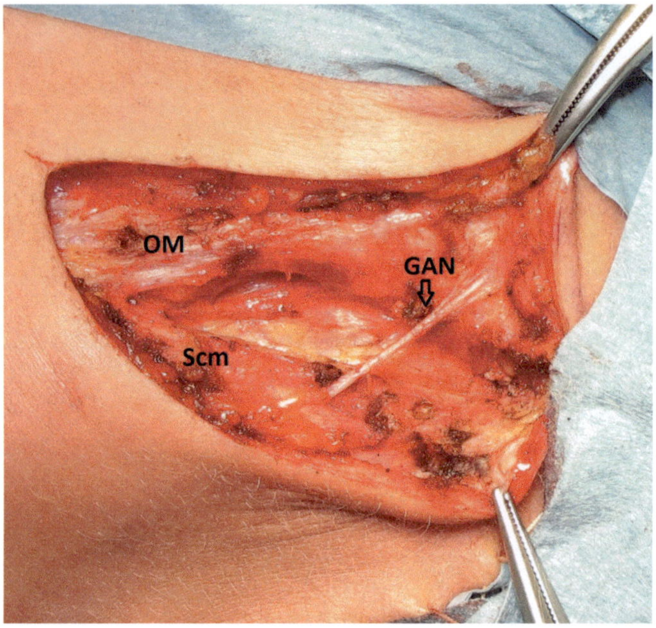

Fig. 10.11 Left side: sternocleidomastoid muscle (*SCM*), great auricular nerve (*GAN*), and omohyoid muscle (*OM*)

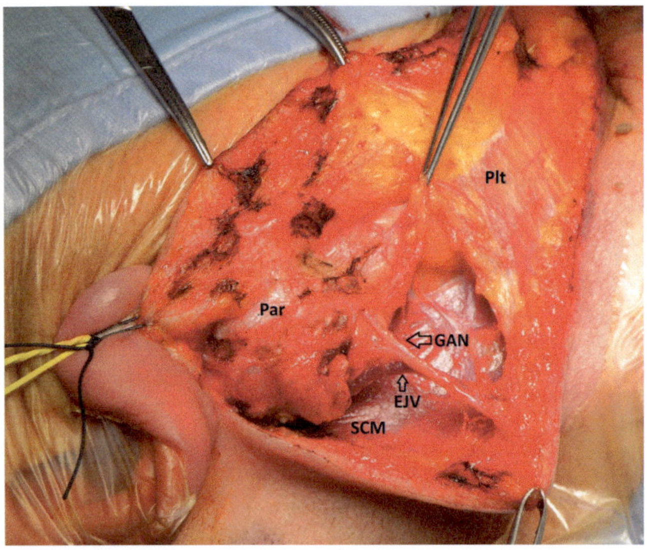

Fig. 10.10 Right side: Exposure of the anatomical landmarks. In brachytype patients, the parotid gland is lower. External jugular vein (*EJV*), great auricular nerve (*GAN*), sternocleidomastoid muscle (*SCM*), and last part of platysma (*Plt*)

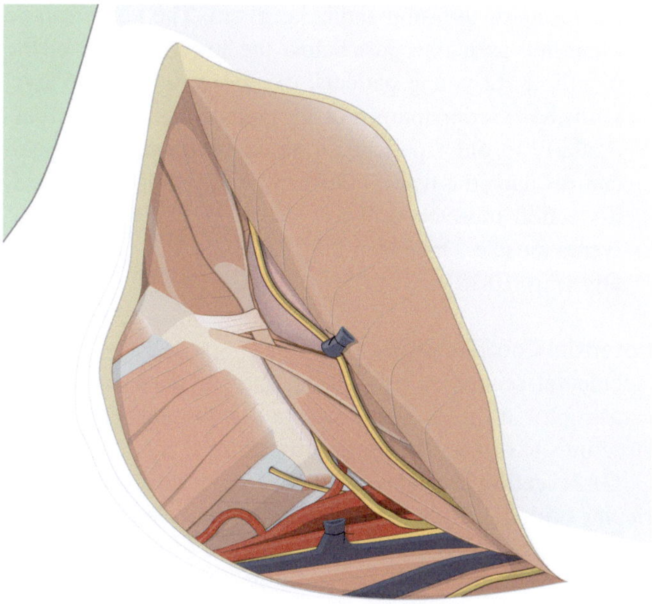

Scheme 10.4 The scheme shows the anatomical landmarks. In particular it underlines the ligation of the facial vein and its overturning in order to protect the mandibular branch of the facial nerve (Hayes-Martin maneuver)

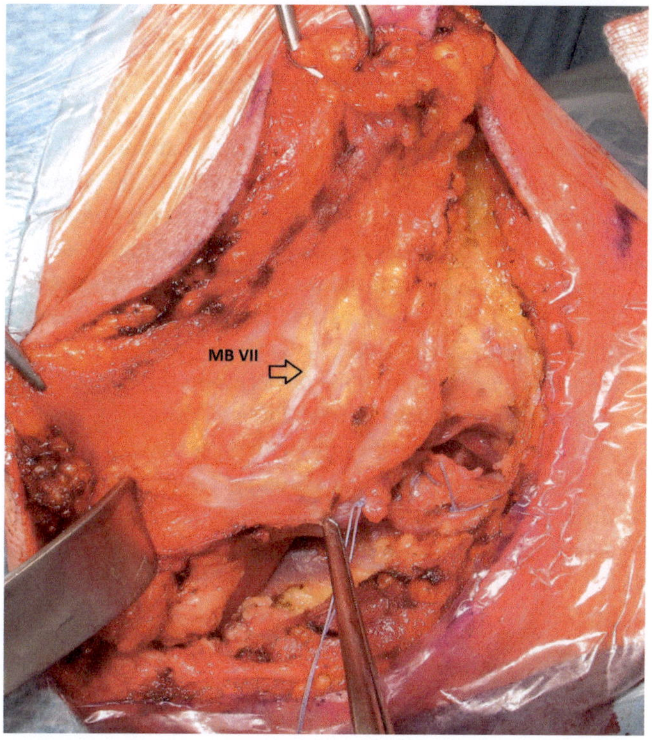

Fig. 10.12 Evidence of mandibular branch of the facial nerve (*MB VII*) lying into the superficial cervical fascia

Pearls

1. Define the inferior margin of the surgical field.
2. Identification of the EJV on the skin, before skin incision, is useful for better identification and preservation of the anatomical structures. The external jugular vein emerges in the parotid loggia from the confluence of the posterior auricular vein and the posterior division of retromandibular vein and then penetrates into the SCM region on the level with the angle of the mandible and crosses the lateral face of the muscle askew. During the dissection, it may be necessary to ligate it.
3. The GAN emerges from the posterior border of the SCM, at Erb's point. It runs parallel and superior to the EJV on the superficial face of the muscle. It gives sensory branches to the skin of parotid and auricular regions. During the dissection, the surgeon usually finds the EJV first and then, about 1 cm laterally and cranially to it, the GAN.
4. The identification of the submandibular is an important step. This salivary gland is a fundamental landmark to recognize a safe plane where preserving the marginal mandibular branch of the facial nerve. The SG lies into the SLDCF division, while the nerve stays below the platysma and above this fascia.
5. The FV lies above the digastric muscle. To carry out a safe dissection, the FV has to be identified, ligated, and retracted cranially (*Hayes-Martin maneuver*). This trick allows protecting the marginal mandibular nerve since in more than 90 % of cases, it passes above the FV, and, in the 10 % left, it is medial to the vein (Fig. 10.13a, b and Scheme 10.3).
6. The surgeon must remain aware of the location of the marginal mandibular branches, which run in the plane of the fascia of the submandibular gland. The nerve can be identified as it dips just below the lower margin of the mandible. In young patients, the nerve runs on the level with the inferior margin of the mandible; however, particularly in older patients, a ptotic submandibular gland can displace the nerve inferiorly. As the flap is elevated, dissection must be kept as close as possible to the platysma muscle. Once identified, it can be transposed cranially (Fig. 10.12).

Potential Complications

Accidental section of the EJV or the FV does not lead to catastrophic bleeding but creates a nonoptimal view of the structures to dissect during the procedure.

GAN section produces an ailment hypoesthesia of periauricular skin; thus, this nerve should be preserved.

The marginal mandibular branch of the facial nerve could be damaged during the elevation of subplatysmal flap. In case of nerve palsy, the patient shows an incomplete smile with very subtle flattening of omolateral nasolabial fold.

10.1.2.3 Third Surgical Step: Dissection of the SLDCF Anteriorly to the SCM

Under cover of the platysma, there is "the investing layer of the deep cervical fascia" or superficial layer of the deep cervical fascia (SLDCF). It surrounds the neck like a collar but splits around the SCM and trapezius to enclose them. Posteriorly, it joins the ligamentum nuchae.

The SLDCF has to be opened along the anterior border of SCM muscle. Lateral retraction of the SCM muscle exposes some of the structures below:

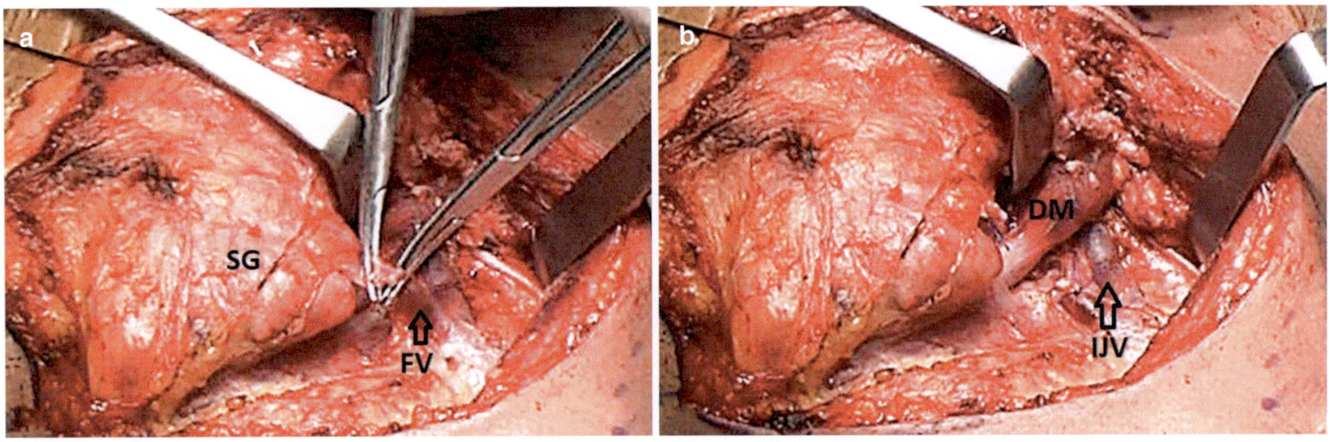

Fig 10.13 Left side: (**a**) ligation of the facial vein (*FV*), that passes over the submandibular gland (*SG*). (**b**) *Hayes-Martin maneuver* and exposure of digastric muscle (*DM*) under the SG

Anatomical Key Points (Figs. 10.14a, 10.15, and 10.16; Scheme 10.5)

1. Digastric muscle (*DM*)
2. Thyrolingual-facial venous trunk (*TLF*) (Fig. 10.14b)
3. Carotid sheath or neurovascular bundle
4. Spinal accessory nerve (*SAN*)
5. Hyoid bone (*HB*)
6. Stylohyoid muscle (*SHM*)
7. Hypoglossal nerve (*HN*)
8. Superior thyroid artery (*STA*)
9. Superior laryngeal nerve (*SLN*)
10. Lingual artery (*LA*)
11. Facial artery (*FA*)
12. External carotid artery (*ECA*)
13. Internal carotid artery (*ICA*)

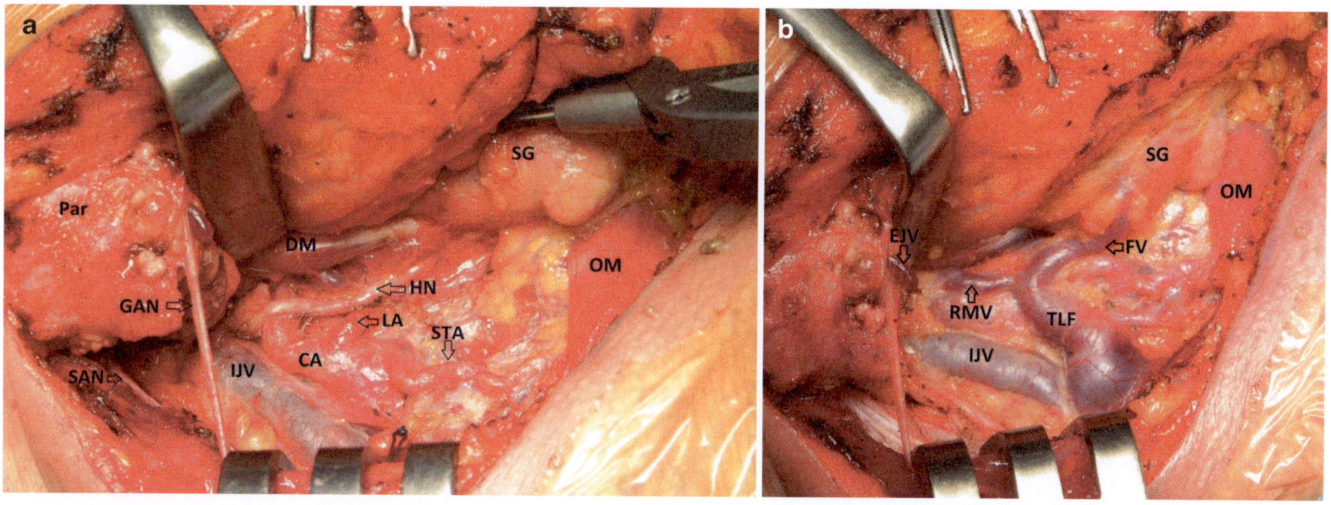

Fig. 10.14 Right side: (**a**) spinal accessory nerve (*SAN*), great auricular nerve (*GAN*), posterior belly of the digastric muscle (*DM*), omohyoid muscle (*OM*), submandibular gland (*SG*), internal jugular vein (*IJV*), external carotid artery (*CA*), hypoglossal nerve (*HN*), superior thyroid artery (*STA*), and lingual artery (*LA*). (**b**) Evidence of external jugular vein ligated (*EJV*), retromandibular vein (*RMV*), facial vein (*FV*), their union into thyrolingual-facial trunk (*TLF*), internal jugular vein (*IJV*), submandibular gland (*SG*), omohyoid muscle (*OM*)

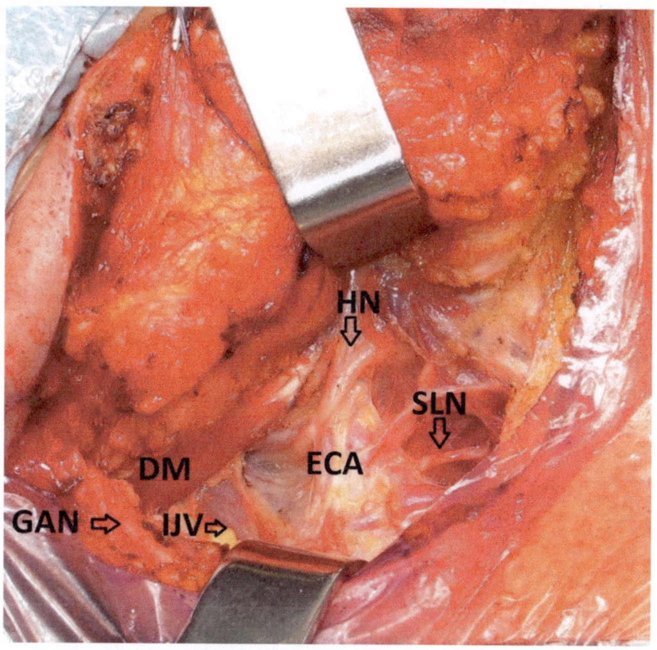

Fig. 10.15 Right side: posterior belly of the digastric muscle (*DM*), hypoglossal nerve (*HN*), superior laryngeal nerve (*SLN*), great auricular nerve (*GAN*), internal jugular vein (*IJV*), and external carotid artery (*ECA*)

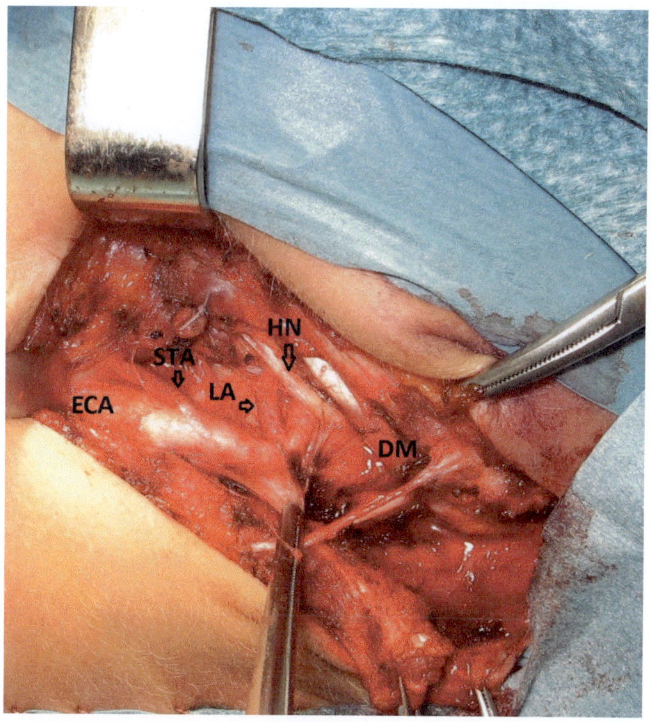

Fig. 10.16 Left side: posterior belly of the digastric muscle (*DM*), external carotid artery (ECA), hypoglossal nerve (*HN*), superior thyroid artery (*STA*), and lingual artery (*LA*)

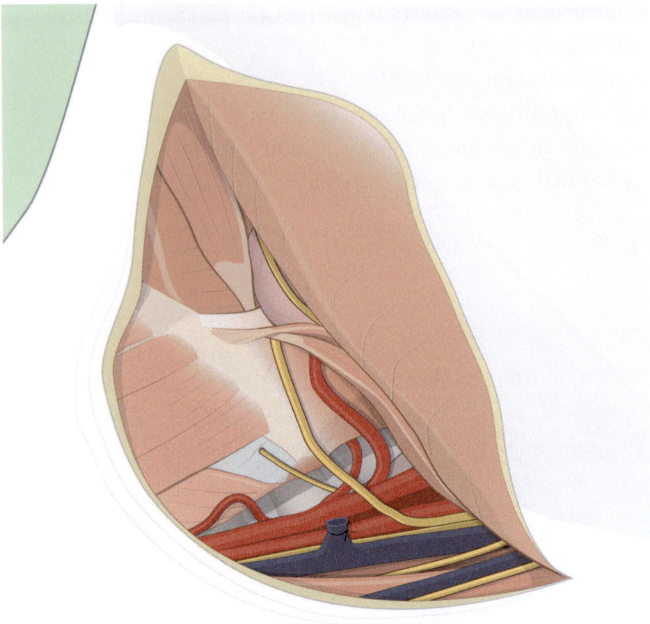

Scheme 10.5 The scheme shows all anatomical landmarks of the surgical step in a simplified way. In the upper portion we find the SG, MG VII and DM retracted. Below the DM there are: the facial artery , the lingual artery, branches of the external carotid artery and the hypoglossal nerve that runs through the field above the external carotid artery. Below the hyoid bone there are the superior thyroid artery and the superior laryngeal nerve. The TLF has been ligated at the origin of the internal jugular vein. More laterally we see the spinal accessory nerve (SAN), EJV and GAN

Pearls

The goal is to identify the three main structures: (Figs. 10.17 and 10.18; Scheme 10.6)

- *Digastric muscle*
- *Hypoglossal nerve*
- *Superior laryngeal nerve*

Once you have the control over these three landmarks, the approach to the vertebral plane can be carried out safely:

1. We find the posterior belly of digastric muscle, which runs obliquely forward and downward and goes to the hyoid bone, where it is fixed with its intermediate tendon to the body and the great horn; then it continues with its anterior belly. It is known as "the resident's friend" because once it is identified, below it is possible to identify important structures like the neurovascular bundle and the TLF.
2. The TLF trunk passes deeply to the SCM and above the digastric muscle to empty into the internal jugular vein. The TLF trunk is formed by the thyroid, pharyngeal, lingual, and facial veins which merge together before opening as a unique vein into the IJV. Only occasionally these veins reach the IJV independently. Under and just before the angle of the mandible, the facial vein receives the blood from the retromandibular vein, before reaching the IJV. The ligation of the TLF trunk should be done near the confluence with the IJV; alternatively the facial vein only can be ligated (Fig. 10.14b).
3. It is formed by the internal jugular vein laterally, by the carotid artery medially, and, in the dihedral angle formed by the two vessels, by the vagus nerve posteriorly. The first vascular structure found during the dissection is the IJV, which is more lateral: it can be identified going just under the digastric muscle.
4. This nerve usually runs below, from the posterior belly of the digastric muscle before piercing the deep surface of the SCM muscle. It passes through this muscle and supplies it and finally emerges from the midpoint of the posterior border of the SCM.
5. The hyoid bone is a fundamental landmark in neck surgical anatomy. Using this approach, below the hyoid, we have to find STA and SLN. The region above shows a high complexity, with several structures that are described in the following paragraphs.
6. This muscle, which is innervated by the facial nerve, arises from the styloid process of the temporal bone and inserts onto the hyoid bone. This muscle is on the same plane of the digastric muscle. When the approach is extended from C2 to CVJ, it is necessary to dissect it.
7. The HN emerges from the hypoglossal canal just medially to the jugular foramen, passes laterally between the internal jugular vein and the internal carotid artery, and goes superficial to the internal carotid artery to reach the posterior edge of the stylopharyngeus muscle. The nerve curves below the lingual artery and disappears beneath the posterior edge of hyoglossus muscle. The triangular space designed by the digastric muscle superiorly, the internal jugular vein laterally, and the TLF inferiorly is useful to identify the HN.
8. The STA arises from the external carotid artery at the point of the hyoid bone and, after providing the upper laryngeal artery, goes down to the upper pole of the thyroid gland. The STA is the first branch of the external carotid artery. To allow the access to the prevertebral space, it is mandatory to bind and dissect this artery. The STA is always under the hyoid bone. To safely bind the artery, we should stay near its origin from the external carotid artery. This maneuver reduces the risk of superior laryngeal nerve injury. During this approach, the STA may not be properly in the surgical field, but without its ligation, it would be difficult and dangerous to obtain a good exposure of the vertebral plane. The surgeons should keep in mind, the possible anatomical variations of the origin of the STA like from thyrolingual (2 %) or thyrolinguofacial (1 %) trunck (Figs. 10.19 and 10.20).
9. The SLN runs medially to the carotid sheath, forming an oblique angle with it. The nerve moves medially toward the thyrohyoid membrane. After the ligation of the STA, it is possible to identify this nerve.
10. The lingual artery is the second branch of the external carotid artery and its first branch above the hyoid bone. Almost immediately after its origin, accompanying the middle constrictor of the pharynx, it meets the posterior margin of the hypoglossal muscle, which takes a horizontal, parallel route to the greater horn of the hyoid bone, approximately half a centimeter above it. Piragoff's triangle is formed by the intermediate tendon of the digastric muscle, the hypoglossal nerve, and the posterior margin of the mylohyoid muscle: this triangle allows surgeons to find the artery, which is usually bound in the space between the digastric superiorly and the HN inferiorly; alternatively, it can also be bound near its origin (Figs. 10.21, 10.22, and 10.23).
11. It's the third branch of the external carotid artery. It emerges behind the posterior belly of the digastric muscle, posteriorly skimming the submandibular gland; running backward to forward, and upward to downward, it becomes superficial to surround the inferior margin of the mandible, immediately anterior to the facial vein. Ligation of the FA should be carried out at its origin. Section of DM should be necessary to approach UCS and to correctly binding of FA (Figs. 10.24 and 10.25).
12. The internal carotid artery passes deep to the posterior suspensory muscles of the pharynx to terminate medial to the styloid process. The ICA continues posteromedially to enter the carotid canal at the base of the skull not giving out any branches.

13. The ICA leaves the carotid sheath passing anterosuperiorly within the carotid triangle. It is useful to keep in mind that the external carotid artery divides the posterior suspensory complex of the larynx by passing between the digastric and stylohyoid muscles laterally and the styloglossus and stylopharyngeus muscles medially. It emits eight branches: five below the posterior belly of the digastric within the carotid triangle and three above it. Surgeons have to follow this artery in order to identify the right plane to approach cervical spine, which lies medially to the vascular structures and laterally to visceral ones (Fig. 10.26).

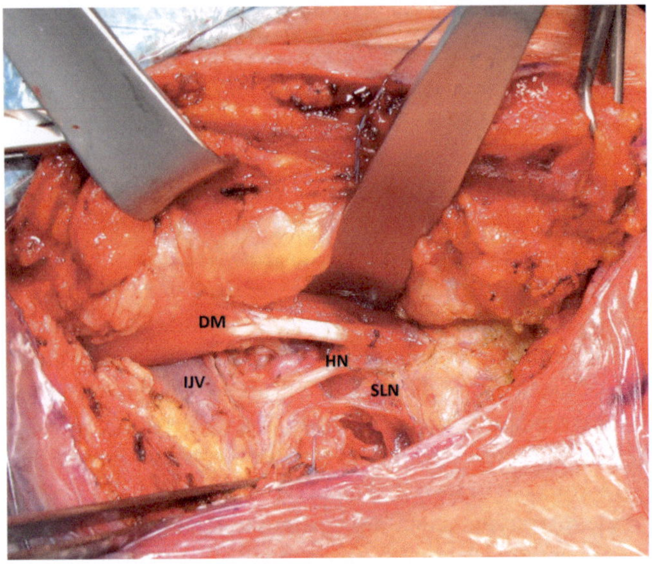

Fig. 10.17 Right side. Evidence of three fundamental landmarks: digastric muscle (*DM*), hypoglossal nerve (*HN*), superior laryngeal nerve (*SLN*), Internal jugular vein (*IJV*)

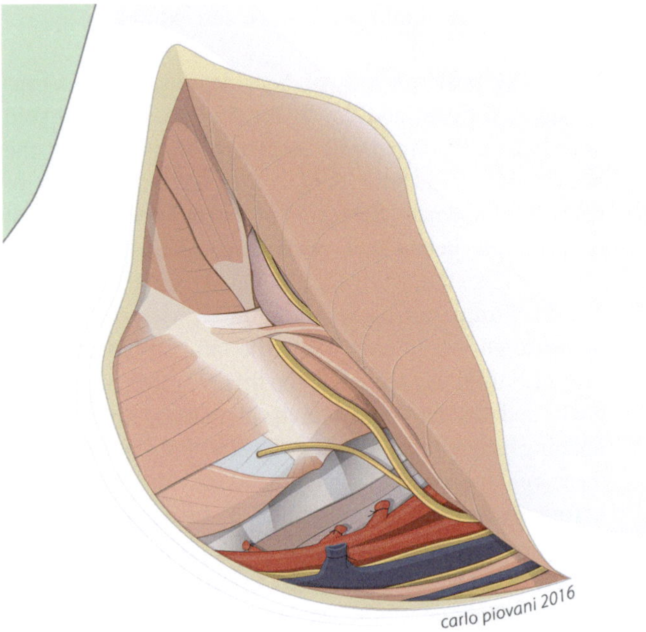

Scheme 10.6 Once you have identified and preserved the following three structures: DM, HN and SLN, the cervical spine can be approached safely

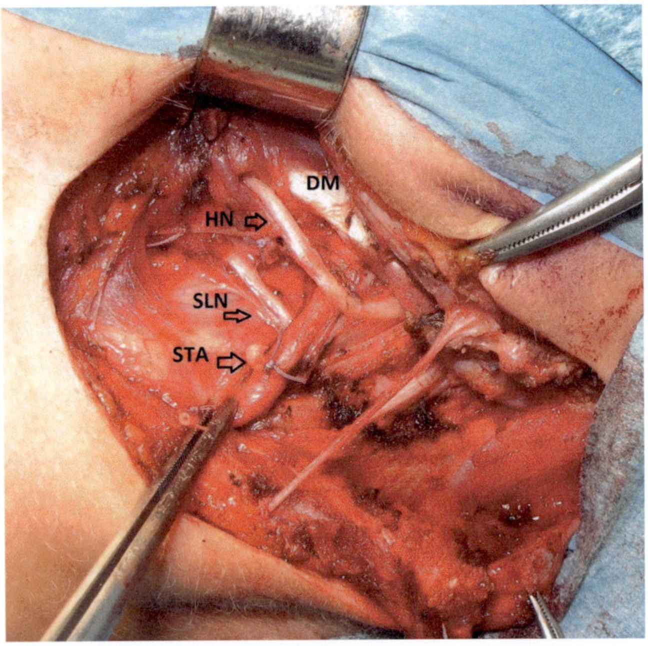

Fig. 10.18 Left side Evidence of three fundamental landmarks. From upper to downer: digastric muscle (*DM*), hypoglossal nerve (*HN*), superior laryngeal nerve (*SLN*), superior thyroid artery (*STA*) ligated

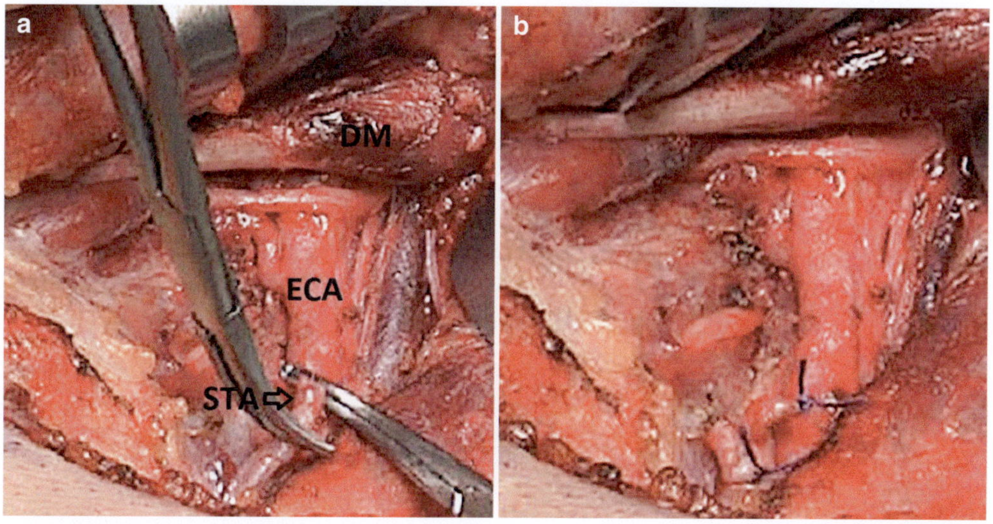

Fig. 10.19 Left side: ligation of superior thyroid artery (*STA*). Evidence of: the posterior belly of digastric muscle (*DM*), external carotid artery (*ECA*), hypoglossal nerve (*HN*)

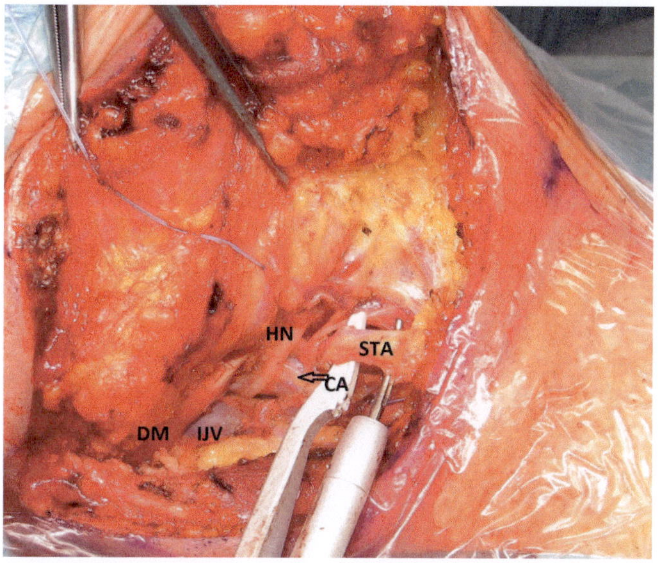

Fig. 10.20 Right side: Ligation of superior thyroid artery (*STA*). Evidence of: the posterior belly of digastric muscle (*DM*), jugular vein (*IJV*), external carotid artery (*CA*), hypoglossal nerve (*HN*)

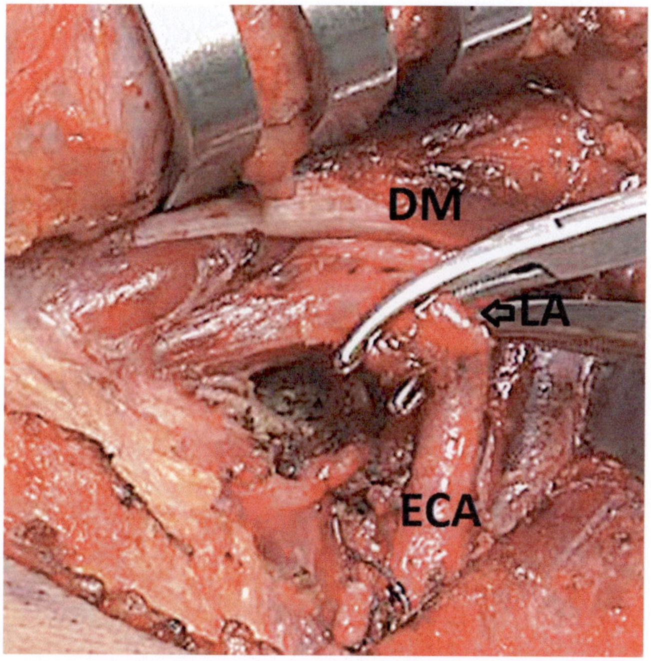

Fig. 10.21 Left side: Ligation of lingual artery (*LA*), The posterior belly of digastic muscle (*DM*), external carotid artery (*ECA*)

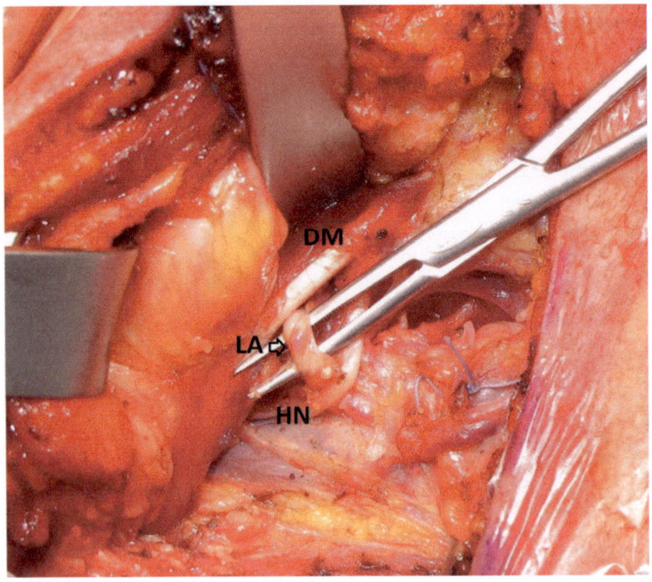

Fig. 10.22 Right side: Ligation of lingual artery (*LA*) between hypoglossal nerve (*HN*) and digastric muscle (*DM*)

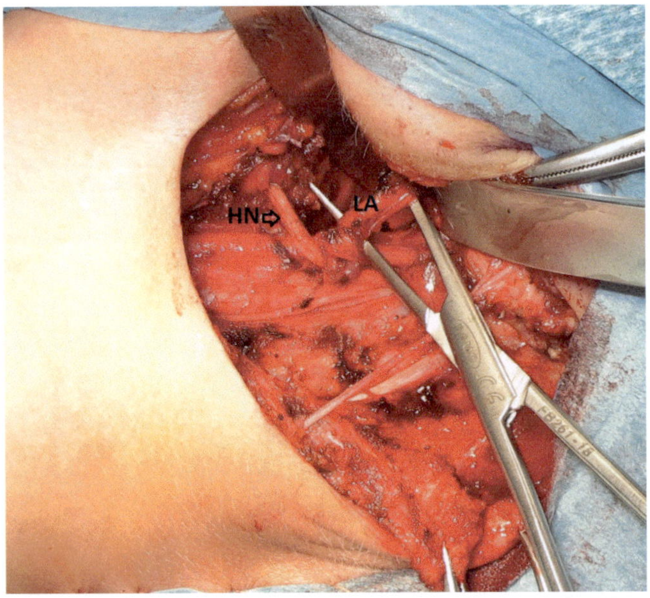

Fig. 10.23 Left side: Ligation of lingual artery (*LA*) between hypoglossal nerve (*HN*)

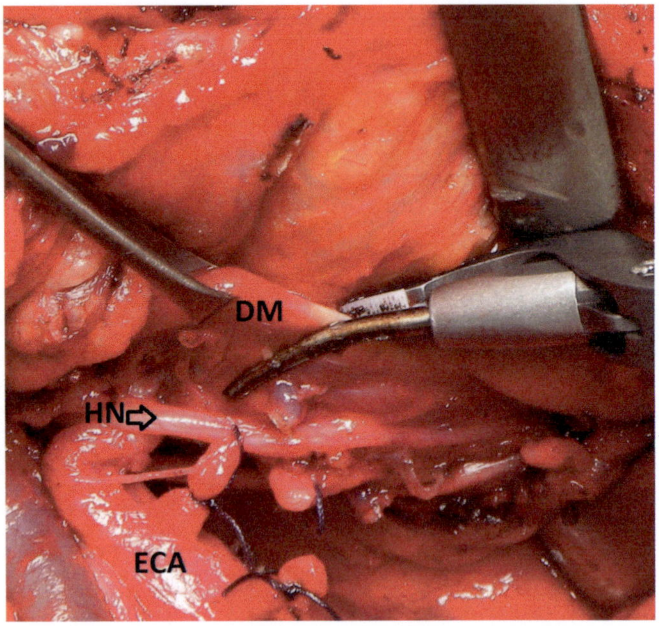

Fig. 10.24 Section of posterior belly of digastric muscle (*DM*) in order to obtain better exposure of upper cervical spine region. External carotid artery (*ECA*) and hypoglossal nerve (*HN*)

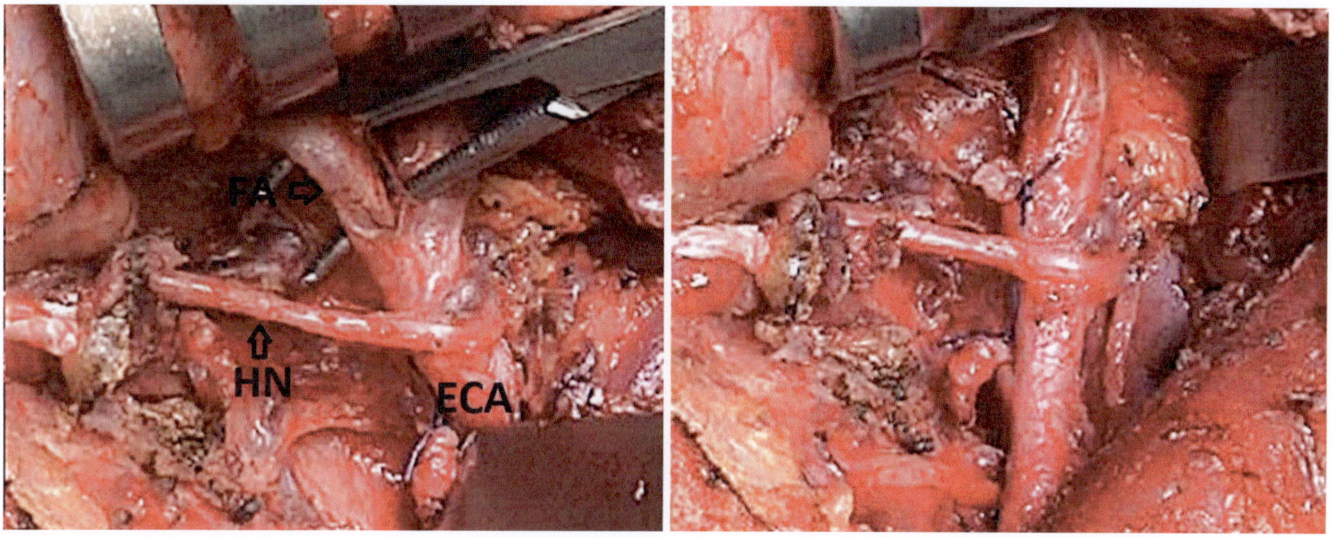

Fig. 10.25 Left side. Ligation of the facial artery(*FA*) after section of digastric muscle

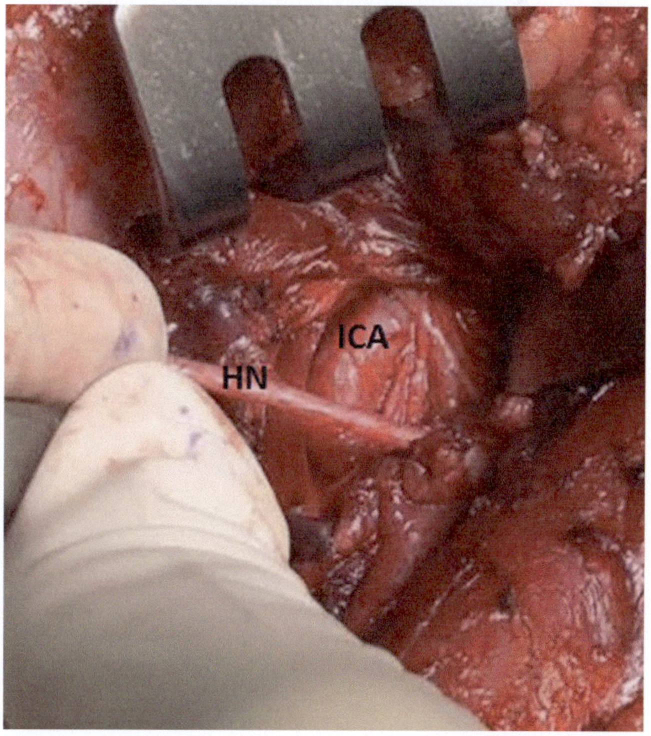

Fig. 10.26 Evidence of internal carotid artery (*ICA*)

Potential Complications

To avoid important bleeding during vascular structure dissection, the surgeon has to isolate veins and arteries plane by plane.

Most common complications are related to nerve injury.

Accidental section and/or palsy of the SAN could be caused by the retraction of SCM muscle. SAN lesions cause weakness of trapezius muscle on involved side and present with mild shoulder droop. The patient might report weakness in shoulder elevation and scapular winging most noticeable in arm abduction.

HN injury is a possible complication during the ligation of lingual and facial artery and in all the dissections above the hyoid bone. HN palsy causes an omolateral deviation of the tongue and difficulty in swallowing and speaking. The tongue often deviates toward the side of the lesion when it's protruded (due to imbalance of genioglossus contraction).

Ligation and dissection of STA implicate high risk of iatrogenic injury of SLN. During these maneuvers, surgeons should focus on the relationship between STA and this nerve. Palsy of SLN causes impaired cough reflex, hoarseness, voice fatigability, and impaired high-pitch phonation. Citare articolo laringeo superiore.

10.1.2.4 Fourth Surgical Step: Dissection of the MLDCF between the Carotid Sheath and the Visceral Column

The middle layer of the deep cervical fascia is contiguous to the carotid sheaths and the muscular-visceral column. It should be dissected just medial to and along the carotid sheath toward the spine. After the complete dissection of the MLDCF, the neurovascular bundle can be retracted laterally, and the midline structures (represented by pharynx and larynx) can be retracted medially. At this point, the vertebral bodies are identified by palpation and should not be confused with the anterior tubercles of the transverse processes.

Anatomical Key Points

1. Retropharyngeal space

Pearls

1. It's a virtual space that lies between the buccopharyngeal fascia anteriorly, which surrounds the pharyngeal muscles, and the alar fascia posteriorly. This space is the upper part of the retrovisceral space and continues down with the retroesophageal space. The whole retrovisceral space extends from the skull base to the mediastinum. The caudal limit is between the level of sixth cervical vertebra and the fourth thoracic vertebra. This space contains the retropharyngeal lymph nodes. Normally, the dissection of this space is carried out in an easy way.

Only if all the structures are precisely identified, it is possible to minimize the risk of complications (Figs. 10.27 and 10.28).

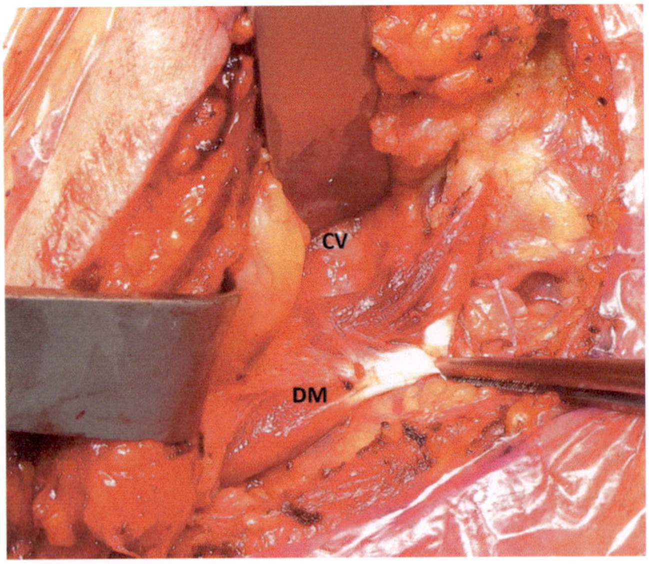

Fig. 10.27 Right side: cervical vertebra C3 (*CV*) and digastric muscle (*DM*)

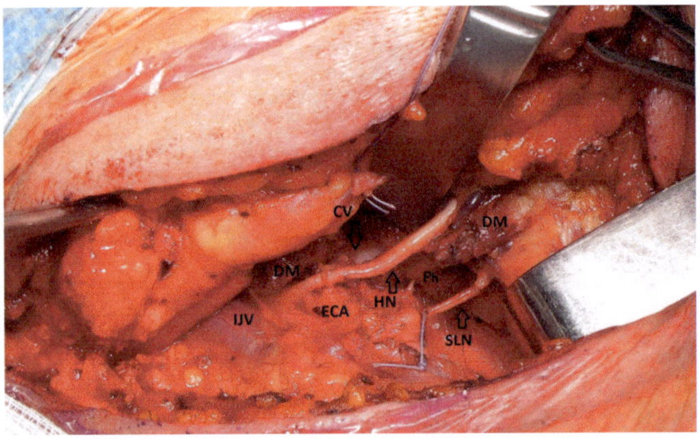

Fig. 10.28 Right side: exposure of the upper cervical spine on the right side. Evidence of internal jugular vein (*IJV*), digastric muscle sectioned (*DM*), external carotid artery (*ECA*), hypoglossal nerve (*HN*), superior laryngeal nerve (*SLN*), pharynx (*Ph*), and cervical vertebra C3 (*CV*)

Potential Complications

Major complications are related to massive traction of the visceral column (like pharyngeal fistula) or of the neurovascular bundle.

During this phase of the dissection, the ORL together with the orthopedic has to focus on the SLN and the HN in order to prevent any injury during traction or instruments positioning.

If the digastric or the stylohyoid muscles prevent an optimal exposure of the vertebral plane, it is possible to dissect them.

10.1.2.5 Fifth Surgical Step: Dissection of Prevertebral Structures and Exposure of the Vertebral Plane (Scheme 10.6)

The retropharyngeal area is detached from the prevertebral fascia, and the contents of the visceral fascia are retracted medially. The prevertebral fascia should be dissected on the median line, thus exposing the prevertebral muscles. The hypoglossal nerve and the superior laryngeal nerve are gently retracted to obtain a good surgical field visualization: in this way, C1, C2, and C3 are exposed. The superior laryngeal nerve crosses the field on the level with C2–C3 (Figs. 10.29 and 10.30).

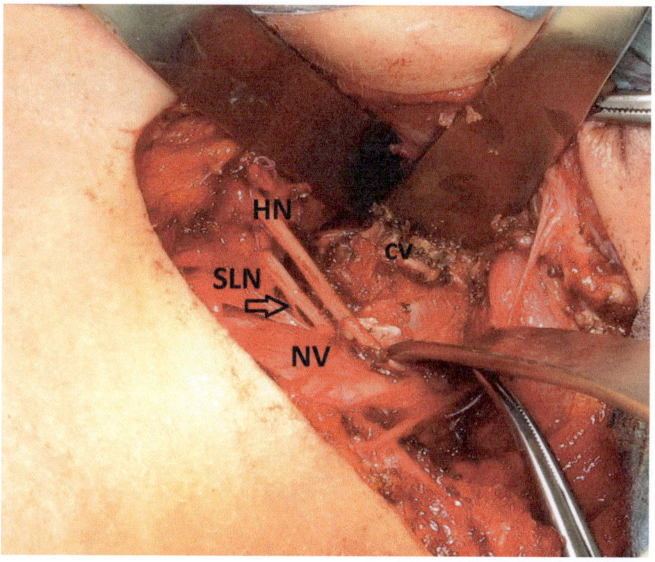

Fig. 10.29 Left side: pathological cervical vertebra (*C3*), hypoglossal nerve (*HN*), and superior laryngeal nerve (*SLN*)

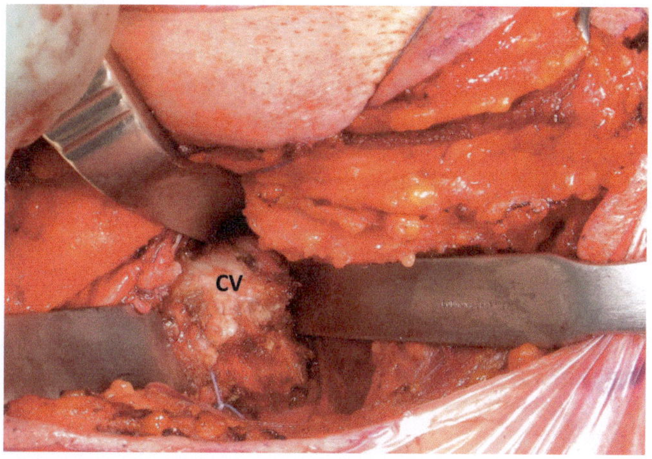

Fig. 10.30 Right side: evidence of a pathological cervical vertebra (*CV*)

Anatomical Key Points

1. Prevertebral muscles
2. Vertebral bodies and intervertebral disks

Pearls

1. The "longus capitis" is located more laterally than the longus colli and originates from the transverse process of C3-C6 and has its insertion at the basilar portion of the occipital bone. Longus colli is loceted on the lateral side of the vertebral bodies, from the atlas to T3. If th diseases does not extend to the lateral portion of vetrabrali bodies, the surgeon has to identify only the longus colli. The cervical midline is defined between right and left longus colli muscles and lies lateral to vertebral bodies, which are bounded across the midline by the prevertebral layer of deep cervical fascia. Prevertebral muscles are dissected in a subperiosteal plane, the contralateral longus colli is retracted medially with the visceral column, and the ipsilateral is retracted laterally with the carotid sheath. During dissection of the long muscles of the neck, the cervical sympathetic chain is easily recognizable compared to the vagus because it has a lower mobility and it also presents ganglion formations.

Potential Complications

Extended lateral retraction of prevertebral muscles might expose the vertebral artery at V2–V3 portions, where it passes through the transverse processes. In case of anatomical variation of the vertebral artery, it is possible to find the VA out of the transverse processes.

Massive traction of the neurovascular bundle could determine palsy of the cervical sympathetic chain leading to the so-called Claude-Bernard-Horner syndrome. The patient shows ptosis (droopy eyelid), miosis (constricted pupil), and facial anhidrosis (impaired facial sweating), with or without enophthalmos.

10.1.3 Midcervical Spine (C3–C6) (Scheme 10.7) and Fig. 10.3

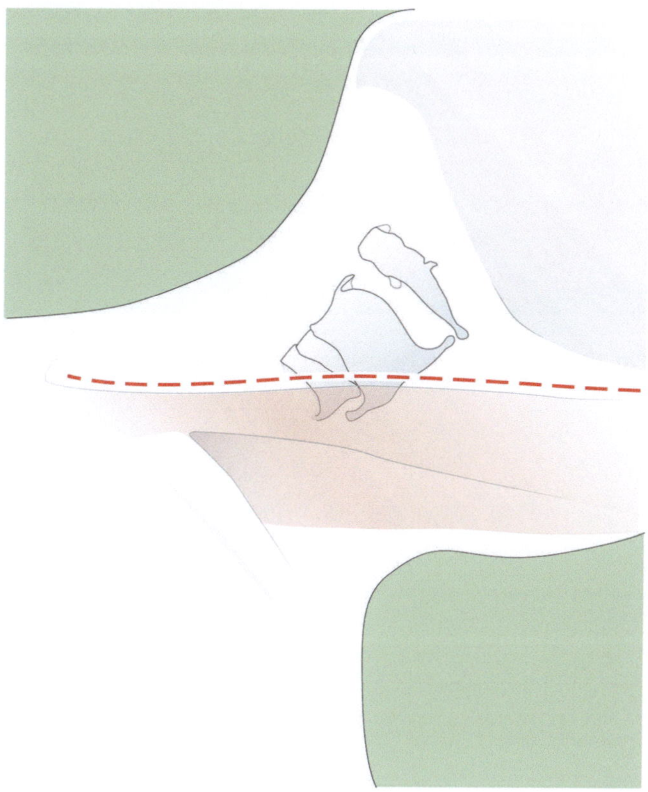

Scheme 10.7 Midcervical spine approach. SCM is a fundamental landmark for skin incision

10.1.3.1 First Surgical Step: Incision of the Skin

The skin incision is performed along the anterior border of SCM muscle, from the tip of the mastoid process to the suprasternal notch. The extent of the incision depends on the pathology's characteristics and the cervical level to treat. Skin and radiology landmarks are assessed during the preliminary surgical step (Fig. 10.31; Scheme 10.8).

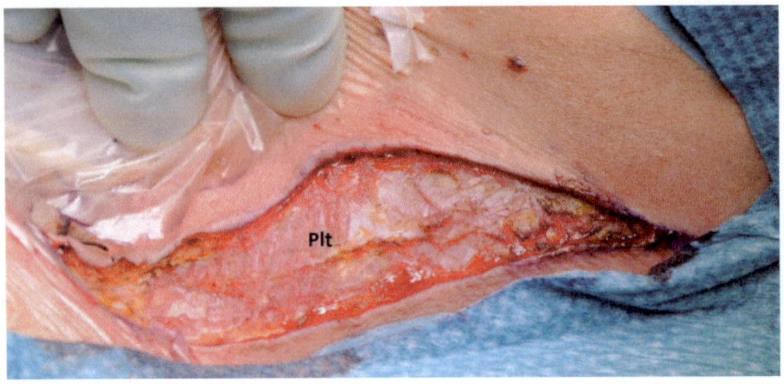

Fig. 10.31 Left side: incision and exposure of the platysma (*Plt*)

Scheme 10.8 After incision of subcutaneous tissue, evidence of the platysma

10.1.3.2 Second Surgical Step: Dissection of the Subplatysmal Flap

Following the skin incision, a dissection of *superficial cervical fascia* (*SCF*) – formed by the subcutaneous tissue and the platysma muscle – is performed. Subplatysmal flap is then retracted medially to expose the SLDCF (superficial layer of deep cervical fascia) (Figs. 10.32, 10.33, 10.34, and 10.35; Scheme 10.9).

Anatomical Key Points

1. Sternocleidomastoid muscle (*SCM*)
2. External jugular vein (*EJV*)
3. Great auricular nerve (*GAN*)
4. Transverse cervical nerves (Scheme 10.9).

Pearls

1. Define the inferior margin of the surgical field.
2. Identification of the EJV on the skin, before skin incision, is useful for better identification and preservation of the anatomical structures. The external jugular vein emerges in the parotid loggia from the confluence of the posterior auricular vein and the posterior branch of retromandibular vein and then penetrates into the SCM region on the level with the angle of the mandible and crosses the lateral face of the muscle askew. During the dissection, it may be necessary to ligate it.
3. The great auricular nerve emerges from the posterior border of the SCM muscle, at Erb's point. It runs parallel and superior to the external jugular vein on the superficial face of the muscle. It gives sensory branches to the skin of parotid and auricular regions. During the dissection, the surgeon usually finds first the EJV and then, about 1 cm laterally and cranially to it, the GAN.
4. The transverse nerve origins from the cervical plexus, emerges at Erb's point, and crosses the sternocleidomastoid muscle transversely. It gives sensory branches to the skin. It can be preserved or sacrificed without postsurgical complications.

Potential Complications

Accidental section of the external jugular vein does not lead to a catastrophic bleeding but creates a nonoptimal visualization of the structures to dissect during the procedure.

Great auricular nerve section produces an ailment hypoesthesia of periauricular skin; thus, this nerve should be preserved.

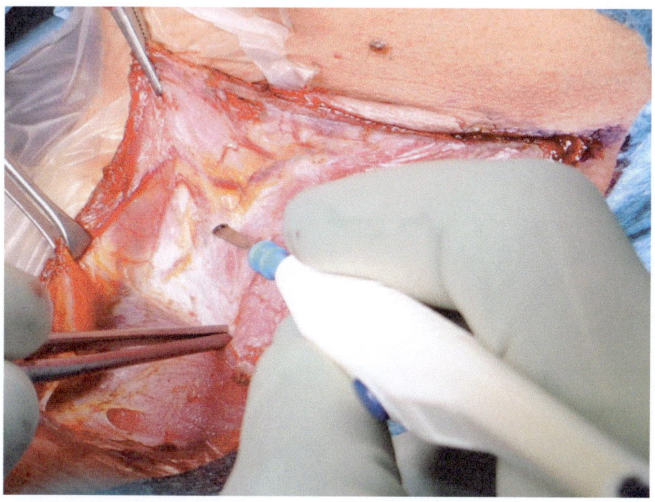

Fig. 10.32 Dissection of the subplatysmal flap

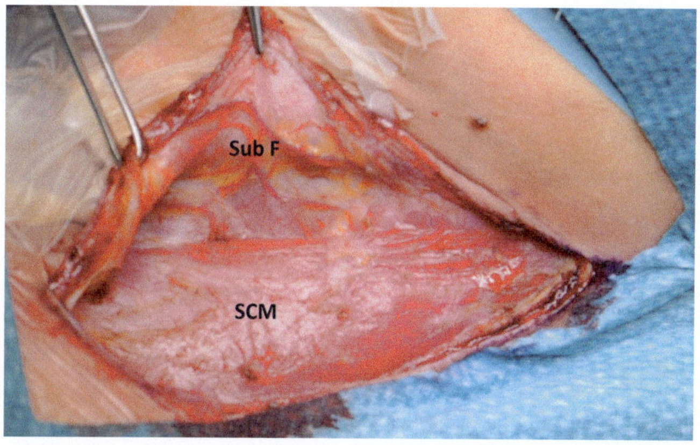

Fig. 10.33 Left side: dissection of subplatysmal flap (*Sub F*) and the sternocleidomastoid muscle (*SCM*)

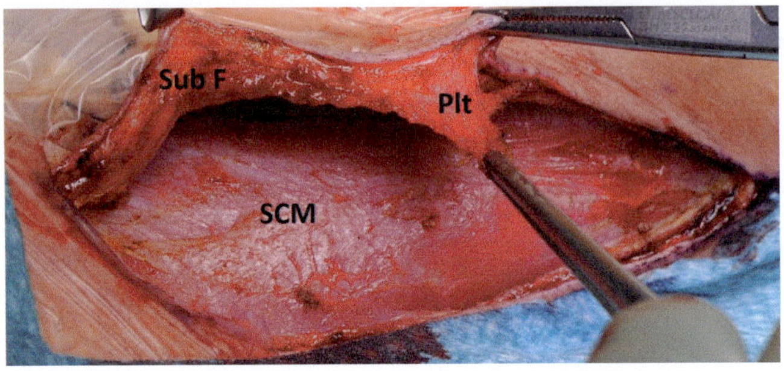

Fig. 10.34 Left side: evidence of platysma in the context of the subplatysmal flap and sternocleidomastoid muscle (*SCM*)

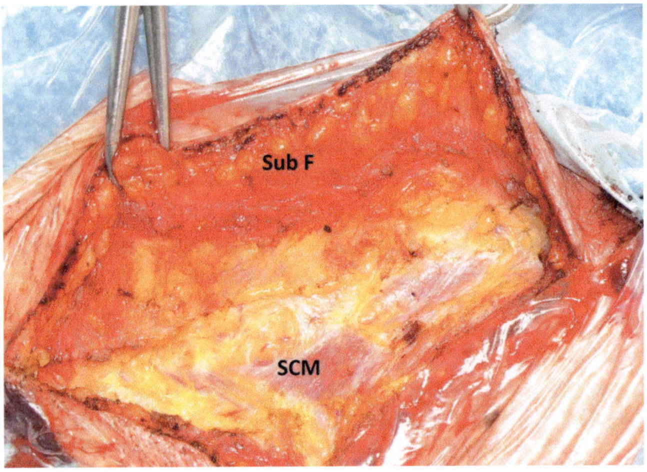

Fig. 10.35 Right side: dissection of the subplatysmal flap (*Sub F*) and sternocleidomastoid muscle (*SCM*)

Scheme 10.9 Elevated of Subplatysmal flap. Evidence of SCM, EJV and GAN. SLDCF covers the omohyoid muscle, that crosses the field

10.1.3.3 Third Surgical Step: Dissection of the SLDCF Anteriorly to the SCM

Under cover of the platysma, there is "the investing layer of the deep cervical fascia" or superficial layer of the deep cervical fascia (SLDCF). It surrounds the neck like a collar but splits around the sternocleidomastoid and trapezius muscles to enclose them. Posteriorly, it joins the ligamentum nuchae. The SLDCF is opened along the anterior border of SCM muscle. Lateral retraction of the SCM muscle exposes some of the structures below: (Fig. 10.36)

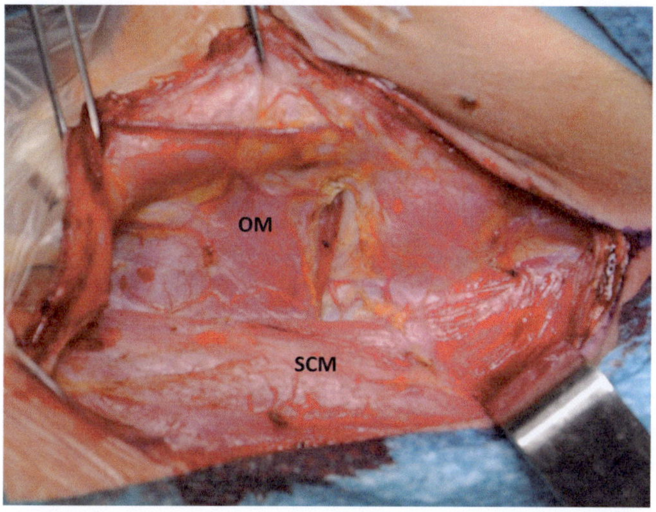

Fig. 10.36 Left side: sternocleidomastoid muscle (*SCM*) and omohyoid muscle (*OM*)

Anatomical Key Points (Scheme 10.10)

1. Omohyoid muscle (*OM*) (Figs. 10.37 and 10.38)
2. Digastric muscle (*DM*)
3. Thyrolingual-facial venous trunk (*TLF*)
4. Carotid sheath or neurovascular bundle
5. Hypoglossal nerve (*HN*)
6. Superior thyroid artery (*STA*)
7. Superior laryngeal nerve (*SLN*)
8. Recurent laryngeal nerve (*RLN*)

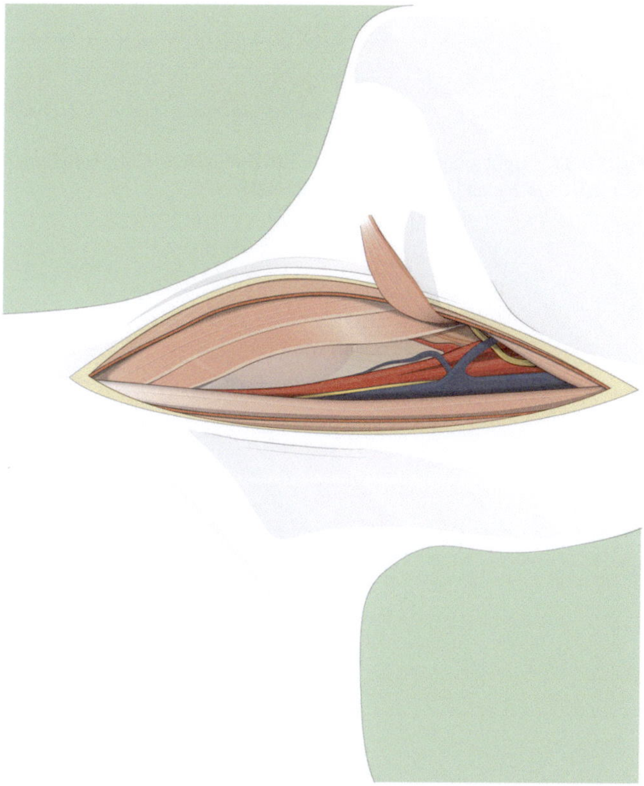

Scheme 10.10 Digastric muscle near upper part of the incision, just below TLF and HN. The image shows the first branch of the external carotid artery, the superior thyroid artery. The omohyoid muscle is dissected in order to simplify approach explanation

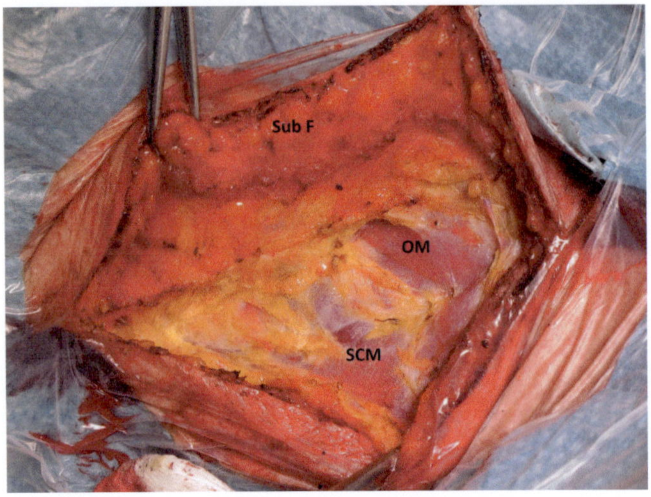

Fig. 10.37 Right side: sternocleidomastoid muscle (*SCM*), subplatysmal flap (*Sub F*), and omohyoid muscle (*OM*)

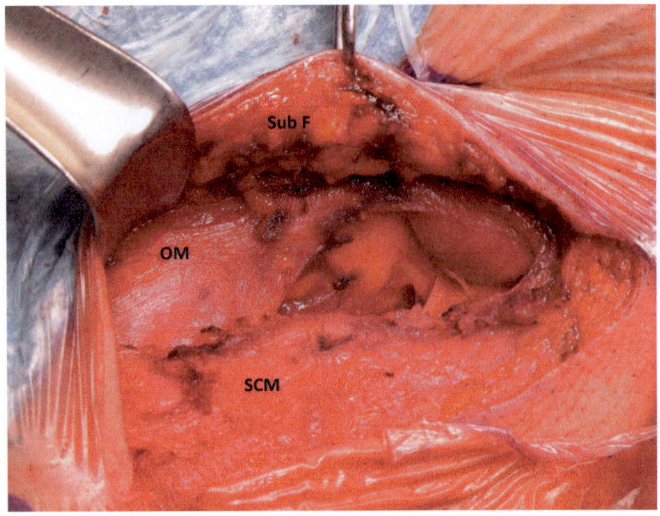

Fig. 10.38 Left side: sternocleidomastoid muscle (*SCM*), subplatysmal flap (*Sub F*), and omohyoid muscle (*OM*)

The goal is to identify the three main structures: (Fig. 10.39)

- *Digastric muscle*
- *Hypoglossal nerve*
- *Superior laryngeal nerve*

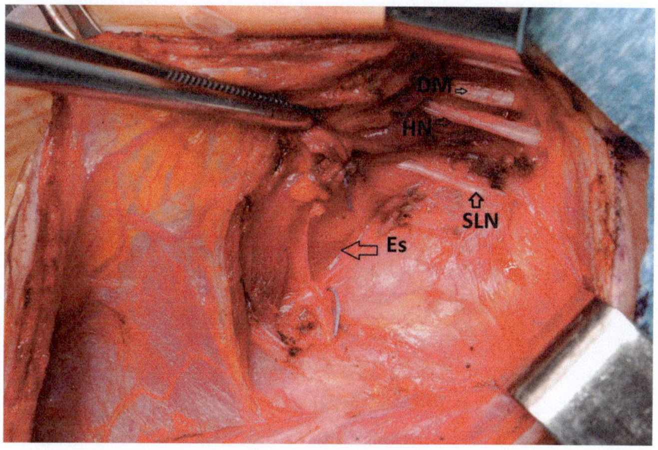

Fig 10.39 Left side: evidence of three fundamental landmarks. From upper to lower: digastric muscle (*DM*), hypoglossal nerve (*HN*), superior laryngeal nerve (*SLN*), and esophagus (*ES*)

Once you manage to control these landmarks, the approach to the vertebral plane is carried out safely (Scheme 10.11).

Scheme 10.11 Superior thyroid artery dissected. At this time, three fundamental landmarks, digastric muscle, HN, and SLA, are under control in order to perform safe cervical spine approach

Pearls

1. The omohyoid muscle is a landmark: the C5–C6 vertebral disks are approximately on the level with the intersection of the omohyoid and the SCM muscle. During the dissection, it is possible to appreciate the acute angle formed by the intersection of the OM and the SCM. The OM can be pulled down or disconnected for an easier access to the plane below. The neurovascular bundle, specifically the IJV, lies under this muscle.
2. On the upper part of this dissection plane, we find the posterior belly of digastric muscle, which runs obliquely forward and downward and goes to the hyoid bone, where it is fixed with its intermediate tendon to the body and the great horn and then continues with its anterior belly. It is defined "the resident's friend" because below it is possible to identify important structures like the neurovascular bundle and the TLF.
3. This vein passes deep to the sternocleidomastoid muscle and above the digastric muscle to empty into the internal jugular vein. It is formed by retromandibular and facial vein. Ligation of this structure should be near the confluence with the IJV.
4. It is formed by the internal jugular vein laterally, by the carotid artery medially and, in the dihedral angle formed by the two vessels, by the vagus nerve posteriorly. The first vascular structure found during the dissection is the IJV, which is just under the digastric muscle or following the TLF.
5. The HN emerges from the hypoglossal canal just medially to the jugular foramen, passes laterally between the internal jugular vein and the internal carotid artery, and goes superficial to the internal carotid artery to reach the posterior edge of the stylopharyngeus muscle. The nerve curves below the lingual artery and disappears beneath the posterior edge of hyoglossus muscle. The triangular space designed by the digastric muscle superiorly, the internal jugular vein laterally, and the TLF inferiorly is useful to identify the HN.
6. The STA arises from the external carotid artery at the point of the hyoid bone and, after providing the upper laryngeal artery, goes down to the upper pole of the thyroid gland. The STA is the first branch of the external carotid artery. To allow the access to the prevertebral space, it is mandatory to bind and dissect this artery. The STA is always under the hyoid bone. The surgeons should keep in mind, the possible anatomical variations of the origin of the STA like from thyrolingual (2 %) or thyrolinguofacial (1 %) trunck. To safely bind the artery, we should stay near its origin from the external carotid artery. This maneuver reduces the risk of superior laryngeal nerve injury. During this approach, the STA may not be properly in the surgical field, but without its ligation, it would be difficult and dangerous to obtain a good exposition of the vertebral plane (Figs. 10.40, 10.41, 10.42, and 10.43).
7. The SLN runs medially to the carotid sheath, forming an oblique angle with it. The nerve moves medially toward the thyrohyoid membrane. After the ligation of the STA, it is possible to identify this nerve.
8. On the left side, the recurrent laryngeal nerve arises from the vagus on the level with the aortic arch, and after having surrounded it, it assumes a vertical course. On the right side, the nerve emerges from the vagus where it crosses the subclavian artery; then it surrounds this artery and ascends obliquely, approaching the esophageal margin. In case of nonrecurrent laryngeal nerve, it is mandatory to convert this approach to the lateral one. In the cervical region, the ILN usually lies within the tracheoesophageal groove.

Potential Complications

The dissection of the omohyoid muscle has to be performed gently; the surgeon has to focus on the IJV lying below.

Most common complications are related to nerve injuries.

HN injury is a possible complication during the ligation of lingual and facial artery and in all the dissections above the hyoid bone. HN palsy causes an omolateral deviation of the tongue and difficulty in swallowing and speaking. The tongue often deviates toward the side of the lesion when it's protruded (due to imbalance of genioglossus contraction).

Ligation and dissection of the STA implicate high risk of iatrogenic injury of the SLN. During these maneuvers, surgeons should focus on the relationship between STA and this nerve. Palsy of th4 SLN causes impaired cough reflex, hoarseness, voice fatigability, and impaired high-pitch phonation.

RLN palsy determines paramedian vocal chords, with poor or no movement, voice fatigability, accumulation of saliva in the piriform fossa, and inability to swallow.

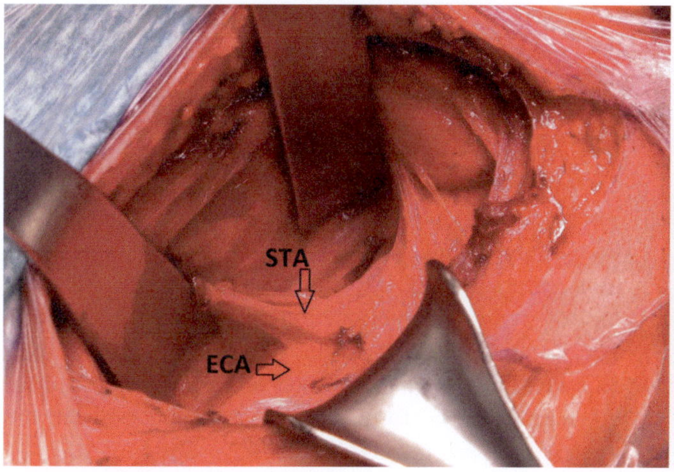

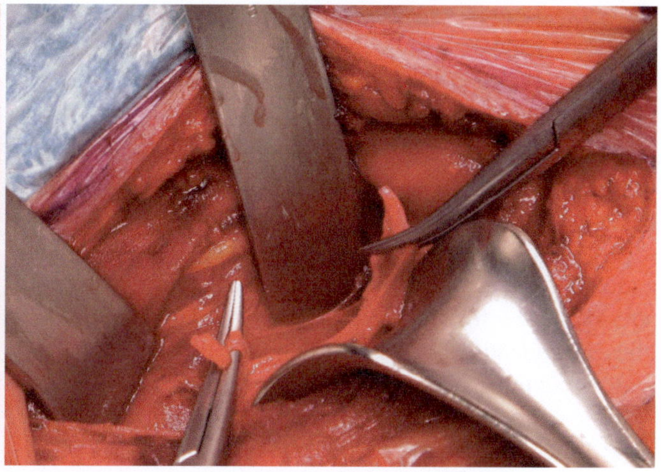

Figs. 10.40 and 10.41 Left side: identification and ligation of the superior thyroid artery (*STA*) at its origin from the external carotid artery (*ECA*)

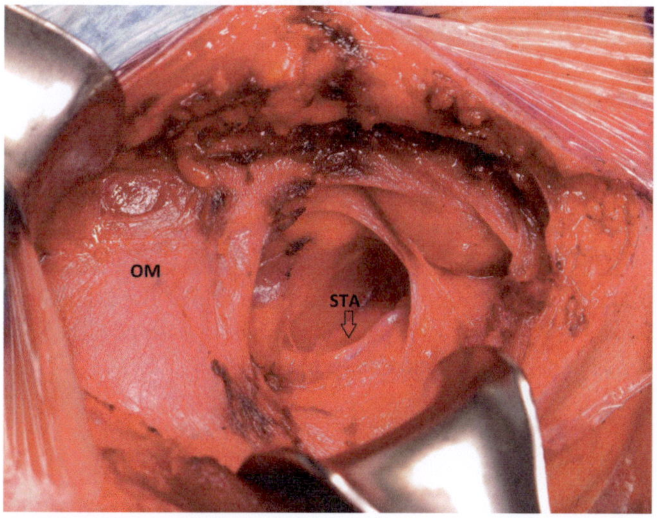

Fig. 10.42 Left side. Identification of superior thyroid artery (*STA*). Omohyoid muscle (*OM*)

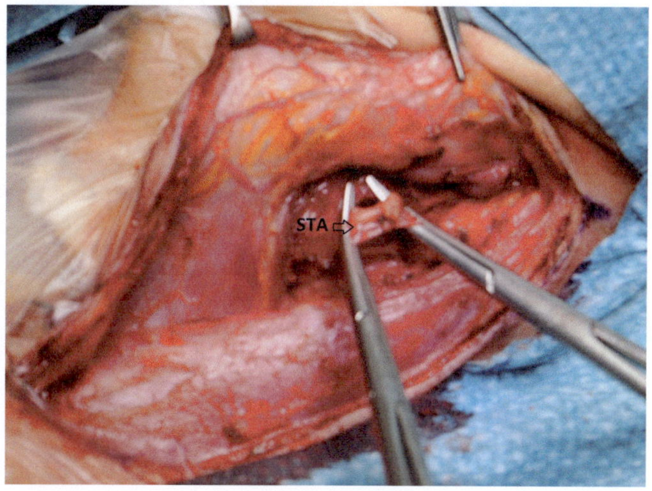

Fig. 10.43 Left side: identification of superior thyroid artery (*STA*)

10.1.3.4 Fourth Surgical Step: Dissection of the MLDCF between the Carotid Sheath and the Visceral Column

The middle layer of deep cervical fascia is contiguous to the carotid sheath and the muscular-visceral column. It should be dissected just medial to and along the carotid sheath toward the spine. After complete dissection of the MLDCF, the neurovascular bundle can be retracted laterally, and the midline structures (represented by larynx, thyroid, trachea, and esophagus) can be retracted medially. At this point, the vertebral bodies are identified by palpation and should not be confused with the anterior tubercles of the transverse processes.

Anatomical Key Points (Figs. 10.44 and 10.45)

1. Larynx,thyroid gland, and trachea
2. Esophagus
3. Vagus nerve
4. Retrovisceral space

Pearls

Only if all the structures are precisely identified, it is possible to minimize the risk of complications.

1. All these structures are covered by the visceral fascia. A palpatory identification of cartilaginous structures like the thyroid cartilage, the cricoid ring, and the hyoid bone usually allows identifying a safe plane to continue the dissection.
2. A right-side approach reduces the risk of iatrogenic esophageal injury, because the physiological curvature to the left of the esophageal tube promotes a better view of the prevertebral plan. Look for the nasogastric tube, preoperatively positioned by an anesthesiologist, by palpation of the posterior portion of the visceral column: this may allow the identification of the esophagus during dissection.
3. The vagus nerve lies posteriorly in the groove between the internal jugular vein and the internal and later common carotid artery. Occasionally, the vagus nerve is located in a more anterior position, related to the carotid artery at the base of the neck.
4. It is identified after the section of the MLDCF. It is an aponeurotic space formed by the buccopharyngeal, the prevertebral, and the alar fascia.

The whole retrovisceral space extends from the skull base to the mediastinum. The caudal limit is between C6 and T4. It contains the retropharyngeal lymph nodes. This space is normally easy to identify, and its dissection is made by finger or with the help of a cottonoid.

Potential Complications

Only if all the structures are precisely identified, it is possible to minimize the risk of complications.

Major risks during this step are related to the esophageal fistula.

RLN palsy is a potential complication during the traction of the visceral column. A correct retraction of the esophagus allows the surgeon to protect the nerve. In case of NRLN, it is necessary to choose another approach.

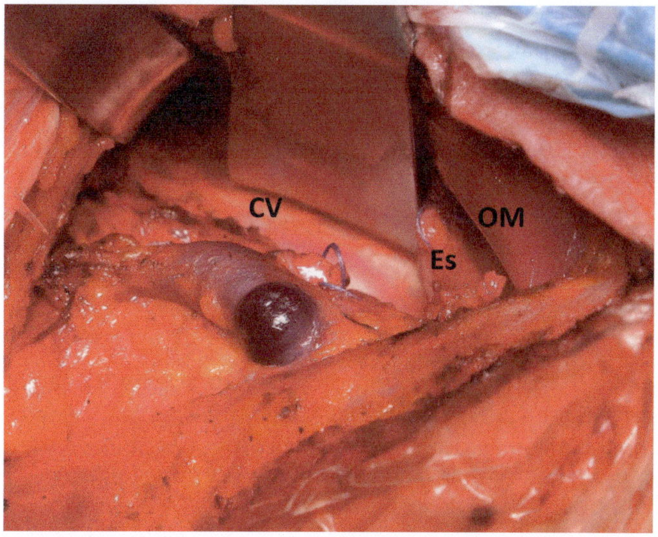

Fig. 10.44 Right side: evidence of cervical vertebra (*CV*). Retraction of esophagus (*Es*), omohyoid muscle (*OM*)

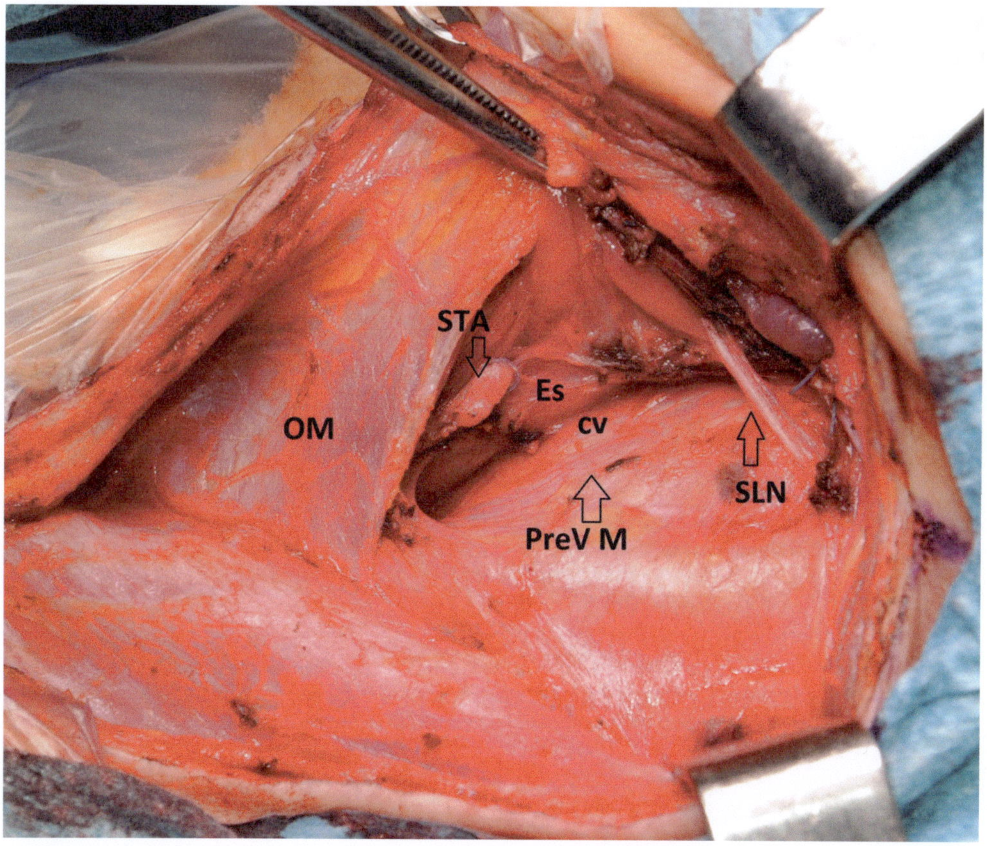

Fig. 10.45 Left side: omohyoid muscle (*OM*), superior thyroid artery (*STA*) dissected, superior laryngeal nerve (*SLN*), esophagus (*ES*), prevertebral muscle (*PreV M*), and cervical vertebra (*CV*)

10.1.3.5 Fifth Surgical Step: Dissection of Prevertebral Structures and Exposure of the Vertebral Plane

The prevertebral layer of the deep cervical fascia origins laterally from the transverse processes and continues with the fascia of the scalene muscles. The prevertebral layer and the anterior longitudinal ligament are incised longitudinally along the midline before their cauterization.

Anatomical Key Points (Figs. 10.46 and 10.47)

1. Prevertebral muscles
2. Vertebral bodies and intervertebral disks (Scheme 10.11)

Pearls

1. The "longus capitis" is located more laterally than the longus colli and originates from the transverse process of C3-C6 and has its insertion at the basilar portion of the occipital bone. Longus colli is loceted on the lateral side of the vertebral bodies, from the atlas to T3. Ðf th diseases does not extend to the lateral portion of vetrabrali bodies, the surgeon has to identify only the longus colli. The cervical midline is defined between right and left longus colli muscles and lies lateral to vertebral bodies, which are bounded across the midline by the prevertebral layer of the deep cervical fascia. Prevertebral muscles are dissected in a subperiosteal plane; the contralateral longus colli is retracted medially with the visceral column, and the ipsilateral is retracted laterally with the carotid sheath. During dissection of the long muscles of the neck, the cervical sympathetic chain is easily recognizable compared to the vagus nerve because it has a lower mobility and it also presents ganglion formations.

Potential Complications

Extended lateral retraction of prevertebral muscles might expose the vertebral artery at V2 portions, where it passes through the transverse processes. In case of anatomical variation of the vertebral artery, it is possible to find the VA out of the transverse processes.

Massive traction of the neurovascular bundle could determine palsy of the cervical sympathetic chain leading to the so-called Claude-Bernard-Horner syndrome. The patient shows ptosis (droopy eyelid), miosis (constricted pupil), and facial anhidrosis (impaired facial sweating), with or without enophthalmos.

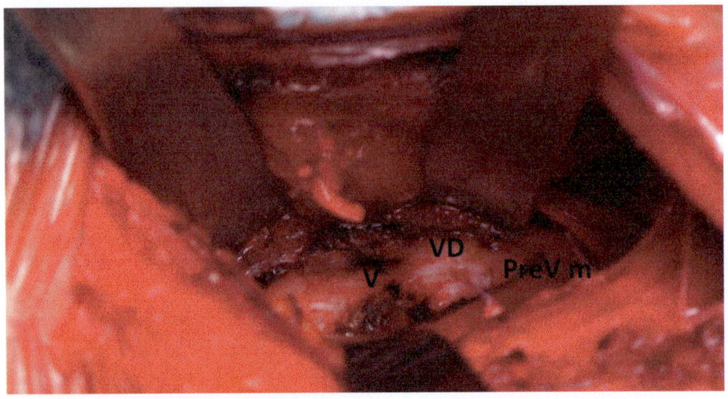

Fig. 10.46 Expose of vertebral plane. Cervical vertebra (*V*), vertebral disk (*VD*), and prevertebral muscles (*PreV M*)

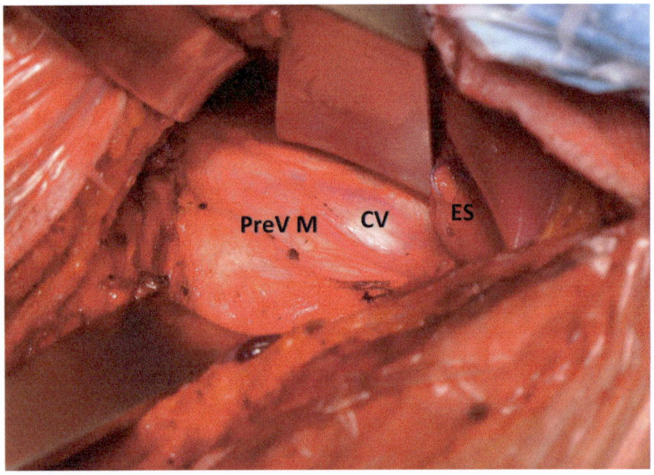

Fig. 10.47 Expose of vertebral plane. Cervical vertebra (*CV*) and prevertebral muscles (*PreV M*). Esophagus (*Es*) retracted laterally

10.1.4 Cervical Mini-invasive Approach

A correct analysis of the risk points during plane-by-plane dissection allows performing a mini-invasive approach to a specific cervical level.

A limited exposure of anatomical structures reduces the possibility of landmarks identification. Surgeons have to focus on variable courses of the structures and perform a plane-by-plane dissection.

Major anatomical details have already been described.

10.1.4.1 First Surgical Step: Incision of the Skin (Figs. 10.48 and 10.49)

A horizontal incision is performed about 4–5 cm from the neck midline to the anterior border of SCM.

The surgeons have to correlate the level to treat to landmarks on the skin (hyoid bone, thyroid cartilage, and cricoid cartilage) and customize the skin incision following the same landmarks.

Heller et al. [9] describe the correlations between cervical spine levels and potentially at risk structures from C2–C6

Vertebral level	Structure (s) at risk	Clinical correlate
C2	HN	Dysphagia, dysarthria, ipsilateral tongue deviation
C2–C3	HN	Dysphagia, dysarthria, ipsilateral tongue deviation
C3	HN, SLN, STA	Dysphagia, dysarthria, ipsilateral tongue deviation, impaired cough reflex
C3–C4	SLN, STA	Impaired cough reflex
C4	SLN, STA, SLA	Impaired cough reflex, hoarseness, voice fatigability, impaired high-pitch phonation
C4–C5	SLN, STA, SLA	Hoarseness, voice fatigability, impaired high-pitch phonation
C5	SLN, STA, SLA	Hoarseness, voice fatigability, impaired high-pitch phonation
C5–C6	SLN, STA	Hoarseness, voice fatigability, impaired high-pitch phonation
C6	SLN, STA	Hoarseness, voice fatigability, impaired high-pitch phonation

From Heller et al. [9] article

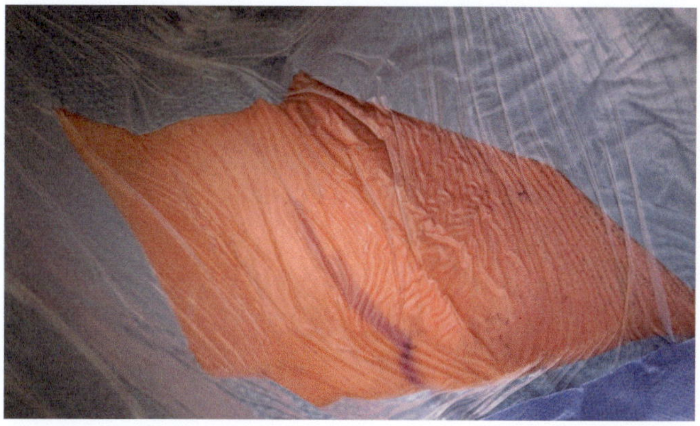

Fig. 10.48 Left side: incision line

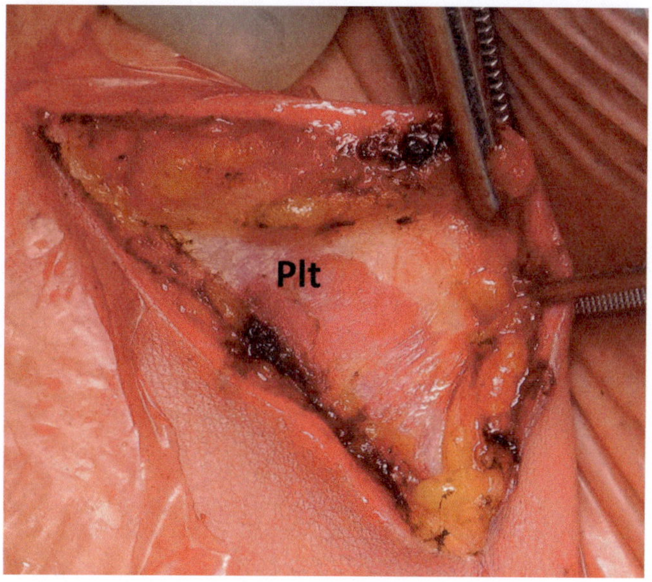

Fig. 10.49 Left side: exposure of platysma muscle (*Plt*)

10.1.4.2 Second Surgical Step: Dissection of the Subplatysmal Flap

Following the skin incision, a dissection of *superficial cervical fascia (SCF)* – formed by the subcutaneous tissue and the platysma muscle – is performed. The flap is then elevated above the inferior border of the mandible to expose the *superficial layer of the deep cervical fascia (SLDCF)* (Fig. 10.50).

Anatomical Key Points

1. Sternocleidomastoid muscle (*SCM*)
2. Anterior jugular vein (*AJV*)
3. Transverse cervical nerve (*TCN*) *C4–C5 levels*

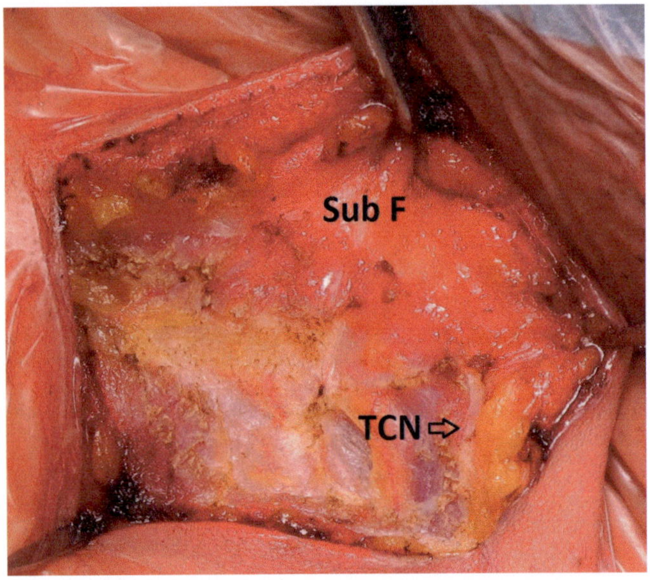

Fig. 10.50 Left side: subplatysmal flap (*Sub F*) and transverse cervical nerve (*TCN*)

Pearls

This vein could be preserved or dissected without postoperative complications. The transverse nerve origins from the cervical plexus, emerges at Erb's point, and crosses the SCM muscle transversely. It gives sensory branches to the skin. It can be preserved or sacrificed without postsurgical complications.

Potential Complications

Accidental section of the AJV jugular vein or communicated vein between AJV and EJV vein does not lead to a catastrophic bleeding but creates a nonoptimal visualization of the structures to dissect during the procedure.

The transverse cervical nerve could be preserved or dissected without postsurgical complications.

10.1.4.3 Third Surgical Step: Dissection of the SLDCF Anteriorly to the SCM

In this approach, the identification of the digastric muscle, the HN, and the SLN is impossible, but it is still mandatory to know which structures could be found during plane-by-plane dissection (Figs. 10.51, 10.52, 10.53, and 10.54).

Anatomical Key Points

1. Omohyoid muscle (*C4–C6 levels*) (*OM*)
2. Thyrolingual-facial venous trunk (*C3–C4 levels*) (*TLF*)
3. Carotid sheath or neurovascular bundle
4. Thyroid gland (*Thy*)
5. Medial thyroid vein (*C4–C6 levels*) (*MTV*)
6. Superior thyroid artery (*STA*) *C3–C6 levels*
7. Superior laryngeal nerve (*SLN*) *C3–C6 levels*
8. Inferior thyroid artery (*ITA*) *C5–C6 levels*
9. Recurrent laryngeal nerve (*RLN*) *C5–C6 levels*

Pearls

The omohyoid muscle allows recognizing the neurovascular bundle; from this point, surgeons have to find the carotid artery and the thyroid gland.

A gentle retraction of these structures simplifies STA, MTV, and ITA dissection and ligation near the neurovascular bundle.

The SLN and the RLN have to be recognized and preserved. It is important to remember RLN's different courses on the left and right side, as previously described.

Potential Complications

Complications are related to nerve and vascular injury. All the clinical presentations are described in the extended approach.

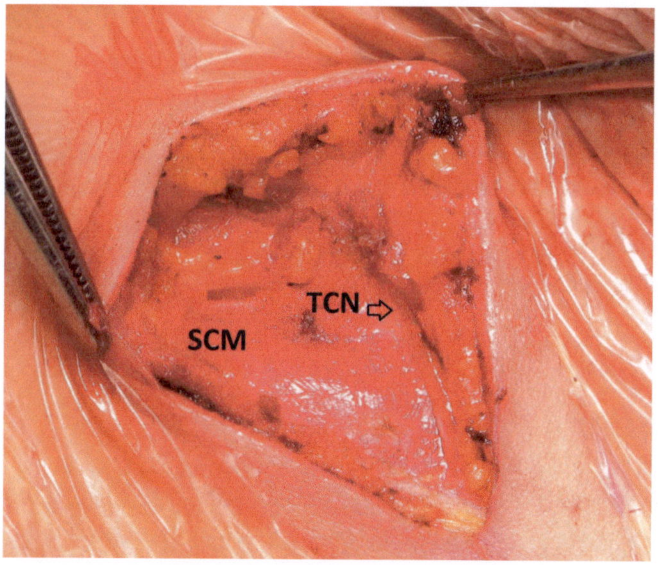

Fig. 10.51 Left side: transverse cervical nerve (*TCN*) and sternocleidomastoid muscle (*SCM*)

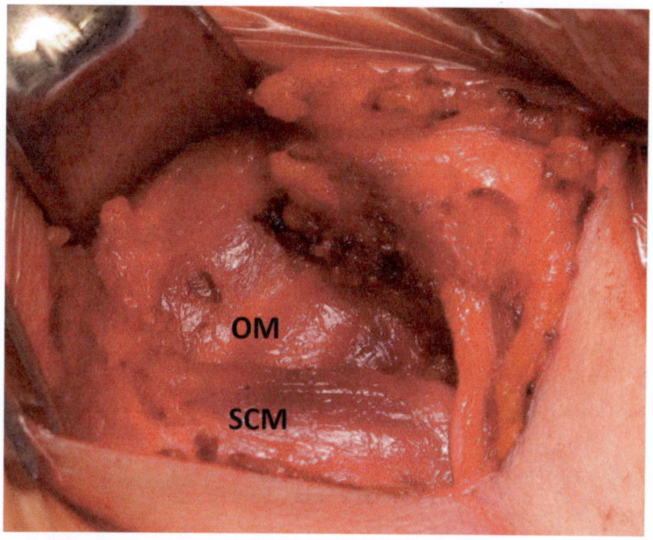

Fig. 10.52 Left side: sternocleidomastoid muscle (*SCM*) and omohyoid muscle (*OM*)

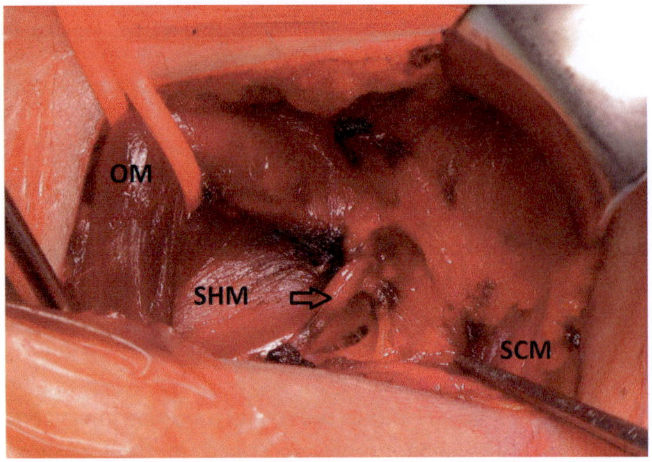

Fig. 10.53 Left side: sternocleidomastoid muscle (*SCM*), omohyoid muscle (*OM*), sternohyoid muscle (*SHM*), and anastomosis between anterior jugular vein and external jugular vein (*arrow*)

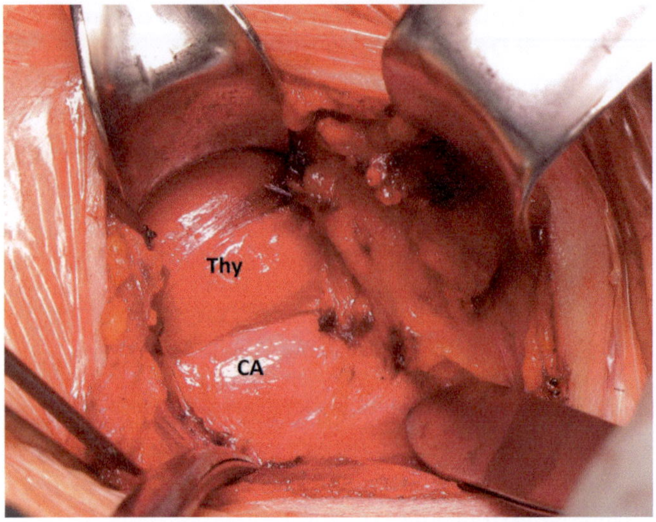

Fig. 10.54 Left side: thyroid gland (*Thy*) and carotid artery (*CA*)

10.1.4.4 Fourth Surgical Step: Dissection of the MLDCF between the Carotid Sheath and the Visceral Column

Anatomical Key Points (Fig. 10.55)

1. Larynx and trachea
2. Esophagus
3. Vagus nerve
4. Retrovisceral space

Pearls and Potential Complications

A right-side approach reduces the risk of iatrogenic esophageal injury, because of its physiological curvature. Major risks during this step are related to the esophagus retraction. The most important maneuver is the correct visualization of the esophagus; the surgeon could touch the gastric tube inside the lumen, preoperatively positioned by the anesthesiologist, and gently retract it medially.

RLN palsy is a potential complication during the traction of the visceral column. A correct retraction of the esophagus allows the surgeon to protect the nerve. In NRLN, it is necessary to choose another approach.

Claude-Bernard-Horner syndrome is the consequence of cervical sympathetic chain palsy due to massive traction of neurovascular bundle.

10.1.4.5 Fifth Surgical Step: Dissection of Prevertebral Structures and Exposure of the Vertebral Plane

Anatomical Key Points (Fig. 10.56)

1. Prevertebral muscles
2. Vertebral bodies and intervertebral disks

Pearls and Potential Complications

Prevertebral muscles are dissected in a subperiosteal plane, contralateral longus colli is retracted medially with the visceral column, and the ipsilateral one is retracted laterally with the carotid sheath.

Extended lateral retraction of prevertebral muscles might expose the vertebral artery at V2 portions, where it passes through the transverse processes. In case of anatomical variation of the vertebral artery, it is possible to find the VA out of the transverse processes.

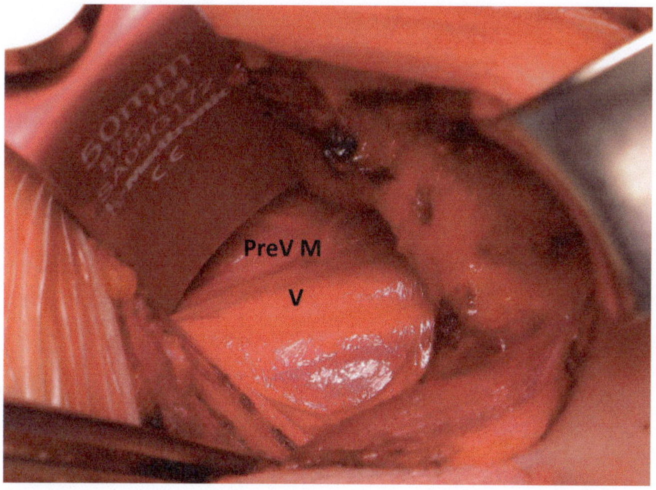

Fig. 10.55 Left side: vertebral plane (*V*) and prevertebral muscle (*PreV M*)

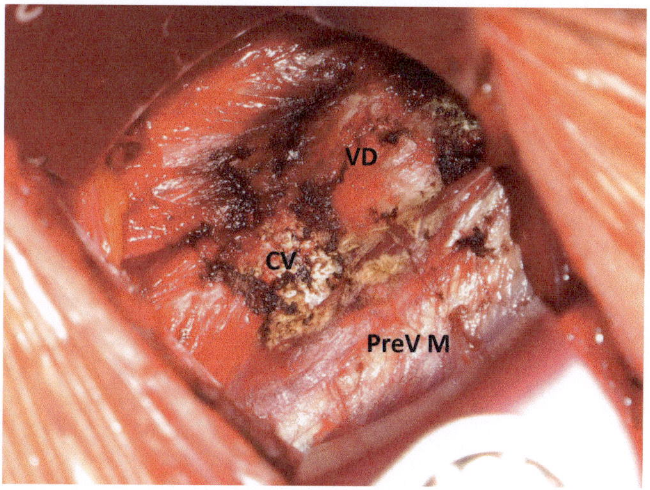

Fig. 10.56 Left side: fifth cervical vertebra (*CV*), vertebral disk between C4 and C5 (*VD*), and prevertebral muscles (*PreV M*)

10.1.5 Caudal Extension: Cervical Thoracic Junction

The approach to the cervical thoracic junction can be considered a lower extension of the midcervical spine approach (Scheme 10.12).

10.1.5.1 Second Surgical Step: Dissection of the Subplatysmal Flap

Anatomical Key Points (Scheme 10.13)
1. Sternocleidomastoid muscle (*SCM*)
2. Anterior jugular vein (*AJV*)
3. Transverse cervical nerve
4. Supraclavicular nerve

Pearls
1. Define the inferior margin of the surgical field.
2. The AJV derives by the joining of several superficial veins in the submental region. It descends anteriorly to the SCM muscle and passes deep to the muscle before draining into the external jugular vein or the subclavian vein. During dissection, it may be necessary to ligate it.
3. The transverse nerve origins from cervical plexus, emerges at Erb's point, and crosses the sternocleidomastoid muscle transversely. It gives sensory branches to the skin. It can be preserved or dissected without postsurgical complications.
4. It passes posteriorly to the SCM muscle at Erb's point. It travels inferiorly in an oblique direction through the posterior triangle of the neck.

Potential Complications
Accidental section of the anterior jugular vein does not determine a catastrophic bleeding but creates a nonoptimal visualization of the structures to dissect during the procedure.

Dissection of the transverse or supraclavicular nerves does not lead to any functional impairment, so surgeons can decide to dissect them.

Scheme 10.12 Incision of the subcutaneous tissue and exposure of the platysma

Scheme 10.13 Subplatysmal flap elevation. The main structures below are: SCM, omohyoid muscle, and transverse cervical nerve

10.1.5.2 TThird Surgical Step: Dissection of the SLDCF Anteriorly to the SCM

Anatomical Key Points (Scheme 10.14)

1. Omohyoid muscle (*OM*)
2. Carotid sheath or neurovascular bundle
3. Thyroid gland (*Thy*)
4. Medial thyroid vein (*MTV*)
5. Inferior thyroid artery (*ITA*)
6. Recurrent laryngeal nerve (*RLN*)

Pearls

1. The omohyoid muscle is a landmark: the C5–C6 vertebral disks are approximately on the level with the intersection of the omohyoid and the SCM muscle. OM can be pulled down or disconnected for an easier access to the plane below. The neurovascular bundle, specifically the IJV, lies under this muscle.
2. It is formed by the internal jugular vein laterally, by the carotid artery medially and, in the dihedral angle formed by the two vessels, by the vagus nerve posteriorly. The first vascular structure found during the dissection is the IJV, which is just under the digastric muscle or following the TLF.
3. Surgeons have to recognize it, and in order to avoid bleeding, it is useful to preserve its fascia.
4. The middle thyroid vein collects the blood from the lower part of the thyroid gland, and after receiving some veins from the larynx and the trachea, it ends in the lower part of the internal jugular vein. Its ligation is normally carried out near the IJV.
5. It arises from the thyrocervical trunk and goes upward, in front of the vertebral artery and the longus colli muscle. Then it turns behind and medially to the carotid sheath and its contents and stays also behind the sympathetic trunk. When it reaches the lower border of the thyroid gland, it divides into two branches, which supply the posteroinferior parts of the gland.
6. On the left side, the recurrent laryngeal nerve arises from the vagus on the level with the aortic arch, and after having surrounded it, it assumes a vertical course. On the right side, the nerve emerges from the vagus where it crosses the subclavian artery, and then it surrounds this artery and ascends obliquely, approaching the esophageal margin. In case of nonrecurrent laryngeal nerve, it is mandatory to convert this approach to the lateral one. In the cervical region, the ILN usually lies within the tracheoesophageal groove. Lore's triangle is useful to find the RLN. It is formed by the trachea medially, the common carotid artery laterally, and the under surface of the retracted inferior thyroid pole superiorly, the apex of the triangle being directed toward the thoracic inlet.

Potential Complications

The dissection of the omohyoid muscle has to be performed gently; the surgeon has to focus on the IJV lying below.

Most common complications are related to nerve injuries.

HN injury is a possible complication during the ligation of lingual and facial artery and in all the dissections above the hyoid bone. HN palsy causes an omolateral deviation of the tongue and difficulty in swallowing and speaking. The tongue often deviates toward the side of the lesion when it's protruded (due to imbalance of genioglossus contraction).

Ligation and dissection of STA implicate high risk of iatrogenic injury of SLN. During these maneuvers, surgeons should focus on the relationship between STA and this nerve. Palsy of SLN causes impaired cough reflex, hoarseness, voice fatigability, and impaired high-pitch phonation.

RLN palsy determines paramedian vocal chords, with poor or no movement, voice fatigability, accumulation of saliva in the piriform fossa, and inability to swallow.

Scheme 10.14 Three fundamental landmarks are not always recognizable. Main structures are the superior thyroid artery and sometimes the superior laryngeal nerve:

10.1.5.3 Fourth Surgical Step: Dissection of the MLDCF between the Carotid Sheath and the Visceral Column

Anatomical Key Points (Scheme 10.15)

1. Trachea and retrovisceral space
2. Esophagus
3. Subclavian artery and vertebral artery (*VA*)
4. Cervical sympathetic chain
5. Thoracic duct on the left side

Pearls

Only if all the structures are precisely identified, it is possible to minimize the risk of complications:

1. It is identified after the section of the MLDCF. It is an aponeurotic space formed by the buccopharyngeal, the prevertebral, and the alar fascia. The whole retrovisceral space extends from the skull base to the mediastinum. The caudal limit is between C6 and T4. It contains the retropharyngeal lymph nodes. This space is normally easy to identify, and its dissection is made by finger or with the help of a cottonoid.
2. A right-side approach reduces the risk of iatrogenic esophageal injury, because the physiological curvature to the left of the esophageal tube promotes a better view of the prevertebral plan. Look for the nasogastric tube, preoperatively positioned by the anesthesiologist, by palpation of the posterior portion of the visceral column: this may allow the identification of the esophagus during dissection.
3. The subclavian artery travels lateral to the trachea into the root of the neck, passing between the anterior and middle scalene muscles. In its first part, it gives origin to the thyrocervical trunk and the vertebral artery. During the lateral retraction of the neurovascular structures, it is important to consider its course because in its V1 portion, the artery goes more superficial and is without bone protection (Fig. 10.57)
4. The stellate ganglion (or cervicothoracic ganglion) is a sympathetic ganglion formed by the fusion of the inferior cervical ganglion and the first thoracic ganglion, which exists in 80 % of cases. The stellate ganglion is located at C7 level (7th cervical vertebrae), anterior to the transverse process of C7, superior to the neck of the first rib, and just below the subclavian artery.
5. The thoracic duct is located on the left side within a triangle bounded medially by the longus colli muscle and the esophagus, laterally by the anterior scalene muscle, and inferiorly by the first rib. Although it may ascend as high as C6, it is most often found between C7 and T1, before it descends to empty into a variable termination at the jugulo-subclavian junction. The rostral extension of the thoracic duct appears to vary by gender, as in patients who have a narrow thoracic inlet, as most women do; the duct may ascend as high as the level of the C6 vertebral body. Many authors have cited the increased possibility of injuring this structure in the left upper thorax.

Potential Complications

Injury of the subclavian artery or the vertebral artery produces a catastrophic bleeding and can determine neurological defects.

Major risks during this step are related to the esophagus retraction. The most important maneuver is the correct visualization of the esophagus that has to be gently retracted medially.

Massive traction of the neurovascular bundle could determine palsy of the cervical sympathetic chain leading to the so-called Claude-Bernard-Horner syndrome. The patient shows ptosis (droopy eyelid), miosis (constricted pupil), and facial anhidrosis (impaired facial sweating), with or without enophthalmos.

Lymphatic fistula or pneumothorax can happen as a consequence of an accidental pleuric lesion.

Scheme 10.15 In order to show CTJ, thyroid gland dissection is mandatory. Superior and inferior thyroid arteries have to be dissected. The RLN lies into the tracheal-esophageal groove and it must be preserved

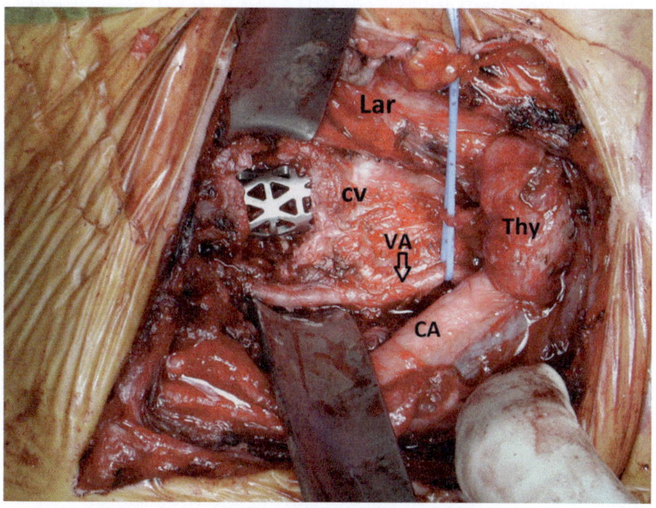

Fig. 10.57 Right side: right lobe of thyroid gland (*Thy*), larynx (*Lar*), carotid artery (*CA*), and vertebral artery (*VA*)

10.1.5.4 Fifth Surgical Step: Dissection of Prevertebral Structures and Exposure of the Vertebral Plane

The prevertebral layer of the deep cervical fascia origins laterally from the transverse processes and continues with the fascia of the scalene muscles. The prevertebral layer and the anterior longitudinal ligament are incised longitudinally along the midline before their cauterization.

Anatomical Key Points (Scheme 10.15)

1. Prevertebral muscles
2. Vertebral bodies and intervertebral disks

Pearl

1. The cervical midline is defined between right and left longus colli muscles and lies lateral to vertebral bodies, which are bounded across the midline by the prevertebral layer of the deep cervical fascia. Prevertebral muscles are dissected in a subperiosteal plane, the contralateral longus colli is retracted medially with the visceral column, and the ipsilateral is retracted laterally with the carotid sheath.

Potential Complications

Extended lateral retraction of prevertebral muscles might expose the vertebral artery at V1 or V2 portions. In case of anatomical variation, it is possible to find the VA in its V2 tract out of transverse processes. Gentle retraction of prevertebral muscles is mandatory.

10.2 Lateral Approach to the Cervical Spine

The lateral approach allows exposing the anterolateral portion of the vertebral bodies and the vertebral artery through a retrovascular way.

The indications to this type of approach are limited to those cases in which you must dominate a lateral mass having an optimum control of the vertebral artery [10].

The approach, as the front prevascular, can be performed for the UCS and MCS. In our experience you can plan an anterior approach and, if necessary, convert it into lateral approach.

This occurrence is related to anatomical abnormalities such as nonrecurrent laryngeal nerve (NRLN), or the need to better control diseases,that arise from the median line and have a lateral extension. NRLN occurs between 0.4 and 2.4 % of people on the right side, while it is extremely rare on the left side.

This anatomical variation is associated with an anomaly in the embryological development. The fourth right aortic arch and the proximal dorsal aorta are occasionally obliterated, and the subclavian artery can originate from the right side of the aortic arch, being called "lusoria" artery.

The course patterns of non-RLN are:

- *Type Ia* – the nerve has a straight course on the level with the upper thyroid pole.
- *Type Ib* – the most common. The nerve runs transversely at the level of the thyroid isthmus.
- *Type II* – the nerve curves downward, eventually reaching the lower pole of the thyroid gland.

In a case report by Yang et al., the non-RLN courses are described in more detail, distinguishing four types:

- In the descending type, the non-RLN descends after originating from the vagus nerve trunk.
- In the vertical type, the non-RLN runs vertically to the cricothyroid joint.
- In the ascending type, the non-RLN runs upward to the cricothyroid joint.
- In the V-shaped type, the non-RLN takes a downward direction and runs upward to the cricothyroid joint.

10.2.1 First Surgical Step: Incision of the Skin

The skin incision is performed along the anterior border of SCM muscle, from the tip of the mastoid process to the suprasternal notch. The extent of the incision depends on the pathology's characteristics and the cervical level to treat. Skin and radiology landmarks are assessed during the preliminary surgical step.

10.2.2 Second Surgical Step: Dissection of the Subplatysmal Flap.

Following the skin incision, a dissection of *superficial cervical fascia* (*SCF*) – formed by the subcutaneous tissue and the platysma muscle – is performed. Subplatysmal flap is then retracted medially to expose the SLDCF.

10.2.2.1 Anatomical Key Points

1. Sternocleidomastoid muscle (*SCM*)
2. External jugular vein (*EXJ*)
3. Great auricular nerve (*GAN*)
4. Transverse cervical nerve

10.2.2.2 Pearls

1. The SCM can be sectioned at its insertion to the tip of the mastoid to allow optimal exposure during extended approach. If not dissected it is displaced laterally, as in the anterior approach.
2. Identification of EJV on the skin, before skin incision, is useful for better identification and preservation of the anatomical structures. The external jugular vein emerges in the parotid loggia from the confluence of the superficial temporal vein and the internal maxillary vein, penetrates into the sternocleidomastoid region on the level with the angle of the mandible, then crosses the lateral face of the muscle obliquely. During dissection it may be necessary to ligate it.
3. The great auricular nerve emerges from the posterior border of the SCM muscle, at Erb's point. It runs parallel and superior to the external jugular vein on the superficial face of the muscle. It gives sensory branches to the skin of parotid and auricular regions. During the dissection, the surgeon usually finds first the EJV and then, about 1 cm laterally and cranially to it, the GAN.
4. The transverse nerve origins from the cervical plexus, emerges at the Erb's point, and crosses the sternocleidomastoid muscle transversely. It gives sensory branches to the skin. It can be preserved or sacrificed without postsurgical complications.

10.2.2.3 Potential Complications

Accidental section of the external jugular vein does not lead to a catastrophic bleeding but creates a nonoptimal visualization of the structures to dissect during the procedure.

Great auricular nerve section produces an ailment hypoesthesia of periauricular skin; thus, this nerve should be preserved.

10.2.3 Third Surgical Step: Dissection of SLDCF and Identification of the Carotid Sheath

Under cover of the platysma, there is "the investing layer of the deep cervical fascia" or superficial layer of the deep cervical fascia (SLDCF). It surrounds the neck like a collar but splits around the sternocleidomastoid and trapezius muscles to enclose them. Posteriorly, it joins the ligamentum nuchae. The SLDCF is opened along the anterior border of SCM muscle. Lateral retraction of the SCM muscle exposes some of the structures below:

10.2.3.1 Anatomical Key Points (Fig. 10.58)

1. Omohyoid muscle (*OM*)
2. Digastric muscle (*DM*)
3. Thyrolingual-facial venous trunk (*TLF*)
4. Spinal accessory nerve (*SAN*)
5. Carotid sheath or neurovascular bundle

10.2.3.2 Pearls

1. The omohyoid muscle is a landmark: the C5–C6 vertebral disks are approximately on the level with the intersection of the omohyoid and the SCM muscle. During the dissection, it is possible to appreciate the acute angle formed by the intersection of the OM and SCM. OM can be pulled down or disconnected for an easier access to the plane below. The neurovascular bundle, specifically the IJV, lies under this muscle.
2. On the upper part of this dissection plane, we find the posterior belly of digastric muscle, which runs obliquely forward and downward and goes to the hyoid bone, where it is fixed with its intermediate tendon to the body and the great horn and then continues with its anterior belly. It is defined "the resident's friend" because below it is possible to identify important structures like the neurovascular bundle and the TLF.
3. This vein passes deep to the SCM muscle and above the digastric muscle to empty into the internal jugular vein. It is formed by the retromandibular and the facial vein. Ligation of this structure should be near the confluence with the IJV.
4. This nerve usually runs below, from the posterior belly of the digastric muscle before piercing the deep surface of the SCM muscle. It passes through this muscle and supplies it and finally emerges from the midpoint of the posterior sternocleidomastoid border.
5. It is formed by the internal jugular vein laterally, by the carotid artery medially, and, in the dihedral angle formed by the two vessels, by the vagus nerve posteriorly. The first vascular structure found during the dissection is the IJV, which is just under the digastric muscle or following the TLF.

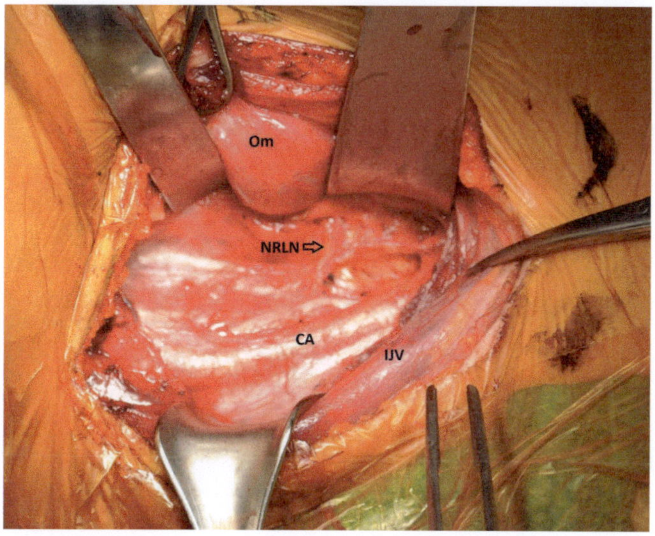

Fig. 10.58 Right side: evidence of nonrecurrent laryngeal nerve (*NRLN*), omohyoid muscle (*OM*) sectioned, carotid artery (*CA*), and internal jugular vein (*IJV*)

10.2.3.3 Potential Complications

6. HN injury is a possible complication during the ligation of lingual and facial artery and in all the dissections above the hyoid bone. HN palsy causes an omolateral deviation of the tongue and difficulty in swallowing and speaking. The tongue often deviates toward the side of the lesion when it's protruded (due to imbalance of genioglossus contraction).
7. Accidental section and/or palsy of the SAN could be caused by the retraction of SCM muscle. SAN lesions cause weakness of trapezius muscle on involved side and present with mild shoulder droop. The patient might report weakness in shoulder elevation and scapular winging most noticeable in arm abduction.
8. Ligation and dissection of the STA implicate high risk of iatrogenic injury of the SLN. During these maneuvers surgeons should focus on the relationship between the STA and this nerve. Palsy of SLN causes impaired cough reflex, hoarseness, voice fatigability, and impaired high-pitch phonation.

10.2.4 Fourth Surgical Step: Dissection of the MLDCF Posteriorly to the Carotid Sheath

During dissection of the IJV it could be necessary to remove the cervical lymph nodes. Once the dissection is completed, it is possible to gently retract all the neurovascular components medially

Behind carotid sheath there are some neurovascular structures: the vagus nerve ,the cervical plexus and the cervical sympathetic chain, that go from skull base to C6. In the UCS approach, behind neurovascular bundle there are ,also, the hypoglossal nerve (HN), glossopharyngeal nerve (GPN) and spinal accessory nerve (SAN).

From C3 to C6 it is possible to identify the phrenic nerve,that lies over the muscular plane (anterior scalene muscle).

Medial retraction of neurovascular bundle allows to exposure of the paravertebral muscles: the scalene muscles and the levator scapulae (Fig. 10.59).

10.2.4.1 Anatomical Key Points

1. Phrenic nerve
2. Cervical sympathetic chain
3. Hypoglossal nerve (HN), Glossopharyngeal nerve (GPN) Vagus nerve and Spinal accessory nerve (SAN)
4. Vertebral artery (VA)

10.2.4.2 Pearls

1. The phrenic nerve runs from ramus of C3-C4 and C5. It can be easily identified on the external surface of the anterior scalene muscle.
2. The cervical sympathetic chain is set against the prevertebral fascia and is contained in its split, medially to the anterior tubercles of the transverse processes. It descends diagonally across the anterior scalene muscles, passes lateral to the thyroid-cervical trunks, and dips between the subclavian artery and vein to enter the chest. To avoid Horner's syndrome, the sympathetic chain and the stellate ganglia should be first recognized and retracted laterally, and then the longus colli muscle and the prevertebral fascia should be longitudinally divided medially to the sympathetic chain.
3. All nerves, except from the HN that passes into the hypoglossal canal, pass through the jugular foramen. GPN, extending anteriorly and describing an anterosuperiorly concave curve, initially runs between the internal jugular vein and the internal carotid artery. Then it passes between the styloglossus and stylopharyngeus and reaches the pharynx and the base of the tongue. Other nerves are have already been described.
4. For a number of reasons, the V2 segment of the vertebral artery is more at risk during decompression of more cephalad vertebrae. First, the diameter of the artery decreases from C2–C3 to C6–C7 (4.88 ± 0.63 mm at C2–C3 to 4.27 ± 0.63 mm at C6–C7), and the anteroposterior diameters of the transverse foramina decrease from C6 to C3 (5.4 ± 1.1 mm at C6 to 4.7 ± 0.7 mm at C3). The amount of the inter-transverse space occupied by the artery, therefore, increases at more rostral levels. Second, the artery ascends medially from C6 to C3 at an angle of approximately 4° compared to the midline, making it more likely to cross the surgical field at higher cervical levels.

10.2.4.3 Potential Complications

Major complications are correlated with neurovascular injuries.

Lesion of the cervical sympathetic chain, leads to Claude-Bernard-Horner syndrome characterized by ptosis (droopy eyelid), miosis (constricted pupil), and facial anhidrosis (impaired facial sweating) with or without enophthalmos.

Emi-diaphragm palsy is subsequent to phrenic nerve palsy.

Lesions of the glossopharyngeal nerve rarely occur in alone and usually are associated with vagus and spinal accessory nerve. Main symptoms are dysphagia, ipsilateral palatal weakness and loss of gag reflex.

Vagus nerve palsy lead to dysphonia and dysphagia, that are a consequence of the homolateral vocal cord paralysis.

A potentially catastrophic bleeding with brain injury can result from vertebral or carotid artery injury.

HN and SAN injuries have already been described.

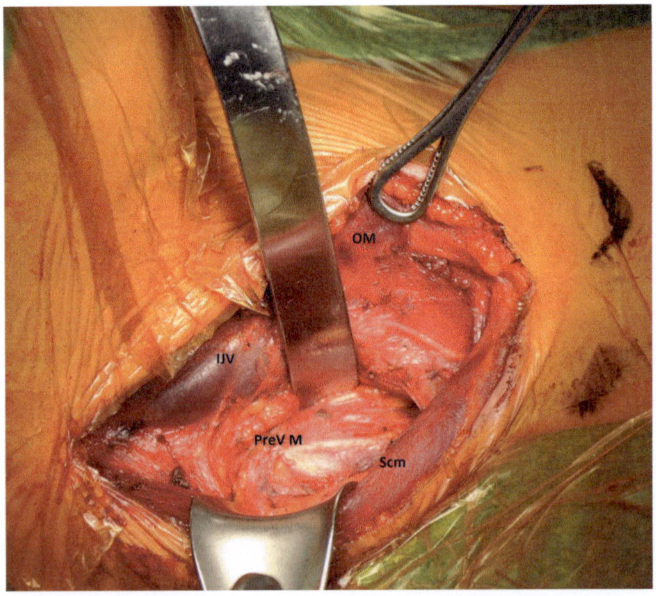

Fig. 10.59 Right side: prevertebral muscles (*PreV M*), internal jugular vein (*IJV*), and omohyoid muscle (*OM*) sectioned

10.2.5 Fifth Surgical Step: Dissection of Prevertebral Structures and Exposure of the Vertebral Plane

The prevertebral layer of the deep cervical fascia origins laterally from the transverse processes and continues with the fascia of the scalene muscles. The prevertebral layer and the anterior longitudinal ligament are incised longitudinally along the midline, before their cauterization (Figs. 10.60 and 10.61).

10.2.5.1 Anatomical Key Points

1. Prevertebral Muscles
2. Vertebral bodies and intervertebral disk

10.2.5.2 Pearls

The "longus capitis" is located more laterally than the longus colli and originates from the transverse process of C3-C6 and has its insertion at the basilar portion of the occipital bone. Longus colli is loceted on the lateral side of the vertebral bodies, from the atlas to T3. In this phase of the operation, hemostasis must be accurately done using bipolar forceps and wax. The cervical midline is defined between right and left longus colli muscles and lies lateral to vertebral bodies, which are bounded across the midline by the prevertebral layer of deep cervical fascia. Prevertebral muscles are dissected in a subperiosteal plane, than are retracted medially with the carotid sheath. During dissection of the long muscles of the neck, the cervical sympathetic chain is easily recognizable compared to the vagus because it has a lower mobility and it also presents ganglion formations.

10.2.5.3 Potential Complication

The complications are related to the possible vertebral artery injury, or to traction of the nerves described in the previous surgical steps.

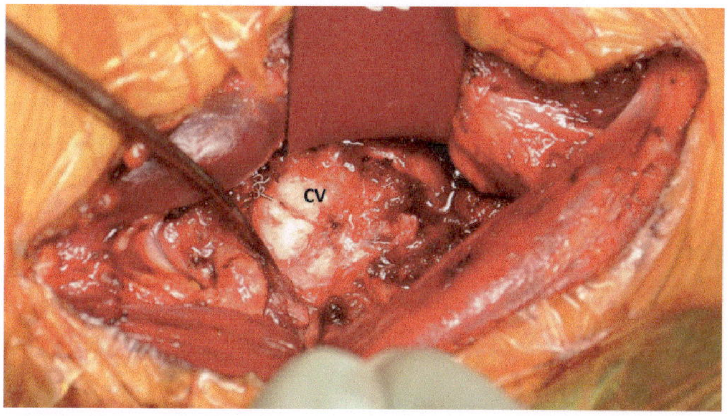

Fig. 10.60 Right side: lateral approach and exposure of cervical vertebra (*CV*)

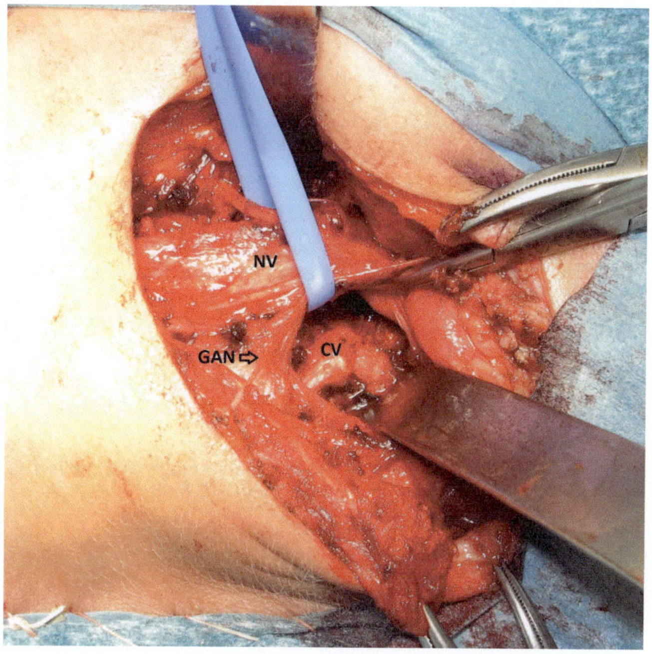

Fig. 10.61 Left side: neurovascular bundle (*NV*) retraction and cervical vertebra (*CV*) exposure. Great auricular nerve (*GAN*)

Bibliography

1. Smith GW, Robinson RA (1958) The treatment of certain cervical-spine disorders by anterior removal of the intervertebral disc and inter- body fusion. J Bone Joint Surg Am 40-A:607–623
2. Southwick W, Robinson RA (1957) Surgical approaches to the vertebral bodies in the cervical and lumbar regions. J Bone Joint Surg Am 39:631–644
3. Cloward R (1958) The anterior approach for removal of ruptured cervical disks. J Neurosurg 15:602–617
4. McAfee PC, Bohlman HH, Riley LH et al (1987) The anterior retropharyngeal approach to the upper part of the cervical spine. J Bone Joint Surg Am 69:1371–1383
5. Fang HS, Ong GB (1962) Direct anterior approach to the upper cervical spine. J Bone Joint Surg Am 44:1588–1604
6. Whitesides TE, Kelly RP (1966) Lateral approach to upper cervical spine for anterior fusion. South Med J 59:879–883
7. Whitesides TE, McDonald AP (1978) Lateral retropharyngeal approach to the upper cervical spine. Orthop Clin North Am 9:1115–1127
8. Russo A, Albanese E, Quiroga M et al (2009) Submandibular approach to the C2–3 disc level: microsurgical anatomy with clinical application. J Neurosurg Spine 10:380–389
9. Haller JM, Iwanik M, Shen FH (2011) Clinically relevant anatomy of high anterior. Spine 36(25):2116–2121
10. Laus M, Alfonso C, Ferrari D et al (1995) Lateral retrovascular approach to the upper cervical spine. Chir Organi Mov 80(1):65–75

Posterior Approach to Cervical Spine

11

R. Ghermandi, M. Girolami, A. Gasbarrini, and S. Boriani

In the neck, the spine takes up the mid-part of the cross-sectional area, and its posterior aspect is surrounded by a thick layer of powerful muscles (see §2. Anatomy).

The posterior muscular coverage is subdivided in two symmetrical compartments, which extend longitudinally in continuous with the adjacent thoracolumbar extensor muscles.

The standard posterior approach to the cervical spine is carried through these two posterior muscular compartments whose innervation and vascular supply is segmental from the dorsal routes of cervical spinal nerves (Fig. 11.1).

Thus, ligamentum nuchae, the septum between the two continuous muscular compartments, is a true internervous plane, and no particular structure is at risk if dissection is carefully conducted through it. Moreover, this minimizes blood loss and maintains a stout tissue layer for closure of the wound.

The posterior midline approach allows surgical exposure of the posterior arches of the vertebrae up to the occiput.

After positioning of the patient (Fig. 11.2), prior to any incision, an accurate trichotomy of the nuchal area might be necessary depending on the segments to be exposed.

The main superficial landmarks are as follows:

- External occipital prominence of the occiput (inion)
- First palpable spinous process → C2 (axis)
- Prominent spinous process of the cervicothoracic junction → C7 (vertebra prominens)
- Mastoid processes

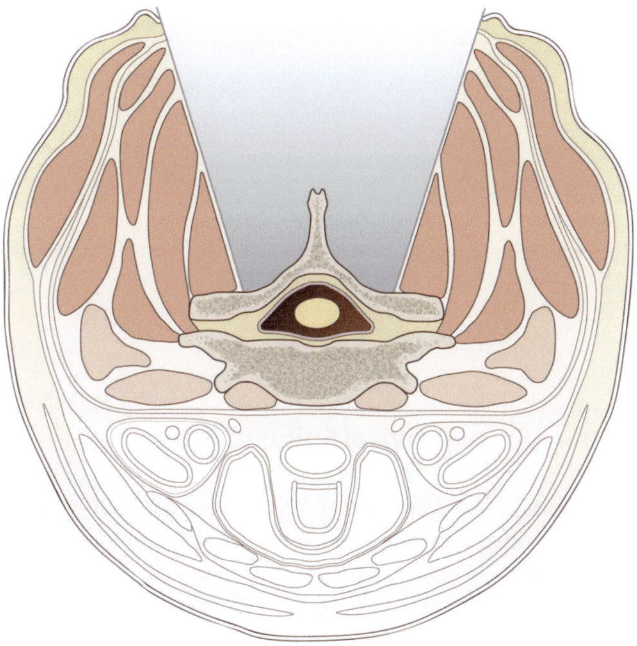

Fig. 11.1 The standard midline posterior approach to the cervical spine is carried through the two symmetrical extensor muscle compartments down to the dorsal surface of the neural arches

R. Ghermandi (✉) • M. Girolami • A. Gasbarrini • S. Boriani
Oncologic and Degenerative Spine Surgery Department, Rizzoli Orthopedic Institute, Bologna, Italy
e-mail: riccardoghermandi@gmail.com

S. Boriani et al. (eds.), *Atlas of Craniocervical Junction and Cervical Spine Surgery*, DOI 10.1007/978-3-319-42737-9_11

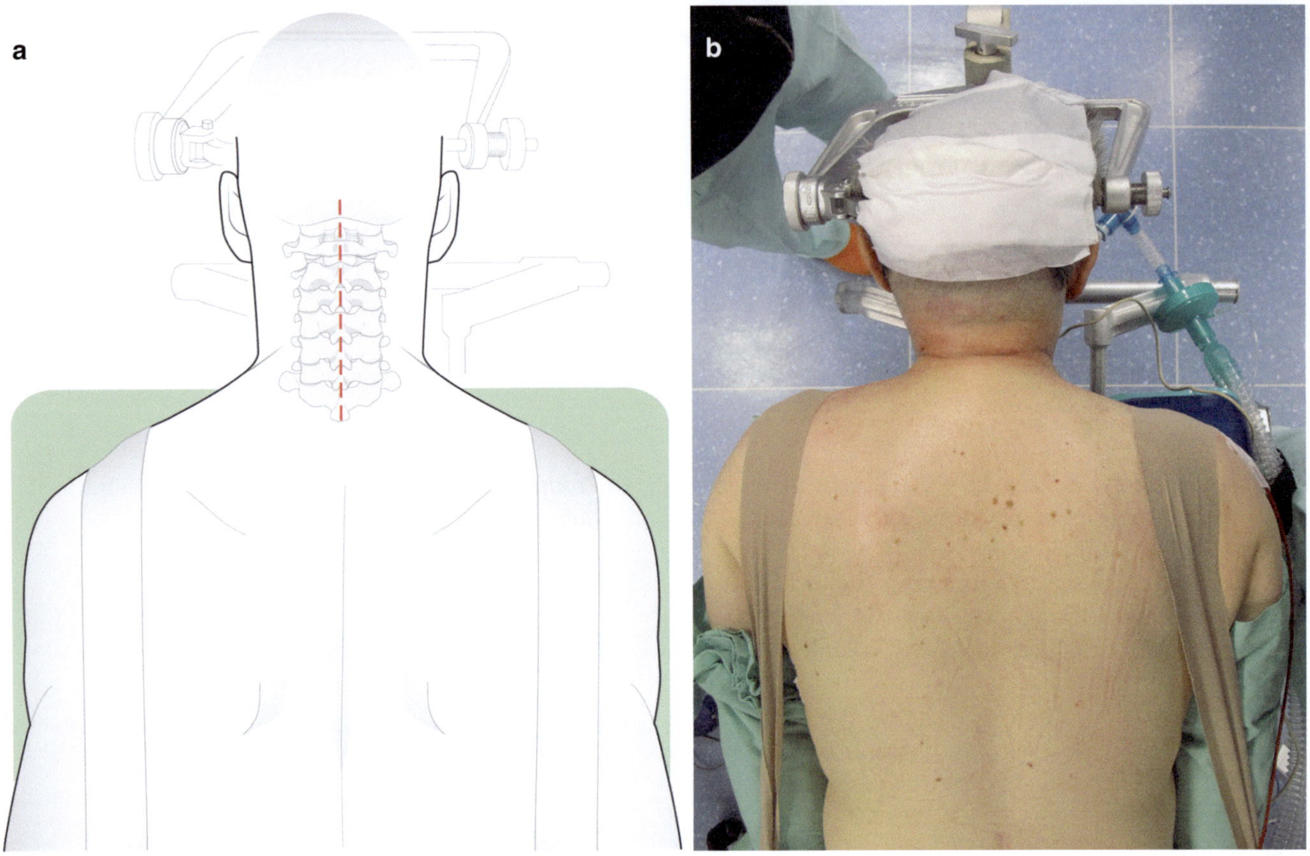

Fig. 11.2 Prior to incision, identification of the superficial landmarks is crucial. Shoulders are retracted in order to allow intraoperative fluoroscopic control at the cervicothoracic junction

11.1 Upper Cervical Spine (C1-C2)

The width of the spinal canal at the craniocervical junction (CVJ) is wider than that in the lower segments of the cervical spine. Thus, instability is a more frequent indication than cord compression to assess the CVJ from posterior.

After midline skin incision, the tip of the spinous process of C2 can be palpated under the inferior aspect of the occipital bone. Immediately above, on the midline and 1–2 cm anteriorly, lies the posterior tubercle of C1 under the thick layer of suboccipital muscles (rectus capitis posterior – major and minor; oblique – superior and inferior, Fig. 11.3).

The bony topography between the lamina and the lateral mass of C2 (axis) is indistinct. Thus, dissection of these muscles must be carried gently and special care must be taken beyond the first 12–15 mm in order not to damage the following neurovascular structures:

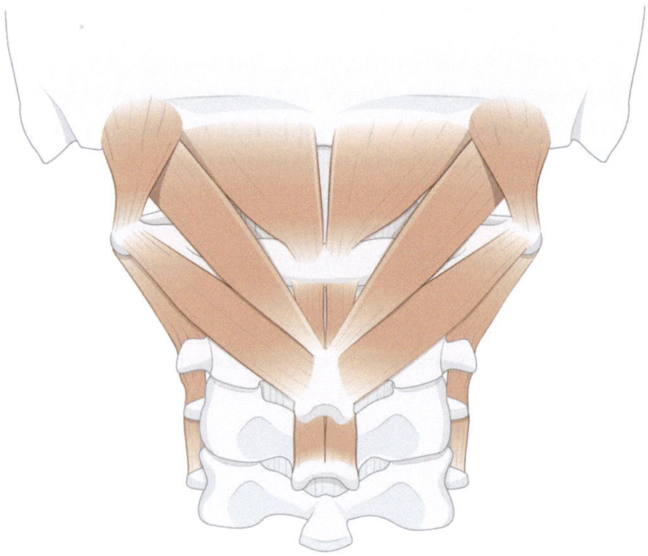

Fig. 11.3 Muscles of the occipitocervical region (rectus capitis posterior – major and minor; oblique – superior and inferior)

- Vertebral artery
- First cervical nerve
- Second cervical nerve (Arnold nerve)
- Suboccipital venous plexus (behind C1-C2 interspace)

11.1.1 Anatomical Key Points

1. Posterior tubercle of C1
2. Transverse processes of C1
3. Spinous process of C2 (bifidus)

11.1.2 Focus On: Vertebral Artery

Surgical anatomy of the vertebral artery (VA) can be subdivided in four segments (Fig. 11.4):

V1, from its origin from the subclavian artery (anterior to the C7 transverse process) to the entry point of C6 foramen transversarium
V2, within C6-C1 transverse foramina
V3, from the superior aspect of the arch of the atlas to the foramen magnum (C0)
V4, intradural course from the foramen magnum to unite with the contralateral VA to form the basilar artery

The highest risk for injury during posterior approaches is in the V3 tract while dissecting the occipitocervical junction (4.2–8.1%). Ebraheim et al. suggest that dissection on the posterior aspect of the posterior ring should remain within 12 mm lateral to the midline, and dissection on the superior aspect of the posterior ring should remain within 8 mm of the midline.

However, depending on the procedures to be performed, different tracts of the VA can be at risk (Fig. 11.5).

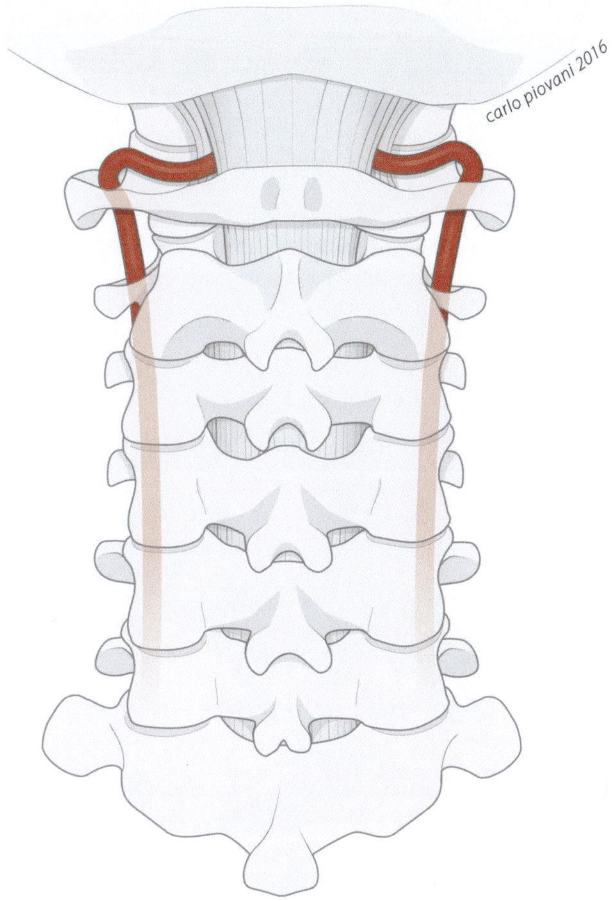

Fig. 11.4 Course of the vertebral artery (VA)

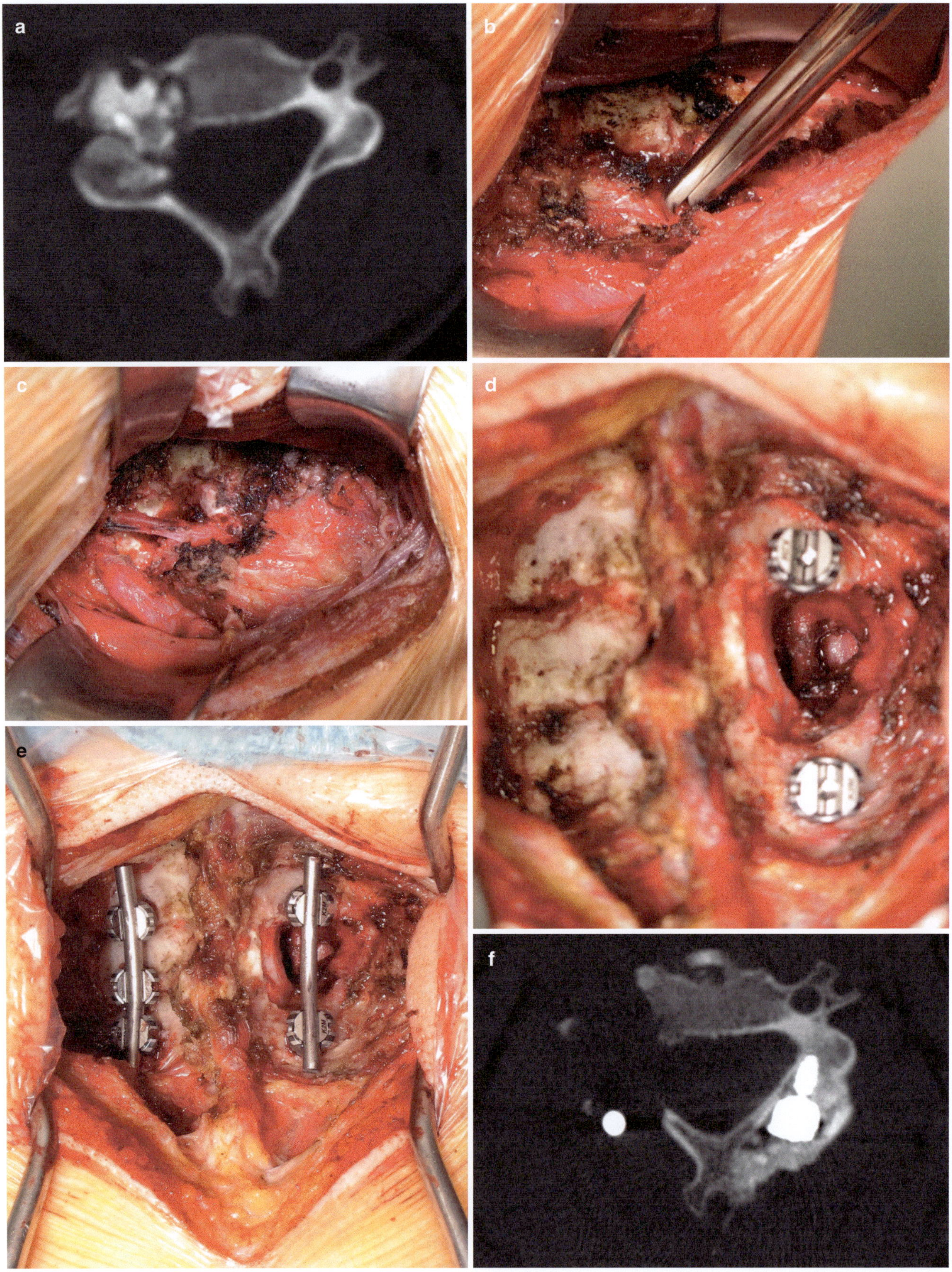

Fig. 11.5 Case report. A 19-year-old female, C4 right lateral mass osteoblastoma (OBL, WBB 8–10/A-D) complaining of neck pain with occasional numbness to the right arm (**a**). Right dominant VA symptomatic at occlusion test prior to embolization. Double-approach procedure, A: pre-SCM approach, VA release (**b**) and intralesional excision (**c**); P: posterior approach, stabilization with lateral mass screws and complete excision (**d**). Unlike most posterior procedures, VA is at risk at V2 crossing the deep surgical field (**e**). Postoperative CT scan shows complete excision (**f**)

11.2 Subaxial Cervical Spine (C3-C7)

The most common approach is carried through a midline skin incision along the line connecting the external occipital prominence and the tip of the spinous process of C7 (Fig. 11.6).

The incision must be centered on the level of the pathology, extending it cranially and caudally depending on the levels to be exposed.

Intraoperative imaging is important for level confirmation: prior to skin incision, so as should be used, after dissection, prior to any procedure on the bone.

Once the tips of the spinous processes and/or the ligamentum nuchae are identified (Fig. 11.7), subperiosteal dissection of the paravertebral muscles from the posterior arches is carried out in a mediolateral direction up to the lateral margin of the lateral masses (Fig. 11.8).

11.2.1 Anatomical Key Points

1. Ligamentum nuchae
2. Spinous processes (C2, C7-T1) and supraspinous ligament
3. Lateral masses
4. First thoracic vertebra and ribs

11.2.2 Pearls

1. Muscular bleeding can be particularly annoying and hard to dominate completely, impairing vision on the surgical field. Moreover, limiting the dissection without worthless exposure of levels preserves innervation and blood supply to the paravertebral muscles.
2. The superior articular surfaces of the lateral masses are directed dorsally and cranially, and the inferior surfaces ventrally and caudally. The slope increases in craniocaudal

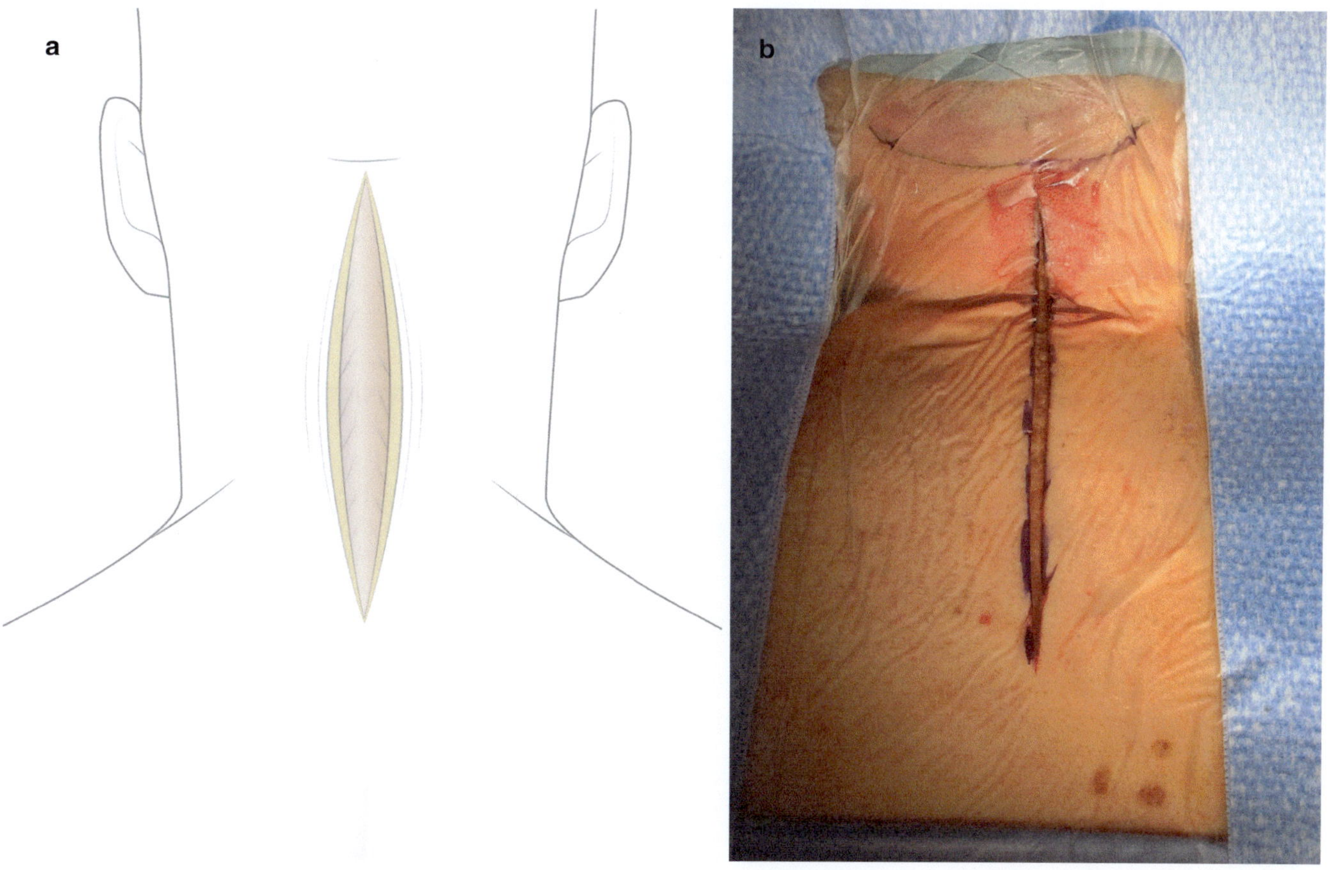

Fig. 11.6 Skin incision and dissection through the subcutaneous tissue

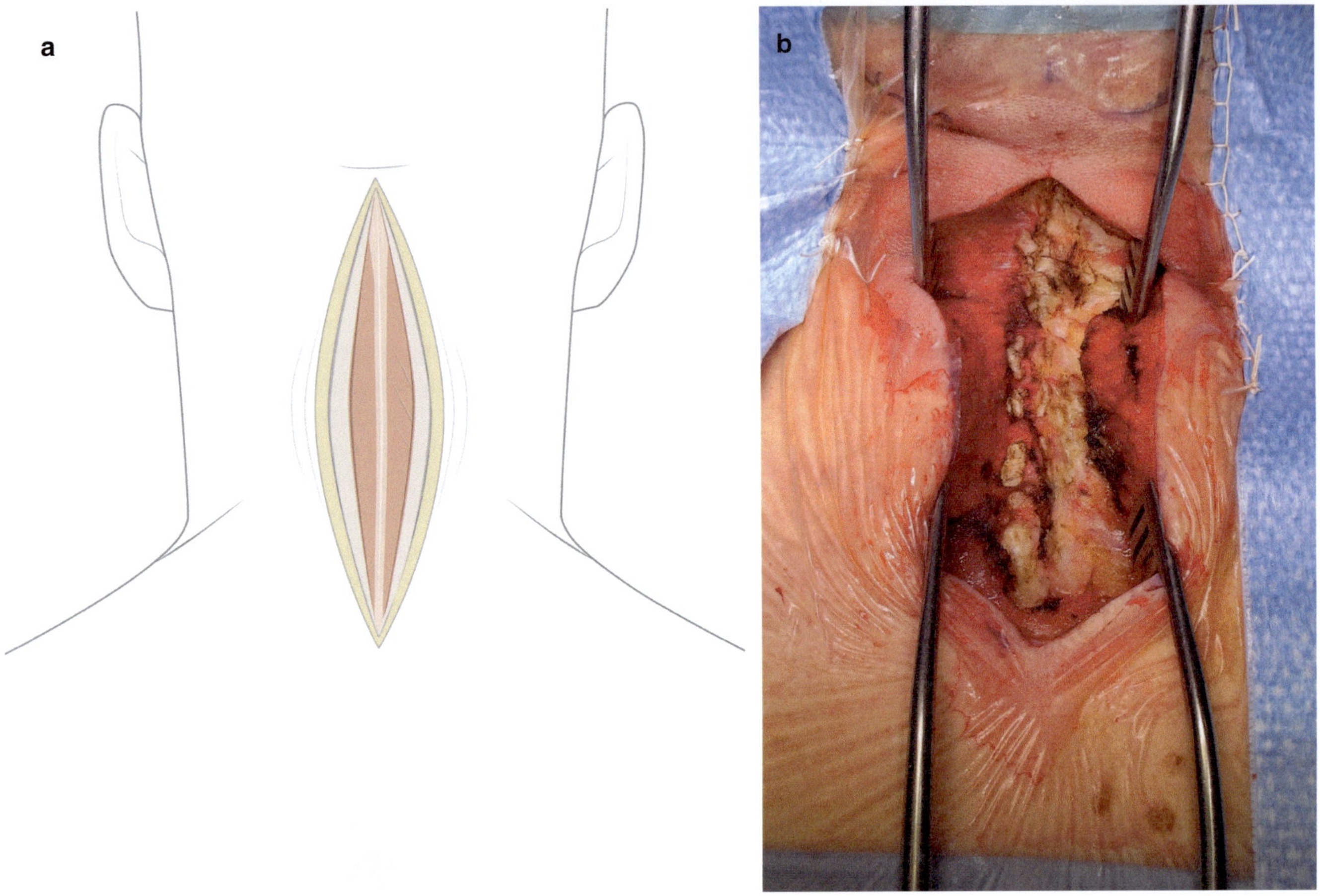

Fig. 11.7 Incision of the fascia in correspondence to the ligamentum nuchae

direction up to the thoracic spine where the articular surfaces are almost vertical.

3. The spinous process of the cervical vertebrae is bifid, except for C7.
4. The first thoracic vertebra has a transverse process that articulates with the first rib.

11.3 Caudal Extension: Cervicothoracic Junction

The posterior approach to the cervical spine can be extended downwards to the cervicothoracic junction.

The anatomical plane through which dissection is carried out is continuous until the caudal end of the spine in the lumbar spine and sacrum, so dissection could be ideally extended downwards up to the surgeon's preference.

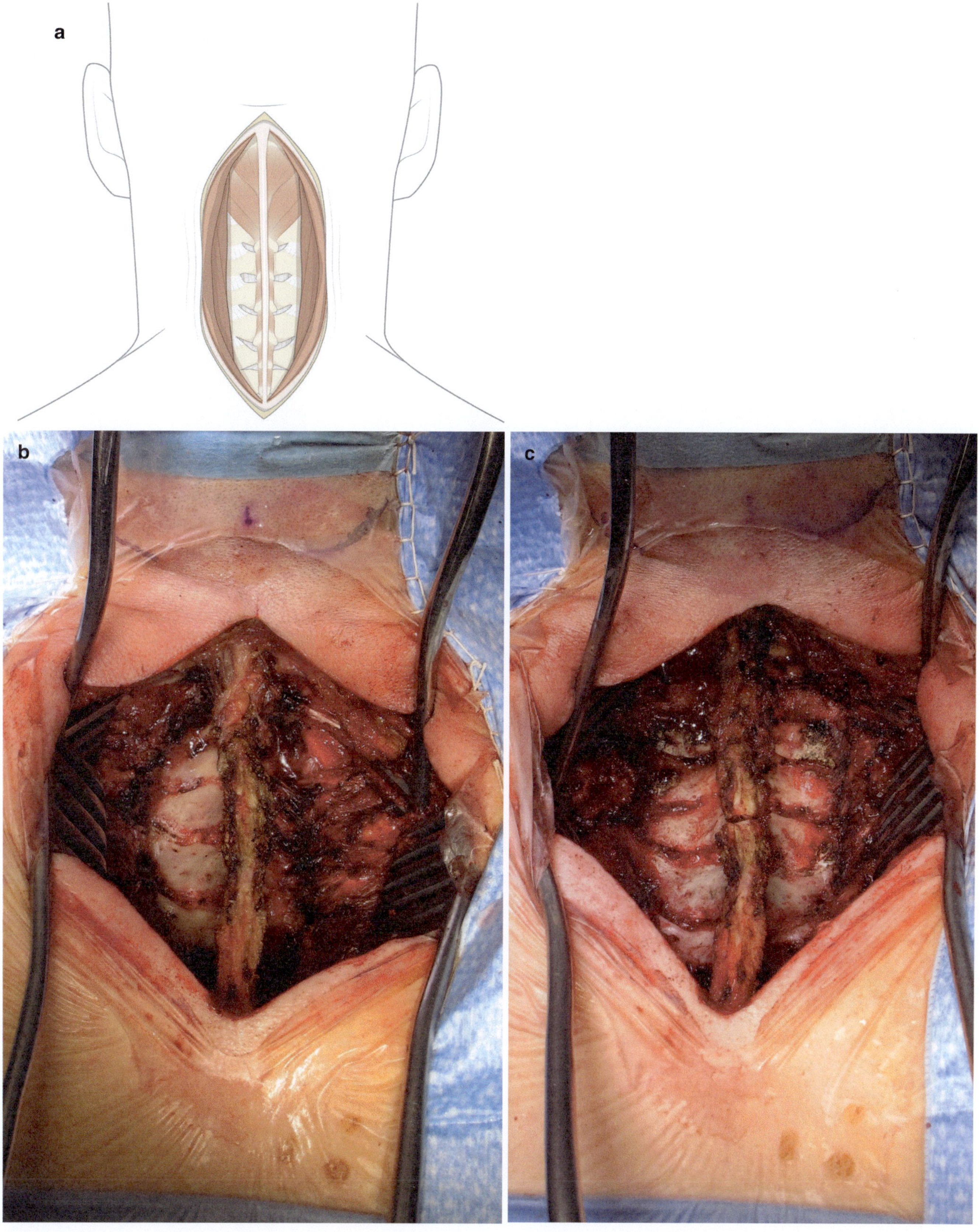

Fig. 11.8 Dissection is carried out along the internervous plane of the ligamentum nuchae (**a**). It is extended laterally until the lateral margin of the lateral masses. Identification of the correct dissection plane (**b**) allows for less intraoperative bleeding – thus, avascular surgical field – and ease of identification of the anatomical key structures (i.e., entry point for screw insertion, **c**)

11.4 Dissection of Spinal Canal

The spinal canal is an anatomical compartment circumscribed anteriorly by the discs and the vertebral bodies and posterolaterally by the neural arch, formed by the pedicles and the laminae. Preliminary to its dissection, from a standard posterior approach, subperiosteal detachment of the paraspinal muscles is required to get to the posterior surface of the laminae.

In order to enter the canal, laminae and yellow ligament must be removed mono- or bilaterally (laminectomy and flavectomy, Fig. 11.9). Flexion of the head opens the interspinous space increasing the working space available.

Removal of the posterior arch exposes the content of the canal: neural structures and venous epidural plexus immersed in the epidural fat (Fig. 11.10). The dorsal aspect of the vertebral body can be reached by gentle mobilization of the dural sac and displacement of the correspondent spinal nerve. Care must be taken during these maneuvers since the ventral surface of the dural sac might be linked to the dorsal surface of the anterior column by the inconstant presence of the Hoffmann ligament (uncommon in the cervical spine).

The sac itself is structured into layers: dura mater and arachnoid. Dura mater, the outer layer, is whitish, smooth, and more thick, while the arachnoid, the inner layer, is colorless and extremely thin (spider weblike). Superficial injuries might leave the arachnoid intact; thus, it might herniate

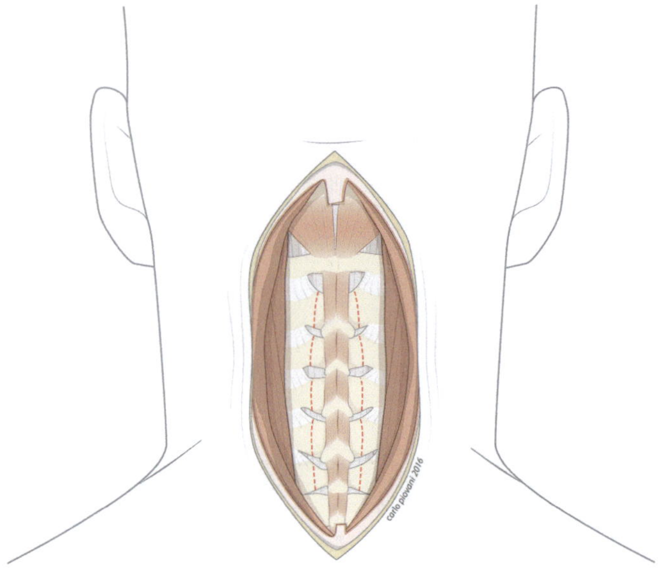

Fig. 11.9 In order to access the canal, laminae are removed mono- or bilaterally

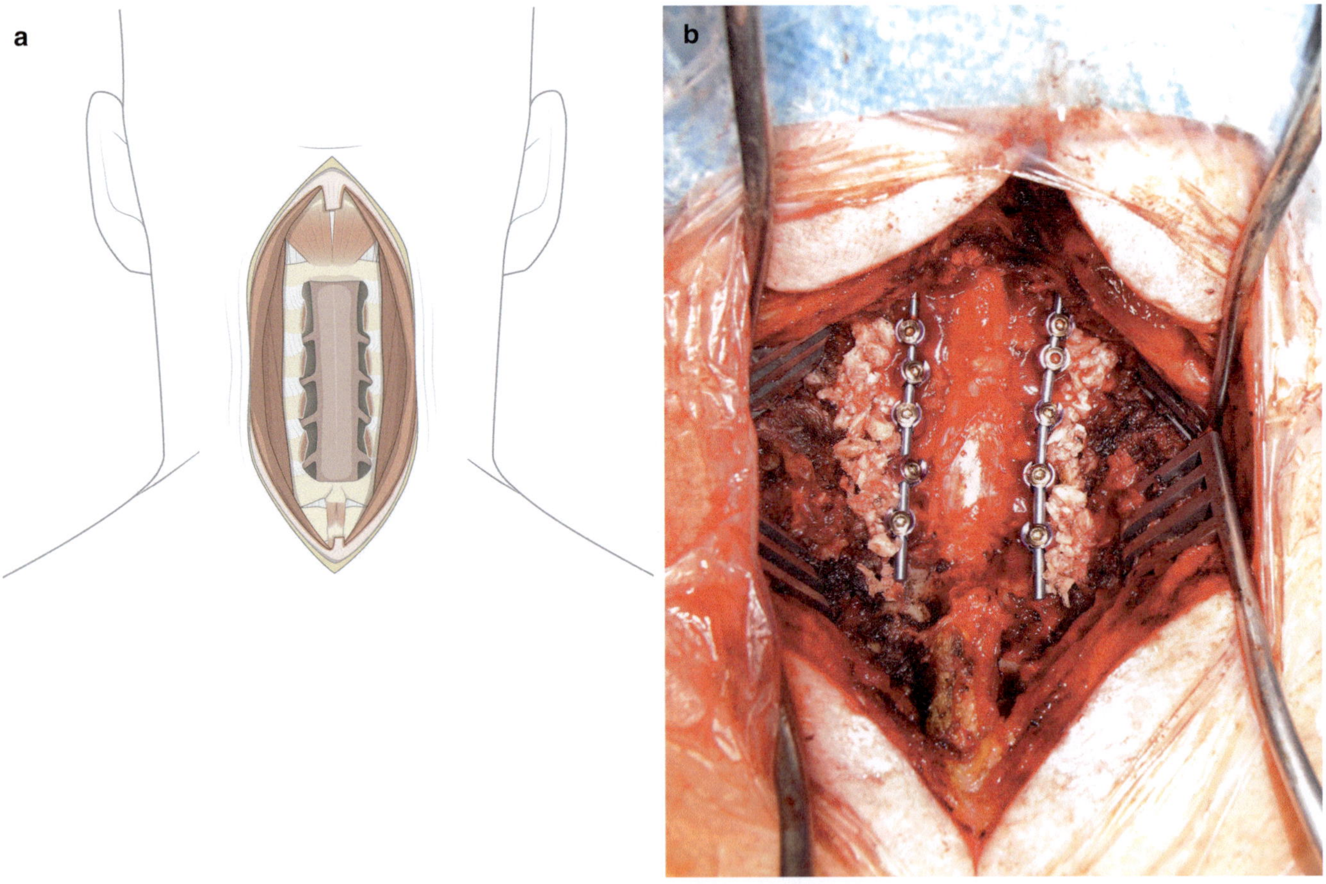

Fig. 11.10 After laminectomy is carried out, the dural sac and the emergence of the spinal nerves can be identified and explored

through the leak in the dura mater, appearing like a drop on the dural sac. Since this lesion is extremely likely to break, it must be reinforced.

References

1. Abumi K (2015) Cervical spondylotic myelopathy: posterior decompression and pedicle screw fixation. Eur Spine J 24(Suppl 2):186–196
2. Bauer R, Kerschbaumer E (1987) Operative approaches in orthopedic surgery and traumatology. Thieme, New York
3. Ebraheim NA, Xu R, Ahmad M, Heck B (1998) The quantitative anatomy of the vertebral artery groove of the atlas and its relation to the posterior atlantoaxial approach. Spine (Phila pa 1976) 23(3):320–323; Epstein NE (2015) Open laminoforaminotomy: a lost art? Surg Neurol Int 6(Suppl 24):S600–S607
4. Harel R, Stylianou P, Knoller N (2016) Cervical spine surgery: approach-related complications. World Neurosurg 94:1–5
5. Jalai CM, Worley N, Marascalchi BJ, Challier V, Vira S, Yang S, Boniello AJ, Bendo JA, Lafage V, Passias PG (2016) The impact of advanced age on peri-operative outcomes in the surgical treatment of cervical spondylotic myelopathy: a nationwide study between 2001 and 2010. Spine (Phila Pa 1976) 41(3):E139–E147
6. Phan K, Scherman DB, Xu J, Leung V, Virk S, Mobbs RJ (2017) Laminectomy and fusion vs laminoplasty for multi-level cervical myelopathy: a systematic review and meta-analysis. Eur Spine J 26(1):94–103
7. Schroeder GD, Hsu WK (2013) Vertebral artery injuries in cervical spine surgery. Surg Neurol Int 4(Suppl 5):S362–S367
8. Seng C, Tow BP, Siddiqui MA, Srivastava A, Wang L, Yew AK, Yeo W, Khoo SH, Balakrishnan NM, Bin Abd Razak HR, Chen JL, Guo CM, Tan SB, Yue WM (2013) Surgically treated cervical myelopathy: a functional outcome comparison study between multilevel anterior cervical decompression fusion with instrumentation and posterior laminoplasty. Spine J 13(7):723–731
9. Smith JS, Ramchandran S, Lafage V, Shaffrey CI, Ailon T, Klineberg E, Protopsaltis T, Schwab FJ, O'Brien M, Hostin R, Gupta M, Mundis G, Hart R, Kim HJ, Passias PG, Scheer JK, Deviren V, Burton DC, Eastlack R, Bess S, Albert TJ, Riew KD, Ames CP; International Spine Study Group (2016) Prospective multicenter assessment of early complication rates associated with adult cervical deformity surgery in 78 patients. Neurosurgery 79(3):378–388

12 Exemplificative Cases in Cervical Spine

A. Gasbarrini, M. Girolami, R. Ghermandi,
Y.E. Akman, and S. Boriani

12.1 Upper Cervical Spine

Fourteen-year-old female, presenting with progressive neck pain in the nuchal area radiating to the head with painful contracture of the right paraspinal musculature. Pain was unrelated to weight-bearing and was increasing during the night. Physical examination revealed mild hyperreflexia, gait disturbance, positive Hoffmann sign, and clonus.

AP and LL radiographs with transoral projection show an osteoproductive lesion replacing the right lateral mass and extending into the vertebral body.

Subsequent CT scan provides more details on the features of the lesion, host bone reaction, and extent of bony involvement. Sagittal scan shows the lesion completely invading the lateral mass, breaching into the posterior articular space and the ipsilateral foraminal space. Axial scan shows disruption of the posterolateral cortex limiting the spinal canal revealing epidural space invasion.

Angio-MRI shows hypervascular nature of the lesion and the close relationship with the vertebral artery. Thus, angiography and embolization of the vertebral artery are mandatory prior to surgical attempt of removing the lesion.

Double-approach procedure, first anterior than posterior, was planned in order to excise, reconstruct, and restore segmental spinal stability. Through a lateral approach, which allowed full view on the right lateral mass, piecemeal removal of the tumor was performed.

Harms cages filled with bone graft were used for reconstruction, along with posterior instrumentation.

Four-month follow-up shows signs of posterior fusion and osteointegration of the graft. No recurrence occurred and patient is NED with solid arthrodesis at 3-year follow-up (Figs. 12.1, 12.2, 12.3, 12.4, 12.5, 12.6, 12.7, and 12.8).

A. Gasbarrini (✉) • M. Girolami • R. Ghermandi • S. Boriani
Oncologic and Degenerative Spine Surgery Department, Rizzoli Orthopedic Institute, Bologna, Italy
e-mail: gasbarrini@me.com

Y.E. Akman
Orthopaedics and Traumatology Department, Metin Sabanci Baltalimani Bone Diseases Training and Research Hospital, Istanbul, Turkey

S. Boriani et al. (eds.), *Atlas of Craniocervical Junction and Cervical Spine Surgery*, DOI 10.1007/978-3-319-42737-9_12

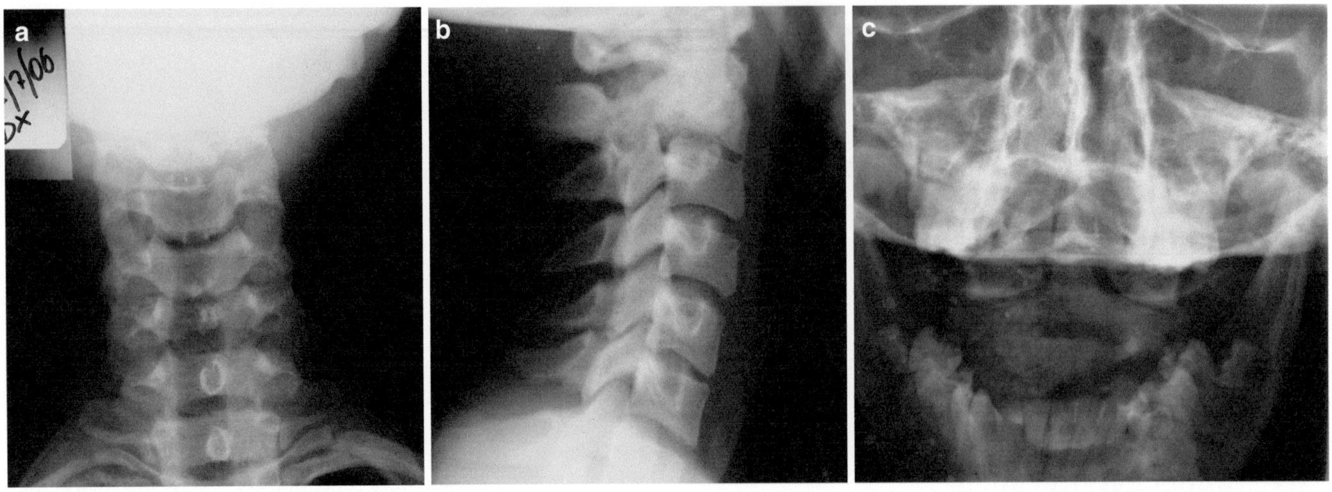

Fig. 12.1 Pre-operative radiographs in anteroposterior (**a**), lateral (**b**) and trans-oral (**c**) projetions

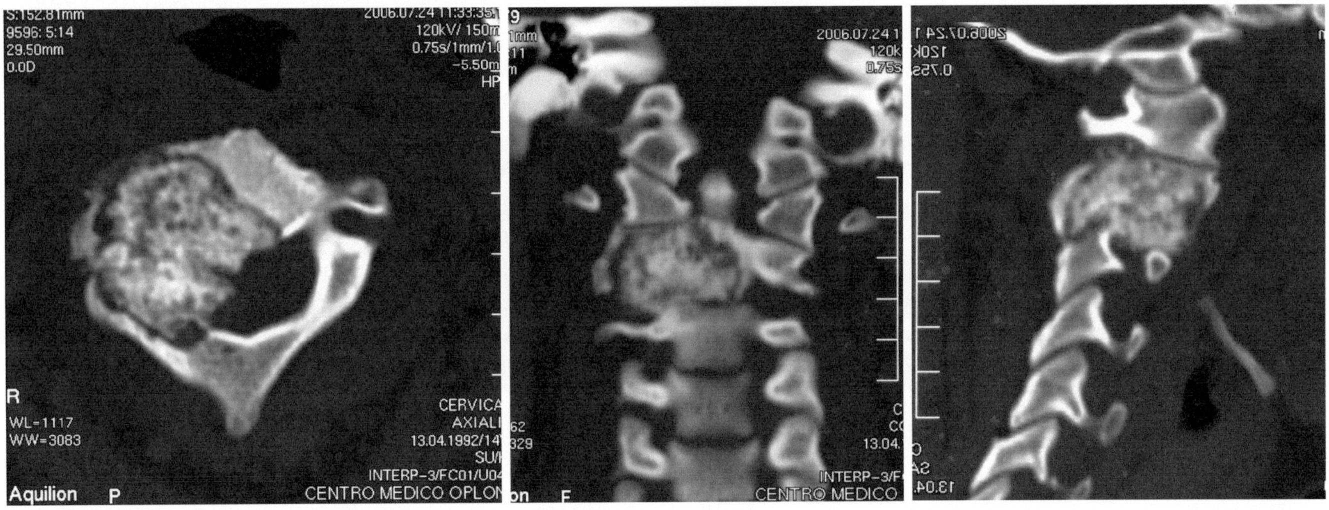

Fig. 12.2 Pre-operative CT scans of the C2 lesion, pathognomonic for osteoblastoma

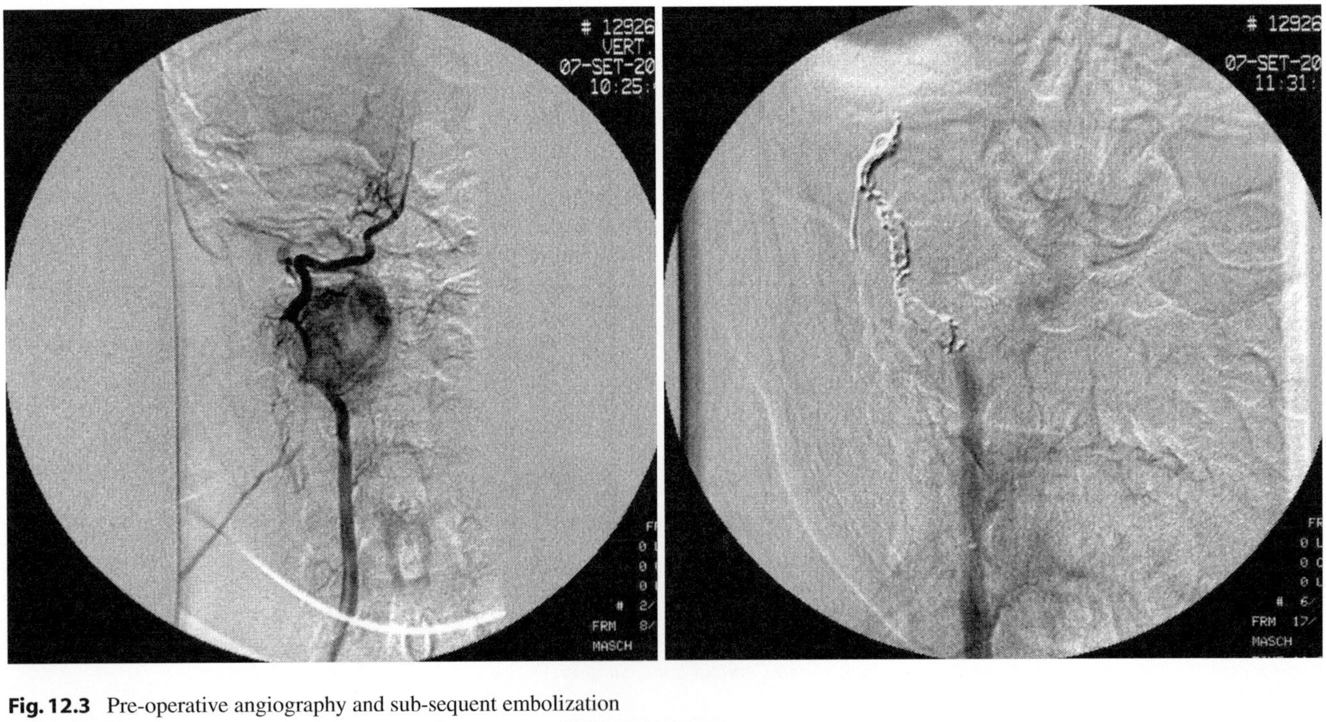

Fig. 12.3 Pre-operative angiography and sub-sequent embolization

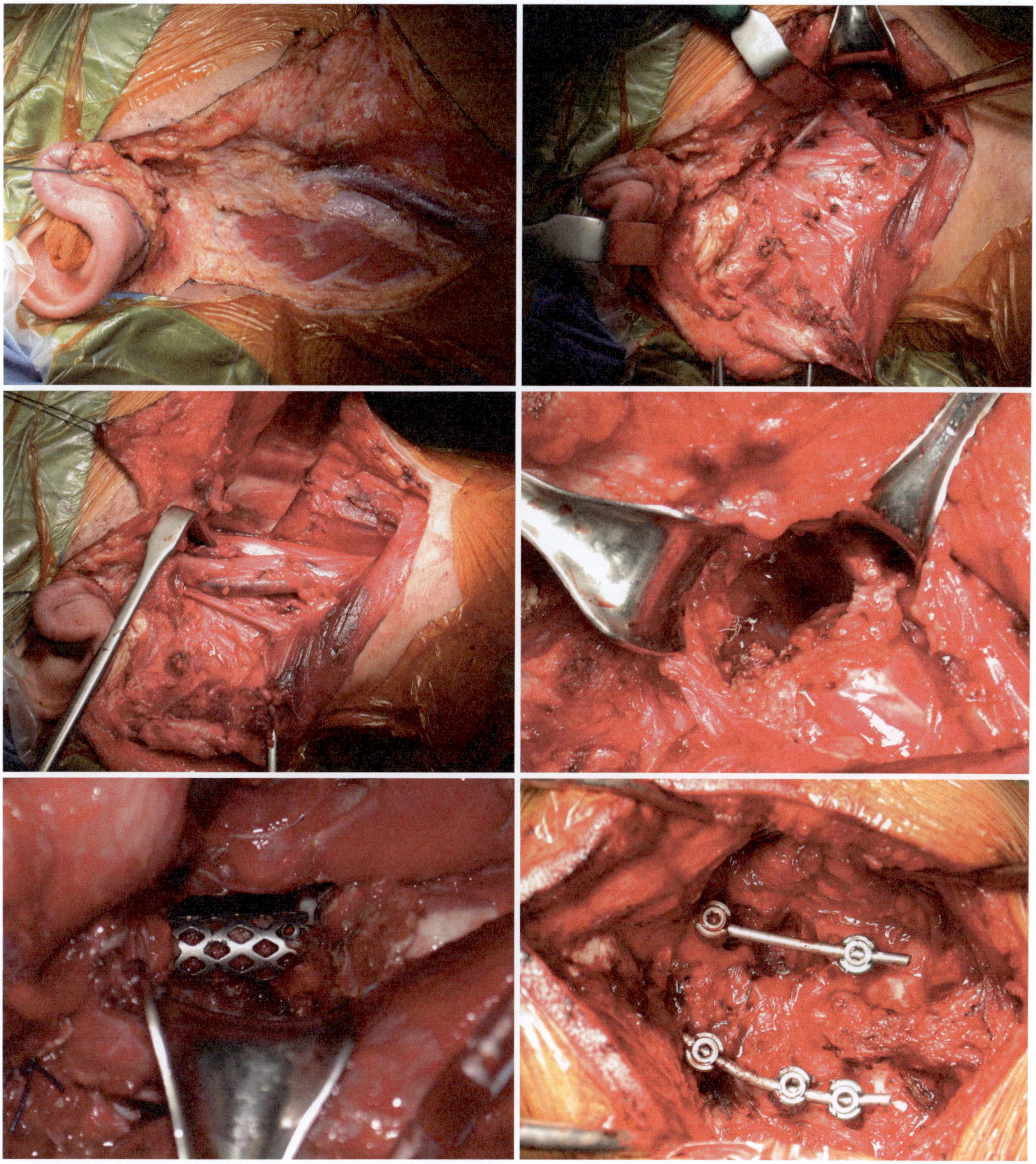

Fig. 12.4 Intra-operative images, lateral and posterior approach

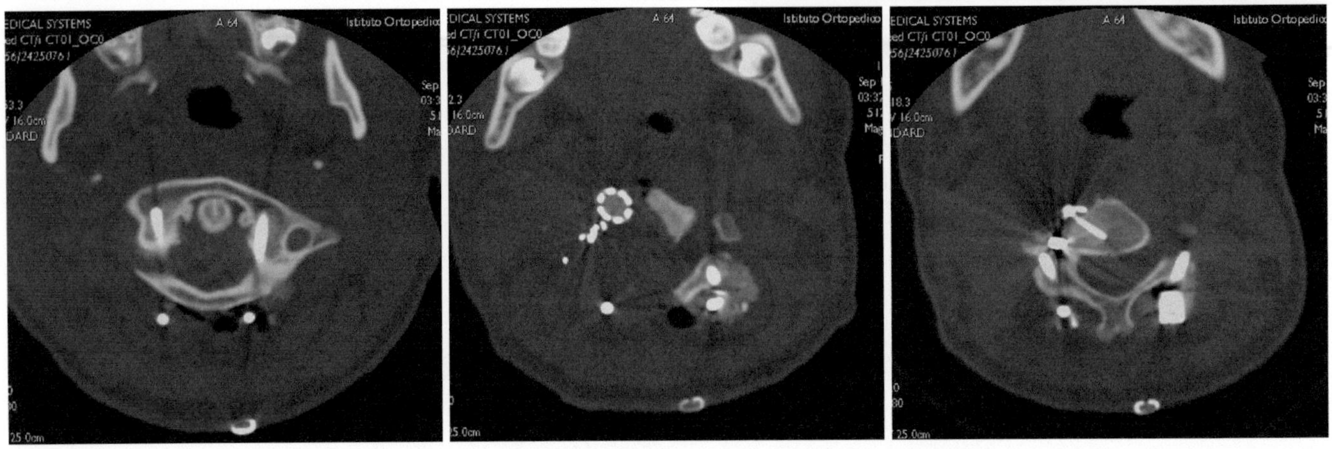

Fig. 12.5 Post-operative CT scan

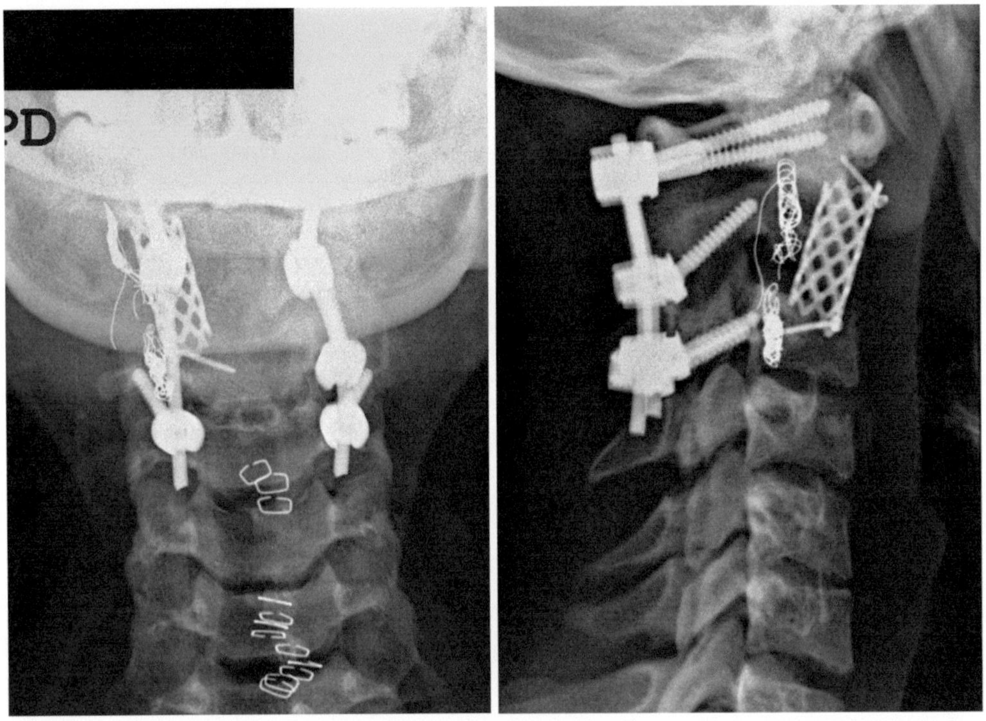

Fig. 12.6 Post-operative standing plain radiographs

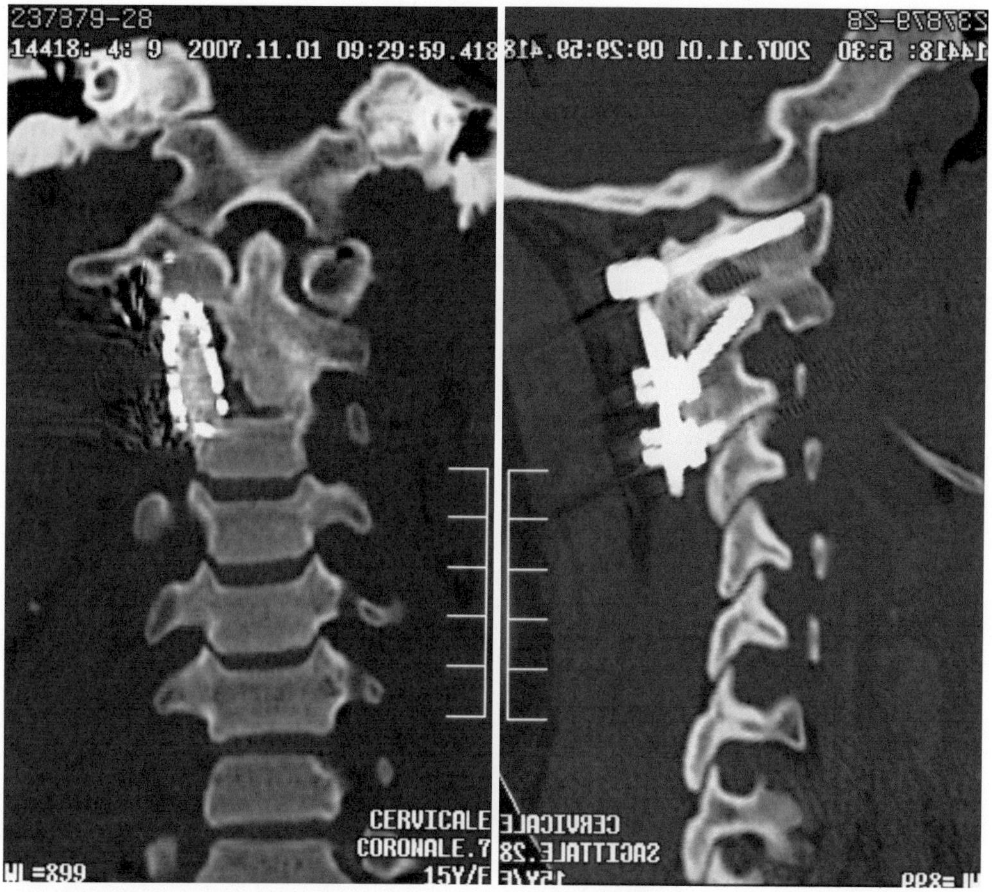

Fig. 12.7 4-months follow-up CT-scan

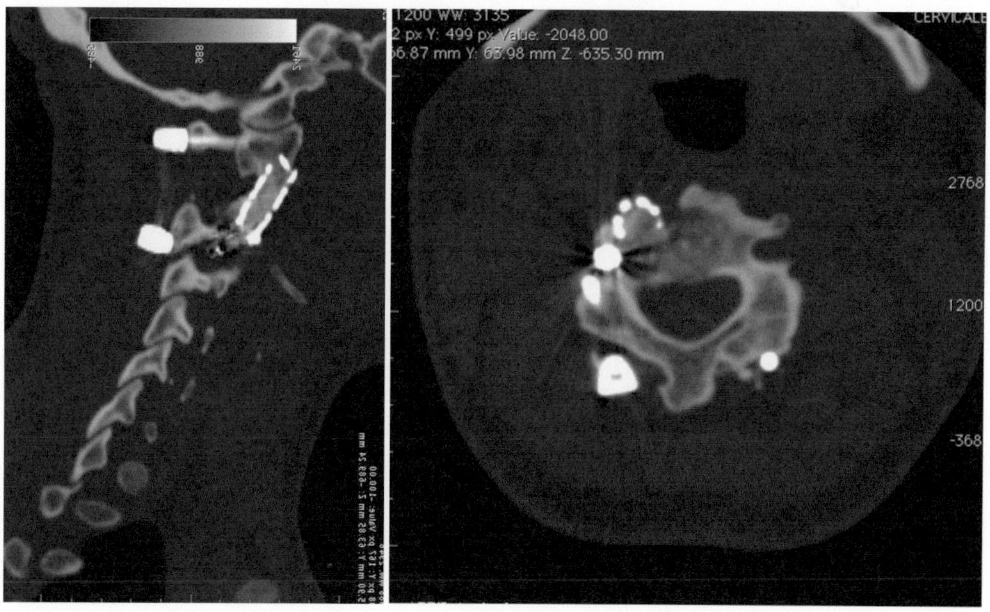

Fig. 12.8 3-years follow-up CT-scan

12.2 Subaxial Cervical Spine

12.2.1 Anterior Procedures

12.2.1.1 Anterior Cervical Discectomy and Fusion (ACDF)

Forty-three-year-old female, complaining of neck pain radiating to the right arm in territory of C7, resistant to a six-week course of conservative treatments. She referred paresthesias and numbness, and physical examination revealed a mild (4/5) motor deficit in wrist flexors and finger extensors and hyporeflexia in triceps reflex.

Sagittal and axial MRI shows herniated C6–C7 disk, obliteration of the anterior liquoral column, and cord compression.

Anterior left prevascular approach was performed. After intraoperative fluoroscopic control, C6–C7 discectomy was completed in order to achieve decompression of the canal and the root. Distraction is applied to adjacent levels during end plate preparation. Wire locking PEEK and tantalum interbody cage was placed stand-alone, and mild compression is applied before final locking (3-month follow-up plain radiographs and CT scan) (Figs. 12.9, 12.10, and 12.11).

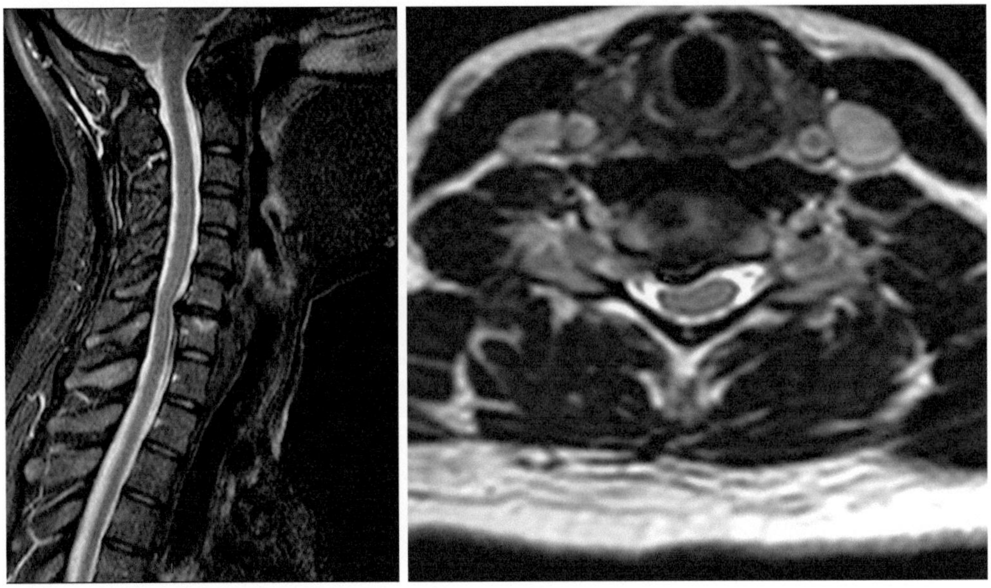

Fig. 12.9 Pre-operative MRI

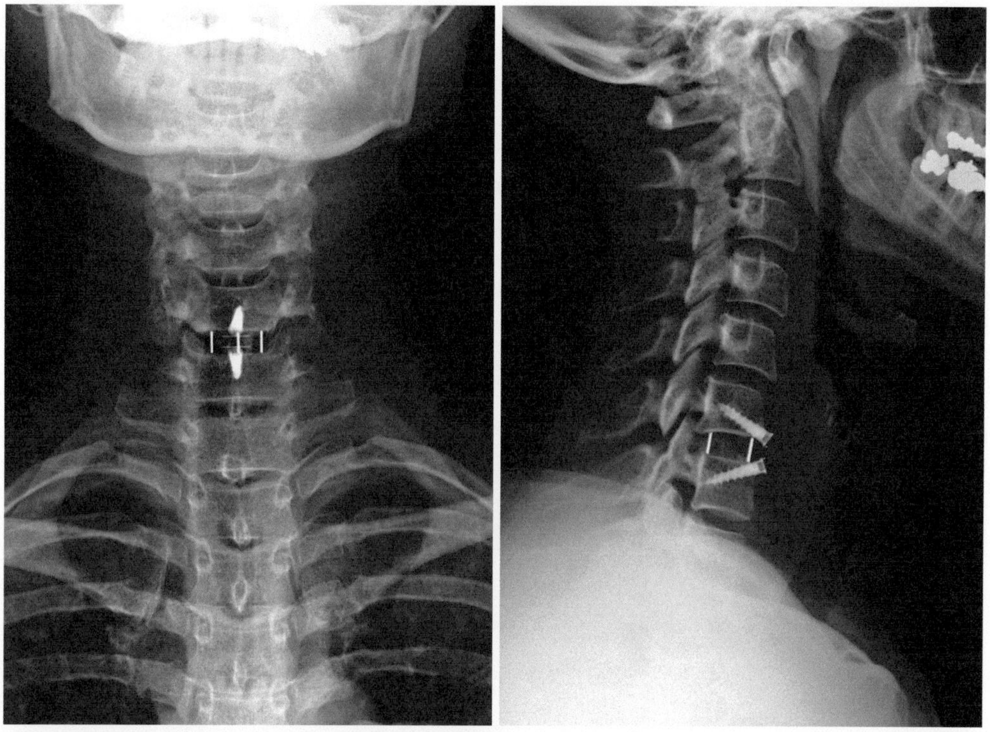

Fig. 12.10 Post-operative standing radiographs

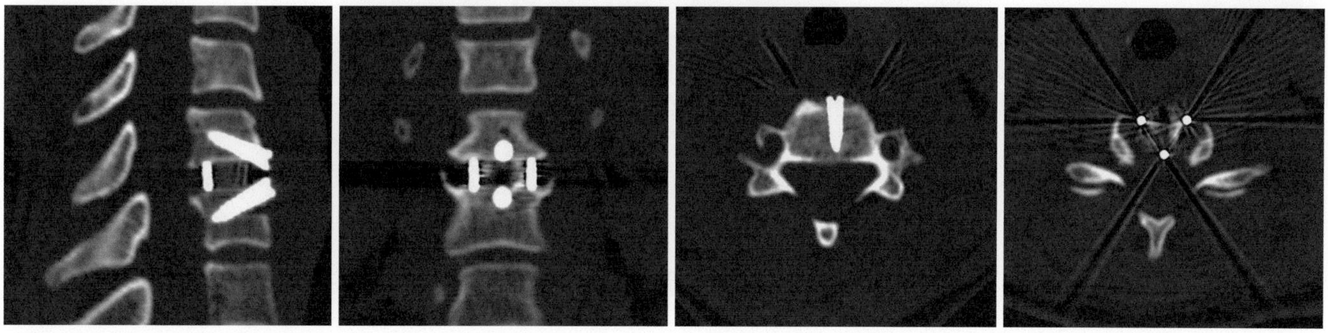

Fig. 12.11 Post-operative CT-scan

12.2.1.2 Cervical Disk Prosthesis

Forty-eight-year-old female, complaining of neck pain radiating to the right arm in territory of C6, resistant to six-week course of conservative treatments. She referred paresthesias and numbness, and physical examination revealed a mild (4/5) motor deficit in wrist extensors and hyporeflexia in brachioradialis reflex.

Sagittal and axial MRI shows herniated C6–C7 disk, obliteration of the anterior liquoral column, and cord compression.

Anterior left prevascular approach was performed. After intraoperative fluoroscopic control, C5–C6 discectomy was completed in order to achieve decompression of the canal and the root. Distraction is applied to adjacent levels during end plate preparation. C5–C6 Bryan Cervical Disc System was placed to restore disk height and allow motion at the treated level (Figs. 12.12, 12.13, 12.14, 12.15, and 12.16).

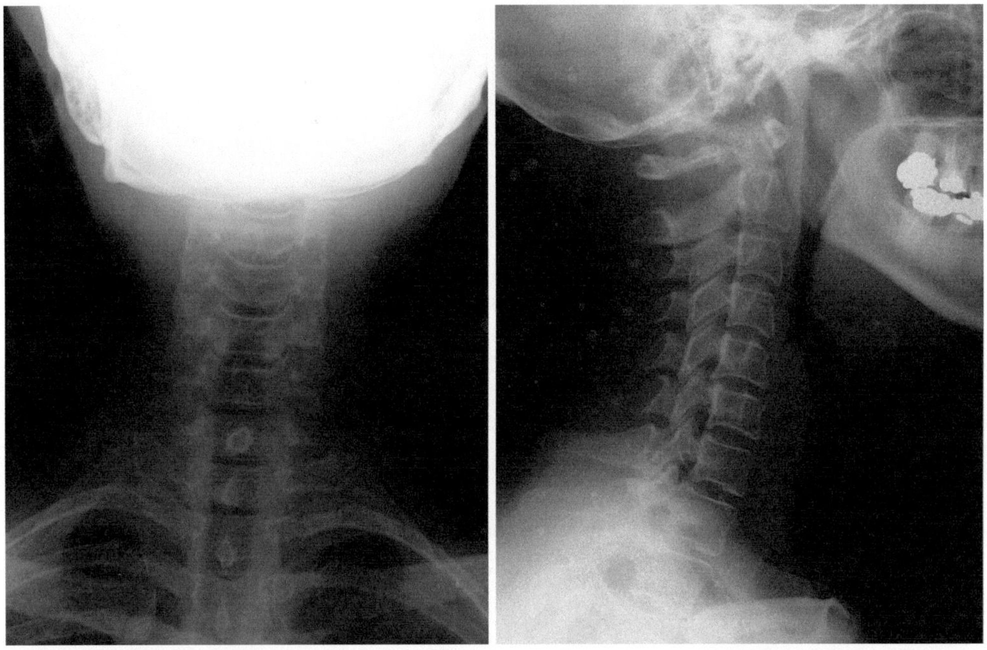

Fig. 12.12 Pre-operative standing radiographs

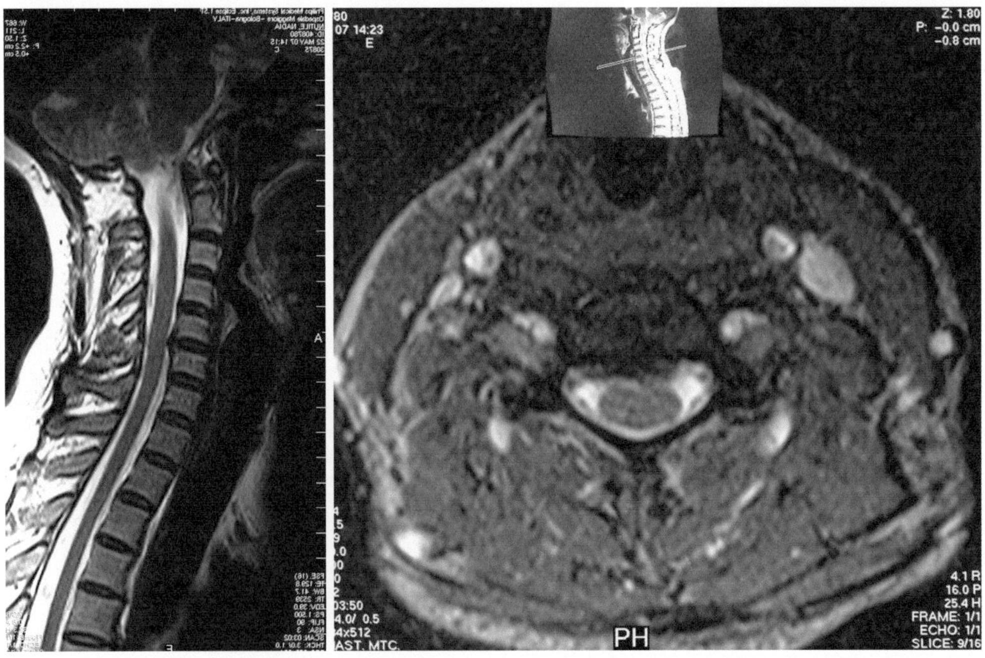

Fig. 12.13 Pre-operative MRI

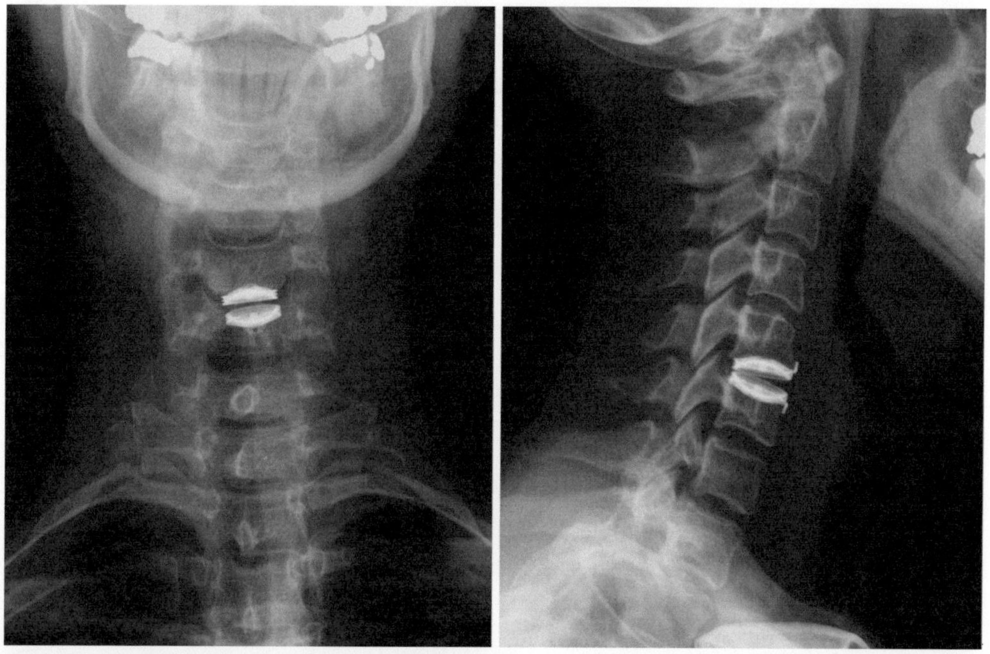

Fig. 12.14 Post-operative standing radiographs

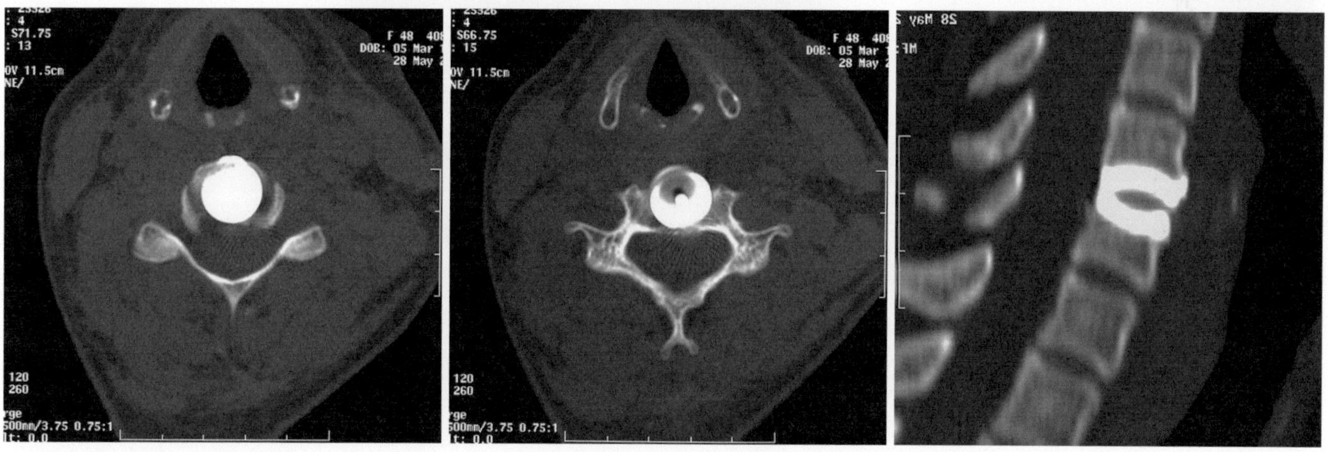

Fig. 12.15 Post-operative CT-scan

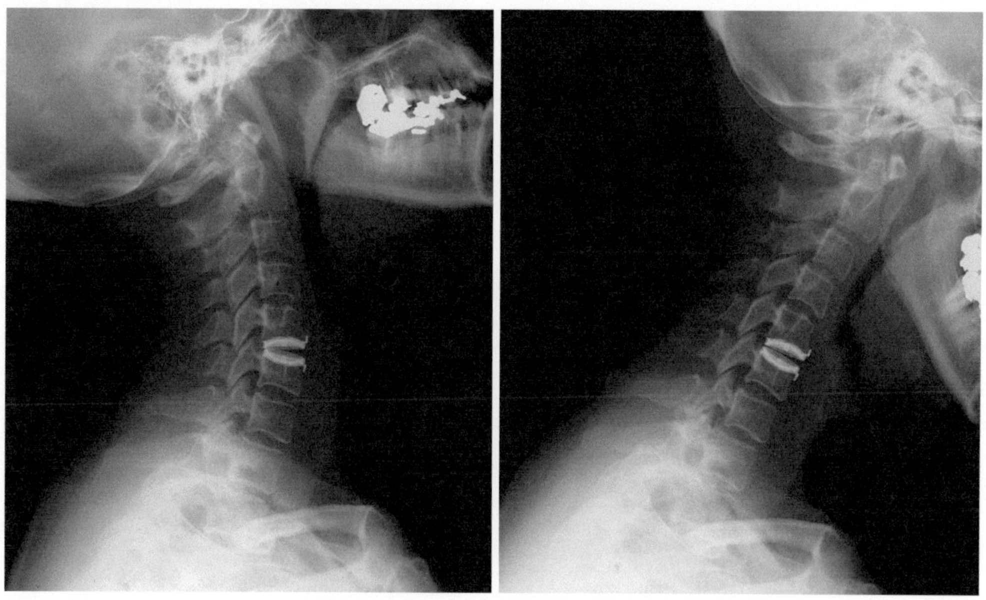

Fig. 12.16 Post-operative flexion-extension radiographs

12.2.1.3 ACDF for Revision of Cervical Disk Prosthesis Failure

Fifty-six-year-old female, presenting with recurrency of progressive neck pain 10 months after C6–C7 disk arthroplasty. AP and LL radiographs show posterior dislocation of the implant; thus, dynamic exam did not reveal segmental instability.

Intraoperative picture shows removal of the device, and carbon fiber-reinforced polymer (CFRP) cage was placed as interbody device to restore segmental stability and provide anterior column support. Additional plate was placed to restore the anterior tension band (Figs. 12.17, 12.18, 12.19, 12.20, 12.21, 12.22, and 12.23).

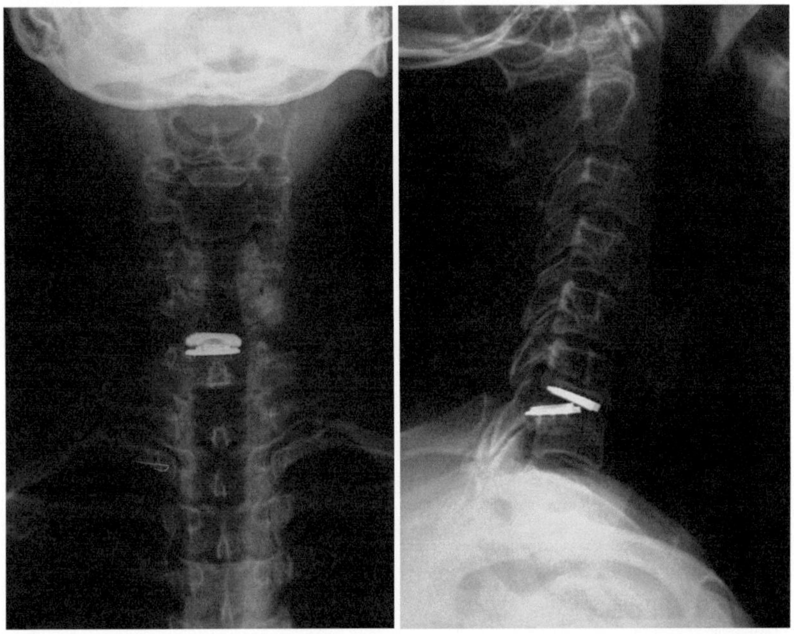

Fig. 12.17 Pre-operative standing radiographs

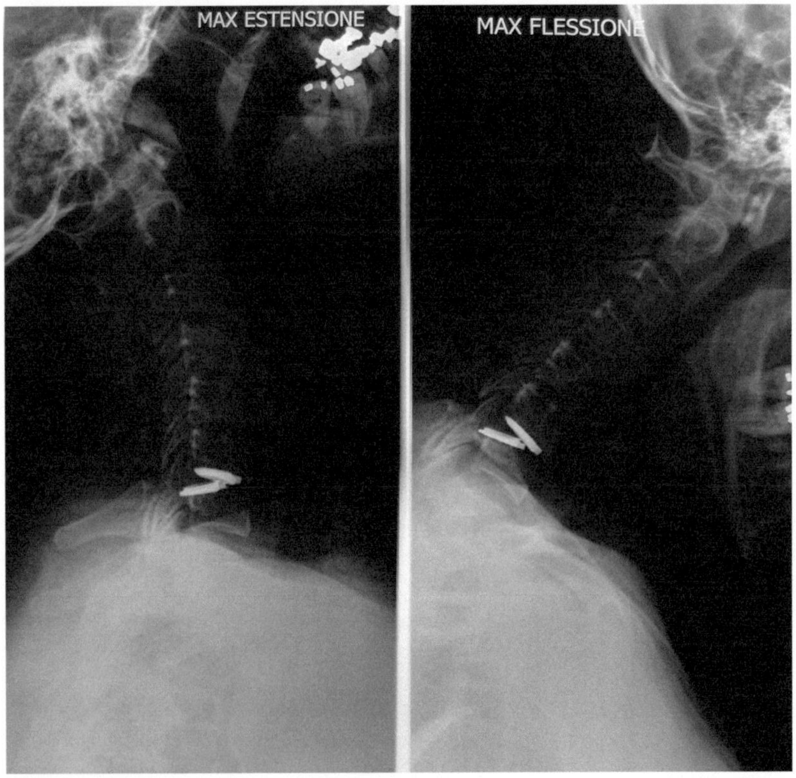

Fig. 12.18 Pre-operative flexion-extension radiographs

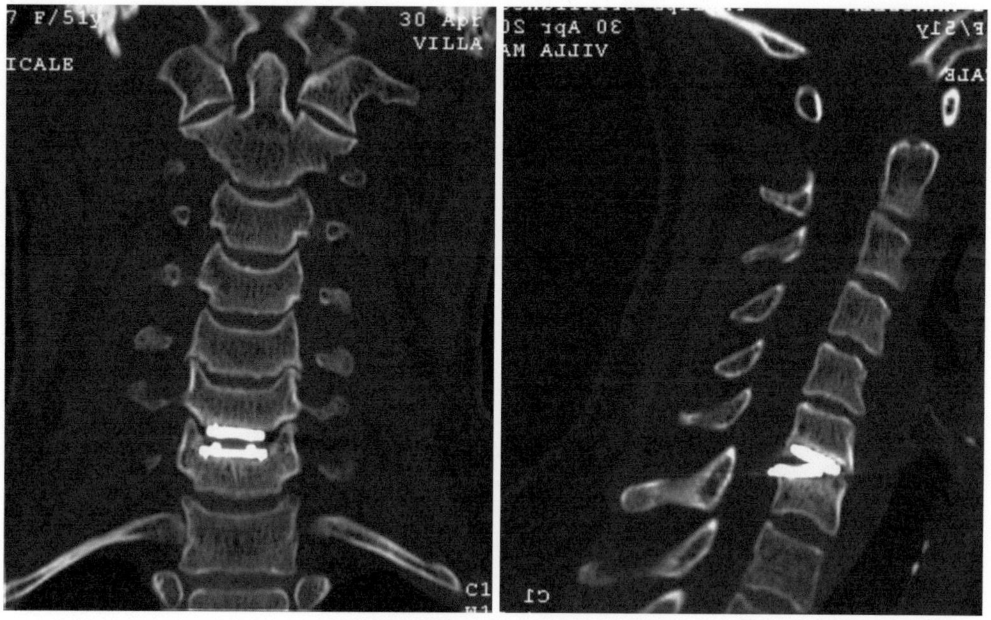

Fig. 12.19 Pre-operative CT-scans

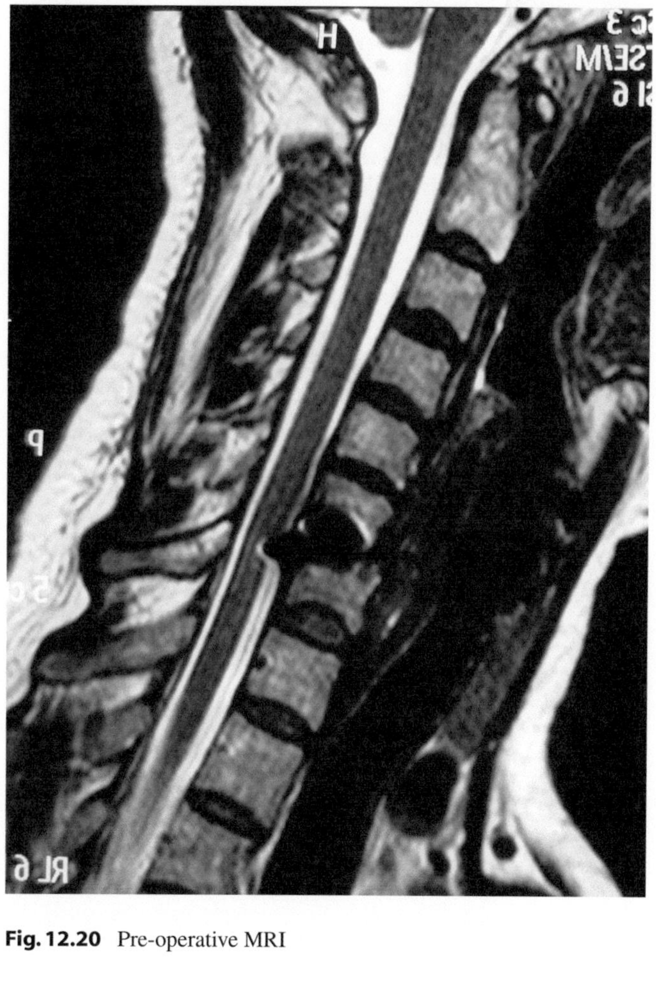

Fig. 12.20 Pre-operative MRI

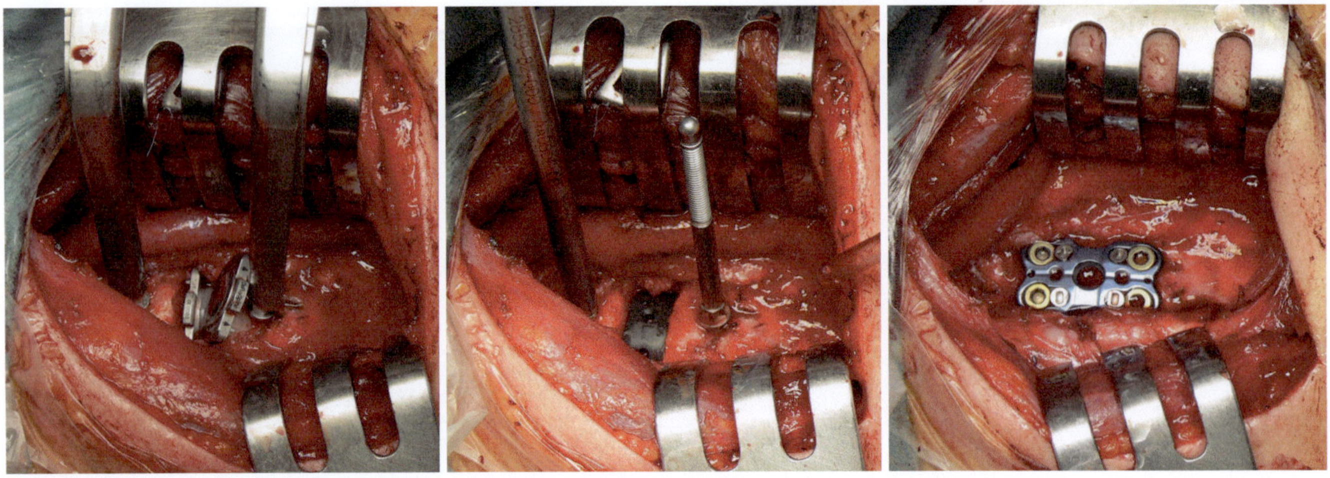

Fig. 12.21 Intra-operative images

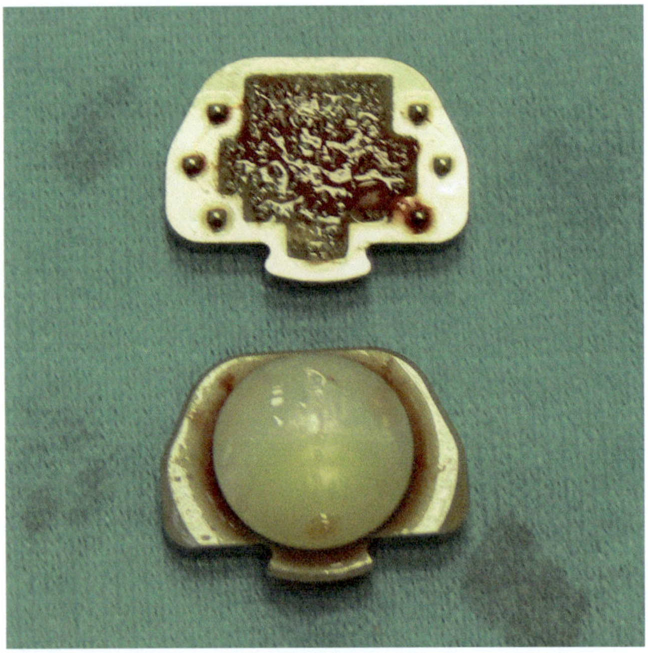

Fig. 12.22 Post-operative radiologic control

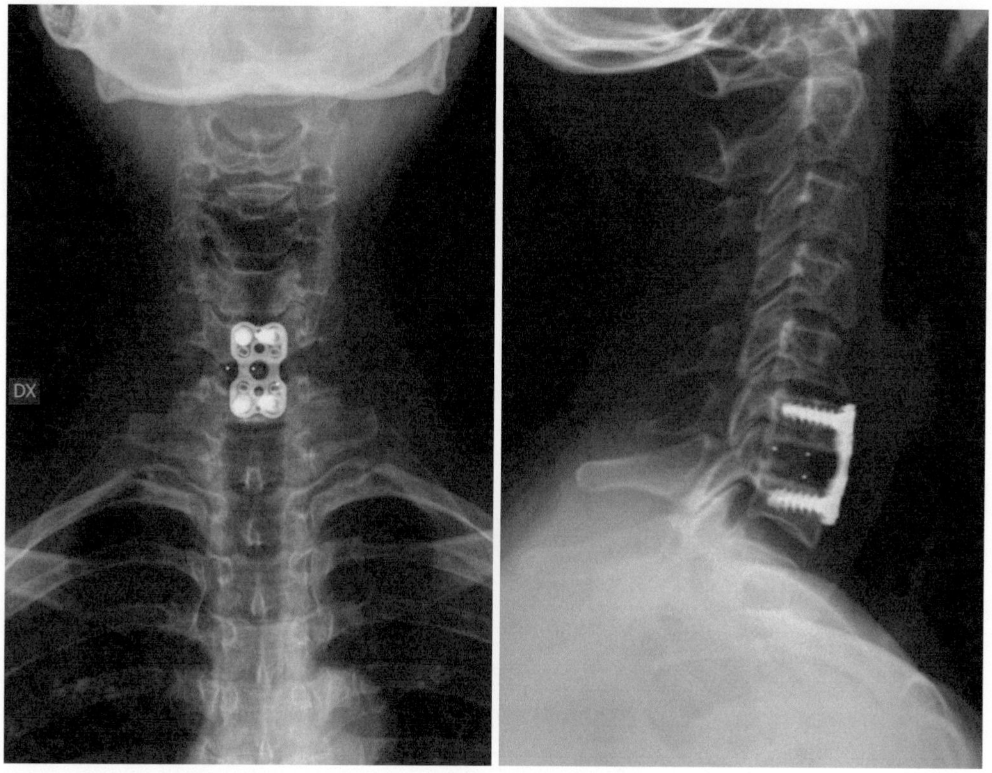

Fig. 12.23 Post-operative standing radiographs

12.2.1.4 Anterior Cervical Corpectomy and Fusion (ACCF)

Fifty-three-year-old male, complaining of lower extremity numbness and gait disturbance. Patient revealed occasional neck pain with seldom irradiation to the arms with no particular dermatomal distribution. Physical examination revealed diffuse increase of muscle tone, hyperreflexia, positive Hoffmann sign, and clonus.

AP and LL radiographs show cervical spondylosis with loss of cervical lordosis.

Sagittal MRI shows C5–C6 and C6–C7 disk protrusion with cord hyperintensity and sign of myelopathy. Dynamic radiographs do not show segmental instability; thus, posterior tension band is preserved.

Intraoperative pictures show anterior approach, C6 corpectomy, and reconstruction using a titanium cage (Pyramesh, Medtronic) filled with mixture of auto- and allograft (from the registered tissue bank of the hospital).

Anterior plate was used to restore the anterior tension band.

Postoperative plain radiographs and CT scan show correct placement of the hardware (Figs. 12.24, 12.25, 12.26, 12.27, and 12.28).

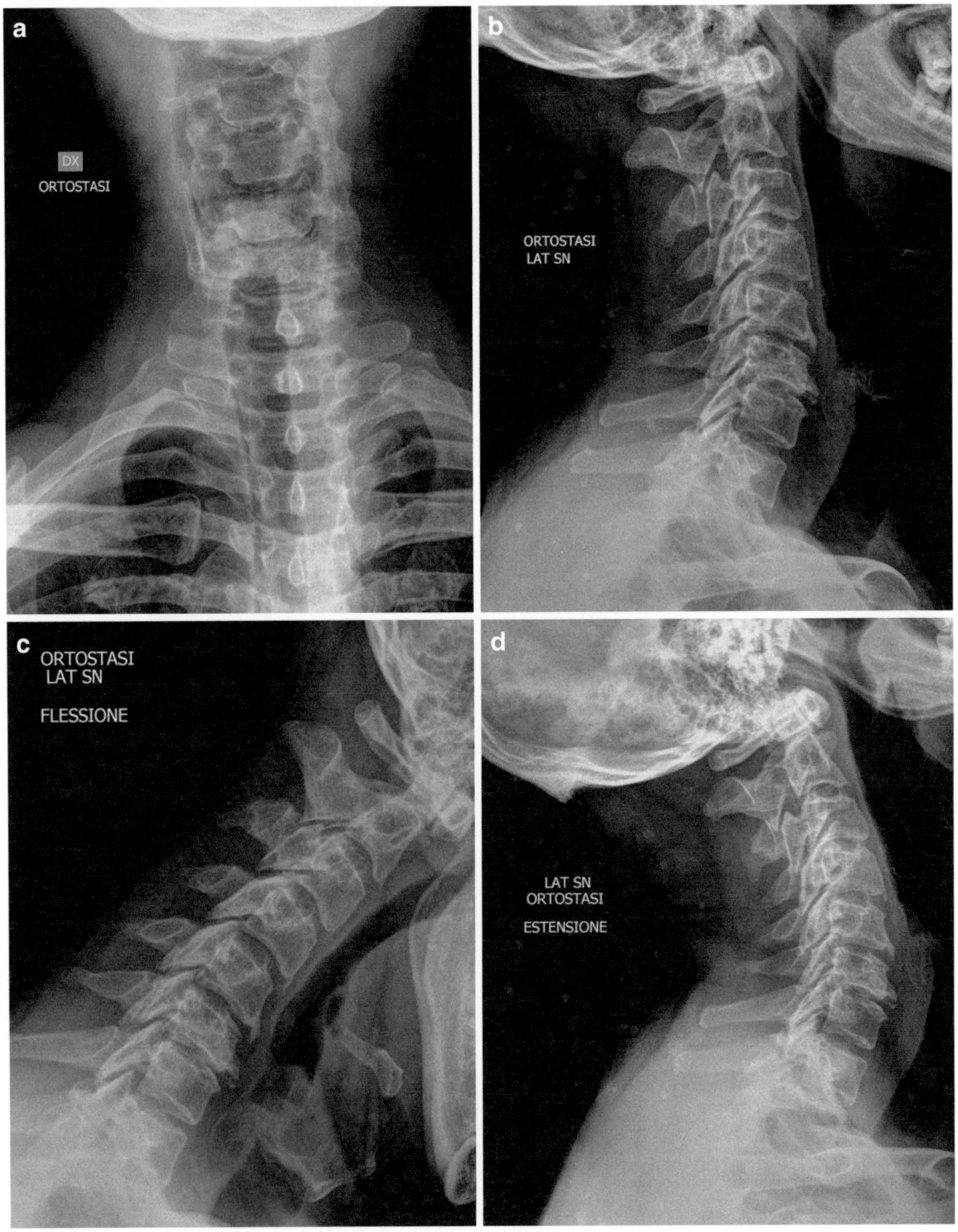

Fig. 12.24 Pre-operative standing radiographs (a, b) and flexion-extension radiographs (c, d)

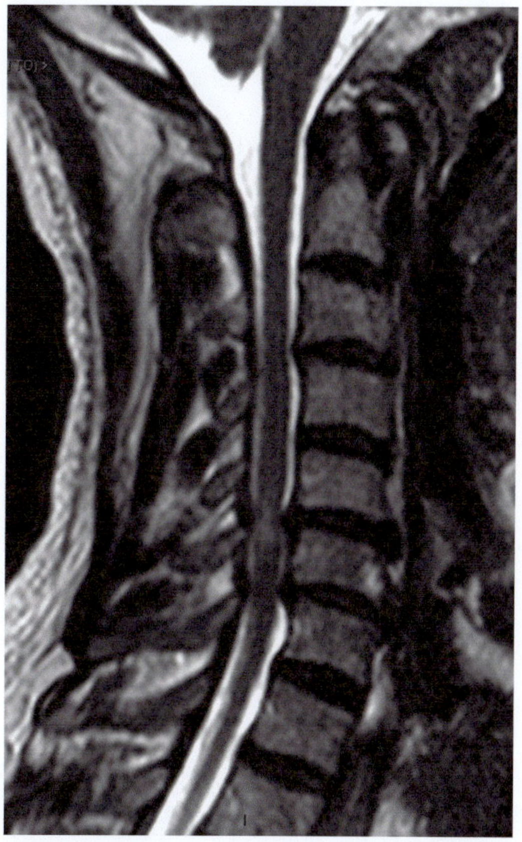

Fig. 12.25 Pre-operative MRI

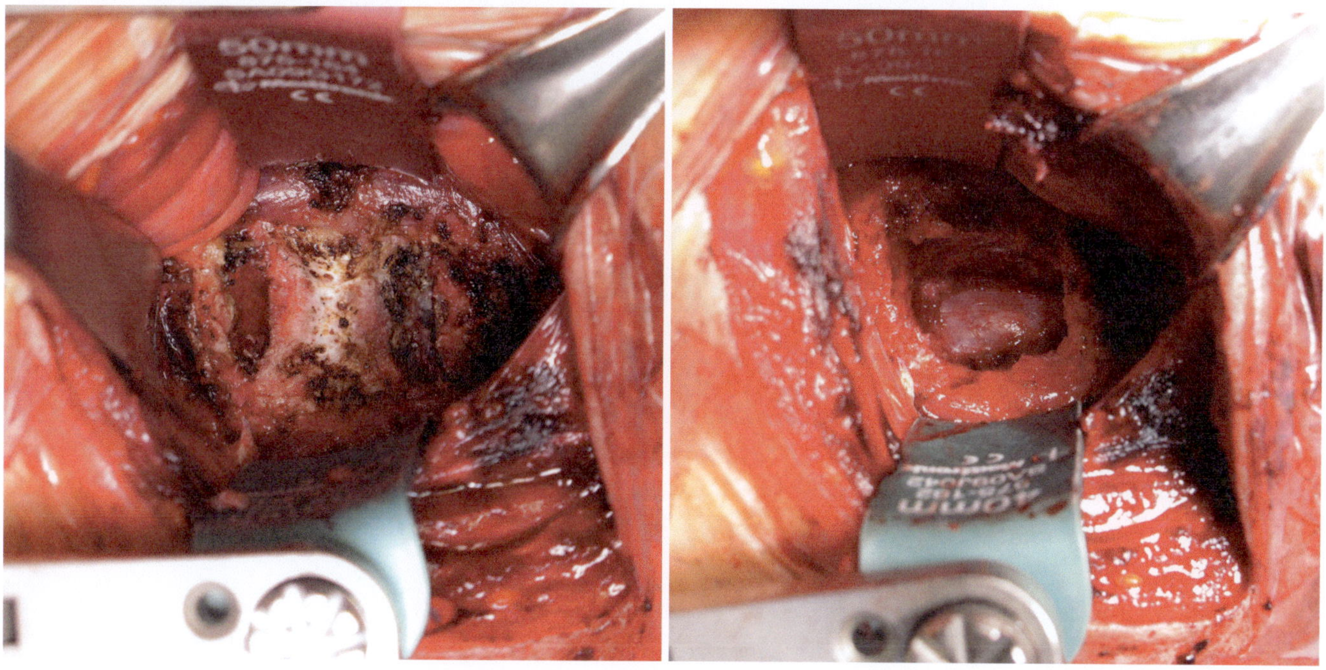

Fig. 12.26 Intra-operative images

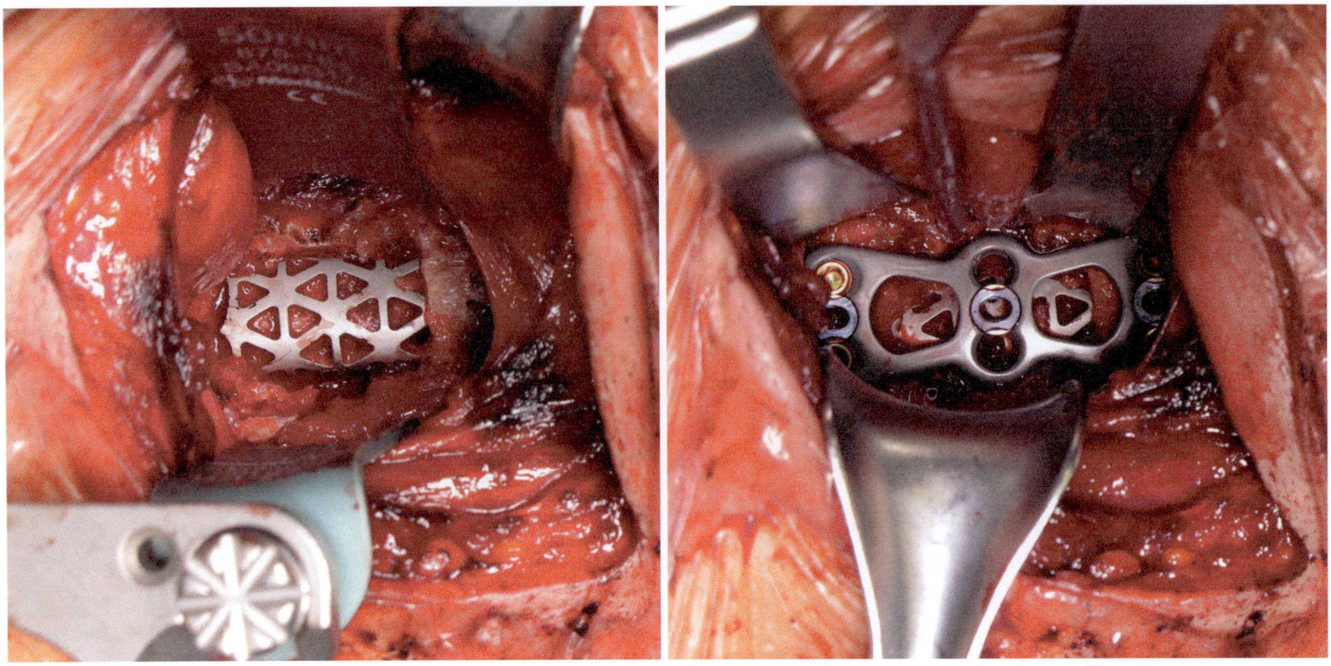

Fig. 12.26 (continued)

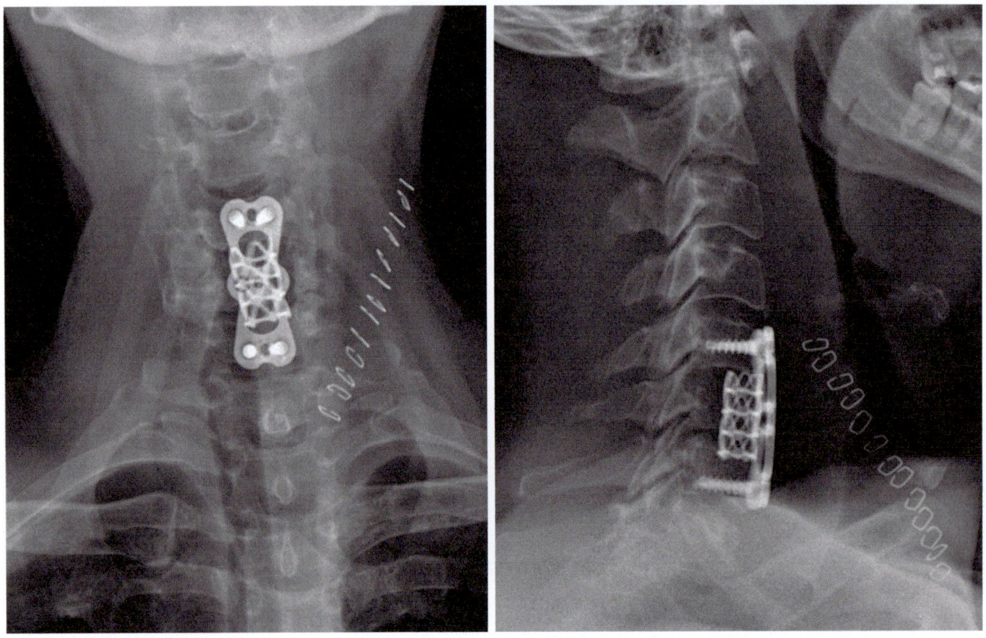

Fig. 12.27 Post-operative standing radiographs

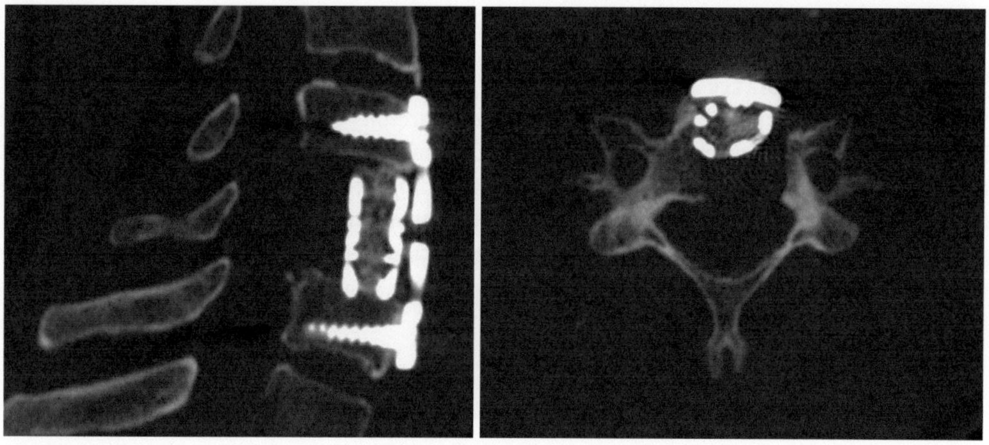

Fig. 12.28 Post-operative CT-scan

12.2.2 Posterior Procedures

12.2.2.1 Laminoplasty

Sixty-three-year-old male, complaining of lower extremity numbness and gait disturbance. Patient revealed occasional neck pain with seldom irradiation to the arms with no particular dermatomal distribution. Physical examination revealed diffuse increase of muscle tone, hyperreflexia, positive Hoffmann sign, and clonus.

AP and LL radiographs show cervical spondylosis, and sagittal MRI shows multilevel circumferential compression and narrowing of the canal.

Through a posterior approach, laminoplasty with Tomita technique was performed, with allograft to stabilize posterior arch widening at C3–C4–C5–C6 (Figs. 12.29, 12.30, 12.31, 12.32, 12.33, and 12.34).

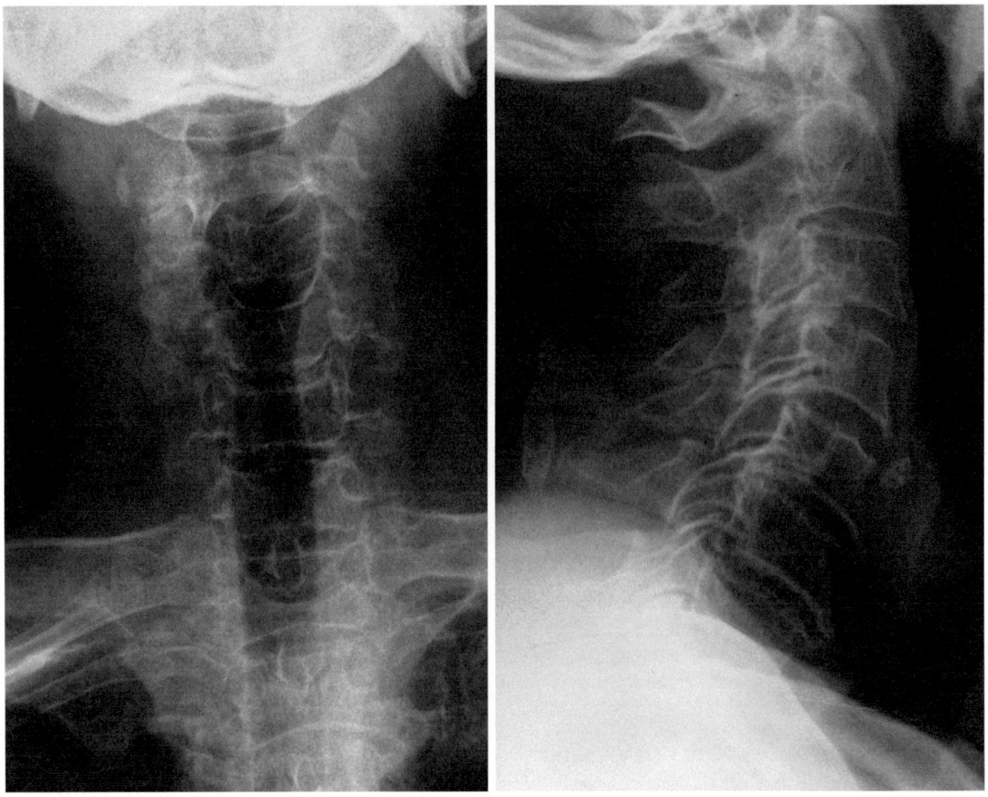

Fig. 12.29 Pre-operative standing radiographs

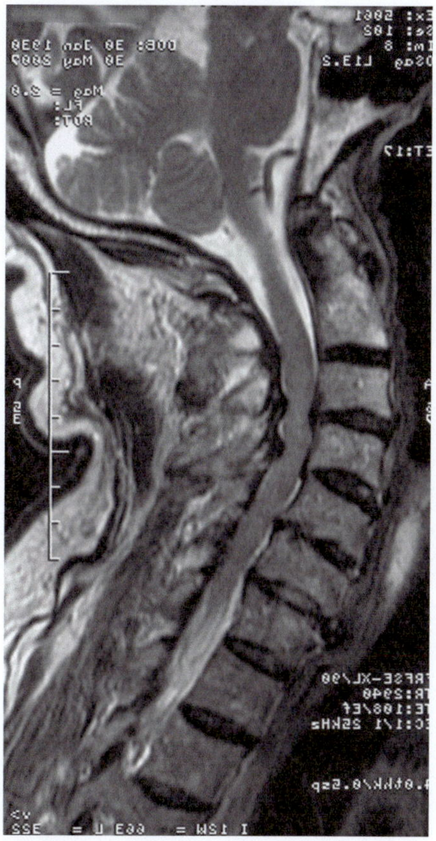

Fig. 12.30 Pre-operative MRI

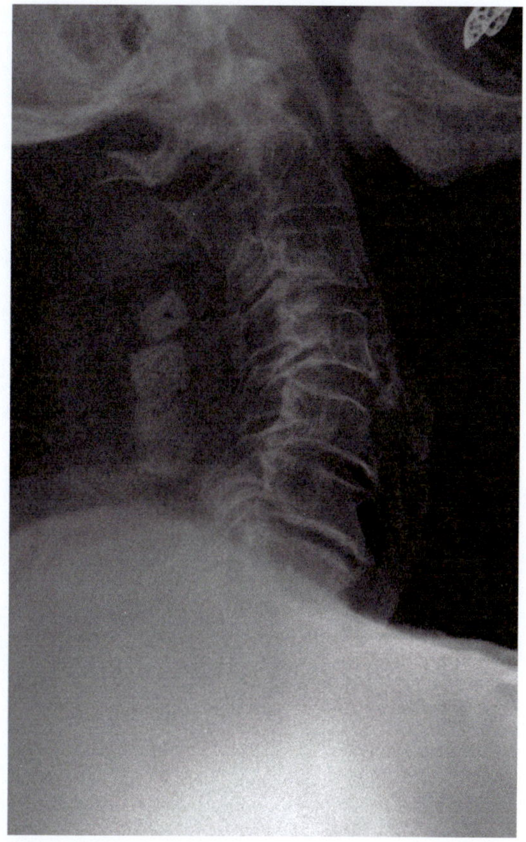

Fig. 12.32 Post-operative standing radiographs

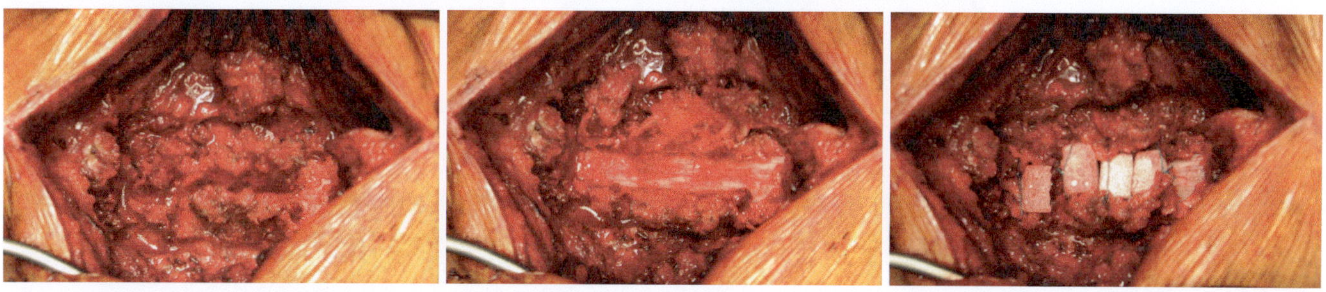

Fig. 12.31 Intra-operative images

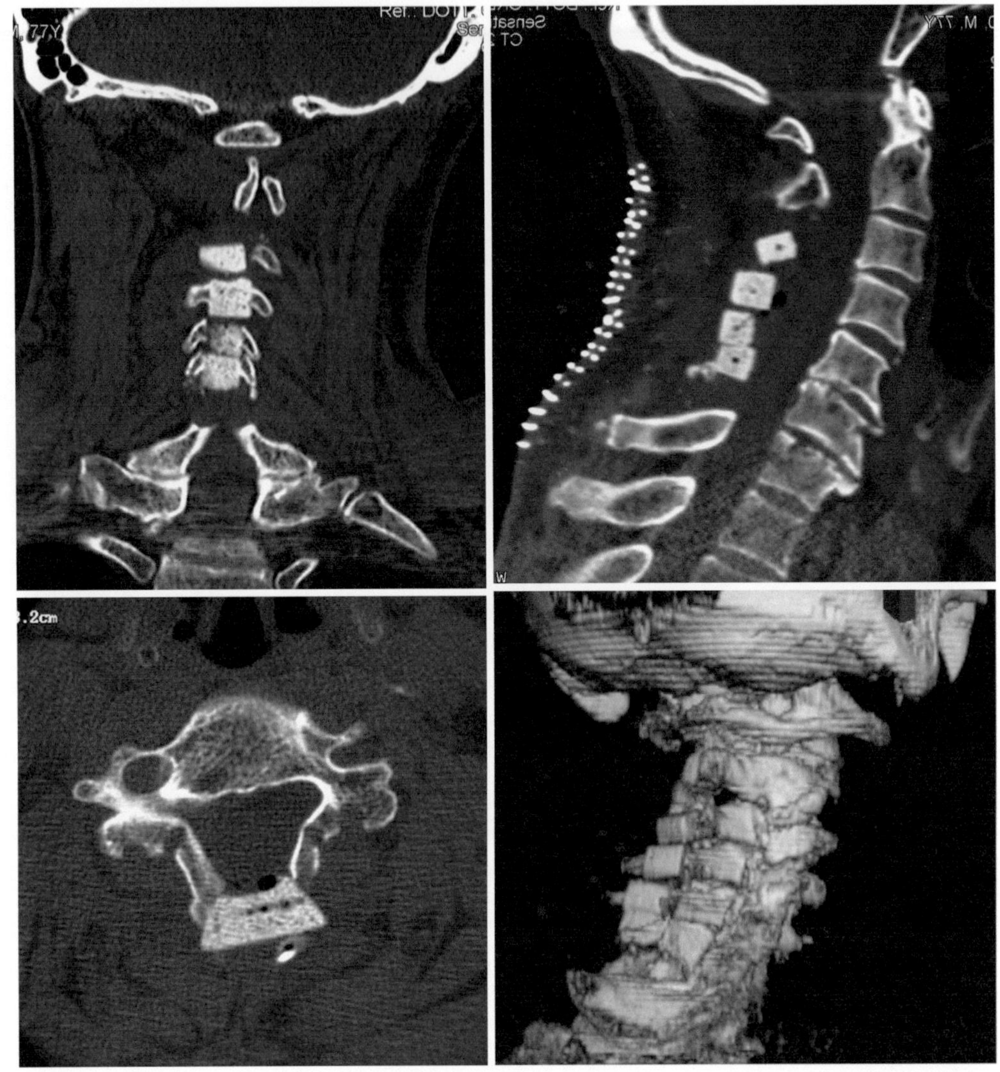

Fig. 12.33 Post-operative CT-scans

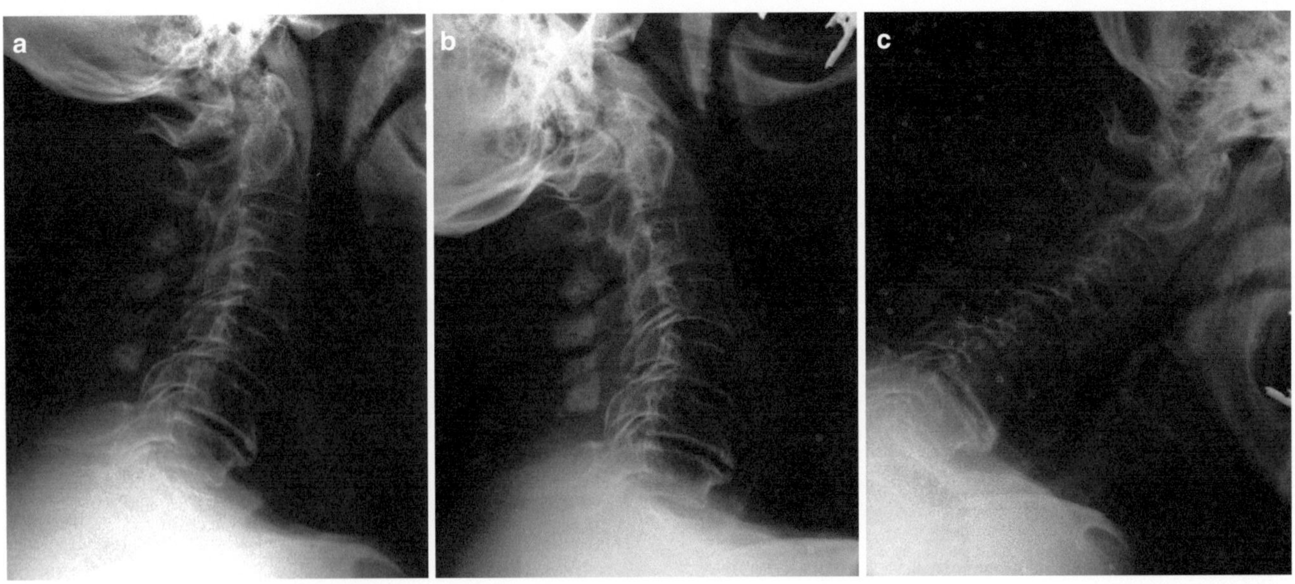

Fig. 12.34 Post-operative flexion-extension (A, C) and standing (B) radiographs

12.2.2.2 Posterior Instrumented Fusion

Thirty-two-year-old female complaining of progressive neck pain since several months and recent onset of a hard swelling in the right paravertebral muscle compartment. Neurologic examination did not reveal any pathological findings. Previous attempts to control symptoms with physical therapy resulted ineffective.

Plain radiographs show osteolytic areas in the posterior elements and a shadow in the posterior soft tissues. MRI shows lesion arising in the right lateral mass of C7 extending into the paravertebral muscle compartment. CT scan shows intralesional "popcorn" ossifications, typical of cartilage tumors.

Biopsy was performed and histopathological examination on the specimen was diagnostic for chondrosarcoma grade 1.

Through a posterior-only approach, en bloc resection of posterior elements from C4 to C7 was performed with wide margins.

Reconstruction of spinal stability was achieved via posterior screw-rods instrumentation and massive cortical bone allograft for spinal canal reconstruction (roof technique). Pedicle screws were placed with both transpedicular route and into the lateral masses.

Cortical bone allograft was contoured in an H shape and placed bridging the posterior arch of C3 to that of T1 and secured to the posterior instrumentation. This technique promotes posterior arthrodesis due to the osteoconductive and osteoinductive properties of the graft. Moreover it provides a posterior covering of the spinal canal, protecting the neurostructures and providing immediate stability to the posterior column of the spine (Figs. 12.35, 12.36, 12.37, and 12.38).

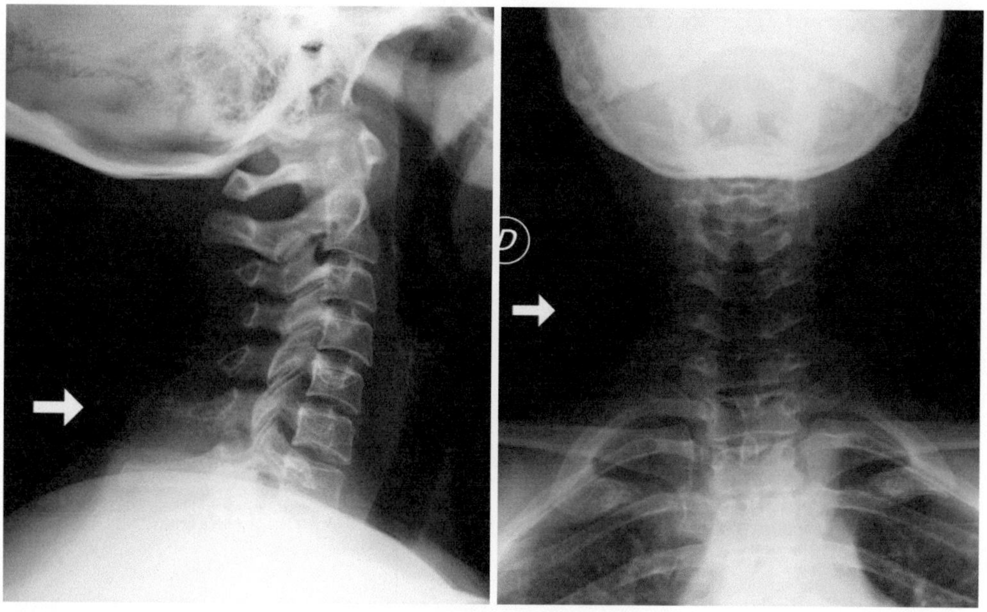

Fig. 12.35 Pre-operative standing radiographs. *Arrow* indicates the shadow of the lesion extending into the paravertebral muscular compartment

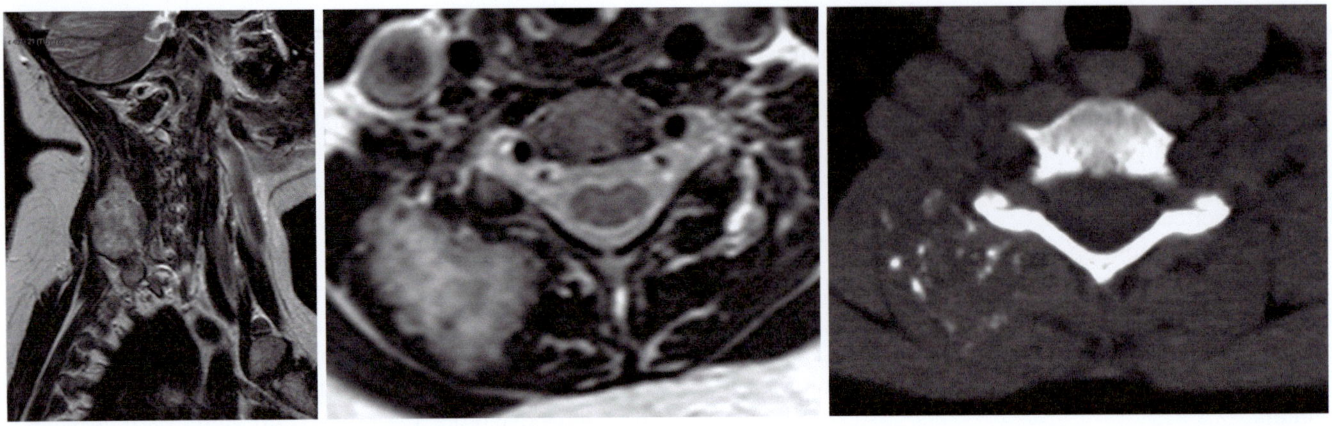

Fig. 12.36 Pre-operative MRI and trasverse CT-scan

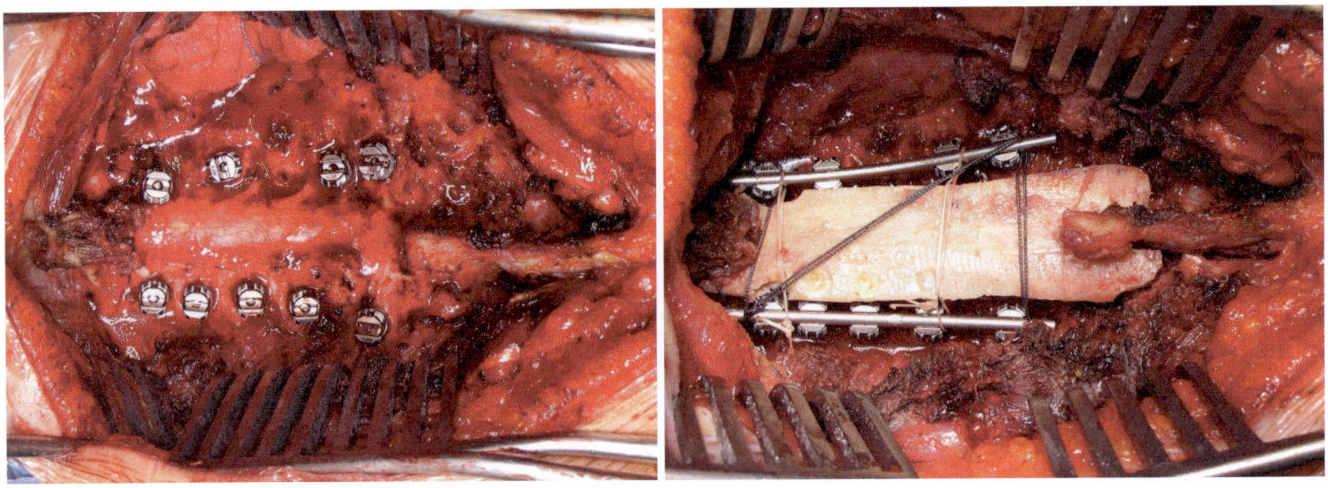

Fig. 12.37 Intra-operative images. Posterior reconstruction with allograft (Roof technique)

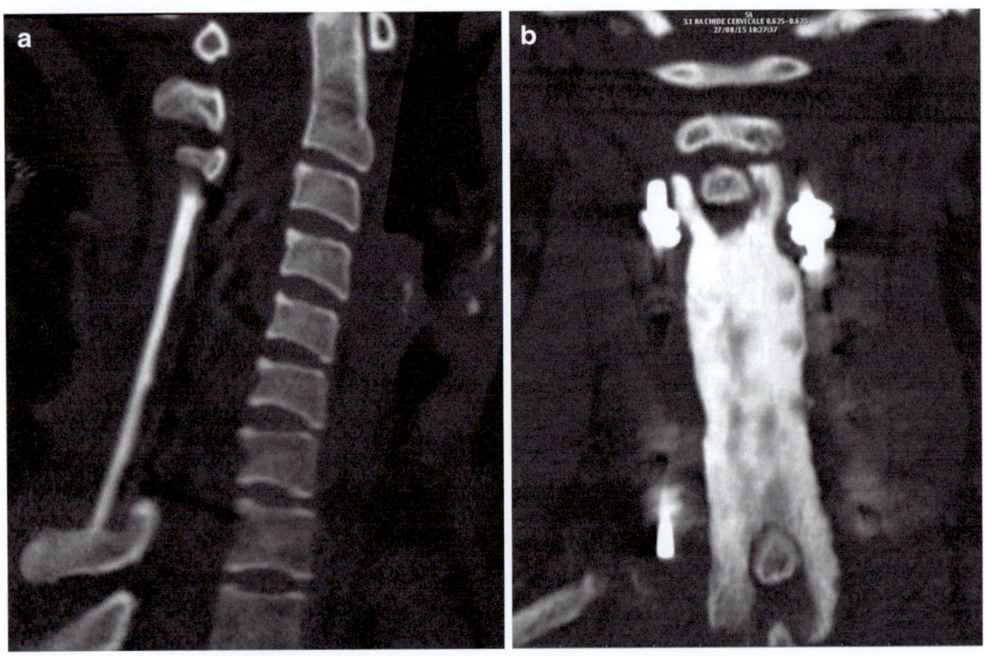

Fig. 12.38 Post-operative CT-scan

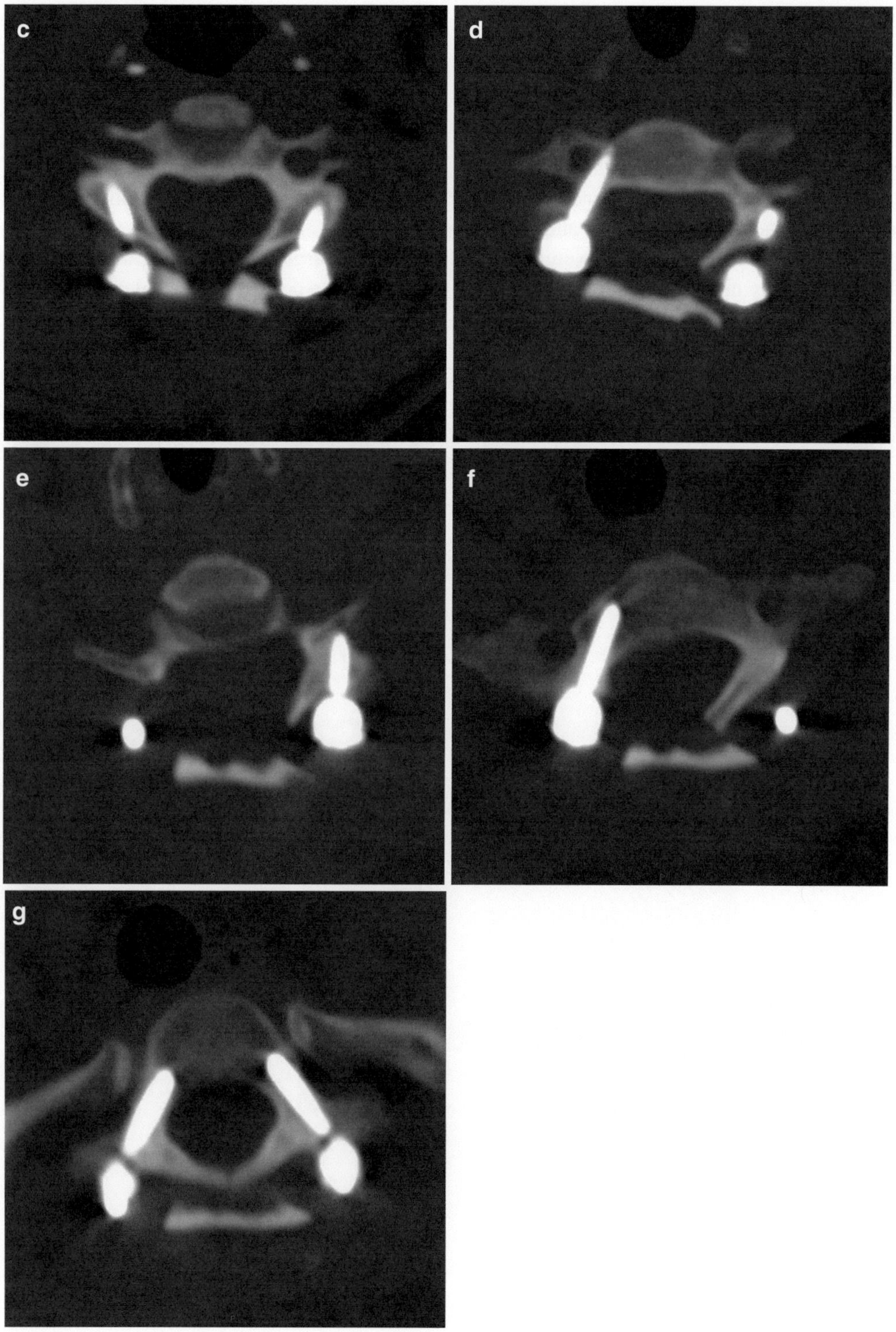

Fig. 12.38 (continued)

12.2.3 A + P Procedures

12.2.3.1 Single-Level

Fifty-six-year-old female complaining of sudden onset of neck pain.

She had been treated for colon cancer 4 years before and was in an oncologic follow-up program ever since. Physical examination revealed incapacitating neck pain and painful contracture of the paravertebral muscles. Neurological examination did not reveal any neurological deficit.

Plain radiographs show C6 vertebral body fracture, and sagittal MRI shows signal alteration of the bone at that level but no signs of cord compression.

Further CT scan is mandatory in order to clarify the pattern of the fracture.

Biopsy was performed and histopathological examination on the specimen was diagnostic for colon cancer metastasis.

Bone scintigraphy does not show any other location of the disease. Thus, being it a solitary bone lesion, excisional surgery is indicated.

Through an anterior prevascular approach, the vertebral body is exposed, and extracapsular excision of the tumor was performed.

PMMA-armed bone allograft was used to fill the bone defect and reconstruct the anterior column. Anterior plate was used to restore the anterior tension band.

Laminectomy with posterior instrumentation was used to circumferentially decompress the cord and restore spinal stability.

Postoperative CT scan shows correct placement of the hardware (Figs. 12.39, 12.40, 12.41, 12.42, 12.43, 12.44, 12.45, and 12.46).

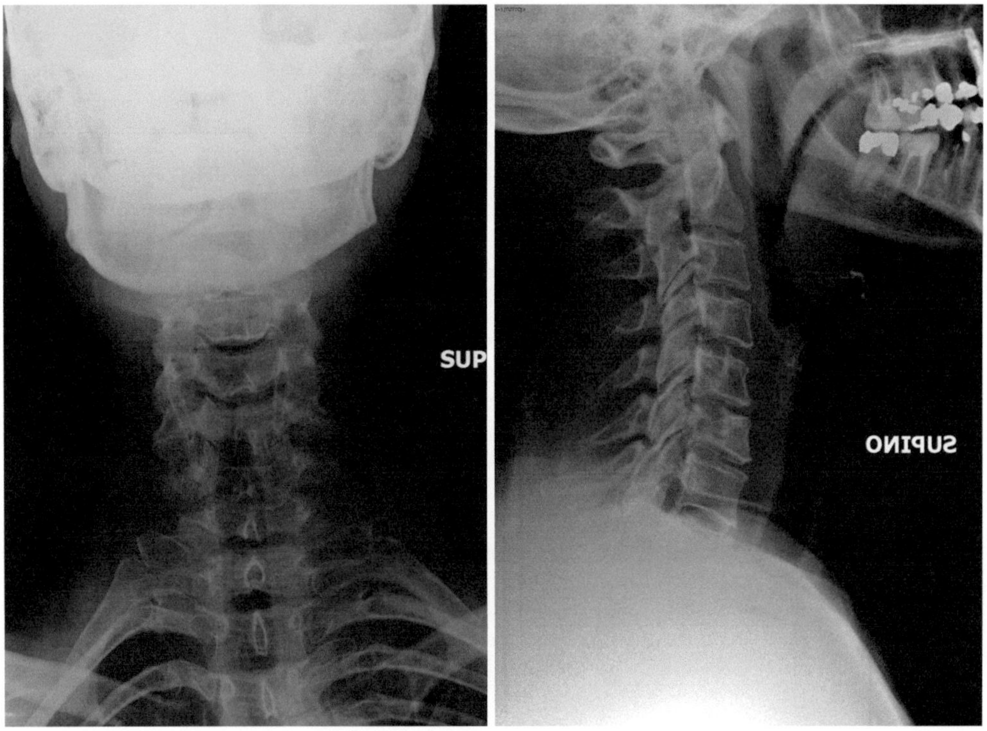

Fig. 12.39 Pre-operative plain radiographs

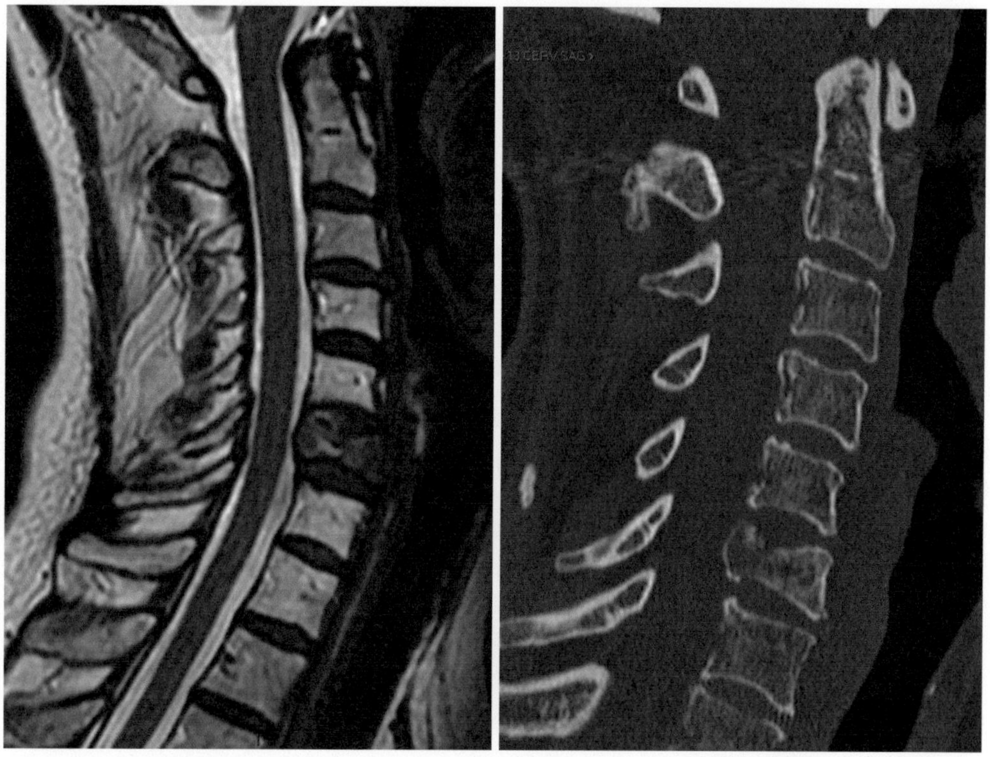

Fig. 12.40 Pre-operative MRI and CT scan

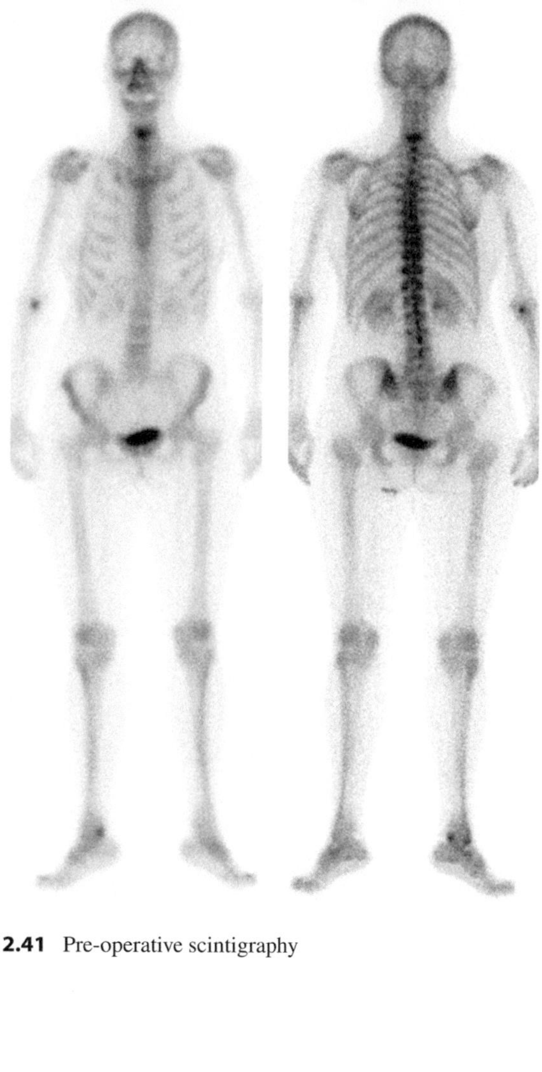

Fig. 12.41 Pre-operative scintigraphy

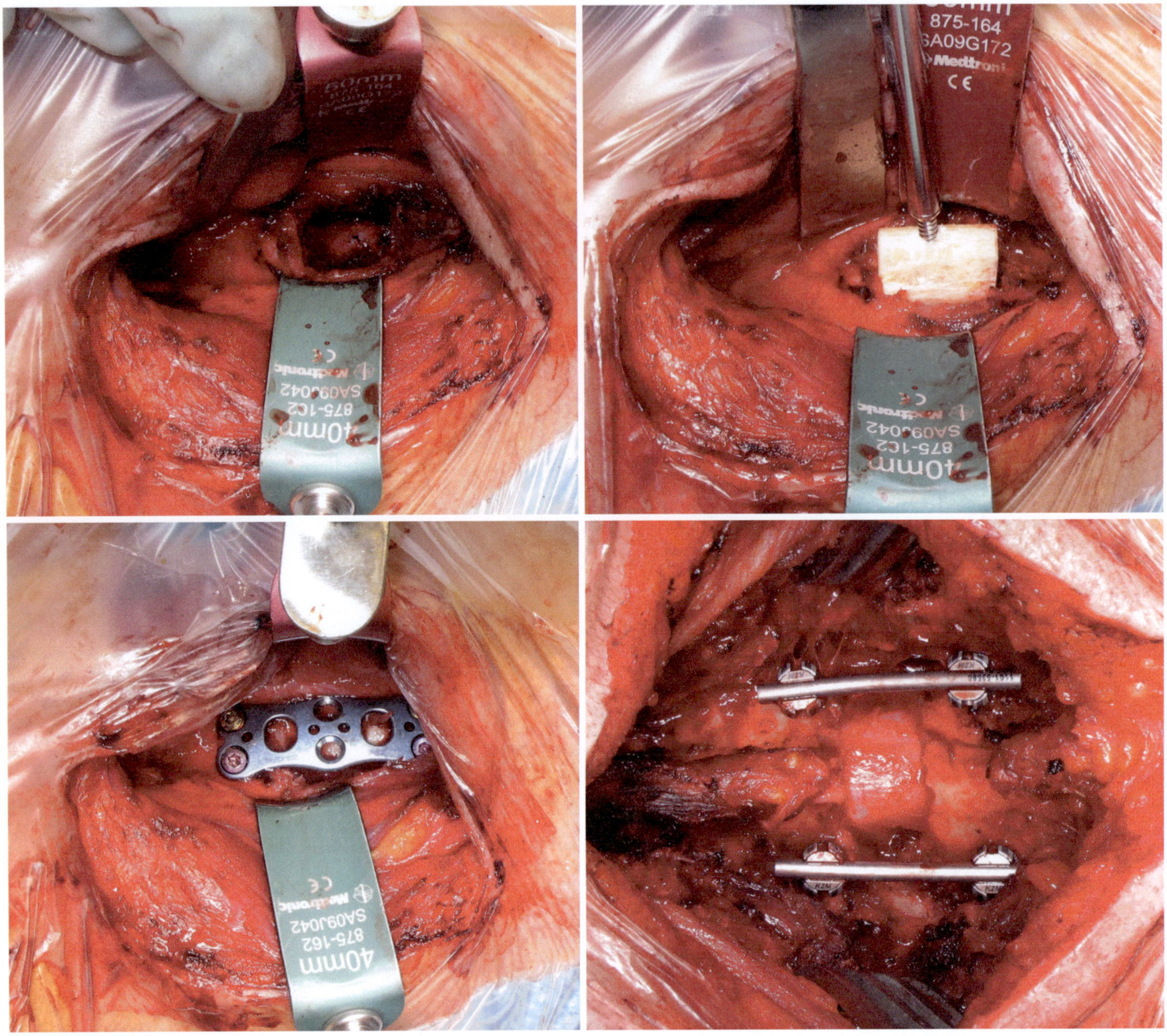

Fig. 12.42 Intra-operative images

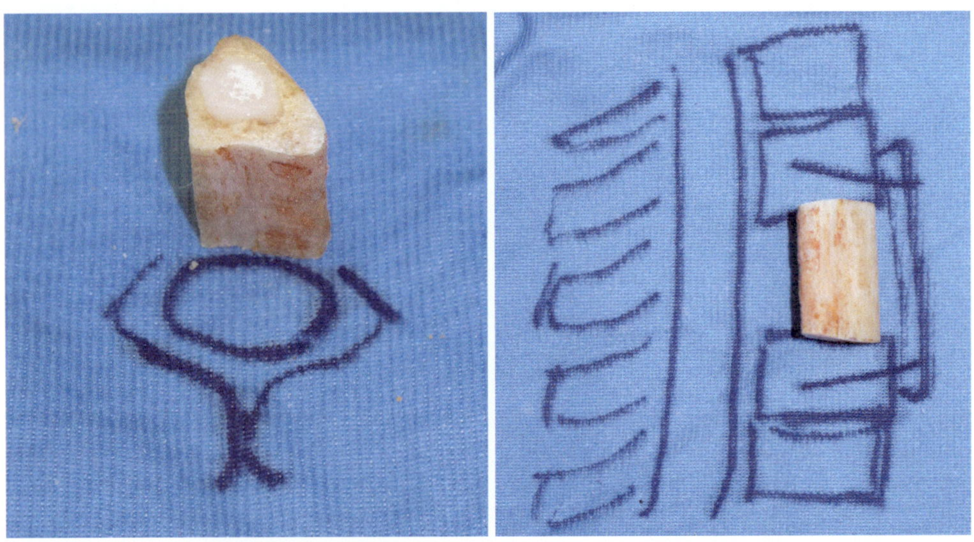

Fig. 12.43 Reconstruction of vertebral body using PMMA-armed allograft

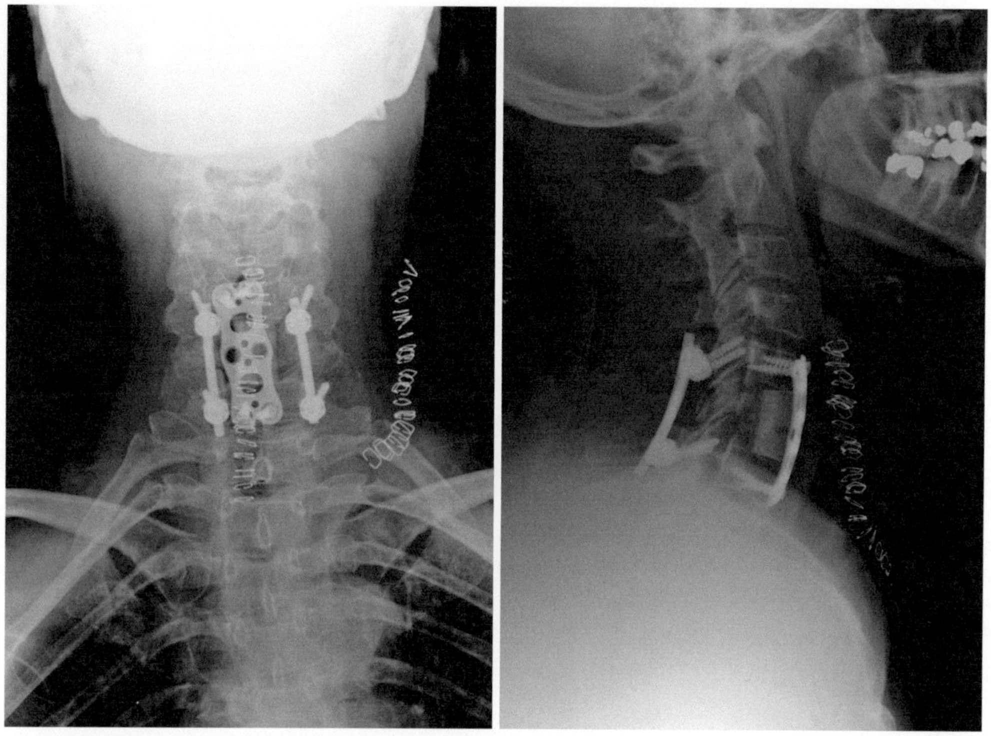

Fig. 12.44 Post-operative standing radiographs

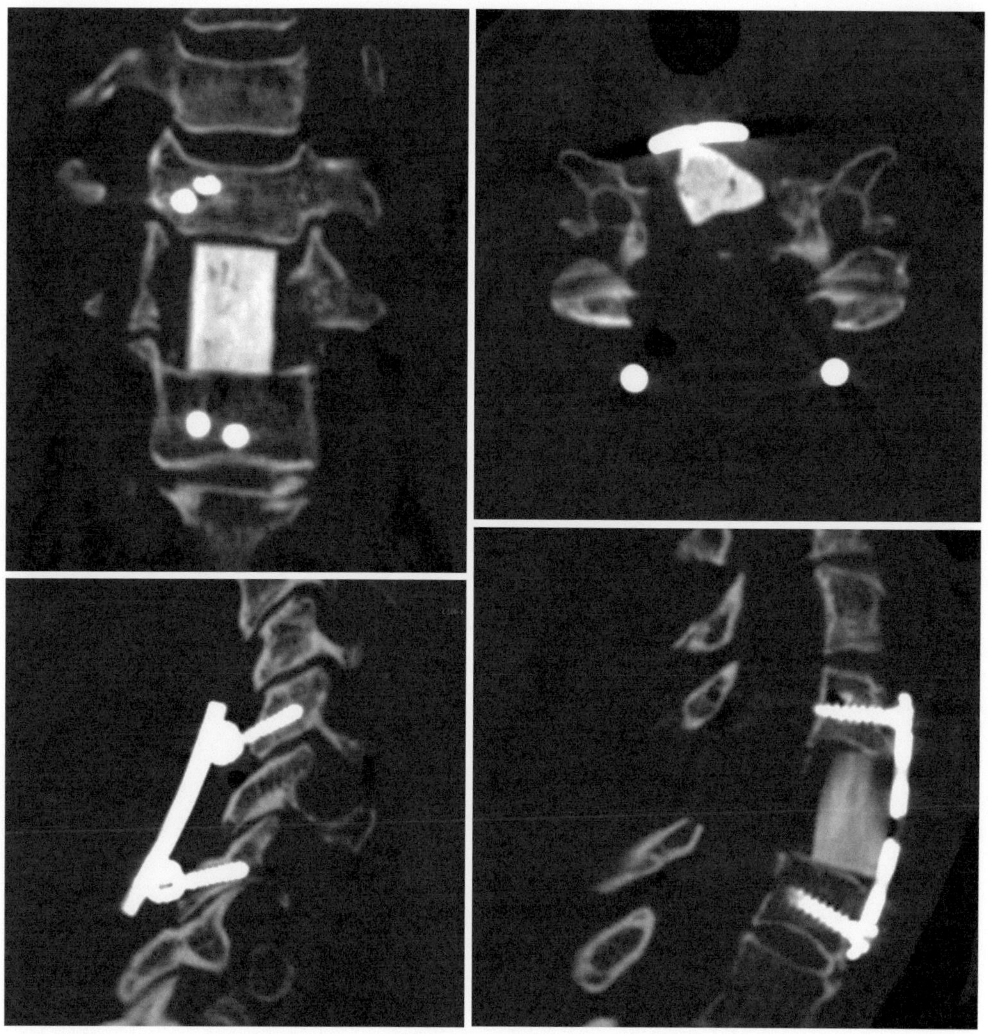

Fig. 12.45 Post-operative CT-scan

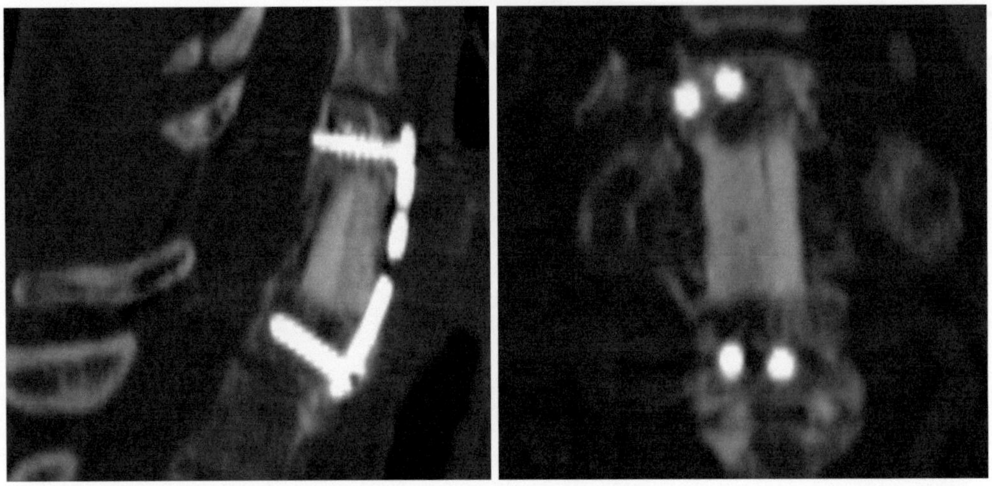

Fig. 12.46 2-years follow-up CT-scan

12.2.3.2 Multiple Levels

Thirty-nine-year-old male complaining of progressive onset of incapacitating neck pain since several months. Previous attempts to control symptoms with physical therapy resulted ineffective. Past medical history was negative for oncologic disease. Physical examination revealed painful contracture of the paravertebral muscles, but neurological examination did not reveal any neurological deficit.

MRI shows a lesion replacing the left lateral mass of C4, protruding in the spinal canal and extending into the prevertebral layer. CT scan provides further information regarding the lesion and host bone reaction to it.

Biopsy was performed and histopathological examination on the specimen was diagnostic for chondrosarcoma grade 2 with dedifferentiated areas of osteogenic and chondroblastic sarcoma grade 3.

Double approach is mandatory to resect the lesion, reconstruct the spinal column, and restore segmental stability.

Through a posterior approach, screw-rods instrumentation was placed, and debulking was performed after laminectomy.

Anterior prevascular approach allowed control of the vertebral artery, mobilization, and complete excision of the tumor.

Reconstruction of the anterior column was performed using tricortical autogenous iliac bone graft.

Anterior plate was used to restore the anterior tension band.

Postoperative plain radiographs and CT scan show balanced spine and correct placement of the hardware.

Two-year follow-up shows fusion of the allograft to the host bone (Figs. 12.47, 12.48, 12.49, 12.50, 12.51, 12.52, and 12.53).

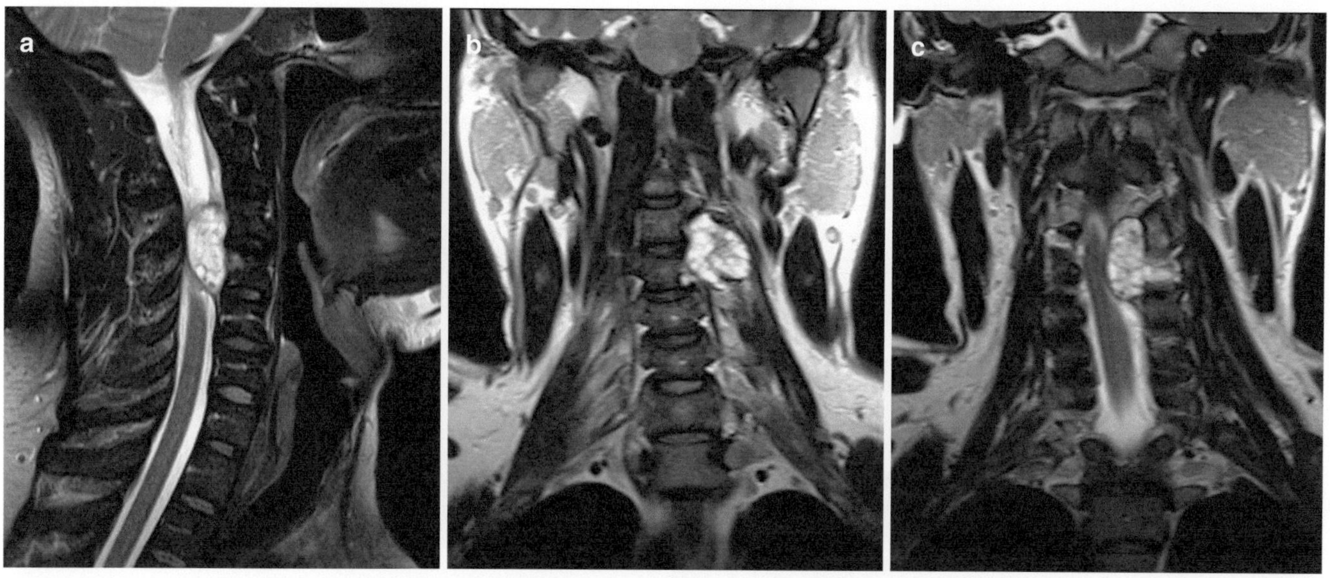

Fig. 12.47 Pre-operative MRI

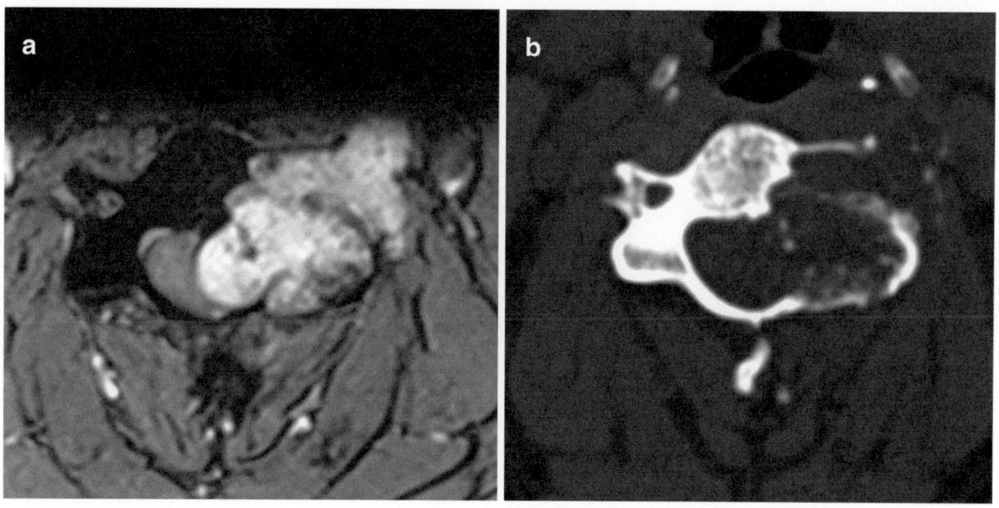

Fig. 12.48 Pre-operative axial MRI and CT-scans

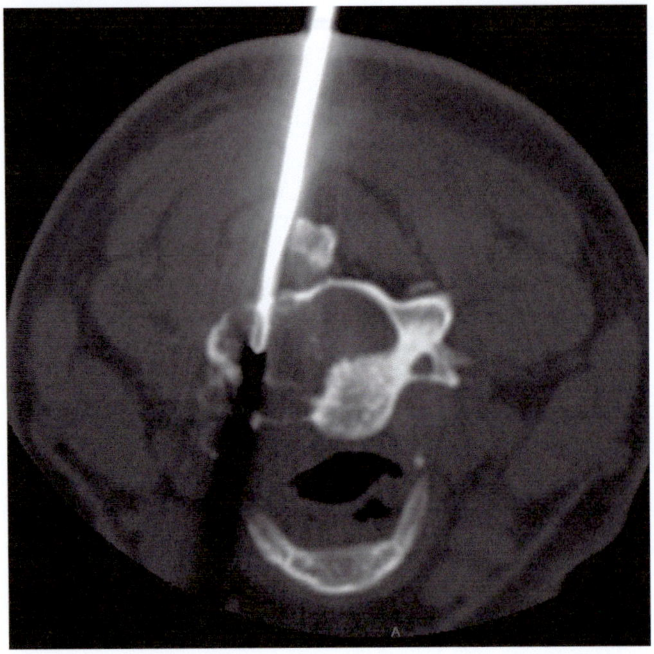

Fig. 12.49 CT-guided trocar biopsy

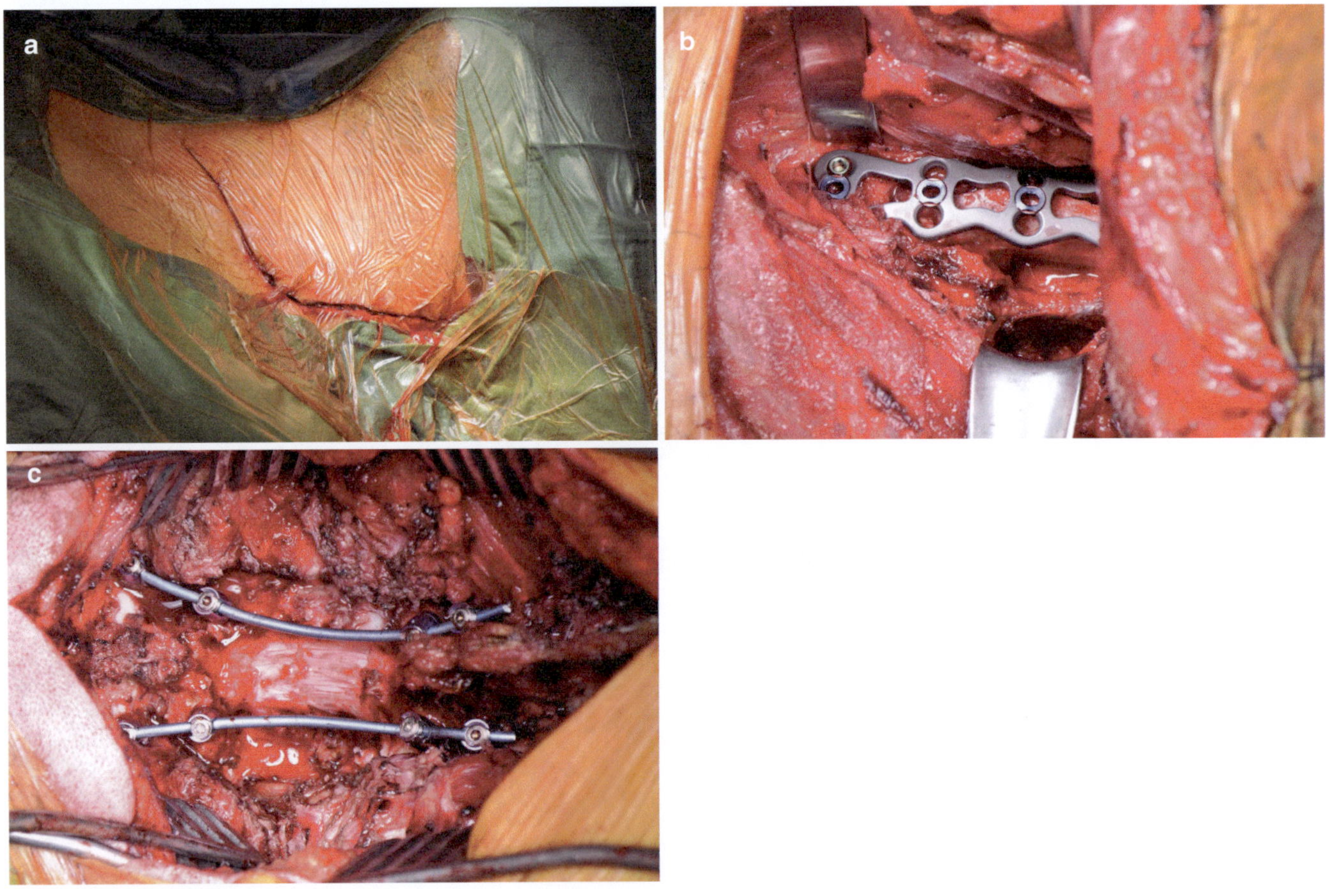

Fig. 12.50 Intra-operative images

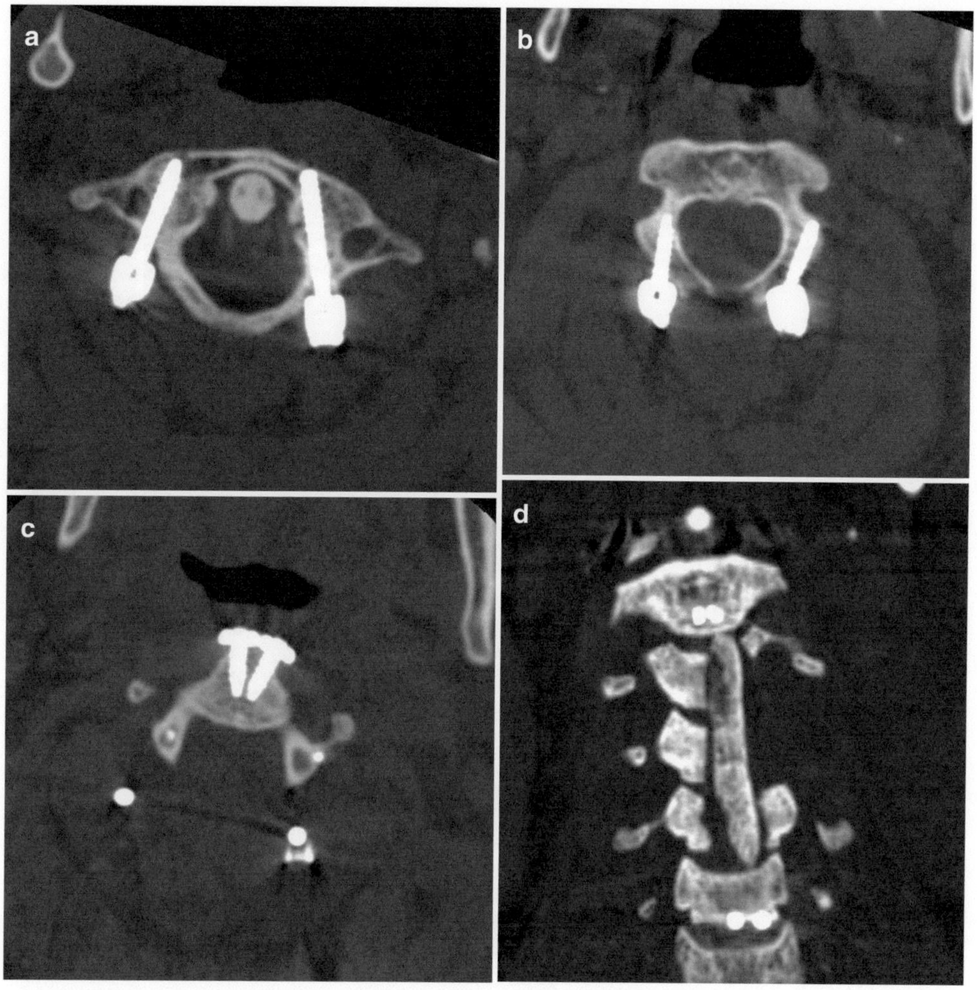

Fig. 12.51 Post-operative CT-scan

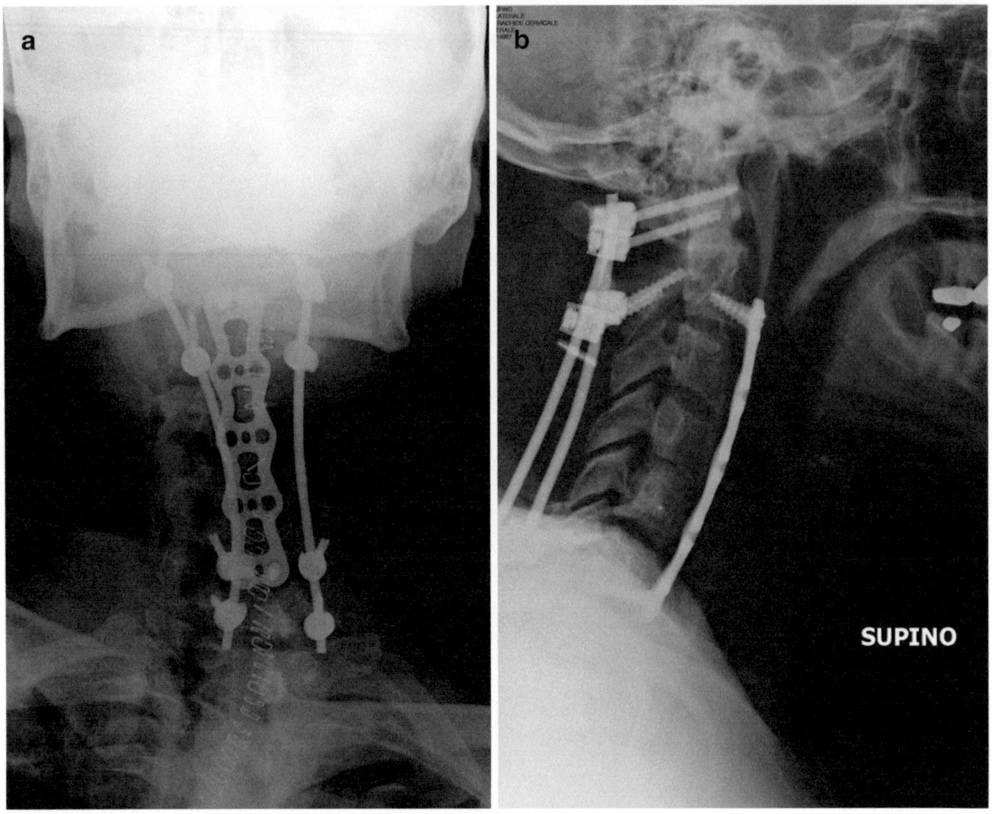

Fig. 12.52 Post-operative plain radiographs.

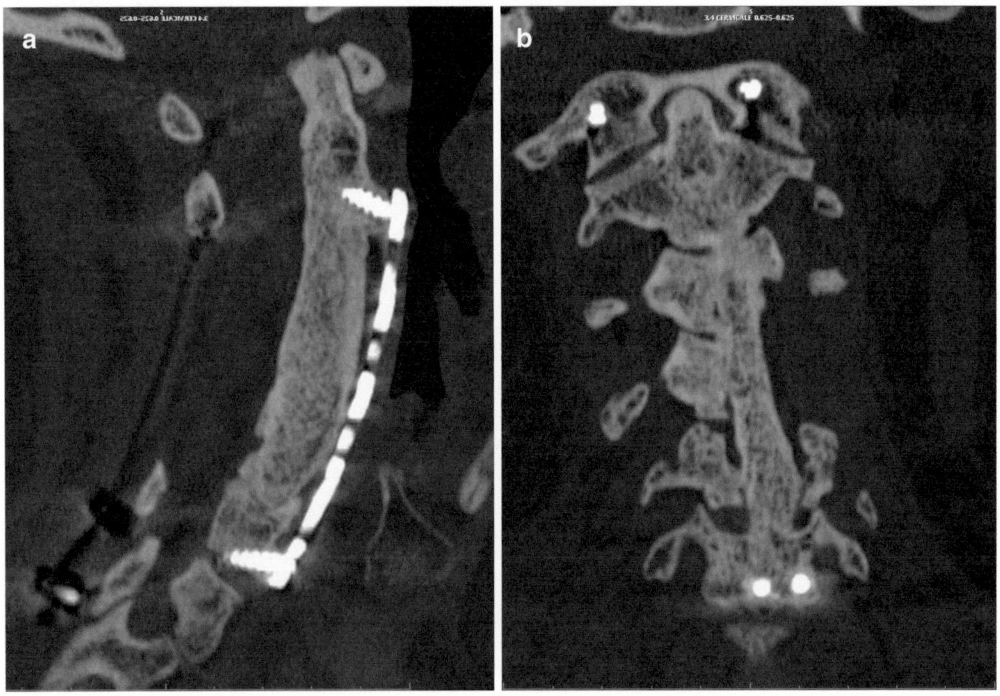

Fig. 12.53 2-years follow-up CT-scan

Complications of Cervical Spine Surgery

13

Gabriele Molteni, Marco Giuseppe Greco, and Pierre Guarino

The surgical approach to the cervical spine requires, in addition to a good surgical experience, a thorough anatomical knowledge of the cervical region and skull base. Indeed this region is crossed by many vascular, nervous, respiratory and digestive structures. Surgical techniques have been widely described in the previous chapters. We now focus our attention on the main complications related to this kind of surgery.

In cases of cervical spine surgery for degenerative diseases in the United States, an incidence rate of complications of 3.93 % during the decade 1992–2001 was reported. More in detail, the highest incidence of complications was found in patients with a primary diagnosis of cervical spondylosis with myelopathy (6.5 %) or with a posterior fusion (10.5 %) or combined anterior and posterior fusion (9.02 %).

The number of complications is also reported to be related with a longer length of hospital stay.

In the same decade, the incidence of in-hospital mortality was 0.14 % overall and still closely related to the age of patients: from 0.03 % for patients ages 20–34 years to 1.33 % for patients ages 75 years and older. From the analysis carried out by C. Wang and colleagues, the incidence of complications appears to be significantly higher when using a posterior approach (still in surgery for degenerative diseases). This partially explains the lowest number of complications in the approach described earlier. In literature, there are therefore conflicting data regarding the number of complications encountered in relation to the type of approach [1].

The incidence of complications in cervical spine surgery over the years was not reduced significantly; this is clear in a multicenter study published in 1997 by Zeidman and colleagues conducted on 4589 patients operated on by 35 different surgeons between 1989 and 1993. The rate of complications in elective surgery, regardless of the approach used, was found to be equal to about 5 %. Comparing the number of complications arising during the 5 years, the rate remained fairly steady (65 % of patients had used an anterior approach and the remaining 35 % a posterior approach).

It was also found that the use of steroids in the perioperative period does not modify the complication rate in a negative or positive way; steroids had only partially reduced the length of hospital stay. Death occurred in 0.8 % of cases and occurred mainly in patients undergoing surgery for trauma associated with quadriplegia. Death cases were very rare in elective surgery [2].

Complications in cervical spine surgical approaches can be schematically divided in generic complications (not correlated with surgical approach) and specific complications. Specific complications, which are related to the site and to the type of surgical approach, can be distinguished in intraoperative complications and postoperative ones, which can be in turn further subdivided in short-term complications and long-term complications. As reported in literature, the risk of severe complications is significantly higher in the anterior approach than in the posterior one, and this is related to the anatomical structures encountered. Complications like tracheal-oesophageal injury, graft dislodgement and injury to large neck vessels are more frequent in the anterior approach [3].

As in all surgical operations, complications can be also related to anaesthesiologic procedures: malignant hyperpyrexia (rare occurrence), anaphylactoid reactions to anaesthetic drugs, trauma of the vocal cords with risk of tracheal or laryngeal stenosis (related with the length of the hospital stay) or respiratory failure from glottic oedema or trauma of the endotracheal tube. Among generic complications, those related to bladder catheterization (urethritis, cystitis, pyelonephritis, globe bladder) have to be kept in mind. The inci-

G. Molteni, MD, PhD, FEBORL-HNS (✉)
Department of Otolaryngology-Head and Neck Surgery, Azienda Ospedaliera Universitaria Integrata, University Hospital of Verona, Modena, Italy
e-mail: gabbomolteni@gmail.com

M.G. Greco, MD • P. Guarino, MD
Head and Neck Surgery Department, University Hospital of Modena, Modena, Italy

S. Boriani et al. (eds.), *Atlas of Craniocervical Junction and Cervical Spine Surgery*, DOI 10.1007/978-3-319-42737-9_13

dence rate of postoperative urinary retention (POUR) after cervical spine surgery with anterior approach for degenerative diseases of discs is 11.1 % as described by Jung et al. Other complications are related to venous-arterial catheterization (phlebitis, septic shock), thromboembolic risks and bedsores [4, 5].

Most important complications are certainly those specifically related to the surgical procedure. In this contest we distinguish intraoperative complications from postoperative ones.

13.1 Intraoperative Complications

Intraoperative complications in soft tissues are more frequently related to anterior surgical approach, and they may occur as lesions of carotid artery, jugular vein, vertebral artery, thoracic duct, pneumothorax, oesophageal and/or tracheal perforation, wound infection and haematoma. It is important to mention intraoperative complications affecting the nervous structures: spinal cord injury and nerve roots, dural fistula (common complications to both approaches), recurrent laryngeal nerve (RLN) palsy and Horner syndrome due to injury to the cervical sympathetic chain (typically related to the anterior approach). Among bone complications for the anterior approach, the most frequent are inadequate positioning of the graft resulting in compression of the spinal canal and improper placement of the screws in the fixation of the plate. For the posterior approach wire pull-out, a fusion of the wrong level and inadequate positioning of plate and screws are the most common.

13.2 Postoperative Complications

After an anterior approach, the most common complications that may occur are oedema; tracheal, oesophageal, or tracheal-oesophageal fistula; mediastinitis; and dysphonia.

Related to nervous structures, spinal injuries may result in monoparesis or paraparesis, Brown-Sequard syndrome, quadriparesis, central cord syndrome and anterior cord syndrome, dural fistula and meningitis.

Also bone-related complications may arise: pseudoarthrosis, aseptic necrosis and discitis, graft dislodgement into the canal, angulation deformity and fracture of bone graft (rather rare occurrence).

Rear approach can frequently cause non-union, infection, spinal cord and/or nerve root lesions and complications related to instrument failure (such as wire or screw breakage or pull-out).

Also halo-related complications may arise such as dislodgement, infection and penetration [3].

Surgical complications can be classified in many ways; we consider minor complications as those that resolve spontaneously within few days and major complications as those that require prolonged hospital stay and/or vigorous treatment according to Zeidman and Ducker.

Minor complications are dysphagia, hoarseness transient sore throat, haematoma and abscess. Major complications include tracheal-oesophageal injury, root injury, recurrent laryngeal nerve injury, spinal cord injury, CSF leak and graft extrusion.

One or more of these complications may be associated with each other. The most common complications are recurrent laryngeal nerve injury (range 0.07–11 %) and dysphagia (2–60). Spanu et al. reported complication rate of ten-year experience: dysphagia (5.6 %), hoarseness (5.6 %), transient sore throat (4.8 %), worsening of pre-existing myelopathy (3 %), graft extrusion in (1.7 %) and root injury, haematoma and wound infection (0.87 %). They describe one case of oesophageal injury (0.43 %) and no deaths related to surgical approach.

Death indeed is very uncommon in elective cases. Almost all cases of deaths reported occurred in trauma with severe neurologic complications and are associated with pulmonary insufficiency.

Hoarseness and temporary dysphagia resulting from oedema of the pharynx, oesophagus and trachea should not be contemplated as complication, but as a condition connected to the surgical management, similar to wound pain after any surgical procedure [2, 6].

13.3 Dysphagia

It is a common early finding after anterior spine surgery that usually recovers at 6 months after surgery. Incorrect positioning of self-retaining retractor and/or a vigorous retraction of the medial musculovisceral column are the commonest causes of this complaint. Another well-known aetiology of postoperative dysphagia may be caused by the presence of cervical plate. Causes that may not be identified immediately after surgery can be related to local haematoma and/or tissue oedema, superior laryngeal nerve lesion and denervation of pharyngeal plexus [6].

13.4 Osteophytes-Related Symptoms

Cervical osteophytes formation after multilevel ACDF (anterior cervical discectomy and fusion) or other cervical spine surgical procedures develop in approximately 20–30 % of cases, and most of them are asymptomatic. Sometimes, however, osteophytes may cause a delayed presentation of functional dysphagia and dysphonia through a mass effect on the anterior structures such as the larynx and cervical oesophagus. Persistent cough due to irritation of the vagus nerve by osteophytes after cervical spine surgery is a very rare but described condition. Surgical untethering can alleviate symptoms [7, 8].

13.5 Recurrent Laryngeal Nerve Injury

This complication is related to the anterior cervical spine approach. It may occur contemporarily to transient sore throat and dysphagia or as the only complaint. When unilateral it is associated to voice hoarseness. Although controversial, it seems there are no differences between anterior right side and left side approach to the cervical spine in order to avoid RNL injury. The nonrecurrent laryngeal nerve, which arises directly from the main trunk of vagus nerve, is associated on the right side to an aberrant subclavian artery with an incidence of 1 %. On the left side, it is very rare and associated with a right-sided aortic arch.

RLN injury is due to a sharp dissection or a direct vigorous retraction, with angled or sharp retractors, along the medial musculovisceral column [6]. It can be permanent or temporary and in the latter it recovers from few days to some month after surgical procedure.

13.6 Transient Sore Throat

It is a very common complaint following anterior cervical spine approach that fully disappeared in few days. It may occur alone or associated with RLN injury, dysphagia and graft extrusion [6].

13.7 Deep Wound Haematoma

It is usually related to a muscular bleeding of the sternocleidomastoid muscle or longus colli muscle within 12–14 h after surgery. In literature the association with dysphagia ranges between 1 and 11 %. To reduce haematoma collection is mandatory to identify the medial border of the sternocleidomastoid gently opening the middle cervical fascia and approaching through the avascular plan medial to the SCM and then proceeding with blunt fingertip dissection directly to the spine. Deep wound haematoma recognition generally may be delayed by late onset (it is not palpable and not externally visible). When associated with tracheal obstruction, a rapid evacuation is required and an anterior cervicotomic approach is performed [6].

13.8 Wound Infections and Abscess

They are uncommon and range between 0.3 and 3%. Retropharyngeal abscess can also be a delayed complication after anterior spinal surgery. It may be associated with graft extrusion, screw pull-out and dysphagia. CT scan could be helpful (Fig. 13.1). Immediate drainage and antibiotic coverage usually did not lead to serious consequences. These patients are going to have some degree of retropharyngeal fibrosis with persistent swallowing problems [2, 6].

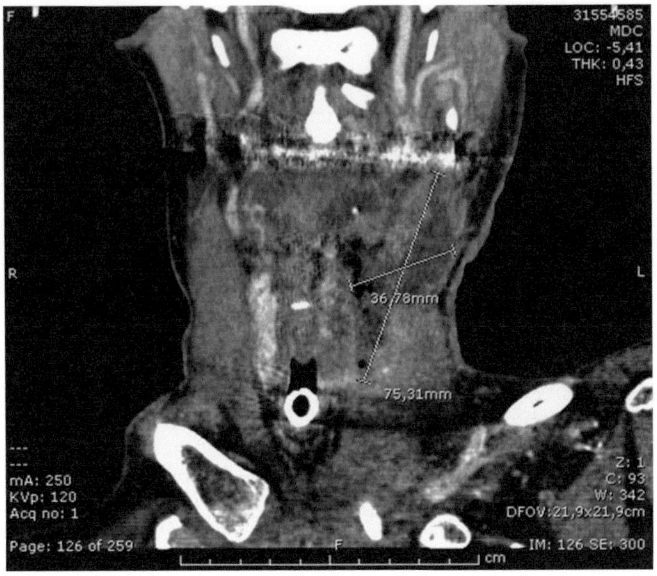

Fig. 13.1 Left lateral cervical abscess following anterior approach in cervical spine stabilization

13.9 Spinal Cord Damage

Spinal cord injury can be an expression of worsening of preexisting myelopathy or associated to RLN injury, graft extrusion and hoarseness. It is described in literature as a fat graft migration causing cervical cord compression after dural tear repairs following anterior discectomy [9].

13.10 Graft Extrusion

Anterior graft extrusion occurs in 2.1 % of cases and can be associated with RLN injury, spinal cord damage retropharyngeal abscess and consequent dysphagia. It may appear if there is no appropriate distraction in the disc space to hold the graft in a correct position. Approaching the disc space to have just the proper amount of body distraction is not always easy. Anterior graft extrusion may be avoided using the correct amount of interbody distraction to place the graft and modelling it to the containing site.

Posterior graft failure occurs in 0.3 % of cases, and it is associated with wire breakage, inferior screw pull-out or fracture of the graft with a fusion failure. This is common in rheumatoid patients [2, 6].

13.11 Root Lesion (With CSF Fistula)

It may occur after ACD and it can be associated to dysphagia and dysphonia. In these cases, a quick reoperation with CSF fistula repair and a wide-spectrum antibiotic therapy is required [6].

13.12 CSF Leaks

They occur in 0.4 % of cases and often after corpectomies associated with ossification of the posterior longitudinal ligament (OPLL) or very severe spondylosis [2]. Fistula repair is generally required.

13.13 Meningitis

Meningitis is a rare but potentially life-threatening complication that could happen after cervical spine surgery. A high index of suspicion for meningitis should be maintained in patients with the clinical triad of fever, neck stiffness and consciousness disturbance after spinal surgery. Intraoperative incidental durotomy is the most important predictor. An early diagnosis, also through the imaging (Fig. 13.2) and appropriate antibiotic treatment, can lead to a good outcome [10].

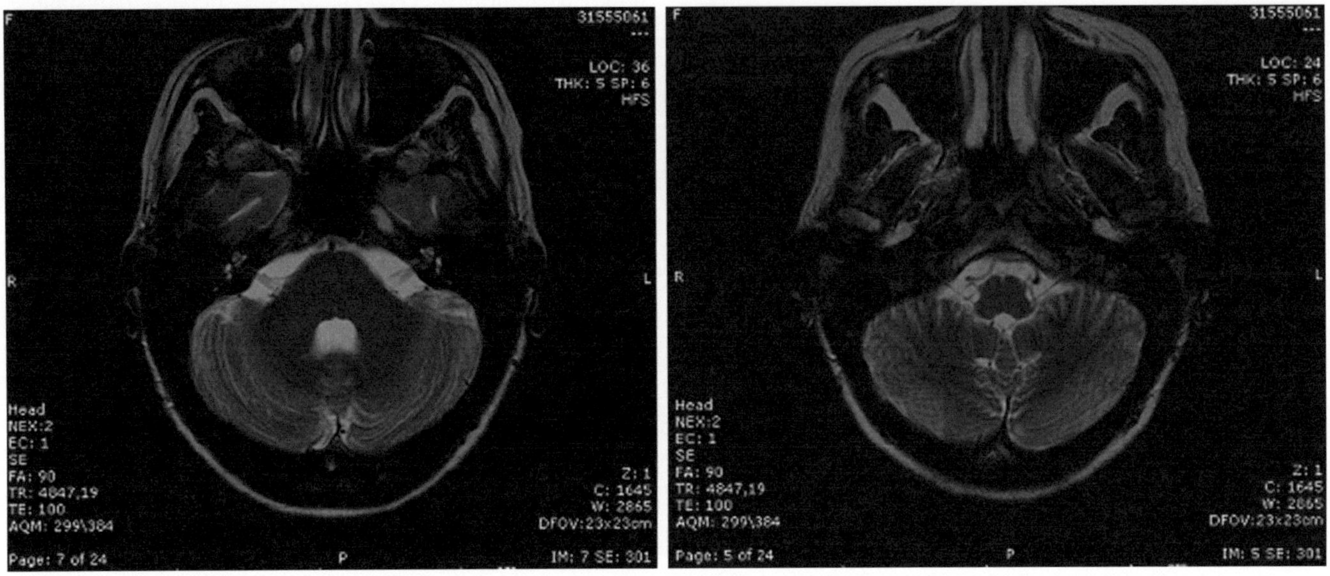

Fig. 13.2 Axial T2 MRI showing meningeal phlogosis

13.14 Oesophageal Perforation

Oesophagus perforation following anterior cervical spine surgery is a rare but potentially life-threatening lesion which can lead to death in 6–34 % of cases. Diagnosis is difficult and the treatment often delayed, resulting in cervical abscess, mediastinitis, septic shock and even death. The proper treatment should be surgical repair as soon as possible. Cervical oesophagus perforations are considered to be less critical than the intrathoracic and intra-abdominal ones due to slow formation of descending mediastinitis, but they may be still fatal in up to 16 % of cases.

It usually reveals within few hours after surgical procedure, but a case of delayed oesophageal perforation after cervical spine plating with a protruding screw is described. When a perforation is suspected (neck pain, dysphagia, odynophagia, fever, cervical emphysema), a CT scan with oral administration of a water-soluble contrast should be employed (Figs. 13.3, 13.4 and 13.5).

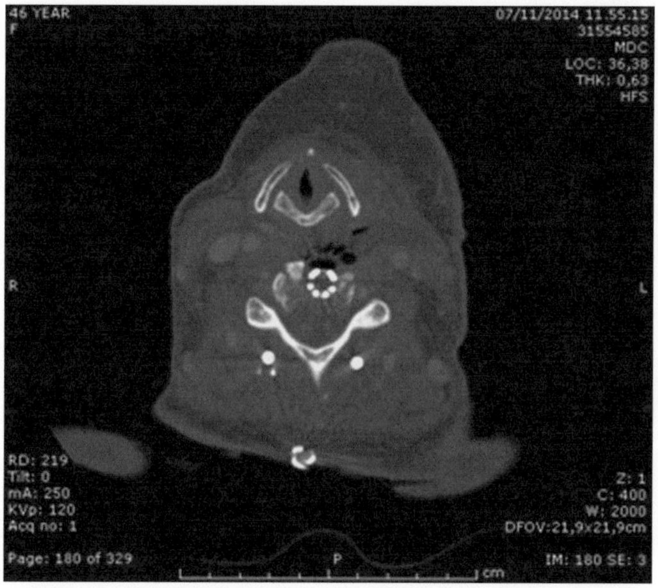

Fig. 13.3 Oesophageal fistula in C5–C6 stabilization with implants (axial CT)

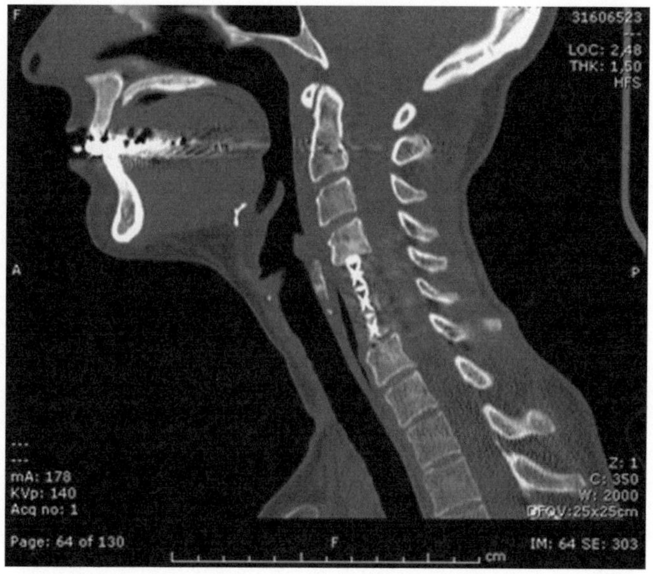

Fig. 13.5 Oesophageal recanalization after tracheal-oesophageal fistula repair

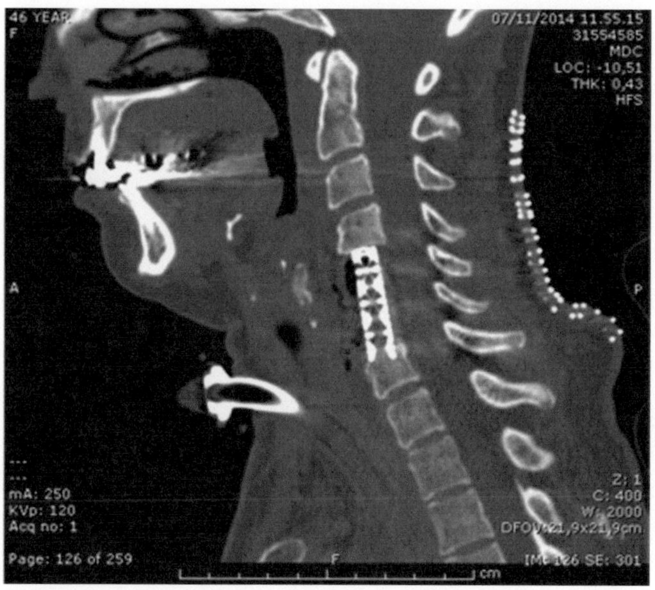

Fig. 13.4 Anterior paravertebral abscess after cervical spine stabilization (sagittal CT)

13.15 Pharyngocutaneous Fistula (PEF)

It is a rare but serious complication of anterolateral approach to the cervical spine with an incidence of 0–1.62 %. Proximity to the vertebral column and thin walls makes the upper digestive tract vulnerable to injury in cervical trauma. Clinical presentation and management are very similar to the oesophageal perforation.

A CT scan or MRI is necessary to seek the presence of any concurrent abscesses or vertebral stabilization failures (Fig. 13.6).

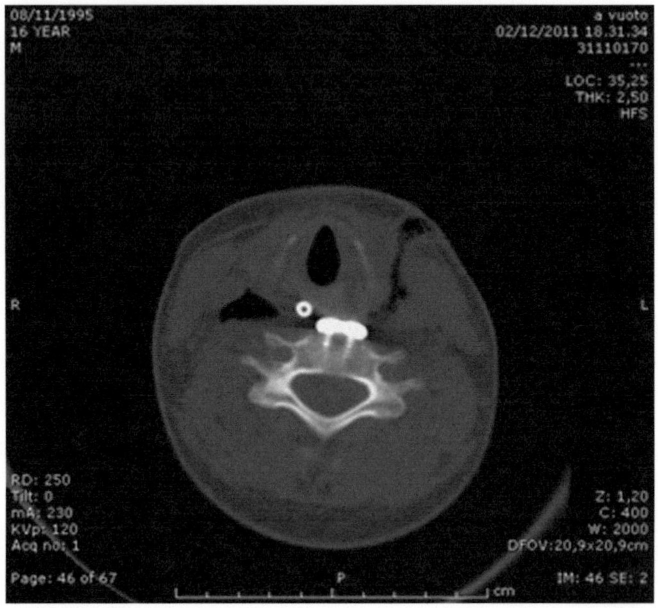

Fig. 13.6 Axial CT scan: left oesophagocutaneous fistula; concurrent abscessual collections at C6 level; vertebral stabilization

13.16 Osteomyelitis

Osteomyelitis following cervical spine surgery is a serious complication that can overlap other scenarios like abscess and oesophageal and/or tracheo-oesophageal fistula [11].

CT scan is always needed for diagnosis (Figs. 13.7, 13.8 and 13.9); also bone scintigraphy may be helpful.

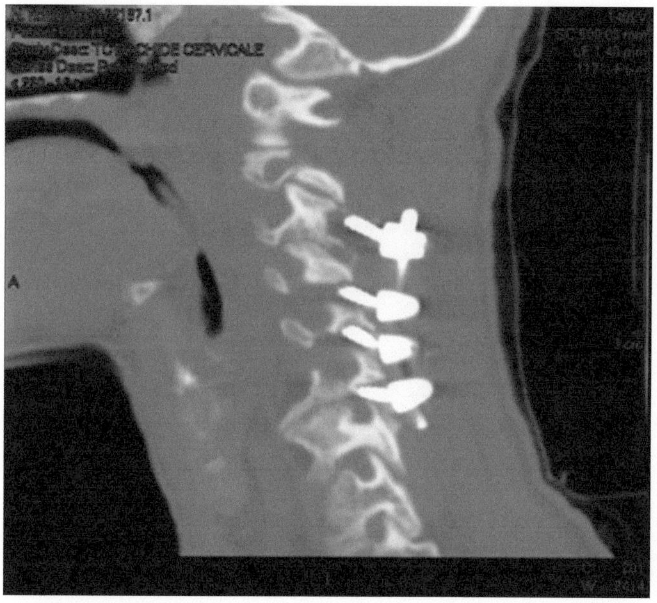

Fig. 13.7 Osteomyelitis with posterior cervical plate dislodgement after post-traumatic fixation

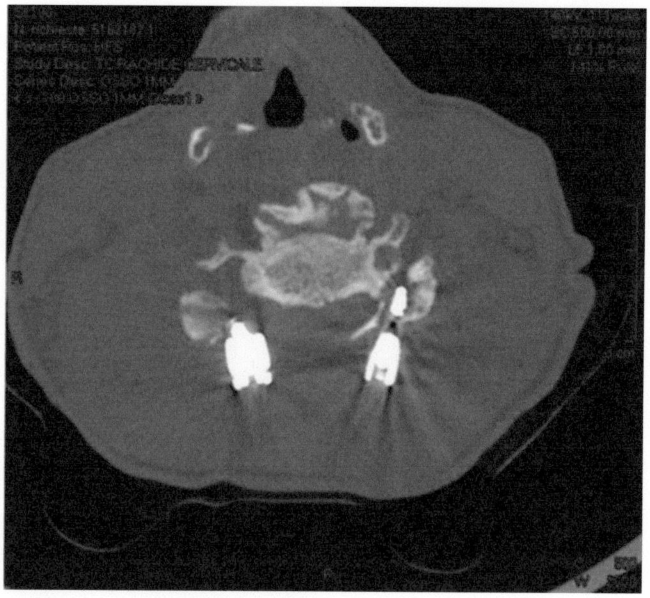

Fig. 13.9 Osteomyelitis after transpedicular fixation

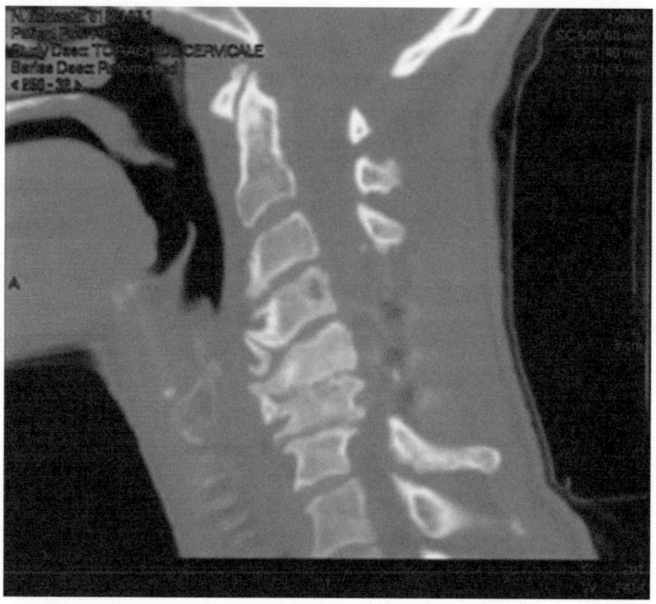

Fig. 13.8 Osteomyelitis with bony reabsorption of the spinose process

13.17 Complications Treatment

Treatment of complications after anterior cervical spine approach is often surgical, especially for major complications. Treatment options may vary and sometimes surgery is needed for more than one complication in the same patient.

Physical examination and blood test are needed, but imaging plays a key role in the evaluation and mostly in the planning of surgical approach. In case of infections, intravenous broad-spectrum antibiotic therapy is provided, and in case of serious dysphagia and oesophageal fistula, oral feeding is interrupted and nasogastric tube is used for feeding.

In the presence of a pharyngocutaneous fistula, an SCM muscle flap interposition may be performed; sometimes a cricopharyngeal myotomy can be useful in order to reduce intrapharyngeal pressure and prevent fistula recurrence, especially for defects above the upper oesophageal sphincter [12].

In case of intraoperative oesophageal perforation, primary double-layer suture (mucosal layer and muscular layer) is performed with absorbable stitches in a tension-free way to avoid secondary stricture of the oesophagus.

In case of postoperative perforations, a wound debridement and direct suture of the fistula with/without interposition of a muscular flap are mandatory. When related to early failure of internal fixation, the implant should be removed, and the spine should be maintained by external or internal fixation. Furthermore, free foreign bodies like pull-out screws are removed (Fig. 13.10).

For large perforations, reinforcement of primary suture with a muscle flap repair (SCM muscle flap, pectoralis major muscle flap and longus colli muscle flap) may be helpful [6, 13, 14].

A similar management can be employed in tracheal-oesophageal fistula (Figs. 13.11, 13.12, 13.13 and 13.14) even if in these cases a muscular flap either pedicle or microvascular is required [15]. For large TEFs, especially in patients who underwent previous radiation therapy, radial forearm and anterolateral thigh free flap demonstrated their advantages [16], mostly due to the amount of vascularized tissue transferred.

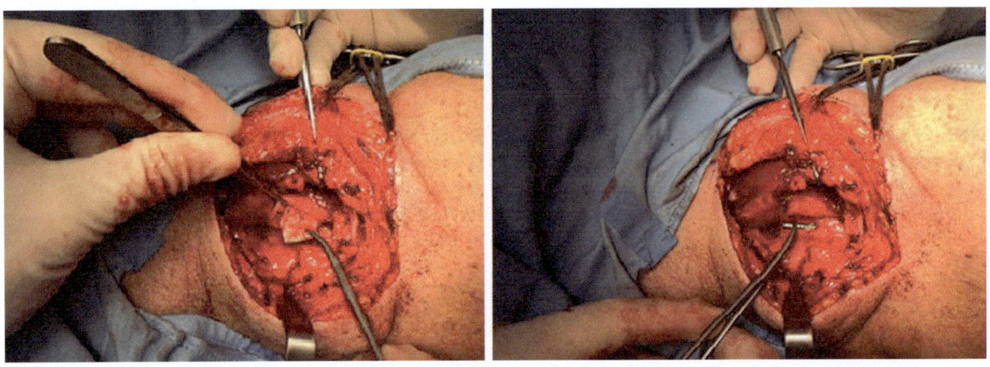

Fig. 13.10 A dislocated screw is removed from the neck

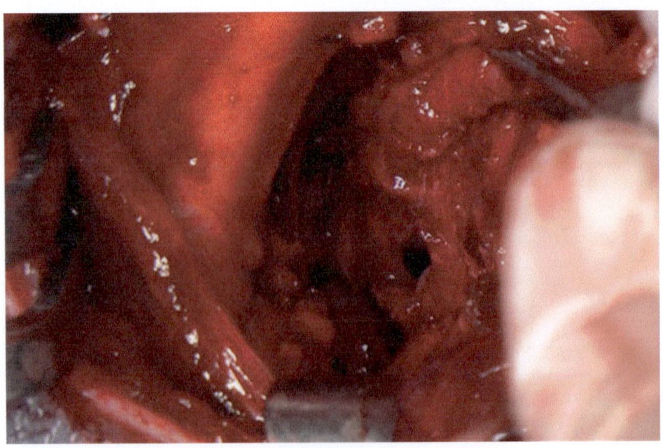

Fig. 13.11 Tracheal-oesophageal fistula

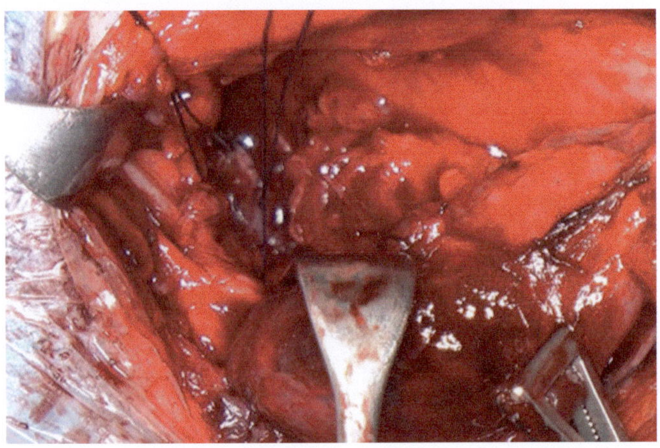

Fig. 13.12 Tracheo-oesophageal fistula repaired

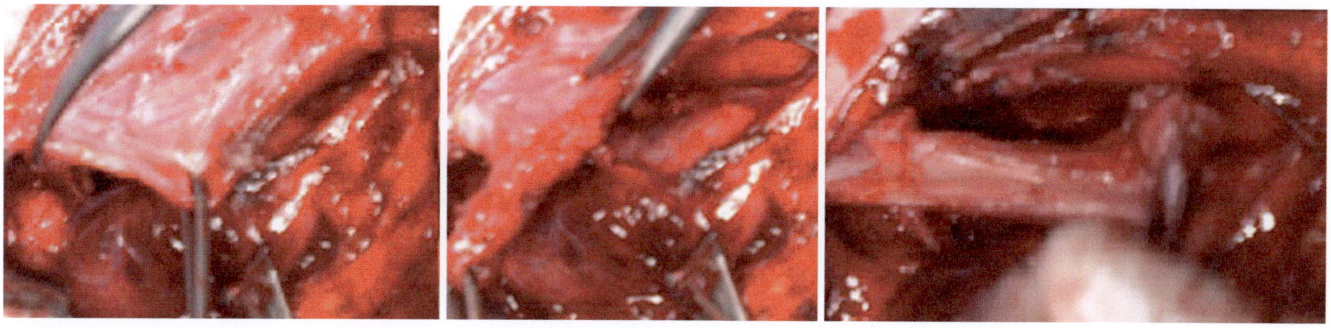

Fig. 13.13 Reinforcement by an SCM flap interposition between oesophagus and trachea

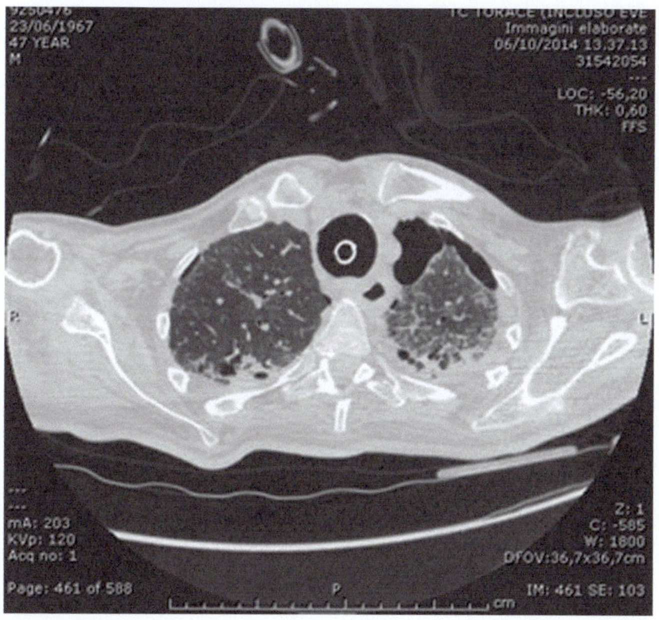

Fig. 13.14 Postoperative CT scan showing the good outcome of the TEF repair

References

1. Wang MC, Chan L, Maiman DJ, Kreuter W, Deyo RA (2007) Complications and mortality associated with cervical spine surgery for degenerative disease in the United States. Spine (Phila Pa 1976) 32(3):342–347
2. Zeidman SM, Ducker TB, Raycroft J (1997) Trends and complications in cervical spine surgery: 1989–1993. J Spinal Disord 10(6):523–526
3. Graham JJ (1989) Complications of cervical spine surgery. A five-year report on a survey of the membership of the Cervical Spine Research Society by the Morbidity and Mortality Committee. Spine (Phila Pa 1976) 14(10):1046–1050
4. Rampersaud YR, Moro ER, Neary MA, White K, Lewis SJ, Massicotte EM, Fehlings MG (2006) Intraoperative adverse events and related postoperative complications in spine surgery: implications for enhancing patient safety founded on evidence-based protocols. Spine (Phila Pa 1976) 31(13):1503–1510
5. Jung HJ, Park JB, Kong CG, Kim YY, Park J, Kim JB (2013) Postoperative urinary retention following anterior cervical spine surgery for degenerative cervical disc diseases. Clin Orthop Surg 5(2):134–137. doi:10.4055/cios.2013.5.2.134, Epub 2013 May 15
6. Spanu G, Marchionni M, Adinolfi D, Knerich R (2005) Complications following anterior cervical spine surgery for disc diseases: an analysis of ten years experience. Chir Organi Mov 90(3):229–240
7. Shih P, Simon PE, Pelzer HJ, Liu JC (2010) Osteophyte formation after multilevel anterior cervical discectomy and fusion causing a delayed presentation of functional dysphagia. Spine J 10(7):e1–e5. doi:10.1016/j.spinee.2010.04.014, Epub 2010 May 20
8. Orhan KS, Acar S, Ulusan M, Aydoseli A, Güldiken Y (2013) Persistent cough associated with osteophyte formation and vagus nerve impingement following cervical spinal surgery: case report. J Neurosurg Spine 19(2):167–169. doi:10.3171/2013.4.SPINE12428, Epub 2013 May 24
9. Kansal R, Nama S, Mahore A, Dange N, Kukreja S (2012) Fat graft migration causing recurrent cervical cord compression. Turk Neurosurg 22(4):502–505. doi:10.5137/1019-5149.JTN.3916-10.1
10. Lin TY, Chen WJ, Hsieh MK, Lu ML, Tsai TT, Lai PL, Fu TS, Niu CC, Chen LH (2014) Postoperative meningitis after spinal surgery: a review of 21 cases from 20,178 patients. BMC Infect Dis 14:220. doi:10.1186/1471-2334-14-220
11. Zairi F, Tetard MC, Thines L, Assaker R (2012) Management of delayed oesophagus perforation and osteomyelitis after cervical spine surgery: review of the literature. Br J Neurosurg 26(2):185–188. doi:10.3109/02688697.2011.609604, Epub 2011 Oct 4
12. Iyoob VA (2013) Postoperative pharyngocutaneous fistula: treated by sternocleidomastoid flap repair and cricopharyngeus myotomy. Eur Spine J 22(1):107–112. doi:10.1007/s00586-012-2451-4, Epub 2012 Sept 19
13. Lu X, Guo Q, Ni B (2012) Esophagus perforation complicating anterior cervical spine surgery. Eur Spine J 21(1):172–177. doi:10.1007/s00586-011-1982-4, Epub 2011 Aug 27
14. Kim SJ, Ju CI, Kim DM, Kim SW (2013) Delayed esophageal perforation after cervical spine plating. Korean J Spine 10(3):174–176. doi:10.14245/kjs.2013.10.3.174, Epub 2013 Sep 30
15. Jaiswal D, Yadav P, Shankhdhar VK, Gujjalanavar RS, Puranik P (2015) Tracheoesophageal puncture site closure with sternocleidomastoid musculocutaneous transposition flap. Indian J Plast Surg 48(3):278–282. doi:10.4103/0970-0358.173125
16. Wreesmann VB, Smeele LE, Hilgers FJ, Lohuis PJ (2009) Closure of tracheoesophageal fistula with prefabricated revascularized bilaminar radial forearm free flap. Head Neck 31(6):838–842. doi:10.1002/hed.20971

MIX
Papier aus verantwortungsvollen Quellen
Paper from responsible sources
FSC® C105338

If you have any concerns about our products,
you can contact us on
ProductSafety@springernature.com

In case Publisher is established outside the EU,
the EU authorized representative is:
Springer Nature Customer Service Center GmbH
Europaplatz 3, 69115 Heidelberg, Germany

Printed by Libri Plureos GmbH
in Hamburg, Germany